Ashwani K. Gupta · Hukam C. Mongia ·
Pankaj Chandna · Gulshan Sachdeva
Editors

Advances in IC Engines and Combustion Technology

Select Proceedings of NCICEC 2019

Springer

Editors
Ashwani K. Gupta
Department of Mechanical Engineering
University of Maryland
College Park, MD, USA

Hukam C. Mongia
LLC
CSTI Associates
Yardley, PA, USA

Pankaj Chandna
Department of Mechanical Engineering
National Institute of Technology
Kurukshetra
Kurukshetra, Haryana, India

Gulshan Sachdeva
Department of Mechanical Engineering
National Institute of Technology
Kurukshetra
Kurukshetra, Haryana, India

ISSN 2195-4356 ISSN 2195-4364 (electronic)
Lecture Notes in Mechanical Engineering
ISBN 978-981-15-5995-2 ISBN 978-981-15-5996-9 (eBook)
https://doi.org/10.1007/978-981-15-5996-9

This Springer imprint is published by the registered company Springer Nature Singapore Pte Ltd.
The registered company address is: 152 Beach Road, #21-01/04 Gateway East, Singapore 189721, Singapore

Lecture Notes in Mechanical Engineering

Lecture Notes in Mechanical Engineering (LNME) publishes the latest developments in Mechanical Engineering—quickly, informally and with high quality. Original research reported in proceedings and post-proceedings represents the core of LNME. Volumes published in LNME embrace all aspects, subfields and new challenges of mechanical engineering. Topics in the series include:

- Engineering Design
- Machinery and Machine Elements
- Mechanical Structures and Stress Analysis
- Automotive Engineering
- Engine Technology
- Aerospace Technology and Astronautics
- Nanotechnology and Microengineering
- Control, Robotics, Mechatronics
- MEMS
- Theoretical and Applied Mechanics
- Dynamical Systems, Control
- Fluid Mechanics
- Engineering Thermodynamics, Heat and Mass Transfer
- Manufacturing
- Precision Engineering, Instrumentation, Measurement
- Materials Engineering
- Tribology and Surface Technology

To submit a proposal or request further information, please contact the Springer Editor of your location:

China: Dr. Mengchu Huang at mengchu.huang@springer.com
India: Priya Vyas at priya.vyas@springer.com
Rest of Asia, Australia, New Zealand: Swati Meherishi at swati.meherishi@springer.com
All other countries: Dr. Leontina Di Cecco at Leontina.dicecco@springer.com

To submit a proposal for a monograph, please check our Springer Tracts in Mechanical Engineering at http://www.springer.com/series/11693 or contact Leontina.dicecco@springer.com

Indexed by SCOPUS. The books of the series are submitted for indexing to Web of Science.

More information about this series at http://www.springer.com/series/11236

Preface

This book contains high-quality research papers selected and presented at the 26th National Conference on Internal Combustion Engines and Combustion (NCICEC 2019), held at the National Institute of Technology Kurukshetra, Haryana, under the aegis of The Combustion Institute—Indian Section (CIIS). The CIIS has been working in the domains of combustion science with exciting projects that include the development of new energy sources with a low carbon footprint and the use of methanol as a fuel for both household cooking and transportation sector applications. This book covers many facets of combustion, fuels, and emissions, including numerical analysis and experimental analysis of the combustion phenomenon, fossil fuels, alternative fuels, internal combustion engines, gas turbines, emission, and flame propagation.

Each chapter in this book was peer-reviewed by renowned domain experts in the field. All authors revised their papers based on the comments provided, which improved the quality of the technical content with an appeal to a wider audience in the field from the new wealth of knowledge provided.

The editors of this book extend their warm and sincere appreciation to all the chairmen of the technical sessions, including Prof. S. Chakravarthy, IIT Madras; Prof. R. V. Ravikrishna, IISC Bengaluru; Dr. Debasis Chakraborthy, DRDL Hyderabad; Prof. C. G. Sarvanan, Annamalai University, Tamil Nadu; Dr. G. N. Kumar, NIT Suratkal; Dr. Ramanujachari, IIT Madras; Dr. Porpatham, VIT Vellore; Dr. N. K. Gupta, ISRO; and worthy members of the technical review committee for their valuable contributions to the success of the conference. Our sincere thanks also extend to several international supporters of the event, including Dr. Kailasanath, U.S. Naval Research Laboratory, USA; Distinguished University Professor, Dr. Ashwani K. Gupta, University of Maryland, USA; Dr. Hukam C. Mongia, CSTI Associates, USA; Dr. Kamil Ekstein, University of West Bohemia, Czech Republic; Dr. Bhupendra Khandelwal, University of Alabama, USA; and Dr. Keiichi Okai, JAXA, Japan, for their invaluable support and contributions to this conference. The organizers gratefully acknowledge the financial contributions provided by BrahMos Aerospace, NTPC, Toyota Kirloskar, India Oil R&D, CSIR-NAL, EduTech, and LPS Bossard. We also extend our sincere appreciation to all the participants of the conference as well as all the authors with their contributed chapters that resulted in the production of this book after peer review of each chapter. We also acknowledge the support provided by Springer to produce the high-quality book that will serve as a reference book for researchers and engineers working in the field of study.

College Park, USA — Ashwani K. Gupta
Yardley, USA — Hukam C. Mongia
Kurukshetra, India — Pankaj Chandna
Kurukshetra, India — Gulshan Sachdeva

Contents

About the Editors

Dr. Ashwani K. Gupta is a distinguished university professor at the University of Maryland, College Park. He obtained his PhD and higher doctorate (DSc) from the University of Sheffield, UK, and also DSc from the University of Southampton, UK. He received honorary doctorates from the University of Wisconsin Milwaukee, the University of Derby, UK, and KMUTNB, Thailand, bestowed by the Princess of Thailand. His research interests include swirl flows, combustion, air pollution, sprays, high intensity distributed combustion, fuel reforming, waste and biomass to clean fuels, pyrolysis and gasification, acid gas treatment for sulfur recovery, and laser diagnostics for high speed flows. He is honorary fellow of American Society of Mechanical Engineers (ASME) and Fellow of American Institute of Aeronautics and Astronautics (AIAA) and Society of Automotive Engineers (SAE), American Association for the Advancement of Science (AAAS) and Royal Aeronautical Society (RAeS), UK. He is the founding co-editor of the Energy Engineering and Environment Series published by CRC Publishers. He is Associate Editor of J. Propulsion & Power, Intl. J. Sprays & Combustion Dynamics, and J. Applied Energy. He is the recipient of AIAA Energy Systems, Propellants & Combustion, Air Breathing Propulsion, and Pendray awards; ASME George Westinghouse Gold, James Harry Potter Gold, James Landis, Worcester Reed Warner, Holley Medal, Honda Medal, and Melville Medal awards; ASME-AIM Percy Nicholls award; ASEE Ralph Coats Roe award. At the University of Maryland, he received President Kirwan Research award and College of Eng. Research award. He has co-authored over 270 journal papers, 500 plus conference papers, 3 books, and 15 edited books.

Dr. Hukam C. Mongia continues to provide services in gas turbine combustion science and technology innovation since May 2011. During his 7-year stay at Purdue University, he collaborated with his colleagues and students on research activities relevant to next-generation energy efficient fuel-flexible ultra-low emissions gas turbine engines for propulsion and power generation. These included collaborative works with IISc, IITM, UCONN/HAPRI-ACAE, Spectral Energies, NASA, Woodward, Nexus, FAA/UT and Dresser Rand. During his 37-year career with three engine design and manufacturing companies (GE Aviation, Allison now Rolls Royce of North America, and Garrett now Honeywell Aerospace), he has contributed significantly in developing combustion technology, design methodology and tools in addition to transition of technology into products; e.g., TAPS for GEnx, LEAP-X, GE9X and other future GE propulsion and aero derivative industrial engines. Dr. Mongia is credited with more than 300 publications, 30 patents and several significant awards and recognition, including Fellow of AIAA and ASME, recipient of AIAA Air Breathing award, Best Paper award by AIAA Terrestrial Energy Committee, "125 Alumni to Watch", University of Massachusetts 125th Anniversary Celebration in 1988, and 1st Edison Award "to annually recognize key technical innovators from General Electric".

Dr. Pankaj Chandna is currently working as a Professor in the Department of Mechanical Engineering, National Institute of Technology Kurukshetra, Haryana. He has more than twenty-six years of professional experience, with a blend of teaching, research and education administration and an excellent academic record. He has the privilege to serve as the President (Vice Chancellor), Mewar University, Rajasthan. His academic contributions include publishing of about ninety-two research papers in international and national journals/conferences, supervision of 14 PhD scholars and 38 M.Tech. dissertations in the varied areas of industrial engineering including scheduling, inventory management, ergonomics and GSCM, CAE etc. In addition to the above, he has been serving as member of board of governors/academic councils/board of studies of various universities and engineering institutions; member of various evaluation/selection committees & regulatory bodies; member of editorial boards & reviewer of journals. He has participated, presented papers/talks and chaired technical sessions in more than thirty-two conferences/workshops in Australia, Canada, Dubai, France, India, Malaysia, Singapore and USA. He has been the recipient of many international and national awards, appreciations including ENVIROENERGY Technocrat Appreciation Award- 2010 (WFO, UK); Technocrat Par Excellence Scroll of Honour-2009 (MSME, Govt. of India); appreciation by Hon'ble Board of Governors, NIT Kurukshetra-2009; appreciation by Hon'ble Chairman, Board of Governors, NIT Kurukshetra-2007 etc. He has been very actively involved in the administrative/managerial activities of the institute in terms of creating various industry supported centers of excellence; organization of twenty international/national conferences, short term courses, workshops, symposium etc. He has also worked as Vice-Chairman with South Asia forum for Energy Efficiency (SAFEE), a joint Venture of World Energy Council & WFO UK. Currently, he also shoulders the responsibilities of the Dean-Industry & International Relations and the President, NIT Kurukshetra Alumni Association (NITKAA).

Dr. Gulshan Sachdeva is an Assistant Professor in the Department of Mechanical Engineering of National Institute of Technology Kurukshetra, India. He obtained his PhD from NIT Kurukshetra in 2010. His research interests include: Computational Fluid Dynamics, Heat Transfer Enhancement Techniques,

Heat Exchangers Design, Ejector Refrigeration System, Vapor Absorption System, Hybrid Cooling Techniques, Optimization of the HVAC Systems, Solar Assisted Cooling Methods, and Bio-gasifiers, Toroidal Gasifier etc. He is a life member of the Institution of Engineers and Indian Society of Heat and Mass Transfer. He is actively engaged in providing consultancy to various industries/organizations on cooling load estimation, duct design and air intake systems. He has 18 years of teaching and research experience. He has guided 3 PhD and 35 M. Tech students. He is presently working on DST sponsored research project on "Performance Enhancement of Vapor Compression System of Small Capacity using Ejector as an Expansion Device". He has delivered invited talks on emerging areas, a few are: Ejector Cooling System, Heat Transfer Enhancement using Vortex Generators, Solar Thermal Systems in reputed technical institutions of India. He had organized 9th and 10th iteration of a series of International Workshop on Energy, Power and Propulsion which were held at NIT Kurukshetra in association with University of Maryland, University of Illinois, IIT Kanpur and ACRi USA. He has presented his research work in reputed International Conferences held in Paris, Melbourne, Miami and Venice. He has co-authored over 25 research papers in reputed International Journals, and 35 conference papers.

Plenary Invited Contributions

Impact of Flowfield on Pollutants' Emission from a Swirl-Assisted Distributed Combustor

Joseph S. Feser, Serhat Karyeyen, and Ashwani K. Gupta

1 Introduction

According to Annual Energy Outlook, the share of energy coming from the consumption of petroleum and other liquid fuels is expected to decrease slightly during the coming years and then increase gradually again, while for natural gas is forecasted to continuously increase [1]. Due to high energy density of fossil fuels, it is favorable to continue using fossil fuels to meet the rising energy demand; however, there are environmental concerns arising from the emission of pollutants produced from the combustion of fossil fuels. Awareness of environmental impacts and depletion of fossil fuel resources has motivated combustion engineers to seek out novel combustion methods for enhanced efficiency and reduced pollutants' emission. Around the turn of the last century, various research groups and engineers began to develop new combustion techniques such as: flameless oxidation (FLOX) [2–4], moderate or intense low oxygen dilution (MILD) [5–7], colorless distributed combustion (CDC) [8, 9], and high-temperature air combustion (HiTAC) [10].

CDC and FLOX combustion require (a) temperature of reactants must be higher than the self-ignition temperature and (b) entrainment must be enough to reduce the final product gas temperature. For HiTAC combustion, the inlet reactant gas temperature must be higher than the auto-ignition temperature. Parameters for achieving MILD are similar to HiTAC, but there is an additional constraint on the temperature difference between the inlet and maximum temperature, which should be smaller than the self-ignition temperature [5]. CDC is derived from the concept of high-temperature air combustion (HiTAC) [10], wherein preheated air along with

J. S. Feser · S. Karyeyen · A. K. Gupta (✉)
Department of Mechanical Engineering, University of Maryland, College Park, MD 20742, USA
e-mail: akgupta@umd.edu

S. Karyeyen
Department of Energy Systems Engineering, Gazi University, 06500 Ankara, Turkey

A. K. Gupta et al. (eds.), *Advances in IC Engines and Combustion Technology*, Lecture Notes in Mechanical Engineering,
https://doi.org/10.1007/978-981-15-5996-9_1

a reduced oxygen concentration in the reactant stream is used to achieve colorless distributed condition (or green combustion) through internal entrainment of hot and reactive gases. Adequate entrainment of the hot reactive species occurring prior to ignition is key to promote mixing while increasing the chemical time scale in order to volumetrically broaden the reaction zone throughout the combustor. In order to promote entrainment, the oxygen concentration in the oxidizer is reduced, slowing the reaction rate allowing for enhanced mixing prior to ignition. The production of distributed reaction zone eliminates the local hot spots which are normally present under traditional combustion condition. Distributed combustion helps in mitigating peak temperatures to provide ultra-low pollutants' emission [11–18].

In this study, the non-reacting flowfield was examined along with OH* chemiluminescence flame signatures and pollutants' emission from a swirl-stabilized flame under traditional and CDC combustion conditions. In order to obtain flowfield information, PIV, a non-intrusive flow diagnostic technique, was used to obtain information on the velocity field and eddy size under simulated conditions of traditional air combustion and simulated CDC condition. OH* chemiluminescence was used as a marker for the onset of distributed combustion condition. Pollutants' emission was measured and analyzed to determine the effect of flowfield on emissions.

2 Experimental Facility

A swirl-stabilized burner with a swirl angle of 45° and a calculated swirl number of 0.77 was used in the present study. Details of the burner can be found elsewhere [19]. The fuel line was situated along the longitudinal central axis of the combustor with the surrounding coaxial oxidizer stream supplied around the fuel line. Fuel was injected radially into the oxidizer stream immediately after the swirl vane blades. Propane was used as the fuel, while the oxidizer consisted of air, the air/diluent mixtures that simulated the entrainment of hot reactive gases for seeking distributed combustion condition. Either N_2 or CO_2 was used as the diluent gas in order to simulate entrainment of reactive combustion gases. Laminar flow controllers having an accuracy of ±0.8% of the reading and ±0.2% of full scale were used to control flow rates of air and nitrogen, leading to an overall accuracy of about 1.5% of the reading. Propane and carbon dioxide flow rates were controlled using gravimetric flow controllers having accuracies of 1.5% of full scale. Chemiluminescence of OH* was recorded via an ICCD that had an interference filter centered at 307 nm with a FWHM of ±10 nm in order to determine the oxygen concentration at transition to CDC. The analyzer was used to measure concentrations of NO/NO_x, CO, and O_2. The O_2 concentration determined by the galvanic cell method allowed for corrections of the NO and CO emission to standard 15% O_2 concentration. NO and CO concentrations were measured with estimated accuracies of 1% of full scale. High-frequency (3 kHz) 2-D particle image velocimetry (PIV) was used to determine the velocity field under non-reacting conditions to provide critical information on the role of flowfield and turbulent mixing as well as resulting Kolmogorov length scales.

A fluidized bed seeder was used to seed alumina (Al_3O_2) particles (nominal diameter of 2 μm) in the flow. The interrogation window used consisted of an adaptive 24 × 24 pixels grid. Further processing for calculation of the mean velocity, root mean square velocity, and Kolmogorov length scale profiles was conducted using MATLAB.

3 Experimental Condition

The experimental investigations reported here were primarily focused on obtaining velocity fields, turbulence quantities, and Kolmogorov length scales along with OH* flame signatures and pollutants' emission for the cases examined. For the oxidizer mixture, oxygen concentration in the oxidizer was gradually reduced in 1% increments. Reduced oxygen concentration was achieved through either N_2 or CO_2 dilution, which were introduced to simulate entrainment of hot reactive species to seek colorless distributed combustion condition. Propane was used as the fuel in this study. For all conditions, heat load, heat release intensity, mixture temperature, pressure, and equivalence ratio were kept constant at 3 kW, 5.72 MW/m^3.atm, 300 K (room temperature), 1 atm., and 0.9, respectively.

4 Results and Discussion

4.1 OH Chemiluminescence*

The OH* chemiluminescence flame signatures with various dilution amounts and diluent type are presented in Fig. 1. The results show the effect of reduced oxygen concentration on flame signature using differing diluents (N_2 or CO_2) entrained into the oxidizer. Reduction of oxygen concentration provided considerable decrease in peak OH* flame signatures for both the diluents examined. However, the effect of specific diluent on OH* flame signature was different in terms of oxygen concentration for transition to CDC. The results of OH* flame signatures showed that transition to colorless distributed condition occurred at oxygen concentration of approximately

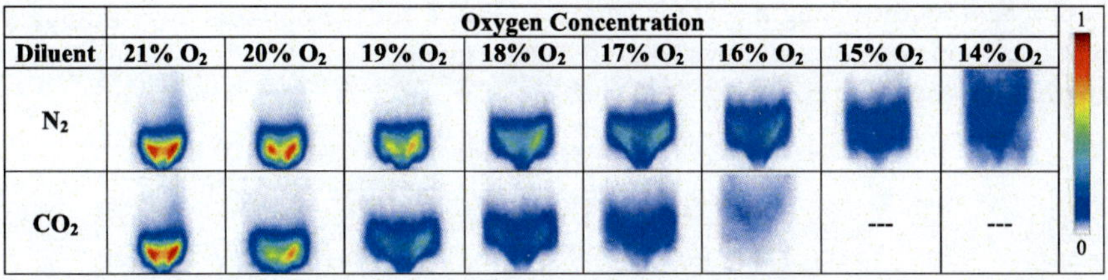

Fig. 1 OH* chemiluminescence signatures using either N_2 or CO_2 dilution

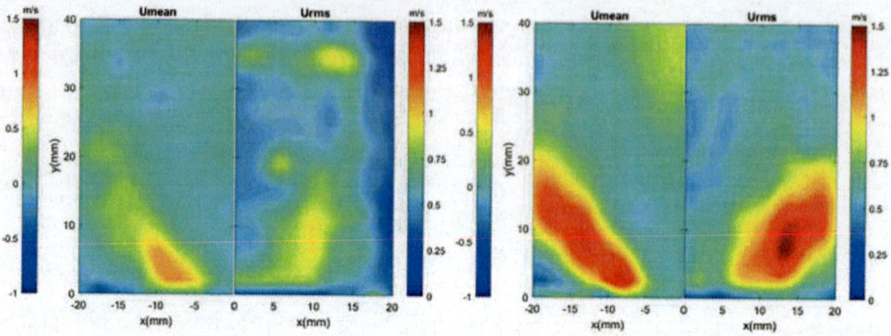

Fig. 2 Radial mean (left of frame) and rms (right of frame) velocity for normal air combustion case (left) and 15% oxygen concentration with N_2 dilution (right) under non-reacting condition

15% for N_2 and 17% for CO_2 diluents. The variation of CDC transition point for each diluent is attributed to different chemical properties of the diluents used. Higher heat capacity for CO_2 than N_2 caused considerable reduction in the reaction rate to cause in the transition to CDC at a higher oxygen concentration. Oxygen concentrations of 14 and 16% were indicative of the blow off limits of propane when N_2 or CO_2, respectively, were introduced as the diluents. The flames at these conditions were not stable so that any further increase in diluents resulted in flame blow off.

4.2 *Non-reacting Flow Field*

The mean and rms velocity profiles under non-reacting condition were determined using PIV. The mean and root mean square (rms) radial velocity profiles for the normal air case and simulated CDC case using N_2 dilution are shown in Fig. 2. The addition of dilution increased the mean velocity due to increase in flow rate. Additionally, the rms fluctuating velocity values increased with increase in dilution that enhanced turbulent mixing. Figure 4 shows the mean and rms axial velocity profiles. Both the mean and fluctuating axial velocity increased, similar to that found for radial velocity. The presence of an enhanced inner entrainment zone under simulated CDC condition along the centerline can be seen in Fig. 3. This enhanced entrainment allows for greater mixing with hot reactive gases enabling CDC.

The Kolmogorov length scale is shown in Fig. 4 for the non-reacting simulated normal air and non-reacting simulated CDC case using N_2 dilution. The Kolmogorov length is given by:

$$\eta = \left(\frac{\upsilon^3}{\varepsilon}\right)^{1/4} \tag{1}$$

where η is the Kolmogorov length scale,

$$\upsilon$$

is the kinematic viscosity, and ε is the dissipation rate. The dissipation rate is equal to the kinetic energy production rate, which is proportional to the velocity magnitude $(V_{\mathrm{el}})^3$ normalized by the characteristic length (L), which is the diameter of the burner outlet, and is given by:

$$\varepsilon = \frac{V_{\mathrm{el}}^3}{L} \tag{2}$$

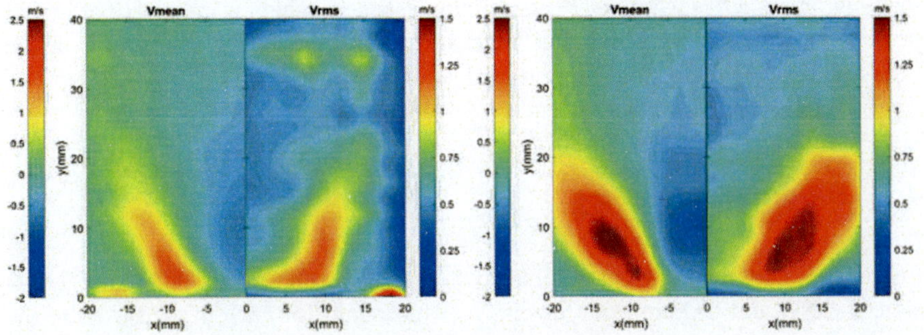

Fig. 3 Axial mean (left of frame) and rms (right of frame) velocities for normal air combustion case (left) and 15% oxygen concentration with N_2 dilution (right) under non-reacting condition

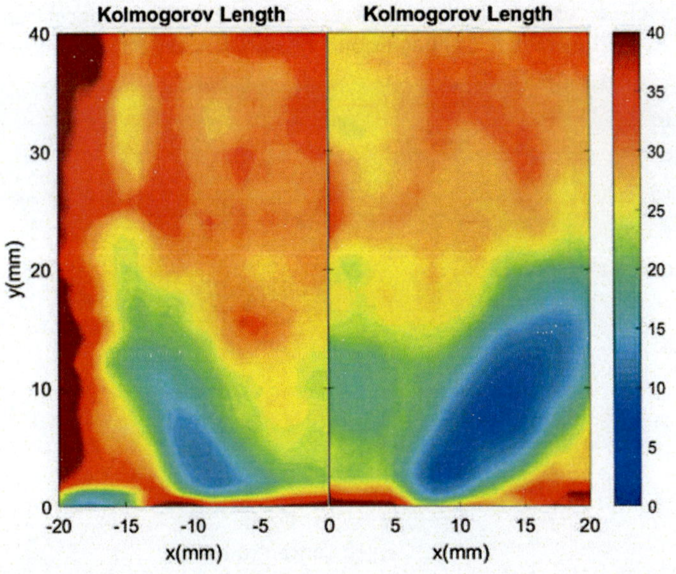

Fig. 4 Kolmogorov length scale for normal air combustion case (left frame) and 15% oxygen concentration with N_2 dilution (right frame)

The mixture viscosity was determined using Herning and Zipperer [20] equation for partial viscosities given by:

$$\mu_{\text{mixture}} = \frac{\Sigma\left(\nu_i x_i \sqrt{M_i}\right)}{\Sigma\left(x_i \sqrt{M_i}\right)} \tag{3}$$

where x is the mole fraction and M is the molecular weight of each component i.

Figure 4 shows that increased dilution decreased the eddy size. This suggests smaller scale mixing which provides more rapid mixing and higher turbulent dissipation rates. In particular the eddy size became significantly smaller in the swirl lobe region, which is where OH* chemiluminescence (which is indicative of higher peak temperature zones) signal is highest, to result in better dissipation in the high-temperature region and support mitigation of hot spots.

4.3 *Pollutants' Emission*

In order to examine the effect of reduced oxygen concentration on pollutants' emission, the NO and CO emission levels were measured for each diluent used at the combustor exit under conventional flame and near distributed combustion condition. Results on the emission of NO corrected to 15% O_2 concentration are shown in Fig. 5 for the specific conditions reported here. The maximum NO levels were measured at oxygen concentration of 21% representing the normal air combustion condition. The NO levels decreased significantly with decrease in oxygen concentration in the oxidizer. At the favorable CDC conditions, NO levels were below 2 ppm for N_2 as

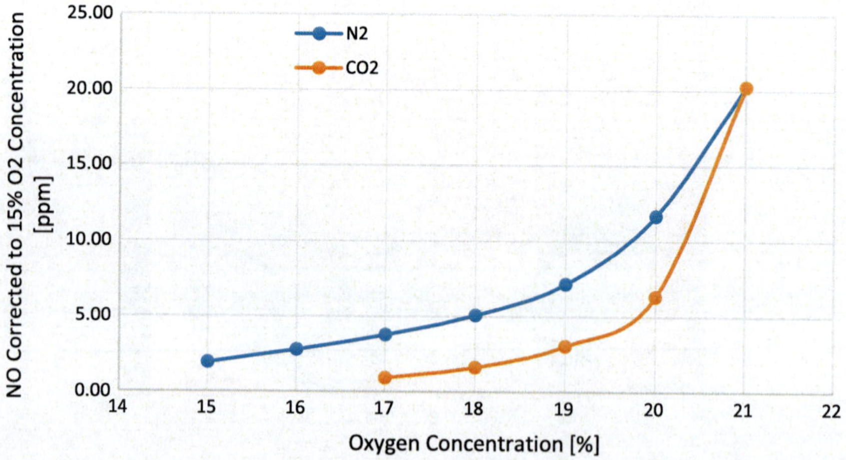

Fig. 5 NO emission corrected to 15% O_2 conc. under various oxygen concentrations

the diluent and 1 ppm for CO_2 as the diluent. Therefore, ultra-low NO emissions were achieved under CDC. Carbon dioxide has been shown to have a greater effect on mitigating NO, which is attributed to higher heat capacity than that of nitrogen.

The effect of reduced oxygen concentration on CO emission is presented in Fig. 6. Carbon monoxide levels decreased considerably when N_2 was used as the diluent. At the favorable CDC condition, single-digit CO levels were achieved under N_2 dilution condition. However, with CO_2 dilution, CO emission first decreased and then increased as residence time prohibited further oxidation of CO. Carbon monoxide emission was approximately 30 ppm under the CDC case. Due to CO_2 dissociation to CO, the carbon monoxide emission was somewhat higher than the N_2 dilution case. The results show both diluents provide ultra-low CO emission under CDC condition.

Figure 7 shows the corrected NO over the average eddy size (Kolmogorov length) for N_2 and CO_2 dilution cases. The eddy size reduced as dilution increased resulting

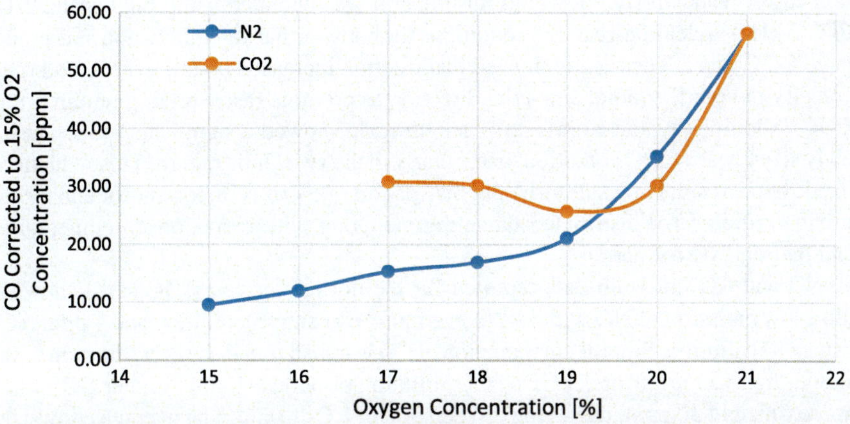

Fig. 6 CO emission corrected to 15% O_2 conc. under various oxygen concentrations

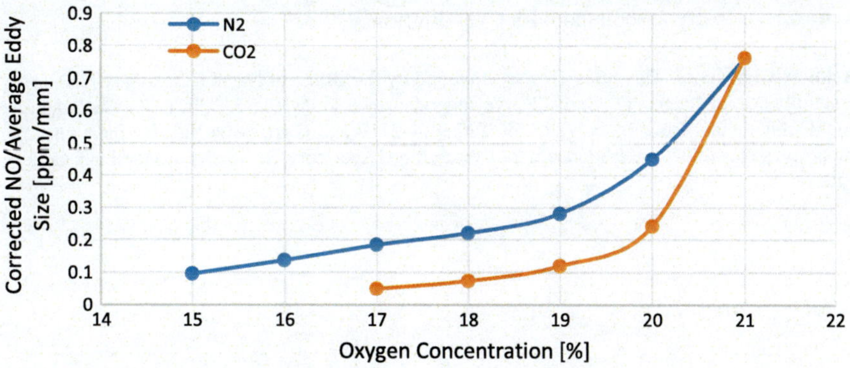

Fig. 7 Effect of eddy size on NO emission

in higher turbulent dissipation, to result in lower peak temperatures and reduction of thermal NO_x. Due to a much lower kinematic

viscosity of CO_2 than N_2, the eddy size under carbon dioxide dilution was further reduced resulting in enhanced turbulent mixing and lower NO emission.

5 Conclusions

Colorless distributed combustion conditions were examined in a swirl-assisted propane–air combustor that incorporated dilution with either N_2 or CO_2 as the simulated entrained reactive gases. OH* chemiluminescence signatures were used to determine the onset of conditions for a volumetrically distributed reaction zone. The CDC conditions were reached at 15 and 17% O_2 concentration for N_2 and CO_2 dilution cases, respectively. Flowfield information was obtained using high speed 2-D PIV (3 kHz) under non-reacting conditions for both normal air combustion and simulated CDC cases. The results showed that dilution increased both the mean and rms velocity in both the radial and axial direction to promote faster mixing required for CDC. Additionally, the Kolmogorov length scale showed a significant reduction in eddy size toward CDC condition, particularly in the swirl lobe region. This reduction in eddy size results in faster turbulent dissipation rate that is important for enhanced mixing required for distributed flame regime along with reduction in temperature and thermal NO formation.

NO and CO emission data reported for the normal air and CDC cases showed ultra-low emission of NO with less than 2 ppm NO using N_2 dilution and 1 ppm NO using CO_2 dilution. Significant reduction in CO was also found using N_2 dilution with single-digit CO levels under CDC. CO_2 dilution provided slightly higher CO level, approximately 30 ppm, due to the dissociation of CO_2 to CO. The average eddy size (Kolmogorov) was determined to examine the effect of flowfield on NO emission. Increase in dilution decreased the eddy size which in turn promoted mixing to reduce NO levels. CO_2 dilution provided a much stronger impact on NO-level reduction due to lower viscosity associated with CO_2 dilution.

Acknowledgements This research was supported by the Office of Naval Research (ONR) and is gratefully acknowledged. Dr. Serhat Karyeyen gratefully acknowledges TUBITAK (The Scientific and Technological Research Council of Turkey)—2219 and Gazi University for their financial supports as a post-doctoral associate at the Combustion Laboratory, University of Maryland, College Park.

References

1. U.S. Energy Information Administration. Annual Energy Outlook 2019 with projections to 2050. AEO 2019. Washington DC, USA: U.S. Department of Energy
2. Wunning JA, Wunning JG (1997) Flameless oxidation o reduce thermal NO formation. Prog Energy Combust Sci 23:81–94
3. Lammel O, Schutz H, Schmitz G, Luckerath R, Stohr M, Noll B, Aigner M, Hase M, Krebs W (2010) FLOX combustion at high power density and high flame temperature. J Eng Gas Turbines Power 132(12):121503
4. Zornek T, Monz T, Aigner M (2015) Performance analysis of the micro gas turbine turbec T100 with a new FLOX-combustion system for low calorific values. Appl Energy 159:276–284
5. Cavaliere A, de Joannon M (2004) MILD combustion. Prog Energy Combust Sci 30(4):329–366
6. Weber R, Smart JP, vd Kamp W (2005) On the (MILD) combustion of gaseous, liquid, and solid fuels in high temperature preheated air. Proc Combust Inst 30(2):2623–2639
7. Ozdemir IB, Peters N (2001) Characteristics of the reaction zone in a combustor operating at MILD combustion. Exp Fluids 30:683–695
8. Arghode VK, Gupta AK (2010) Effect of flowfield for colorless distributed combustion (CDC) for gas turbine combustion. Appl Energy 78:1631–1640
9. Khalil AEE, Gupta AK (2011) Swirling distributed combustion for clean energy conversion in gas turbine applications. Appl Energy 88:3685–3693
10. Tsuji H, Gupta AK, Hasegawa T, Katsuki M, Kishimoto K, Morita M (2003) High temperature air combustion: from energy conservation to pollution reduction. CRC Press, Boca Raton
11. Khalil AEE, Gupta AK (2018) Fostering distributed combustion in a swirl burner using prevaporized liquid fuels. Appl Energy 211:513–522
12. Correa SM (1992) A review of NO_x formation under gas turbine combustion conditions. Combust Sci Technol 87:329–362
13. Khalil AEE, Gupta AK (2017) Acoustic and heat release signatures for swirl assisted distributed combustion. Appl Energy 193:125–138
14. Khalil AEE, Gupta AK (2016) Fuel property effects on distributed combustion. Fuel 171:116–124
15. Khalil AEE, Gupta AK (2015) Impact of internal entrainment on high intensity distributed combustion. Appl Energy 156:241–250
16. Khalil AEE, Gupta AK (2015) Thermal field investigation under distributed combustion conditions. Appl Energy 160:477–488
17. Khalil AEE, Gupta AK (2017) Towards colorless distributed combustion regime. Fuel 195:113–122
18. Khalil AEE, Gupta AK (2016) On the flame-flow interaction under distributed combustion conditions. Fuel 182:17–26
19. Kim HS, Arghode VK, Linck MB, Gupta AK (2009) Hydrogen addition effects in a confined swirl-stabilized methane-air flame. Int J Hydrogen Energy 34(2):1054–1062
20. Herning F, Zipperer L (1936) Calculation of the viscosity of technical gas mixtures from the viscosity of individual gases. Gas-und Wasserfach 79:69–73

Emerging Engine Technologies for Reducing Fuel Consumption and Emissions

K. Kailasanath

1 Introduction

Two major societal problems that we, combustion scientists and engineers, can address are as follows: (a) How to deal with the depletion of worldwide petroleum-based fuel resources? and (b) how to reduce the harmful emissions from burning these hydrocarbon fuels? Of course, the answers to these questions are complex and involve both technical and non-technical entities throughout the world. The perspective of this presentation is from the point of view of the combustion scientist or engineer. In this paper, two emerging technologies that can significantly reduce: (a) the fuel consumption and (b) the exhaust gas emissions from combustion systems are highlighted. Both these technologies are emerging from fundamental research and require some more development before "practical" implementation. The first technology has the potential to reduce the fuel consumption of gas turbine and other engines by moving to a more efficient energy conversion process. In this new class of hybrid engines, the standard "constant pressure" combustion process is replaced by a "pressure gain" combustion process, such as "constant volume" combustion or "detonation". This results in an improvement in the thermodynamic energy-conversion efficiency by more than 20%. In this paper, key information available in the open literature is presented and used to assess the state of the art of these hybrid engines under research and development. Reducing fuel consumption will of course reduce the exhaust gas emissions during a given mission or task. But there are other options for more directly reducing the exhaust gas emissions from engines. The second technology called "high-temperature air combustion" [1] is discussed in this context.

Engines based on a "constant-volume" thermodynamic cycle have the potential to significantly reduce fuel consumption when compared to engines based on the traditional "constant-pressure" (Brayton) thermodynamic cycle. The detonation process

K. Kailasanath (✉)
22079 Lorton, VA, USA

A. K. Gupta et al. (eds.), *Advances in IC Engines and Combustion Technology*,
Lecture Notes in Mechanical Engineering,
https://doi.org/10.1007/978-981-15-5996-9_2

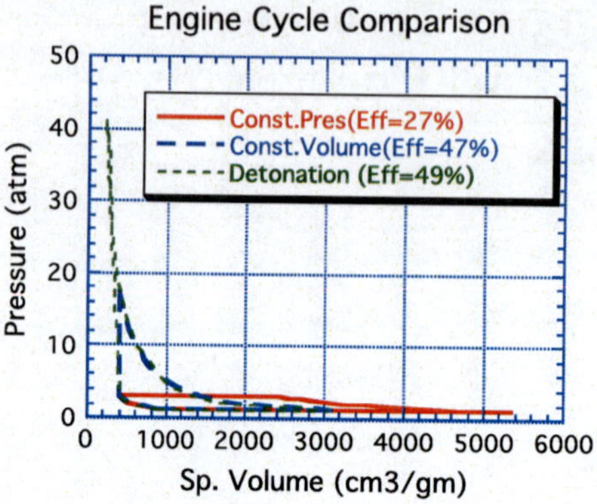

Fig. 1 Relative thermodynamic energy conversion efficiencies of three engine cycles

provides a practical way to attain near constant-volume combustion. Let us consider the relative efficiency of the three underlying engine cycles for a specific operating condition as shown in Fig. 1. For purposes of comparison, the only process that is different in the three cycles is the mode of energy conversion or heat addition. For the three cases, heat is added at constant pressure, constant volume or in a detonation. Hence, the three cycles have been referred to as "constant pressure", "constant volume" and "detonation" cycle, respectively. For ease of comparison, the amount of heat added is kept the same at 50 kcal/mole (a value typical of hydrocarbon fuels) for the three cycles. In all cases, the fuel–air mixture is initially compressed adiabatically from 1 to 3 atm. before heat addition. After heat addition, the products of combustion are expanded adiabatically back to 1 atm. For the efficiency, the work output is divided by the heat input, which was set to be the same for all three cycles. The calculated thermodynamic efficiencies for the three cycles are: 27% for constant pressure, 47% for constant volume and 49% for detonation. From the figure and the above numbers, we see that the thermodynamic efficiency of the detonation cycle is close to that of the constant volume cycle and significantly higher than that of the constant-pressure cycle that is typically used today.

Although the detonation-wave engine concept has been around for more than sixty years [2, 3], the past two decades of extensive research and development activities on the pulsed-detonation engine (PDE) were needed before the demonstration of a PDE powered aircraft flight [4]. The history of the development of the PDE concept has been discussed in References [2, 3] and elsewhere. There is the potential to increase further the performance gains made by the pulsed-detonation engine (PDE) by moving from "pulsed" or "intermittent" detonations to a more "continuous" mode of operation with detonations [5]. A schematic of a rotating detonation wave engine is shown below in Fig. 2.

An RDE is a type of continuous detonation wave engine where a detonation is initiated once and then remains within the engine, while the engine is running. RDEs use an annular ring combustion chamber, where a detonable gas mixture is injected

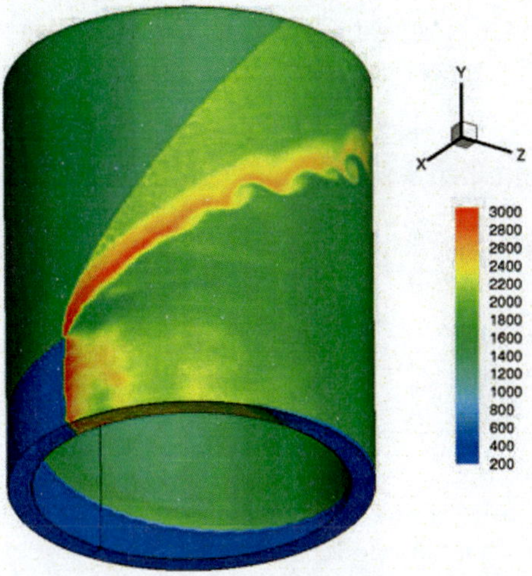

Fig. 2 A schematic of a rotating detonation-wave engine (RDE)

axially at the head end of the combustion chamber, and a detonation propagates circumferentially around the annular ring near the injection plane. Product gases are expanded and exhausted out the far end of the combustion chamber. The RDE concept does, however, have its own technical challenges. Because the detonable mixture is injected axially and the detonation wave runs circumferentially around the combustion chamber, the flowfield within an RDE has both very strong axial and azimuthal components, which makes analysis of the engine more complex and an efficient design more challenging. Further details on the flow field and operation of the RDE will be presented.

While decreasing the fuel consumption will indirectly reduce the amount of pollutants emitted by the engine during a given operation, there are other methods for directly addressing the reduction of unwanted pollutants. One such technique, which is still in the research and development stage, is "high-temperature air (HiTAC)", sometimes also called "flameless" or "distributed" combustion [1, 6–12]. It is a technique based on operating with the oxidizer or air at temperatures higher than the auto-ignition temperature of the fuel–air mixture. There are several practical ways of raising the temperature of the oxidizer or air including using exhaust gas recirculation. A traditional flame is not visible in the combustor and hence the term "flameless" combustion. Typically, the turbulence level is also high in this combustor. Because the conditions are close to that of a well-stirred reactor, the burning is more distributed and yields high efficiency with low emissions. Although a general understanding of this combustion concept is present, there are many aspects still under research. A brief summary of the state of the art of this technology for reducing unwanted pollutant emissions is discussed next, and an up-to-date status can be found in Ref. [1].

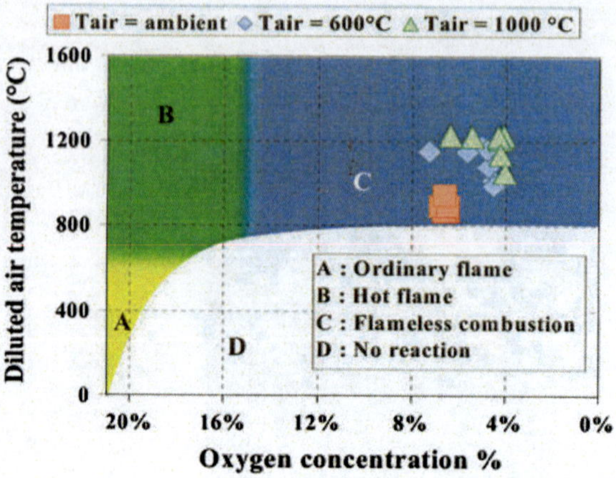

Fig. 3 A characterization of the flameless combustion mode

Traditionally, improvements in combustion efficiency are typically achieved by moving towards higher flame temperatures. However, one of the consequences of just raising the combustion temperature is to increase the amount of NO_x produced by the "thermal" mechanism. Thus, a more judicious approach is needed to balance improved efficiency without increasing pollutants such as NO_x. Combustion using highly preheated air and exhaust gas recirculation is one such approach [1]. In this approach, high internal recirculation leads to a significant dilution of the air by the combustion products, thus reducing the local concentration of oxygen. Furthermore, peaks in temperature are avoided due to the higher turbulence and more uniform distributed combustion. The net result is a potential for reduction in the amount of NO_x produced. A characterization of this mode of combustion with respect to an "ordinary flame" as well as a "hot flame" is shown in Fig. 3 (adapted from Reference 11).

Further research on this topic is taking place in a number of countries including Germany, France, Japan and the USA [6–12]. The state of the art of this technology is much lower than that of the detonation-wave engines. More research needs to be done in the application of this technique for gas-turbine engines. It is quite possible that pure high-temperature air or flameless combustion throughout the operating range of a typical gas-turbine engine is not feasible. However, the beneficial characteristics of HiTAC or flameless combustion could be used to varying degrees if the physics and chemistry behind this concept were better understood.

References

1. Tsuji H, Gupta AK, Hasegawa T, Katsuki, K, Kishimoto K, Morita M (2019) High temperature air combustion—from energy conservation to pollution reduction. CRC Press. ISBN 9780367395643
2. Bussing T, Pappas G (1996) Pulse detonation engine theory and concepts. In: Murthy SNB, Curran ET (eds) Developments in high-speed-vehicle propulsion systems, Progress in astronautics and aeronautics, vol 165, pp 421–472
3. Kailasanath K (2000) Review of propulsion applications of detonation waves. AIAA J 38:1698–1708
4. Kailasanath K (2003) Recent developments in the research on pulse detonation engines. AIAA J 41:145–159
5. Kailasanath K (2011) The rotating-detonation-wave engine concept: a brief status report. AIAA Paper 2011-0580, American Institute of Aeronautics and Astronautics, Reston, VA
6. Wunning JA, Wunning JG (1997) Flameless Oxidation to Reduce Thermal NO Formation. Prog Energy Combust Sci 23:81–94
7. Gupta A (2000) Flame characteristics and challenge with high temperature air combustion. In: Proceedings of the 2000 international joint power generation conference, Miami Beach, Fl
8. Choi GM, Katsuki M (2001) Advanced low NO_x combustion using preheated air. Energy Convers Manag 42:639–652
9. Flamme M (2004) New combustion systems for gas-turbines (NGT). Appl Therm Eng 24:1551–1559
10. Delacroix F (2005) The flameless oxidation mode: an efficient combustion device leading also to very low NO_x emission levels. In: Producing more with less: efficiency in power generation
11. Guillou E, Cornwell M, Gutmark E (2009) Application of "Flameless" Combustion for gas turbine engines. AIAA paper 2009-0225, American Institute of Aeronautics and Astronautics, Reston, VA
12. Johnson RF, Schwer D, Kercher A, Corrigan A, Kailasanath K, Kessler D, Gutmark E (2018) Flow characteristics of a recirculating flameless combustor configuration. In: AIAA paper 2018-0136, American Institute of Aeronautics and Astronautics, Reston, VA

Synopsis of Propulsion Engine Combustion Technology/Product Development and Analysis Substantiated During Last 47 Years

Hukam C. Mongia

1 Introduction

The author considers himself privileged working for 37 years with excellent combustion teams of coworkers at Garrett (now Honeywell Aerospace), Allison (now Rolls-Royce North America), GE Aviation at Cincinnati and Bangalore that made possible pulling together majority of this manuscript. Readers interested in details may want to browse through [1–22]. During this period of 37 years and since retirement in 2009, considerable progress has been made in combustion technology and products when expressed in terms of the following parameters:

- Takeoff pressure ratios have increased from 10 to 47 with potential capability increasing to 60 in GE9X currently underway.
- Combustor overall fuel/air ratio (FAR) has increased from 0.015 to 0.05+ with potentials demonstrated for near-stoichiometric FAR operation with combustion efficiency essentially comparable to the equilibrium values.
- Mean times between shop visits have increased from less than 1000 h to more than 20,000 h.
- Combustor exit hot streak characteristics expressed as pattern factor have decreased from 0.35 to less than 0.2.
- Sea-level idle combustion efficiencies have increased from 90 to 99%, and cruise combustion efficiency from 98 to 99.9%.
- Altitude relight capabilities have increased from 7 to 5 psi with technology potentials demonstrated to 2 psi.
- Combustor volumetric heat release intensity has gone up by a factor of 5.
- Engine exhaust emissions of smoke and gaseous emissions compared to engines certified before 1995 have improved considerably, as discussed in Sect. 3.

H. C. Mongia (✉)
CSTI Associates, LLC, Yardley, PA, USA
e-mail: hmongia43@hotmail.com

A. K. Gupta et al. (eds.), *Advances in IC Engines and Combustion Technology*, Lecture Notes in Mechanical Engineering,
https://doi.org/10.1007/978-981-15-5996-9_3

When a manuscript subject to page limitation is pulled together, the author has to make tough decision on which topics need to be covered. Section 2 gives a very brief description of a hypotheses-based combustion system technology and product development methodology [23] that has been used successfully for 36 combustors. This section gives very brief summary of the six advanced combustors the author is very proud of. An overview of low emission technology is given in Sect. 3 that shows back-to-back comparison between the GEnx and the rich-dome combustion product emissions certified before 1995 at the two original equipment manufacturers (OEMs), CFM and GE Aviation, followed by in Sect. 4 comparing LEAP-X with GEnx. After an introduction to emission technologies in Sect. 5, we give an overview of dynamics in Sect. 6 followed by extensive discussion on rich-dome low-emission combustors (LEC), lean combusting dual-annular combustors (DAC), second-generation lean-dome known popularly as twin-annular partially premixing swirl (TAPS) stabilized concentric flames, the second- and third-generation single venturi lean direct injection SV-LDI-2 and SV-LDI-3 in Sect. 7. A brief summary is provided in Sect. 8.

2 Selected List of Advanced Combustors

All the gas turbine combustors whether technology or products in which Mongia can claim significant technical participation [1–22] have relied on the methodology described schematically in Fig. 1; see [23] for details. It differs from conventional design approach because of two areas of process activities, namely (1) a set of hypotheses and (2) its continuing verification via specialized rig tests, design specific engineering correlations supported by multi-dimensional CFD simulations that are used qualitatively or quantitatively in helping decide the next combustor modification. This approach has worked for all the programs the author has been

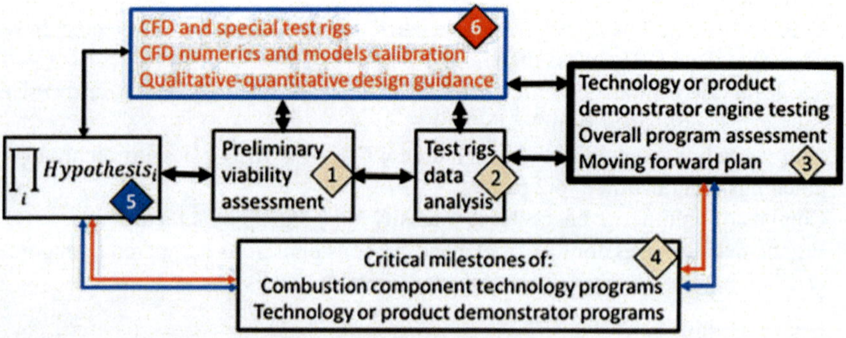

Fig. 1 Hypotheses-based approach comprises conventional combustion technology, or product development process comprised of boxes 1 through 4 complemented by boxes 5 and 6 has worked well since the early 1970s

Table 1 Range of takeoff pressure ratios, characteristic LTO HC, characteristic SAE smoke number, characteristic LTO values of CO and NO_x for all pre-LEC engine models of the two selected OEMs, CFM and GE Aviation compared to GEnx engine models that use a second-generation lean-dome combustion technology

Engine models	P3, PR	HC	Smoke	CO	NO_x
CFM56 Pre-LEC	21.3–31.3	3.5–10.2	3.9–21.6	30.8–105	45.2–71.3
GE Pre-LEC	24.7–35.7	5.1–44.2	4.3–16.1	35.4–108	39.7–68.9
GEnx	35.5–47.5	0.6–1.2	0.75	17.5–28.5	28.9–57.3
Increase or decrease %	33–67	83–97	81–96	43–74	19–27

involved since 1972; we can call it 'good luck' or 'good Karma.' Refer to Table 1 in [24] for a complete list of 36 technologies and product combustors that have used this methodology. This approach has worked for high temperature rise combustors, some approaching near-stoichiometric conditions, regenerative turbine cycle engines, small to large engine combustors, for achieving significant advances in combustion technology including lean blowout, altitude ignition, pattern factor, durability, alternative fuels covering even coal combustion producing only 25 PPM, low emissions for industrial and propulsion engines. Figures 2, 3 and 4 present schematic representation of the selected six combustors, and details of these are given in [14–16, 18].

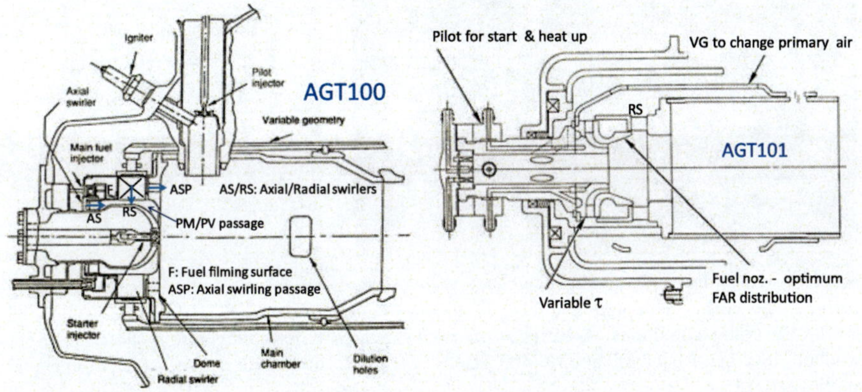

Fig. 2 Two regeneratively cooled automotive gas turbine combustion systems described in [15]; AGT101 idle operating condition combustor inlet temperature T3 and fuel/air ratio are 1941 °F and 0.0032, respectively; the corresponding max power conditions are 1168 °F and 0.012 with the attendant combustor exit temperature T4 of 2500 °F. Technology's emission goals, g/mile, were 0.41, 3.38 and 0.41 for NO_x, CO and HC, respectively

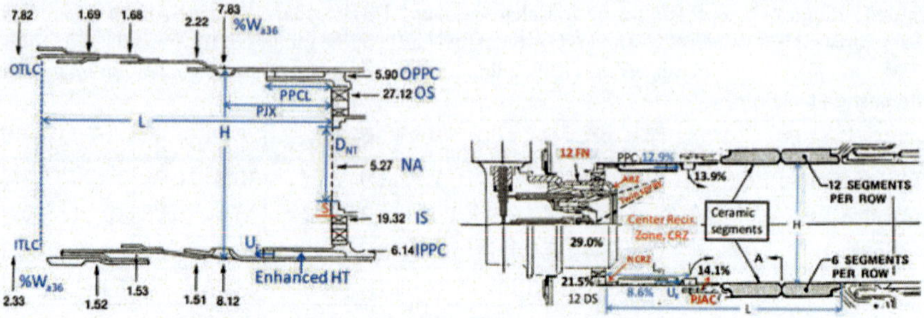

Fig. 3 Schematic cross sections of two of the several high temperature rise combustors described in [14]. Combustor temperature rise, $\Delta T = 2400$ °F and $\Delta T = 3400$ °F

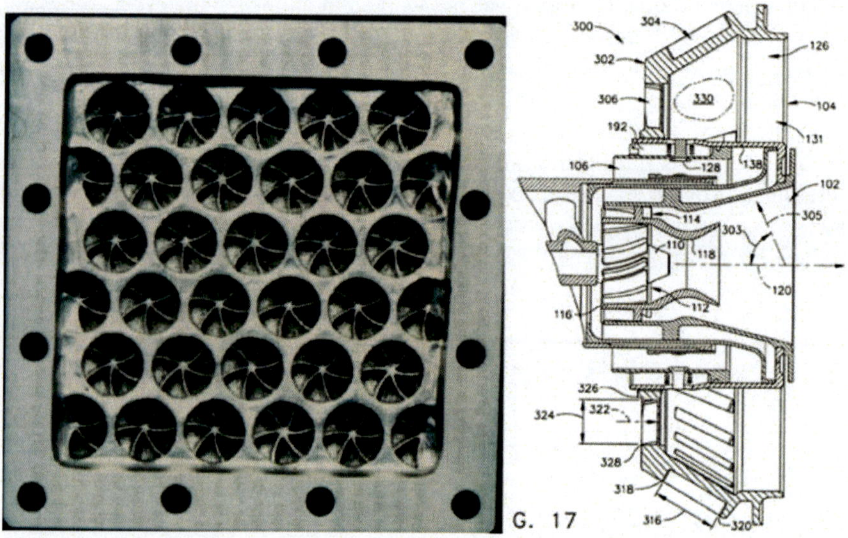

Fig. 4 Two ultra-low-NO_x combustion concepts demonstrated in flame tube test rigs. Accelerating swirl passage cluster with premixed Jet-A fuel shown on left produced NOxEI of 1.0 at simulated supersonic cruise conditions. A partially premixing multi-swirler configuration shown on right produced less than 1.0 NOxEI at P3 = 400 psi, T3 = 1000 °F and fuel/air ratio of 0.027. For details, refer to [16, 18]

3 From GE Pre-LEC to TAPS GEnx

As explained very well on https://www.easa.europa.eu/easa-and-you/environment/icao-aircraft-engine-emissions-databank, commercial propulsion engines' smoke and gaseous emission regulatory standards are written in terms of characteristic landing takeoff values based on engine emission data measured at the specified engine operation at idle, approach, takeoff and climbout. The following regulatory levels for smoke, HC and CO have not changed:

$$\text{SAE Smoke number} = \text{Minimum}\left[50, 83.6/F_{00}^{0.274}\right]$$

Here, F_{00} is takeoff thrust in kN.

$$\text{LTO unburned hydrocarbon (HC)} = 19.6\text{gCH}_4/\text{kN}$$

$$\text{LTO CO} = 118\text{gCO/kN}$$

The LTO NO_x regulatory standard started with a simple correlation, LTO NO_x = $40 + 2 \times \text{PR}$ which in its latest form for engines of a type or model of which the date of manufacture of the first individual production model was after January 1, 2014, has been divided into engines with takeoff pressure ratio divided into three groups, $\text{PR} \leq 30$, $30 < \text{PR}\langle 104.7$ and $\text{PR}\rangle 104.7$ and takeoff thrust F_{00} also into two groups, namely $26.7 < F_{00}\langle 89$ and $F_{00}\rangle 89$. Here, PR is takeoff pressure ratio. The resulting expression relevant to GEnx is LTO $NO_x = -9.88 + 2 \times \text{PR}$, known popularly as CAEP-8 as opposed to the base regulation as CAEP-1. Therefore, compared to CAEP-1 regulatory standards of 112 and 135 at the takeoff pressure ratios of 35.5 and 47.5, respectively, the corresponding CAEP-8 regulatory values are 61 and 85 with the corresponding increased stringency of 45 and 37%. Therefore, the GEnx has 53% margin at PR = 35.5 and 33% margin at PR = 47.5 from the CAEP-8 regulatory requirement.

Figure 5 shows an illustration for summarizing engine emission reductions from the earlier generation of rich-dome combustor emissions certified before 1995, identified as pre-low emission combustors (LECs) or more popularly familiar rich–quench–lean (RQL), to the first successfully introduced second generation of lean-dome combustion products GEnx. We prefer to use LEC over RQL because it implies simultaneous reduction in HC, CO, NO_x and smoke when required; refer to Table 1 for more discussion that follows on this topic. Figure 5 shows data from two OEMs, GE and CFM, selected basically because of the author's familiarity with their products. It is a common practice to plot characteristic values of LTO data because the regulatory standards are applicable to the characteristic values. Since the exhaust smoke number regulations are in terms of takeoff thrust F_{00}, we have added parts (a) and (b); the former shows smoke number as function of takeoff pressure ratio and the latter as a function of F_{00}. Data shown here can be interpreted in many ways. The summary presented in Table 1 leads to very interesting observation.

Calculating range for any variable does not require any clarification; but % increase or decrease needs to be defined as shown below:

$$\text{Range of pressure ratio increase} = \left[\frac{\text{min(GEnx)}}{\text{min(PreLEC)}} - 1, \frac{\text{max(GEnx)}}{\text{max(PreLEC)}} - 1\right]$$

$$\text{Range of emissions decrease} = \left[1 - \frac{\text{min(GEnx)}}{\text{min(PreLEC)}}, 1 - \frac{\text{max(GEnx)}}{\text{max(PreLEC)}}\right]_{\text{HC,Smoke,CO,NOx}}$$

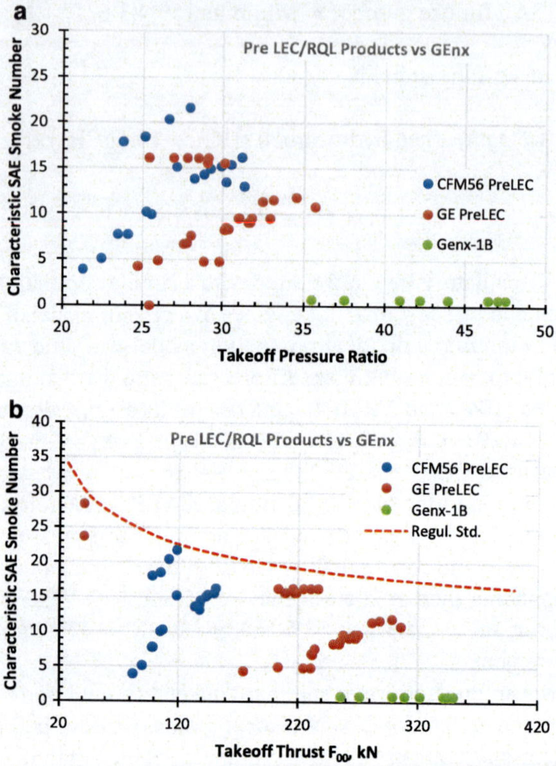

Fig. 5 Landing takeoff emission characteristics of the GEnx lean-dome TAPS combustion system compared to GE and the CFM engine model emissions tested before 1995. **a** Characteristic SAE smoke number during landing takeoff (LTO) cycle comprised of taxi idle, approach, takeoff and climbout. GEnx exhaust smoke number is less than 1.0 compared to the regulatory requirement between 16.8 and 19.3. **b** Characteristic SAE smoke number versus takeoff thrust F_{00}, kN. Some pre-LECs too close to the smoke regulatory standard SAE Smoke number $=$ Minimum$\left[50, 83.6/F_{00}^{0.274}\right]$ shown by dotted red curve. **c** Characteristic LTO unburned hydrocarbon (HC) emission, g of CH_4/kN. GEnx exhaust HC emission is less than 1.2 compared to the regulatory requirement of 19.6. **d** Characteristic LTO CO emission, g of CO/kN. GEnx exhaust CO emission is between 17.5 and 28.5 g/kN compared to the regulatory requirement of 118. **e** Characteristic LTO NO_x emissions, g NO_2/kN versus takeoff pressure ratio PR. Baseline CAEP regulatory NO_x standard is given by CAEP $= 40 + 2 \times$ PR. GEnx has 53% margin at PR $= 35.5$ and 33% margin at PR $= 47.5$ from the CAEP-8 regulatory requirement, LTO $NO_x = -9.88 + 2 \times$ PR. If required, future GEnx combustion products can be further reduced by approximately 50% as shown by dotted green line. GEnx LTO is represented by LTO $NO_x = 2.483 \times \text{PR} - 59.726$. **f** Bypass ratio versus takeoff pressure ratio of pre-LEC/RQL engines compared with GEnx, LEAP-1A and LEAP-1B engines

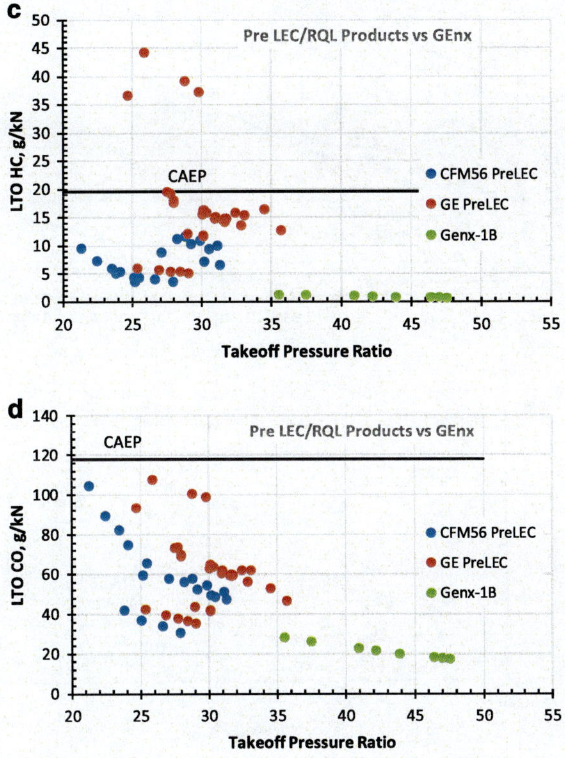

Fig. 5 (continued)

The readers have freedom to interpret the results in different ways. But if we go with the definition used here, we get very interesting results. The GEnx engine takeoff pressure ratios have increased by 33– 67% compared to pre-LEC engines; the corresponding reductions in HC, smoke, CO and NO_x are 83–97%, 81–96%, 43–74% and 19–27%. Getting the lowest level of NO_x reduction compared to other pollutants (HC, smoke and CO) is not surprising because increasing pressure ratio (combined with bypass ratio; see Fig. 5f and reduced parasitic losses) results in reduced fuel consumption but attendant with higher temperature regions and NO_x production rates.

Since emission regulatory standards are written in terms of landing takeoff emissions as they should be as obvious from the definitions given below, in Fig. 5 parts (a) through (e) do not directly relate to fundamentals of pollutant formation and the guidance technologists and designers need in order to move forward.

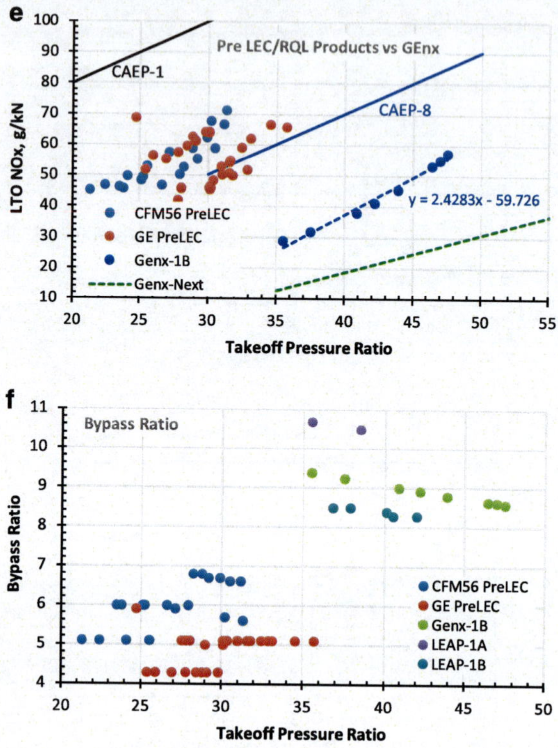

Fig. 5 (continued)

LTO emissions of unburned hydrocarbons, CO and NO_x, are given by the following expression:

$$\text{LTO Emissions}_{\text{HC, CO, NO}_x} = \frac{1}{F_{00}}\left[26 \times \dot{m}_{\text{idle}}\text{EI}_{\text{Idle}} + 4 \times \dot{m}_{\text{approach}}\text{EI}_{\text{approach}} + 2.2 \times \dot{m}_{\text{climb}}\text{EI}_{\text{climb}} + 0.7 \times \dot{m}_{\text{takeoff}}\text{EI}_{\text{takeoff}}\right]_{\text{HC,CO,NO}_x}$$

Here, $\dot{m}$ represents fuel flow rate, kg/min for each of the LTO cycles comprised of 26 min at idle, 4 min at approach, 2.2 min at climbout and 0.7 min at takeoff power setting. Characteristic LTO value is popularly represented as Dp/Foo for HC, CO and NO_x. It is simply LTO values divided by fraction less than 1.0 that for each of the four pollutants is a function of the number of engines tested. The author prefers to use the nomenclature LTO in his figures instead of Dp/Foo.

4 LEAP-X Compared with GEnx

There are many topics that one can choose in order to show that empirical gas turbine combustion technology and design experience backed by semi-analytical models, simple to complex CFD-based design tools, are useful tools, not dissimilar to machines in modern factories; however, it is ingenuity of the user that makes the difference. If combustion technology development process is controlled by continually evolving database backed by commonsense correlations and simple to complex CFD tools, it is destined to achieve the design objectives that are consistent with the '2nd law of thermodynamics.' The data shown in Fig. 6 motivated us to select the topic on successful demonstration of hypotheses-based approach for design and development of low emission gas turbine combustion technology and products.

If we assume that the GEnx combustion team approach for scaling from existing database was adopted by the LEAP-X team, then the data presented in Fig. 6a raises one's eyebrows. Why did we get higher LTO NO_x with LEAP-X? Perhaps one can try to simply attribute it to the differences in the engine size. Or simply similar to a common run combustion engineer, one could go along with the marketing folks who can still claim 37% reduction in LTO NO_x compared to current product, namely the CFM56-5B Tech Insertion engine that uses a rich-dome LEC/RQL combustor as shown in Fig. 6b. But LTO NO_x represents contribution made from the engine fuel burn and NO_x emission index; a combustion person can claim credit for emission reduction and not for reduction in fuel burn.

The results presented in Fig. 6, parts a–c, can be summarized in Table 2 similar to Table 1.

From Table 2, the bottom-line conclusion is that even though the marketing person can claim 37% reduction in Fig. 6b, reality check says that LEAP-1A can claim 25% NO_x reduction in the CFM56 Tech Insertion engine when compared to minimum LTO values of the both engines. But when compared to their maximum values, LEAP-1A gives 17% increase in NO_x emissions. On the other hand, the LEAP-1B results in 30–36% increase in LTO NO_x. This is somewhat discouraging because all the three engines, the CFM56-5BTI, LEAP-1A and LEAP-1B, fall within a very narrow band of rated takeoff thrust levels. Additionally, the LEAP-X design and development effort had at its disposal a broad knowledge base on TAPS in regard to empirical know-how, CFD simulations and fundamental investigations which we will discuss later in Sects. 6 and 7. It should be pointed out that normally OEMs feel uncomfortable when LTO NO_x margin from the regulatory standard, here CAEP-8, falls below 20%. LEAP-1A engine models with different F_{00} ratings have NO_x margin that varies between 11 and 52%, average being 33%; the corresponding margin values for LEAP-1B fall between 6 and 34% with average being 21%. Therefore, for now LEAP-X is okay in regard to meeting LTO NO_x requirement.

We discovered in 1975 [25] that proper metric for evaluating progress on engines' gaseous emissions should be done by plotting takeoff NOxEI as a function of takeoff pressure ratio, idle COEI as a function of takeoff NOxEI and idle COEI as a function of idle HCEI. This progress evaluation approach has worked since 1975 and provides

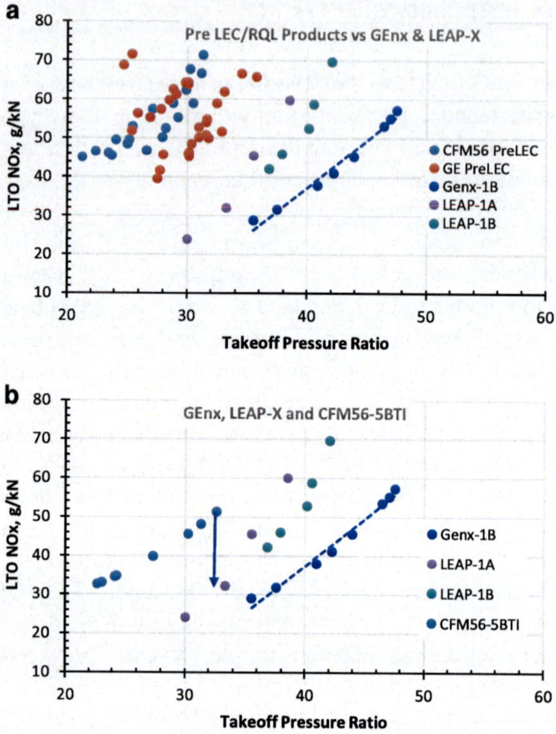

Fig. 6 Comparison of NO_x emissions of pre-LEC rich dome, LEC/RQL dome of the CFM56-5B Tech Insertion and TAPS-based GEnx, LEAP-1A and LEAP-1B. **a** LEAP-1A and LEAP-1B produce higher values of LTO NO_x compared to the GEnx-1B. **b** Marketing folks still can sell LEAP-X because it has 37% lower LTO NO_x compared to the CFM56-5B Tech Insertion rich-dome combustor at the same takeoff pressure ratio as shown by arrow. But is this acceptable? No, see the following figure. **c** LTO NO_x versus takeoff thrust. All three engines (CFM56-5CTI, LEAP-1A and LEAP-1B) fall within a very narrow band of F_{00}, 96-142 kN; the latter two have higher LTO NO_x but with significant reduction in fuel burn. LTO NO_x margin from CAEP-8 is between 11 and 52% for LEAP-1A, which reduces to 6 and 34% for LEAP-1B. Maybe, the CFM is working on further reducing NO_x emissions of LEAP-X. **d** Takeoff NOxEI versus takeoff pressure ratio of the CFM56-5B Tech Insertion, LEAP-1A, LEAP-1B and GEnx. Combustion engineer should not like trendlines of LEAP-X compared to GEnx

the insight required for advancing low emission technology and products. We will pick up this topic later in Sect. 7; here, we will summarize the following from Fig. 6d.

Takeoff NOxEI data summarized in Fig. 6d shows the following for LEAP-1A engine model:

- 32% lower than the rich-dome at 30 PR, but NOxEI comparable at 32.6 PR.
- Comparable with lean DAC GE90 at 35.2 PR; at PR > 35.2, it is higher than GE90.
- At 42 PR, takeoff NOxEI of LEAP-1B is more than double of GEnx.

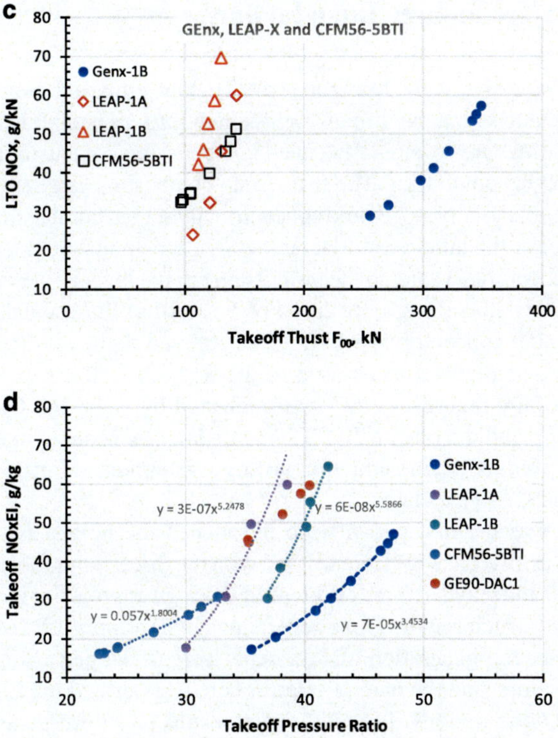

Fig. 6 (continued)

Table 2 Range of increase in LTO NO_x of LEAP-1A and LEAP-1B compared to their predecessor the CFM56-5B

	CFM TI	LEAP-1A	LEAP-1B
PR	22.7–32.6	30–38.5	36.8–42
LTO NOx	32.5–51.2	24.1–59.9	42.2–69.7
F_{00}, kN	96–142	107–143	111–130
LTO NOx, % increase		17 to −25	30–36

GE Aviation was very proud of GEnx TAPS combustion system and created an interesting YouTube https://www.youtube.com/watch?v=Y8QqX6q6RSk&list=PLB40438A15A6B2501&index=37. YouTube https://www.youtube.com/watch?v=XEiWwRyq_9E&feature=youtu.be&list=PLB40438A15A6B2501 shows that GE9X uses a SAC TAPS combustion system unlike AST TAPS which used DAC TAPS as described in [11].

5 Emission Technology Introduction

As discussed extensively by Mongia [16–22], low emission propulsion engine combustion products can be broadly classified into rich and lean combustion concepts; the former are colloquially called low emission combustor (LEC) or rich–quench–lean (RQL) combustor. The lean-dome combustors can be broadly divided into axially staged, radially staged and concentrically staged combustion concepts. So far, only two out of the hundreds of laboratory-scale-demonstrated lean combustion technology concepts have entered the propulsion engine revenue service, namely the dual-annular combustors (DACs) for the CFM-56 and the GE90, and the twin-annular partially premixing concentric swirl (TAPS) stabilized flames for the LEAP-X and the GEnx. TAPS is planned to be used in the upgraded 60 pressure ratio GE9X. The first technology report on DAC was published by Bahr and Gleason in 1975 described in [26], and the one on TAPS 29 years later by Mongia in 2004; see [11]. The first DAC product engine emission testing was conducted in 1995 and that of the TAPS in 2009, 14 years later.

Bob Tacina was the first person who introduced the nomenclature lean direct injection (LDI) in 1990; see [27]. He worked with the three major fuel nozzle vendors in the early 2000 and pursued three different LDI design approaches including the one discussed in [28] which is called first-generation single venturi LDI. Approximately ten years later, second-generation LDI concepts were developed again with the same three nozzle vendors, and the one for interest here is described in [29] that included three different configurations. The most recent results on a third-generation SV-LDI are reported by Tacina et al. in [30]. Because of page limitations, two topics are covered here. The first very relevant to low emission combustion technology and product combustion dynamics is covered in Sect. 6 followed by a rather extensive assessment of low emission combustion products of the two OEMs, CFM and GE Aviation engines in Sect. 7.

6 Combustion Dynamics

Combustion instability and dynamics topics were very popular when the author was a graduate student in 1967–1971, and are pursued vigorously even today. Many of the combustion dynamics titans, Professors Fred Culick, Vigor Yang and Ben Zinn, are my close friends; their works are simply amazing. Several dynamics publications are presented in every important combustion conference including the most recent ASME Turbo Expo 2019 Combustion sessions which had 37 papers dealing with different aspects of dynamics that included GT2019 papers numbered 90321, 90438, 90447, 90546, 90600, 90793, 90814, 90834, 90852, 90870, 90878, 90924, 91100, 91152, 91182, 91246, 91473, 91547, 91656, 91835 and many of my friends whose papers' numbers are included in parentheses, Sebastien Candel (90738), S.R. Chakravarthy (91252, 91981, 92022), Matthias Ihme (92052), Tom Lieuwen

(90954, 91209, 92096), Wolfgang Polifke (90140, 90732, 91604, 91784), Domenic Santavicca (91577) and Thomas Sattelmayer (90238, 90239, 90240, 90241). These papers and thousands before have taught us 'all the fundamentals' that we know about gas turbine combustion dynamics.

Unfortunately, dynamics is a boundary value problem, and we have yet to learn to duplicate in the laboratory the boundary conditions combustion system encounters in real-life gas turbine engine system of systems. Therefore, dynamics can appear or may not during the last phase of combustion system demonstration on an engine. I was worried about dynamics in a CFM56 demonstrator engine tested in January 2001, wherein we started with a new lean-dome technology combustor, twin-annular partially premixing swirlers (TAPSs) producing concentric flames. Everyone was happy when the engine ran smoothly without any dynamics; I thanked God of Combustion. The same injector/mixer produced dynamics, and attendant unsteady flashback and liftoff of the flame base caused perhaps because of its cylindrical test section, the rig installation and the test conditions; see [31, 32]. However, the mixer did reproduce the flow characteristics we were hoping to expect as illustrated in Fig. 7. Another TAPS mixer tested in a rectangular test section at higher operating conditions did not produce any dynamics, albeit its flow characteristics (Fig. 8) were very similar to those shown in Fig. 7.

The CFM56 TAPS single-annular combustor did not produce any dynamics in full-scale annular test rig and its demonstrator engine that went through extensive testing as described in Ref. [11]. Similarly, the AST TAPS dual-annular combustor tested in a full-scale annular test rig did not produce any dynamics either. The GEnx TAPS combustor did not create any noticeable dynamics in the test rigs and the prototype engine until the later part of the engine durability testing phase. We will discuss emission characteristics of TAPS engine later. Here, we want to stay focused on the challenges of encountering dynamics issues that generally crop up late in the engine development phases as described in [34, 35]. All major OEMs have developed know-hows to deal with it; outline of our approach is given in [34–36].

Dynamic studies for the second-generation LDI as reported in [37] did not raise any concerns other than for an overly fuel-rich operation at simulated idle which may not be relevant in an LDI-based engine product. A parametric dynamic study was investigated (see [38]) on a dynamically robust single-element injector/mixer configuration of [28] in order to document effect of boundary conditions and combustor length on amplitude and frequency of combustion dynamics for the nominal test conditions listed in Table 3; schematic of the experimental rig setup is shown in Fig. 9 with more details presented in Fig. 10.

As intended, the LDI-1 robust pilot configuration that is not expected to produce unacceptable dynamics in real gas turbine application gave different magnitudes of pressure oscillations and frequencies as a function of equivalence ratios as summarized in Fig. 11. Amplitude of these oscillations ranged between 200 kPa and less than 5 kPa (20–0.5% P3, here P3, combustor operating pressure is equal to 1 MPa) along with frequencies varying between approximately 100 and 7500 Hz. An injector that would normally be selected for low-power operation by combustion engineers is now fraught with dynamics with attendant hesitancy in using it. Perhaps, one would

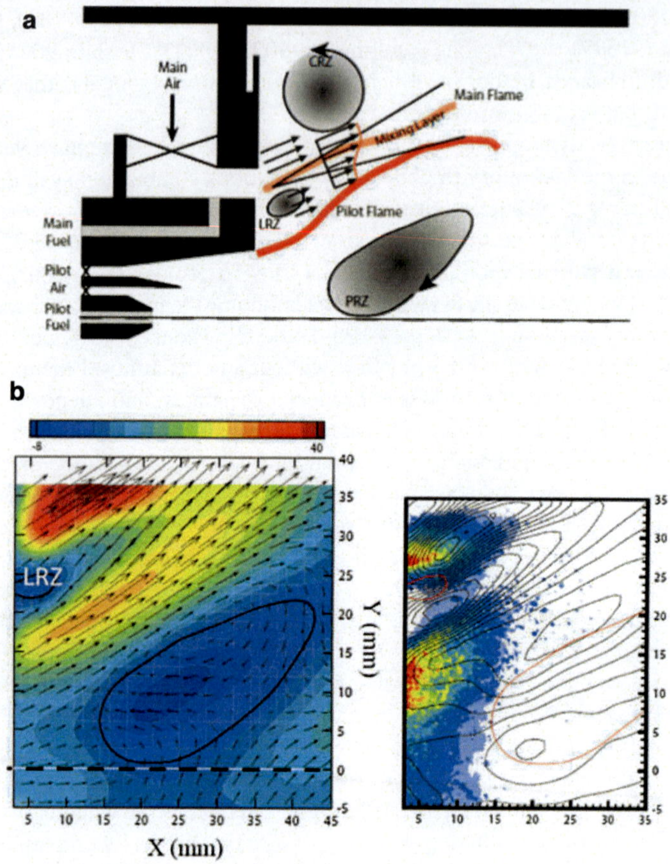

Fig. 7 Schematic representation of a TAPS injector mixer and flow characteristics in a laboratory test rig environment; modified figures from [31]. **a** Schematic representation of TAPS mixer and flow characteristics; here, RZ means recirculation zone caused by pilot P, pilot lip L and sudden expansion corner C formed between the TAPS mixer and combustor dome and liner wall, identified, respectively, as PRZ, LRZ and CRZ. **b** Mean axial velocity vectors (left) and PLIF image of CH_2O on right

not like to sustain more than 1% amplitude of pressure oscillations for a durable combustion system.

Clearly, dynamic characteristics of the selected LDI injector are strongly influenced by boundary conditions and equivalence ratio; the former is reproduced only in an actual engine environment. In conclusion, the LDI configuration selected for the dynamic test reported in [38] with enormous levels of dynamics is not expected to produce any dynamics concerns in real gas turbine environment. Similar studies on a cylindrical can combustor with continuously varying length afforded by a moving piston, colloquially called tunable combustor acoustic (TCA) at high T3 and P3 along with a range of overall fuel/air ratios, gave us a trove of engineering data similar to

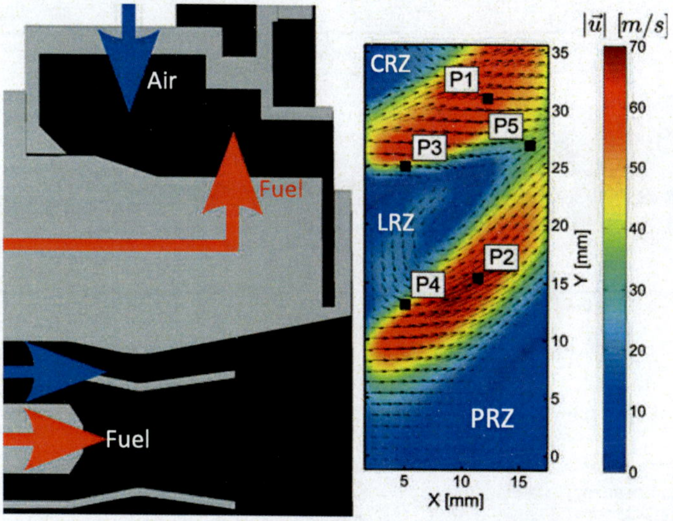

Fig. 8 TAPS flow characteristics measured in a rectangular test section at higher operating conditions, modified figure from [33]. Schematic representation of TAPS air mixer and fuel injection shown on left side of figure comprises of concentric multi-swirling air streams identified here as blue arrows along with LDI pilot and partially premixing discrete fuel jets in swirling cross-flow. On the right side of the figure are shown mean velocity vectors, P1, P2, – P5 and spatial locations where time series investigations are reported in [33]

Table 3 Summary of design envelope and nominal operation parameters for the single venturi swirler LDI investigated in [38]

Combustor nominal inlet pressure	1 MPa
Fuel	Jet-A and FT-SPK
Inlet air temperature	650–800 K
Equivalence ratio	0.37–0.7

Inlet boundary condition: constant mass inflow from a choked orifice
Exit boundary condition: choked nozzle
Diameter of combustor: 50.8 mm
Diameter of air plenum: 25.4 mm
See Fig. 10 for lengths of air plenum (A.P.) and combustion chamber C.C

Fig. 11. The so-called best injector/mixer selected from the TCA rig did not necessarily give acceptable dynamic characteristics in full-scale annular rig (FAR) tests. Similarly, the best dynamically accepted combustor configuration developed on FAR did not necessarily meet the dynamics limits on the prototype engines.

The author who believes in the gas turbine combustion technology and product development methodology shown schematically in Fig. 1 that has worked well since the middle 1970s on 36 gas turbine combustors for meeting all the critical design

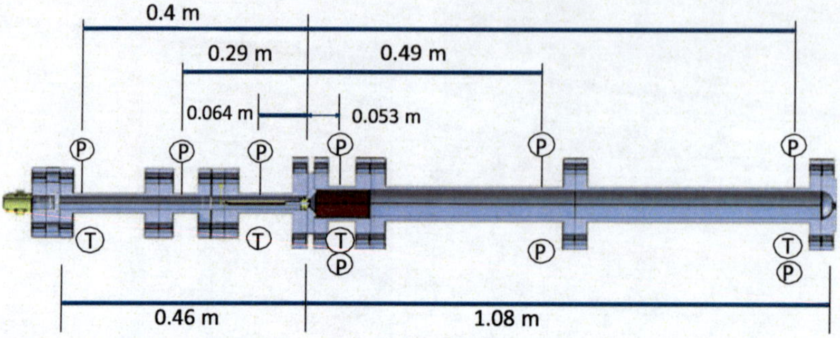

P: Pressure transducers 100 kHz; mounted in a recessed cavity w/ Helmholtz resonant frequency of 14 kHz
T: Type K T/C installed in 3/8th air plenum and ½ wave combustion chamber.

Fig. 9 Experimental schematic for dynamic studies on LDI-1 configuration investigated in [38]. P: Pressure transducers 100 kHz; mounted in a recessed cavity w/Helmholtz resonant frequency of 14 kHz T: type K T/C installed in 3/8th air plenum and ½ wave combustion chamber

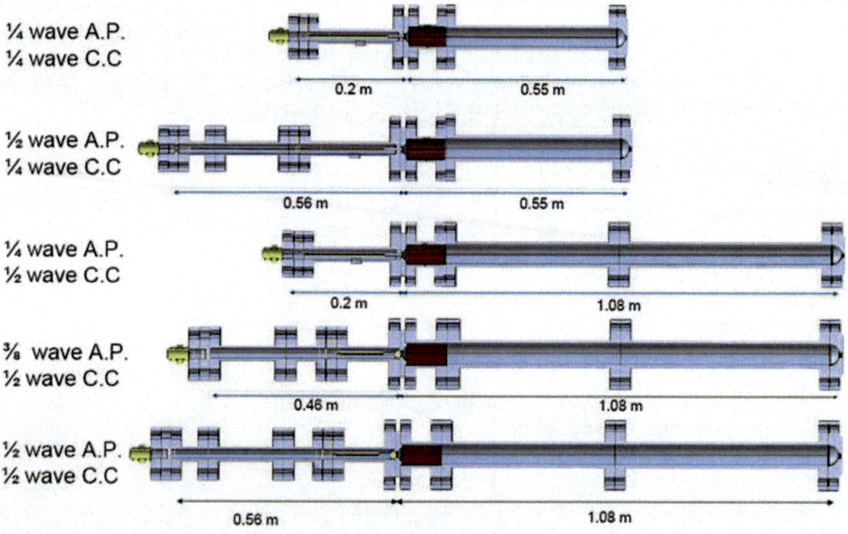

Fig. 10 Five different configurations were tested including combinations of ¼, 3/8 and ½ wave air plenum (A.P.) tube lengths, ¼ and ½ wave combustion chamber (C.C) lengths

requirements hesitates in extending it for solving combustion dynamics-related challenges. He makes fun of his approach by calling it 'my gut feeling' or 'my way or the highway'; but when it comes to dynamics, he follows the philosophy of 'design by committee' in order to avoid getting blamed for the numerous failures that are

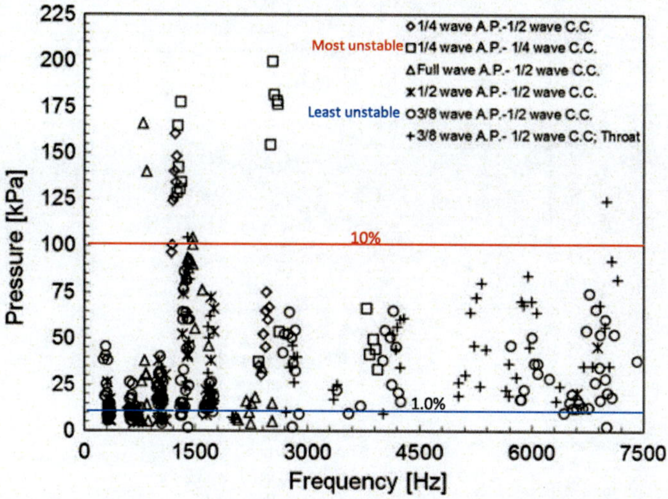

Fig. 11 Summary of pressure fluctuation amplitudes as a function of frequency for all test configurations and nominal test conditions listed in Table 3

bound to occur before we hit an acceptable solution for keeping combustion dynamics within limits required to achieve durability objectives.

7 Evolution of Low Emission Combustion Products

Development experience on low emission propulsion engine combustion technology programs since the early 1970s has led to a protocol for conducting comparative assessment favored by the author. Part of this process includes Figs. 12, 13 and 14 shown here for the three selected technologies specifically for the CFM and GE propulsion engines. The source of data presented in these figures is the ICAO Engine Emissions Database published on May 27, 2019; https://www.easa.europa.eu/easa-and-you/environment/icao-aircraft-engine-emissions-databank. LDI-3 data is from [30].

From data summarized in Fig. 12a, the GE rich-dome LEC/RQL takeoff NOxEI characteristics represented by NOxEI $= 0.052 \times \text{PR}^{1.82}$ are comparable with products from other original equipment manufacturers (OEMs). Similarly, from Fig. 12b the relationship between idle CO emissions versus takeoff NOxEI, Idle COEI $=$ $427.54/(\text{Takeoff NOxEI})^{0.843}$, can be regarded as typical characteristic value for LEC/RQL product combustors from different OEMs. Unlike Fig. 12a where all well-developed LECs fall very tightly along the trendline given by NOXEI $=$ $0.052 \times \text{PR}^{1.82}$, there is a wide scatter in the idle COEI versus takeoff NOxEI which

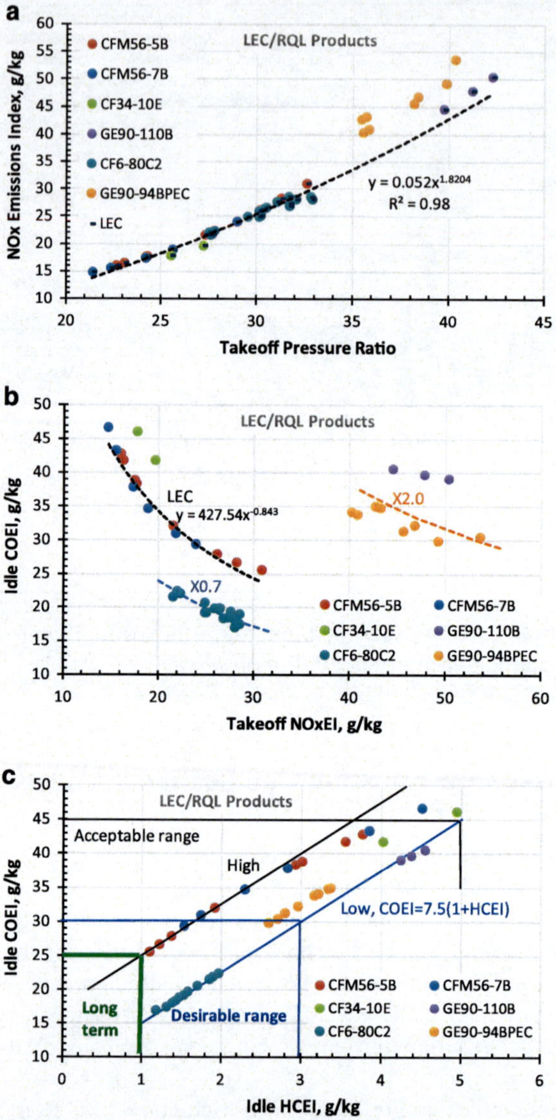

Fig. 12 Gaseous emission characteristics of low emission combustors (LECs) or popularly called rich–quench–lean (RQL) of the CFM56-5B, CFM56-7B, CF34-10E, GE90-110/115B, CF6-80C2 and GE90-76/94B PEC engine models, the latter initially emissions certified in 1997 has not been updated to the LEC capability level. Therefore, it is not included as part of GE and CFM family of LEC/RQL combustors. **a** Takeoff NO_x emission index (NOxEI) versus ICAO takeoff pressure ratio. **b** Idle CO emission index (COEI) versus takeoff NOxEI. **c** Idle COEI versus unburned hydrocarbon emission index (HCEI)

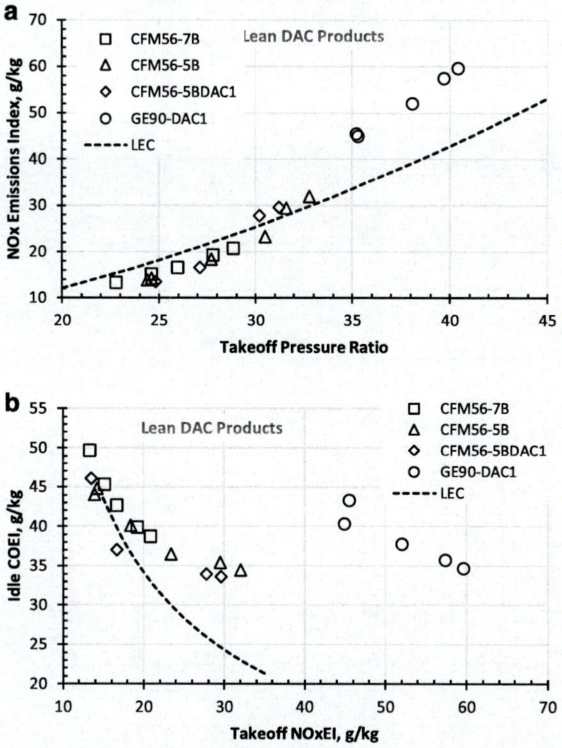

Fig. 13 Gaseous emission characteristics of the lean-dome dual-annular combustors (DACs) of the CFM56-7B, CFM56-5B, DAC1 generation for the CFM56-5B and GE90 engine models compared with the rich-dome LEC product capabilities. Part d has been added to illustrate that DAC products performed better than the first DAC demonstrator tested in 1978. **a** Takeoff NO_x emission index (NOxEI) versus ICAO takeoff pressure ratio. **b** Idle CO emission index (COEI) versus takeoff NOxEI. **c** Idle COEI versus unburned hydrocarbon emission index (HCEI). **d** Dual-annular combustion (DAC) product emission capabilities have improved considerably from the first experimental clean combustor program (ECCP) demonstrator engine tested in 1978 at different thrust level settings; see [39]. High smoke at idle and approach, high HCEI at approach and higher power setting

can be attributed to many reasons. The most common causes are poor fuel spray atomization resulting in high concentration of unburned hydrocarbons and resulting in CO quenching or quenching of CO from the cooling film. The latter could be due to tradeoff between structural durability and idle CO emissions, especially for higher pressure ratio engines. The data sets for the CF6-80C2 and the GE90-94BPEC can be represented well by Idle COEI $= 427.54 \times F/(\text{Takeoff NOxEI})^{0.843}$; here, $F =$ 0.7 and 2.0, respectively, give good agreement with the data from the CF6-80C2 and GE90-94B PEC.

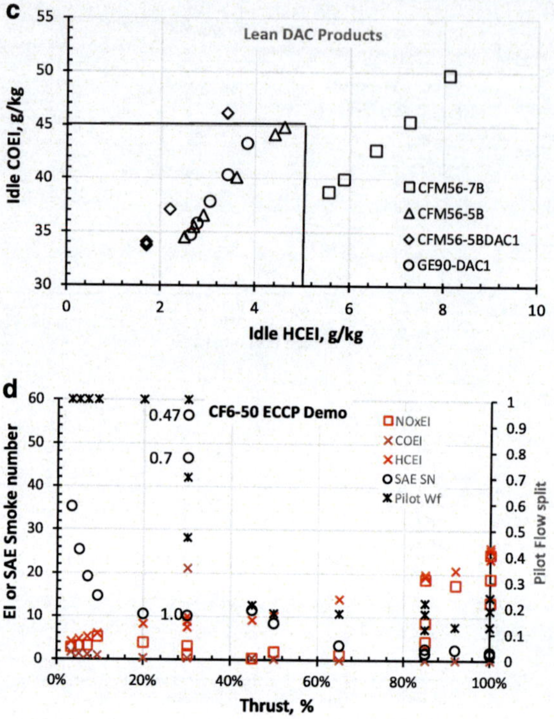

Fig. 13 (continued)

From Fig. 12c, it is reasonable to assume a linear relationship between idle COEI and idle HCEI, namely Idle COEI $= a \times (b + \text{Idle HCEI})$. For the selected six rich-dome GE engine combustors, the low emission levels of idle COEI and HCEI are represented by Idle COEI $= 7.5 \times (1 + \text{HCEI})$. On the other hand, the highest idle COEI combustor can be represented by another parallel straight line drawn in black color. In summary, idle COEI and HCEI characteristics of LEC/RQL combustion products can be divided into three groups of combustors, namely the ones having desirable range of 30 COEI and 3 HCEI; acceptable range of not exceeding 45 COEI and 5 HCEI; and finally, to have a long-term goal of not exceeding 25 COEI and 1.0 HCEI. The latter was set as the goal for the TAPS combustion products.

Looking back at the low emission technology development activities of the past 50 years, one can see a common mistake that almost every novice technologist can be accused of making: Lowering high-power NO_x is only ~5% of the effort, and everything else takes 95% of the effort for implementation of low NO_x technology into product. Other frequent occurrence is neglecting the probable progress the state-of-the-art technology might make in 10–15 years, the time it normally takes to transition new combustion technology into product. Let us illustrate this point by comparing LEC takeoff NO_x technology curve, namely NOxEI $= 0.052 \times PR^{1.82}$

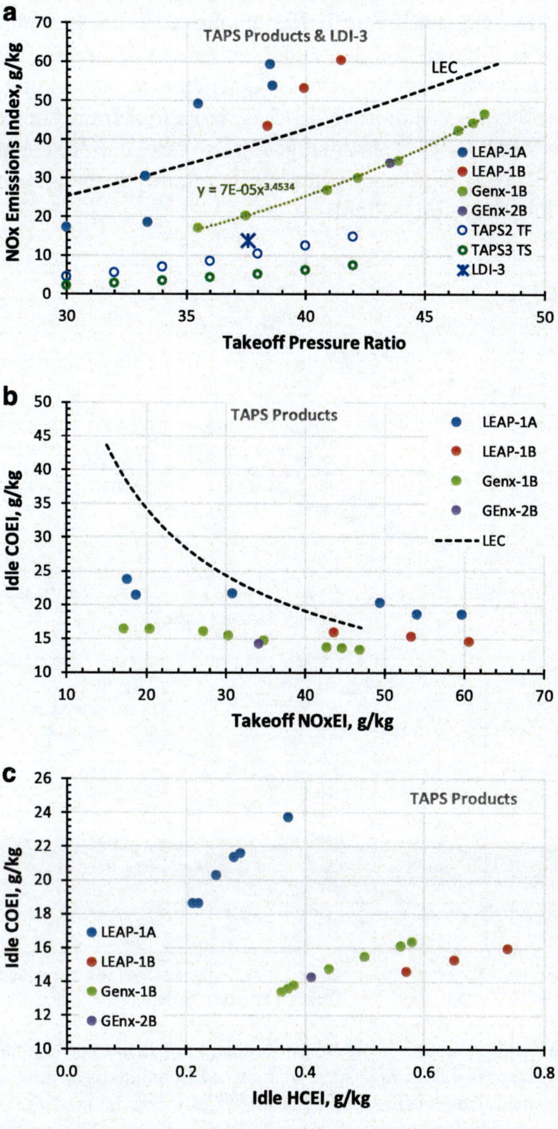

Fig. 14 Gaseous emission characteristics of the twin-annular partially premixing swirl (TAPS) stabilized concentric flame combustors of the LEAP-1A, LEAP-1B, GEnx-1B and GEnx-2B engine models. TAPS is expected to be used in GE9X engine model and thereby extending its takeoff pressure ratio up to 60. **a** Takeoff NO_x emission index (NOxEI) versus ICAO takeoff pressure ratio. **b** Idle CO emission index (COEI) versus takeoff NOxEI. **c** Idle COEI versus unburned hydrocarbon emission index (HCEI)

with GE's rich-dome engines before 1995 summarized as a scatter plot in Fig. 15a for the CFM56 class of engines and the remaining being GE engines. LEC technology brought in 20% reduction from the pre-LEC. Therefore, the minimum expectation for lean-dome technology should be 40–50% reduction from the LEC technology curve $NOxEI = 0.052 \times PR^{1.82}$. From Fig. 15b and c, we can see the rationale behind making the CAEP NO_x regulatory rules but it certainly does not provide guidance for developing low emission technology.

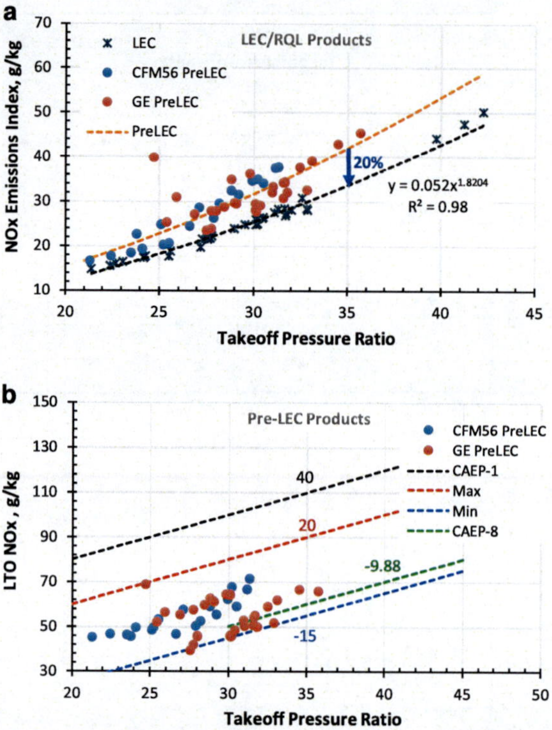

Fig. 15 Evolutionary continuous takeoff NOxEI reduction of rich-dome combustors led to only ~20% reduction from the pre-LEC combustion products. **a** LEC combustion product developed falls narrowly along the trend line $NOXEI = 0.052 \times PR^{1.82}$, whereas the pre-LEC products showing a large scatter band are indicative of different degrees of rich-dome technology development level. Approximately 20% reduction in takeoff NO_x emission index. **b** LTO NO_x is the right criteria for ranking different engines, and there was 'gentlemen's agreement' among the combustion community not to use low emissions as a marketing tool, and NO compromises with safety. Pre-LEC engine LTO data with significant scatter showed a consistent slope of 2.0, and the data fell within the following two boundaries, max and min, Max LTO $NO_x = 20 + 2 \times PR$ and Min LTO $NO_x = -15 + 2 \times PR$ compared to CAEP − 1LTO $NO_x = 20 + 2 \times PR$ and CAEP − 8LTO $NO_x = -9.88 + 2 \times PR$. All four lines are shown in this figure. **c** LEC technology for LTO data for $PR \leq 33$ scatters around best fit data of the CFM56-7B Tech Insertion given by LTO $NO_x = 1.68 \times PR - 3.41$ compared to that of the GE90-110/115B given by LTO $NO_x = 2.66 \times PR - 33.81$

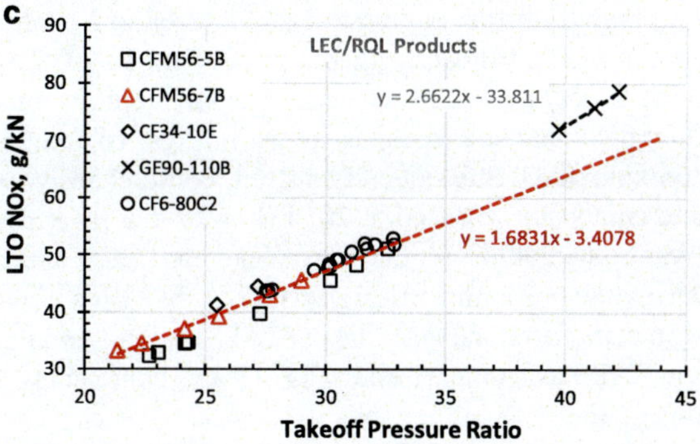

Fig. 15 (continued)

Consequently, as summarized in Fig. 13a, DAC technology combustors for the CFM56 class of engines did not achieve any NO_x reduction compared to the LEC rich-dome products. On the other hand, for the high-pressure ratio GE90, DAC gave significantly higher NO_x than LEC, namely 36% increase for a 40.4 pressure ratio engine. DAC, a significantly more complex, heavier and expensive system with significantly higher idle and approach CO and HC emission levels (see Figs. 13b, c and later Figs. 18 and 21), loss of durability, higher pattern factor and radial profile challenges with attendant impact on mission fuel burn, simply lost its appeal compared to continuously evolving rich-dome SACs.

However, the DAC technology development and transition into product experience gave a trove of lessons learned leading to a more well-balanced TAPS technology development effort as summarized in [11], a process extended to its transition into the GEnx product given in [13] that resulted in the desired low emission product as summarized in Fig. 14. GEnx achieved significant reduction in takeoff NO_x compared to LEC, namely 50% reduction for 36 pressure ratio (PR) engine, but only 20% reduction at 48 PR. The latter can be addressed by applying TAPS2 and TAPS3 technology with resulting potential NO_x reduction of 60% and 80%, respectively. GEnx also achieved idle COEI levels lower than or comparable with the LEC technology. As summarized in Fig. 14c, the GEnx achieved the long-term GE LEC goal of COEI $\leq$ 25 and HCEI $\leq$ 1.0.

GEnx TAPS Combustor Design and Development Process

Let us digress here how the design of the GEnx combustor evolved and how further gains in TAPS technology and fundamentals were expected to provide guidance during the product development of LEAP-X limited to 42 takeoff pressure ratio and ongoing GE9X TAPS combustor operating at pressure ratio of 60.

Limitations of the rich-dome anchored design methodology [8] for lean domes as pointed out in [40, 41] kicked off intensive development of a new set of CFD models in 2001. These models, an overview of these models given in [42], were not ready for application during the conceptual, preliminary and detailed design phases of the GEnx; e.g., the GEnx first engine to test occurred in March 2006. The first comprehensive set of model validation shared in 2007 for two TAPS design configurations in pilot mode only [43] showed clearly that apparently similar designs have very different flow field characteristics, as shown in Figs. 16 and 17 with potential serious impact on emissions. Subsequently shared, comprehensive analysis results on one rich-dome combustor released in 2014 showed that its emission prediction capability [44] was comparable with that of the anchored CCD approach [8].

Therefore, during the preliminary design phase of the GEnx combustion system we had the CFM56 demonstrator engine data of TAPS and full-annular rig data of the AST dual-annular TAPS as summarized in Fig. 18 and nonreacting comprehensive analysis capability [42].

Let us assume that we knew TAPS combustor product specifications for the four engines, GEnx-1B, LEAP-1A, LEAP-1B and GE9X when the preliminary design effort for the GEnx-1B started, as summarized in Table 4.

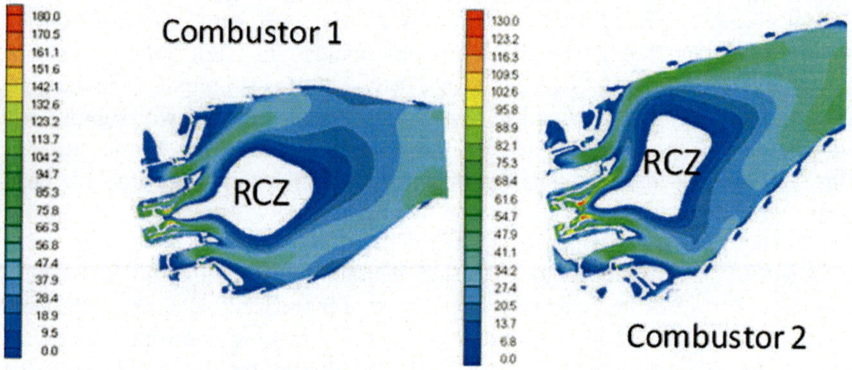

Fig. 16 Reacting flow mean axial velocity contours (m/s) predicted by comprehensive combustion system analytical modeling approach from [43]

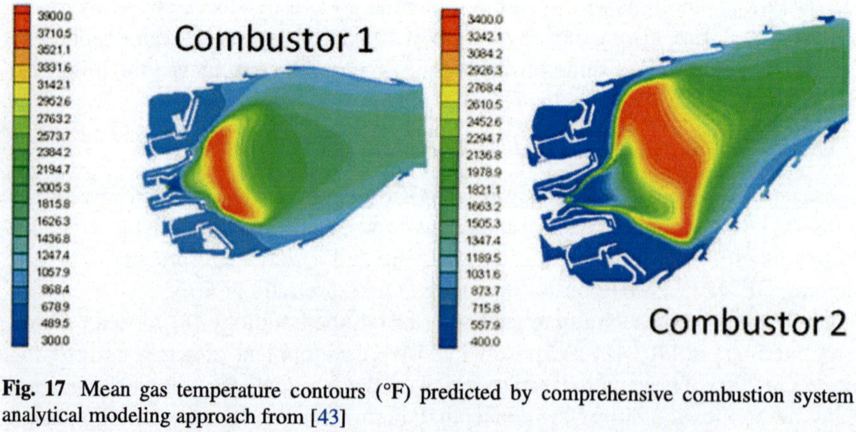

Fig. 17 Mean gas temperature contours (°F) predicted by comprehensive combustion system analytical modeling approach from [43]

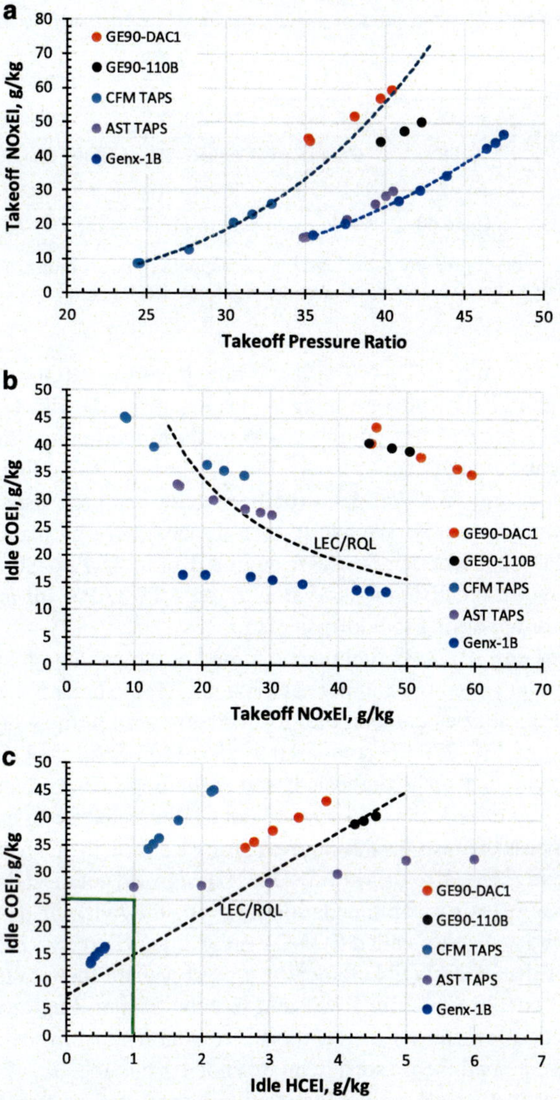

Fig. 18 Emission characteristics of LEC/RQL, lean DAC and TAPS represented, respectively, by the GE90-110/115B, GE90-DAC1 and CFM TAPS, AST TAPS and GEnx. **a** Takeoff NOxEI versus takeoff pressure ratio. **b** Idle COEI versus takeoff NOxEI. **c** Idle COEI versus idle HCEI

As summarized in Fig. 18a, we had takeoff NOxEI data from the CFM TAPS engine demonstrator and full-annular rig of the AST TAPS. The former's trendline when extended to the lean-dome GE90-DAC1 pressure ratios reminded us about a similar after the fact situation we faced when extending nominal 30 PR DAC technology to 40 PR GE90; see Fig. 13a. We wanted to have GEnx SAC TAPS with

Table 4 Four TAPS combustion product engine specifications compared to existing four LEC combustion products

TAPS Combustors	PR	F_{00}, kN	LEC Combustors	PR	F_{00}, kN
GEnx-1B	36–48	255–349	CF6–80C2	28–33	231–272
LEAP-1A	30–39	107–143	CFM56–5B	23–33	96–142
LEAP-1B	37–42	111–130	CFM56–7B	21–28	87–117
GE9X	Up to 60	500±	GE90–110B	40–42	493–514

Note ICAO engine takeoff pressure ratios and the corresponding thrust ratings F_{00} for GE9X are not publicly available; therefore, we have assumed F_{00} to be 500± kN

NOxEI capability similar to DAC TAPS and not above LEC/RQL GE90-110/115B as shown in Fig. 12a. The ongoing diagnostics effort for the CFM TAPS at lower P3 and T3 conditions was not going to help us on this dilemma; the rig had also encountered combustion dynamics issues, Fig. 7. The high P_3/T_3 diagnostics, Fig. 8, was not even planned in 2004. Validation of the reacting comprehensive CFD analysis had not even started in 2004. We had to take the challenge of scaling 27 pressure ratio CFM TAPS ~100 kN demonstrator engine data to 45 PR ~300 kN GEnx-1B on delivering takeoff NOxEI comparable with the AST TAPS full-annular rig data without raising doubt during the downselect process.

The estimated cost of TAPS fuel nozzle turned out to be more than 3–4 times of conventional product injectors compared to goal of 1.5; the latter forced us to set our design target of 22 nozzles compared to 30 currently used in the CF6-80C2 engine with rated thrust of 231–272 kN compared to 255–349 kN for the GEnx.

As per the guideline on hypothesis-based technology or product development process, Fig. 1, you miss 3 hits, and you are out as illustrated by an example here in regard to the process that was used for moving forward from the CFM56-5BDAC1 to the CFM56-5B, Fig. 13a. The CFM56-5BDAC1 design did not meet design objective leading to my design team recommendation for trying the swirl cup design with corotating swirlers and no venturi called CONOVEN. I did not agree with my own team's interpretation of the CONOVEN element test rig. I wanted them to take more time to reassess their conclusions. But I was told that the System Engineering Manager (SEM) would not allow more time. When SEM went against my recommendation even after I presented all the reasons in an extended group meeting, I had no option except to tell him that it will not get the results as predicted by my team. It did not go well with my team; but I had to go with my gut feeling. Many years later, spray diagnostics proved that I was right about the spray quality at ambient conditions; but we needed at the high-power point in order to show its impact on high-power NO_x.

The results summarized in Fig. 13 showed that we did not get the results the team was expecting. However, it helped me to build a trust with the SEM leading to getting more empowerment in meeting the customer's expectation on the next project, namely the CFM56-7B DAC combustion product. Even though we barely met the customer's expectation and delivered the DAC engines as per the contract, I was not happy with how DACs turned out compared to marketing binge. A representative from the same customer in 2006 made it very clear how he felt about DAC.

For moving forward on the GEnx TAPS design, had the author followed the conventional component design organizational practice of assigning a senior combustion engineer as team leader subject to the chief engineer's office representatives to oversee the design process, it would have turned into a commonly used practice of 'design by committee.' I never bought into this practice which I blame partially for the results summarized in Fig. 13. Instead, I preferred alliance with key experts similar to what was done for the TAPS technology development and the LEC GE90-110/115B combustion product which gave 23% reduction in takeoff NOxEI and considerably improved durability and altitude relight capability compared to the GE90-DAC1. Therefore, I appointed myself as the team leader on GEnx until through the full-annular combustor rig demonstration because of the limited time we had in addition to the following.

When preliminary design on the GEnx TAPS combustor started, we had at our disposal comprehensive nonreacting combustion system analysis fully calibrated with existing engines in regard to predicting accurately combustion system overall pressure drop, pressure and airflow distribution around the combustor [43]. Therefore, in order to save time and resources, unlike previous rich-dome and lean-dome combustion system development process, we decided not to design and conduct optimization studies by using combustor diffusion system or high-pressure sector test rigs. Only limited testing on single cup could be afforded again because of time constraints. The planned supporting activities on the CFM TAPS engine did not come through in time to make critical design decisions on GEnx.

Because of the concerns about the scaled GEnx mixer falling along the extended trendline of the CFM TAPS (Fig. 18a), we for the first time went with four different mixer configurations tested concurrently in a 22-nozzle full-annular test rig using so-called pizza pie sector approach; one sector comprised 7 TAPS1 mixers, and the remaining 5-nozzle sectors with three different TAPS2 mixer designs. In addition to engineering guidelines for these mixers, we made qualitative use of nonreactive scalar mixing CFD simulations in downselecting the four mixer designs. One senior engineer from the chief engineer made an interesting observation on our process of mixer selection: I see you making use of CFD in the process, but I don't understand how you made the decision. He was right. But the results turned out good. Instead of showing the full-annular rig data in Fig. 18, we chose to summarize engine results showing that by good luck wc achieved all of our objectives, here discussion limited only to takeoff NOxEI, idle HCEI and COEI. Takeoff NOxEI falls along the AST TAPS full-annular rig data, a combustor with 60 TAPS mixers riding on 30 DAC nozzles, compared to 22 injectors/mixers on the GEnx. Idle COEI versus takeoff NOxEI characteristics fall below the LEC trendline; these values are considerably reduced compared to the GE90 DAC1 and the GE90-110/115B LEC DAC combustor. Idle COEI and HCEI data fall very safely within the long-term box comprised of COEI ≤ 25 and HCEI ≤ 1.0.

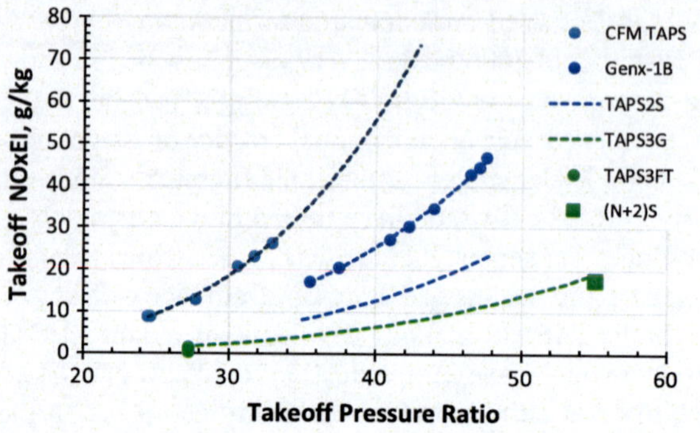

Fig. 19 GEnx TAPS1 technology backed by TAPS2 and TAPS3 along with the CFM TAPS when extrapolated to 40 pressure ratio provides combustion design engineers a wide range of TAPS-based design options

The TAPS2 configurations tested in the pizza pie annular test rig achieved the intended objective, namely 50% reduction compared to TAPS1, shown approximately in Fig. 19 as TAPS2S compared to two TAPS engines, namely the CFM TAPS and the GEnx. A TAPS3 configuration was tested in a flame tube at NASA GRC at fixed condition of P3 = 400 psi, T3 = 1000 °F and fuel/air ratio of 0.027 with pilot fuel flow split as a design variable. The corresponding measured NOxEI varied between 1.39 and 0.29 as reported in [42]; these results are identified as TAPS3FT. Dr. John Hebron, GE Aviation, Cincinnati, undertook a very impressive TAPS technology development effort for a N + 2 cycle with takeoff pressure ratio of 55; see [45]; this included a sector rig testing at pressures approaching takeoff condition. His takeoff NOxEI result shown in Fig. 19 is identified as (N + 2) S. It is interesting to see that the TAPS3 goal line (TAP3G) defined as GEnx takeoff NOxEI divided by 4 goes through these two data points.

The author was very pleased to see the data summarized in Fig. 19 as it provides almost unlimited choices for undertaking fundamental investigations. The experienced GE combustion team enriched by the TAPS relevant fundamental research activities conducted at the GE Global Research Center, the University of Cincinnati, the University of Michigan, Georgia Tech and Purdue University combined with continuously improving comprehensive combustion models can confidently undertake downscaling for the LEAP-X and upscaling for the GE9X TAPS.

Should we see the emission results for LEAP-X summarized in Fig. 14a for takeoff NOx as failure or a warning for being alert for scaling? However, idle COEI and HCEI results are consistent with GEnx. Therefore, in spite of all we have tried to accumulate the knowledge base in TAPS combustion fundamentals and design, the author was surprised to see these results showing continuing challenges in dealing with lean combustion systems.

The results for the third-generation swirl–venturi LDI combustor reported in [30] are summarized in Figs. 20, 21 and 22 along with comparison with emission characteristics of TAPS and DAC combustion products. Readers should refer to [30] for

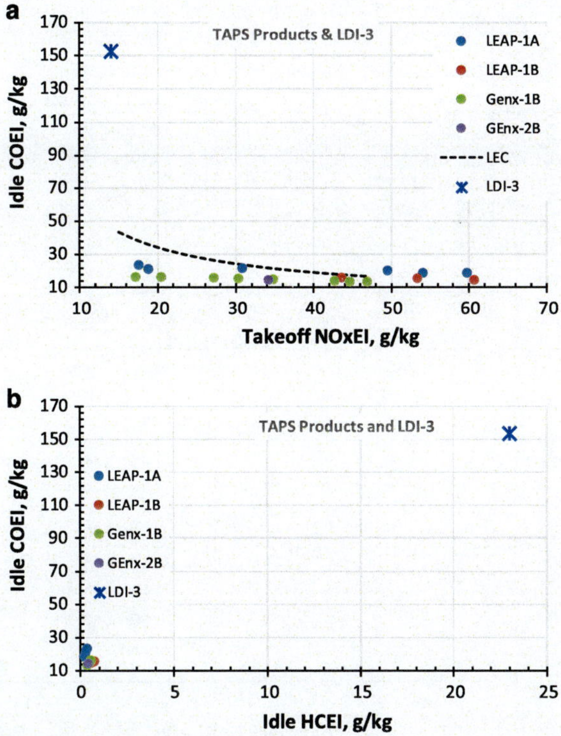

Fig. 20 LDI-3 technology compared with TAPS products at idle. **a** Idle COEI versus takeoff NOxEI. **b** Idle COEI versus idle HCEI

details on this design configurations and attendant idealized ceramic lined combustor without dome or liner cooling air. Intermittent effort since 1990, this work represents the best state-of-the-art LDI technology in a multi-nozzle sector simulating the NASA N + 3 small-core cycle for an engine with 30,000 lb_f (133 kN) rated thrust.

For the LDI-3 at the takeoff pressure ratio of 37.6, we assigned takeoff NOxEI of 14.08 compared to the range of values calculated in [30] being between 31.2 and 13.6. Therefore, compared to GEnx, LDI-3 takeoff NOxEI is 30% lower which when we compare with idle COEI and HCEI emissions at idle (see Fig. 20) and approach (Fig. 21) shows LDI-3 (1) is significantly inferior to TAPS in regard to idle CO and HC, Fig. 20; (2) is comparable to DAC in regard to approach CO and HC, Fig. 21; and (3) may require very different design configurations for the low-power operation in order to eventually get a configuration that matches approach CO and HC levels comparable to TAPS as summarized in Fig. 22.

Figure 23 and accompanying Table 5 provide summary of two LEC products, GE90-110/115B and the CFM56-5B Tech Insertion, emissions tested in 2003 and 2005, and three TAPS products, namely GEnx tested in 2009, and LEAP-1A and

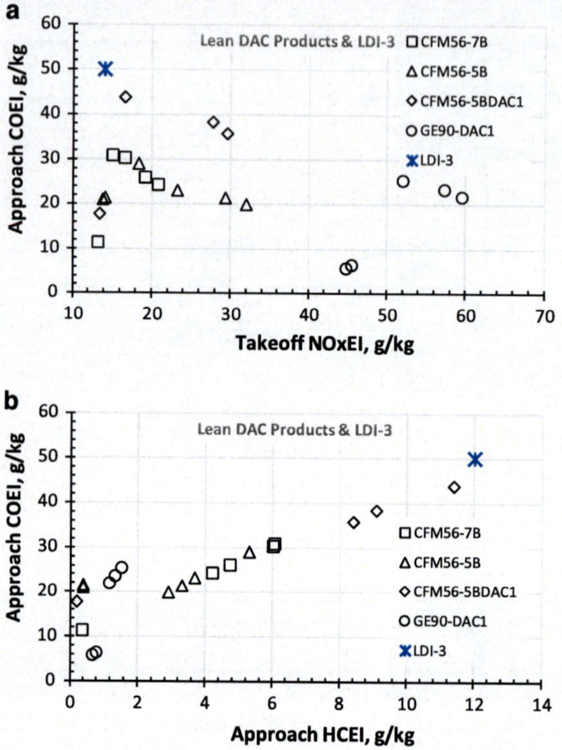

Fig. 21 LDI-3 technology comparable with lean DAC products at approach. But LDI-3 technology is long way off from TAPS products as shown in Fig. 6. **a** Approach COEI versus takeoff NOxEI. **b** Approach COEI versus approach HCEI

LEAP-1B tested in 2016 along with the corresponding applicable CAEP NO_x regulatory standards for engines produced on or after 1996, 2004, 2008 and 2014, respectively. Normally, the margin from the regulatory standards decreases with takeoff pressure. In other words, the margin values at the highest takeoff pressure ratios are of concern to the OEMs as well as customers. These values are listed in Table 5 along with the corresponding LTO NO_x and NO_x margin. Among these products, GEnx has the highest NO_x margin of 39% compared to only 6% for LEAP-1B. This table also includes the characteristic value of LDI-3 calculated by the average of the LTO values reported in [30] divided by 0.8627 that is required in order to calculate characteristic value from a single engine test. The resulting LTO NO_x of 14.46 and the corresponding NO_x margin of 78% are reasonable if we assume the technical feasibility of producing next low-NO_x TAPS-based product 50% lower than the current GEnx.

In concluding section of this manuscript, Fig. 24 shows an interesting look-back thought-provoking image on the process of LTO NO_x regulatory stringency in regard to pre-LEC combustion products of GE and CFM. This data set can be covered by the

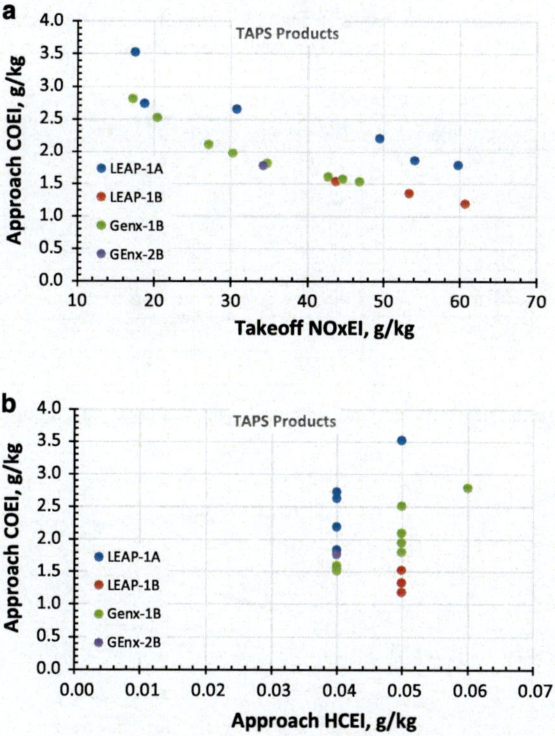

Fig. 22 TAPS products' COEI and HCEI at approach comparable with LEC products. **a** Approach COEI versus takeoff NOxEI. **b** Approach COEI versus approach COEI

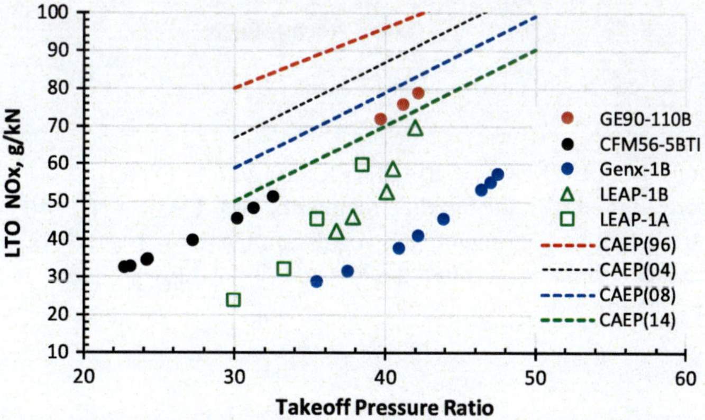

Fig. 23 Summary figure for the GE Aviation and CFM engines' characteristic LTO NO_x versus takeoff pressure ratio compared to applicable CAEP-2, CAEP-4, CAEP-6 and CAEP-8 NO_x regulatory standards. Here, the CAEP standards have been identified with the start year of application, namely CAEP (96), CAEP (04), CAEP (08) and CAEP (14) color coordinated with data symbols being brown, black, blue and green, respectively

Table 5 Summary of engine data shown in Fig. 23 in regard to engine model, year of emission certification, maximum takeoff pressure ratio PR, and the corresponding characteristic LTO NO_x and margin from the applicable regulatory NO_x standards. Year is color coordinated with the applicable NO_x regulatory standards shown in Fig. 23

Engine model	Year	Max PR	LTO NO_x	NO_x Margin
GE90-110B	2003	42.24	78.7	21%
CFM56-5BTI	2005	32.6	51.2	29%
GEnx	2012	47.5	57.27	39%
LEAP-1A	2016	38.5	59.9	11%
LEAP-1B	2016	42	69.66	6%
LDI-3*	Future	37.60	14.46	78%
GEnx Next**	Future	37.60	15.97	76%

[a]Flame tube data from [30]; [b]arbitrary projection by assuming 50% reduction from GEnx

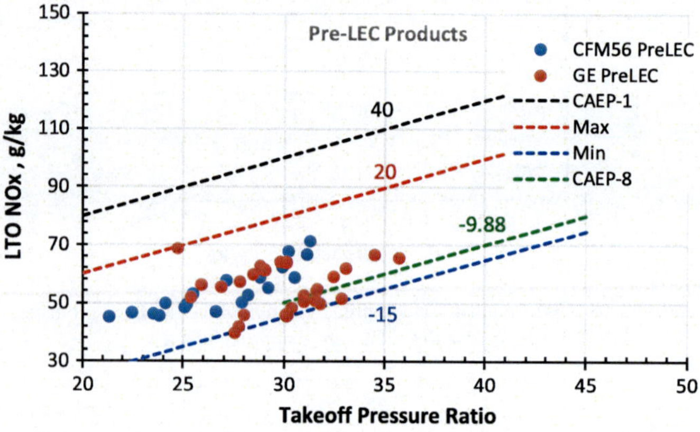

Fig. 24 Pre-LEC combustion products of the CFM and GE Aviation

two equations: Max LTO $NO_x = 20 + 2 \times PR$; and Min LTO NOx $= -15 + 2 \times PR$. We, therefore, started with the baseline regulatory NO_x CAEP $- 1 = 40 + 2 \times PR$, in 2014 introduced more stringent NO_x regulation CAEP $- 1 = -9.88 + 2 \times PR$, a green line that still passes through the low end of pre-LEC products, and proudly claimed 49.9% reduction.

8 Summary

An overview is given of hypothesis-based combustion technology and product design and development practiced since the early 1970s applied successfully for 36 combustion systems including description of author's six most favorite advanced combustion schematics. The challenges of developing the three technologies and products, low emission combustors (LECs), dual-annular combustors (DACs) and TAPS at GE Aviation and CFM, and the lessons learned to further continue making continuing progress have been summarized. While the landing takeoff NO_x of the third-generation single venturi lean direct injection demonstrated in an idealized multi-injector test rig shows potentials for achieving 50% reduction compared to current lowest NO_x GEnx TAPS combustor, its future activities should be focused on reducing low-power CO and HC emissions in addition to commonly encountered operability challenges associated with lean-dome combustors.

References

1. Mongia HC, Smith KF (1978) An empirical/analytical design methodology for gas turbine combustor. In: AIAA, p 998
2. Mongia HC, Coleman EB, Bruce, TW (1981) Design and testing of two variable geometry combustors for high altitude propulsion engines. In: AIAA, p 1389
3. Sanborn JW, Mongia HC, Kidwell JR (1983) Design of a low-emission combustor for an automotive gas turbine. In: AIAA, p 0338
4. Sanborn JW, Scheiling PE, Coleman EB, Johnson KP, Davis FG (1984) Design and performance evaluation of a two-position variable-geometry turbofan combustor. In: AIAA, p 1171
5. Mongia HC, Reynolds RS, Srinivasan R (1986) Multidimensional gas turbine combustion modeling: applications and limitations. AIAA J 24(6):890–904
6. Sanborn, JW, Lenertz, JE, Johnson, JD (1987) Design and test verification of a combustion system for an advanced turbofan engine. In: AIAA, p 1826
7. Mongia HC (1993) Application of CFD in combustor design technology. AGARD CP-536 pp 12–1/12–18
8. Danis AM, Burrus DL, Mongia HC (1997) Anchored CCD for gas turbine combustor design and data correlation. J Eng. Gas Turbine Power 119:535–545
9. Mongia HC (1997) Recent progress in low-emissions gas turbine combustors. ISABE
10. Joshi ND, Mongia HC, Leonard G, Steggmaier JW, Vickers EC (1998) Dry low emissions combustor development. ASME1998-GT-310
11. Mongia HC (2003) TAPS—a 4th generation propulsion combustor technology for low emissions. In: AIAA, p 2657
12. Mongia HC (2009) GE Aviation low emissions combustion technology evolution. Combustion Science and Technology Recent Advances, pp 79–99. ISBN 978-81-8487-014-5
13. Foust MJ, Thomsen D, Stickles Cooper RC, Dodds W (2012) Development of the GE Aviation low emissions TAPS combustor for next generation aircraft engines. In: AIAA, p 0936
14. Mongia HC (2011) Engineering aspects of complex gas turbine combustion mixers Part I: high ΔT. In: AIAA, p 0107
15. Mongia HC (2011) Engineering aspects of complex gas turbine combustion mixers Part II: high T3. In: AIAA, p 0106
16. Mongia HC (2011) Engineering aspects of complex gas turbine combustion mixers Part III: 30 OPR. In: AIAA, p 5525
17. Mongia HC (2011) Engineering aspects of complex gas turbine combustion mixers part IV: swirl Cup. In: AIAA pp 2011–5526
18. Mongia HC (2011) Engineering aspects of complex gas turbine combustion mixers Part V: 40 OPR. In: AIAA, p 5527
19. Mongia HC (2013) N+3 and N + 4 generation aeropropulsion engine combustors Part 1: large engines' emissions. GT, p 94570

20. Mongia HC (2013) N+3 and N+4 generation aeropropulsion engine combustors part 2: medium size rich-dome engines and lean-Domes. In: GT p 94571
21. Mongia HC (2013) N + 3 and N + 4 generation aeropropulsion engine combustors Part 3: Small engine emissions and axial staging combustion technology. GT, p 94572
22. Mongia HC (2014) Future trends in commercial aviation engines' combustion. In: Agarwal A, Pandey A, Gupta A, Aggarwal S, Kushari A (eds) Novel combustion concepts for sustainable energy development. Springer, New Delhi
23. Mongia, HC, Ajmani, K, Sung, CJ (2020) Hypotheses driven combustion technology and design development approach pursued since early 1970s. In: Akshai K (ed) Runchal 50 years of CFD in engineering sciences: a commemorative volume in memory of D. Brian Spalding. Springer, Singapore
24. Mongia HC, Ajmani K, Sung CF (2020) Fundamental combustion research challenged to meet designers' expectations. Under preparation
25. Bruce TW, Davis FG, Kuhn TE, Mongia HC (1977) Pollution reduction technology program small jet engines Phase I Final report. NASA CR-135214
26. Bahr DW, Gleason CC (1975) Experimental clean combustor program, Phase I—final report. NASA CR-134737
27. Tacina RR (1990) Low-NOx potential of gas turbine engines. In: AIAA, p 0550
28. Tacina R, Lee P, Wey C (2005) A lean-direct injection combustor using a 9-point swirl-venturi fuel injector. In: ISABE, p 1106
29. Tacina KM, Podboy DP, He ZJ, Lee P, Dam B, Mongia H (2016) A comparison of three second-generation swirl-venturi lean direct injection combustor concepts. In: AIAA, p 4891
30. Tacina KM, Podboy DP, He ZJ, Lee P, Dam B (2019) A third-generation swirl-venturi lean Direct injection combustor with a prefilming pilot injector. In: GT, p 90484
31. Dhanuka SK, Driscoll JF, Mongia HC (2008) Instantaneous flow structures in a reacting gas turbine combustor. In: AIAA, p 4683
32. Dhanuka SK, Temme JE, Driscoll JF, Mongia HC (2009) Vortex-shedding and mixing layer effects on periodic flashback in a lean premixed prevaporized gas turbine combustor. Proc Combust Inst 32:2901–2908
33. Slabaugh CD, Pratt AC, Lucht RP (2015) Simultaneous 5 kHz OH-PLIF/PIV for the study of turbulent combustion at engine conditions. Appl Phys B 118:109–130
34. Mongia HC, Held TJ, Hsiao GC, Pandalai RP (2003) Challenges and progress in controlling dynamics in gas turbine combustors. J. Propul Power 19:822–829
35. Mongia HC, Held TJ, Hsiao GC Pandalai RP (2005) Incorporation of combustion instability issues into design process: GE aeroderivative and aero engines experience. In: Lieuwen TC, Yang V (eds) Progress in astronautics and aeronautics, vol 210, pp 43–64
36. Pandalai RP, Hsiao GC, Mongia HC (1999) Empirical and anchored methodologies for controlling combustion dynamics. In: RTO meeting proceedings, vol 14, pp 5–1 through 5–14
37. Tacina KM, Chang CT, Lee P, Mongia H (2015) An assessment of combustion dynamics in a low-NO_x, second-generation swirl-venturi lean direct injection combustion concept. In: ISABE, p 20249
38. Anderson, WE, Lucht, RP, Mongia, H (2015) Integrated physics-based modeling and experiments for improved prediction of combustion dynamics in low-emission systems. NASA/CR-2015-NNX11AI62A
39. Bahr DW, Gleason CC (1979) Experimental clean combustor program, Phase III—Final report. NASA CR-135384
40. Hura HS, Joshi ND, Mongia HC (1998) Dry low emissions premixer CCD modeling and validation. ASME1998-GT-444
41. Hura HS, Mongia HC (1998) Prediction of NO emissions from a lean dome gas turbine combustor. In AIAA, p 3375
42. Mongia HC (2008) Recent progress in comprehensive modeling of gas turbine combustion. In: AIAA, 1445
43. Mongia, H, Krishnaswami, S, Sreedhar, PSVS (2007) Comprehensive gas turbine combustion modeling methodology. In: Fluent's International Aerospace CFD Conference, June 18, 2007, Paris

44. Sripathi M, Krishnaswami S, Danis AM, Hsieh SY (2014) Laminar flamelet based NO_x predictions for gas turbine combustors. In: GT, p 27258
45. Lee CM, Chang C, Kramer S, Hebron JT (2013) NASA project develops next generation low-emissions combustor technologies. In: AIAA, p 0540

Emission Characteristics on Combustion of HEFA Alternative-Aviation Fuel Under In-Flight Conditions

Hitoshi Fujiwara and Keiichi Okai

1 Introduction

Global climate change, due to the rapid increase in CO_2 emissions, especially caused by aviation, is one of the critical issues to be solved through international collaborations. Though the amount of CO_2 emission of aviation is only around 2% of the total CO_2 emission, it is important to start any possible measures now to suppress CO_2 emission from aviation. This is particularly important considering that there has been the rapid growth of aviation transportation in recent times. In contrast sincere efforts are on the rise in the other fields, such as automotive and power generation, where electric and hybrid vehicles are prevailing and carbon dioxide capture and storage (CCS) have already been installed in some stationary power plants. In aviation, more than 90% of CO_2 emissions from commercial aircraft operations are generated by large aircraft, which indicates that one must pursue fundamental and applied research to reduce emissions from commercial aircraft with a focus on technology applicable to these large commercial aircraft [1].

In aviation, alternative fuels are considered one of the important options to reduce CO_2 emission. The specifications of alternative and biofuels are strictly defined in ASTM D7566 "Standard Specification for Aviation Turbine Fuel Containing Synthesized Hydrocarbons" [2]. The Annexes in this document defines not only the chemical and physical properties of the fuels but also the manufacturing process of those fuels, which is of crucial importance from the aviation safety point of view.

H. Fujiwara (✉) · K. Okai
Japan Aerospace Exploration Agency, Chofu, Tokyo 182-8522, Japan
e-mail: fujiwara.hitoshi@jaxa.jp

A. K. Gupta et al. (eds.), *Advances in IC Engines and Combustion Technology*,
Lecture Notes in Mechanical Engineering,
https://doi.org/10.1007/978-981-15-5996-9_4

2 Background

In Japan, discussions and investigations of bio-derived aviation alternative fuels started around 2008. At JAXA, the effect of biofuels has been investigated [3] and combustion testing began with available (not-certified and general) biofuel (fatty acid methyl-ester (FAME) fuel) in 2011 [4]. As expected, the fuel had higher levels of CO emissions at lower load condition due to low flame temperature [5]. In 2012, the University of Tokyo, Airlines and Airports in Japan and related institutes formed the Initiatives for Next-generation of Aviation Fuels (INAF) and published a report in 2015 [6]. In the same year, the Ministry of Economy, Trade and Industry of Japan (METI) established a process study committee for clarifying problems that should be considered and also for planning the future process in realizing flights using bio-jet fuel during the 2020 Supper Olympic Games and Paralympic Games in Tokyo [7]. Inspired from these discussions and proposals, three major active projects focusing on the production of bio-derived aviation fuels were evolved. In 2015, Euglena and their partners announced their "Made-in-Japan Biofuels Project" to produce and supply bio-jet/diesel fuels by 2020 [8]. In 2017, IHI and Kobe University started a project to commercialize bio-jet fuel production by developing integrated production process of microalgae-based biofuel supported by New Energy Industrial Technology Development Organization (NEDO) [9]. Also, in 2017, Mitsubishi-Hitachi Power System (MHPS) with partners (Toyo Engineering Cooperation (TEC), Chubu Electric Power (now JERA) and JAXA) started a project to conduct a pilot-scale plant testing for a FT-SPK fuel production derived from lignocellulosic biomass that was supported by NEDO [10]. For this project, JAXA is presently participating in conducting the final combustion tests of the product.

This chapter provides an overview of our experimental results obtained on the investigations of aviation biofuels that provides an understanding on the limitations and potentials of bio-derived aviation fuels for more environmentally friendly aviation. The alternative turbine fuel used is hydro-treated ester and fatty acid (HEFA) made from tallow fat. The manufacturing process is specified in Annex 2 of ASTM D7566. The HEFA fuel is available for commercial flights, provided the blending ratio does not exceed 50 volume % as specified in Annex 2.

Specifications for HEFA are provided in ASTM D7566 Annex 2 [2]. To understand the similarities and differences of HEFA fuel compared to the baseline jet fuel, the properties of HEFA fuel were analyzed. Several considerations on the analysis are summarized below:

1. HEFA showed a very low freezing point, considered to be caused by branched paraffins.
2. Dynamic viscosity and surface tension of HEFA and baseline jet fuel differ only a little from each other. The slight differences might affect the atomization behavior of HEFA fuels as compared to the baseline jet fuel.
3. The density of HEFA is lower than that of baseline jet fuel. The density of baseline jet fuel refined in Japan tends to be lower than those of other countries, possibly causing the density of the fuel blend to be outside of the required specification.

To meet the specification, slightly lower ratio of HEFA would be acceptable in Japan.

4. The lubricity of HEFA is lower than that of baseline jet fuel due to lack of sulfur content. The refined baseline jet fuel in Japan tends to have smaller amount of sulfur content which might cause insufficient lubricity of the 50:50 blend fuel specified in D7566.
5. The most remarkable difference of the two fuels (baseline jet fuel and HEFA) lies in its aromatics content. HEFA contains less aromatic compounds while baseline jet fuel contains about 20% of aromatic compounds by volume, which consists of mainly benzenes and its derivatives with low polycyclic aromatic hydrocarbons (PAHs).

3 Experimental Apparatus and Conditions

A single burner combustion rig test was performed at JAXA AP7 high-pressure test rig (Fig. 1) and medium pressure test rig. This test rig is usually used for the demonstration of a new combustor concept and for the development of some innovative measurement technologies. An exhaust gas sample probe with eight ($\varphi = 0.8$ mm) sampling holes was located at the exit of the combustor liner.

Exhaust sample gas was connected to the measurement instruments through a stainless-steel pipe with a valve to control the mass flow and temperature of the sampled gas. NO_x was measured using a chemiluminescence detector (CLD), CO and CO_2 were measured using nondispersive infrared (NDIR) detectors and THC was measured using flame ionization detector (FID), respectively, Horiba MEXA ONE. Non-volatile PM mass (nvPM) was measured through photoacoustic soot sensor, (PASS) AVL MSS 483.

Fig. 1 JAXA AP7 high-pressure combustion test rig

Table 1 Inlet air conditions (high-load condition)

Inlet temperature (K)	Inlet pressure (kPa)	Pressure loss ratio (%)
$T_{A1} = 644$	$P_A = 1350$	4.6
$T_{A2} = 739$	$P_A = 1350$	4.1
$T_{A3} = 803$	$P_A = 1350$	4.2
$T_B = 803$	$P_B = 2000$	4.2

Inlet air conditions shown in Table 1 were tested for high-load condition (simulating actual high-temperature and high-pressure operational condition) with the total air/fuel ratio (AFR) ranging from around 38–130.

4 Results

Figure 2 shows the RQL combustor [11] used in this study, wherein 10% of total air flow enters through the upstream Parker-Hannifin type air blast fuel nozzle (shown in Fig. 5), while 90% of total air enters through the air holes on the liner (shown in Fig. 2 as combustion/dilution air holes). The fuel nozzle has only one fuel inlet, so that no fuel staging was performed.

Inlet air conditions were determined so that the emissions for the simulated operational conditions can be investigated. Total air/fuel ratio ranged from around 38–150.

Results obtained from the experiments are given below.

- A combustion rig test of an RQL combustor for an aero-engine was performed with both baseline jet fuel and hydro-treated ester and fatty acid (HEFA) alternative jet fuel.
- Both of non-volatile PM mass and number were reduced with HEFA, while NO_x, THC and CO emissions were similar to each other for the two fuels.

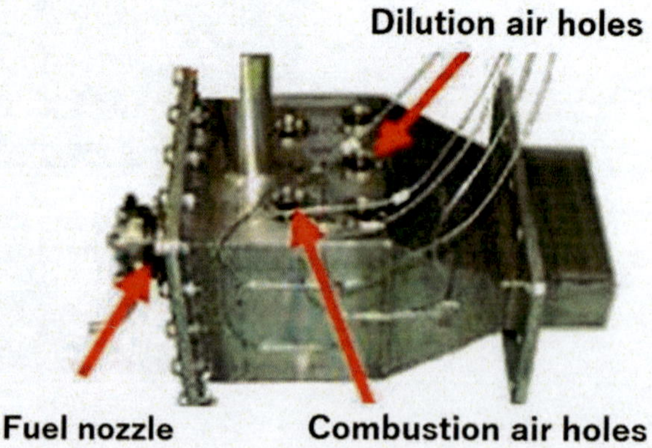

Fig. 2 RQL combustor

- The reduction of nvPM mass with HEFA is more critical than the number of nvPM indicating that large size nvPMs are removed with HEFA and the average size of nvPM with HEFA was smaller than that with kerosene fuel.

Figure 3 shows a comparison of the direct flame images at low load with HEFA and baseline jet fuel. The images show that HEFA had less flame brightness than that from baseline jet fuel combustion; this is consistent with the results of the reduced nvPM emission (both on mass and number basis) when using HEFA. Figure 4 shows the spatial distribution of soot showing the KL factor determined using Hottel and Broughton two-color radiometry [12]. KL factor in baseline jet fuel flame showed a region having large value of KL factor around luminous flame. In contrast, KL factor in HEFA flame showed smaller value than baseline jet fuel case.

Figure 5a–b shows pressure and temperature effects on the change of CO emission characteristics with the rig air to fuel ratio (AFR). The values are much smaller than low-load result (near idle condition) (not shown in the figure). For the baseline jet

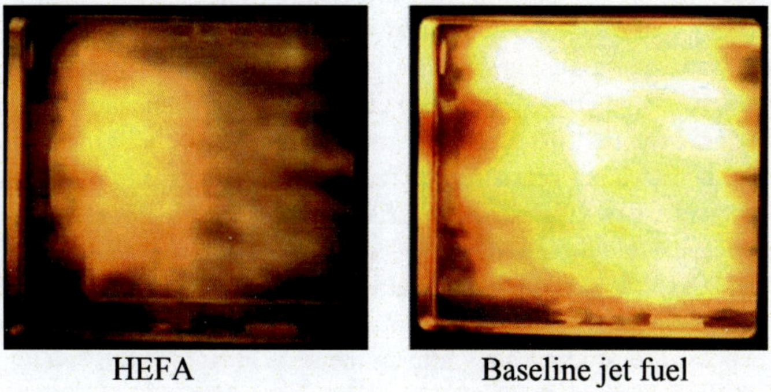

Fig. 3 Comparison of the flame visualization (low-load condition)

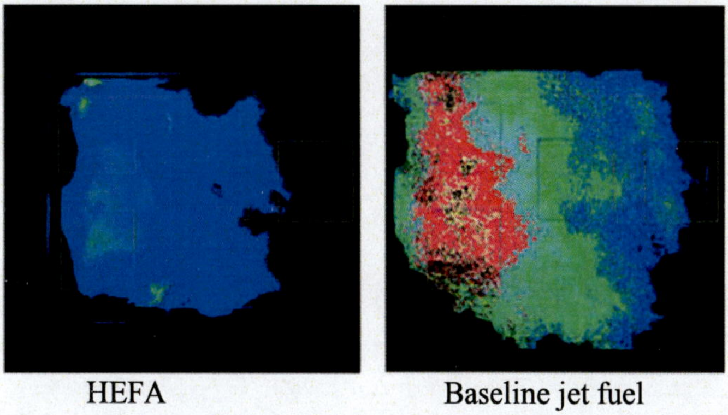

Fig. 4 Comparison of the distribution of KL factor obtained with Hottel and Broughton two-color radiometry (at low-load condition)

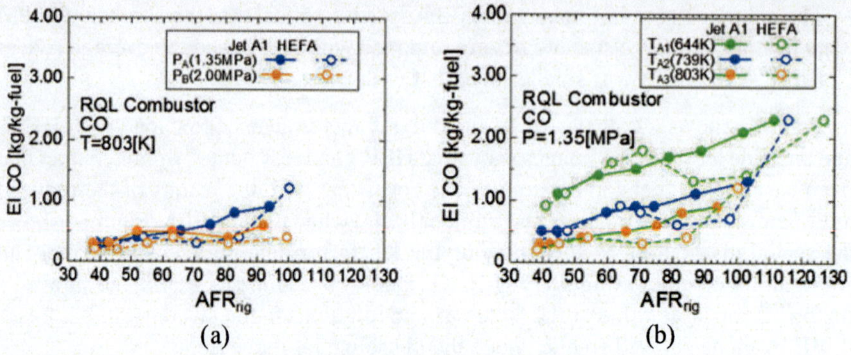

Fig. 5 Pressure and temperature effects on emission characteristics of CO (at high-load condition). **a** Pressure effects (at $T = 803$[K]), **b** temperature effects (at $P = 1.35$[MPa])

fuel case, increase in combustion pressure reduces CO emission possibly due to suppressed chemical dissociation as shown in Fig. 5a [5]. A strong air temperature effect on CO was observed as shown in Fig. 5b. This is similar to the description given by Lefebvre AH, Ballal DR [5]. For each condition, the CO emission characteristics were similar between the two fuels examined.

Figure 6a, b shows the pressure and temperature effects on the change in NO_x emission characteristics with the rig air to fuel ratio (AFR). For conventional combustors, it is generally found that $NO_x \propto p^n$, where n has values ranging from around 0.5 to around 0.8 [5]. Based on the typical value of $n = 0.5$, a correlated value on P_B is plotted in Fig. 6a showing general similarity. An increase in inlet air temperature would be expected to produce a significant increase in NO [5] (and NO_2) as shown in Fig. 6b. The results show that for each condition, the NO_x emission characteristics were similar for the two fuel cases examined here.

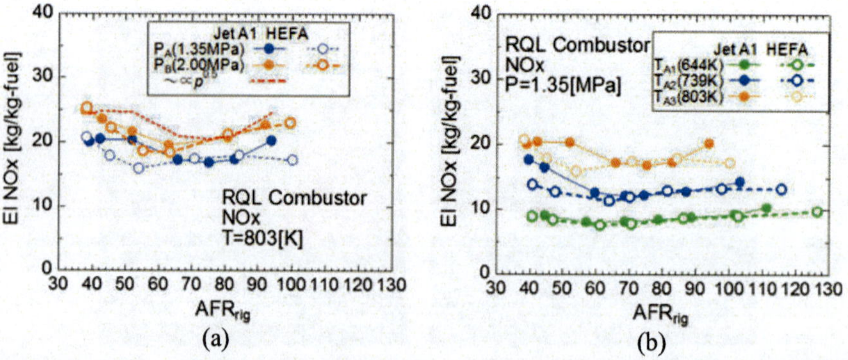

Fig. 6 Pressure and temperature effects on NO_x emission characteristics (under high-load condition). **a** Pressure effects (at $T = 803$[K]), **b** temperature effects (at $P = 1.35$[MPa])

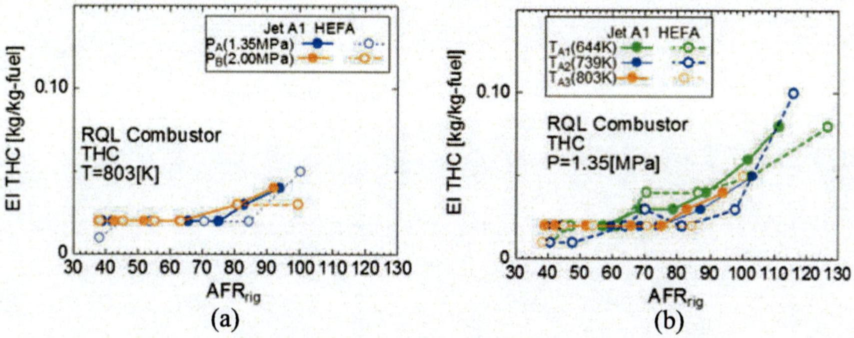

Fig. 7 Pressure and temperature effects on emission characteristics of THC (at high-load condition). **a** Pressure effects (at $T = 803$[K]), **b** temperature effects (at $P = 1.35$[MPa])

Figure 7a–b shows the pressure and temperature effects on the change of THC emission characteristics with the rig air to fuel ratio (AFR). For the high-load condition presented in the figures, the effect of pressure and temperature seems small, though these values are much smaller than those found at low-loads (representing near idle condition) (not shown in the figure). For each condition reported here, the NO_x emission characteristics were similar for the two fuel cases reported here.

From the above figures, it can be said that NO_x, CO and THC emissions with HEFA are quite similar to those from normal jet fuels.

Figure 8a–b shows the effect of pressure on PM emission characteristics (both mass and number) at constant temperature condition (803 K). These figures show increase in pressure decreases both mass and number of PM emission; however, the pressure effect is less pronounced for the PM number case.

A correlation for soot formation is given by Döpelheuer (DLR) [13]. From the equation, soot concentration (in mg/m_3) is proportional to $p^{1.35}$. Based on this dependency, a correlated value on P_B is plotted in Fig. 8a that showed good general agreement.

Figure 8c–d shows the effect of temperature on PM emission characteristics (mass and number) at constant pressure condition (1350 kPa). In the case, temperature effect on PM mass is not clear, but clearly the PM number decreases with increase in temperature.

5 Conclusions

The effect of fuel change from baseline jet fuel to HEFA is remarkable, especially in soot formation and soot emission phenomena. The reduction of emission from HEFA compared to baseline jet fuel is attributed largely to the reduced aromatic concentration in this fuel. Other combustion properties tested included the emission of CO, NO_x and THC. They showed almost no effect from the change of fuel. This

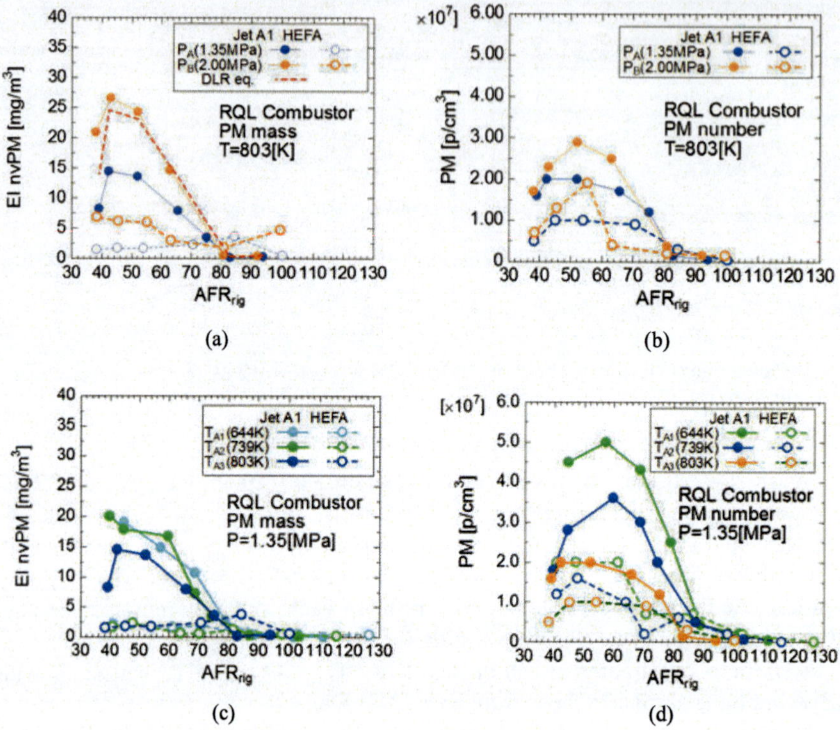

Fig. 8 Pressure and temperature effects on emission characteristics of PM mass and number (at high-load condition). **a** Pressure effects on PM mass (at $T = 803$[K]), **b** pressure effects on PM number (at $T = 803$[K]), **c** temperature effects on PM mass, **d** temperature effects on PM number [at $P = 1.35$[MPa] for (c) and (d)]

information is aimed to provide helpful guidelines for total environmental impact of changing jet fuel to the new fuels along with the actual operational conditions.

Acknowledgements The authors thank Professor Ashwani K. Gupta of the University of Maryland for his comments and revisions.

References

1. National Academies of Sciences, Engineering, and Medicine (2016) Commercial aircraft propulsion and energy systems research: reducing global carbon emissions. The National Academies Press, Washington DC. https://doi.org/10.17226/23490
2. ASTM D7566-18 (2018) Standard specification for aviation turbine fuels containing synthesized hydrocarbons, ASTM International, West Conshohocken, PA

3. Fukuyama Y, Fujiwara H, Okai K, Alternative fuels and their impact on the turbofan engine design and performance under the realistic flight conditions, ACGT paper No. 51, Asian Joint Congress on Gas Turbines 2009 (ACGT 2009), Tokyo, Japan 2009 (ACGT2009-051)
4. Okai K, Fujiwara H, Hongoh M, Shimodaira K (2012) Application of a Bio-fuel to a single sector combustor for an experimental small aero-engine, Asian Joint Congress on Gas Turbines 2012 (ACGT 2012), Shanghai, China (ACGT 2012-2121)
5. Lefebvre AH, Ballal DR (2010) Gas turbine combustion: alternative fuels and emissions, 3rd edn. CRC press, Taylor & Francis
6. Initiatives for next-generation aviation fuels, report of the initiatives for next-generation aviation fuels (English translation of the original Japanese version) (2015) https://urldefense.proofpoint.com/v2/url?u=http-3A__aviation.u-2Dtokyo.ac.jp_inaf_roadmap-5Fen.pdf&d=DwIFJg&c=vh6FgFnduejNhPPD0fl_yRaSfZy8CWbWnIf4XJhSqx8&r=eX-g8gqe3d-GoWycz_7rrFqNjbM5XPbKrfS6HBeL60XuS2ovalzEMTtidqJEphSc&m=QAcgkidUX2FbqCIFXX90FcMspD02ZtK_ljfACAXNeK0&s=v6HBureYBA0U46R9_UzLyTP5EuPsdo-CYZOzv-D6e08&e=. Accessed 18 June 2020
7. Ministry of Economy, Trade and energy press release, July 2 2015, Agency for natural resources and energy, establishment of a committee for the study of a process leading to introduction of bio jet fuel for the 2020 summer olympic games and paralympic games in Tokyo, http://www.meti.go.jp/english/press/2015/0702_01.html. Accessed 18 June 2020
8. Terasaki NH (2017) Road to Tokyo 2020 and beyond; Japan's initiatives. In: IATA Alternative Fuel Symposium, 16–17 November 2017
9. IHI press release, IHI to implement pilot scale experiment of algae mass cultivation for biofuel in Thailand, November 6 2017, https://www.ihi.co.jp/en/all_news/2017/other/2017-11-06/index.html. Accessed 18 June 2020
10. Mitsubishi-Hitachi Power Systems, An activity of R&D and verification of bio fuel production -a pilot-scaled testing of throughflow production process, 2017 (in Japanese). http://www.nedo.go.jp/content/100870869.pdf. Accessed 18 June 2020
11. Makida M, Yamada H, Shimodaira K (2014) Detailed research on rich-lean type single sector and full annular combustor for small air-craft engine. In: 29th congress of the international council of the aero-nautical sciences (ICAS 2014), St. Petersburg, Russia (ICAS2014-0628)
12. Hottel HC, Broughton FP (1932) Ind Eng Chem 4(2):166–174
13. Döpelheuer A (2001) Quantities, characteristics and reduction potential of aircraft engine emissions. SAE 2001-01-3008

What Artificial Intelligence Can Do for You

Kamil Ekštein

1 Extended Abstract

Artificial intelligence (AI) is nowadays an immensely vast research area set in a borderland among many branches of knowledge, particularly among computer science, cybernetics, mathematics, statistics, and others.

Some courses of AI studies are medialized and popularized way faster than the others as the world's most influential Internet companies (e.g. Google) use them in their marketing as display windows demonstrating the companies' capabilities. Therefore, the medial image of AI is unfortunately narrowed to self-driving vehicles and voice interaction-enabled smartphones.

In fact, today's AI offers many more applications based on techniques that are less attractive and eye-catching. However, they are very useful, practical, efficient and reliable. Some few examples of how these AI techniques may serve in the area of combustion (or chemical process modelling) are described below.

2 Computer Vision

Computer vision (CV) is a branch of AI that studies mathematical techniques and algorithms enabling computers to (at least partially) understand the content of scenes captured in digital images and their sequences (videos). From an application point of view, CV automates tasks that otherwise need a human operator using his/her

K. Ekštein (✉)
Department of Computer Science and Engineering, Faculty of Applied Sciences,
University of West Bohemia, Plzeň, Czech Republic
e-mail: kekstein@kiv.zcu.cz

A. K. Gupta et al. (eds.), *Advances in IC Engines and Combustion Technology*,
Lecture Notes in Mechanical Engineering,
https://doi.org/10.1007/978-981-15-5996-9_5

eyesight to accomplish certain assignment (e.g., count some items, tell whether the item is damaged or not, etc.).

Generally, artificial intelligence techniques always perform with certain accuracy (or reliability) which is desired to reach up to 100% (meaning that the solution provided by an algorithm for the given task is flawless and in accord with a human-provided solution no matter how many times the task was executed) but it actually happens only rarely—most of the techniques usually prove accuracy between 90 and 100% (in order to be considered excellent).

However, on certain specific tasks, the contemporary CV techniques offer practically 100% reliability. For instance, shape detection techniques based on *Hough transform* [1] are nowadays capable of identifying and locating shapes (especially lines and circles) in digital images with reliability approaching 100%. Figure 1 depicts how *two-stage circular Hough transform* can be applied on a high-speed video sequence of isooctane ignition: the CV technique automatically detects the boundary of the spherical flame, thus enabling, e.g., to compute the velocity of the flame propagation or the gaseous product inflation, etc.

Computer vision techniques are very useful also in the task of image structure analysis: Fig. 2 shows results provided by the *Canny edge detector* [2] in the task of assessing non-homogeneity of the spherical flame. The number of white pixels in the processed images normalized to the areas of the analysed circular image clips gives an objective measure of the roughness of the spherical flame surface.

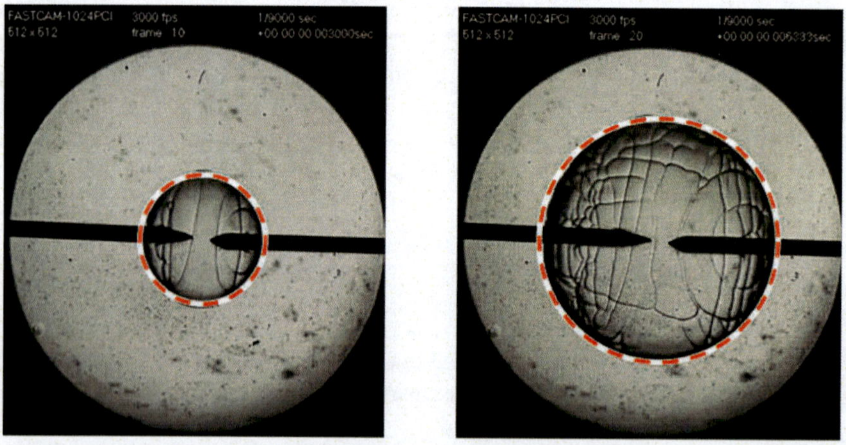

(a) Frame #10, i.e. $t = 0.0033$ sec (b) Frame #20, i.e. $t = 0.0066$ sec

Fig. 1 Ignition of isooctane captured by Photron FASTCAM 1024 PCI high-speed camera at 3000 fps. A computer vision algorithm (based on *two-stage circular Hough transform*) enables automatic detection of the spherical flame boundary, and thus, e.g., computation of the temporal derivatives of the volume, flame propagation velocity, surface non-homogeneity through time, etc.

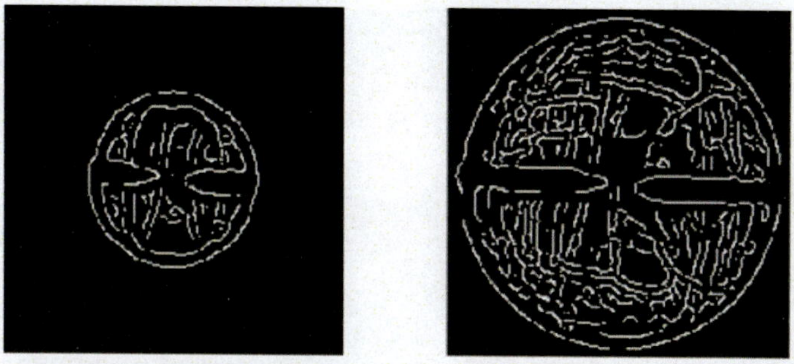

(a) Frame #10 (b) Frame #20

Fig. 2 Ignition of isooctane captured by Photron FASTCAM 1024 PCI high-speed camera at 3000 fps. *Canny edge detector* makes it possible to objectively assess the coarseness of the spherical flame as each "bubble" boundary is marked with white pixels

3 Cellular Automata

A *cellular automaton* (CA) [3, 4] is a discrete computational model of spatio-temporal development of a certain physical entity, feature, or reality, e.g., forest fire, air pollution/contamination, or contagious disease spreading, etc.

Fundamental principles of cellular automata were first introduced in works of Ulam and von Neumann in 1940s. Later, some attention was paid to CA in 1950s and 1960s but the real sound interest has appeared since 1970s.

The very basic principle of the *cellular automaton* is that the state space of the given task is discretized into a regular *n*-dimensional grid of uniform arbitrarily[1] small volumes, acting like volume differentials of Newtonian integration. These small volumes are called *cells*. Each cell has a definite state (usually expressed by different colours as shown in Fig. 3). Each cell has also a set of neighbouring cells that influence the respective cell. The influence is precisely described and quantized by a set of rules that are applied to all cells of the CA in each step of the discretized time.

As the task time advances from t_i to t_{i+1}, the automaton changes states of all cells according to the available (and applicable) rules. As such, the CAs are well-suited to the tasks investigating system dynamics: Even though the simple rule-based changing of cell states might in certain cases lead to lower simulation precision, this disadvantage is generously balanced by feasibility—most of the models can be run in (nearly) real time even on lower computational power devices.

Later on, certain specific types of CA addressing fluid dynamics appeared: *lattice gas automata* (LGA) or *lattice gas cellular automata* [5] are a type of cellular

[1]The discretization step is in fact limited only by the available memory and computational power of the used computer.

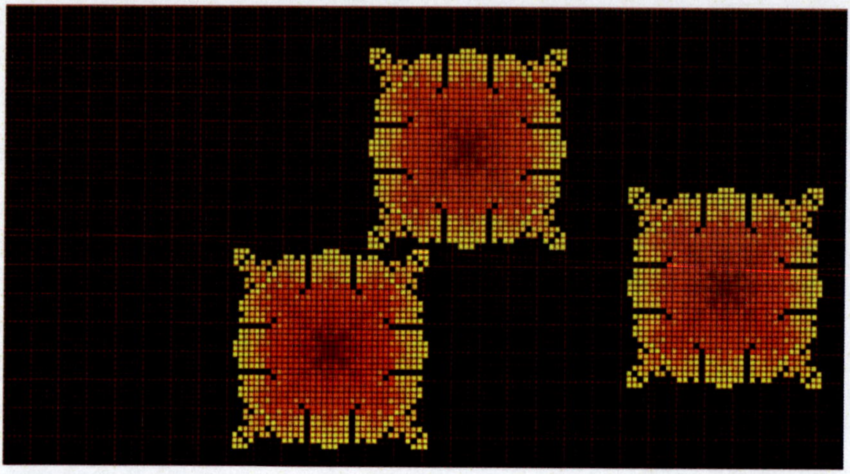

Fig. 3 Nearly real-time mathematical model of fireworks explosions performed by a cellular automaton. The software used for the model computation and visualization was MCell 4.20 [7]

automaton used to simulate fluid flows. They were the precursor to the lattice Boltzmann methods. From lattice gas automata, it is possible to derive the macroscopic Navier–Stokes equations [6].

Obviously, the ideas behind CAs and LGAs gave birth to more complex computational models on a discretized continuum, such as FEM or CFD[2] that are, however, significantly more demanding when it comes to computational power and usually cannot deliver the simulation results in real time.

Figure 3 depicts how a CA implemented in a free CA software MCell [7] effectively models fireworks explosions.

4 Evolutionary Algorithms

An *evolutionary algorithm* (EA) [8] is a generic population-based metaheuristic optimization algorithm. An EA uses mechanisms inspired by biological evolution, such as *reproduction*, *mutation*, *recombination*, and *selection*. Candidate solutions to the optimization problem play the role of individuals in a population, and the *fitness function* determines the quality of the solutions [9].

The final product of an evolutionary optimization (performed by the evolutionary algorithm) can be understood as the best (via the fitness function) individual "cultivated" from crossbreeding its predecessors (e.g. Fig. 5).

A well-known example of this process is the work of Lohn et al. [10]. NASA commissioned a research team to develop an X-band antenna for a small satellite

[2]Finite element method, Computational fluid dynamics.

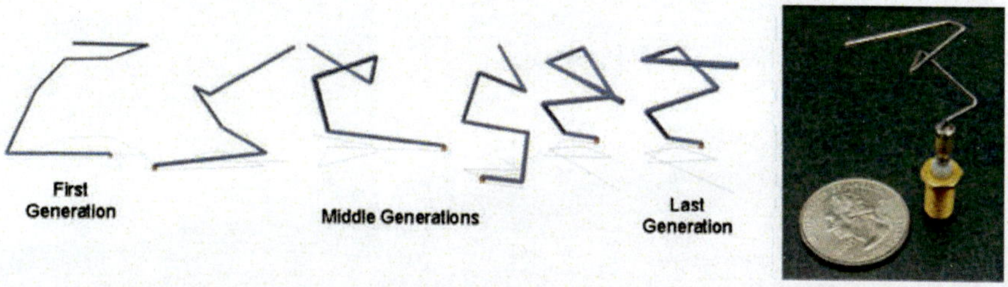

(a) Sequence of evolved antennas (b) Final product

Fig. 4 Interesting application of an *evolutionary algorithm* to design an X-band antenna for the 2006 NASA mission "Space Technology 5". The algorithm worked in the so-called multi-objective mode: the goal was to design antenna which is (i) as small as possible while (ii) it has the desired radiation pattern and (iii) the maximum possible EM radiation output power. Images from [10]

with a given radiation pattern and desired output power and other specifications. The antenna should have been also as small as possible. Lohn et al. suggested to use an evolutionary algorithm to "cultivate" an antenna with the requested properties and designed it. The result (successfully tested during the NASA ST5 mission) is depicted in Fig. 4.

Evolutionary algorithms are traditionally used in AI to find (sub)optimal solution(s) in high-dimensional feature spaces where the problem objective function is very complex and not differentiable at all or with difficulties. Even in situations where the traditional approaches (e.g., gradient descent optimization) fail, the EAs are capable to find a suitable solution within an acceptable time.

In mechanical engineering, EAs can be used to find, e.g., optimal structure from the point of view of, let us say, rigidity or compactness or stability, etc. The only requirements are (i) to provide the *fitness function* which assesses how the newfound solution is good/acceptable from the point of view of the optimization criteria and (ii) to be able to somehow "map" (or project) the features of the design to form a searchable space, i.e., to unambiguously project the design into a point (n-component vector) in an n-dimensional space.

5 Neural Networks

Artificial neural networks (ANN) [11] are by far the most noticeable product of research in the field of AI. Recently, there has been a lot of attention paid to ANNs also in media worldwide, namely thanks to research outcomes of the world's most influential Internet companies. Currently, the theory behind ANNs is vastly extensive and there exist thousands of different types and architectures of ANNs.

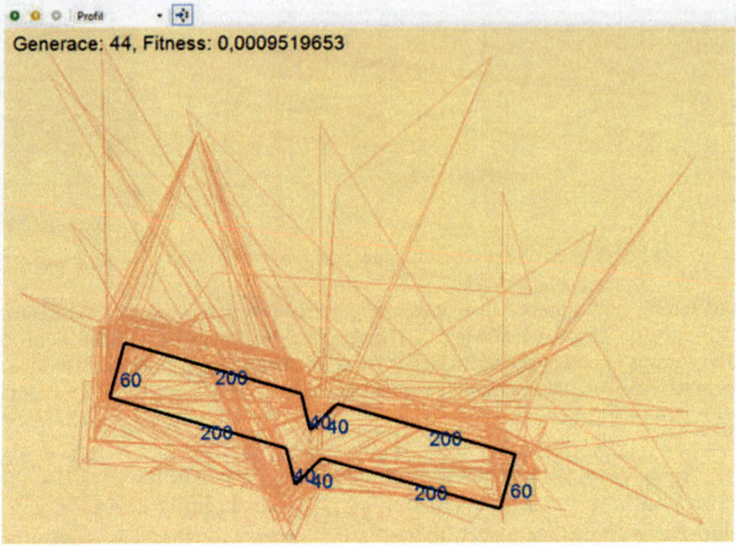

Fig. 5 EA-based shape optimizer developed for the Department of Machine Design, Faculty of Mechanical Engineering, UWB

Generally speaking, the ANN is usually a *classifier,*[3] i.e., a function $f(\mathbf{x})$: $\Re^N \rightarrow \aleph^1$. The function value is the so-called *class label* that identifies into which class the input feature vector (a projection of a real-world object into the task state space) most probably belongs.

The analytic description of $f(\mathbf{x})$ by a formula is initially unknown: it is derived from data during an iterative process called *training*. The network is successively exposed to samples of input feature vectors and correct output class labels (in the case of the so-called *supervised learning*). Stepwise execution of a *backpropagation*[4] algorithm causes the ANN to modify its internal structure to improve the classification accuracy.

Based on the above-mentioned properties, the engineering applications of ANNs vary almost endlessly. Whenever there is a need for an explication of the measured data (by means of building a model that provides sound predictions), an ANN should be considered.

[3]However, it can play nowadays many different roles as well.

[4]Currently, there exist a large number of modifications of the original backpropagation. Thus, the term backpropagation denotes rather a class of algorithms than a specific one.

References

1. Duda RO, Hart PE (1972) Use of the Hough transformation to detect lines and curves in pictures. Commun ACM. 15(1):11–15. https://doi.org/10.1145/361237.361242
2. Canny J (1986) A computational approach to edge detection. IEEE Trans Pattern Anal Mach Intell 8(6):679–698
3. Ulam S (1952) Random processes and transformations. Proc Int Congr Math 2:264–275
4. Preston K Jr, Duff MJB (1984) Modern cellular automata—theory and applications. Springer US. https://doi.org/10.1007/978-1-4899-0393-8. ISBN: 978-0-306-41737-5
5. Hardy J, de Pazzis O, Pomeau Y (1976) Molecular dynamics of a classical lattice gas: transport properties and time correlation functions. Phys Rev A 13(5):1949–1961. https://doi.org/10.1103/physreva.13.1949
6. Wikipedia: Lattice gas automaton (2019) https://en.wikipedia.org/wiki/Lattice_gas_automaton
7. Wojtowicz M (2005) MCell 4.20 http://www.mirekw.com/ca/index.html
8. Ashlock D (2006) Evolutionary computation for modeling and optimization. Springer. ISBN: 0-387-22196-4
9. Wikipedia: Evolutionary algorithm (2019) https://en.wikipedia.org/wiki/Evolutionary_algorithm
10. Lohn JD, Linden DS, Hornby G (2008) Advanced antenna design for a NASA small satellite mission. In: Proceeding of 2nd annual AIAA/USU conference on small satellites
11. Bishop CM (1995) Neural networks for pattern recognition. Clarendon Press. ISBN: 978-0198538493. OCLC 33101074

Experimental Investigation of Aromatic Blended Binary Fuel on Pollutant Emissions from Compression Ignition Engine

Paramvir Singh, Saurabh Sharma, Bandar Awadh Almohammadi, Sudarshan Kumar, and Bhupendra Khandelwal

1 Introduction

The stringent emission standards on a global scale for lower pollutant emissions, especially PM from compression ignition engines demand a significant improvement in existing engine technology and alternative fuels with lower emissions from these combustion engines. Automobile companies are continuously working on the numerous engine technologies that would help in reducing the emissions from compression ignition engines like flexible fuel injection systems, exhaust gas recirculation, variable compression ratio, and innovative combustion approaches like low-temperature combustion [1–5]. At the same time, fuel companies are focusing on alternative renewable fuels, alcohol blended fuels, and other novel fuels [6, 7]. It is also interesting to assess the effect of diesel fuel components on the performance of the engine and their exhaust emissions.

Diesel is a complex mixture of hydrocarbons, which usually contains n-alkanes, iso-alkanes, cyclo-alkanes, and aromatics [8]. Aromatics have continuously gained attention because of their higher content (up to 30% by mass) in diesel and high tendency to emit particulate matters during combustion due to the higher degree of unsaturation [9]. PM emissions from diesel-fueled engine exhaust mainly constitute elemental and organic carbon [10]. The proportion of PM emissions in the exhaust

P. Singh (✉) · S. Sharma · S. Kumar
Combustion Research Laboratory, Department of Aerospace Engineering, Indian Institute of Technology Bombay, Powai, Mumbai 400076, India
e-mail: param016@gmail.com

B. A. Almohammadi
Low Carbon Combustion Centre, University of Sheffield, Unit 2, Crown Works Industrial Estate, Rotherham Road, Sheffield S20 1AH, UK

B. Khandelwal
Department of Mechanical Engineering, University of Alabama, Tuscaloosa, AL 35487, USA

A. K. Gupta et al. (eds.), *Advances in IC Engines and Combustion Technology*,
Lecture Notes in Mechanical Engineering,
https://doi.org/10.1007/978-981-15-5996-9_6

of diesel engines mainly be influenced by the pyrolysis of fuel droplets, combustion byproducts, and unburned fuel [11]. Numerous mutagenic and carcinogenic compounds have been found in the particulate matters emitted from diesel engine exhaust [12].

The diesel operated engines are the main contributor for the elemental carbon emissions present in the atmosphere [13]. Ryan and Waytulonis [14] reviewed the consequence of aromatics content on tailpipe emissions of the diesel engine. The study concluded that the emissions mainly PM reduces with the decrement in the content of aromatics from fuel blends. The main concluding remark from the extensive review of various studies is that emission formation capability of aromatics with high flash point and low distillation endpoint is lower than the hydrocarbon fuel alone. The authors mainly concentrated on the impact of aromatic contents on PM emissions. Some researchers indicated that PM emissions increase with the aromatic contents [8, 15]. However, some authors mentioned that the change in PM emissions with varying aromatic content was not statistically significant [16]. Not all aromatics contribute to the same levels of emissions at all conditions. Still, there is no particular study present in the literature that systematically analyzes the consequences of different aromatics on diesel engine emissions and their effects on fuel properties. Moreover, the explanation of aromatic properties and its structure on emissions is still missing. It is, therefore, imperative to study the effects of aromatics content on base fuel properties that affect the tailpipe emissions and to analyze the facts from several perspectives to advance the quality fuel for future applications.

The effects of aromatics structure and their properties like H/C ratio, boiling point, calorific value, density, and cetane number on regulated emissions (CO, HC, NO_x, and PM) are studied. Toluene, indene, and 1-methylnaphthalene were blended with base fuel (de-aromatized hydrocarbon) in three blending ratios (15–20–25% by mass). The chemical composition and structure of aromatics are shown in Fig. 1. A straight-chain de-aromatized hydrocarbon solvent containing the mixture higher single-chain alkanes is considered as base fuel.

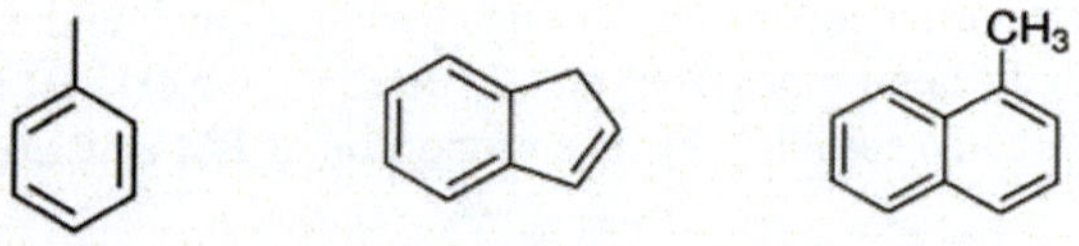

C_7H_8 Toluene — C_9H_8 Indene — $C_{10}H_7CH_3$ Methylnaphthalene

Monocyclic aromatics **Polycyclic aromatics**

Fig. 1 Structure and chemical composition of monocyclic and polycyclic aromatics

2 Experimental Setup and Methodology

The experiments were executed on a diesel engine test rig in the laboratory (see Fig. 2). The water-cooled single-cylinder four-stroke diesel engine was used for the present study. The engine was coupled with a dynamometer and control panel was connected for electric loading. The engine was started on diesel fuel initially, until it reaches the steady-state conditions. For each fuel blend, the values were taken at both low and high load conditions. The engine specification used for experimentation is tabulated in Table 1.

The emission testers were fixed to measure the CO, HC, NO_x, and PM. CO, HC and NO_x, were assessed by using Testo 350 gas analyzer. The measurement range and accuracy are provided in Table 2. The values of the PM emissions were measured with Atmos air quality meter. The effect of these aromatics on properties of the blend on PM emissions from a diesel engine is investigated.

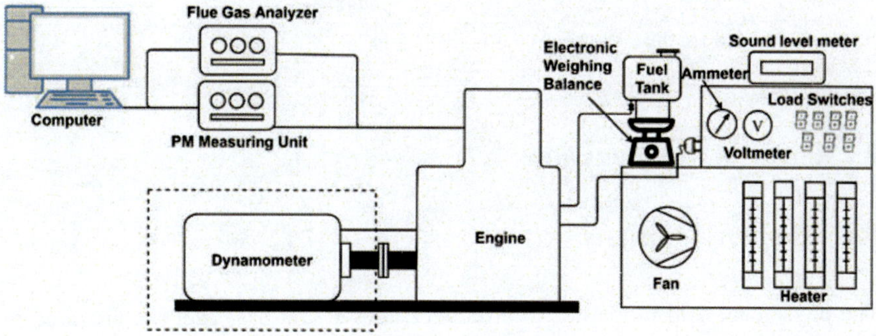

Fig. 2 Schematic diagram of the engine set up

Table 1 Specifications of the CI engine used for experimentation

Specification	Compression ignition engine
No of cylinders	Single
Cooling system	Water-cooled
Fuel pump type	Mechanical (In-line fuel injection)
Type of injection	Direct
Engine displacement	481 cm^3
Stroke	8 cm
Bore	8.75 cm
Rated engine power	6 KW
Rated speed	1500 rpm
Compression ratio	16.7:1

Table 2 Measurement range and accuracy of gas analyzers

Measurement	Measurement range	Accuracy	Resolution
CO	0–10,000 ppm CO	±5 ppm CO (0–199 ppm CO) ±5% of mv (200–2000 ppm to 2000 ppm CO) ± 10% of mv (2001–10,000 ppm CO)	1 ppm CO (0–10,000 ppm CO)
NO	0–4000 ppm NO	±5 ppm NO (0–99 ppm NO) ±5% of mv (100–1999.9 ppm NO) ± 10% of mv (2000–4000 ppm NO)	1 ppm NO (0–3000 ppm NO)
UHC	100–40,000 ppm	<400 ppm (100–4000 ppm) <10% of mv. (>4000 ppm)	10 ppm
PM1, PM2.5, and PM10	0–1000 $\mu g/m^3$	±1 $\mu g/m^3$	–

3 Results and Discussion

3.1 *Property Analysis*

The physiochemical properties of aromatics, diesel, and base fuel are provided in Table 3. The density was measured according to ASTM D1298 methods.

The assumption of linear variation in cetane number was taken to estimate the cetane number for the fuel blends [17]. Bomb calorimeter was used to measure the calorific value of the fuel blends as per ASTM D240. The densities of various fuels were measured as per the ASTM D1298 standard. There was no mechanical failure reported during the testing of different fuel blends. Diesel was run periodically

Table 3 Physiochemical properties of aromatics, diesel, and base fuel

Aromatics	Molecular mass	Density (kg/m^3)	H/C ratio	Specific energy (kJ/kg)	Cetane number	Flash point (°C)
Toluene	92.14	865	1.143	42,439	10	63
Indene	116.16	996	0.78	28,090	7	78.3
Methylnaphthalene	142.2	1001	1	36,700	0	82
De-aromatize hydrocarbon solvent	168	810	1.91	43,600	80.2	75
Diesel	167	835	1.91	42,700	51	66

throughout the testing period to check and verify the working of the research engine along with the coupled equipment.

3.2 *CO and HC Emissions*

CO variation as a function of aromatic contents and loads in fuel blends are presented in Fig. 3a, b. Both CO and HC emissions reduced at high load compared to low load conditions due to the complete CO oxidation at high load conditions with higher in-cylinder temperature. CO emissions increase with an increase in aromatic contents in the fuel blends. Higher C-H ratio in aromatics as compared to de-aromatized hydrocarbon solvent, which could enhance the oxidation of carbon atoms relative to that of hydrogen at a particular energy release [18], hence, leading to higher CO emissions. CO emissions are higher for polycyclic aromatics as compared to monocyclic aromatics. The higher degree of unsaturation in polycyclic aromatics that promote the production of polycyclic hydrocarbon, and hence deteriorate the combustion of binary mixtures in the engine cylinder [8]. 1-methylnaphthalene has double-ring polycyclic structure and the complexity in the structure adversely affects the burning of fuel, hence CO emission increases. Moreover, the higher boiling point of methylnaphthalene deteriorates the vaporization, and hence combustion of the methylnaphthalene blended fuels. Despite having a single ring structure, indene blended fuel emitted more CO emission as compared to other single-ring monocyclic aromatics due to the polycyclic structure of these aromatics that increases the degree of unsaturation.

HC variation as a function of aromatic contents and loads in fuel blends are presented in Fig. 4a, b. The HC emissions increase with the increasing fraction of aromatic contents in the fuel. When the aromatic fraction of a particular aromatic component increases from 5 to 10%, the incremental percentage for HC is lower as

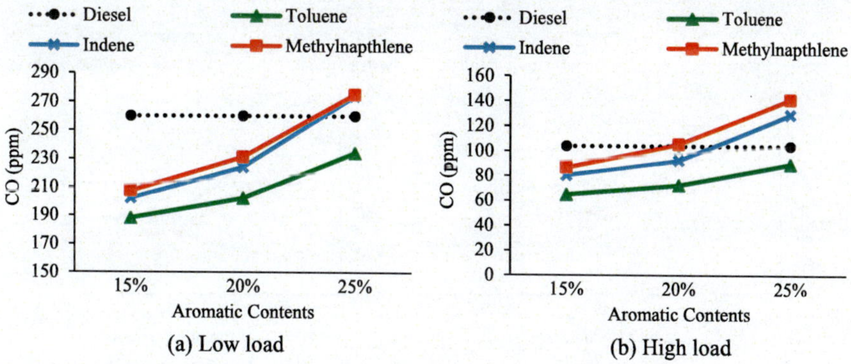

Fig. 3 CO emission variation with aromatic contents

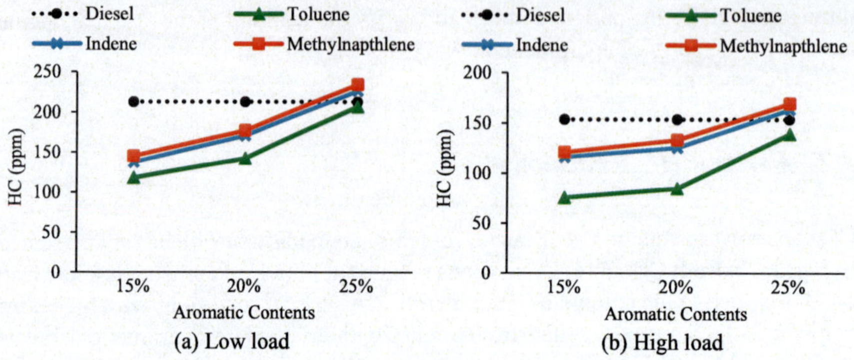

Fig. 4 HC emission variation with aromatic contents

compared to the increment from 10 to 15%. The percentage increment in HC emissions is higher at low load conditions compared to high load conditions due to the efficient oxidation of HC at higher in-cylinder temperatures attained for high load conditions. The fuel fraction that goes toward the cylinder boundary layer inside the combustion chamber influences the overall HC emissions [19]. Higher HC with higher aromatic content in fuel composition occurs with the deterioration in combustion characteristics of fuel blends in the engine chamber due to lower cetane number and lower calorific value of aromatics. The HC emissions are more influenced by the base fuel having polycyclic aromatics instead of monocyclic aromatics.

3.3 NO_x Emissions

The variation in NO_x emissions for different fuel blends is presented in Fig. 5a–b.

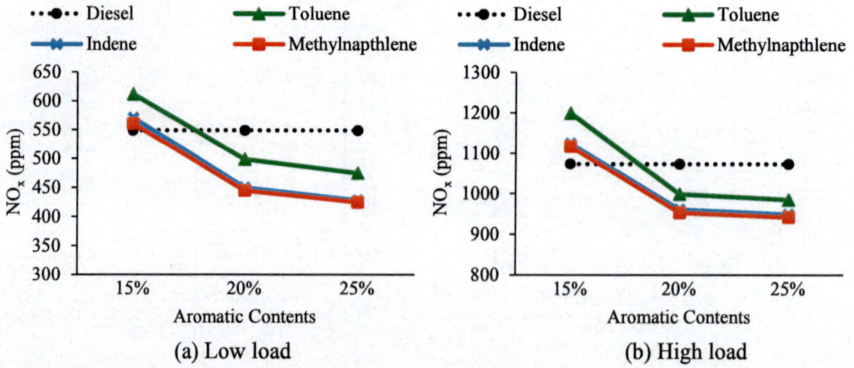

Fig. 5 NO_x emission variation with aromatic contents

NO_x emissions from diesel engine exhaust comprised of NO_x with a small quantity of NO_2 [20]. As the aromatic contents increase in the base fuel, NO_x emissions reduce. This could be expected as the NO_x emissions from a diesel engine depending on the combustion temperature. As the aromatic content increases in the fuel blends, the cetane number and the calorific value reduce, resulting in lower in-cylinder temperature. NO_x emissions are higher for lower aromatic contents in fuel blends. Higher cetane number and calorific value favored by better combustion for 20% aromatic blend case help in better combustion with reduced ignition delay. The percentage reduction in NO_x emissions with an increase in the aromatic fraction from 20–25% from base fuel is not significant.

3.4 PM Emissions

PM emission variation with aromatic contents is presented in Fig. 6. The fuel blends significantly alter PM emissions with 15, 20, and 25% of aromatic contents. Binary mixture with 15 and 20% emit fewer particulates compared to 25% aromatic content and diesel fuel. Higher aromatic contents in base fuel promote the formation of PM. As the global hydrogen to carbon ratio decreases with higher aromatic contents, PM emission increases. Heavy aromatics such as 1-methylnaphthalene also affected the PM emissions due to the poor thermophysical properties and poor atomization of the fuel blend that promotes the high flame temperature in rich fuel region. Due to a higher degree of unsaturation, polycyclic single and double-ring aromatics (indene and 1-methylnaphthalene) produce higher PM and NO_x emissions compared to single-ring monocyclic aromatics (toluene).

The polycyclic exhibits higher boiling points compared with monocyclic single-ring aromatics leading to reduced fuel vaporization rate, which adversely affects the combustion of the binary mixture and results in increased PM emissions. Based on the present studies, it can be concluded that the variation in the aromatics content and

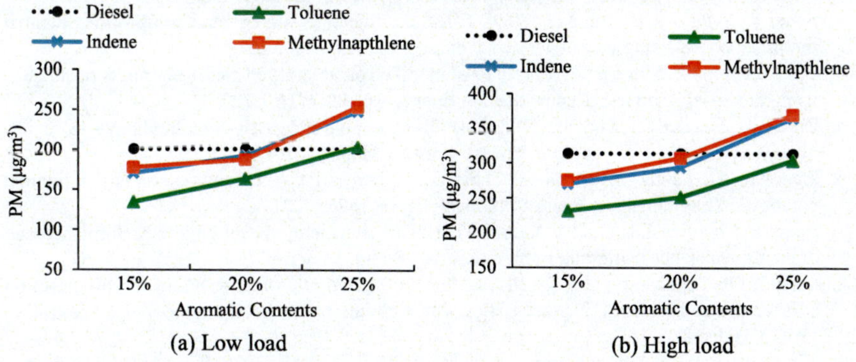

Fig. 6 PM emission variation with aromatic contents

its structure profoundly impact the PM emission formation in compression ignition engines.

The blends with 15% aromatic content show lower PM emissions, however, acoustic emissions (combustion noise and mechanical noise) were found to be higher for these blends. This may be due to a higher cylinder pressure curve for the fuel with lower content of aromatic and higher cetane number of base fuel.

4 Conclusions

Aromatics content in the fuel forms a significant component of fossil-based diesel fuel, and its compatibility with the engine is important to be investigated. Single-ring monocyclic aromatics show promising results as compared to single-ring polycyclic and double-ring polycyclic aromatics due to the high degree of unsaturation of these aromatics. 20% content of toluene in base fuel is an optimum fuel for lower CO, HC, NO_x, and PM emissions compared to 15 and 25% aromatic contents and other tested aromatic. Similarly, 1-methylnaphthalene and indene emitted higher emissions and are not suggested to be part of formulated fuel. Further reduction in aromatic is possible by adding some renewable lubricants in fuel.

Acknowledgements Support for this work by UKIERI (No.16UGC001) is gratefully acknowledged. The authors also acknowledged the contribution of teams at Combustion Research Laboratory, IIT Bombay, India and Low Carbon Combustion Centre, University of Sheffield, UK.

References

1. Varun SP, Tiwari SK, Singh R, Kumar N (2017) Modification in combustion chamber geometry of CI engines for suitability of biodiesel: a review. Renew Sustain Energ Rev 79:1016–1033
2. Knothe G, Razon LF (2017) Biodiesel fuels. Prog Energy Combust Sci 58:36–59
3. Sener R, Yangaz MU, Gul MZ (2020) Effects of injection strategy and combustion chamber modification on a single-cylinder diesel engine. Fuel 266:117122
4. Goel V, Kumar N, Singh P (2018) Impact of modified parameters on diesel engine characteristics using biodiesel: a review. Renew Sustain Energy Rev 82:2716–2729
5. Zheng M, Reader GT, Hawley JG (2004) Diesel engine exhaust gas recirculation—a review on advanced and novel concepts. Energy Convers Manag 45(6):883–900
6. Agarwal AK (2007) Biofuels (alcohols and biodiesel) applications as fuels for internal combustion engines. Prog Energy Combust Sci 33(3):233–271
7. Singh P, Chauhan SR, Goel V, Gupta AK (2020) Enhancing diesel engine performance and reducing emissions using binary biodiesel fuel Blend. J Energy Resour Technol 142(1)
8. Qian Y, Qiu Y, Zhang Y, Lu X (2017) Effects of different aromatics blended with diesel on combustion and emission characteristics with a common rail diesel engine. Appl Therm Eng 125:1530–1538
9. Almohammadi BA, Singh P, Sharma S, Kumar S, Khandelwal B (2020) Impact of alkylbenzenes in formulated surrogate fuel on characteristics of compression ignition engine. Fuel 266:116981

10. Shah SD, Cocker DR, Miller JW, Norbeck JM (2004) Emission rates of particulate matter and elemental and organic carbon from in-use diesel engines. Environ Sci Technol 38(9):2544–2550
11. Gysel NR, Russell RL, Welch WA, Cocker DR III (2016) Impact of aftertreatment technologies on the in-use gaseous and particulate matter emissions from a tugboat. Energy & Fuels 30(1):684–689
12. Mohankumar S, Senthilkumar P (2017) Particulate matter formation and its control methodologies for diesel engine: a comprehensive review. Renew Sustain Energy Rev 80:1227–1238
13. Jayaram V, Agrawal H, Welch WA, Miller JW, Cocker DR III (2011) Real-time gaseous, PM and ultrafine particle emissions from a modern marine engine operating on biodiesel. Environ Sci Technol 45(6):2286–2292
14. Ryan TW, Storment JO, Wright BR, Waytulonis R (1981) The effects of fuel properties and composition on diesel engine exhaust emissions—a Review. SAE technical paper 810953
15. Fukuda M, Tree DR, Foster DE, Suhre BR (1992) The effect of fuel aromatic structure and content on direct injection diesel engine particulates. SAE technical paper 920110
16. Van Beckhoven LC (19991) Effects of fuel properties on diesel engine emissions-a review of information available to the EEC-MVEG Group. SAE technical paper 910608
17. Yanowitz J, Ratcliff MA, McCormick RL, Taylor JD, Murphy MJ (2017) Compendium of experimental cetane numbers. National Renewable Energy Lab (NREL), Golden, CO (United States)
18. Hellier P, Ladommatos N, Allan R, Rogerson J (2013) Combustion and emissions characteristics of toluene/n-heptane and 1-octene/n-octane binary mixtures in a direct injection compression ignition engine. Combust Flame 160(10):2141–2158
19. Verma P, Sharma MP (2015) Performance and emission characteristics of biodiesel fuelled diesel engines. Int J Renew Energy Res 5(1):245–250
20. Hoekman SK, Robbins C (2012) Review of the effects of biodiesel on NO_x emissions. Fuel Process Technol 96:237–249

Fuel Development and Alternative Fuels

Experimental Investigation on a Compression Ignition Engine with Blends of Plastic Oil and Diesel as Fuel

Manoj Kumar and J. M. Mallikarjuna

1 Introduction

As we all know, the carbon dioxide is a greenhouse gas and plays in important role in global warming. The fossil fuels are also ceasing at a rapid rate [1], and in future, there will be scarcity of fossil fuels. So in future to meet energy demand, we have to look over on developing new alternate sources of energy [2]. The plastic oil can be used to obtain energy [3, 4]. It can be obtained by pyrolysis, catalytic cracking, etc [5, 6]. The most commonly used process is the pyrolysis process. Pyrolysis occurs in presence of no oxygen. The plastic waste can be converted into liquid, gas and solid by pyrolysis technique. Pyrolysis plastic oil possesses similar properties as that of petroleum products so can be used in diesel engines [7, 8]. The objective of this research work was to determine whether the use of pyrolysis plastic oil as an alternative diesel engine fuel would be feasible or not and estimate the optimum operational conditions.

Manoj Kumar · J. M. Mallikarjuna (✉)
Indian Institute of Technology Madras, Chennai 600036, India
e-mail: jmmallik@iitm.ac.in

Manoj Kumar
e-mail: mech.er.manoj@gmail.com

A. K. Gupta et al. (eds.), *Advances in IC Engines and Combustion Technology*, Lecture Notes in Mechanical Engineering,
https://doi.org/10.1007/978-981-15-5996-9_7

2 Methods and Materials

Pyrolysis Process

Pyrolysis is a technique of thermal degradation of matter in absence of oxygen. Conversion of waste plastic in oil is shown in Fig. 1. The gasses formed are condensed in condenser to obtain a hydrocarbon distillate, and the fluid is separated using refining to get the oil [9–11].

Comparison of properties of plastic oil and Diesel

#	Properties	Plastic oil	Diesel
1	Density(kg/m^3)	792	850
2	Ash content (%)	<1.02%wt	0.044
3	Calorific value(kJ/kg)	41,800	42,000
4	Kinematic viscosity @40C	2.15	3.04
5	Cetane number	51	55
6	Flash point °C	40	50
7	Fire point °C	45	56
8	Carbon residue (%)	0.01%wt	0.2
9	Pour point °C	−4	3–15
10	Sulphur content	<0.02	<0.034

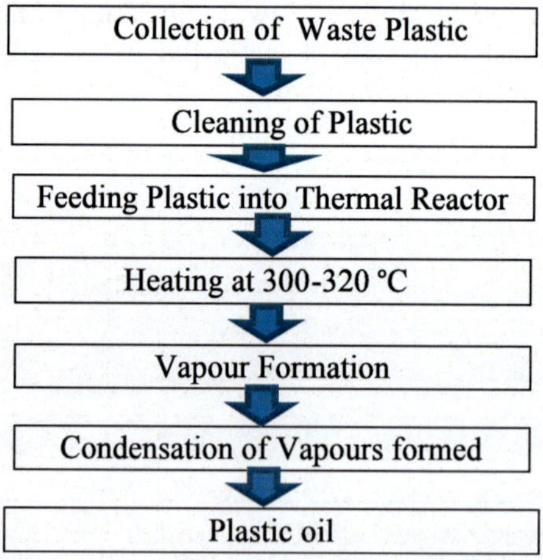

Fig. 1 Conversion of waste plastic into oil through pyrolysis method

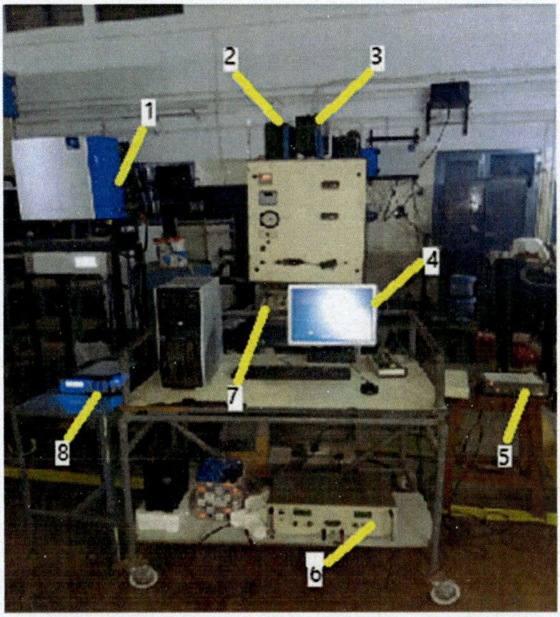

Fig. 2 Front view of experimental setup. 1 Smoke meter, 2 diesel tank, 3 blend tank, 4 DAQ, 5 pressure pick up, 6 voltage regulator, 7 dynamometer controller, and 8 DI GAS analyser

Test Engine

The technical specifications of the engine are given in Appendix A. The engine setup is shown in Fig. 2. The cylinder pressure was measured by a piezoelectric pressure transducer. Two fuel tanks were installed so that diesel and blends can be used separately for the experiments. The mechanical pump was plunger type. For each reading, pressure and TDC data were recorded by a piezoelectric transducer and a TDC sensor using data acquisition system (DDAS). Figure 2 shows front view of the setup used in this work. Cooling water pump was used to provide cooling to dynamometer and pressure pick up.

Performance, Emission and Combustion Data measurement devices

Appendix A Specifications of the engine

Type	Four stroke, naturally aspirated
Manufacturer	KIRLOSKAR TAFI
No. of cylinders	1
Bore × Stroke	87.5 mm × 110 mm
Connecting rod length	220 mm
Rated power	4.4 kw @ 1500 rpm

(continued)

(continued)

Type	Four stroke, naturally aspirated
Compression ratio	17.5: 1

3 Results and Discussion

Combustion characteristics

3.1 *Effect of Blend Ratio and Operating Parameters on Peak in-Cylinder Pressure*

The variation in peak pressure for various loads for blends is represented in Fig. 3. It was found that low blending ratios make little effect on maximum pressure irrespective of the operating conditions, but the peak pressure is increased for higher blending ratios. It may be due to lower viscosity and cetane number of plastic oil as compared to diesel [12]. The spray atomisation and evaporation of the fuel are improved due to lower viscosity of PPO [13], which increased the premixed combustion phase duration. The combustion was delayed because of low cetane number thus increased the premixed burning phase period.

Engine Performance Parameters

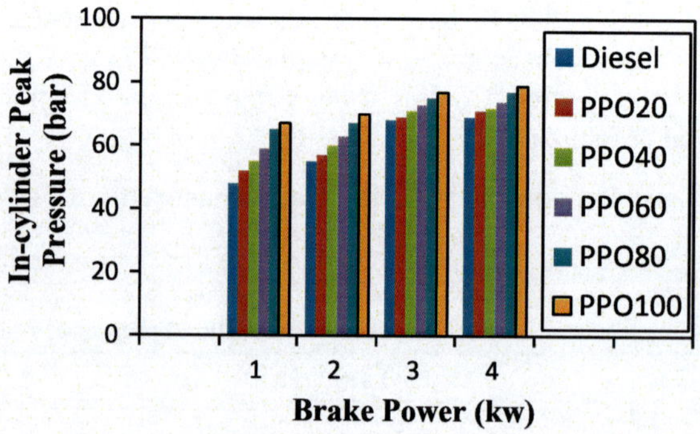

Fig. 3 Effect of blend ratio and operating parameters on maximum in-cylinder pressure

3.2 *Effect of Blend Ratio and Brake Power on Brake Specific Fuel Consumption (BSFC) and Thermal Efficiency (BTE)*

The variation of the BSFC and thermal efficiency for blends and diesel for various loading conditions is shown in Fig. 4a, b, respectively. Due to more fuel demanded, the BSFC for blends is increased for all operating conditions even with the lowest PPO proportions. The lower heating value of PPO fuel was the reason for this. The lower BTE in case of blends may be due to more aromatics contents present in PPO as more energy is needed to breakdown the aromatics bonds [14]. The high heat transfer losses occurred may be because of more burning temperature in case of blends.

3.3 *Effect of Blend Ratio and Brake Power on Volumetric Efficiency*

The variation of volumetric efficiency with brake power for blends and diesel at constant speed is plotted in Fig. 5. The decrease in volumetric efficiency with increasing the blend ratios as shown in Fig. 5 is due to availability of oxygen during combustion (Fig. 6).

Exhaust emissions

3.4 *Effect of Blend Ratio and Operating Parameters on Nitrogen Oxides*

The NO_x mechanisms occur in three different ways, the thermal, prompt and fuel bound NO mechanism [13–17]. The thermal mechanism plays a major role for the formation of NO_x because of the high combustion temperature and more O_2 availability. The longer ignition delay duration caused in large portion of premixed burning, thus more heat and combustion temperature obtained and hence the concentration of NO_x emissions increased parallely as the proportion of blends increased. The more nitrogen availability in the fuel also aggravates the NO_x formation by fuel mechanism.

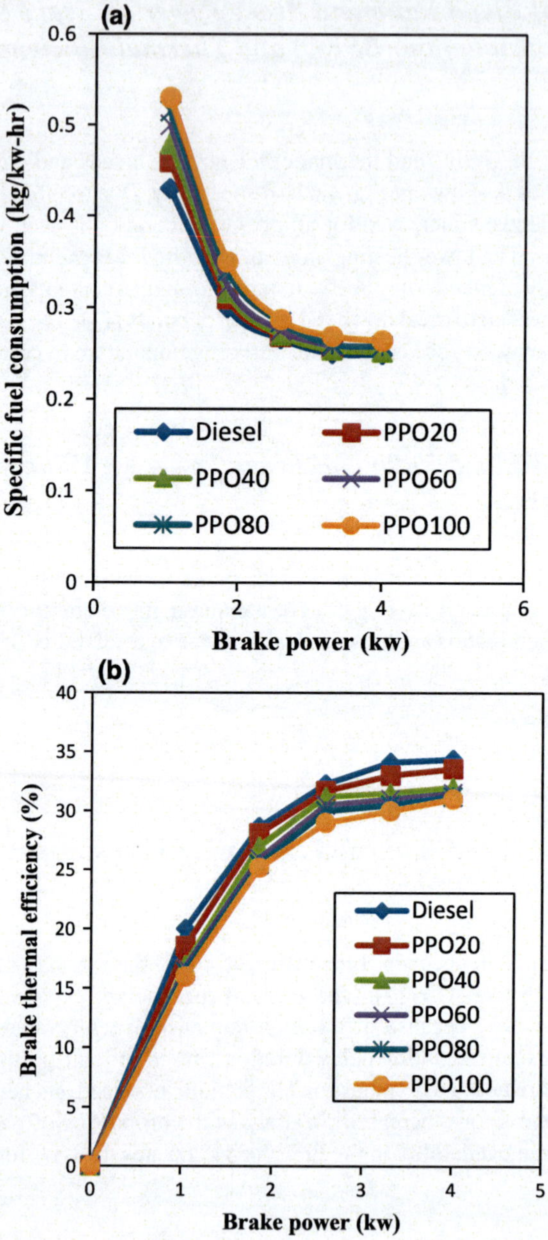

Fig. 4 **a** Effect of blend ratios and BP on BSFC at constant speed (1500 rpm) and **b** effect of the blend ratios and BP on BTE at constant speed

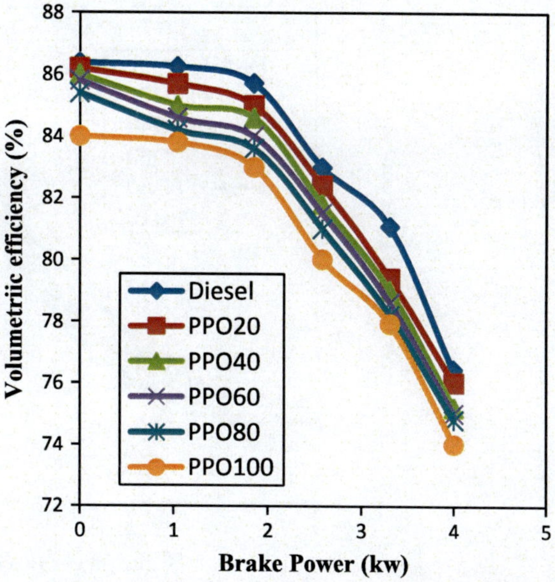

Fig. 5 Effect of blend ratio and BP on volumetric efficiency at constant speed (1500 rpm)

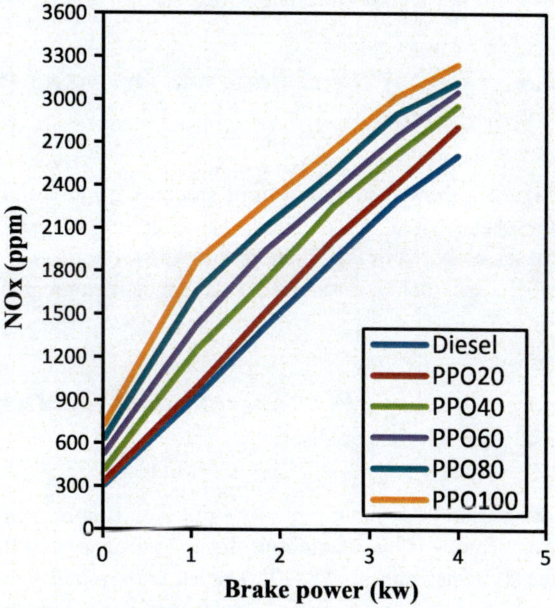

Fig. 6 Variation of NO_x with BP at constant speed (1500 rpm)

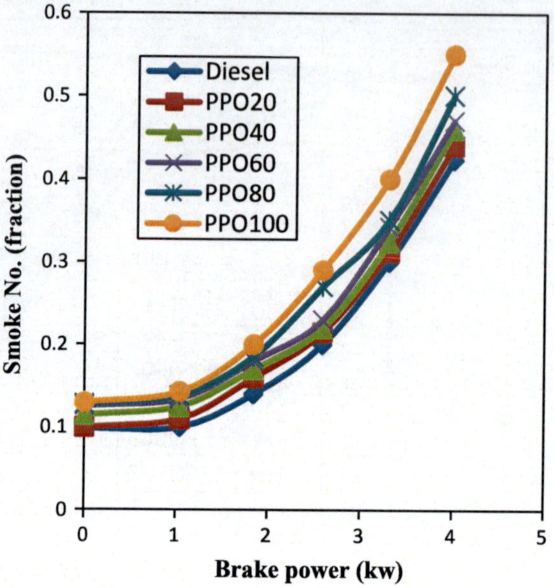

Fig. 7 Effect of brake power on smoke at constant speed (1500 rpm)

3.5 *Effect of Blend Ratio and Operating Parameters on Smoke*

Figure 7 shows the variation of smoke with brake power for different blends and diesel at constant speed. The smoke emitted is more in case of blends as compared to diesel as shown in Fig. 7. Increasing the proportion of pyrolysis plastic oil in blends with diesel increases the smoke as can be seen clearly from Fig. 7.

3.6 *Effect of PPO Concentrations and Operating Parameters on Hydrocarbons*

The variation of unburned HC emissions is shown in Fig. 8. The under-mixing or over-leaning (bulk-quenching) local zones and wall flame quenching mainly affect the HC emissions [15, 18, 19]. The HC emission in case of PPO blends is significantly more than that of diesel, and the HC emissions also increase with more proportions of PPO in blending ratios. For the same PPO blend, HC emissions decrease with load.

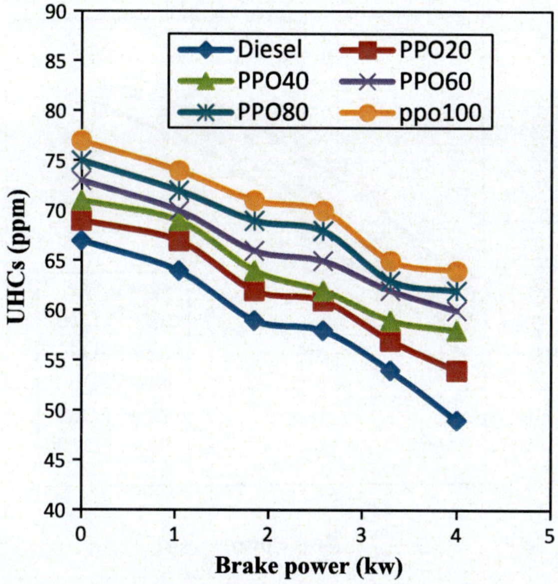

Fig. 8 Variation of HC with BP at constant speed (1500 rpm)

3.7 Effect of Blend Ratio and Operating Parameters on Carbon Dioxide

The carbon dioxide is emitted due to complete combustion of the carbon atoms. Since it is a greenhouse gas so CO_2 emissions should be reduced. The carbon dioxide emitted is more in case of PPO blends as can be seen clearly from Fig. 9. This has bad effect on the environment, nevertheless, this fuel is obtained from wastes which is also non bio-degradable and more CO_2 would be released in future.

3.8 Effect of Blend Ratio and Operating Parameters on Carbon Monoxide (CO)

The CO emissions are mainly formed due to incomplete burning of carbon contained in the fuel, and the CO is mainly affected by the mixture strength and temperature. The tests results show that the CO emissions decrease as load increases irrespective of the fuel as can be seen clearly from Fig. 10. In general, more quantity of plastic oil in blends produces more CO emissions. The incomplete burning at these operating parameters is more due to low cetane number of PPO.

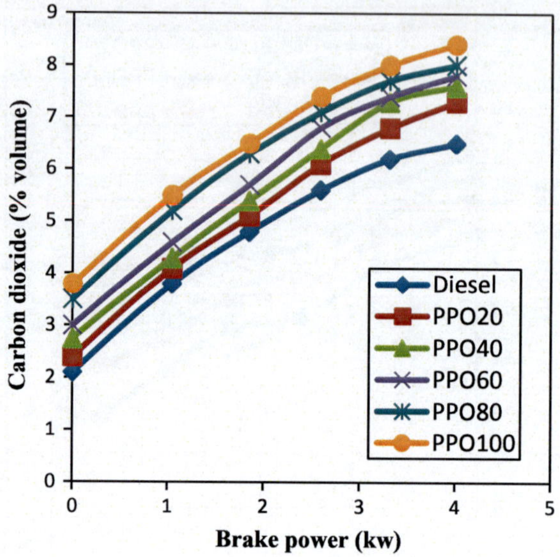

Fig. 9 Effect of blend ratio and BP on CO_2 at constant speed (1500 rpm)

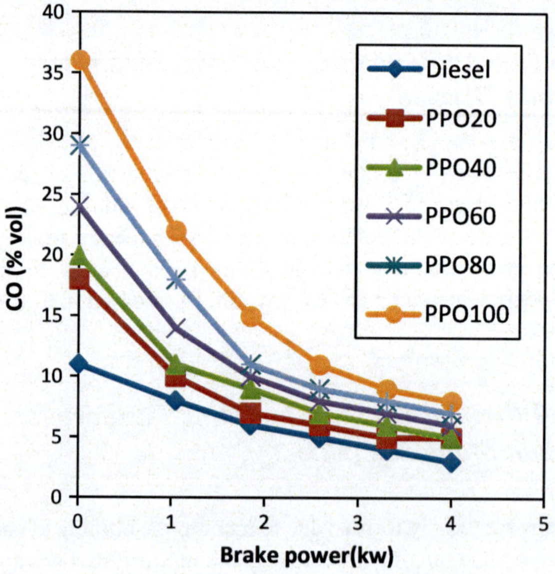

Fig. 10 Effect of blend ratios and BP on CO at constant speed (1500 rpm)

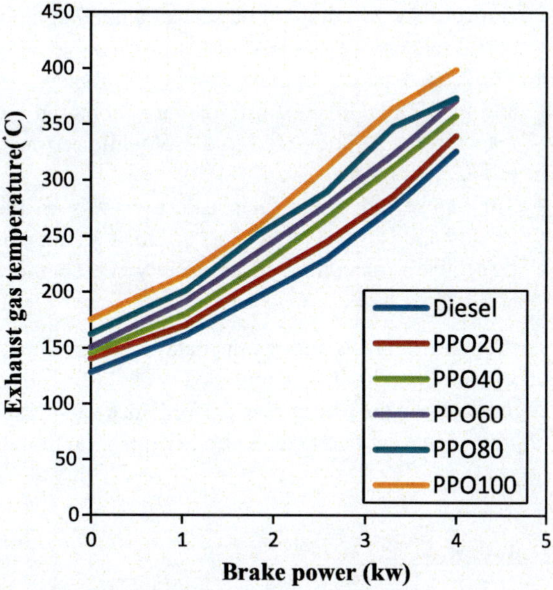

Fig. 11 Effect of BP on EGT at constant speed (1500 rpm)

3.9 The Variation in Exhaust Gas Temperature (EGT) with Different Blend Ratios and Operating Parameters

The EGT in case of blends are increased at all loads as compared to that of diesels as can be clearly seen from Fig. 11. If the quantity of plastic oil is increased in blends, then EGT is also increased. This variation in EGT can be explained by the longer ignition duration.

4 Conclusions

The following conclusions drawn about the emissions are:

- The NO_x emissions for plastic pyrolysis oil blends PPO20, PPO40, PPO60, PPO80 and PPO100 at 90% of full load are increased by 7.65, 13, 17, 23 and 24.2%, respectively, as compared to diesel while engine was running in constant speed (1500 rpm) mode.
- The UHC emissions for plastic pyrolysis oil blends PPO20, PPO40, PPO60, PPO80 and PPO100 at 90% of full load are decreased by 6.5, 17.4, 26, 30 and 39%, respectively, as compared to diesel while engine was running in constant speed (1500 rpm) mode.
- The CO_2 emissions for plastic pyrolysis oil blends PPO20, PPO40, PPO60, PPO80 and PPO100 at 90% of full load are increased by 12.3, 16.9, 20, 23 and 29.2%,

respectively, as compared to diesel while engine was running in constant speed (1500 rpm) mode.

- The CO emissions for plastic pyrolysis oil blends PPO20, PPO40, PPO60, PPO80 and PPO100 at 90% of full load are decreased by 6.5, 17.4, 26, 30.4 and 7.6%, respectively, as compared to diesel while engine was running in constant speed (1500 rpm) mode
- The smoke emissions for plastic pyrolysis oil blends PPO20, PPO40, PPO60, PPO80 and PPO100 at 90% of full load are increased by 4, 8.7, 11, 18 and 30%, respectively, as compared to diesel while engine was running in constant speed (1500 rpm) mode.

From the analysis and results data, it is found that the blends of 60–80% plastic oil operating at 80–90% engine loads could be utilised in CI engine such as diesel engine to attain the best engine performance and reduce exhaust emissions with some improvements such as cetane number injection timing and pressure.

References

1. Mani M, Nagarajan G, Sampath S (2010) An experimental investigation on a DI diesel engine using waste plastic oil with exhaust gas recirculation. Fuel 89:1826–1832
2. Wongkhorsub C, Chindaprasert N (2013) A comparison of the use of pyrolysis oils in diesel engine. Energy and Power Eng 5:350–355
3. Guntur R, Kumar D, Reddy V (2011) Experimental evaluation of a diesel engine with blends of diesel-plastic pyrolysis oil. Int J Eng Sci Technol 6:5033–5040
4. Mani M, Nagarajan G, Sampath S (2011) Characterisation and effect of using waste plastic oil and diesel fuel blends in compression ignition engine. Energy 36:212–219
5. Kumar S, Singh RK (2011) Recovery of hydrocarbon liquid from waste high density polyethylene by thermal pyrolysis. Braz J Chem Eng 28(04):659–667
6. Kumar S, Prakash R, Murugan S, Singh R (2013) Performance and emission analysis of blends of waste plastic oil obtained by catalytic pyrolysis of waste HDPE with diesel in a CI engine. Energy Convers Manage 74:323–331
7. Mani M, Subash C, Nagarajan G (2009) Performance, emission and combustion characteristics of a DI diesel engine using waste plastic oil. Appl Therm Eng 29:2738–2744
8. Pratoomyod J, Laohalidanond K (2013) Performance and emission evaluation of blends of diesel fuel with waste plastic oil in a diesel engine. Int J Eng Sci Technol 2:57–63
9. Consea JA, Font R, Marcilla A, Garcia AN (1994) Pyrolysis of polyethylene in a fluidized bed reactor. Energy Fuel 8:1238–1246
10. Panda AK, Singh RK (2011) Catalytic performances of kaoline and silica alumina in the thermal degradation of polypropylene. J Fuel Chem Technol 39(3):198–202
11. Park JW, Kim JH, Seo G (2002) The effect of pore shape on the catalytic performance of zeolites in the liquid-phase degradation of HPDE. Polym Degrad Stability 76(3):495–501
12. Frigo S, Seggiani M, Puccini M, Vitolo S (2014) Liquid fuel production from waste tyre pyrolysis and its utilisation in a diesel engine. Fuel 116:399–408
13. Murugan S, Ramaswamy M, Nagarajan G (2008) The use of tyre pyrolysis oil in diesel engines. Waste Manage 28:2743–2749
14. Kaimal VK, Vijayabalan P (2015) A detailed study of combustion characteristics of a di diesel engine using waste plastic oil and its blends. Energy Convers Manage 105:951–956
15. Heywood JB (1988) Internal combustion engine fundamentals. McGraw-Hill Inc

16. Kidoguchi Y, Yang C, Kato R, Miwa K (2000) Effects of fuel cetane number and aromatics on combustion process and emissions of a direct-injection diesel engine. JSAE Rev 21:469–475
17. Semerjian H, Vranos A (1977) NO_x formation in premixed turbulent flames. Symp (International) Combustion 16(1):166–179
18. Greeves G, Khan IM, Wang CHT, Fenne I. Origins of hydrocarbons emissions from diesel engines. SAE Technical Paper (770259)
19. Mendez S, Kashdan JT, Bruneaux G, Thirouard B, Vangraefschepe F. Formation of unburned hydrocarbons in low temperature diesel combustion. SAE Technical Paper (2009-01-2729)

Combustion Characteristics of Conventional Diesel Engine and Low Heat Rejection Diesel Engine with Biodiesel Blends

Sharad P. Jagtap, Anand N. Pawar, Subhash Lahane, and D. B. Lata

1 Introduction

There is overwhelming evidence that enhanced greenhouse effect from human activates is changing the global climate. To protect the environment, energy efficiency needs to be adopted and alternative to petroleum/fossil fuel for sustainable development is to be developed. Biodiesel is one of the options for motivating renewable and lower pollution making fuel for internal combustion engine. Hence, it is studied worldwide by long ago as a fuel for compression ignition engines. Biodiesel can substitute or blended with diesel fuel and can be used with some or no changes in CE operation. It is non-toxic, biodegradable and sulfur, aromatics is almost negligible [1]. Oil derived by crushing of oil seeds is not suitable as engine fuel. It has lower heating value, high viscosity, and low volatility. Transesterification is mostly used to reduce viscosity and improve other properties [2, 3]. The engine performance and combustion characteristics can be improved by applying thermal barrier coating (TBC) on engine cylinder head, piston crown, and both valves that causes low heat rejection passing through engine surfaces into jacket cooling water. A wide range of

S. P. Jagtap (✉)
Department of Mechanical Engineering, Jawaharlala Darda Institute of Engineering and Technology, Yavatmal, Maharashtraa, India
e-mail: jagtapsp@gmail.com

A. N. Pawar
Maharashtra State Board of Technical Education, Regional Office, Aurangabad, Maharashtra, India

S. Lahane
Department of Mechanical Engineering, Marathwada Institute of Technology, Aurangabad, Maharashtra, India

D. B. Lata
Department of Energy, Central University of Jharkhand, Ranchi, India

A. K. Gupta et al. (eds.), *Advances in IC Engines and Combustion Technology*, Lecture Notes in Mechanical Engineering,
https://doi.org/10.1007/978-981-15-5996-9_8

TBC materials like PSZ [4] YSZ [5, 6], ZrO [7], mullite [8], $CaZrO_3$, $MgZrO_3$ [9], etc., is used as TBC material on engine components. Here for the experimentation investigation, Oerlikon Metco made METCO 6150 (mullite material) is considered as a TBC material [10] with 250 μm thickness and NiCrAlY as a bond coat with 75 μm is considered.

1.1 Biodiesel Influences on Combustion Characteristics During CE

Biodiesel includes fuel quality like negligible sulfur content, higher cetane number, and lesser aromatic [2]. As higher the cetane number, better will be the cold starting ability. It also provides lower noisy operation and extended engine life. As per Laboratory of National Renewable Energy report, higher bulk modulus is the reason for automatic advance in fuel injection timing. It is the front-runners to quickflow of pressure wave from fuel pump linear injector needle. It is instigating slightly higher fuel line pressure as compared to conventional diesel and causes advance fuel injection time (AIT) [11]. Further start of combustion is also earlier while biodiesel operation because of excess oxygen available in fuel than normal diesel [12]. As biodiesel has low carbon content (10%) and more oxygen (11%) than diesel fuel [13], ID is reduced while fueled with biodiesel, as it has AIT during fuel injection. Larger ignition delay hints to higher value in maximum rate of pressure rise (MRPR) which leads to knocking and noise. As discussed above, dissolved oxygen available in biodiesel fuel and lower C/H ratio [12] lead to improved combustion and result in reduction of carbon monoxide (CO), hydrocarbon (HC), and smoke except NO_x [14]. Peak pressure rises in premixed phase with biodiesel operation which is mostly answerable to greater NO_x emission [15]. Total combustion timing span is also increased due to lower heat value of biodiesel [1].

1.2 Biodiesel Influences on Combustion Characteristics During LHRE

The TBC on combustion chamber components reduces engine coolant heat loss; hence, this unloosed heat may be utilized as the extra heat to enhance engine performance. Higher in-cylinder temperature during injection is observed due to TBC coating, which leads to advances in start of combustion. Hence, ID further reduced and consecutively higher peak pressure and rate of heat release during premixed period [16]. More amount of fuel burned in premixed combustion phase and hence because of the early start with higher in-cylinder temperature results in marginal increased premixed combustion duration [17] with biodiesel blend. Musthafa et al.

[18] reported that because of the longer ignition delay in CE, combustion is introduced at late than LHRE. Hence, the peak point is shifted further away from TDC and causes lower peak cylinder gas pressure during CE. Krishnamoorthi [19] reported that CO emission is further reduced in LHR engine. This is because of 10% higher dissolved oxygen in the biodiesel and higher in-cylinder temperature due to TBC. HC emission is also low in the LHRE because additional heat available due to TBC. Higher heat after combustion causes the burning of volatile HC. NO_x emissions in LHR engine are higher than that for CE [20]. This is because of excess dissolved oxygen available in biodiesel and also higher temperature during fuel combustion due to TBC.

1.3 Effect of Ethanol on Combustion Characteristics

With the aim to reduce NO_x emission and specific fuel consumption with both CE and LHRE, ethanol is blended in biodiesel blend. Alcohols provide higher oxygen level in burning fuel but it increases the energy need for fuel evaporation. Thus, it lowers the combustion chamber temperature and creates scope for dropping NO_x and PM. Low cetane number (CN) and the phase parting from ethanol–diesel mixture are the two major disputes with ethanol use. But if it is used in association with biodiesel and with limited quantity (to form ethanol–biodiesel–diesel blend), the problems could be reduced. Ethanol has lowermost cetane number among diesel and biodiesel causes extended ID period and hence start of combustion timing is increased. How et al. [21] considered two blend fuels, B20 [esterified coconut oil (20%) and diesel (80%)] and B20E05 [esterified coconut oil (20%), ethanol (5%), and diesel(75%)]. The characteristics from experimentation on four cylinders, higher pressure, and common-rail DI engine with B20 and B20E05 are compared with baseline diesel. The extended ignition delay time, lesser viscosity, and better volatility are contributed by the ethanol fraction in DBE blends. That also leads to higher peak HRR with B20 blend. During B20E05 fuel operation, the peak cylinder gas pressure decreases than B20 fuel. This is due to combustion that initiates late and peak pressures point that shifted far away from the top dead center during the expansion stroke. The cooling effect of ethanol could decrease in NO_x emission marginally [22]. In present experimentation, various combustion characteristics of biodiesel blend (B10) through CE and LHRE are compared with baseline diesel at CE operation. To predict further improvement in performance and emission, fraction of ethanol is added in biodiesel blend (to form E05B10 fuel) and again its characteristics are compared with baseline diesel to evaluate scope for effective biodiesel utilization.

2 Investigational Setup

For the investigation of characteristics, single cylinder, four strokes, kirlosker make, water-cooled TV1 model, with eddy current dynamometer attached engine is used. Specifications of the test rig are mentioned in Table 1. Schematic diagram of the test rig is displayed in Fig. 1. The test rig has attached with various instruments for measurement of rate of fuel, adjusted weight, CGP, CGT, crank angle, etc. Two sets of rotameters connected to water pump line are used to maintain water flow for engine and calorimeter. Fuel consumption rate and air flow rate can be measured by using graduated burette and U-tube manometer, respectively. K-type thermocouples are connected to digital indicators to measure temperatures at various essential points. A pressure (piezo) sensor attached to the cylinder head to measure CGP for one degree crank angle position of the cycle time. A rotary encoder is attached on main shaft for crank angle signal record. Both the pressure and crank angle signals are simultaneously measured and communicated to computer through data acquisition unit.

The computer interface engine test rig has externally attachable facility to measure all emission parameters by exhaust gas analyzer. Specifications of exhaust gas analyzers are shown in Table 2. Crude jatropha oil has high viscosity, lower volatility, and lower heating value characteristics that may cause some problems (such as injector chocking, sticky piston rings, thickening lubricating oil, filter gumming with extensive use) during fuel injection and its combustion. The problems are limited by doing transesterification on crude jatropha oil to form monoesters (or biodiesel). The thermo-physical properties of diesel, esterified jatropha oil, and ethanol are shown is Table 3. The transesterification of crude jatropha oil is carried out at M/s Parshuram Bio-energy Pvt. Ltd., Pune. Three types of fuel, baseline diesel, 10% esterified jatropha oil (EJO) blended with diesel (B10) and 5% ethanol (99.8% pure) and 10% EJO blended with pure diesel (E05B10), are considered to evaluate of combustion characteristics. The characteristics for B10 and E05B10 fuels are compared with pure diesel. The comparative characteristics are studied in both CE as well as LHR diesel engine operation. For converting CE to LHRE, 250 μm thick, METCO 6150 (mullite) TBC material is sprayed over cylinder head, piston crown, and crown surface of valves. Initially bond coat with 75 μm thickness was provided to increase

Table 1 Specifications of diesel engine

Rated output	5 HP (3.67 kW)
Speed	1500 RPM
Compression ratio	18:01
Injection pressure	210 bar
Dynamometer	Eddy current, water cooled with loading unit

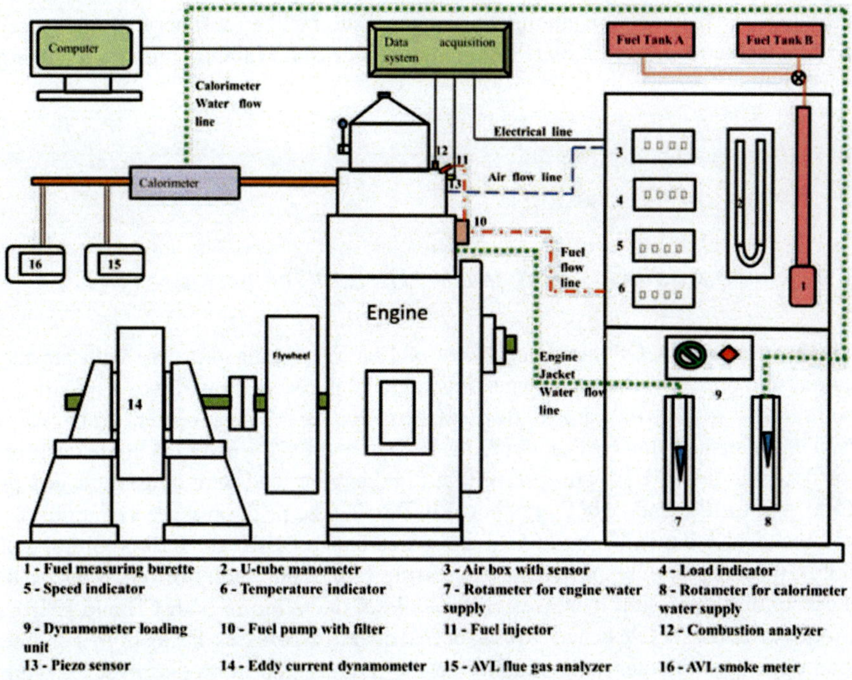

Fig. 1 Representation of experimental setup

Table 2 Emission characteristics gauging equipment

Equipment:	Exhaust gas analyzer
Make and model	AVL 444 N (di-gas analyzer)
Measuring parameters	CO, HC, NO_x
Range	CO (0–10% of exhaust volume), HC (0–2000 ppm), NO_x (0–5000 ppm)
Equipment:	Smoke meter
Make and model	AVL 437
Measuring parameters	Smoke opacity
Range	0–100% capacity in %

Table 3 Properties of base fuels

Property	Diesel	Ethanol	Esterified jatropha oil (EJO)
Density (kg/m^3)	846	789	870
Kinematic viscosity (cSt)	2.51	1.09	12.408
Calorific value; CV (kJ/kg)	42,600	26,700	35,500
Flash point (°C)	64	13	180
Cetane number	46	7	51.3

the adhesive characteristic among top coats and subtract. Atmospheric plasma spray coating technique is used at m/s Plasma Spray Processor, Mumbai (India), for coating operation.

3 Result and Discussion

3.1 *Combustion Gas Pressure (CGP) and Temperature (CGT)*

Characteristics of CGP show capability of fuel for mixing properly with air and combusted properly. During expansion stroke, CGP reached rapidly to a maximum point in few degrees of crank angle displacement after top dead center (a TDC) and then it diminished slowly. The variations in CGP and CGT with crank angle position for baseline diesel, B10, and E05B10 at rated engine load with constant speed of 1500 rpm for CE and LHRE are shown in Fig. 2. The peak pressure variations are 61.79 and 61.4 bar in CE and 63.24 and 62.75 bar in LHRE for B10 biodiesel and E05B10 EBD blend, respectively. Comparing to baseline diesel of 60.16 bar peak pressure, peak pressure increases with biodiesel operation in both CE and LHRE. Biodiesel has lesser CV, hence burning of fuel quantity increases during higher engine load operation. Biodiesels has higher cetane number and more dissolved oxygen, causes improved combustion and higher heat release rate observed during premixed combustion (discussed in latter phase). As temperature of combustion chamber gas is proportionate to pressure, CGT also found increased in biodiesel operation. Also, it is noticed that compared to biodiesel blend fuel operation, CGP and CGT of EBD blend is marginally reduced. This is mainly due to ethanol that has higher latent heat for evaporation and hence during combustion of atomized fuel, it gains heat

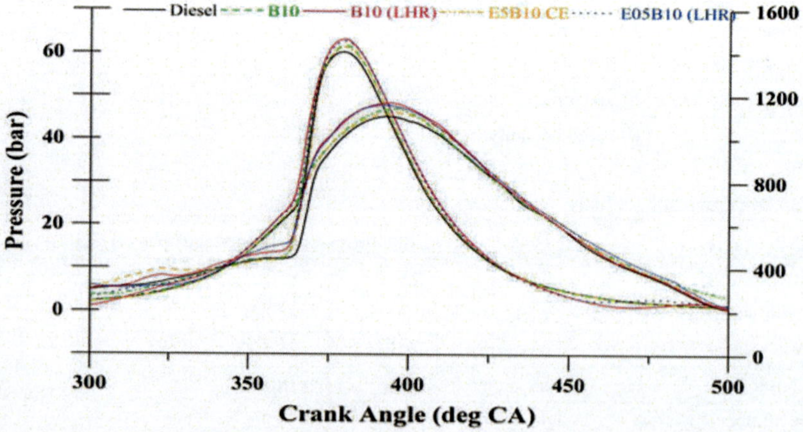

Fig. 2 Variations in combustion chamber pressure and temperature with crank angle

from combustion chamber itself and reduces CGT [21] accordingly CGP also found reduced.

3.2 *Change in CGP Per Degree Crank Angle (ROPR)*

Favorable ROPR variation indicates the smoothness in engine operation. The variations in ROPR with respect to crank angle for baseline diesel, EJO blend (B10), and EBD blend (E5B10) at maximum engine load condition are shown in Fig. 3. It is observed that during the compression, ROPR increased initially due to the increased force over the cylinder piston and reaches maximum value near to peak pressure condition and then it is reduced. During expansion, CGP started retarding, indicating negative. The maximum ROPR variations are 5.48 and 5.61 bar/°CA in CE and 6.1 and 6.16 bar/°CA in LHRE for B10 biodiesel and E05B10 EBD blend, respectively. Compared to baseline diesel (5.83 bar/°CA), the maximum ROPR with biodiesel operation reduces by 0.35 bar/ bar/°CA in CE and increases by 0.18 bar/°CA LHRE operation. The reduction is mainly because of higher cetane number of biodiesel in CE and increment is because of higher in-cylinder temperature (due to TBC on engine components) and early start of combustion in LHRE. The maximum ROPR is reduced by 0.13 and 0.15 bar/°CA in CE and LHRE, respectively, during EBD operation. This is because of lower cetane number of ethanol. The maximum ROPR variations are lower to 9 bar/°CA for immobile engine and indicate that CE as well as LHRE operation is smoother [1].

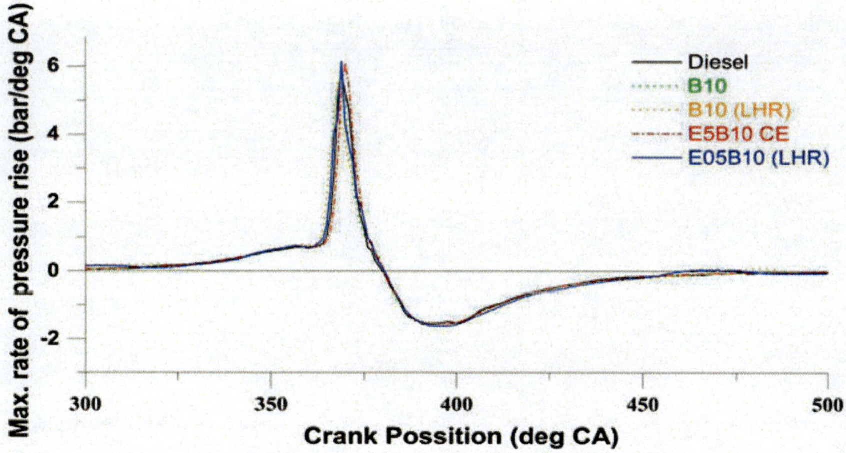

Fig. 3 Variation of rate of pressure rise

3.3 Net Heat Release Per Degree Crank Angle (NHRR)

The NHRR plot informs the maximum available energy which may be converted into useful form. The variations in NHRR for baseline diesel, EJO blend (B10), and EBD blend (E5B10) at maximum engine load condition are shown in Fig. 4. NHRR is an important tool to evaluate combustion characteristics like ignition delay, premixed, diffusion, and afterburning combustion phases for each fuel, which are used to optimize the thermal efficiency and pollutant of diesel engine. NHRR can be calculated using first law of thermodynamics [23, 24]. NHRR ($\frac{dQ_t}{d\theta}$) can be estimated with the relation as shown in Eq. (1). For precise estimation of sudden heat transfer, the cycle-averaged coefficient of heat transfer for each surface among cylinder geometry with varying surface temperatures is to be considered. Heat removal through wall [24] is can be known from Eq. (2)

$$\frac{dQ_t}{d\theta} = \frac{\gamma}{(\gamma - 1)} P \frac{dv}{d\theta} + \frac{1}{(\gamma - 1)} V \frac{dP}{d\theta} + \frac{dQ_{ht}}{d\theta} \quad (1)$$

and

$$\frac{dQ_{ht}}{d\theta} = \frac{h_c \cdot A_c \cdot (T_g - T_w)}{6 \cdot N} \quad (2)$$

{enlargethispage24ptThe instantaneous heat transfer coefficient through cylinder wall [15] in diesel engine and the mean piston velocity (V_p) can be evaluated by Eqs. (3) and (4)

$$h_c = \frac{130 P_c^{0.8} (V_p + 1.4)^{0.8}}{V^{0.06} \cdot T_g^{0.4}} \quad (3)$$

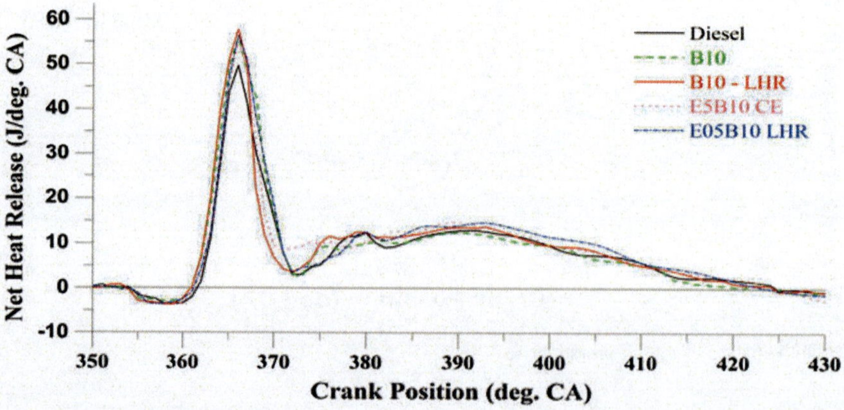

Fig. 4 Net heat release rate

and

$$V_p = \frac{2 \cdot N \cdot S}{60} \tag{4}$$

γ	Specific heat ratio (ranges from 1.3–1.35)	$\frac{dQ_{ht}}{d\theta}$	Heat rejection through cylinder wall (J/°CA)
V_p	Average piston speed (m/s)	V	Cylinder volume at crank position (m^3)
P	Instantaneous pressure (N/m^2)	T_g	Cylinder charge temperature (k)
V	Instantaneous Volume (m^3)	N	Engine rpm
h_c	Coefficient of heat transfer (W/m^2K)	T_w	Temperature of engine wall (k)
P_c	Instantaneous cylinder pressure (bar)	A_c	Surface area for heat transfer (m^2)

3.4 *Ignition Delay (ID)*

Both physical and chemical delay constitutes total ignition delay of injected fuel. It is the time delay before substantial part of the injected fuel energy released. ID is more valuable tool for predicting engine operating characteristics like smooth running, smoke emissions, and comfort in engine start. The physical delay consists of atomization, vaporization, and mixing of fuel droplets with air before the combustion initiate. Chemical delay is crank duration needs for air–fuel precombustion reaction. The discussed delay parameters are altered by fuel characteristics, design of engine, and working variables. Start of injection through nozzle advances automatically due to more bulk modulus of B10 fuel than conventional diesel fuel. The pressure waves are generated in fuel line thereby fuel (biodiesel blends) propagates more rapidly toward the injector and hence start of injection is advanced. Biodiesel has more bulk modulus (1500 MPa) than diesel (1350 MPa) [12]. Also, biodiesel has more dissolved oxygen at the core of fuel molecule causing earlier and improved combustion. It advances start of combustion and hence effectively ID time is reduced with B10 fuel. This ID reduction effect is quite less in case of EBD fuel, because ethanol required more hidden energy for vaporization (changing phase) of injected fuel blend and hence during combustion of atomized fuel, it gains heat from combustion chamber itself and makes slightly delay in initiation in combustion. ID is measured from negative NHRR loop before initiation of fuel burning as shown in Fig. 5. ID variations are 7.14 and 7.26 °CA in CE and 7.08 and 7.24 °CA in LHRE for B10 biodiesel and E05B10 blend, respectively, at maximum engine load operation. Compared to baseline diesel of 7.35 °CA, the ID is reduced by 0.21 °CA with biodiesel and 0.09 °CA with EBD blend during CE operation. ID is further reduced by 0.27 and 0.11

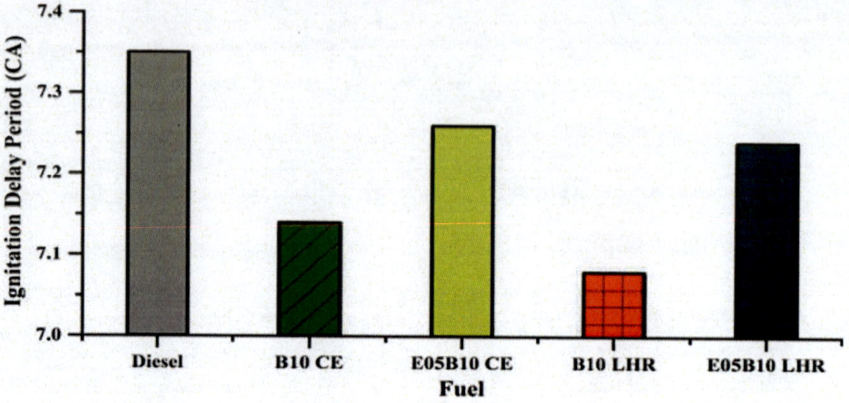

Fig. 5 Ignition delay

°CA with B10 and E05B10 fuels, respectively, during LHRE operation. This further reduction of ID in LHRE is mainly because of more CGT (because of TBC on engine components), automatic advance in injection, and early initiation of fuel burning in LHRE.

3.5 Combustion Phases

From NHRR, the premixed combustion, diffusion combustion, and afterburning combustion duration (in CA) for B10, E05B10, and for baseline diesel are evaluated as stated in Fig. 6. Premixed combustion variations are 11.48 and 10.84 °CA in CE and 11.84 and 11.02 °CA in LHRE for B10 biodiesel and E05B10 blend, respectively, at maximum engine load operation. Compared to baseline diesel of 10.54 °CA, the premixed combustion is increased by 0.94 and 0.30 °CA with B10 and E05B10 blend during CE operation, respectively. Premixed combustion duration is further increased by 1.30 and 0.48 °CA with B10 and E05B10 fuels, respectively, during LHRE operation. Rise in premixed results in greater NO_x formation. Hence, the addition of 5% ethanol in existing B10 biodiesel blend results in reduction of NO_x emission. The diffused combustion variations are 37.74 and 33.92 °CA in CE and 37.60 and 33.52 °CA in LHRE for B10 biodiesel and E05B10 blend, respectively, at maximum engine load operation. Compared to baseline diesel of 39.27 °CA, the diffused combustion is decreased by 1.53 and 5.35 °CA with B10 and E05B10 blend during CE operation, respectively. Diffused combustion duration is further decreased by 1.67 and 5.65 °CA with B10 and E05B10 fuels, respectively, during LHRE operation. The decrease in diffused combustion phase results in drop of CO and HC pollutants. Also, with addition of 05% ethanol in existing B10 fuel, there is further decrease in CO and HC pollutants. EBD blend may be advantageous to suppress all CO, HC, and NO_x generation from biodiesel blend operation. The afterburning

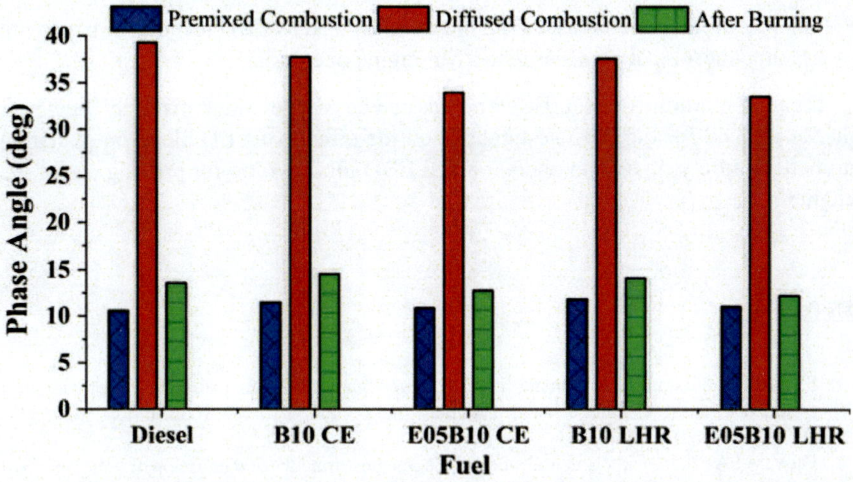

Fig. 6 Combustion phases

combustion variations are 14.54 and 12.76 °CA in CE and 14.10 and 12.22 °CA in LHRE for B10 biodiesel and E05B10 blend, respectively, at maximum engine load operation. Compared to baseline diesel of 13.54 °CA, the afterburning combustion is increased by 1.0 and decreased by 0.78 °CA with B10 and E05B10 blend during CE operation, respectively. Afterburning combustion duration is marginally increased by 0.56 and decreased by 1.32 °CA with B10 and E05B10 fuels, respectively, during LHRE operation. The decrease in afterburning combustion phase due to 5% ethanol indicates overall reduction in combustion duration.

4 Conclusions

Esterified jatropha biodiesel–diesel blend (B10) and ethanol–biodiesel–diesel blend (E05B10) fuels are used in four strokes, single cylinder, direct injection-type diesel engine and mullite material insulated LHRE. The combustion characteristics from both CE and LHRE are compared with baseline diesel of CE. The following conclusions are drawn:

- Higher CGP and CGT with biodiesel blend (B10) may enhance NOx formation. The CGP and CGT can be slightly reduced by adding ethanol in the biodiesel blend (to form E5B10) which could be reduced NO_x emission.
- Premixed combustion phase increases in both B10 and E05B10 fuel blends, may results in increased specific fuel consumption and also improved combustion.
- The decrease in diffused combustion phases with B10 as well as E05B10 results in more mass fraction burned in work generating premixed combustion results in less CO formation.

- The MROP rise is decreased with biodiesel as well as EBD blend compared with baseline diesel that provides smoother engine operation.

Ethanol–biodiesel–diesel (EBD) blends can be a better alternative fuel; however, there is a need for further investigations on the effects of EBD blend on oxidation properties, safety on engine operation and EBD mixing ratio for prolong use of the engine.

References

1. Shelke PS, Sakhare NM, Lahane S (2016) Investigation of combustion characteristics of a cottonseed biodiesel fuelled Diesel Engine. Proc Technol 25:1049–1055. https://doi.org/10.1016/j.protcy.2016.08.205
2. Patwary SA (2014) Experimental study of Engine performance and emission of palm, mustard and calophyllum biodiesel blends in a Diesel Engine Sanjid Ahmed Patwary Dissertation Submitted in Fulfillment of the Requirements for the Degree of Master of Engineering Science Facu 2014
3. Sahoo PK, Das LM, Babu MKG, Naik SN (2007) Biodiesel development from high acid value polanga seed oil and performance evaluation in a CI engine. Fuel 86:448–454. https://doi.org/10.1016/j.fuel.2006.07.025
4. Rajendra Prasath B, Tamilporai P, Shabir MF (2010) Analysis of combustion, performance and emission characteristics of low heat rejection engine using biodiesel. Int J Therm Sci 49:2483–2490. https://doi.org/10.1016/j.ijthermalsci.2010.07.010
5. Mittal N, Athony RL, Bansal R, Ramesh Kumar C (2013) Study of performance and emission characteristics of a partially coated LHR SI engine blended with n-butanol and gasoline. Alexandria Eng J 52:285–293. https://doi.org/10.1016/j.aej.2013.06.005
6. Vamsi Krishna K, Sastry GRK, Murali Krishna MVS, Deb Barma J (2018) Investigation on performance and emission characteristics of EGR coupled semi adiabatic diesel engine fuelled by DEE blended rubber seed biodiesel. Eng Sci Technol Int J 21:122–129. https://doi.org/10.1016/j.jestch.2018.02.010
7. Işcan B, Aydin H (2012) Improving the usability of vegetable oils as a fuel in a low heat rejection diesel engine. Fuel Process Technol 98:59–64. https://doi.org/10.1016/j.fuproc.2012.02.001
8. Rao BN, Prem Kumar BS, Kumar Reddy KV (2016) Effect of CNG flow rate on the performance and emissions of a Mullite-coated diesel engine under dual-fuel mode. Int J Ambient Energy 37:589–596. https://doi.org/10.1080/01430750.2015.1023835
9. Taymaz I (2006) An experimental study of energy balance in low heat rejection diesel engine. Energy 31:364–371. https://doi.org/10.1016/j.energy.2005.02.004
10. Spray T, Products P, Metco I, Coating EB, Ebcs S, Composites CM et al (2014) Material product data sheet premium EBC-Grade Aluminum Silicate (Mullite) Powder, 3–5
11. Tat ME, Van Gerpen JH (2003) Measurement of biodiesel speed of sound and its impact on injection timing: final report; Report 4 in a Series of 6. Other Inf PBD 1 Feb 2003; Relat Inf Work Performed by Iowa State Univ Ames, Iowa 2003: Medium: ED; Size: 120 pages
12. Lahane S, Subramanian KA (2015) Effect of different percentages of biodiesel-diesel blends on injection, spray, combustion, performance, and emission characteristics of a diesel engine. Fuel 139:537–545. https://doi.org/10.1016/j.fuel.2014.09.036
13. Dwivedi G, Jain S, Sharma MP (2013) Diesel engine performance and emission analysis using biodiesel from various oil sources—review. J Mater Environ Sci 4:434–447. https://doi.org/10.5897/AJB12.2931

14. Nabi MN, Rasul MG, Anwar M, Mullins BJ (2019) Energy, exergy, performance, emission and combustion characteristics of diesel engine using new series of non-edible biodiesels. Renew Energy. https://doi.org/10.1016/j.renene.2019.03.066
15. Subramanian KA, Lahane S (2012) Comparative assessment of injection, combustion, performance and emission characteristics of a diesel engine for biodiesel-diesel blends. Int J Renew Energy Technol 3:410. https://doi.org/10.1504/ijret.2012.049526
16. Abedin MJ, Masjuki HH, Kalam MA, Sanjid A, Ashraful AM (2014) Combustion, performance, and emission characteristics of low heat rejection engine operating on various biodiesels and vegetable oils. Energy Convers Manage 85:173–189. https://doi.org/10.1016/j.enconman.2014.05.065
17. Parlak A, Yasar H, Sahin B (2003) Performance and exhaust emission characteristics of a lower compression ratio LHR Diesel engine 44:163–175
18. MohamedMusthafa M, Sivapirakasam SP, Udayakumar M (2011) Comparative studies on fly ash coated low heat rejection diesel engine on performance and emission characteristics fueled by rice bran and pongamia methyl ester and their blend with diesel. Energy 36:2343–2351. https://doi.org/10.1016/j.energy.2010.12.047
19. Krishnamoorthi T (2019) Performance and emission characteristics analysis of thermal barrier coated diesel engine using palm biodiesel
20. Karthickeyan V (2018) Experimental analysis on thermally coated diesel engine with neem oil methyl ester and its blends. Heat Mass Transf Und Stoffuebertragung 54:1961–1974. https://doi.org/10.1007/s00231-018-2286-6
21. How HG, Masjuki HH, Kalam MA, Teoh YH (2014) Engine performance, emission and combustion characteristics of a common-rail diesel engine fuelled with bioethanol as a fuel additive in coconut oil biodiesel blends. Energy Proc 61:1655–1659. https://doi.org/10.1016/j.egypro.2014.12.185
22. Zhu L, Cheung CS, Zhang WG, Huang Z (2010) Science of the total environment emissions characteristics of a diesel engine operating on biodiesel and biodiesel blended with ethanol and methanol. Sci Total Environ 408:914–921. https://doi.org/10.1016/j.scitotenv.2009.10.078
23. Lahane S, Subramanian KA (2014) Impact of nozzle holes con figuration on fuel spray, wall impingement and NO_x emission of a diesel engine for biodiesel e diesel blend (B20), vol 64, pp 307–314
24. Sahoo PK, Das LM (2009) Combustion analysis of Jatropha, Karanja and Polanga based biodiesel as fuel in a diesel engine. Fuel 88:994–999. https://doi.org/10.1016/j.fuel.2008.11.012

Performance and Emission Analysis of RCCI Engine Fuelled with Acetylene Gas

M. Sonachalam and V. Manieniyan

1 Introduction

Fossil fuels powered engines are used as electrical generators, automobiles, and ships for its unique workability and thermal efficiency. Although by using diesel engines, it will emit enormous amount of pollutants like carbon monoxide (CO), hydrocarbon (HC), nitrogen oxides (NO_x), particulate matters, and sulfur oxides (SO_x). For saving the fuel utilization and reducing pollutant emission for environmental protection, the research is carried out toward the efficiency improvement and emission reduction [4]. To resolve this issue researchers looks for the fuel management Diesel engines is evolved. The promising secondary gaseous fuels for CI engines are LPG, CNG, biogas, producer gas, hydrogen, and acetylene. The advantage of gaseous fuel usage in CI engines is not only effectively utilized by the CI engine and also it will emit low levels of pollutants. The gaseous fuels are normally high level of self-ignition temperature [12].

With regard to fuel consumption and environmental safety, the research on engine envelopes the increase in efficiency and in emission deduction. With an end goal to diminish NO_x and soot emissions in-cylinder, while keeping up the high thermal efficiency, numerous compression ignition combustion methodologies have been carried out in the literature. Reactivity-controlled compression ignition (RCCI) combustion is one of the least complicated approaches to achieve low soot emissions and NO_x in a CI engine. RCCI engine is most capable of achieving greater heat effectiveness and reduced NO_x emissions [13]. The reactivity-controlled combustion is accomplished by creating chamber reactivity stratification characterized by specifically diverse

M. Sonachalam (✉) · V. Manieniyan
Annamalai University, Annamalainagar, India
e-mail: altimatesona@gmail.com

V. Manieniyan
e-mail: manieniyan78@gmail.com

A. K. Gupta et al. (eds.), *Advances in IC Engines and Combustion Technology*, Lecture Notes in Mechanical Engineering,
https://doi.org/10.1007/978-981-15-5996-9_9

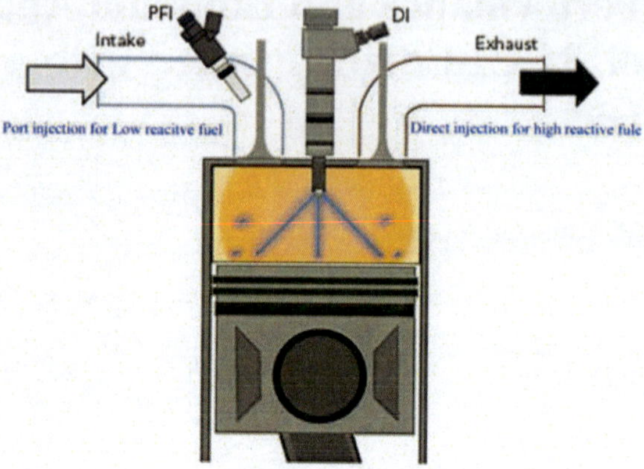

Fig. 1 RCCI engine line diagram

cetane figures through the use of two fuels. Initially, the intake manifold is premixed by the small cetane amount of gas and air, after which a direct injector injects the elevated fuel of cetane number. The reactivity stratification is subsequently generated. By intentionally altering the ratio of two fuels and the timing of the injection, the reactivity gradient generated can track the phasing of combustion and mitigate the rate of pressure increase and the rate of thermal release [7].

Figure 1 shows the RCCI engine line diagram. The intake manifold consists of a port fuel injector (PFI). The low reactivity fuel (LRF) is gasoline. During the intake stroke of combustion process, gasoline is passed through PFI, and then, it is premixed with air in the cylinder. The high reactivity fuel (HRF) is diesel; it is passed through the direct injector (DI) into the cylinder. During compression stroke, the injection strategy adopted is single, double, or triple. The HRF injected in the early hours focuses at the squish region while the comparatively late injected HRF stages an ignition source job. This results in RCCI combustion. In this work, the low reactive fuel used is acetylene gas and high reactive fuel is diesel. Lime stone-produced acetylene gas ($CaCO_3$) is in nature renewable and has comparable characteristics to hydrogen (Fig. 2).

Benajes et al. [2] conducted the test in RCCI engine with diesel. They reported that emission of CO is reduced 90% compared to standard diesel engine. Kakoee et al. [9] investigated the compressed natural gas and hydrogen gas with diesel in RCCI engine. The emission of CO and HC emission was reduced 35.5% and 29.8% in natural gas compared to diesel blends, but higher in NO_x emission. The results conclude that both performance and emission were superior in natural gas and hydrogen gas with pre-chamber. Li et al. [10] carried the investigation in RCCI with gasoline and biodiesel blends. The double injection strategy technique were introduced one is dwell and other one is split fraction. The results were obtained longer ignition delay due to double injection strategy. The results were reported that the BTH was higher in biodiesel blends with gasoline while compared to biodiesel, and emission was also reduced except for oxides of nitrogen. Liu et al. [11] conducted the different parameters of shape of combustion chamber and injection diesel and natural gas engine to

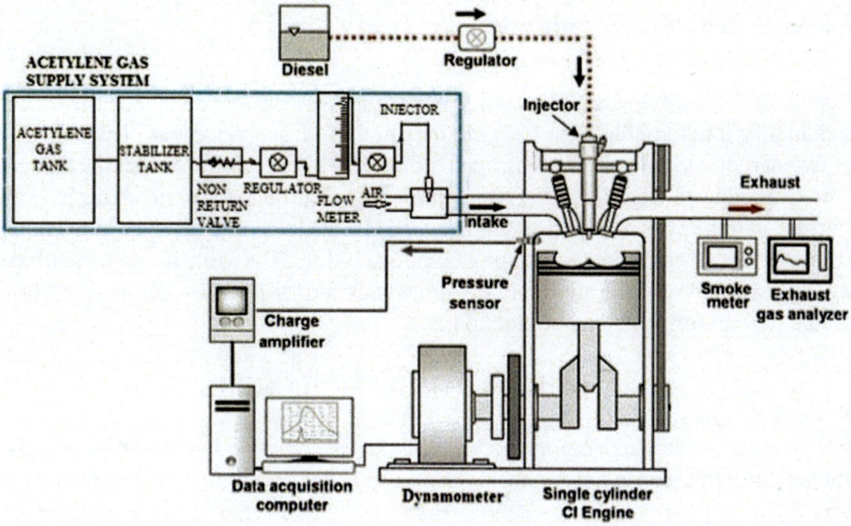

Fig. 2 Experimental arrangement

obtain reduced emissions and fuel consumption. Results show that standard combustion chamber is efficient in reducing CH_4 emissions and improving consumption of fuel, and BTE reaches 50.2%. Ilhak et al. [8] conducted research in a spark ignition engine with acetylene blends with gasoline, and this research provides experimental outcomes of engine efficiency, combustion, and emission. Results were shown that acetylene to gasoline reduced the BTE at wealthy and stoichiometric circumstances owing to reduced ignition timing to avoid knocking and increased CO emissions. Sasikumar and Sankaranarayanan [15] analyzed in a single cylinder diesel engine with biodiesel and acetylene gas in diesel. Their results were shown with improved performance and decreased emission compared to that of biodiesel with acetylene gas compared to diesel. Srivastava et al. [17] conducted the impact of the compression ratio on the emission combustion and performance features of the stationary CI engine fuelled by diesel acetylene. Emissions of CO, HC, and smoke were also evaluated and discovered to be smaller, whereas NO_x emissions were greater in dual-fuel mode at optimized values compared to diesel.

The idea of RCCI also has the benefit of being able to use various fuels with distinct features of reactivity. In particular, it was discovered that RCCI engine combustion is capable of running at an acceptable pressure increase rate over a broad spectrum of engine loads with a negligible quantity of NO_x and soot. An engine should minimize the need for decrease of emissions after therapy in order to achieve maximum engine efficiency. Generally speaking, RCCI motors are advances in dual-fuel combustion technology that utilizes at least two distinct reactivity fuels mixed together in-cylinder blends.

2 Acetylene Gas Production

Acetylene is a colorless and distinctive odor gas. When acetylene is mixed with air, it is highly inflammable. The C_2H_2 is the symbol of acetylene gas. This consists of two carbon atoms and two hydrogen atoms. There are several processes that are used to produce acetylene gas. The simplest way to produce acetylene gas is from calcium carbide (CaC_2) reacting with water [8]. Calcium carbide is an industrially produced by carbo-thermal reduction of lime at 2000 °C. The form of calcium carbide structure in room temperature is a distorted rock structure. The chemical reaction can be written for acetylene production is,

$$CaC_2 + 2H_2O \rightarrow C_2H_2 + Ca(OH)Ca(OH)_2$$

The by-product which comes out at the end of the reaction is calcium hydroxide. The calcium hydroxide can be back again in to water and lime at a temperature of 512 °C.

3 Experimental Setup

Four stroke, water cooled, and single cylinder Kirloskar TV 1 diesel engine was used in this experiments to measure the performance and emission. The engine was connected to the dynamometer. A control panel was interfaced with the dynamometer. The engine was run at a 1500 rpm for entire test. Emissions such as CO, HC, and NO_x were evaluated in the AVL DI gas analyzer, and AVL smoke meter was used to measure smoke density. Thermocouple was used to calculate the exhaust gas temperature. The combustion parameter such as cylinder pressure was calculated using AVL combustion analyzer. The fuel diesel as a high reactivity fuel (HRF) which is injecting to combustion chamber direct injection (DI) and acetylene as a low reactivity fuel (LRF) which is injected along with air through the inlet manifold of the engine in the range of 1 LPM to 5 LPM (liter per minute) with equal increment of 1 LPM. The engine specifications are given in Table 1.

Table 1 Physical properties of fuels

Properties	Diesel	Acetylene
Density kg/m^3	825	1.092
Self-ignition temperature (°C)	210	305
Calorific value (kJ/kg)	42,500	48,225
Critical temperature (°C)	446.85	35.5

4 Results and Discussion

Figure 3 shows the difference of brake thermal efficiency (BTE) in diesel and acetylene fuel. As the quantity of acetylene increases, the BTE increases in RCCI combustion in entire loads. The BTE is increased about 10.5% by injecting 3 LPM acetylene compared to other acetylene blends in RCCI mode combustion. The enhancement of the air flow rate shall be decreased because of the heavy acetylene pressures in the combustion chamber, which is to be increased. When the charge is high, the burning at full load is normal, and the additional acetylene can be concluded to be active in the cylinder [15].

Figure 4 shows that the difference of specific fuel consumption (SFC) for conventional diesel fuel combustion and RCCI engine combustion. The SFC slowly increases with higher percentage of acetylene. The graph shows that the 3 LPM RCCI combustion is having lesser SFC when compared to all other RCCI combustion. The reason behind the acetylene having higher heat content enables the combustion more efficient [5]. The engine based in acetylene also shows a combination of pre-combustion and diffusion. However, for a short time, the pre-premixed combustion is the noticeable fact that differs from the baseline diesel engine really [17].

Figure 5 illustrates that the oxides of nitrogen (NO_x) for conventional diesel fuel combustion and RCCI engine combustion. The trend shows that there is an increment in NOx emission of each increment of acetylene quantity in LPM. NO_x formation is due to the temperature of the sample, the length of the response, and the accessibility of oxygen [16], according to the Zeldovich system. In this case, the rise in NO_x

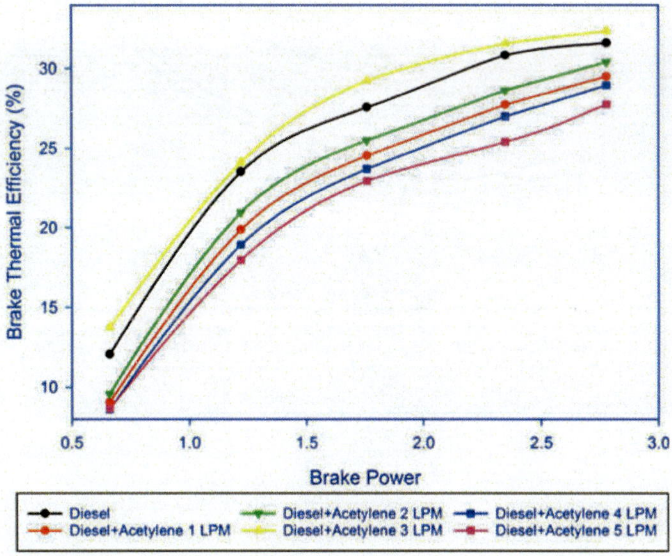

Fig. 3 Brake thermal efficiency with brake power

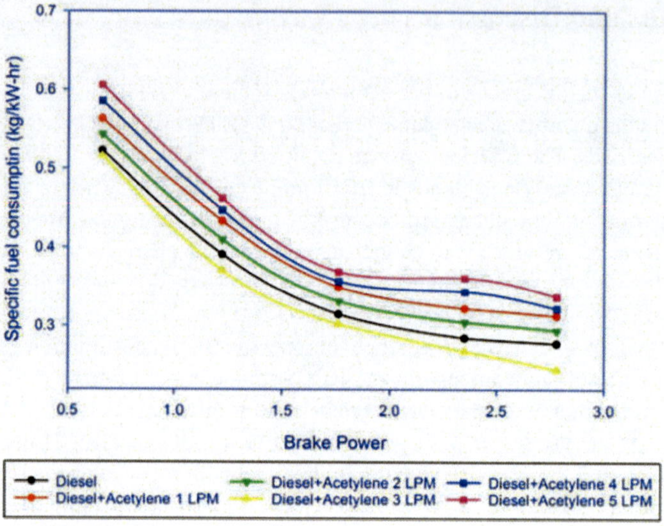

Fig. 4 Specific fuel consumption with brake power

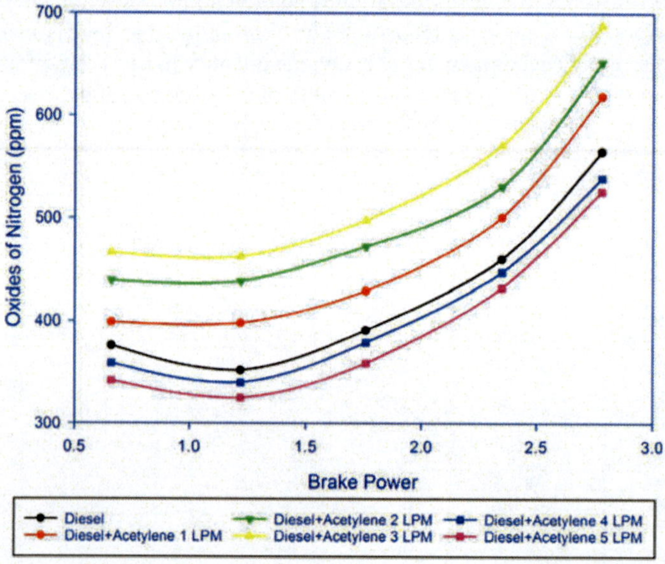

Fig. 5 Oxides of nitrogen with brake power

Table 2 Engine specifications

Type	Four stroke, vertical inline diesel, water cooled engine
No of cylinder	Single cylinder
Compression ratio	17.5:1
Stroke, bore	110 mm, 87.5 mm
Dynamometer	Eddy current
Brake power	5.2 kW
Speed	1500 rpm
Ignition system	Compression ignition
Ignition timing	23° BTDC (rated)
Injection pressure	220 kgf/cm^2
No of cylinder	Single cylinder

may be due to the enhanced higher temperature due to quicker discharge of energy, which is verified by the enhanced higher heat release rate, when inducing acetylene. The increment of oxides of nitrogen (NO_x) emission is about 17% in 3 LPM RCCI combustion mode when compared to diesel (Table 2).

Figure 6 illustrates the smoke density (HSU) with brake power (BP) for conventional diesel fuel combustion and RCCI engine combustion. The trend shows there will be a decrease in smoke density due to the increase in acetylene amount. Generally, the amount of smoke rises as the rate of diesel flow rises. The amount of smoke

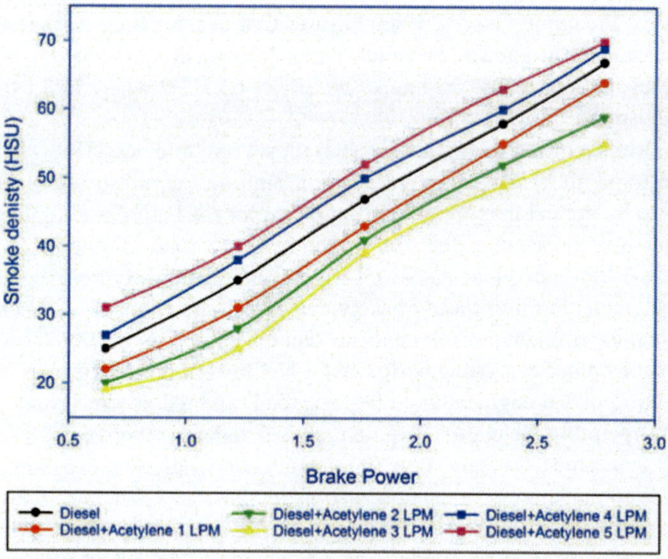

Fig. 6 Smoke density with brake power

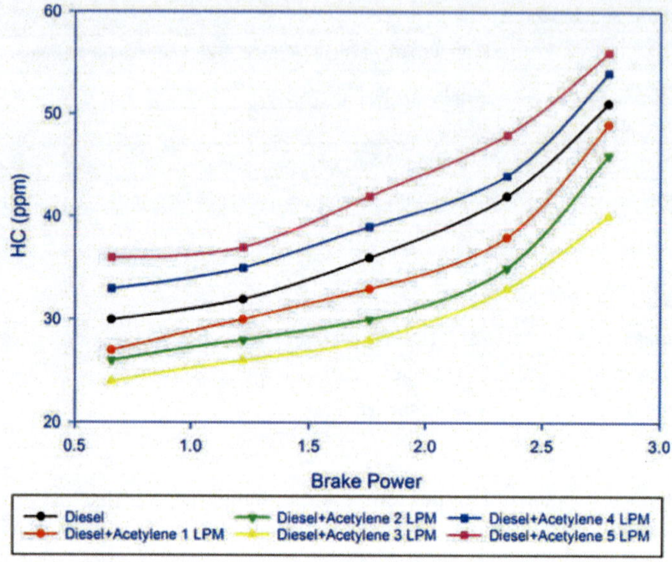

Fig. 7 HC with brake power

density in the exhaust gas is based on how the mixture is formed, the combustion procedures, and the quantity of fuel injected prior to ignition [18]. In RCCI combustion, the air is well mixed with acetylene gas which is before entering into the combustion chamber. In heterogeneous mixtures close to the mixture state used in CI engines, the engine works which ensures that charges are consumed because of the discussion flame and smoke which forms during a process like this [8]. At full load in diesel fuel operation, the smoke density is 67 HSU where in 3 LPM RCCI mode combustion is 55 HSU.

Figure 7 shows the of hydrocarbon (HC) diesel fuel and acetylene gas in RCCI mode. The trend shows that the hydrocarbon emission contrasts from 30 ppm at lower load to 51 ppm at maximum load condition for the base diesel fuel operation. The hydrocarbon emission differs from 24 ppm at low load to 40 ppm at full load condition for 3 LPM acetylene gas RCCI engine combustion. Hydrocarbon decrease is due to the greater burning speed of acetylene, which increases the combustion rate [6]. This clear experimental result confirms that the entire combustion of the load in the combustion chamber is being performed. The limits of ignition temperatures and slight operation of the engine should be expanded, and the specific energy content released for operating the engine is quite practical and constructive [1].

Figure 8 shows the CO (% by volume) for conventional diesel fuel combustion and RCCI engine combustion. A graph observation demonstrates that the variation of CO emissions indicates a comparable trend to that of hydrocarbon. The CO emissions of 3 LPM acetylene gas RCCI combustion is lowered by 35% when compared to the diesel. It is very obvious that the CO emission for rich mixtures is noted to be relative higher due to insufficient availability of oxygen for carbon dioxide combustion. In

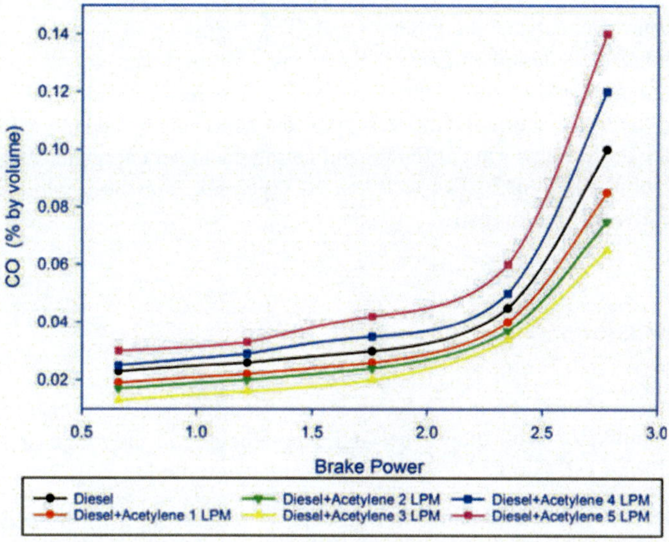

Fig. 8 CO with brake power

case of acetylene-enriched motor operation, CO emissions are a little lower; this might be due to operation on a lean mixture of a RCCI engine [14].

Figure 9 depicts the exhaust gas temperature (EGT) for diesel and acetylene combustion. The trend shows that the increment in quantity of acetylene will lead

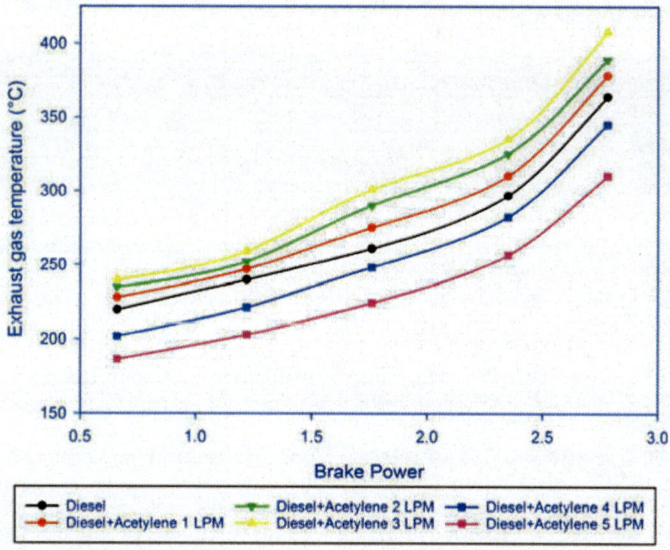

Fig. 9 EGT with brake power

to decrement in exhaust gas temperature (EGT). Induction of acetylene gas along with air reduced the temperature of exhaust gas at entire loads suggesting progress of energy liberate and greater flame velocity [3]. The reason behind the decrement in EGT is due to the thermal conductivity of the acetylene gas which is high when compared to the air. Lowering of EGT is indicating the lower energy loss. The EGT is increased about 44 °C for 3 LPM acetylene RCCI combustion mode while compared to diesel at the maximum load.

5 Conclusions

Experiments were carried in DI diesel engine, and the following conclusions are made in comparison with diesel fuel and acetylene gas in the range of 1 LPM, 2 LPM, 3 LPM, 4 LPM and 5 LPM in RCCI combustion mode.

- The specific fuel consumption is significantly reduced in 3 LPM acetylene gas RCCI combustion mode due to higher calorific value of acetylene than diesel.
- The BTE is found to be higher about 10.5% by injecting 3 LPM acetylene in RCCI mode combustion than diesel fuel combustion.
- The EGT is increased about 44 °C for 3 LPM acetylene RCCI combustion mode when compared with diesel at the maximum load condition.
- There is a considerable reduction in smoke density, HC, and CO emissions which are observed, while compared to diesel.
- With the implementation of acetylene gas, the emission of NOx is considerably reduced. The peak decreases with 3 LPM acetylene gas that was approximately 17 percent at peak load.

References

1. Balasubramanian K, Purushothaman K (2019) Effect of acetylene addition on performance, emission and combustion characteristics of neem biodiesel and corn biodiesel-fueled CI engine. J Thermal Anal Calorimetry 138(2):1405–1414
2. Benajes J, Garcia A, Monsalve-Serrano J, Sari RL (2018) Experimental investigation on the efficiency of a diesel oxidation catalyst in a medium-duty multi-cylinder RCCI engine. Energy Conversion Manage 176:1–10
3. Charitha V, Thirumalini S, Prasad M, Srihari S (2019) Investigation on performance and emissions of RCCI dual fuel combustion on diesel-bio diesel in a light duty engine. Renew Energy 134:1081–1088
4. Deivajothi P, Manieniyan V, Sivaprakasam S (2019) Experimental investigation on DI diesel engine with fatty acid oil from by-product of vegetable oil refinery. Ain Shams Eng J 10(1):77–82
5. Dhas AG, Anderson, Devarajan Y, Nagappan B (2018) Analysis of emission reduction in ethyne–biodiesel-aspirated diesel engine. Int J Green Energy 15(7):436–440

6. Gupta K, Suthar K, Jain SK, Agarwal GD, Nayyar A (2018) Design and experimental investigations on six-stroke SI engine using acetylene with water injection. Environ Sci Pollut Res 25(23):23033–23044
7. Harari PA (2018) Comprehensive Review on Enabling Reactivity Controlled Compression Ignition (RCCI) in Diesel Engines. Integr Res Adv 5(1):5–19
8. İlhak MI, Tangöz S, Akansu SO, Kahraman N (2019) An experimental investigation of the use of gasoline-acetylene mixtures at different excess air ratios in an SI engine. Energy 175:434–444
9. Kakoee A, Bakhshan Y, Aval SM, Gharehghani A (2018) An improvement of a lean burning condition of natural gas/diesel RCCI engine with a pre-chamber by using hydrogen. Energy Conversion Manage 166:489–499
10. Li J, Ling X, Liu D, Yang W, Zhou D (2018) Numerical study on double injection techniques in a gasoline and biodiesel fueled RCCI (reactivity controlled compression ignition) engine. Appl Energy 211:382–392
11. Liu J, Wang J, Zhao H (2018) Optimization of the injection parameters and combustion chamber geometries of a diesel/natural gas RCCI engine. Energy 164:837–852
12. Manieniyan V, Senthilkumar R, Sivaprakasam S (2015) Performance, combustion and emission analysis on a Diesel Engine fueled with methyl ester of neem and madhua oil. Int J ChemTech Res 7(5):2355–2360
13. Olmeda P, Garcia A, Monsalve-Serrano J, Sari RL (2018) Experimental investigation on RCCI heat transfer in a light-duty diesel engine with different fuels: comparison versus conventional diesel combustion. Appl Thermal Eng 144:424–436
14. Raman R, Naveen Kumar(2019) Experimental investigation to analyze the effect of induction length of diesel-acetylene dual fuel engine. Energy Sources Part A: Recovery Utilization Environ Effects, 1–15
15. Sasikumar R, Sankaranarayanan G (2019) A study of emission & performance characteristics of Diesel Engine run by dual fuel (Bio Diesel + Acetylene Gas). J Mech Eng Res Dev (JMERD) 42(1):104–108
16. Singh G, Esmaeilpour M, Ratner A (2019) The effect of acetylene black on droplet combustion and flame regime of petrodiesel and soy biodiesel. Fuel 246:108–116
17. Srivastava AK, Soni SL, Sharma D, Sonar D, Jain NL (2017) Effect of compression ratio on performance, emission and combustion characteristics of diesel–acetylene-fuelled single-cylinder stationary CI engine. Clean Technol Environ Policy 19(5):1361–1372
18. Tangoz S, Ilhak MI, Akansu SO, Kahraman N (2018) Experimental investigation of performance and emissions of an SI engine fueled by acetylene-methane and acetylene-hydrogen blends. Fresenius Environ Bull 27(6):4174–4185

Engine Performance and Emission Studies with Cotton Seed—Simarouba and Cotton Seed—Mahua Oil Blends as a Partial Replacement Biofuel

Ramakrishna N. Hegde and B. Jagadeesh

1 Introduction

Owing to the fast depleting petroleum-based fuels of late, many alternate ways are being explored by researchers across the globe to partially or fully substitute them. One of the ways to meet the growing demand found by researchers is use of biodiesel which has multiple advantages and could be used without altering the existing IC engines. India spends quite a higher percent of its GDP on oil imports despite being an agriculture-based country. Millions of tons of agri wastes are generated in India which means partial replacement of conventional petroleum-based fuels is not a tough job. Non-edible oils and seeds like cotton seed, Simarouba, Mahua, etc. are being widely used for this purpose. Even though using the blends is not uncommon, research scope is still available to investigate the double blends along with diesel, which is the main focus of the present paper. After characterization of both oil blends using a heterogeneous catalyst ($MgPO_4$), subsequent trans-esterification, and checking conformity as per ASTM standards, the performance testing was done using three combinations of blends and engine operating conditions.

Some of the literatures available are quoted here. Leenus et al. [1] investigated the using cotton seed oil in a diesel engine and reported the respective brake thermal efficiencies of 32.3 and 28% for diesel and CSO. They further noticed that the preheated (90 °C) 60–40 blend increased brake thermal efficiency and decreased the smoke level. Dilip et al. [2] investigated effect of Simarouba Oil Methyl Ester (SOME) at

R. N. Hegde (✉)
Department of Automobile Engineering, Srinivas Institute of Technology, Mangaluru 574143, India
e-mail: rkhegderk@gmail.com

B. Jagadeesh
Department of Aeronautical Engineering, Srinivas Institute of Technology, Mangalore, India
e-mail: jagadishramb@gmail.com

A. K. Gupta et al. (eds.), *Advances in IC Engines and Combustion Technology*,
Lecture Notes in Mechanical Engineering,
https://doi.org/10.1007/978-981-15-5996-9_10

different fuel injection timing and reported that at 23°before TDC injection timing with blend S20 at 220 bar performance was superior and less polluting. Santhosh B et al. [3] conducted experiment using Mahua oil-based methyl ester blends (15–45%) at injection pressures of 200 bar, compression ratio of 17:5:1 using a single cylinder diesel engine and observed that 25% blend gave better engine performance. Zhang and Van Gerpen [4] conducted experiments using methyl esters—soybean oil blends in diesel in a modified bowl—in-piston, turbocharged, four-cylinder, and DI engine. They reported shorter ignition delay using the blends having similar combustion characteristics as that of diesel.

Ladommatos et al. [5] studied exhaust gas recirculation effect on engine emissions in a diesel engine and reported considerable NO_x emissions reduction. However, their findings pointed increased particulate matter and UBHC emissions. Desantes et al. [6] used a diesel blend of rapeseed oil methyl ester for testing on a one-cylinder diesel engine. They reported considerable decrease in the CO, NO_x, and smoke emissions on increasing the inlet air temperature from the ambient conditions. From the literature survey, it is evident that further scope is available to test and compare the blends of dual blend of cotton seed—Simarouba oil (CSSB) and cotton seed—Mahua oil (CSMB).

2 Preparation, Characterization of CSSB and CSMB Blends

The preparation, characterization of CSSB and CSMB blends were done as per the ASTM and IS standards and the major physiochemical properties of these blends vis-a-vis neat diesel are given in Table 1. In order to establish repeatability, each blend was tested for fuel characterization three times. Both CSSB and CSMB blends were subjected to the standard procedure which involved. As a part of preparation and characterization, a 1:1 ratio by volume of cotton seed and Simarouba (CSSB) and cotton seed and Mahua (CSMB) were mixed and then subjected to gas chromatography to check the fatty acid composition. After knowing the contents of fatty

Table 1 Specific gravity and density of biodiesel blends and diesel

S. No	Physiochemical properties	Standard referred	Standard range	CSSB blend	CSMB blend
1	Calorific value (kJ/kg)	ASTM:D-240	–	35,474	37,613
2	Flash point (°C)	ASTM: D93	>130	154	169
3	ν (m^2/s) at 40 °C	ASTM: D445	1.9×10^{-6} to 6×10^{-6}	5.05×10^{-6}	5.50×10^{-6}
4	Specific gravity	ASTM:D4052	0.87–0.90	0.86	0. 89
5	Pour point	IS:1448 (P10)	−15 to 10 °C	5 °C	6 °C

acids, the free fatty acid contents of the raw oil were checked and it was found that FFA content of CSSB and CSMB blends were, respectively, 4.52 and 4.51, respectively. Based on these values 1.25 ml of H_2SO_4 was added to reduce the FFA in esterification process. The resulting FFA after esterification process was 1.97. Further, the biodiesel was produced by trans-esterification using calcined Di-magnesium phosphate ($Mg_2P_2O_7$), with a yield of 91%. The trans-esterification was done with the following process parameters: Methanol to oil molar ratio: 9:1; amount of calcined magnesium phosphate catalysts: 2%(w/v); reaction temperature: 70 °C; reaction time: 3 h. Figure 1 shows the trans-esterified blends of CSSB and CSMB.

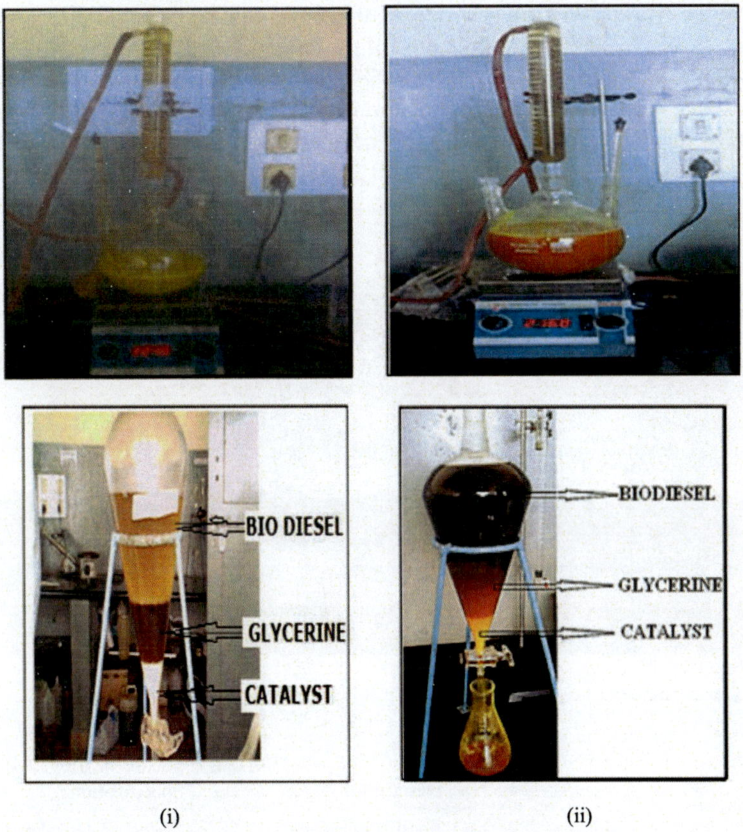

Fig. 1 Transesterified biodiesel with three layers (i) CSSB blend (ii) CSMB blend

3 Experimental Setup

The experiments were conducted on Kirloskar diesel engine TAF-1 using cotton seed—Simarouba (CSSB) and cotton seed—Mahua (CSMB) oil blends with neat diesel viz. in three volume percentages CSSBX and CSMBX where X represents 10, 20 and 30 (percent by volume). The engine was run with net diesel in the beginning at loading conditions, no load to full load, maintaining injection timing of 23° ± 1° (normal condition), injection pressure of 200 bar and a constant compression ratio of 17.5:1. After achieving the stability, readings were taken to evaluate different engine performance parameters. The experiments were repeated with blended diesel prepared from heterogeneous catalyst calcined magnesium phosphate catalyst. Figure 2 shows the photograph of the test rig.

Owing to better performance with 20% blend, engine performance testing was further done by advancing the injection timing to 27° without changing other parameters (CR and injection pressure). The detailed engine specification is given in Table 2.

Fig. 2 Diesel engine test rig

Table 2 Engine specifications

S. No	Description	Specifications
1	Engine make/type	Kirloskar-4-stroke single cylinder
2	Rating and rpm	4400 W at 1500 rpm
3	Bore, stroke, displacement volume	87.5 mm, 110 mm, 661.5 cc
4	Compression ratio	17.5:1

4 Results and Discussion

4.1 Engine Performance

Engine performance was investigated using both CSSBX and CSMBX blends by varying the load in the range of no load to full load. The brake thermal efficiency (BTE), brake specific fuel consumption (BSFC), pressure-crank angle variation, emission and exhaust gas temperature are the parameters discussed hereunder, for both the blends as mentioned above, considering normal and advanced injection timings.

4.1.1 Normal Working Conditions

Variation of brake thermal efficiency and BSFC versus load using the biofuel blends is shown in the following Fig. 3. It is evident that efficiency increases with load in both neat diesel and the blends. At all loads, the brake thermal efficiency of neat Diesel is superior than the blended diesel with lower specific fuel consumption. Among the CSSB and CSMB blends, brake thermal efficiency of CSSB blends is higher and among CSSB blends, CSSB30 gave better performance but with higher specific fuel consumption. The reason for this could be owing to higher viscosity of the blends that result in poor fuel spray formation, coupled with air entrainment affecting the combustion process [7]. Owing to lower volatility, lower calorific value of vegetable oils, more fuel is consumed and higher the blend ratio, fuel consumption is also slightly more. The highest brake thermal efficiencies recorded by the respective blends CSSB30 and CSMB30 28.4 and 29.8% as against 29.1% of neat diesel. Fuel consumption wise, BSFC for all blends at all loads is 4–8%, more compared to neat diesel.

Figure 4 shows the variation of pressure with crank angle at different load conditions obtained using cylinder pressure data for diesel, CSSB and CSMB blends.

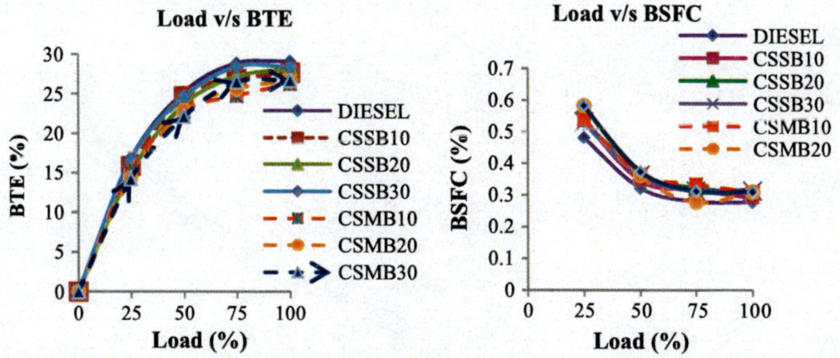

Fig. 3 Performance plots of BTE and BSFC versus load

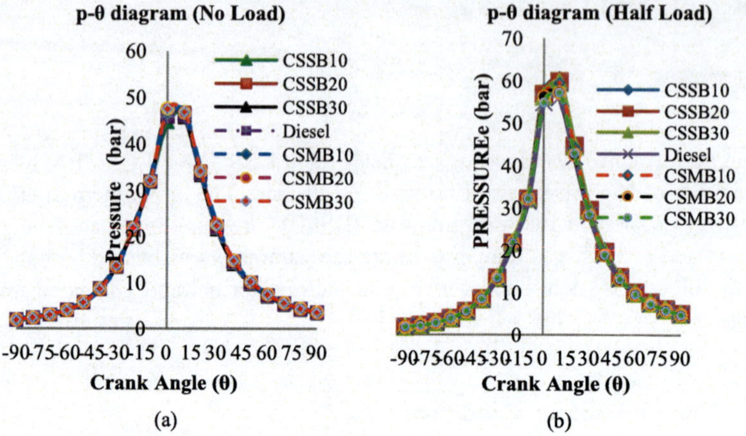

Fig. 4 Combustion characteristic plots: pressure versus CA

Referring Fig. 4, with neat diesel heat release just after TDC is higher, closely followed by the blend CSSB30, CSMB30, CSSB20 and CSMB20, respectively, at no load, 50% load and full load conditions.

During the test run, the peak pressure attained by neat diesel was found to be at 4°–5° after TDC (ATDC) for no load, 6°–7° ATDC for 50% load and 9°–11° ATDC for full load (figure is not shown here). Further, both CSSBX and CSMBX blends recorded higher peak pressures ranging between 4 and 8% compared to neat diesel. It may also be noticed that the peak pressure curves for all the blends offset by 1–2° crank angle toward the right of neat diesel pressure curve, indicating shifting of combustion at the end phase, which could influence heat loss in the exhaust.

Figure 5 shows the exhaust gas temperature versus load profile for all the blends at normal working conditions. The exhaust gas temperature is lower for all the blends compared with neat diesel. At full load, while diesel recorded 485 °C, CSSB20 and CSMB20 blends recorded 65 °C and 47 °C lesser, respectively. Further, higher the blend percent lower is the gas temperature in case of CSSB, which could be due to the

Fig. 5 Exhaust gas temperature (EGT) versus load

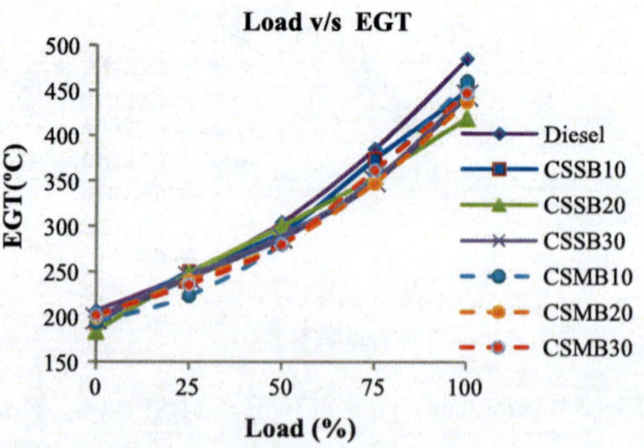

presence of excess oxygen present in blends, expediting more complete combustion. Among CSSB and CSMB blends, the latter recorded lower exhaust gas temperature. However, at higher loads, blends CSSB20 and CSMB20 fare better.

4.1.2 Engine Emission

As it is customary, performance of the engine with regard to emission is also a major field of research focus. The major effluents from the engine are CO, UBHC and NO_x. From Figs. 6a, b, it may be inferred that CSSB and CSMB blends show either equal or the lower CO and UBHC emissions as compared to neat diesel, respectively, between 25% less CO (at half load and with CSSB20) and 2.85% less UBHC (at half load and with CSSB10) to71.42% less CO (at full load and with CSMB30) 57.69% less UBHC (at full load and with CSMB30), respectively, in the load range tested. As discussed earlier, excess or additional oxygen traces in biofuel could be one of the reasons for the reduction in emissions. Among the blends considered, CSMB blends emit nearly 7% less CO compared to CSSB blends at full load conditions. The lower viscosity of the blend could be the reason for this which would result in better fuel injection spray. Also, increased blend percentage of CSSB and CSMB

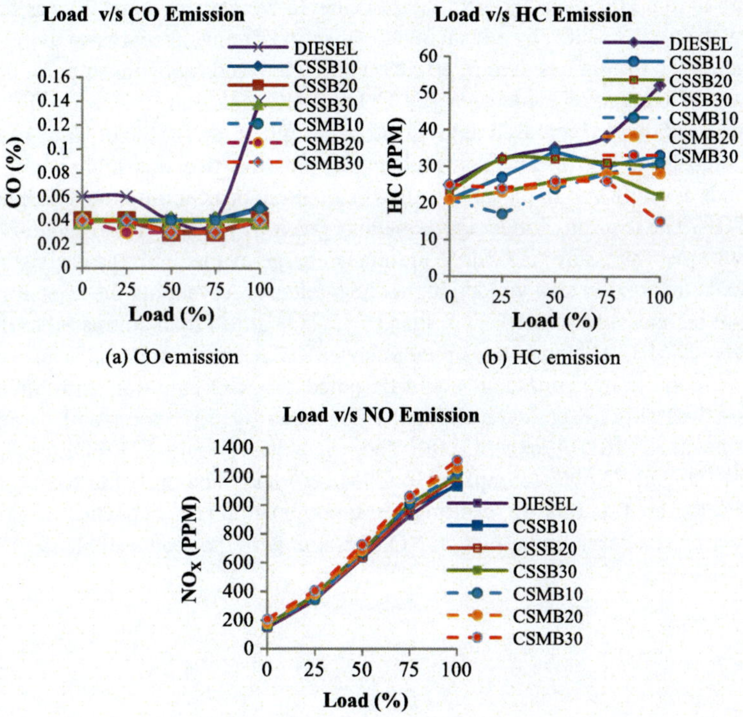

Fig. 6 Engine emission versus load

blends increases the UBHC emissions from 0.03 to 0.05%. This may be reasoned due to poor atomization of the blended fuel because of higher viscosity when compared diesel. Additionally, other influencing factors could be physical properties of fuels such as density and viscosity [8].

NO_x emission is another dominant pollutant from the engine. Figure 6c shows load versus NO_x emission for neat diesel and its blends. The trend here is, increasing load on the engine the NO_x emission also increases, which is quite obvious with higher combustion temperatures. While neat diesel fares better as far as NO emission is concerned, similarly, among the blends, CSSB has lower NO emission levels when compared to CSMB, a maximum of 8.8%. Further, with both blends, increase in the blend percentage NO_x emissions increases. The CSSB blends have nearly the same NO_x emission (in particular CSSB20) as petro diesel with as little higher from 1.43 to 6.78% at all loads except for no load. Available literature [9] says that NO_x emission has multifold ill effect when compared to carbon dioxide as a greenhouse gas.

4.2 Blend 20 Performance Evaluation; Normal Versus Advanced İnjection

Keeping in mind the overall performance centered weightage, blend 20 was chosen to study the performance, by advancing the injection timing, as represented in Fig. 7. Advancing the timing has dual ill effect on efficiency and fuel consumption in both the cases of blends, as evident. While CO and HC emission was reduced (higher pressure peak due to increased delay), NO traces increased for the obvious reasons.

As seen from Fig. 8, advancing injection, pressure rise was higher with both the blends as compared to diesel and also the pressure peak shifted toward right side from TDC. The resulting longer ignition delay given way to larger fuel accumulation and consequent pressure rise. More premixed charge available in the early stage of premixed combustion also yielded higher heat release. Advancing the injection, HC emission increased from 3.5% at ¾ load to 35.48% at full load compared to diesel. However, CSMB20 blend still resulted in lower UBHC emission (42.3% less) under the same conditions. Further, it could be noted that CO emission with advanced injection was still lower (even though it increased by 50% compared to normal injection with CSSB20A) except at full load where diesel emitted 31.94% (compared to CSSB20) and 92.69% (compared to CSMB20) less. The probable reason could be due better air-fuel mixing resulting better combustion [10, 11]. Thus, advancing the injection has cascading effect on NO emission with the biodiesel blends.

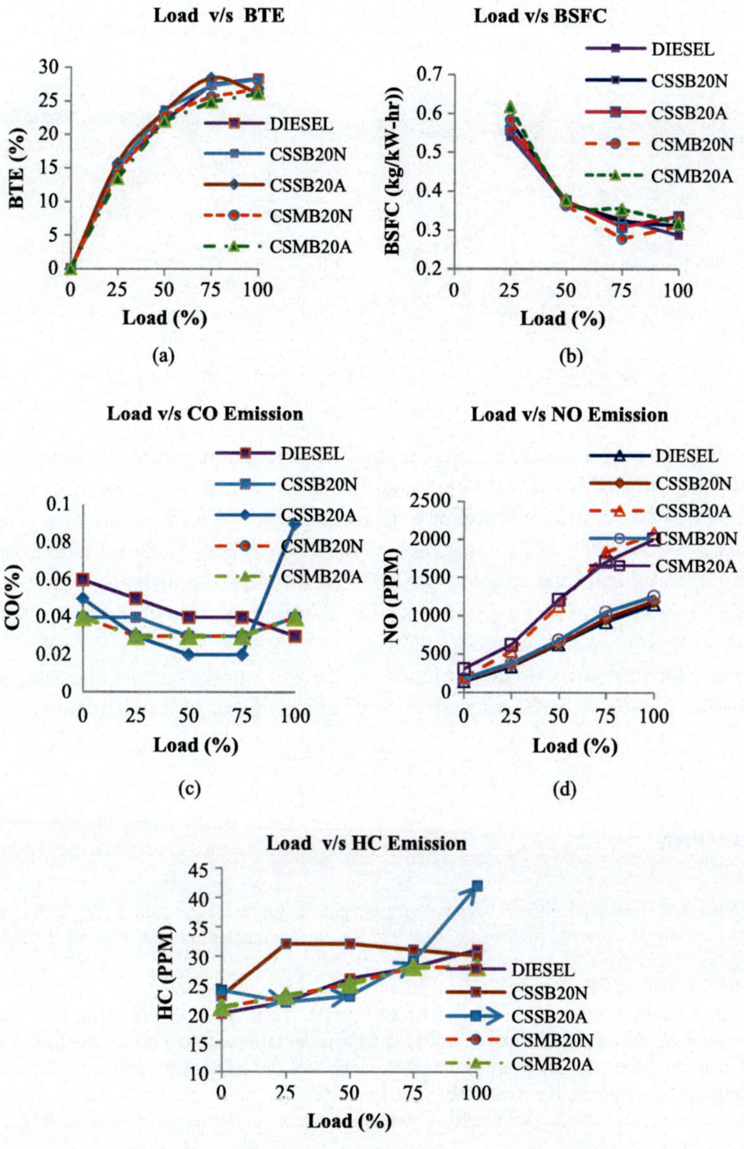

Fig. 7 Blend 20 performance evaluation for normal and advanced injection

5 Conclusions

Based on experimental investigation on engine performance using dual blend combinations of cotton seed—Simarouba, CSSB and cotton seed—Mahua, CSMB blends and subsequent comparison with neat diesel, the following conclusions are drawn.

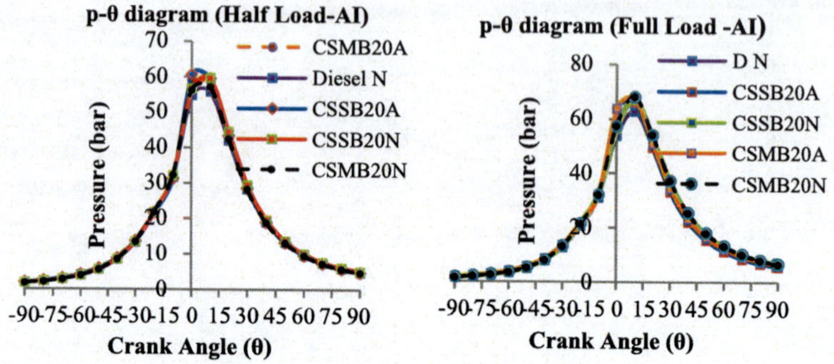

Fig. 8 Blend 20—pressure versus crank angle for normal and advanced injection

- The engine power and efficiency of the blends are comparable with neat diesel. CSSB30 and CSMB30 were 28.4 and 29.8% as against 29.1% of neat diesel.
- The specific fuel consumption of both blends is 4–8% higher than neat diesel.
- The emission levels of CO were lesser by 25% (at half load and with CSSB20) & 71.42% (at full load and with CSMB30) and UBHC emissions were lesser by 57.69% (at full load and with CSMB30) & 2.85% (at half load and with CSSB10). However, NO_x emissions are quite high.
- Overall performance of 20% blend is superior compared to other blends and CSMBX blends show better performance compared to CSMBY blends.

References

1. Leenus Jesu Martin M, Edwin Geo V, Kingsly Jeba Singh D, Naglingam B (2012) A comparative analysis of different methods to improve the performance of cotton seed oil fuelled diesel engine. Elsevier J 372–378
2. Dilip Sutraway YUB, Katti VV (2014) Effect of fuel injection time on performance of ci engine using (Simarouba) biodiesel as fuel. J Mech Eng Res Technol 2(1):1741–1746
3. Santhosh B, Suresh R, Yathish KV (2013) Experimental investigations on the use of mahua oil methyl ester produced using magnesium phosphate catalyst as fuel in a compression international. J Innovative Res Sci Eng Technol 2(8)
4. Zhang Y, Jon H, Van Gerpen (1996) Combustion analysis of esters of soybean oil in a diesel engine, SAE paper 960765
5. Ladommatos N, Abdelhalim SM, Zhao H, Hu Z (1998) The effects of carbon dioxide in exhaust gas recirculation on diesel engine emissions, J Automobile Eng 212:25–42
6. Arregle J, Ruiz S, Desantes JM, Delage A (1999) Characterization of the injection combustion process in a D.I. Diesel engine running with rape oil methyl ester, SAE paper, pp 01–1497
7. Misran RD, Murthy MS (2009) Performance, emission and combustion evaluation of soapnut oil–diesel blends in a compression ignition engine. Fuel 90, 2514–2518
8. Humke AL, Barsic NJ (1981) Performance and emissions characteristics of a naturally aspirated diesel engine with vegetable oil fuels, SAE paper No. 810955
9. Devan PK, Mahalakshmi NV (2009) Performance, emission and combustion characteristics of poon oil and its diesel blends in a DI diesel engine fuel 88:861–867

10. Gangadhara Rao G, Kumar N, Herbert M (2018) Effect of injection pressure on the performance and emission characteristics of the CI engine using Vateria indica biodiesel. Int J Ambient Energy. https://doi.org/10.1080/01430750.2017.1421575
11. Syed Aalam C, Saravanan CG, Prem Anand B (2016) Impact of high fuel injection pressure on the characteristics of CRDI diesel engine powered by mahua methyl ester blend. Appl Therm Eng S1359–4311(16):30855–30859

Effect of Bio-Additive Blends with Diesel Fuel Utilization in a Diesel Engine

R. Senthilkumar, V. Sukumar, V. Manieniyan, and S. Sivaprakasam

1 Introduction

Vegetable oil offers many benefits, including sustainability, reduction of greenhouse gas emission, territorial advancement, and improvement in farming. The compound creation of vegetable oil helps in lessening the emission of undesirable parts when they are scorched. A few strategies like impeded fuel infusion timing, reused to exhaust gas, and after treatment gadgets are utilized for reducing nitric oxide discharge from an engine. It is seen that these methods while decreasing nitric oxide from the fumes of an unmodified engine experiences at least one impediment as a result of the natural exchange off as for particulate issue or cost [4]. Any post-treatment technique requires certain extra plans for their execution. From powerful and straightforwardness stance, the methodologies on fuel alterations to reducing nitric oxide discharges stay alluring. The standard necessity of powers for rapid diesel engines is great start quality. Poor start quality fuel can prompt broadened delay and the exhaust in diesel knock. To alter the fuel quality, added substances can be included little amount either to improve motor execution or to decrease the

R. Senthilkumar (✉)
Mechanical Engineering, Srinivasa Subbaraya Polytechnic College, Puthur, Tamil Nadu, India
e-mail: srsau@yahoo.com

V. Sukumar
Mechanical Engineering, Government Engineering College, Bodinayakanur, Tamil Nadu, India
e-mail: sukumsr.rp87@gmail.com

V. Manieniyan
Mechanical Engineering, Government Engineering College, Srirangam, Tamil Nadu, India
e-mail: manieniyan78@yahoo.com

S. Sivaprakasam
Mechanical Engineering, Annamalai University, Tamil Nadu, India
e-mail: rgssiva2002@yahoo.co.in

A. K. Gupta et al. (eds.), *Advances in IC Engines and Combustion Technology*,
Lecture Notes in Mechanical Engineering,
https://doi.org/10.1007/978-981-15-5996-9_11

emanation [4, 11]. Many researcher have investigated in the field for past years, some of the relevant study are given below.

Prabu et al. [12] conducted the experiments in diesel engine using butylated hydroxytoluene (BHT) and n-butanol additive. The results reported on the minimum emission were identified with cetane improver additive and commercial smoke abatement additives. Also noted the particulate emission increased in ideal condition. Overall the emission and performance parameter were better in cruise condition with cetane improver additive compare to other sixteen additives. Dueso et al. [6] studied the impact of natural antioxidant additive produced from bio-oil extraction and diesel fuel. The test was conducted in four-stroke diesel engine with bio-additive and diesel fuel separately at rated rpm. The emission was reduced in with additive and compare to diesel. They reported in combustion analysis with additive has been the superior than other fuel. Nayyar et al. [10] experimented was carried in DI diesel engine with cetane improver additive blends of diesel in range of 1 ml, 3 ml, and 5 ml. noted in all emission parameter and performance parameter such as BTE and SFC. The results show cetane value improved was used to reduce the NO_x emission. The additive of 3 ml was selected as a best blend of both performance and emission parameter. They reported cetane improved additive produced higher NO_x emission from low load and full load.

2 Experimental Investigation

2.1 Bio-Additive Preparation Process

The present investigation relates to bio-additive obtained from Mahua oil. The process of preparation is detailed here. The present research provides detoxified and transesterified Mahua oil extracted from the seeds of Mahua by a process which is simple, economical and which gives high yields of bio-additive. The extraction of Mahua oil, its economic benefits along with the favorable emission and pollution parameters are evaluated by the present investigation. The bio-additive so obtained is with high efficiency, and better biodegradability. Mahua oil is highly economic, eco-friendly, and an efficient bio-additive which is economically accessible without any seasonal restriction. By undertaking cultivation on a large scale, the availability can be assured. The seed is collected, dried, and processed to obtain crude oil by standard method. The crude oil may contain the toxic alkaloidssanguinarine and dihydrosanguinarine, four quaternary isoquinoline alkaloids such as dehydrocorydalmine, jatrorrhizine, columbamme, and oxyberberme. The present invention discloses a detoxified and transesterified oil of Mahua. The modified and processed oil is useful as bio-additive to petroleum diesel. In a preferred aspect, the present invention discloses a process of preparation of Mahua oil as bio-additive which is described here, degumming of crude Mahua oil by treating with mineral acid and lower alcohol to remove alkaloid and gums. Detoxifying process is the above-degummed oil by treating with

Table 1 Properties—biofuel additive (2.0 ml BFA)

Test property	Values	Standards
Fire point in °C	51	ASTM D 92
Kinematic viscosity @ 40 °C in cst	2.53	ASTM D 445
Flash point in °C	48	ASTM D 92
Cetane number	60	ASTM D 976
Calorific value kJ/kg	41,670.4	ASTM D 240

0.1% aqueous NaOH solution to obtain alkaloid free or detoxified Mahua oil with high purity. Transesterifying detoxified Mahua oil, by successive treatment with acid and alkali in the presence of methanol to obtain transesterified Mahua oil. The various physical properties of the new bio-additive that was named as biofuel additive (BFA) was ascertained as per international standards and are shown in Table 1.

2.2 *Experimental Procedure*

Experimental work was conducted on Kirloskar TV I diesel engine with rated speed and various loads by using bio-additives blended with diesel. The loading device is an Eddy current dynamometer. The combustion parameters were measured with the help of AVL combustion analyzer. Emission parameters were noticed with help of AVL smoke meter and AVL-di-gas analyzer. All bio-additive blends when compared with diesel. Heat release rate was reduced and maximum cylinder pressure was nearer to the diesel value for all bio-additives. It was seen that the exhaust gas temperature is less for all the bio-additives. The bio-additive is blended with diesel in the rage of 1 ml, 2 ml, 3 ml, 4 ml, and 5 ml. Figure 1 shows as experimental setup.

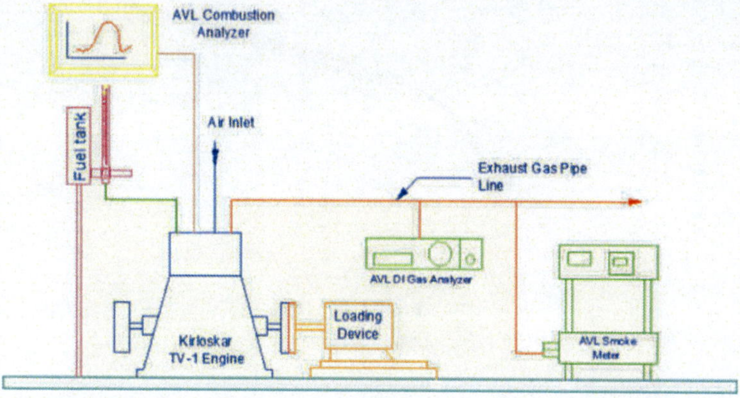

Fig. 1 Experimental setup

3 Results and Discussion

The variation in specific fuel consumption of biofuel additive and diesel with brake power is shown in Fig. 2. Compared to diesel and other BFA blends, it can be noted that the fuel consumption of 2.0 ml BFA is smaller. This is because of BFA's low heating value and low viscosity compared to diesel. Biofuel additive is heterogeneous when blended with air, leading to a reduction in fuel usage [7].

The brake thermal efficiency (BTE) variation is shown in Fig. 3. It is noted from the test outcomes that the biofuel additive brake heat efficiency improves gradually. With diesel, the highest thermal efficiency of 27.71% was accomplished, while the brake thermal efficiency of 2.0 ml BFA mixes was 28.89 percent at the highest brake

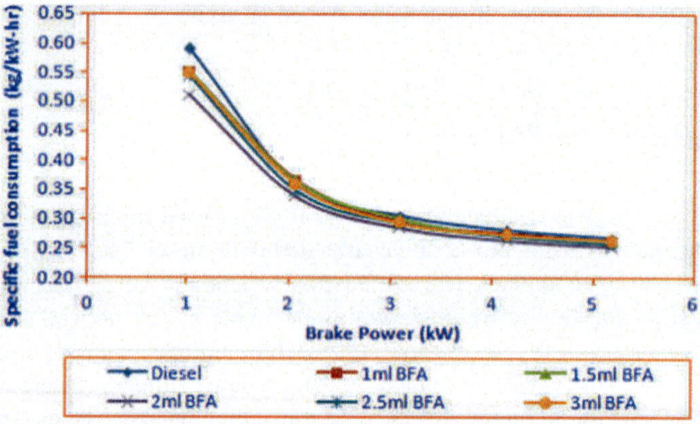

Fig. 2 Specific fuel consumption versus brake power—BFA

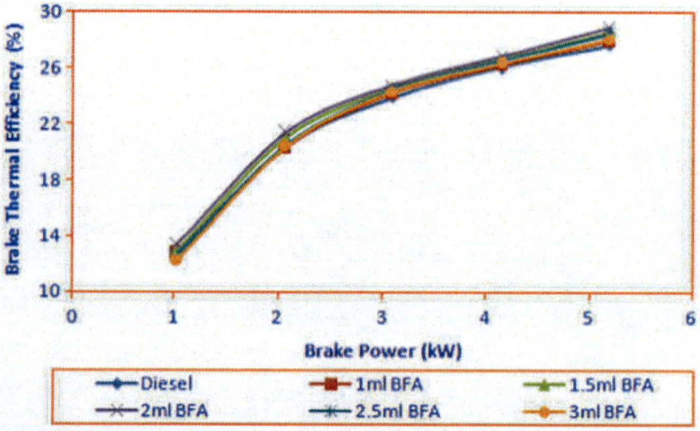

Fig. 3 Brake thermal efficiency versus brake power—BFA

energy. This tiny variation in BTE may be due to BFA's chemical properties that promote the process of combustion [9].

Figure 4 shows the variation on smoke density with brake power. It can be observed that smoke density increases with an increase in the brake power. Maximum emission occurs to maximum load. It can be seen that 2.0 ml BFA blend effectively reduces the smoke density. The oxygen enrichment provided by the bio-additive leads to smoke reduction [13].

For diesel and BFA blends, Fig. 5 demonstrates nitrogen oxides (NO_x) emissions. It is well recognized that a tiny quantity of nitrogen is in vegetable-based fuel. This will contribute to the manufacturing of NOx. The characteristics of oil spray rely on droplet momentum, droplet size, degree of mixing with air, evaporation rate, penetration rate, and transfer rate of radiant heat. A change in any of these

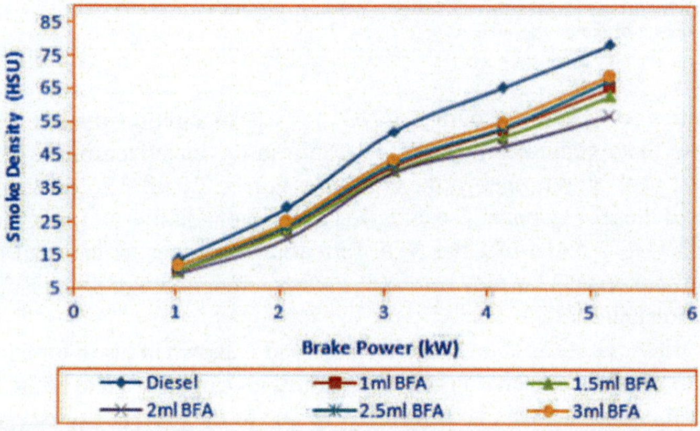

Fig. 4 Smoke density versus brake power—BFA

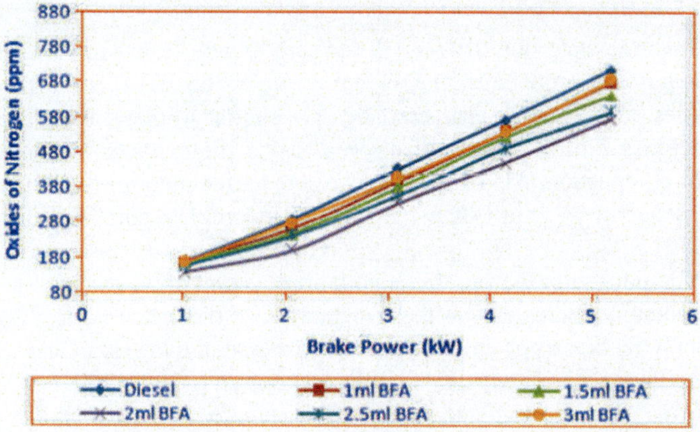

Fig. 5 Oxides of nitrogen versus brake power—BFA

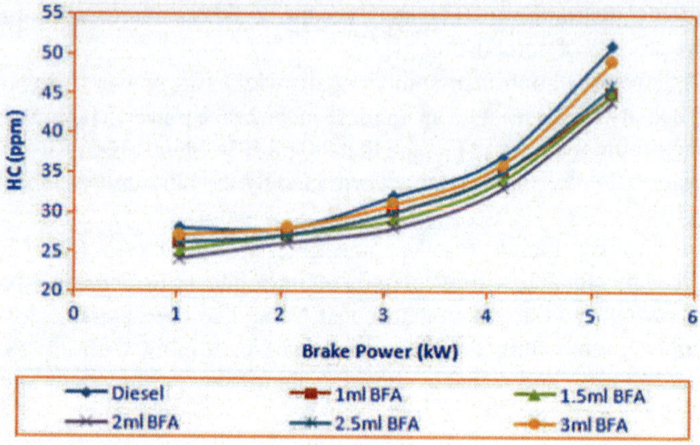

Fig. 6 HC versus brake power—BFA

characteristics could alter the output of NO_x [8, 16]. In addition, the impacts of fuel chemistry in flame region could result in a shift in the manufacturing of NO_x. It is noted that at peak engine energy, the NO_x emission of 2.0 ml BFA mixes is around 576 ppm and diesel 715 ppm. The emission of biofuel additive NO_x is smaller than that of diesel. It is noted that the NOx formation improves as the load rises and reaches the peak load. This may be due to greater temperature of combustion inside the greater load tank [3].

Hydrocarbon (HC) emission variation with load is shown in Fig. 6 for BFA blends and diesel 6. HC for the operation of biofuel additives can be noted to be smaller than that of diesel operation. This is believed to be due to the presence of oxygen in the fuel, which promotes the combustion process. Because the biofuel additive includes oxygen in the structure itself, the combustion is supposed to be enhanced [15]. The HC is small at peak brake energy for 2.0 ml BFA mix at 44 ppm and for diesel at 45 ppm.

Figure 7 demonstrates the BFA and diesel carbon monoxide emission with brake power. CO emissions from an engine rely primarily on the fuel's chemical and physical properties, and the BFA characteristics are similar to diesel fuel. The biofuel additive includes a tiny quantity of oxygen inside the cylinder, which acts as a promoter of combustion [5]. This leads to better biofuel additive combustion than diesel. Consequently, CO in exhaust gas due to incomplete combustion is smaller than diesel [2].

Figure 8 demonstrates the variability with brake power of the exhaust gas temperature. The outcome indicates that the temperature of the exhaust gas rises in both instances with the brake power. For 2.0 ml BFA blend, the lowest EGT of 252°C is observed, while for diesel only 277°C is discovered.

Figure 9 indicates the crank angled heat release rate. It is deduced that with the load the heat release rate improves, when compared to other blends, 2.0 ml BFA blend has higher heat release rate. The heat release rate indicates a brief delay period

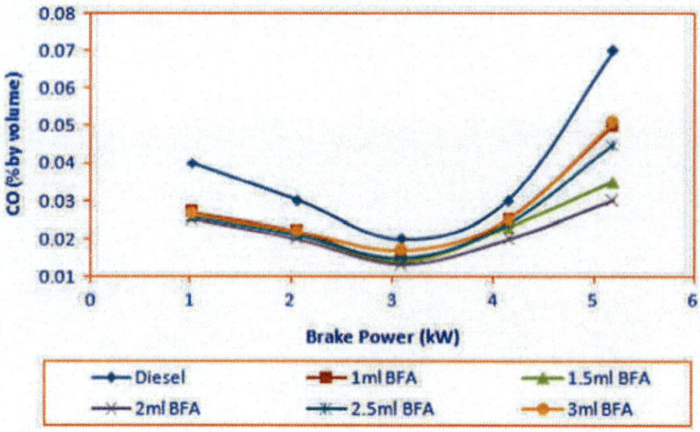

Fig. 7 CO versus brake power—BFA

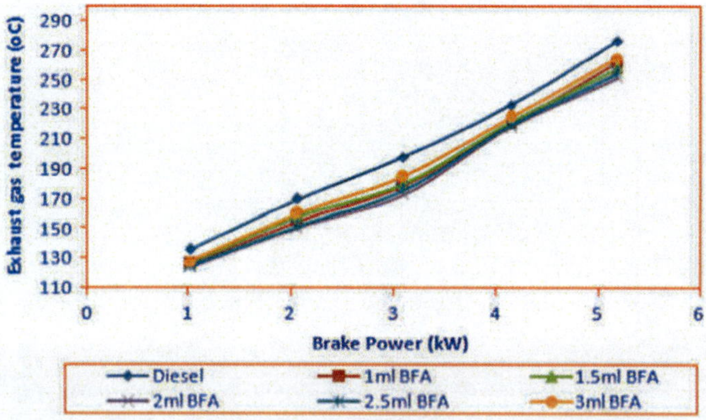

Fig. 8 Exhaust gas temperature versus brake power—BFA

for the bio-additive mixed fuel. However, there is no distinction in the phases of premixed combustion of all BFA blends. But in bio-additive blended engine, the rate of diffusion combustion is much faster than in the diesel operation [14, 17].

The correlation of cylinder pressure with crank angle is as appeared in Fig. 10. The BFA mixes are having higher pressure contrasted with diesel. As the diesel amount in the mixes builds, the measure of fuel partaking in the uncontrolled burning phase of the blend decrease, which results in a higher pressure rise. The cylinder pressure is reliant on consumed fuel portion during the premixed consuming stage. Maximum cylinder pressure for all BFA blends is marginally increased when compared with diesel and other blends (Appavu et al. [1].

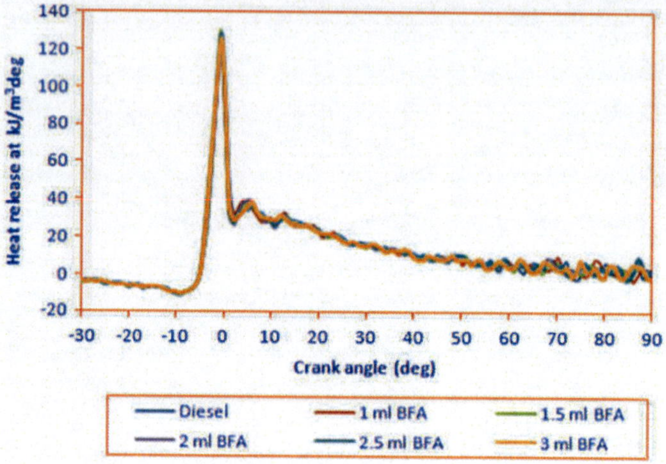

Fig. 9 Cylinder pressure versus crank angle—BFA

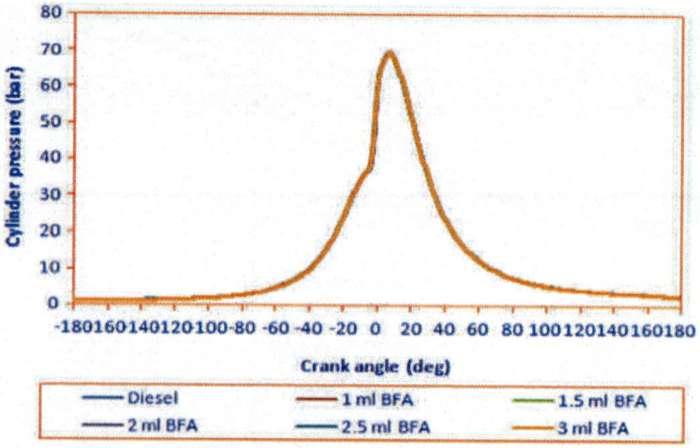

Fig. 10 Heat release rate versus crank angle—BFA

4 Conclusion

The conclusions are drawn on this investigation can be explained as follows.

- It is found that SFC is least for 2.0 ml BFA bio-additive blend compared to diesel.
- At maximum brake power 2.0 ml BFA bio-additive blend is observed optimal with thermal efficiencies at 29.89%.
- Smoke emissions are lower for 2.0 ml BFA bio-additive blend (56.7 HSU).
- NO_x emission is reduced to 576 ppm at maximum brake power with 2.0 ml BFA bio-additive blend.

- The CO emission from the 2.0 ml BFA bio-additive blend is lower compared to diesel fuel. In case of 2.0 ml BFA, bio-additive blend of 3% and diesel of 7% CO emission at full load.
- The lowest value of EGT of 269°C is observed for 2.5 ml OLE bio-additive, whereas for diesel, it is found to be 277 °C only.
- At a maximum load condition, the maximum cylinder pressure is noticed to be 69.62 bar at 7 degree crank angle with diesel, 69.456 bar at 7° crank angle with 2.0 ml BFA bio-additive blend.

References

1. Appavu P, Madhavan VR, Venu H, Mariadoss A (2019) Effect of fuel additives and exhaust gas recirculation in biodiesel fuelled CI engine: a review. Int J Ambient Energy:1–7
2. Baltacioglu, Kaan M, Kenanoglu R, Aydın K (2019) HHO enrichment of bio-diesohol fuel blends in a single cylinder diesel engine. Int J Hydrogen Energy
3. Debnath R, Sastry GRK, Rai RN (2019) An experimental investigation of performance and emission of thumba biodiesel using butanol as an additive in an IDI CI engine and analysis of results using multi-objective fuzzy-based genetic algorithm. Environ Sci Pollut Res 26(3):2281–2296
4. Deivajothi P, Manieniyan V, Sivaprakasam S (2019) Experimental investigation on DI diesel engine with fatty acid oil from by-product of vegetable oil refinery. Ain Shams Eng J 10(1):77–82
5. Devarajan Y (2019) Experimental evaluation of combustion, emission and performance of research diesel engine fuelled Di-methyl-carbonate and biodiesel blends. Atmos Pollut Res 10(3):795–801
6. Dueso C, Muñoz M, Moreno F, Arroyo J, Gil-Lalaguna N, Bautista A, Gonzalo A, Sánchez JL (2018) Performance and emissions of a diesel engine using sunflower biodiesel with a renewable antioxidant additive from bio-oil. Fuel 234:276–285
7. Londhe H, Luo G, Park S, Kelley SS, Fang T (2019) Testing of anisole and methyl acetate as additives to diesel and biodiesel fuels in a compression ignition engine. Fuel 246:79–92
8. Manieniyan V, Senthilkumar R, Sivaprakasam S (2015) Performance, combustion and emission analysis on a diesel engine fueled with methyl ester of neem and madhua oil. Int J ChemTech Res 7(5):2355–2360
9. Musthafa MM, Ashok Kumar T, Mohanraj T, Chandramouli R (2018) A comparative study on performance, combustion and emission characteristics of diesel engine fuelled by biodiesel blends with and without an additive. Fuel 225:343–348
10. Nayyar A, Sharma D, Soni SL, Bhardwaj B, Augustine M (2019) Modeling and experimental investigation for performance and emissions on a diesel engine using bio-oxygenated ternary fuel blends. Energy 168:136–150
11. Padmanabhan S, Ganesan S, Anvesh AMV, Pradeep B (2018) Influence of cerium oxide additive and aloe vera biodiesel on a CI engine. Int J Ambient Energy 39(5):516–520
12. Prabu SS, Asokan MA, Roy R, Francis S, Sreelekh MK (2017) Performance, combustion and emission characteristics of diesel engine fuelled with waste cooking oil bio-diesel/diesel blends with additives. Energy 122:638–648
13. Rajendra P, Kamalesh J, Pranali G, Vishal B, Bhushan S (2018) A comprehensive review on influence of biodiesel and additives on performance and emission of diesel engine. Chem Eng Trans 65:451–456
14. Sachuthananthan B, Krupakaran RL, Balaji G (2019) Exploration on the behaviour pattern of a DI diesel engine using magnesium oxide nano additive with plastic pyrolysis oil as alternate fuel. Int J Ambient Energy: 1–12

15. Senthilkumar R, Manieniyan V, Sivaprakasam S (2015) Experimental investigation on DI diesel engine using bio additive. Int J Appl Eng Res 10(8):6210–6214
16. Shah PR, Gaitonde UN, Ganesh A (2018) Influence of soy-lecithin as bio-additive with straight vegetable oil on CI engine characteristics. Renew Energy 115:685–696
17. Sukumar V, Manieniyan V, Sivaprakasam S (2019) Experimental studies on DI diesel engine fueled in sweet lime pyrolysis oil with biodiesel. Int J Appl Eng Res 14(5):1145–1150

Experimental Investigation of the Influence of Nanoparticle Additive with Biodiesel in Diesel Engine

T. Deepak Kumar, Manjunatha, and D. K. Ramesha

1 Introduction

In our daily life, the CI engine is essentially using everywhere due to flexible operation and low cost. Due to modernization in industries and lifestyle, it causes depletion in conventional fuel causing lack of petroleum-derived fuels, which reflects on its price. Most of the search for suitable alternative fuel for diesel engines in this regard. The biodiesel is as desirable fuels. At present, the research is continued into producing biofuel from food, it has been found that the waste chicken fat provides additional hopeful source for the invention of biodiesel. The straight use of such chicken fat oil in compressive ignition engine is not possible owing to the content of triglycerides. Therefore, the transesterification system in which triglycerides reacts with alcohol in the existence of alkali catalyst for the development of biodiesel. However, biodiesel also has certain drawbacks, such as high molecular weight, high viscosity, lower volatility and high pour point as matched with the diesel. These negatives lead to reduced atomization and cause imperfect combustion. Nanoparticles are the biochemical components used to improve the characteristic of the fuel. They fused with diesel, biodiesel, so on and so forth, to increase the efficiency of the fuel and meeting desired emission control standards.

Vairamuthu G et al. [1] employed CIME combined with sulphur and a combination of CeO_2 additives to examine engine output and exhaust gas releases. They noted that it significantly lower hydrocarbon and oxide of nitrogen emission with enhanced BTE. Nanthagopal K et al. [2] utilized CIME biodiesel, Zno and TiO_2 additives are used to assess the analyses of performance, combustion and release gases characteristics of diesel engine. In the concerned assessment, they found out that the

T. Deepak Kumar (✉) · Manjunatha · D. K. Ramesha
Department of Mechanical Engineering, University Visvesvaraya College of Engineering, Bangalore University, Bangalore 560001, India
e-mail: deepu.doit@gmail.com

A. K. Gupta et al. (eds.), *Advances in IC Engines and Combustion Technology*, Lecture Notes in Mechanical Engineering,
https://doi.org/10.1007/978-981-15-5996-9_12

increase in BTE in the range of 5–17% and NO_x emission was less in comparison with pure Calophyllum inophyllum biodiesel. Ramesh DK et al. [3] conducted an experimental assessment in CI engine by operating poultry litter oil B20 blend and B20 blend with 30 mg/l alumina nanoparticles. They realized noticeable Reduction in UBHC, NO_x and CO emissions with alumina nanoparticle. Seela CR et al. [4] conducted an experimental investigation that the engine analyses of performance and exhaust gas releases mission by addition of CeO_2 nanoparticle additives with mahua methyl ester blends. The experimental results exhibited that BTE of B20+100 CeO_2 was better by 1.8 with an improved 1% in SFC. Releases gases of HC and CO were also noticeably lower than diesel fuel. Jiaqiang E et al. [5] the engine performance and emission characteristics impact of these technologies will be summarized. Using LTC mode is more efficient and viable than other modes when compared. This can simultaneously the both NOx and PM emissions by up to 95% and 98%. Sahar et al. [6] waste cooking oils may, along with reducing pollution, be converted into biodiesel, as an energy source. Debbarma S et al. [7] the presence of INP Shorter ignition delay and advanced heat release was investigated experimentally for the efficiency and emission characteristics of a bio-diesel engine with or without Iron nanoparticles (INP). M. Mohamed Musthafa et al. [8] the purpose of the study is to lower the NOx emission of palm oil methyl ester combined with diesel fuel (B20), use the improved Cetane DTBP number and increase ignition efficiency of the tested fuel. Gerhard Knothe et al. [9] many metals and their oxides will have effects on the oxidative stability of biodiesel. X.J. Man et al. [10] the images TEM and HRTEM show that primary particles are measured by their combustion process size and configuration. The HRTEM images are only slightly affected by the engine speed. Rolvin D'Silva et al. [11, 12] increased brake thermal efficiency of the engine is possible with the introduction of TiO2 nanoparticles. Broad brake fuel consumption decreases at full load. C.S. Cheung et al. [13] B10 diesel fuel (biodiesel 10%), B20, B30 and pure biodiesel. B20 fuel. Five loads of engines have been experimented. Show that biodiesel leads to reduction of HC, CO. From the review, it is established that up to 20% of diesel-powered biodiesel is caused by improved performance, controlled combustion and reduced diesel engine emissions. Similarly, nanoparticles such as nickel oxide also can be used to blend directly to the biodiesel (Abbrevation Table 1). In this experimental investigation, the performance and emission sequences were analysed with the application of nickel oxide nanoparticle on waste chicken fat oil methyl ester.

2 Methodology

2.1 Biodiesel Manufacture Procedures

The raw unused chicken fat was collected from places like restaurant, hostels and fast food centres, and the biodiesel was prepared using transesterification method (Fig. 1).

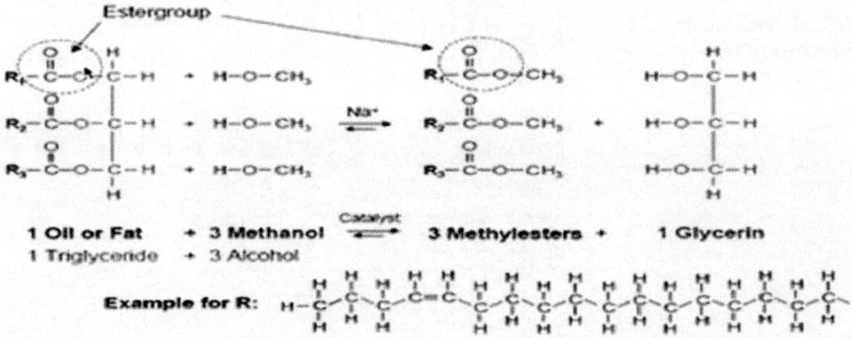

Fig. 1 Schematic representation of transesterification

Table 1 Abbreviation

BP	Brake power
BTE	Brake thermal efficiency
NiO	Nickel oxide
D	Diesel
B20	20% biodiesel
CFME	Chicken fat methyl Ester
UBHC	Unburnt hydrocarbon
NO_x	Oxides of nitrogen
CO	Carbon monoxide
B20CFME	20% chicken fat methyl Ester (CFME) biodiesel + 80% clean diesel
B20CFME+NIO	20% chicken fat methyl ester (CFME) biodiesel + 80% clean diesel + nickel oxide
ppm	Parts per million

In the occurrence of KOH and methyl alcohol, the transesterification methods of the raw chicken fat oil were carried out in a biodiesel reactor machine produces the chicken fat methyl ester (CFME), as in the demonstration below.

2.2 *Preparation of Nanofluids*

In the present experimental investigation, nickel oxide is used as a nanoparticle additive. The morphology of NiO was examined utilizing SEM as shown in Table 2. The Nanoliquids are set up for three diverse grouping of fifty (50 ppm), seventy-five (75 ppm) and hundred (100 ppm). The ultrasonicator assists in compelling the scattering of nanoparticles in biodiesel .

Table 2 Nanoparticles properties of NiO

Parameters	NiO
Manufacture	Nano Research Lab
Chemical name	Nickel oxide
Form colour	Powder grey
Particle size	30–50 nm
Specific surface area	120 m^2/g
Purity	99.9%

Table 3 Properties of fuels used for testing

Properties	Diesel	B20CFME	B20CFME NiO +50	B20CFME + NiO 75	B20CFME+NiO 100
Kinematic viscosity (40 °C, Cst)	3.05	2.2	3.219	3.41	3.92
Heat value (MK/kg)	43,900	42,437	42,697	42,663	43,000
Density (kg/m^3)	850	835	840	845	850
Flash point °C	56	68	71	69	62
Cetane number	51	54	57	56	55

2.3 *Resolving of Fuel Properties*

The properties of diesel collected chicken fat methyl ester and B20 CFME+NIO blends were determined and tested to meet the ASTM standard requirement. The required fuel characters are as orange in Table 3.

2.4 *Experimental Details*

In this investigation, the fuel used was petroleum diesel CFME and CFME + NiO blends with and without nickel oxide. The computerized single cylinder, four strokes 3.5 kw at constant speed static diesel engine was selected for experimentation. The selected engine was empowered with required instrumentation, and i3sys gas analyser and i3sys smoke metre EDM 1601 were employed for measuring amount of smoke emitted (Fig. 2).

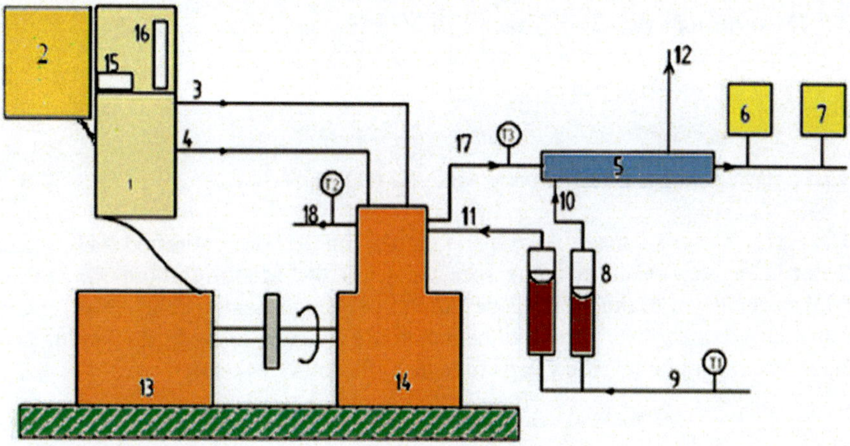

Fig. 2 Engine setup

2.5 *Error Analysis and Uncertainty*

The experimental results will certainly have error and uncertainties, which can rise from the incorrect calibration of instruments due to excessive handling and mishandling, surrounding conditions, experimental test conditions and planning, surveillance and reading. The errors which are gotten become a hindrance to obtain accurate results. Thus, to nullify these errors, tools and methods from mathematics and statistics are used. Generally, the method used is repetition of the taking data from the experimentation (at least three times) and finding the mean in order to minimize the error, which may occur (Kakoee et al. 2018). Measurements of uncertainties were calculated and results are shown in Table 4.

Table 4 Uncertainties percentage

Parameters	Average uncertainties (%)
Air flow rate	1.2
LCV of fuel	1.0
Engine speed	1.3
Liquid fuel flow rate	0.15
Temperature	1.0
Gas flow rate	2.1
Cylinder pressure	0.8

3 Outcomes and Argument

3.1 *Analysis of Performance*

3.1.1 Brake Thermal Efficiency Analysis (BTE)

The BTE (%) versus load (%) graph is plotted Fig. 3. For total tested fuels, brake thermal efficiency enhanced with respect to load. B20CFME+NiO 100 shows a better BTE as compared to other blends because of oxygen present in chicken fat oil and better surface area/volume ratio of nanoparticles assisting to complete burning and higher the rate of combustion leads to higher efficiency [14].

3.1.2 Brake-Specific Fuel Consumption (BSFC)

The change in brake-specific fuel consumption (BSFC) with respect to varying load is shown in this BSFC versus load in Fig. 4. It is evident that BSFC has an inverse relation with increasing load because of lower calorific values and the higher oxygen content present in NiO.

3.2 *Analysis of Emission*

3.2.1 Hydrocarbon (HC)

HC emission of biodiesel reduces as it self-hold oxygen, which increase the combustion rate of the fuel. In the graph Fig. 5, we recognize that as the load

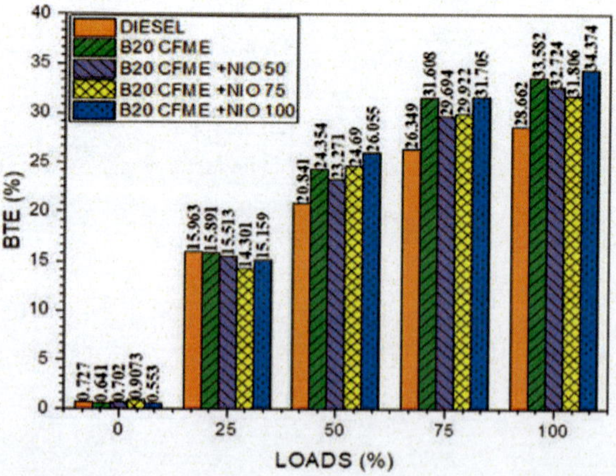

Fig. 3 BTE versus load

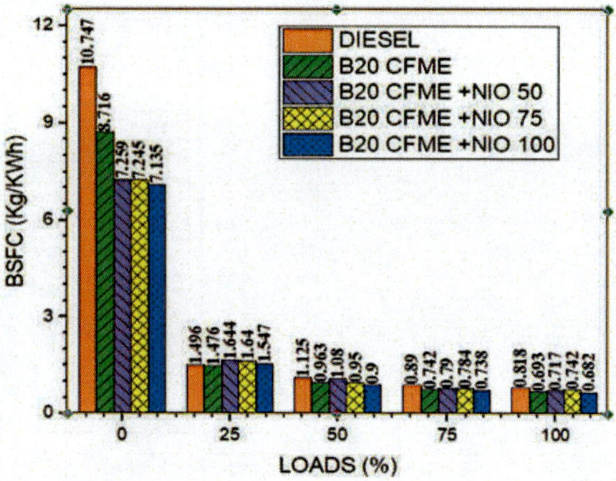

Fig. 4 BSFC versus load

increases the B20CFME+NiO gives low UBHC than the diesel. This indicates B20CFME+NiO100 gives complete combustion.

3.2.2 Carbon Monoxide (CO)

The change in carbon monoxide (CO) with respect to the load is shown in Fig. 6. It is found that biodiesel itself contains oxygen, using biodiesel as a fuel, it improves the combustion of fuel. Improving in combustion leads to reduction in CO emission. Addition of nickel oxide nanoparticle leads to further reduction in carbon monoxide at higher loads [12, 14].

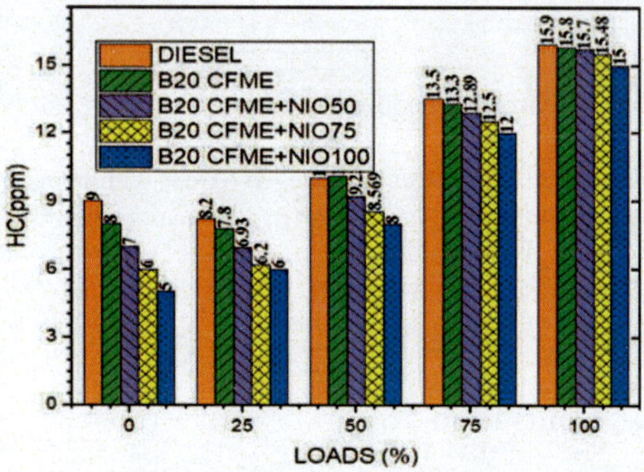

Fig. 5 HC versus load

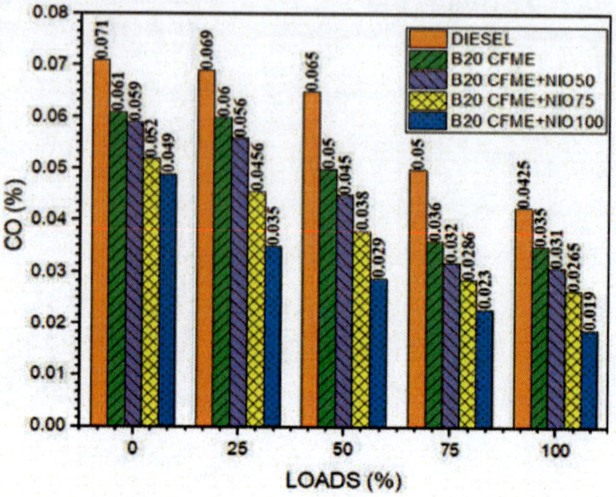

Fig. 6 CO versus load

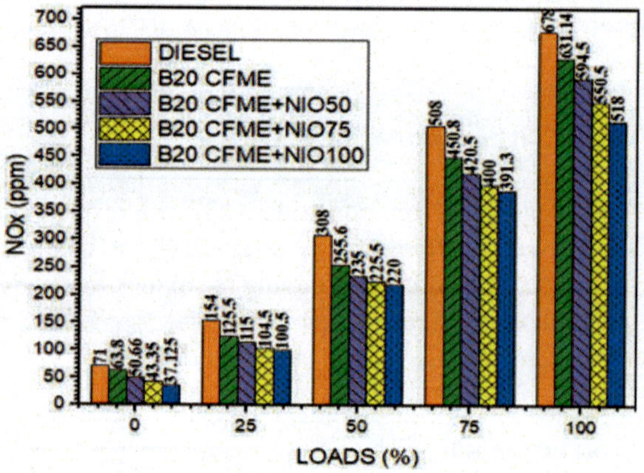

Fig. 7 NO_x versus load

3.2.3 Nitrogen Oxides (NO_x)

From Fig. 7, variation of nitrogen oxides with respect to load, this is due to rise in load, there is an enhancement in combustion chamber temperature. However as the addition of nanoparticle increases the surface area to vulume ratio, at maximum load, B20CFME NiO shows the marginal reduction in NO_x.

4 Conclusion

The present investigation was conducted with the blends of diesel, B20CFME, B20CFME+NIO50, B20CFME+NIO75. B20 CFME+NIO100, for zero to maximum load condition. The tests were conducted at above-mentioned engine specification and parameters,

- The B20 CFME+NIO100 showed enhance BTE with respect to diesel and other blends.
- The nitrogen oxides drastically lowered for the B20 CFME+NIO100 when compared with diesel due to the combined effect of nickel oxide nanoparticles.
- For B20 CFME+NIO100, the emission of HC and CO decreases as compared to diesel signifying a near-complete and an efficient combustion.

Finally, B20 CFME+NIO100 can be a promising fuel for diesel.

References

1. Vairamuthu G, Sundarapandian S, Kailasanathan C, Thangagiri B (2015) Experimental investigation on the effects of cerium oxide nanoparticle on CalophyllumInophyllum (Punnai) biodiesel blended with diesel fuel in DI diesel engine modified by nozzle geometry. J Energy Inst 89(4):668–682
2. Nanthagopal K, Ashok B, Tamilarasu A, Johny A, Mohan A (2017) Influence on the effect of zinc oxide and titanium dioxide nanoparticles as an additive with Calophylluminophyllum methyl ester in a CI engine. Energy Conversion Manage 146:8–19
3. Ramesh DK, Dhananjaya Kumar JL, Hemanth Kumar SG, Namith V, Parashuram (2016) Study on effects of Alumina nanoparticles as additive with Poultry litter biodiesel on performance, combustion and emission characteristic of diesel engine. In: Proceedings PMME (2016) EMT–338
4. Seela CR, Ravi Sankar B, Kishore D, Babu MVS (2017) Experimental analysis on a DI diesel engine with cerium oxide added Mahua Methyl Ester blends. ISSN: 0143-0750 (Print) 2162-8246
5. Jiaqiang E, Pham M (2017) Effect of different technologies on combustion and emissions of the diesel engine fueled with biodiesel. Renew Sustain Energy Rev 80:620–647
6. Sahara SS, Iqbala ACJ, Ullahd I (2018) Biodiesel production from waste cooking oil: an efficient technique to convert waste into biodiesel. Sustain Cities Soc 41:220–226
7. Debbarma S, Misra RD (2017) Effects of iron nanoparticles blended biodiesel on the performance and emission characteristics of a diesel engine. J Energy Resources Technol. ASME, July 2017, vol 139/042212-1
8. Mohamed Musthafa M, Ashok Kumar T, Mohanraj T, Chandramouli R (2018) A comparative study on performance, combustion and emission characteristics of diesel engine fuelled by biodiesel blends with and without an additive. Fuel 225:343–348
9. Knothe G, Steidley KR (2018) The effect of metals and metal oxides on biodiesel oxidative stability from promotion to inhibition. Fuel Process Technol 177:75–80
10. Man XJ, Cheung CS, Ning Z (2015) Effect of diesel engine operating conditions on the particulate size, nanostructure and oxidation properties when using wasting cooking oil biodiesel. Energy Proc 66:37–40
11. D'Silvaa R, Binu KG, Bhat T (2015) Performance and emission characteristics of a C.I. engine fuelled with diesel and TiO_2 nanoparticles as fuel additive. Mater Today Proc 2:3728–3735
12. Ashok B, Nanthagopal K, Mohan A, Johny A, Tamilarasu A (2017) Comparative analysis on the effect of zinc oxide and ethanox as additives with biodiesel in CI engine. EGY 11515. https://doi.org/10.1016/j.energy.2017.09.021
13. Cheung CS, Man XJ, Fong KW, Tsang OK (2015) Effect of waste cooking oil biodiesel on the emissions of a diesel engine. Energy Proc 66:93–96

14. Ashok B, Nanthagopal K, Subbarao R, Johny A, Mohan A (2017) Experimental studies on the effect of metal oxide and antioxidant additives with CalophyllumInophyllum Methyl ester in compression ignition engine. J Cleaner Prod. https://doi.org/10.1016/j.jclepro.2017.08.050 S0959-6526(17)31768-7
15. Attia AMA, Hassaneen AE (2016) Influence of diesel fuel blended with biodiesel produced from waste cooking oil on diesel engine performance. Fuel 167:316–328

Performance, Emission and Combustion Characteristics of Turpentine Cottonseed Oil Ester Blend

P. Udayakumar and G. Kasiraman

1 Introduction

There has been increase in energy demand as the economic growth and population are growing. As there are limited quantity of fossil fuel, researchers are looking at alternative fuels. Specially biodiesel fuel is the recent important field, in which many people are interested to explore the opportunities, due to its renewable source and lesser emissions than diesel. Over the last decade, there has been plethora of researches happened on vegetable oils (edible or non-edible oil), for instance, Jatropha, cottonseed, neem, olive, turpentine, cashew nutshell oil and sunflower oils from which the biodiesel can be prepared. These biodiesels are easy to produce and are having good fuel properties like cetane index, lesser HC, CO emissions due to its oxygen-rich content. Biodiesel is non-toxic, renewable and biodegradable fuel which can be used in combustion ignition engine in pure form or blended form without any requirement of design change of engine. In recent years, many researchers are concentrating on using vegetable oil fuels with low viscous fuel so that combined viscosity turns to be comparatively near to the diesel. Research work carried out on various biodiesel-ethanol and biodiesel-methanol, cashew nutshell oil-camphor oil, biodiesel-eucalyptus oil blends. Many researchers have done investigation on a duel fuel to eliminate the use of diesel completely and they have found that engine can be run with these biodiesels without any modification of engine.

Dubey et al. [1] investigated the Jatropha biodiesel and turpentine oil blend in a CI engine. They found that the BTE of fuel blends was lower than diesel. The nitrogen oxides and carbon monoxides release were lesser than the diesel. Altun et al. [2] have done investigations on usage of cottonseed biodiesel on compression ignition engine. The NO_x, HC and carbon monoxide emissions were lesser for CSO

P. Udayakumar (✉) · G. Kasiraman
SRM Institute of Science and Technology, Kattankulathur, India
e-mail: udayakup@srmist.edu.in

A. K. Gupta et al. (eds.), *Advances in IC Engines and Combustion Technology*,
Lecture Notes in Mechanical Engineering,
https://doi.org/10.1007/978-981-15-5996-9_13

blended fuel in comparison with diesel. Dubey et al. [3] studied the combustion and performance of dual fuel blends (Karanja and turpentine oil) and have observed that there has been decrease in NO_x, CO_2 and carbon monoxide release and rise in unburned hydrocarbon release in the exhaust for fuel blend in comparison with diesel. Vallinayagam et al. [4] studied various blends of pine oil with kapok methyl esters. It was observed that brake thermal efficiency is marginally higher than pure diesel for B50P50 and emissions HC and CO were found to be 8.1% and 18.9%, respectively, lower than diesel, with a comparable NO_x emission with diesel.

Vallinayagam et al. [5] studied various blends of pine oil with diesel its give brake thermal efficiency and maximum heat release rate increase by to 5% and 27% and HC,CO emissions are reduced for all loads. Kasiraman et al. [6] studied that cashew nutshell oil 70% and camphor oil 30% blend gives brake thermal efficiency is 29.1% at a peak load compared to diesel and NO_x emission is 1040 ppm. Prem Anand et al. [7] observed various blends of Turpentine oil fuels with diesel. 30% turpentine oil fuel blend found give high brake power and net heat release rate than diesel and also emission reduces, except CO_2. Karthikeyan et al. [8] studied that the brake thermal efficiency of the diesel fuel engine fueled by turpentine diesel blend is 29%, very close to that of diesel up to 75% of load, beyond which it starts decreasing because of occurrence of knock. Venkatesan et al. [9] tested pine oil-soapnut oil methyl ester blends in twin-cylinder tractor diesel engine. The brake thermal efficiency of the blend 50% of pine and 50% of soapnut gives 8% increased with diesel. Devana et al. [10] studied that 50% paradise oil—50% of eucalyptus oil blend gives 2.4% increase in brake thermal efficiency at full load and emissions also decreased.

In this study, a blend of high CN and low CN fuels is used. The emission and performance characteristics of duel fuel blends, cottonseed oil ester (high CN) and turpentine (low CN) are studied on a 5.2 kW, single cylinder, 1500 rpm constant speed diesel engine. The studies were conducted with various proportions of fuel blend such as 50% by volume of turpentine and 50% of cottonseed oil ester by volume (T50), and 60% turpentine—40% cottonseed oil ester (T60), and similarly 70% turpentine—30% cottonseed oil ester (T70), and 80% turpentine—20% cottonseed oil ester (T80). The experiments were conducted at different loads on the engine from no load up to maximum load. The performance and emission characteristics of the engine were plotted.

Properties of the oils

All the physical and physical properties of diesel, turpentine, cottonseed oil ester and fuel blends have been given in below Table 1. The calorific value of the blends is improving. The cetane number is found to increase from base level for T50 and T60 blends. Viscosity of the blends decreases with increment of turpentine oil as the cetane number of the cottonseed oil ester is higher, the ignition quality and hence the combustion of turpentine oil is expected to improve.

Table 1 Physical and chemical properties of fuels

Fuel	Cottonseed oil ester	Turpentine oil	Diesel	T50	T60	T70	T80
Formula	$C_{20}H_{39}O_2$	$C_{10}H_{16}$	$C_{10}H_{20}$	–	–	–	–
Kinematic viscosity (cst)	5.81	2.51	2.5	4.16	3.83	3.1	2.9
CV (MJ/kg)	39.64	44.4	45	42	42.5	43	43.5
Density (kg/l)	0.91	0.8	0.855	0.85	0.84	0.83	0.82
Cetane number	57	35	55	46	43	41	39
Fire point (C)	160	50	74	120	100	85	70

2 Experimental Setup

The experiments were carried out on a four-stroke single cylinder, water cooled, direct injection engine has rated capacity of 5.2 kW at a constant speed 1500 RPM. Schematic diagram is shown in Fig. 1. It has coupled with Eddy current dynamometer for applying brake load. The output signals from all sensors of load, pressure and temperature sent to control panel's board then connected to a personal computer. To record exhaust gas emissions AVL 5 gas analyzer was attached to the set up for measuring the NO_x, CO_2, HC and CO emissions present in the exhaust gas. Engine technical specifications are shown in Table 2.

The specifications for the exhaust gas analyzer have been given in Table 3.

Fig. 1 Schematic diagram

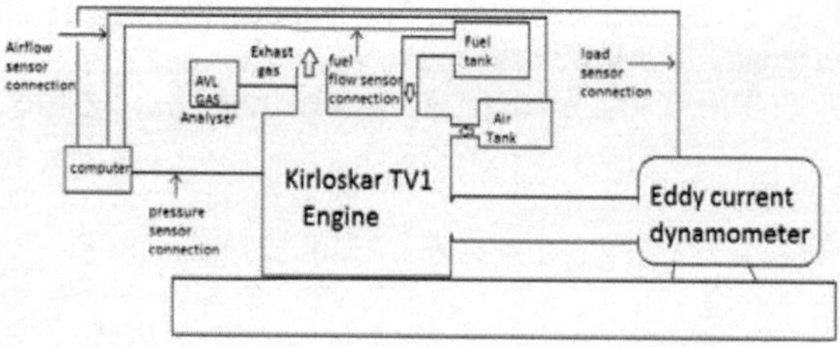

Table 2 Engine specifications

Feature	Description
Engine type	TV1, multifuel
Cylinders	Four-stroke single cylinder
Cooling medium	Water cooled
capacity	5.2 kw @ 1500 rpm
Range of compression ratio	17.5:1

Table 3 Exhaust gas analyzer specifications

Gas emitted in exhaust	Range of measurement	Resolutions	Precision
CO	0 to 10% of volume	0.01% of volume	<0.6% ± 0.03% vol ≥ 0.6% ± 5% vol of ind val
HC	0 to 20,000 ppm	<2000:1 ppm volume > 2000:10 ppm volume	<2000 ppm of vol ± 10 ppm ≥ 2000 ppm of vol ± 5% of individual val
CO2	0 to 20% of vol	0.1% of vol	<10% vol ± 0.5% ≥ 10% vol ± 5% of ind val
NO_x	0 to 5000 ppm	1 ppm vol	<500 ppm of vol ± 50 ppm ≥ 500 ppm vol ± 10% of individual val

3 Results and Discussion

3.1 Performance Characteristics

3.1.1 Brake Thermal Efficiency

The brake thermal efficiency variation with load is shown in Fig. 2. The brake thermal efficiency increases with respect to load. There is rise in the brake thermal efficiency with increasing the concentration of cottonseed oil in the blend. This indicates the better combustion quality of cottonseed oil ester as compared to turpentine oil. But still the all fuel blends are lower thermal efficiency than diesel because of higher viscosity and lower choleric value. The brake thermal efficiencies for diesel, T50, T60, T70, T80 are 30%, 28%, 25%, 27%, and 25.79%, respectively.

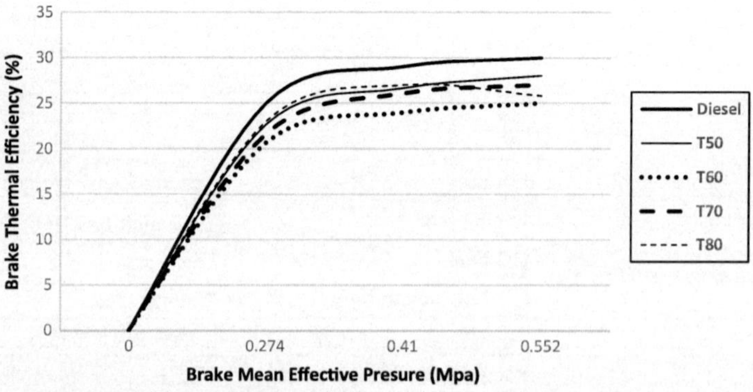

Fig. 2 Brake thermal efficiency graph

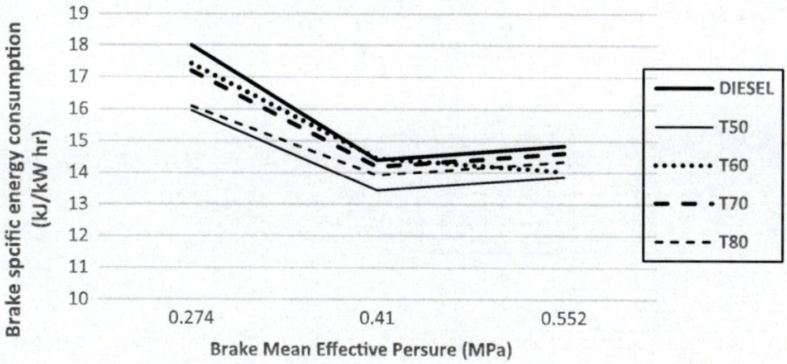

Fig. 3 Brake specific energy consumption

3.1.2 Brake Specific Energy Consumption

The brake specific energy consumption for various loads is shown in Fig. 3. It is observed that T50 blend has the lowest energy consumption among all the blends. It develops 5.2 kW power output with energy consumption of 14 kJ/kW hr. This is due to better combustion of blend and high cetane number. The brake specific energy consumption of other T60, T70, T80 blends are 14.025, 14.62, 14.355 kJ/kWh, respectively, for same power output.

3.2 Emission Characteristics

3.2.1 Carbon Monoxide Emission

The carbon monoxide emission of pure diesel and blend fuel is illustrated in graph for various load conditions is shown in Fig. 4. Carbon monoxide formation is due to incomplete combustion of fuel blend. Due to high viscosity of the blend the combustible mixture formation is slower. The emission of carbon monoxide for blends T50, T60, T70 and T80 is higher than diesel. CO emission at full load condition is found to be 30–40% more than diesel.

3.2.2 Hydrocarbon Emission

The hydrocarbon emission for both the blend fuel and diesel fuel has shown increasing trend with increase in the load is shown in Fig. 5. In spite of fuel bound oxygen in blended fuels, the HC emission is higher than diesel. This indicates the partial combustion of fuel, due to high density and viscosity of cottonseed oil ester in the blend. At full load, the unburned hydrocarbon emission for fuel blends is 20–40% more than diesel.

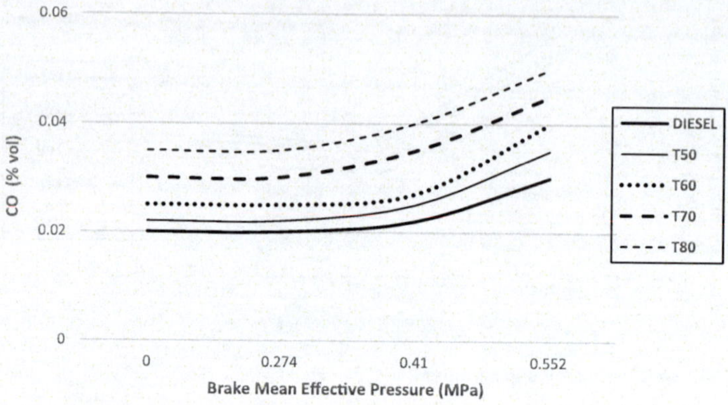

Fig. 4 Carbon monoxide emission

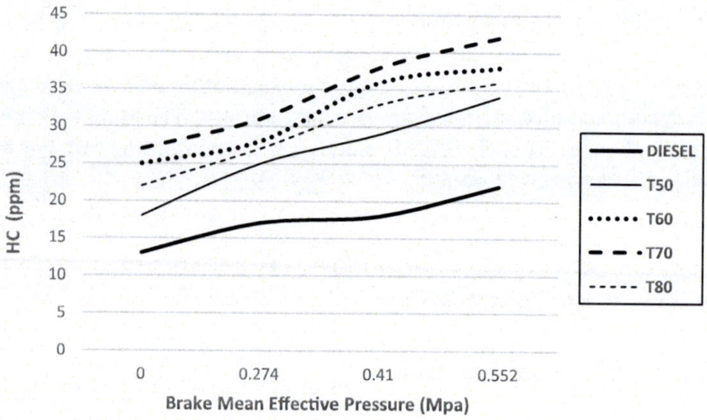

Fig. 5 Hydrocarbon emission

3.2.3 NO_x Emission

The NO_x emission has shown increasing trend as shown in Fig. 6. During all the load conditions, NO_x emission of the fuel blends T50, T60, T70 and T80 were less than diesel emissions. At the full load conditions, the nitrogen oxides emission for fuel blends was about 20%–30% lesser than diesel due to lower calorific value and combustion impairment of the fuel blends the maximum in cylinder temperature should be lesser than that of diesel hence the NO_x emission is lesser than the diesel.

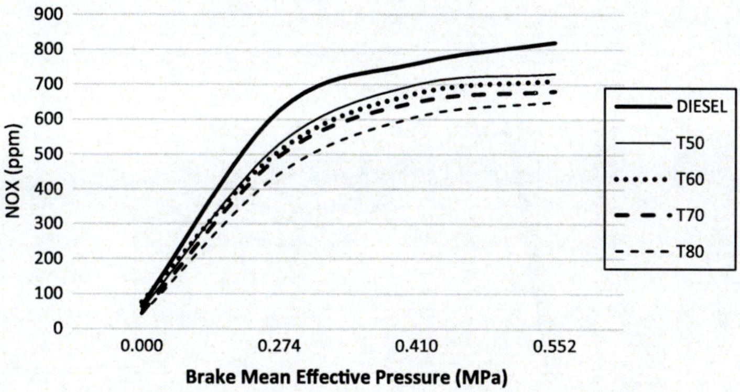

Fig. 6 NO_x emission

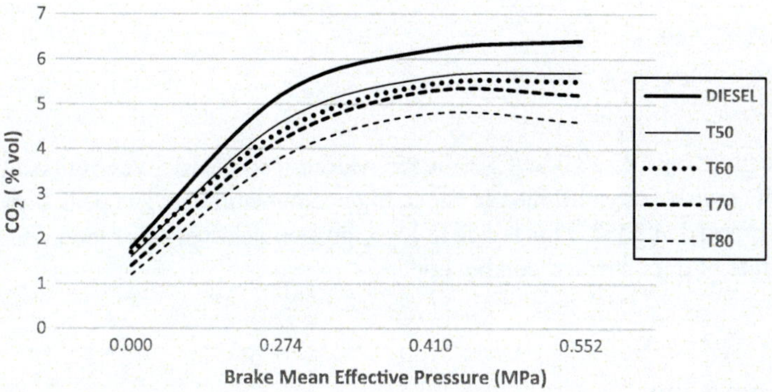

Fig. 7 Carbon dioxide emission

3.2.4 Carbon Dioxide Emission

The CO_2 emission increases with load as shown in Fig. 6. The CO_2 emission indicates the combustion quality of fuel, greater the CO_2 emission, better the fuel burned. The CO_2 emission for fuel blends was less than the diesel emissions at all the loads. This might be because of poor combustion of turpentine. At full load condition, the CO_2 release for fuel blends was 20 30 percent less than the diesel emissions (Fig. 7).

4 Combustion Characteristics

Heat release rates of diesel and blends of turpentine and cottonseed oil ester are shown in Fig. 8. Heat release rate for diesel is 18.95 J/CAD at 353 degree, for T50 21.05 j/CAD at 355 degree, for T60 16.52 J/CAD at 353 degree, for T70 16.83 J/CAD

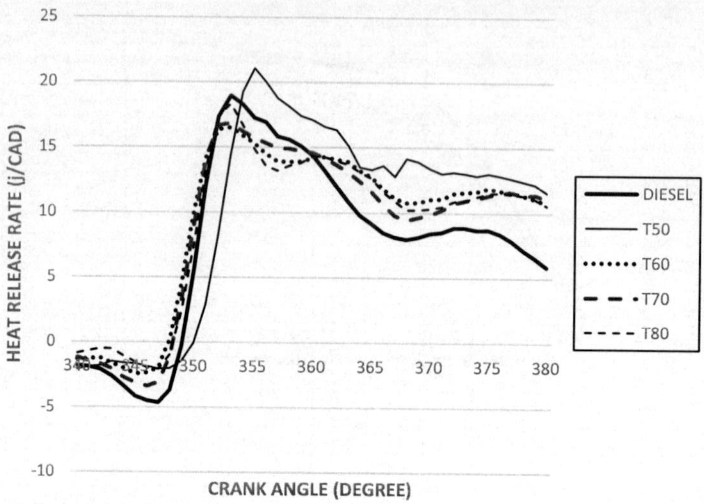

Fig. 8 Heat release rate at full load

at 353 degree, for T80 18.32 J/CAD at 353 degree. T50 blend gives high heat release rate with compare other all blends due to high cetane number. The peak heat release rate is observed at 355°. This is due to high ignition delay of this blend which leads to high rate of uncontrolled combustion.

5 Conclusion

The present work fuel blend of cottonseed oil and turpentine oil was blended in various proportions. The performance, emission and combustion characteristics were studied and compared with that of pure diesel. The following conclusions were drawn.

1. For the fuel blend T50, the engine has shown comparatively better performance as compared to other blends. At maximum load, the BTE for fuel blends was around 28%, next highest after diesel.
2. CO_2 and NO_x emission of the blend T50 is less than that of diesel. T50 emits 730 ppm and 5.7%.
3. The heat release rate of T50 blend is 21.056 J/CAD whereas other T60, T70, T80 blends have heat release rates of 16.52,16.83,18.32 J/CAD, respectively.

T50 blend is found to be most suitable blend among all blends tested. However, CO, HC emissions are higher than diesel. The long-term engine tests are required to assess the engine deposits that may arise due to the high viscous biofuels.

References

1. Dubey P, Gupta R (2017) Effects of dual bio-fuel Jatropha biodiesel and turpentine oil on a single cylinder naturally aspirated diesel engine without EGR. Appl Thermal Eng 115:1137–1147
2. Altun S, Ãu ner C, Yasar F, Adärn H (2011) Biodiesel production from raw cottonseed oil and its performance in a diesel engine. Technology 14(3):95–102
3. Dubey P, Gupta R (2016) Study of the performance and emission characteristics for a dual fuel powered single cylinder diesel engine. Int J Autom Mech Eng 13(2):3373–3388
4. Vallinayagam R, Vedharaj S, Yang WM, Lee PS, Chua KJE, Chou SK (2013) Pine oil–biodiesel blends: a double biofuel strategy to completely eliminate the use of diesel in a diesel engine. Appl Energy
5. Vallinayagam R, Vedharaj S, Yang WM, Lee PS, Chua KJE, Chou SK (2013) Combustion performance and emission characteristics study of pine oil in a diesel engine. Energy 57:344–351
6. Kasiramana G, Nagalingam B, Balakrishnan M (2012) Performance, emission and combustion improvements in a direct injection diesel engine using cashew nut shell oil as fuel with camphor oil blending. Energy 47:116–124
7. Prem Anand B, Saravanan CG, Ananda Srinivasan C (2010) Performance and exhaust emission of turpentine oil powered direct injection diesel engine. Renew Energy 35:1179–1184
8. Karthikeyan, Mahalakshmi NV (2007) Performance and emission characteristics of a turpentine–diesel dual fuel engine. R Energy 32:1202–1209
9. Venkatesan V, Nallusamy N (2019) Pine oil-soapnut oil methyl ester blends: a hybrid biofuel approach to completely eliminate the use of diesel in a twin cylinder off-road tractor diesel engine. Fuel
10. Devana PK, Mahalakshmi NV (2009) A study of the performance, emission and combustion characteristics of a compression ignition engine using methyl ester of paradise oil–eucalyptus oil blends. Appl Energy 86(5):675–680

Experimental Catalyst Optimization and Artificial Neural Network Modeling in Biodiesel Preparation

P. Deivajothi, G. Vinodhini, and V. Manieniyan

1 Introduction

Nowadays, the diesel-fueled vehicles are becoming progressively famous because of its effectiveness and performance. In the near future, the world's transportation prerequisites will depend on the diesel engine. The utilization of vegetable oil as alternate fuel is as aged as the diesel engine. Rudolph Diesel who created the diesel motor, observed energy from various alternative resources which also incorporate a few vegetable oils. In 1912, Rudolph Diesel anticipated and communicated, "the utilization of vegetable oils for engine may appear to be insignificant today, yet such oils may progress in near future as significant as the petrol and coal of our time." Notwithstanding Rudolph Diesel's positive thinking, the utilization of biodiesel as a feasible oil option has over and again been investigated in recent years. Biodiesel has turned out to be thrust area of research because of its natural advantages [6, 9]. The expense of biodiesel, in any case, is the primary hurdle to commercialization of the item. The financial favorable circumstance of biodiesel is because of its capacity to decrease green house emissions, diminishes a nation's dependence on unrefined oil imports, and supports farming by giving new job and market open doors for household crops.

Biodiesel fuel is prepared by transesterification of refined vegetable oil, unwanted cooking oil, and utilized frying oil using alkaline catalysts. The idea of catalyst

P. Deivajothi (✉) · V. Manieniyan
Mechanical Engineering, Annamalai University, Tamil Nadu, India
e-mail: pdeivajothi@gmail.com

V. Manieniyan
e-mail: manieniyan78@gmail.com

G. Vinodhini
Computer Science & Engineering, Annamalai University, Tamil Nadu, India
e-mail: vinodhini.g.t@gmail.com

A. K. Gupta et al. (eds.), *Advances in IC Engines and Combustion Technology*,
Lecture Notes in Mechanical Engineering,
https://doi.org/10.1007/978-981-15-5996-9_14

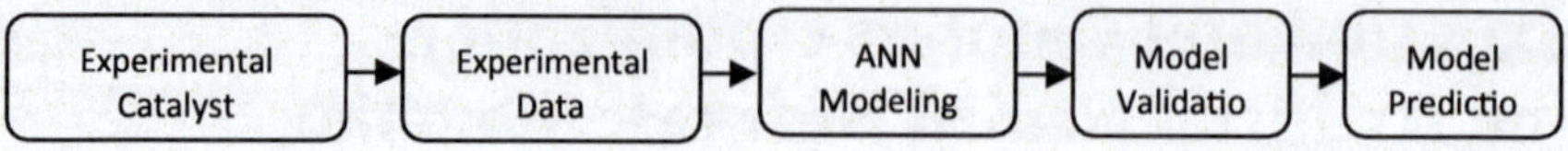

Fig. 1 Problem design

utilized during transesterification response is vital in conversion of triglycerides to biodiesel. Accordingly, various catalysts have being investigated for converting triglycerides to biodiesel fuel. The catalysts generally employed in transesterification method are homogeneous catalysts and heterogeneous catalysts. Ordinarily, homogeneous catalysts, for example, NaOH, KOH, Na_2O_2, and CH_3ONa are all the more regularly utilized in preparing biodiesel than heterogeneous [3, 4]. The capacity of catalyst to perform under moderate conditions has prompted the decision of choosing the optimum catalyst.

Machine learning, a influential method of artificial intelligence (AI), has been widely used for classification, clustering, and forecasting. In the past decades, it was widely utilized in many application fields, such as energy, biology, environment, medicine, and engineering technology [5]. However, in the literature reviewed, there are not many studies that analyze the use of ANN modeling in catalyst optimization of biodiesel preparation. So, this motivated to analyze the capability on application of ANN for optimization of catalysis in biodiesel preparation. The content of this paper comprises of two major sections such as (i) Experimental analysis of catalysts optimization and (ii) Application of ANN modelling for experimental data on catalysis. The problem methodology is as shown in Fig. 1.

2 Experimental Investigation

The alkali-based catalysts for 10 g of potassium hydroxide (KOH), 8 g of sodium hydroxide (NaOH), 11 g of sodium peroxide (Na_2O_2), and 2.5 g of CH_3ONa were utilized. The ethanol was utilized as the alcohol. The optimization of reaction time was conducted in one litter of crude vegetable oil refinery waste (Sunflower acid oil). The optimization work was conducted in various reaction times with constant quantity of catalyst and alcohol. In both stages, the ethanol was changed in the 200 ml. After the finish of response, alcohol catalyst was isolated from upper layer. Next, the transesterified oil is washed with refined water, until the point that the remaining parts of impetus were evacuated and warmed to expel water content. Figure 2 shows the experimental setup for biodiesel plant.

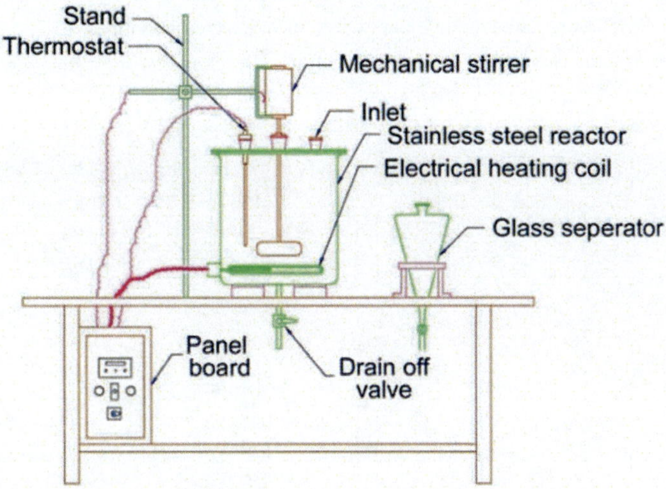

Fig. 2 Experimental setup for biodiesel plant

2.1 Optimization of Catalyst in Experimental Results

The yield of biodiesel for various catalysts with varying reaction time is analyzed to find the optimum reaction of the catalyst. Different reaction times such as 30, 45, 60, 75, and 90 min were chosen for the analysis. The experimental results are tabulated in Table 1. In all the catalyst, when the reaction time was lower certain amount of soap formation is noted in the biodiesel [1, 2]. Soaps were formed along with biodiesel, and during the washing process, the soaps stick to the walls of the glass separator. This soap formation reduces the yield of biodiesel [8]. Among the catalyst used, the KOH and CH_3ONa catalysts were found in optimum yield. The higher biodiesel yield with minimum reaction time is obtained for KOH rather than CH_3ONa. Sodium hydroxide when used in the transesterification process increased the reaction time [3, 4].

3 ANN Modeling

An ANN model contains numerous input factors to predict different output factors. The well-trained ANN is faster in prediction than the traditional mathematical models. In ANN modeling, no extensive iterative computations are needed as in mathematical modeling. However the determination of a optimimum neural network topology and its parameters is a need for best prediction [7, 10]. In this investigation, back propagation neural system (BPN) is utilized. Back propagation neural networks (BPN) are supervised learning method for training multi-layer perceptrons. Figure 3 demonstrates the ANN model applied in this study.

Table 1 Consolidation results for different catalyst in various reaction times

Catalyst	Volume of ethanol in ml	Weight of catalyst in g	Reaction time in min	Biodiesel yield in %
NaOH	200	8	30	81
			45	93
			60	96
			75	96
			90	96
			30	89
			45	92
KOH	200	10	60	97
			75	97
			90	97
			30	49
			45	49
Na_2O_2	200	11	60	51
			75	57
			90	57
			30	93
			45	95
CH_3ONa	200	2.5	60	97
			75	97
			90	97
			30	93.5
			45	95.5

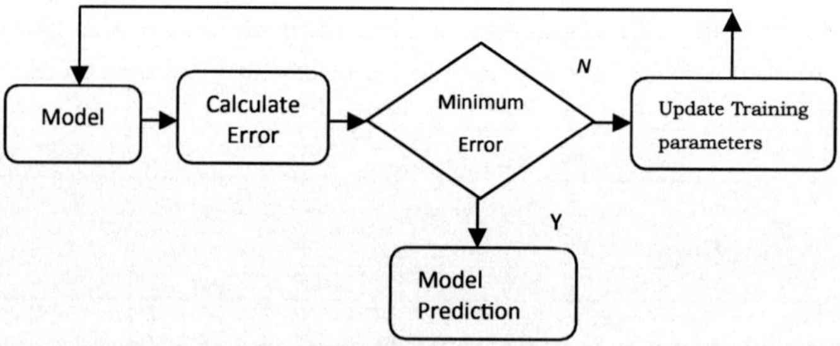

Fig. 3 Back propagation ANN modeling

```
initialize network weights
do
foreach training example (ex)
prediction = neural-net-output(network, ex)
actual = teacher-output(ex)
compute error (prediction - actual)
calculate error from hidden to output layer
calculate error from input to hidden layer
update network weights
until stopping criterion satisfied
return the network
```

Fig. 4 Pseudocode of BPN

The BPN network contains the following layers: inputs, hidden, and output neurons. The steps in BPN Modeling are as given below:

i. In forward propagation, some random values to weight are initialized and propagated forward.
ii. In backward propagation, the network propagates backwards thereby to reduce the error by changing the values of weights and biases.
iii. Finally, all the values are combined together and the updated weight values are calculated. If the error value is minimum, the trained BPN model is complete, else propagate backwards again and update the weight values. Repeat until error becomes minimum. Figure 4 shows the pseudocode for the back propagation algorithm.

3.1 ANN Modeling Results

In this modeling, the number of input neurons is methanol quantity, catalyst quantity, and reaction time. The yield of biodiesel is the output neuron. The number of neurons in one hidden layer is 5. The learning rate used is 0.01. The testing results for biodiesel yield prediction Vs reaction time for catalyst optimization using ANN are obtained. The results are shown in Figs. 5, 6, 7, and 8 for various catalysts used. These figures clearly depict that the yield prediction made by various catalysts is almost similar to the experimental yield obtained for varying reaction times applied during experimental analysis. In Fig. 5, 6, 7, and 8, the minimum deviation between exact values and predicted values of reaction time is because of the better generalization capacity of neural network model employed in this training and testing phases. A slightly higher variation is noted for KOH at the peak yield value and this may be due to variation in the bias of the neural network model.

To validate the performance of the ANN model used, a testing procedure is essential. The model predicted data can be compared with the actual data in the testing set, with the root mean square error (RMSE) calculated by equation

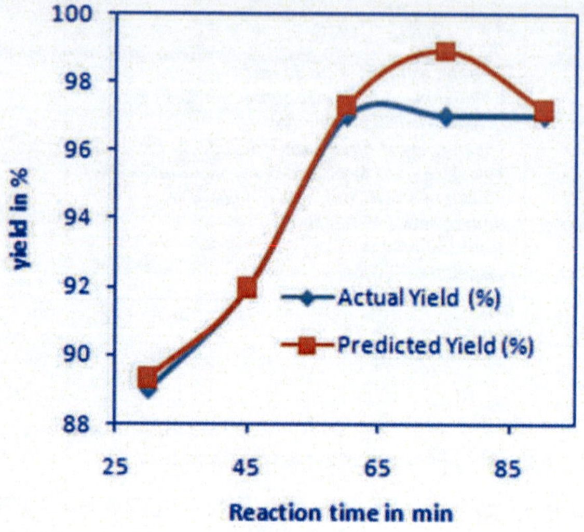

Fig. 5 Biodiesel yields with reaction time (KOH)

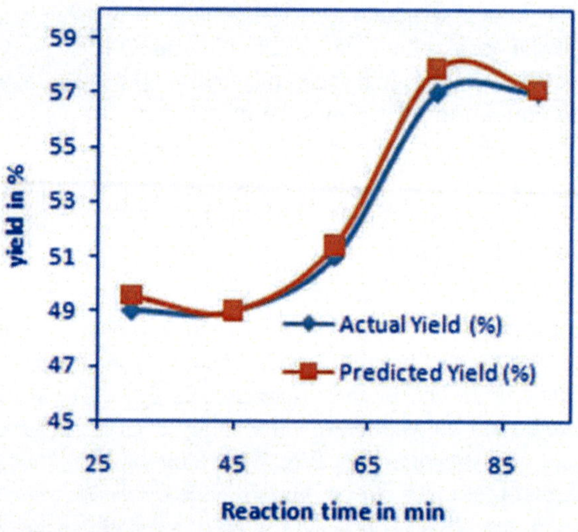

Fig. 6 Biodiesel yield with reaction time (NaOH)

$$\text{RMSE} = \sqrt{\frac{\sum_i^n (P_i - A_i)^2}{a}}$$

where P is the predicted value and A is the actual value. If the calculated RMSE from the testing set is relatively small, it means that the ANN is well trained. The RMSE values calculated are tabulated in Table 2.

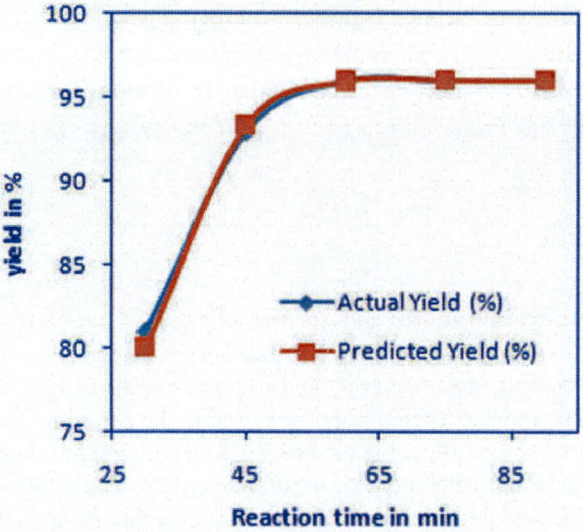

Fig. 7 Biodiesel yield with reaction time (Na_2O_2)

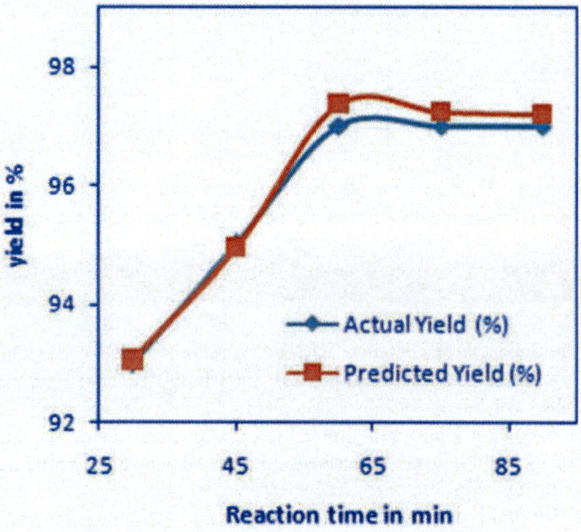

Fig. 8 Biodiesel yield with reaction time (CH_3ONa)

Table 2 RMSE values

Catalyst used	RMSE
NaOH	0.012
KOH	0.009
Na_2O_2	0.015
CH_3ONa	0.011

The RMSE values reveal that for all catalysts employed the BPN modeling result in minimum negligible RMSE values of range 0.009 to 0.015. Thus, it is concluded that ANN modeling can be used as an effective tool for modeling of catalyst optimization in biodiesel production.

4 Conclusion

In this study, the effect of various catalysts such as NaOH, KOH, Na_2O_2, and CH_3ONa was analyzed in biodiesel production. The biodiesel production is carried out by transesterification process using crude vegetable oil refinery waste. In order to identify the optimum catalyst, various reaction times are used in the experimental analysis. The optimum catalyst used in producing biodiesel is KOH. In order to show the effectiveness of ANN modeling, BPN model is used in this study. The results concluded that BPN-based ANN modeling can be used as an effective tool for modeling of catalyst optimization in biodiesel production.

References

1. Baskar G, Aiswarya R (2016) Trends in catalytic production of biodiesel from various feedstocks. Renew Sustain Energy Rev 57:496–504
2. Bharadwaj AS, Singh M, Niju S, Begum KMS, Anantharaman N (2019) Biodiesel production from rubber seed oil using calcium oxide derived from eggshell as catalyst–optimization and modeling studies. Green Process Synthesis 8(1):430–442
3. Deivajothi P, Manieniyan V, Sivaprakasam S (2019) Experimental investigation on DI diesel engine with fatty acid oil from by-product of vegetable oil refinery. Ain Shams Eng J 10(1):77–82
4. Deivajothi P, Sukumar V, Manieniyan V, Senthilkumar R (2019b) Bio Oil from bio waste (Mahua Oil Cake) using fast pyrolysis process. Int J Innov Technol Exploring Eng (IJITEE) 8(10):2525–2528
5. Etim A, Betiku E, Ajala S, Olaniyi P, Ojumu T (2018) Potential of ripe plantain fruit peels as an ecofriendly catalyst for biodiesel synthesis: optimization by artificial neural network integrated with genetic algorithm. Sustainability 10(3):707
6. Manieniyan V, Sivaprakasam S (2013) Artificial neural network based modelling for vibration characteristics of DI diesel engine using bio-diesel. Int J 3(8)
7. Manieniyan V, Vinodhini G, Senthilkumar R, Sivaprakasam S (2016) Wear element analysis using neural networks of a DI diesel engine using biodiesel with exhaust gas recirculation. Energy 114:603–612
8. Mardhiah HH, Ong HC, Masjuki HH, Lim S, Lee HV (2017) A review on latest developments and future prospects of heterogeneous catalyst in biodiesel production from non-edible oils. Renew Sustain Energy Rev 67:1225–1236

9. Sarve A, Sonawane SS, Varma MN (2015) Ultrasound assisted biodiesel production from sesame (*Sesamum indicum* L.) oil using barium hydroxide as a heterogeneous catalyst: comparative assessment of prediction abilities between response surface methodology (RSM) and artificial neural network (ANN). Ultrason Sonochem 26:218–228
10. Soltani S, Rashid U, Roodbar Shojaei T, Nehdi IA, Ibrahim M (2019) Modeling of the nanocrystalline-sized mesoporous zinc oxide catalyst using an artificial neural network for efficient biodiesel production. Chem Eng Commun 206(1):33–47

Performance Evaluation of Hydrogen-Fuelled Internal Combustion Engine

Sauhard Singh, V. K. Bathla, Reji Mathai, and K. A. Subramanian

1 Introduction

Fossil fuels are depleting due to extensive usage in various fields such as transport, industry etc. and their burning degrades the environment. Over a period of time, for advancement in transportation technologies, internal combustion engines became very important to our lifestyle. Oil is that the most significant and therefore the abounding supply of energy in the world on that the IC engines run. World primary energy demand has been increasing by 1.6 percent per year on an average. International fossil fuel oil product consumption continued to grow in 2017 because of increasing demand from transport and petrochemical industry. Oil demand associated with transport sector is in line for development of the worldwide vehicle fleet principally in higher fuel intense vehicles (SUVs) section. Efficiency gain post 2005 in transport sector in USA and Europe is especially offset by this trend and oil demand slightly redoubled in these two markets in 2017. Asian countries, that accounted for many of the international increase in petrochemical oil demand that is especially associated with global economic process, announce a gradual growth in oil consumption, particularly in China, India (above 5% in each countries), Indonesia, Thailand and South Korea [1].

The present energy systems that are getting used in IC engines have severe environmental impacts. Thc historical and present energy systems are predominant by fossil fuels (coal, oil, gas). The output of combustion of these fossil fuels is carbon dioxide and other greenhouse gases which are the foremost driver of the climate and

S. Singh (✉) · V. K. Bathla · R. Mathai
Indian Oil, Research & Development Centre, Faridabad, India
e-mail: singhsauhard@indianoil.in

K. A. Subramanian
IIT Delhi, Delhi, India

A. K. Gupta et al. (eds.), *Advances in IC Engines and Combustion Technology*, Lecture Notes in Mechanical Engineering,
https://doi.org/10.1007/978-981-15-5996-9_15

therefore the environment amendment globally. If we tend to achieve global sustainable and balanced development goals and prevent the appalling depletion of our fossil fuels, the globe needs a stimulating and collective change in its sources of energy.

A nation's general growth and advancement depend on its usable energy basket. The excessive use of fossil fuels for power wishes has resulted in outstanding emission development. Thus, the primary goal was moved to the use of cleaner fuels called alternative fuels to reduce carbon-based emissions to negligible levels [2].

Compressed natural gas (CNG) found application in stationary engines since quite long and also the application of CNG in transport sector advanced throughout the last decade primarily due to the development of high-pressure lightweight storage cylinders [3]. As an alternative fuel, CNG's economic, emission and strategic benefit encouraged many scientists to undertake compressed natural gas research [3–7]. Of the several alternative fuels that are capable of reducing general emissions relative to petrol and diesel, natural gas is regarded predominantly for several reasons [3–5]. Shasby [3] recognized three primary reasons for the ever-increasing use of natural gas in the stationary and transportation sectors as accessibility, compatibility with the environment and usability in standard diesel and gasoline motors. Operating costs are regarded as another reason in accordance with [3]. The argument of the reduced price per natural gas power unit over petroleum-powered cars offers important theoretical benefit for natural gas-powered cars. However, the argument is a little more complicated than this. While it is obvious that natural gas is cheaper than gasoline or diesel in most nations, the assessment is playing out differently. Compressed natural gas (CNG) for five reasons is appealing. (1) It has reduced emissions of greenhouse gases; (2) their use extends the supply of petroleum; (3) they are the only fuel cheaper than petrol or diesel; (4) air pollution emissions are inherently smaller, and (5) enormous amounts of natural gas are accessible in North America.

Vehicle ranges starting from passenger cars to merchandise carrier to heavy duty application, fuel storage and associated dead weight, state-of-the-art infrastructure demand arises because of flammability and high-pressure handling and guaranteeing sufficient provide are the most concern areas with natural gas. The importance of vary as a vehicle characteristics is illustrated in [8]. New gas pipelines and alternative infrastructure demand are coming in huge ways with increasing in the number of CNG vehicles. Though natural gas reserves are immense, still concern is there for doubling the extraction over a few years while not a rise in extraction price. In step with Lave [9], the natural gas vehicles demonstrate capability for lowering the emissions and particulates. However, low-energy volume ratio, knock at high loads and high emission of CH_4 and CO are some of the issues associated with natural gas combustion.

Therefore, due to elevated energy content and zero carbon borne emissions, H_2 is being considered as a sustainable and potential long-term fuel in the midst of all available fuel alternatives [10].

With technologies moving towards guaranteeing sustainable development, new and better fuels are being researched upon. The structure of H_2 is carbonless that makes it appropriate to be used as a future fuel for combustion engines. Hydrogen, thus, being the foremost promising substitute is being studied and tested for its performance, specific fuel consumption, its emissions and comparison of its emissions with those coming back from typical fuels.

The first step towards utilization of H_2 as a fuel in internal combustion engines in Asian nation was taken forward by IndianOil in October 2005.

NHEB: National Hydrogen Energy Board below Ministry of New and Renewable Energy has been constituted for backing of IndainOil (research-oriented institution) and leading vehicle industries like Tata Motors and Mahindra & Mahindra. NHEB had created four subgroups for the aim to specialize in completely different aspects of H_2 energy like H_2 production, H_2 storage and distribution, H_2 applications, H_2 safety standards, security and connected policy problems. However, because of lower-energy costs and taxes, the likelihood of H_2 technology is a lot of favourable in Western Europe compared to India [11].

As per [12], utilization of H_2 in internal combustion engines is taken into account a viable and favourable possibility in the present scenerio. Versatile properties of H_2 fuel make it an ultimate future fuel [13]. H_2 is used as a replacement or addition to CNG and petrol in SI engines due to overall characteristics of H_2 fuel [14]. Backfire, high No_x emission, power drop, etc., measure the number of notable limitation of H_2 fuel that will hamper its performance in internal combustion engines, for which fuel injection techniques are improved over some past years and analysis is underways to more improve it [15]. The properties of H_2 are given in Table 1.

Hydrogen generated through an alternate process has the ability to address both the problem of air pollution as well as of energy security. All facets of hydrogen economy are being worked upon by the researchers globally.

At present, analysis work has been undertaken at IndainOil Research & Development Centre, Faridabad, for the harnessing of hydrogen in genset application.

Table 1 Physio-chemical properties of compressed natural gas and H_2 [16]

Properties	CNG	Hydrogen
Flammability limit in air, vol.%	5–15	4–75
Composition (stoichiometric) in air, vol.%	9.48	29.53
Equivalence ratio	0.7–4	0.1–7.1
Density (kg/m^3)	0.65	0.08
Calorific value (MJ/kg)	50	120
Flame temperature in air, K	2148	2318
Self-ignition temperature, K	813	858
Burning velocity in NTP air, cms^{-1}	45	325
Diffusivity in air, cm^2 s^{-1}	0.2	0.63
Quenching gap in NTP air, cm	0.203	0.064
Normalized flame emissivity	1.7	1.00
Least energy for combustion in air, MJ	0.29	0.02
% of thermal energy radiated	23–33	17–25

The major problem that was faced during this experimentation was of contamination of the engine oil caused due to the formation of water vapour during hydrogen combustion resulting in water ingression in engine oil. This can lead to the serious problems of cavitations, explosion, metal fracture, etc. This may also affect or change the viscosity of the lubricant, thus affecting the reliability of the system. The CNG lubricant differs in several aspects from the hydrogen lubricant. A comparative study suggests that CNG lubricant has a higher total base number (TBN) and is of API CF category. As compared to CNG combustion, lube oil degenerates at a faster rate during hydrogen combustion. Thus, hydrogen lube oil should have a sufficient amount of de-emulsifiers to control the water admittance.

IOCL carried out in-house iterations and therefore developed a lubricant to cater to the problem of moisture ingress in the lubricant. It is observed that, there is no HC, CO emissions in H_2 combustion. However, throughout actual operation, tiny quantities of HC, CO emissions are available. These emissions are variable which may be due to burning of lubricant. To analyse the above issues and to study the effect of moisture ingression in the lubricant, i.e. servo green edge (component-based), a long-period endurance test for establishing the oil drain interval was conducted.

The endurance test was conducted for 100 h. It was divided into six cycles each of 16 h. The various parameters, namely the emissions, equivalence ratio, BSFC, etc., were noted for each cycle at different loads.

2 Experimental Set-up

The existing CNG genset at IOCL, R&D Centre, Faridabad, with a load capacity of 28 kW has been converted into hydrogen genset by making suitable modifications. It consists of an electric load bank panel system having a capacity of 30 kW, an emission measurement system, condensation particle counter, utility like compressed air, exhaust, etc. The facility can be seen in Fig. 1.

2.1 Test Equipment

2.1.1 Electrical Load Bank Panel System

The load bank is mandatory equipment required for genset testing, which is utilized to induct load of different capacities under monitored way. The capacity of load bank is thirty kilowatt having resistive load has been designed to induct artificial loads of different capacities on the hydrogen genset. The load bank is having programmable logic computer and human–machine interface for the desired load to be given. The system is having air-cooled cooling facility which makes this load bank suitable for long-term durability testing. A multifunction meter is incorporated for measuring voltage, current and power.

Fig. 1 Test facility for hydrogen genset testing

2.1.2 Emission Measurement System

Combustion of conventional gasoline fuels results in the emission of exhausts like CO, HC, NO_x, etc. For the measurement of these emissions, motor exhaust gas analyser was coupled for electrical load bank panel system. This was done to measure the intensity of the emitted emission species such as carbon monoxide, carbon dioxide and nitrogen oxides in the tailpipe gas. The major components of the measuring system comprise equipment to heat, divide and control the emissions. It consists of a warmed-up filter, warmed-up line, gas separator, bench governor, etc. The tailpipe gases are quantified in ppm and in percentage for CO and NO_x and for carbon dioxide and oxygen emissions, respectively.

2.1.3 Engine Exhaust Condensation Particle Counter (EECPC)

EECPC could be a device that is employed to sight and count mobile particles of up to 10 nm diameter with high accuracy. Once alcohol condenses on the particles, it leads to the formation of aerosol droplets which might be detected simply exploitation lightweight scattering mensuration technique. Within the condensation particle counter, the flow is saturated once the alcohol evaporates into the sample exhaust stream. The alcohol is then created to withstand a vertical condenser tube that has

Fig. 2 EECPC set-up

been cooled by thermo-electrical apparatus. The vapours of alcohol therefore become saturated and therefore condensed on the particles bigger than 10 nm. The droplets then withstand a skinny ribbon of optical device fall upon exiting the condenser. The sunshine scattered by these particles is then targeted on a lightweight detector, and it gets regenerated to an electrical pulse that is recorded as a particle count [17]. EECPC system can sight nanoparticles within the exhaust sample. Figure 2 shows the EECPC operation.

2.1.4 Experimental Design and Set-up

The experimental set-up consists of various components as shown in Fig. 3.

2.2 Test Engine

A four-cylinder H_2 genset with a capability of nine power unit was accustomed study of the performance of the genset. The genset was coupled with generator for producing electrical lower. The engine characteristics area unit is listed below in Table 2.

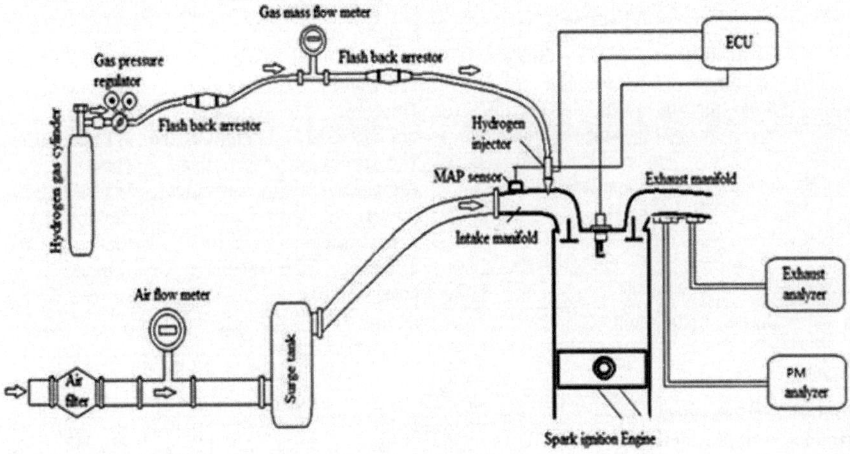

Fig. 3 Schematic experimental set-up

Table 2 Genset characteristics

Generator	Genset Model	Unit	Values
	Rated output	kW	9
	Power factor	P.F.	0.8
	Electrical KVA rating (Standby)	kVA	11.25
Engine	Model		4 R 1040 NA
	Rated output	kW	36
	Aspiration	–	NA
	No. of cylinders	No.	4
	RPM	RPM	1500
	Stroke × Bore	mm	120 × 105
	Compression ratio		12:01
	Governor		Electronic
	Lube oil sump capacity	Litre	9.5
	Lube oil change period	Hrs	800
Alternator	Model		Kirloskar Green
	Type		4 pole brushless
	Enclosure		IP 23
	Voltage regulation	%	< ±1.5
	Range of voltage adjustment	%	± 5
	Alternator efficiency at 100% load	%	89.1

Table 3 Composition of hydrogen fuel

S. No.	Characteristics (minimum mole fraction, %)	Type I—Grade A Gaseous HYDROGEN: ICE for transportation; residential/commercial appliances	Type I—Grade B Gaseous HYDROGEN: Industrial fuel for use, e.g. in power generation or as a heat energy source	Type I—Grade C Gaseous HYDROGEN: Aircraft and space vehicle ground support systems	Type I—Grade D Gaseous HYDROGEN: Gaseous HYDROGEN fuel for PEM fuel cell road vehicle systems
1.	Purity of H2	98.0	99.90	99.995	99.99

Table 4 Performance parameters of the engine oil [18]

S. No.	Properties	Servo green edge (component-based)
1.	Viscosity index	122
2.	Viscosity @ 100 °C, cSt	13.50
3.	Viscosity @ 40 °C, cSt	108.00
4.	TBN, mg/koH gm	5.95

2.3 Test Fuel

The test fuel used is neat hydrogen. The quality of hydrogen fuel as per ISO 14687 is given below in Table 3.

2.4 Test Engine Oil

SAE-40, i.e. servo green edge (component-based) oil was used as engine oil in the genset for studying its performance evaluation. The physio-chemical properties of servo green edge engine oil (component-based) are given in Table 4.

2.5 Test Set-up

Several modifications were made to the CNG genset to use hydrogen as fuel. The following modifications were carried out:

- Exhaust line with a special provision of moisture drainage was fabricated.
- The hydrogen fuel supply line was fitted with flashback arrestor in order to arrest any hydrogen backfire.

- Sampling ports for measurement of emissions and particle no in the exhaust line were incorporated.
- Electronic injection system for hydrogen feed was fitted.
- Genset connection with electrical load bank panel and connected temperature and pressure sensors were carefully made.
- Initial running-in was carried out.

2.6 Test Procedure

The endurance test was carried out for 100 h to establish the drain period and engine oil credentials for hydrogen-powered spark ignition engine.

1. The durability of the engine oil was assessed by measuring the engine oil consumption before and after the endurance test of 100 h. The performance test was conducted after the completion of every three cycles, i.e. at 0, 48 and 96 h as per the following procedure:

- Engine warm-up
- Drain the engine oil in the warm condition through the drain plug and collect it in the tray/jerry can. Weight the engine oil—W1.
- Close the drain plug and charge the complete engine oil in the genset. Run the genset on rated power (9 kW) for 12 h continuously.
- After completing 12 h, drain the engine oil in the warm condition through the drain plug and collect it in the tray/jerry can. Weight the engine oil—W2.
- Engine oil consumption: W1–W2
- Repeat the procedure three times, i.e. on three consecutive days.

Moreover, oil sampling and analysis were undertaken once the completion of each forty eight hours, i.e. after 0, forty eight and ninety six hours. The fulfilment parameters of the servo green edge engine oil like viscosity at 100 °C, viscosity at 40 °C, viscosity index and total base number were measured once completion of every three cycles, i.e. at 0, 48 and 96 h. Hydrogen fuel consumption was measured in kg/hr using Coriolis-based fuel flow meter, and consumption of air was measured using ABB make air flow meter in kg/hr based on hotwire anemometer principle [19, 20].

2. Evaluation of performance of hydrogen-powered SI engine was done by undertaking an endurance test of 100 h, and duty cycle is as per IS 10,000 (Part IX). Test cycle description is shown in Table 5.
3. Performance test of the engine oil, i.e. servo green edge (component-based) was done after completion of every three cycles, i.e. 0 h, 48 and 96 h by measuring the emissions from the engine like CO_2, CO, NO_x, PN, BSFC, etc.
4. Used oil sampling and analysis were also done at 0, 48 and 96 h. The kinetic viscosity of the fresh oil was measured and was compared for used oil in order to estimate the oil drain period. The levels of sulfation, oxidation and nitration were other measured parameters. Wear elements concentration was also measured at 0, 48 and 96 h.

Table 5 Load cycle description

Load (% of Rated Load)	Load (kW)	Time (Hours)
100	9	4
50	4.5	4
110	9.9	1
No load (Idling)	–	0.5
100	9	3
50	4.5	3.5
Duration of 1 cycle		**16**

Thus, endurance test was carried out for hundred hours divided into six cycles each of sixteen hours, i.e. zero to sixteen hours, sixteen to thirty-two hours, thirty-two to forty-eight hours, forty-eight to sixty-four hours, sixty-four to eighty hours and eighty to ninety-six hours, and the tailpipe emissions were measured for each cycle at above loads for respective running time.

Performance of hydrogen-fuelled spark ignition was assessed with hydrogen fuel using servo green edge engine oil (component-based). Many optimizations were created at IOCL and R&D centre for maintaining a similar speed of the engine with totally different loads and throttle mixtures. The spark ignition engine was run at idle till the engine oil reached a temperature of 60 °C. After reaching the desired oil temperature, the SI engine was subjected to increased load through electrical load bank panel system in a step of 1 kW each. While putting the load, the throttle for air supply and hydrogen fuel supply is adjusted in such a manner that the hydrogen-fuelled spark ignition engine speed remains at 1500 rpm. On each load, after stabilizing the value, the fuel flow rate, air flow rate, emissions (CO, CO_2, NO_x, and O_2), particle no., etc., were measured.

3 Result and Discussion

Figures 4, 5, 6, 7 and 8 show the comparison of CO_2, CO, NO_x, PN emissions and brake-specific fuel consumption (BSFC) during the endurance test at 0, 48 and 96 h, respectively.

Figure 4 shows the emission of carbon dioxide from the engine exhaust measured at each load from 0 to 9 kW. From the above figure, it can be inferred that as the hydrogen-powered spark ignition engine starts running, the emission of carbon dioxide from the engine exhaust decreases significantly with time. Further, CO_2 emissions decreased considerably during increase of endurance hours, i.e. lower at 96 h vis-à-vis 0 h, as the wear reduces after running-in completed. It may also be noted that for increase in load on the engine, the emission of carbon dioxide also increases.

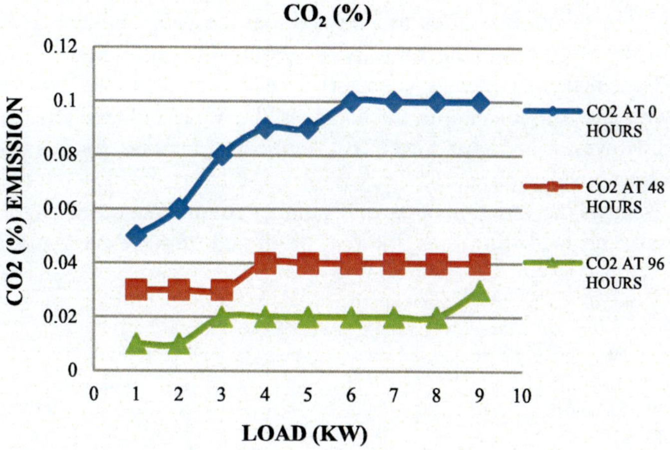

Fig. 4 CO_2 emissions (%)

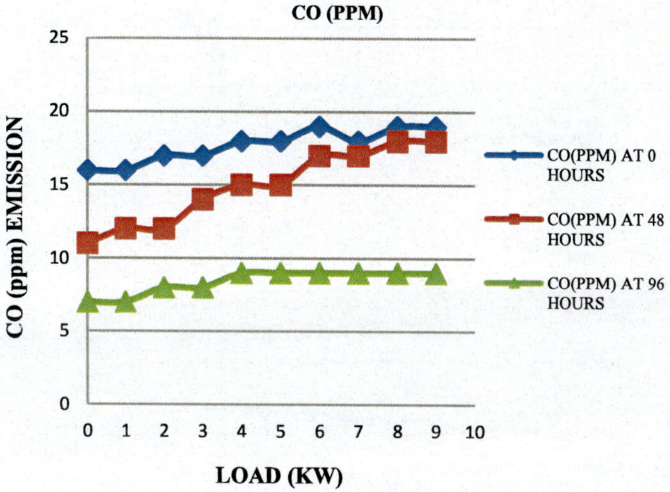

Fig. 5 CO emissions

Figure 5 inferred the variation of carbon monoxide emission from hydrogen-powered spark ignition engine exhaust measured at each load from 0 to 9 kW at 0, 48 and 96 h, respectively during a 100-hour endurance test. From Fig. 5, it may be finalized that the emission of carbon monoxide decreases considerably for increase in an engine running time. Further, CO emissions decreased considerably during increase of endurance hours, i.e. lower at 96 h vis-à-vis 0 h, as the wear reduces after running-in completed. Also, at every instant of time, for increase in engine load, CO emission conjointly increases.

From Figure seven, it is often ascertained that the NO_x emission reduces with increase in the engine running time. Also, throughout the initial loading of the engine, NO_x emission is nearly constant for every time instant. NO_x emissions reduced with increase of endurance hours as the wear reduces after running-in completed. However, at higher loads, NO_x emission increases considerably as can be seen in the above graph.

Figure 7 shows the variation of particle number coming out of the engine exhaust measured at each load from 0 to 9 kW at 0, 48 and 96 h, respectively. It can be

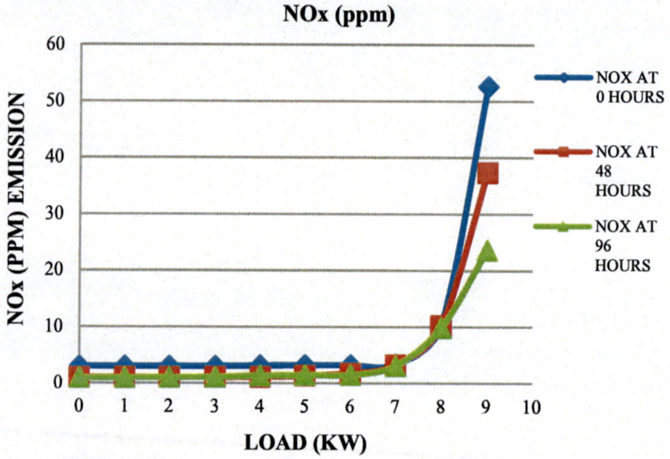

Fig. 6 NO_x emissions

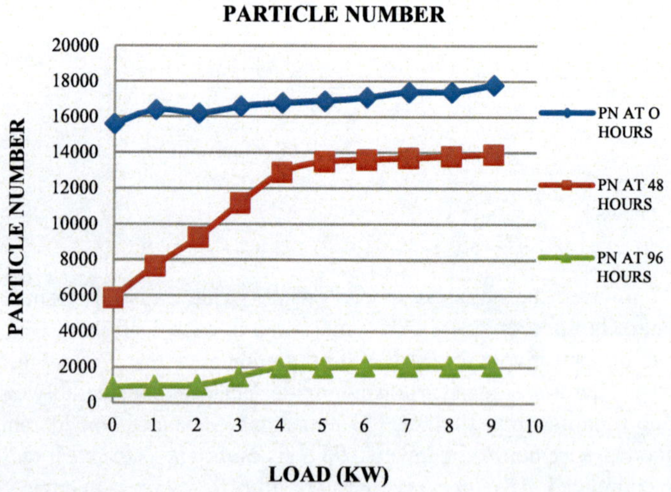

Fig. 7 Particle number

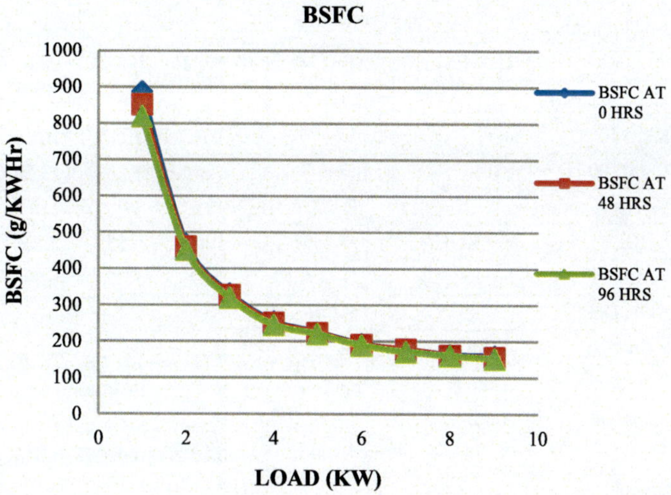

Fig. 8 BSFC of the engine at 0, 48 and 96 h

inferred from the above figure that the number of particles decreases with increase of endurance hours as the wear reduces after running-in completed. Also for every time instant, the particle number increases with a rise in load at lower loads. However, at higher loads, an additional increase in PN is often seen.

Figure nine infers the variation in brake-specific fuel consumption (BSFC) of the engine at zero, forty-eight and ninety-six hours. It can be concluded from the above graph that the BSFC of the engine at zero, forty-eight and ninety-six hours is comparable and reduces with a rise in the load on the engine.

4 Oil Analysis

The emission characteristic analysis of used engine oil for the hydrogen-powered spark ignition engine for servo green edge(component-based) engine oil after completion of running-in of 0, 48 and 96 h is given in Table 6.

It can be concluded from the used oil analysis report that the kinetic viscosity of thc oil at 40 °C and at 100 °C using neat hydrogen fuel decreases after completion of every three cycles, i.e. 48 h as compared to the viscosity at same temperatures for the fresh oil. The viscosity index also deviates from that of fresh oil and shows a decrease from 121 to 119. However, after that, it remains constant after completion of every three cycles, i.e. after 0, 48 and 96 h. The analysis additionally indicates a depletion of TBN and a rise in TAN once completion of each third cycle.

It is inferred from Table 6 that the sulfation, nitration and water ingression of the engine oil increase by a really bit. Also, the wear elements of iron, chromium, copper, nickel, aluminium, lead and tin were increased insignificantly.

Table 6 Used oil analysis

S. No.	Properties	UOM(SI units)	Test Method	Fresh	0 h	48 h	96 h
1	*Kin. viscosity		D-445-17a				
1.1	KV@40 °C	cSt		107.7	105.4	105.6	105.8
1.2	KV@100 °C	cSt		13.48	13.01	13.04	13.07
1.3	Viscosity index	.	D-2270-10	121	119	119	119
2	*TBN	mg KOH/g	D-2896-15	5.9	5.55	5.14	5.05
3	*TAN	mg KOH/g	D-664-17	1.02	1.05	1.09	1.12
4	FTIR Data		E-2412-10				
4.1	Oxidation	A/cm		0	0	0	0
4.2	Sulfation	A/cm		0	0.96	1.2	1.3
4.3	Nitration	A/cm		0	0.5	0.7	0.8
4.4	Water	%		0.0389	0.0108	0.0435	0.0712
4.5	Glycol	%		0	0	0	0
5	Wear elements		D-6595-17				
5.1	Iron	ppm		0	4	6	8
5.2	Chromium	ppm		0	3	5	7
5.3	Copper	ppm		0	3	4	5
5.4	Nickel	ppm		0	0	0	0
5.5	Aluminium	ppm		0	1	2	2
5.6	Lead	ppm		0	1	1	1
5.7	Tin	ppm		0	0	0	0
6	Contaminants		D-6595-17				
6.1	Silicon	ppm		0	0	0	0
6.2	Vanadium	ppm		0	0	1	1

The above table shows no vital degradation within the properties of the SAE-40 engine oil once the completion of ninety-six hours of endurance duty cycle.

5 Outcomes

From the above experimental investigations, the subsequent outcomes can be drawn:

- CNG genset was successfully converted to hydrogen genset and mounted on the test bench.
- Initial running and endurance test was conducted with servo green edge (component-based) engine oil.
- The engine oil successfully completed 96 h of an endurance test.

- CO_2 and CO emissions from the engine decrease with time. These emissions are higher initially at 0 h. They decrease at 48 h by a significant amount and further decrease at 96 h is observed.
- NO_x emissions at 48 and 96 h are lower and comparable as compared to that at 0.
- Particle number from the hydrogen-fuelled internal combustion engine decreases with time.
- The decrease in CO_2, CO, NO_x and PN can be attributed to the actual fact that once the engine starts to run at the start, the friction is high between its numerous elements and consequently the emissions are high too. As the engine continues to run, the frictional loses decrease and so the emissions conjointly decrease.
- The brake-specific fuel consumption of the engine shows insignificant amendment with time. The variation of BSFC with increasing load at zero, forty-eight and ninety-six hours is comparable.
- The engine oil used, i.e. servo green edge (component-based) is appropriate for hydrogen-fuelled genset.
- The performance durability of the lubricant and the performance of hydrogen-fuelled spark ignition engine are good.

From the used engine oil analysis report, it is evident that there is a minor amendment in properties of the engine at zero, forty-eight and ninety-six hours. Considering the engine oil performance (kinematic viscosity at 40 °C, kinematic viscosity at 100 °C, viscosity index, TBN and wear elements), performance and emission data of hydrogen genset, it is often complete that the lubricant analysed in the above research, i.e. servo green edge component-based conjointly referred as SAE-40 oil is appropriate for hydrogen application in spark ignition engine for long drain application.

References

1. Global energy statistical yearbook 2018. https://yearbook.enerdata.net/oil-products/world-oil-domestic-consumption-statistics.html
2. Momirlan M, Veziroglu TN (2005) The properties of hydrogen as fuel tomorrow in sustainable energy system for a cleaner planet. Int J Hydrogen Energy 30(7):795–802
3. Poulton ML (1994) Alternative fuels for road vehicles. Comp. Mechanics Publications, UK
4. Shasby BM (2004) Alternative fuels: incompletely addressing the problems of the automobile, MSc Thesis, Virginia Polytechnic Institute and State University, USA, pp 5–13
5. Brombacher EJ (1997) Flow visualisation of natural gas fuel injection. Master of Applied Science Thesis, University of Toronto, Canada
6. Srinivasan KK (2006) The advanced injection low pilot ignited natural gas engine: a combustion analysis. J Eng Gas Turbines Power 128(1):213–218
7. Hill PG, Douville B (2000) Analysis of combustion in diesel engines fueled by directly injected natural gas. J Eng Gas Turbines Power 122:143–149
8. Aslam MU (2006) An experimental investigation of CNG as an alternative fuel for a retrofitted gasoline vehicle. Fuel 85:717–724, 47

9. Hollnagel C, Neto JAM, Di Nardi ME, Wunderlich C, Muraro W, Miletovic C, Bisetto F (2001) Application of the natural gas engines Mercedes-Benz in moving stage for the carnival 2001 in Salvador City, SAE Technical Paper 2001-01-3824
10. Lave LB, MacLean H, Lankey R, Joshi S, McMichael F, Horvath A, Hendrickson C (2000) Life cycle inventories of conventional and alternative automobile fuel/propulsion systems: summary & conclusions, SAE Technical Paper, 2000-01-1504
11. Negurescu N, Pana C, Cernat A (2012) Aspects of using hydrogen in SI engine, University Politehnica of Bucharest Scientific Bulletin, Series D. Mechanical Engineering 74(1):11–20
12. Van Ruijven B, Hari L, Van Vuuren DP, De Vries B (2008) The potential role of hydrogen energy in India and Western Europe. Energy Policy 36(5):1649–1665
13. Singh S, Jain S, Venkateswaran PS, Tiwari AK, Nouni MR, Pandey JK, Goel S (2015) Hydrogen: a sustainable fuel for future of the transport sector. Alternate Sustain Energy Rev 1(51):623–633
14. Mohammadi A, Shioji M, Nakai Y, Ishikura W, Tabo E (2007) Performance and combustion characteristics of a direct injection SI hydrogen engine. Int J Hydrogen Energy 32(2):296–304
15. MacCarley CA, Van Vorst WD (1980) Electronic fuel injection techniques for hydrogen-powered IC engines. Int J Hydrogen Energy 5(2):179–203
16. Akansua SO, Dulger Z, Kahraman N, Veziroglu TN (2004) Internal combustion engines fuelled by natural gas-hydrogen mixture. Int J Hydrogen Energy 29(14):1527–1539
17. Thipse S, Rairikar D, Kavathekar P, Chitnis P (2009) Development of a six-cylinder HCNG engine using an optimized lean burn concept, SAE Technical Paper 2009-26-031
18. Caldow RC, Palmer MR, Quant FR (1992) Performance of the TSI model 3010 condensation particle counter, paper presented at the American Association for Aerosol Research, 11th annual meeting. California, USA, San Francisco
19. Singh S, Subramanian KA, Mathai R (2019) Development of dedicated lubricant for hydrogen fuelled spark-ignition engine, SAE Technical Paper 2019-28-2511
20. Singh S, Subramanian KA, Mathai R (2018) Setting up of Genset testing facility for hydrogen application. Presented and published in 7th International Hydrogen & Fuel Cell Conference, 2018 (IHFC-2018), December 9–11, Jodhpur, Rajasthan, India

Effects of Methanol Substitution on Performance and Emission in a LPG-Fueled SI Engine

K. Ravi, E. Porpatham, and Jim Alexander

1 Introduction

Growing concern for environmental pollution, the varying trends in the energy scenario and depletion of conventional fuels have led to a search for alternate fuels and efficient combustion technologies. The current rate of demand for fossil fuel exceeds the rate of supply and availability. This has caused an inequality between the supply and demand. Global oil production grew only by 0.4 Million Barrels per Day (MBD) in 2016 whereas the average of consumption growth is around 1.6 MBD [1]. According to the World Oil Outlook 2017 released by OPEC on November 7, 2017, the total primary energy demand is expected to increase by 35% in the period to 2040. The long-term demand growth is the result of growing demand for passenger vehicles in the developing nations which contributes to the increase in road transportation (5.4 MBD) [2]. The change in cost of production of oils and gases from the depleting reserves also explains the fluctuating prices of these resources. As reserves get depleted, the overheads of extraction from them increase. In the period to 2040, the investment required for global oil sector is estimated to be $10.5 trillion [2]. Human interactions and practices undertaken toward the surroundings affect the energy exchange between the Earth, the atmosphere and space. The other foremost threat which endangers the lives of the living population of the Earth is global warming and the increasing rate of pollution. Countries like Canada, India, Brazil, China, the European Union (EU), Australia and Russia comprise the Transport Task Group [TTG] under G20 energy efficiency action plan to prioritize the improvement of energy efficiency and the environmental performance of motor vehicles. The governments and other world organizations are implementing

K. Ravi (✉) · E. Porpatham · J. Alexander
Automotive Research Center, School of Mechanical Engineering, Vellore Institute of Technology, Vellore 632 014, India
e-mail: ravi.krishnaiah@vit.ac.in

A. K. Gupta et al. (eds.), *Advances in IC Engines and Combustion Technology*, Lecture Notes in Mechanical Engineering,
https://doi.org/10.1007/978-981-15-5996-9_16

strict emissions standards that have to be followed in order to control the pollution caused due to vehicle emissions in an aim to improve the air standards. Emissions of carbon monoxide (CO), carbon dioxide (CO_2), hydrocarbon (HC), oxides of nitrogen (NOx) and particulate matter (PM) account for the major air pollutants. Stringent exhaust limits for these emissions are laid down to reduce the health hazards caused by them and to improve the quality of air. The exhaust emission limits of these gases are made stricter with each advancing versions of emissions standards. Though the air quality has improved and there is a reduction of emissions over the past few years, the quality foreseen and recommended by world organizations for much safer environment is yet to be achieved. NOx emissions are of particular concern owing to the significant harm it can cause to health and the environment. Research works toward the emission and reduction of NOx is being carried out extensively in recent times. The reason for this prioritization is the apparent contribution of NOx to the PM formation and occurrence of photochemical smog. Being an irritant gas, it also increases the ozone in the urban air that increases the ultraviolet radiation in the stratosphere which eventually reaches the surface of the Earth [3]. Further the rise in PM and ozone on ground levels can have hostile health effects. Higher concentration of Nitrogen dioxide can lead to swelling of the airways. Inhaling the related particles of nitric acid can worsen the respiratory problems and can also aggravate heart diseases. Being the highest contributor to the emission of NOx, the commercial vehicle emission can be condensed in many possible ways. All these factors have headed to an urge to shift from conventional fuel usage in engines to utilization of alternative fuels for a sustainable development and control.

1.1 Liquefied Petroleum Gas (LPG)

The abundant carbon content in shale gas and progression with advanced extraction processes has made Liquefied Petroleum Gas (LPG) a viable source of alternative fuel. Recent findings of shale gas occurrences and their compositions promises profuse source for LPG extractions. Such rapid supply of LPG from shale gas has transformed the scope of LPG usage in countries like USA. Potential shale gas reserves that are identified in many regions across the world have given a new scope for the utilization of LPG across different fields. LPG has a simple molecular structure which aids in very low tendency of aromatics formation or particulates formation on combustion. LPG's faster vaporization and enhanced mixture is the reason for this reduction in PN emission. It also increases the engine's thermal efficiency [4]. Low-specific fuel consumption (SFC) of LPG compared to that of gasoline means that the engine uses less quantity of LPG compared to that gasoline to yield the same power. LPG is cost-effective and has less energy cost. It costs lesser than gasoline by 47% and diesel by 44% on average [5]. These factors make LPG an energy efficient and cost-effective fuel than gasoline. The primary contents of LPG are propane and butane. Each component has higher octane number than gasoline. Propane has RON value of 112 and n-butane has RON value of 94 compared to that of gasoline's 93.

This property of LPG inhibits detonation and permits the engine to operate on high compression ratios. Being a gas in nature, LPG offers better homogenization. Proper homogenization directly yields to a decrease in NOx emissions.

1.2 Lean-Burn Operation

The SI engines that are fueled by LPG often work at stoichiometric Air/Fuel Ratio (AFR) conditions employing a three-way catalyst to control the NOx emissions [4]. But the low energy density per volume of LPG results in poor fuel economy compared to that of gasoline engines operating at same conditions. Lean-burn engines were developed after World War II as a result of diminishing oil reserves and to reduce exhaust emissions [6]. Lean-burn operation was aimed to improve the fuel efficiency of the engine [7, 8]. Employing lean-burn operation in SI engines fueled by LPG will help in improvement of thermal efficiency. Proper swirl actions and injection ratios along with spark advancements will result in reduction of SFC of the LPG fueled engine by 5–25% [6]. Modern lean-burn engines operate at $\lambda = 1.7$. Lean-burn operation improves brake thermal efficiency and shows lesser emissions. Best-achieved brake thermal efficiency was 28% higher than gasoline-powered engines at stoichiometric conditions (19% best improvement for TBI) [9]. LPG has a broader lean-burn limit than that of gasoline because it is gaseous in nature. Thus, it entirely mixes with air providing proper homogenization and its flame combustion is stable [10]. The lean combustion mode of LPG engines also decreases the exhaust gas temperature (EGT) and hence keeping the engine away from knocking region. LPG as a better fuel and lean-burn combustion contribute to improvement in efficiency [11].

1.3 Methanol

For such operations and to cope up with higher compression ratios, alcohols serve as a better substitutes and additives to fuels. Development of modern methods of extraction of alcohols from shale gas occurrences [12] and production of methanol from coal gasification [13] has offered an energy security and made methanol a suitable alcohol for powering IC engines. Moreover the greater octane number of methanol and its molar expansion will yield a compression without any rise in temperature [14]. Addition of methanol also ensures reliable cold start firing. They are seen as a prominent additives to petrol and diesel engines for the reason that presence of hydroxyl (OH) results in more efficient and clean combustion of fossil fuels [15]. Methanol has a higher latent heat of vaporization than gasoline which helps in enhanced cooling effect of the charge yielding to reduction in NOx emissions. This directly results in the knock resistance of the engine and increases the volumetric efficiency. Several

Table 1 Properties of Methanol, n-Butane and Propane

Properties	Methanol	n-Butane	Propane
Chemical formula	CH_3OH	C_4H_{10}	C_3H_8
Density at NTP (kg/l)	0.79	0.00248	0.002
Energy density (MJ/l)	15.9	26	26
Calorific value (MJ/kg)	23	49.1	50.4
Stoichiometry air/fuel ratio	6.5	15.5	15.5
Boiling point at bar (°C)	65	−0.5	−42
Research Octane Number (RON)	108.7	94	112
Motor Octane Number (MON)	88.6	90.1	97
Latent heat of vaporization (kJ/kg)	1100	386	428
Vapor pressure (bar)	0.25	2.44	9.36

studies and experiments carried out on using alcohols as blend fuels with gasoline have shown that there is an increase in brake power, BMEP and thermal and volumetric efficiencies and decrease in SFC [16–19]. The properties of Methanol, n-Butane and Propane are shown in Table 1.

This paper is focused on the investigation of effects of methanol substitution in a LPG fueled SI engine. The experiment is carried out for LPG and Methanol to study the effects of fuels which are regarded as the alternatives for replacing conventional fuels. The dependency on depleting fossil fuels would be reduced as LPG and Methanol has the ability to completely substitute them as IC engine fuels. Moreover the modifications that need to be done to the engine for these fuels to power it are simple and can be carried out easily. Small amount of methanol is added with LPG to power a modified SI engine. The experiment is carried out under lean-burn condition. The effects methanol addition has on the parameters like brake thermal efficiency (BTE) and emissions of NO*x* are studied. Extensive research on alternative fuels can provide excellent academic resources which can be used in the future for any further development.

2 Experimental Setup

The schematic, photographic view and engine specifications of the experimental setup are shown in Figs. 1, 2 and Table 2. A four stroke, single cylinder diesel engine is modified to operate as a SI engine. This is done because engines fueled by LPG operate at a higher compression ratio than the gasoline engines. The fuel injection pump and the injector of the diesel engine are replaced with venturi type gas carburetor, a spark plug and an ignition system in order for the engine to operate in SI mode. The engine runs on LPG with a displacement volume of 661 cm^3 and is air cooled developing a power of 4.4 kW at 1500 RPM with a compression ratio

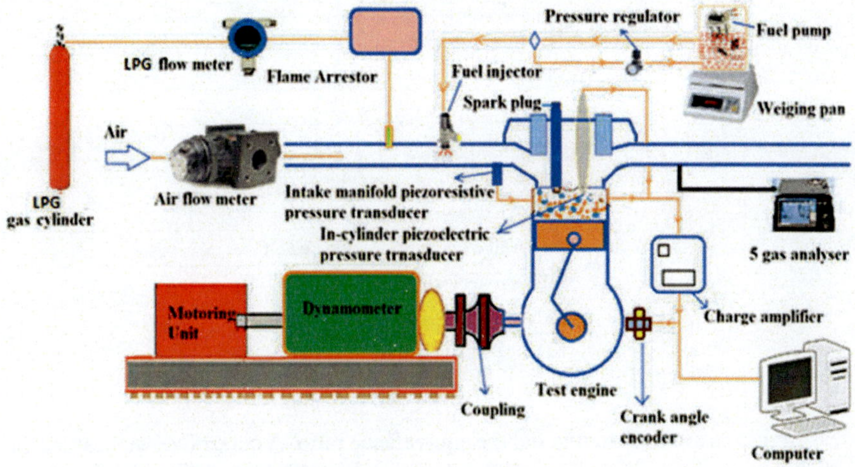

Fig. 1 Schematic diagram of experimental setup

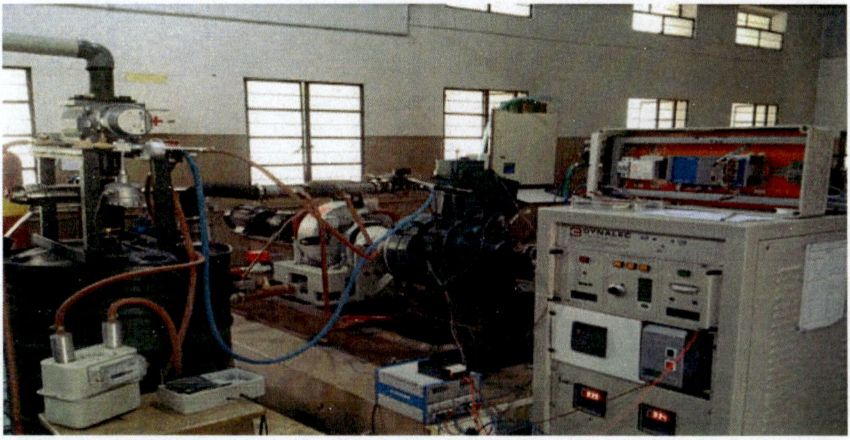

Fig. 2 Photographic view of the experimental setup

of 10.5:1. The clearance volume is changed in order to modify the compression ratio. The diesel engine is chosen for its ability to handle higher compression ratios, withstand knock and accommodate the additional air requirement for lean operation. A water-cooled eddy current dynamometer is coupled with the engine is operated at a constant speed mode at 1500 RPM for the experiment. The LPG storage cylinder is heated using a hot water bath in a constant temperature so that the fuel is available at 1.03 bar (abs). LPG is heated before the intake to avoid any condensation in the flow line. A surge tank is connected to the intake side of the engine. This is done in order to smoothen the air flow. A positive displacement type air flow meter measures the air flow and a positive displacement gas flow meter measures the gas flow. These

Table 2 Engine specifications

Displaced Volume	661 cc
Stroke	110
Bore	87.5
Compression ratio	17.5:1 (CI Version); 10.5:1(SI Version)
Rated power	4.4 kW @ 1500 rpm
No. of valves	2
Inlet valve opening	4.5° bTDC
Inlet valve closing	35.5° aBDC
Exhaust valve opening	35.5° bBDC
Exhaust valve closing	4.5° aTDC

measured rates are used to find out the equivalence ratio. A control valve throttles the LPG flow into the engine. The volume flow rate is multiplied with its actual density to convert it to mass basis. The spark plug is located in the center of the combustion chamber.

A Capacitor Discharge Ignition (CDI) system is fixed to the engine. It is electronically controlled. The spark timing device is developed in the lab. It is microcontroller-based and the reference signals are sent to it from a crank angle encoder. Through this process, the charge is ignited at specified crank angle. In order to record the crank angle values and corresponding pressure, an optical angle encoder is mounted on the cam shaft. It also varies the spark timing according to the crank angle. An in-house electronic device was developed for injecting methanol to the intake manifold of the engine. The device regulates the opening and closing of the injector and varying the period of injection. Methanol is injected at 3 bar pressure and the injection is calibrated. The calibrated pulse width is used to find out the mass flow rate of the methanol. A Non-Dispersive Infra-Red (NDIR) analyzer for HC, CO, CO_2 and electrochemical sensors for the O_2 and NO*x* measurements is used to measure the exhaust gas. A similar setup was used by the author to investigate on the effect of piston squish area on performance and emission characteristics of LPG fueled lean-burn SI engine [20].

3 Results and Discussion

The effect of substitution of methanol on energy basis in a lean-burn LPG fueled SI engine has been studied in this work. Brake thermal efficiency (BTE), nitrogen oxides (NO*x*) emissions, equivalence ratio and Specific Energy Consumption (SEC) were taken into account for this study. A series of experiments were conducted at full throttle condition. The results were studied with reference to Brake Mean Effective Pressure (BMEP). This is done because BMEP serves as a universal yardstick to compare between different parameters under different conditions. BMEP is the equivalent of mean in cylinder pressure throughout the entire cycle responsible for the output brake power. This is calculated by using the following equation.

$$\text{BMEP} = \text{Brake Power}/\text{LAN Watts}$$

where, L = Stroke length, A = Cylinder cross section area, N = Engine revolutions per second.

Parameters like BTE, Nitrogen Oxide emissions (NOx) and SEC for 10, 20 and 30 percentages of methanol on energy basis in the fuel and for pure LPG were plotted against BMEPs of 4, 4.5, 5 and 5.5 bar.

Brake thermal efficiency is the ratio between brake power outputs at the crankshaft to the input chemical energy in fuel. Figure 3 shows the variation of brake thermal efficiency with BMEP for different percentages of methanol and neat LPG at full throttle condition. Results show that as the percentage of methanol in the fuel increase the BTE also increases for all the values of BMEP. The average BTE for neat LPG is 23.7%. The BTE averages are 23.8%, 27.3% and 31.7% for 10%, 20% and 30% of methanol substituted fuel, respectively. This means that the average BTE of 10%, 20% and 30% methanol substituted fuel is better than the average BTE of neat LPG fuel over the four values of BMEP. The peak BTE of 34.71 is attained at 4.5 bar for 30% of methanol in the fuel. The efficiency values observed for methanol substituted fuels are comparable to studies conducted on methanol-fueled IC engines [21]. The values of BTE of neat LPG do not change much over different pressures. But there is a drastic change seen in the variations of BTE of methanol-substituted fuel for different BMEPs. The percentage difference between the peak efficiency of methanol

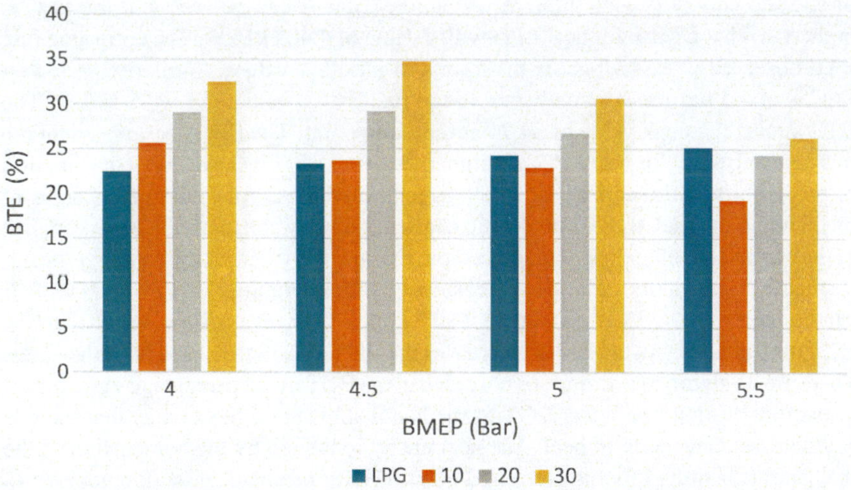

Fig. 3 Variation of brake thermal efficiency with brake mean effective pressure

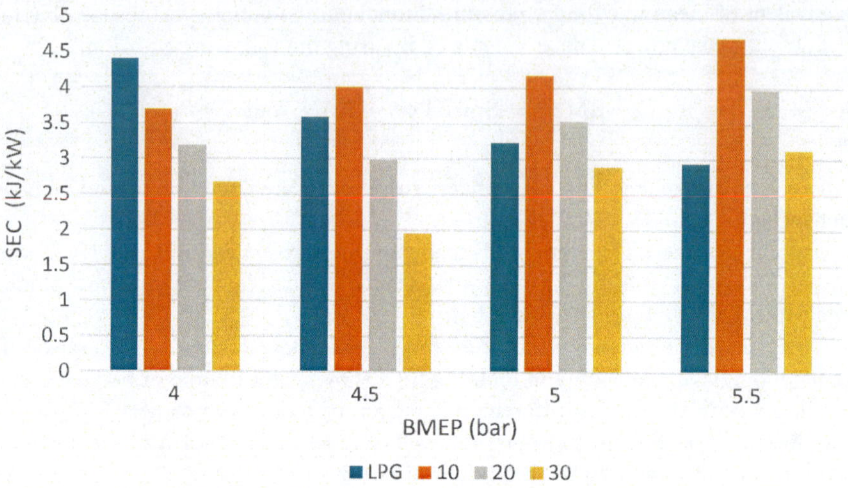

Fig. 4 Variation of specific energy consumption with brake mean effective pressure

substituted fuel which occurs at 4.5 bar for 30% methanol in fuel and that of neat LPG that occurs at 5.5 bar BMEP is 27.65%.

The improvement in the BTE is the direct result of reduction in Specific Energy Consumption (SEC) and increase in brake power. As the energy consumption decreases, the BTE increases. This trend corresponds with the research conducted on Emission and Performance Enhancement of Multi-Cylinder SI Engine Fueled with LPG and Vaporized Water Methanol Induction [22]. Figure 4 shows the variation of specific energy consumption at various BMEPs. It can be seen that the SEC is highest at 4 bar BMEP for neat LPG with a value of 4.409 kJ/kW. The corresponding BTE observed is 22.4 which is the lowest of all other values. Similarly the lowest SEC is at 4.5 bar for 30% methanol in the fuel with a value of 1.9485 kJ/kW. The BTE attains the peak value of 34.71 at the same point. The SEC decreases with the increase in BMEP for neat LPG. At higher values of BMEP, the energy consumption of methanol substituted fuel increases as load increases. For lower percentages of methanol in the fuel, there is no drastic change in energy consumption observed. But as the percentage of methanol increases, the rise in energy consumption is significant.

Figure 5 shows the variation of Exhaust Gas Temperature (EGT) with BMEP. It can be seen that the EGT of methanol substituted fuel always stays lower than the EGT of neat LPG. As the BMEP increases the EGT also increases in both the cases. There are no significant variations observed in the EGT as the percentage of methanol in the fuel differs. The lower EGT of methanol substituted fuels mean that there is an effective conversion of heat. The heat energy obtained by the combustion of the methanol substituted fuel is converted effectively to mechanical work compared to the conversion that takes place in case of neat LPG fuel.

Figure 6 represents the NOx emissions in g/kWh plotted against the four set of BMEPs. The NOx emission decreases with increase in the percentage of methanol in

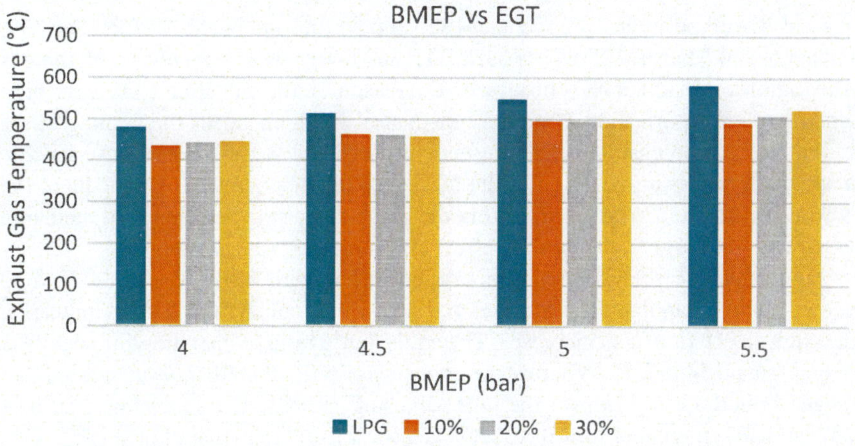

Fig. 5 Variation of exhaust gas temperature with brake mean effective pressure

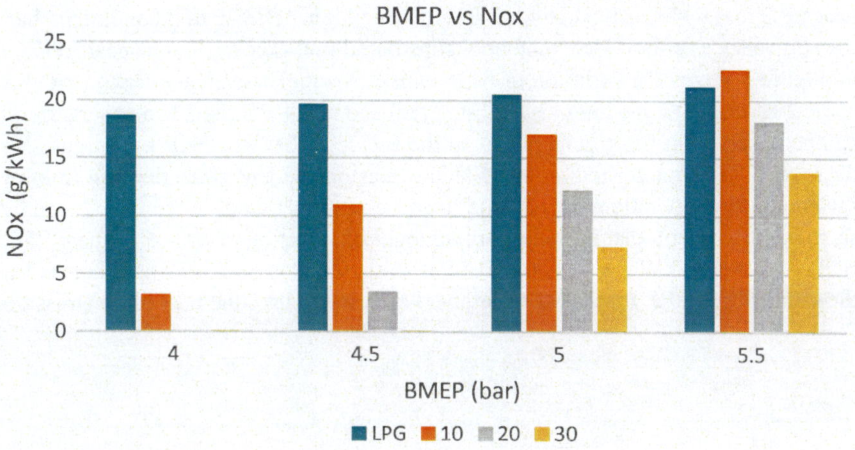

Fig. 6 Variation of NO*x* emissions with brake mean effective pressure

the fuel for each value of BMEP. A similar trend is observed in a research conducted on Emission and Performance Enhancement of Multi-Cylinder SI Engine Fueled with LPG and Vaporized Water Methanol Induction [22] where the NO*x* emission decreases with increase in methanol content in the fuel for various RPMs. As the value of BMEP increases, the emission also increases. With increasing BMEP, there is a drastic rise in the level of NO*x* emission observed for the methanol substituted fuel. The rise in NO*x* emissions for the neat LPG fuel with increasing BMEP is not as significant as that of methanol substituted fuel. The average emission of NO*x* for neat LPG fuel is 19.94 gm/kWh, whereas as the emission of NO*x* for 10% methanol substituted fuel averages to just 13.43 gm/kWh. There is an 83% reduction of NO*x* at 4 bar BMEP between 10% methanol substituted fuel and neat LPG fuel. For 30% methanol at 4.5 bar BMEP where peak BTE and lowest SEC were observed, there is a negligible emission of NO*x* observed. The reason for the reduction NO*x* is the high latent heat of vaporization offered by methanol which enhances the cooling effect of the combustion chamber. The presence of methanol vapor increases the specific heat which results in reduction of the burned charge temperature. This reduces the NO*x* formation. It can be inferred from the graph that as the percentage of methanol in the fuel goes higher, the cooling effect also increases.

Figure 7 shows the variation of equivalence ratio with BMEP for 10, 20, and 30 percentage methanol substituted fuel and neat LPG fuel. As the BMEP increases, the equivalence ratio also increases. The equivalence ratio of operation for neat LPG ranges from 0.54 to 0.75. Whereas the equivalence ratio of methanol substituted fuel ranges from 0.43 to 0.78 over the four values of BMEPs. The equivalence ratio of 20% and 30% methanol substituted fuel is always lower than that of neat LPG. This means that the lean limit of the combustion operation extends when methanol is used along with LPG. The equivalence ratio leans up to 0.43 at 4 bar BMEP for 30% methanol fuel which is 20% leaner than that of lowest equivalence ratio observed for neat LPG fuel which is 0.54 at 4 bar BMEP. At a higher BMEP of 5 bar and 5.5 bar, the equivalence ratio of 10% methanol substituted fuel exceeds that of neat LPG.

Figure 8 shows the variation of NO*x* emission with Equivalence Ratio (ER) for 10%, 20% and 30% methanol substituted fuel and neat LPG fuel. The general trend can be observed in the graphs is that as the ER decreases, the NO*x* emissions also decrease. The scope for reduction of NO*x* emissions widens since the lean limit of the combustion operation extends due to addition of methanol in the fuel. It can be seen from the graph that the minimum amount of emission observed for neat LPG fuel is 18.16 gm./kWh at an ER of 0.54 at least stable operation load. Whereas for the same range of ER for the 30% methanol substituted fuel, the maximum emission

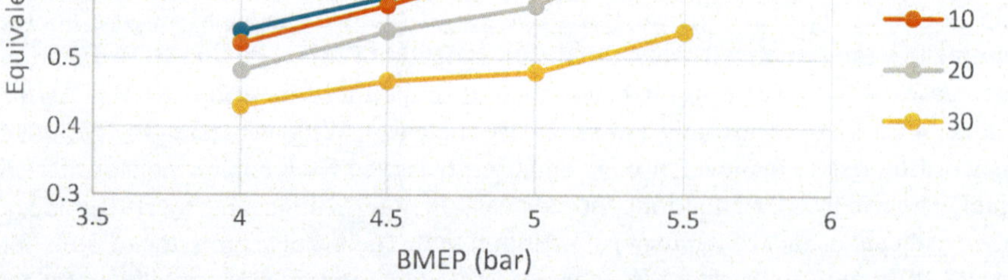

Fig. 7 Variation of equivalence ratio with brake mean effective pressure

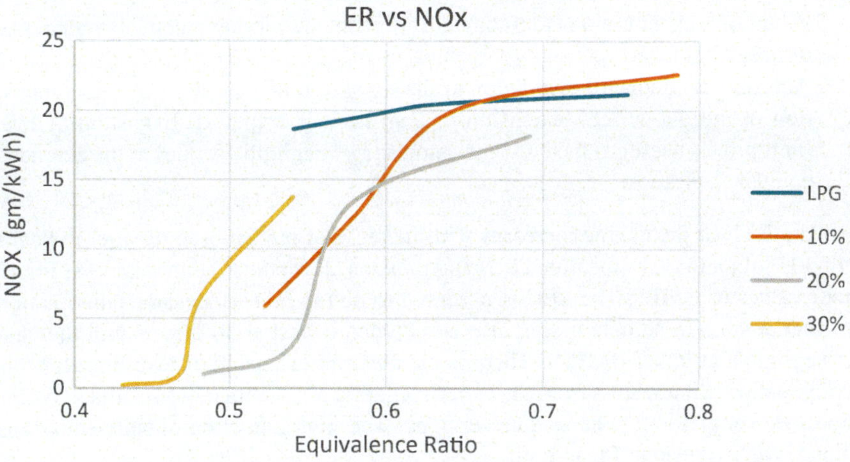

Fig. 8 Variation of NO*x* emissions with equivalence ratio

observed is around 13.711 gm/kWh but for full load condition. As the ER goes below that range for methanol substituted fuel, the emission values also decrease. It reaches a negligible value at the ER range of 0.45. For 30% methanol in fuel, the maximum load conditions are achieved at an ER of 0.54 whereas for neat LPG it is obtained at an ER of 0.75. This means that there is a less usage of fuel in case of 30% methanol in fuel at the same equivalence ratio compared to that of LPG. In spite of presence of higher amount of air in the charge, less combustion temperatures because of dilution and specific heat capacity result in significant NO*x* reduction at higher pressure conditions. The percentage of methanol in the fuel characterizes the lean limit the engine operates on. This in turn affects the emission of NO*x*. Thus, both these factors play a vital role in reducing the NO*x* emissions.

4 Conclusions

1. Brake thermal efficiency increases with increase in methanol percentage in the fuel for a given BMEP. At lower BMEPs of 4 bar and 4.5 bar, the BTE of methanol substituted fuel is higher than that of neat LPG fuel.
2. The SEC values decrease as the percentage of methanol in the fuel increases. At 4 bar BMEP, the value of SEC for 10, 20 and 30 percentage methanol substituted fuel is lower than that of neat LPG fuel. As the BMEP increases, the SEC also increases.
3. There is a massive reduction in emission of NO*x* when methanol used along with LPG to fuel the engine. For certain percentages of methanol substituted fuel the NO*x* emission is negligible at lower BMEPs. For the taken four BMEPs, the

values of NOx emissions for methanol substituted fuel stay much lower than that of neat LPG fuel.

4. Addition of methanol also helps in the extension of lean limit of the combustion operation. A 20% extension of lean limit is achieved. Extension of lean limit proves useful as the NOx emission shows negligible values at the extended equivalence ratios.

The BTE of the engine increases with increase of methanol in the fuel. It peaks for 30% of methanol substituted fuel and remains above the BTE of neat LPG for all four values of BMEP. The SEC is observed to be lowest for the same combination of fuel at 4.5 bar. NOx emission also is recorded lowest at 30% methanol fuel and is negligible at lower BMEPs. There is an apparent extension of lean limit for this combination. So it results from the experiment that 30% methanol substituted fuel is optimal fuel yielding enhanced performance and high reduction in NOx emissions along with extension of lean limit.

Acknowledgements The authors wish to thank the Department of Science and Technology (DST) and Vellore Institute of Technology, Vellore for the financial support extended in carrying out this project.

References

1. Dudley B, Dale S (2017) BP statistical review of world energy, 49, June 2017 http://www.bp.com
2. Lichtblau J (2014) World oil outlook. The leading edge, vol 4. https://doi.org/10.1190/1.1439163
3. Lee D, Goto S, Kim I, Motohashi M (1999) Spectroscopic investigation of the combustion process in an LPG lean-burn SI engine. No. 1999-01-3510. SAE Technical Paper
4. Kim J, Kim K, Oh S (2016) An assessment of the ultra-lean combustion direct-Injection LPG (Liquefied Petroleum Gas) engine for passenger-car applications under the FTP-75 Mode. Fuel Process Technol 154:219–226. https://doi.org/10.1016/j.fuproc.2016.08.036
5. Ugurlu A, Oztuna S (2015) A comparative analysis study of alternative energy sources for automobiles. Int J Hydrogen Energy 40(34:11178–11188. https://doi.org/10.1016/j.ijhydene.2015.02.115
6. Ganesan V (2012) Internal combustion engines. McGraw Hill Education (India) Pvt Ltd
7. Kim TY, Park C, Oh S, Cho G (2016) The effects of stratified lean combustion and exhaust gas recirculation on combustion and emission characteristics of an LPG direct injection engine. Energy 115(x):386–396. https://doi.org/10.1016/j.energy.2016.09.025
8. Danaiah P, Ravi Kumar, Vinay Kumar (2012) Lean combustion technology for internal combustion engines: a review. Sci Technol 2(1):47–50
9. Ali Khan M, Watson H, Baker P, Liew G et al (2006) SI engine lean-limit extension through LPG throttle-body injection for low CO_2 and NOx. SAE Technical Paper 2006-01-0495. https://doi.org/10.4271/2006-01-0495
10. Li L, Wang Z, Wang H, Deng B et al (2002) A study of LPG lean burn for a small SI engine. SAE Technical Paper 2002-01-2844. https://doi.org/10.4271/2002-01-2844
11. Perin M, Achek T (2013) Lean burn engines, No. 2013-36-0402. SAE Technical Paper

12. Zhang C, Jun KW, Kwak G, Lee YJ et al (2016) Efficient utilization of carbon dioxide in a gas-to-methanol process composed of CO_2/steam-mixed reforming and methanol synthesis. J CO_2 Utilization 16:1–7. https://doi.org/10.1016/j.jcou.2016.05.005
13. Bassani A, Bozzano G, Pirola C, Ranzi E et al (2017) Low impact methanol production from sulfur rich coal gasification. Energy Proc 105(0):4519–4524. https://doi.org/10.1016/j.egypro.2017.03.970
14. Svensson E, Li C, Shamun S, Johansson B et al (2016) Potential levels of soot, NO*x*, HC and CO for methanol combustion. SAE Technical Paper 2016-01-0887. https://doi.org/10.4271/2016-01-0887
15. Liu X, Wang H, Zheng Z, Liu J et al (2016) Development of a combined reduced primary reference fuel-alcohols (methanol/ethanol/propanols/butanols/n-pentanol) mechanism for engine applications. Energy 114:542–558
16. Abu-Zaid M, Badran O, Yamin J (2004) Effect of methanol addition on the performance of spark ignition engines. Energy Fuels 18(2):312–315. https://doi.org/10.1021/ef030103d
17. Bilgin A, Sezer I (2008) Effects of methanol addition to gasoline on the performance and fuel cost of a spark ignition engine. Energy Fuels 22(4):2782–2788. https://doi.org/10.1021/ef8001026
18. Hsieh WD, Chen RH, Wu TL, Lin TH (2002) Engine performance and pollutant emission of an SI engine using ethanol-gasoline blended fuels. Atmos Environ 36(3):403–410. https://doi.org/10.1016/S1352-2310(01)00508-8
19. Liu S, Cuty Clemente ER, Hu T, Wei Y (2007) Study of spark ignition engine fueled with methanol/gasoline fuel blends. Appl Thermal Eng 27(11–12):1904–1910
20. Krishnaiah R, Ekambaram P, Jayapaul PB (2016) Investigations on the effect of piston squish area on performance and emission characteristics of LPG fuelled lean burn SI engine, No. 2016-28-0123. SAE Technical Paper
21. Naganuma K, Vancoillie J, Sileghem L, Verhelst S et al (2012) Drive cycle analysis of load control strategies for methanol fuelled ICE vehicle. SAE Technical Paper 2012-01-1606. https://doi.org/10.4271/2012-01-1606
22. Patil B, Nayak V, Padmanabha M (2014) Emission and performance enhancement of multicylinder SI engine fuelled with LPG and vaporized water methanol Induction, No. 2014-01-2764. SAE Technical Paper

Effect of Static Ignition Timing on the Emission and Performance Characteristics of a Four-Cylinder MPFI Engine Fueled by LPG

Vighnesha Nayak, K. S. Shankar, Anusha, P. M. Ashit, Bhushith, and K. L. Vikyath

1 Introduction

Nowadays, alternative fuel plays vital role in the transportation sector. Gaseous fuels are playing important part in automobile industry due to their cost-effective, high calorific values, lower emissions, and high octane rating [1]. Among the gaseous alternative fuels suggested were hydrogen (H_2), methane and LPG (mainly propane). Due to availability and lower price, LPG has been widely used in automobiles recently as an alternative fuel than gasoline. Because of its sufficient reserves in the world, LPG gave promising hopes and expectations to the researchers. Compare to liquid fuels, LPG has positive impact on environment [2]. An engine fueled by LPG generates almost zero particulate matter emissions, very low carbon monoxide emissions, and mild hydrocarbon emissions. As per emission point of view, main disadvantage of using LPG as a fuel is oxides of nitrogen (NO_x) emission which a higher value from liquid fuels during peak loads. From the literature, it has been found that NO_x emission is almost three times that of gasoline-fueled vehicle [3–5].

High octane rate of fuel will indicate the resistance of detonation in the engine cylinder. Because of high octane rating of LPG compared to gasoline, this fuel can be used at higher compression ratio. The flaming velocity of LPG is greater in lean mixture region (equivalence ratio around 0.9 to 1.0) but that of gasoline is promoted in the rich mixtures. Hence, LPG has better combustion characteristics at lean burn

Affiliated to Visvesvaraya Technological University.

V. Nayak (✉) · Anusha · P. M. Ashit · Bhushith · K. L. Vikyath
Department of Mechanical Engineering, A. J. Institute of Engineering and Technology, Mangaluru, Karnataka 575006, India
e-mail: nayakviggu@gmail.com

K. S. Shankar
Department of Mechanical Engineering, Srinivas Institute of Technology, Valachil, Mangalore, Karnataka 574143, India

A. K. Gupta et al. (eds.), *Advances in IC Engines and Combustion Technology*, Lecture Notes in Mechanical Engineering,
https://doi.org/10.1007/978-981-15-5996-9_17

Table 1 Properties of LPG, propane, and butane

Properties	Propane	Butane	LPG
Chemical formulae	C_3H_8	C_4H_{10}	60% C_4H_{10} + 40% C_3H_8
Max. vapor pressure saturated in bar at 65 °C	22.66	6.32	16.87
Gross calorific value in kJ/kg.	49,980	49,560	49,728
Specific gravity (liquid) at 15 °C	0.504	0.582	0.543
ignition temperature (°C)	493–504	482–537	488–502
Max. flame temperature (°C)	1979.44	1990	1985
Boiling point (°C)	−45	−2	−22
Latent heat of vaporization (kJ/kg)	431	444	436.2

engines [6]. The precise structure of LPG differs slightly from one part of the country and refineries to the other [7]. Table 1 shows the properties of propane, butane, and LPG.

1.1 Effect of Ignition Timing

LPG is having higher self-ignition temperature than gasoline which increases the combustion duration. Also LPG is having higher calorific value and latent heat of vaporization than gasoline. When fuels are having different combustion properties, then there is a need of changing the spark ignition timing. Spark ignition timing has a significant impact on the engine performance and emission characteristics when fuels are having different combustion properties. If spark timing is too advanced, there is substantial increase in the cylinder pressure before piston reaches to the top dead center (TDC) point. Hence, there will be more compression work against the exhaust gas and hence net work output will be decreased. Therefore, optimum spark timing will produce the satisfactorily maximum cylinder pressure with peak happening immediately after the TDC. Hence, optimum spark timing will result in minimum work input and maximum power work [8–11].

The current study deals with effect dual fuel mode of engine operation to get better performance and emission characteristics in a modified four-cylinder gasoline engine by changing static ignition timing. The main purpose of this investigation is to optimize the gasoline/LPG proportion and static ignition timing for different speed at full load operation.

2 Experimental Methodology and Procedure

Initially, experiment is conducted with gasoline as a fuel then LPG solenoid is operated to get the mixture of LPG and gasoline fuel (i.e., 25% LPG, 50% LPG, 75% LPG, and 100% LPG) for different advanced ignition timings and speeds. In this experimentation, advanced static ignition timing has taken since self-ignition temperature of LPG is higher hence combustion duration will be more. Engine speed is varied from 2000 to 4500 rpm in steps of 500 rpm. During these experimental investigations, static ignition timings of 5 deg. bTDC, 8 deg. bTDC and 11 deg. bTDC are taken. The experimental procedure for LPG fueled gasoline engine is shown in Fig. 1.

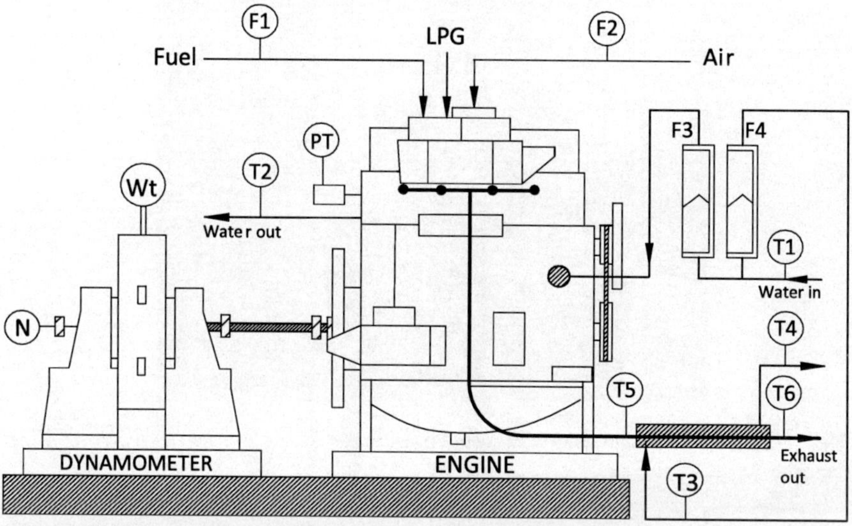

F1	Fuel flow differential pressure unit	F3	Rotameter (Engine)
F2	Air Intake DP unit	F4	Rotameter (Calorimeter)
T1	Engine Cooling water inlet temperature	T2	Cooling water outlet temperature
T3	Calorimeter water inlet temperature	T4	Calorimeter water outlet temperature
T5	Exhaust gas Calorimeter inlet temperature	T6	Exhaust gas outlet Calorimeter temperature
N	RPM decoder	PT	Pressure transducer
Wt	Load on Dynamometer		

Fig. 1 Schematic diagram of the experimental setup

2.1 Experimental Test Rig

The engine configuration comprises of a 4-stroke, 4-cylinder, and Maruti Zen spark ignition (MPFI) engine connecting for loading to the eddy current dynamometer as shown in Figs. 2 and 3 that show photographic view of the experimental setup with control panel. Figure 4 demonstrates line diagram for the experimental setup used for LPG fuel. Near each intake manifold of cylinder, gas injectors have been installed for LPG fuel. Solenoid valves type of gas injectors has been used which can be activated and deactivated from the electrical signal depending upon the requirement of LPG

Fig. 2 Photograph of engine setup with control panel

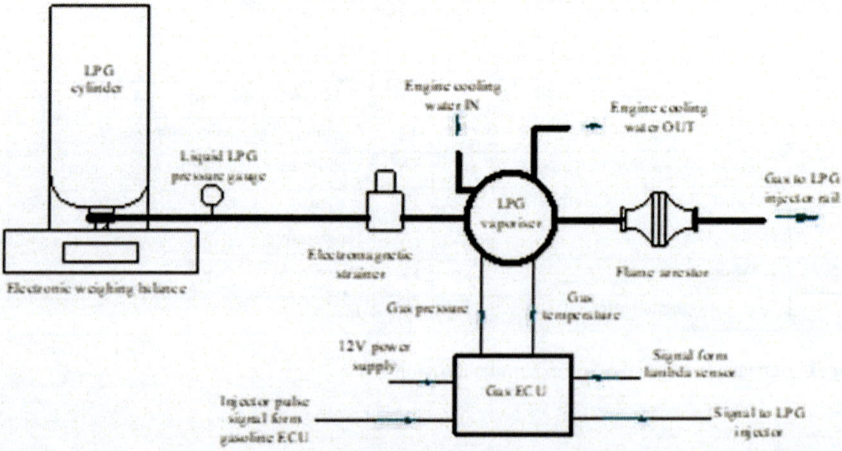

Fig. 3 Schematic diagram of the experimental setup for LPG

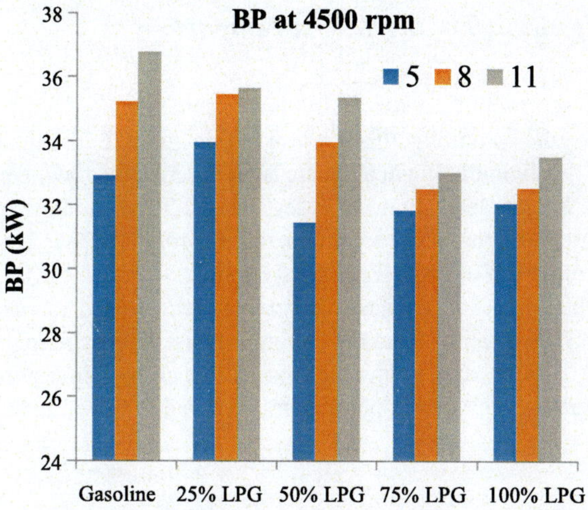

Fig. 4 Comparison of BP w.r.t static ignition timing at 4500 rpm full load condition

flow. A separate gas ECU been designed and installed to operate this gas injectors which controls the activation period of these injectors through the software.

2.2 *LPG Setup*

The engine is modified with to work on LPG fuel. The LPG fuel injectors are controlled by 12 V DC power supply which in turn controls the solenoid. The nozzle diameter of each gas injector is determined based on the power output per cylinder and for the given engine, nozzle diameter of 2 mm is used. The LPG gas ECU is designed and installed by M/s Auto gas, Mangaluru, which includes software Auto gas v 3.1 for controlling the solenoid valve. The schematic diagram of the LPG experimental is demonstrated in Fig. 4. For experimental work, 14 kg LPG cylinder is used with cylinder pressure of 4–7 bar having maximum vapor pressure of 10.5 bar at 40 °C. An unreduced pressure regulator is used in cylinder which reduces the pressure which is just above the atmospheric pressure. Any iron dust particle present in the copper pipe line is absorbed by electromagnetic strainer which is placed after vaporizer. The purpose of vaporizer is to convert liquid LPG into vapor LPG by using thermal energy of engine hot water [6]. The engine cooling water will supply the required thermal energy to vaporizer.

2.3 *Static Ignition Timing Measurements*

The Static ignition timing is measured by the instrument called ignition timing gun. Whenever the engine is running, this ignition timing gun will illuminate the light and manually focus this light on flywheel. Scale is provided in the flywheel for measuring the angle which has the least count of 1 deg. bTDC. To change the ignition timing, the distributor assembly has to be loosened and it is rotated either in the direction or opposite to the direction of rotation of the flywheel.

An AVL makes five gas exhaust analyzer which is used to measure the CO, CO_2, HC, and NO_x emissions from the engine exhaust. The experimental setup does not include any kind of after treatment devices, hence raw emissions readings are presented. Uncertainty of various parameters involved is shown in Table 2 [12].

3 Results and Discussion

A 4 cylinder, MPFI engine is converted to work on LPG-fueled injection system. Experiment has conducted on engine by varying static ignition timing (5, 8, and 11 deg. bTDC) for different fuel conditions. Parameters like engine speed are varied from 2000 to 4500 rpm for different LPG flows at different load conditions. Results are analyzed and compared with baseline fuel gasoline at standard static ignition timing of 5 deg. bTDC. Based on analysis of results, the LPG flow rate and static ignition timing are optimized.

Table 2 Uncertainty of various parameters

Sl. No.	Parameter	Uncertainty (%)
1	Speed	±0.25
2	Torque	±0.32
3	Air flow rate	±1.05
4	Fuel flow rate	±0.81
5	Exhaust gas temperature	±0.50
6	NO_x emission	±5.91
7	HC emission	±5.50
8	CO emission	±3.77
9	Brake power	±0.96
10	Brake thermal efficiency	±0.5

3.1 Performance Characteristics

3.1.1 Brake Power (BP)

Figures 5, 6, 7, 8, 9 and 10 show the variation of BP with respect to engine speed for different advanced static ignition timings at different loading conditions. From the results, it is revealed as the static ignition timing is advanced from 5 deg. bTDC to 11 deg. bTDC, BP increases and it is maximum for 11 deg. bTDC. Since combustion

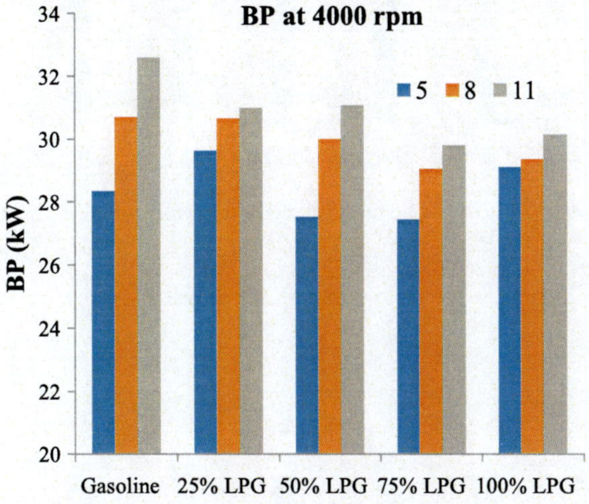

Fig. 5 Comparison of BP w.r.t static ignition timing at 4000 rpm full load condition

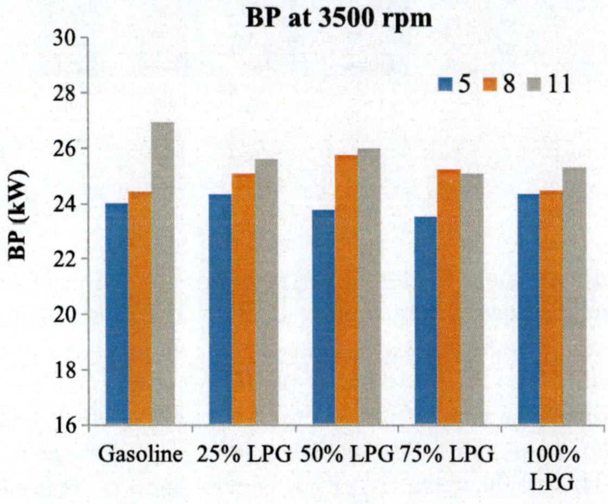

Fig. 6 Comparison of BP w.r.t static ignition timing at 3500 rpm full load condition

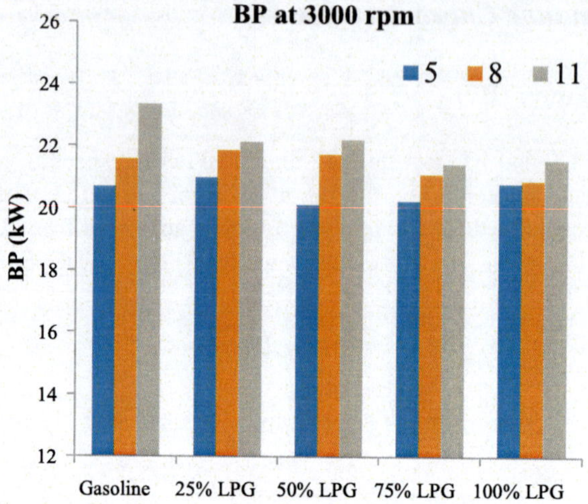

Fig. 7 Comparison of BP w.r.t static ignition timing at 3000 rpm full load condition

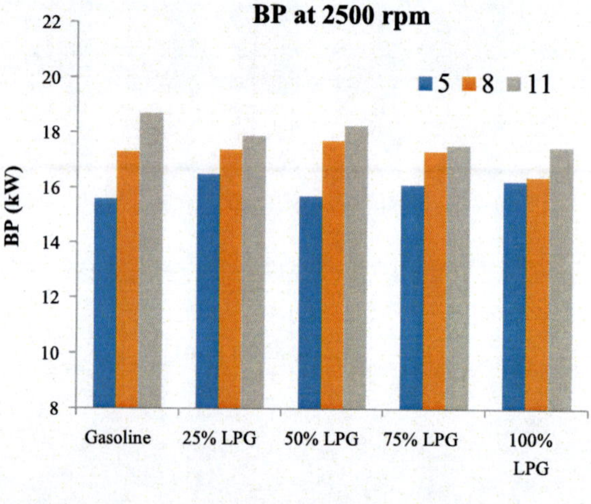

Fig. 8 Comparison of BP w.r.t static ignition timing at 2500 rpm full load condition

process starts earlier in the cycle, torque and power generally increase with the speed and load condition because of more time availability. Therefore, advancing the static ignition timing causes increased power output of an engine. Also, due to the higher flame velocity of LPG compared to that of petrol will reduce the flame propagation time thereby reduces the combustion duration. Hence, net effect will be increased in power output of an engine. So BP increases linearly with engine speeds at higher load condition. This results in the higher power generation in the full load condition for all the fuels. At full load, 4500 rpm, average increase in power output for gasoline is 11.66% at 11 deg. bTDC, 6.92% at 8 deg. bTDC. for 100% LPG is 4.6% at 11 deg. bTDC, 1.8% at 8 deg. bTDC when compared to the 5 deg. bTDC.

3.1.2 Brake Thermal Efficiency (BTE)

The variations of BTE with respect to static ignition timing for gasoline and dual fuel mode at all the engine speed and full load condition are shown in Figs. 11,

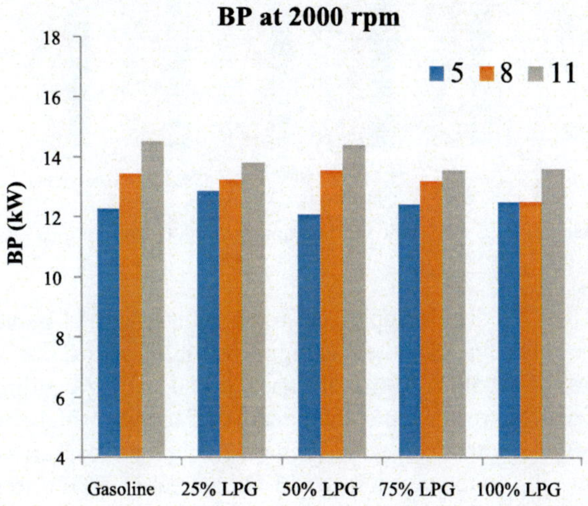

Fig. 9 Comparison of BP w.r.t static ignition timing at 2000 rpm full load condition

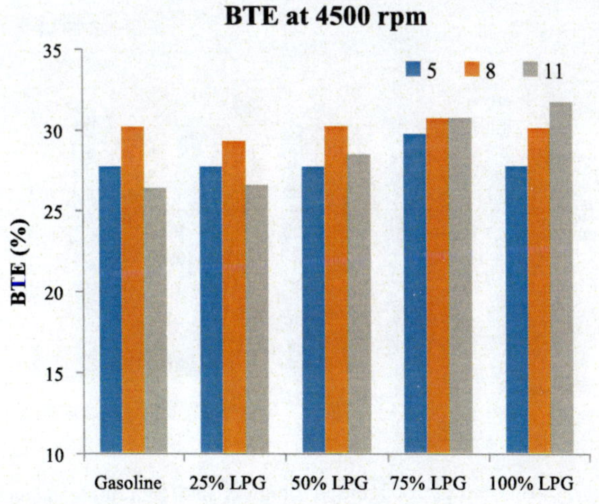

Fig. 10 Comparison of BTE w.r.t static ignition timing at 4500 rpm full load condition

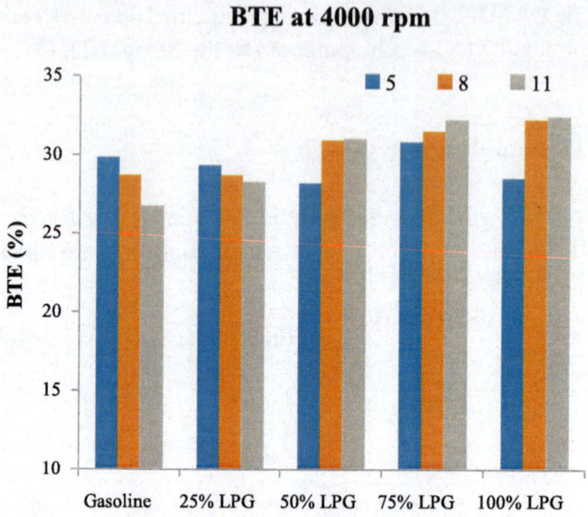

Fig. 11 Comparison of BTE w.r.t static ignition timing at 4000 rpm full load condition

12, 13, 14, 15, and 16. The comparison graph shows that the brake thermal efficiency increases with advancing the static ignition timing up to 8 deg. bTDC, beyond which decreases for gasoline. This is because too advancing the timing will result in higher compression work against the already burned gas which results in higher fuel consumption therefore decrease in thermal efficiency irrespective of increase in the power output. But as the percentage of LPG increases, BTE also increases with advancing the timing and it is maximum for 11 deg. bTDC at higher speeds. This

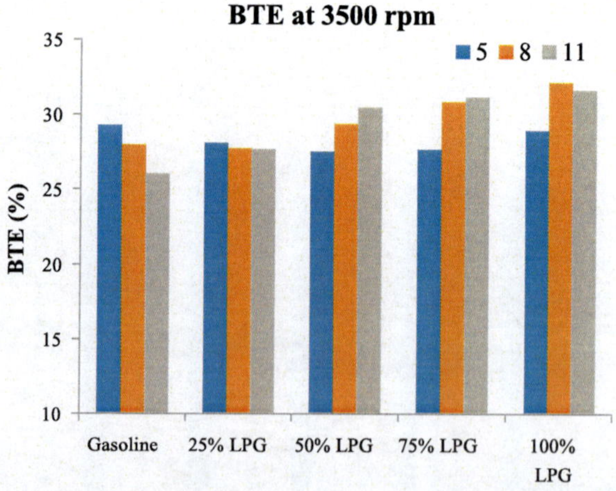

Fig. 12 Comparison of BTE w.r.t static ignition timing at 3500 rpm full load condition

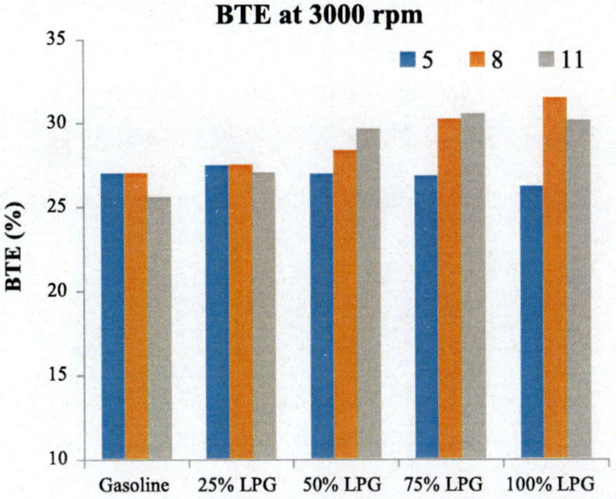

Fig. 13 Comparison of BTE w.r.t static ignition timing at 3000 rpm full load condition

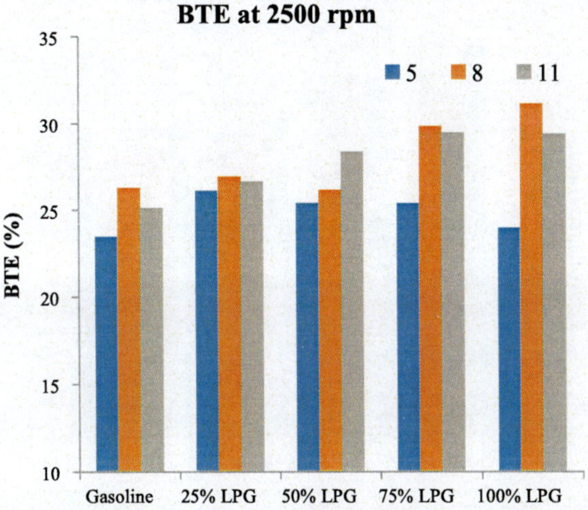

Fig. 14 Comparison of BTE w.r.t static ignition timing at 2500 rpm full load condition

because LPG having higher self-ignition temperature gives more time for combustion hence advancing the timing will result in higher thermal efficiency [13]. But, at lower speed, since combustion duration is more for LPG, too advancing the timing will result higher fuel consumption which leads decrease in BTE. At lower engine speed 8 deg. bTDC results showed better BTE. Average increase in BTE at 4500 rpm for gasoline is 8.89%, −4.79%, and for LPG is 8.56%, 14.36 at 11 deg. bTDC and 8 deg. bTDC, respectively, when compared to the 5 deg. bTDC. Average increase

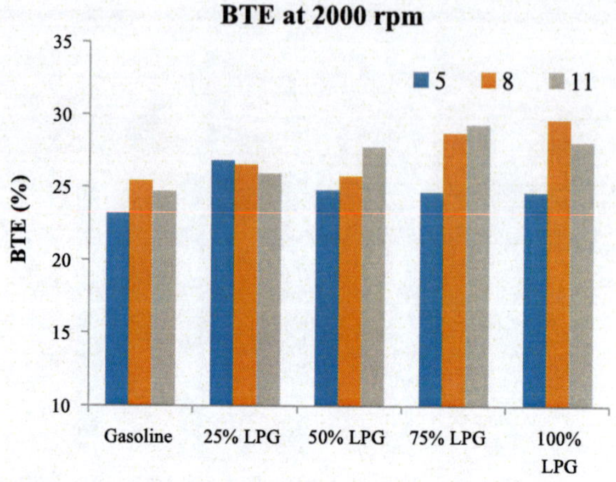

Fig. 15 Comparison of BTE w.r.t static ignition timing at 2000 rpm full load condition

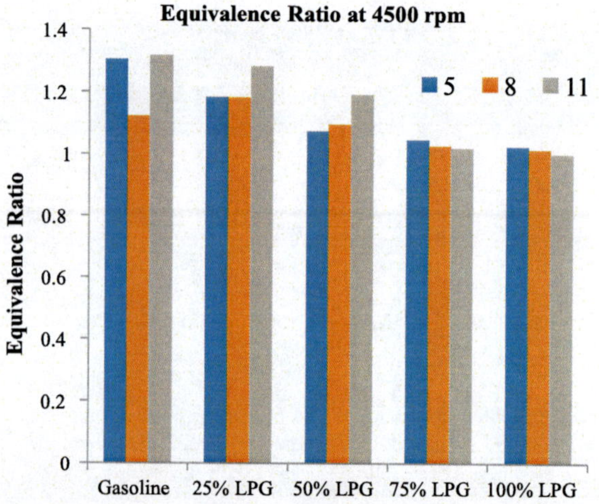

Fig. 16 Comparison of φ w.r.t static ignition timing at 4500 rpm full load condition

in BTE at 2000 rpm for gasoline is 9.55%, 6.37%, and for LPG is 20.3%, 14.07 at 11 deg. bTDC and 8 deg. bTDC, respectively, when compared to the 5 deg. bTDC.

3.1.3 Equivalence Ratio (φ)

Figures 17, 18, 19, 20, 21, and 22 indicate the comparison of equivalence ratio with respect to static ignition timing for different speeds and fuel blends at full load conditions. This equivalence ratio will determine whether the engine is working in

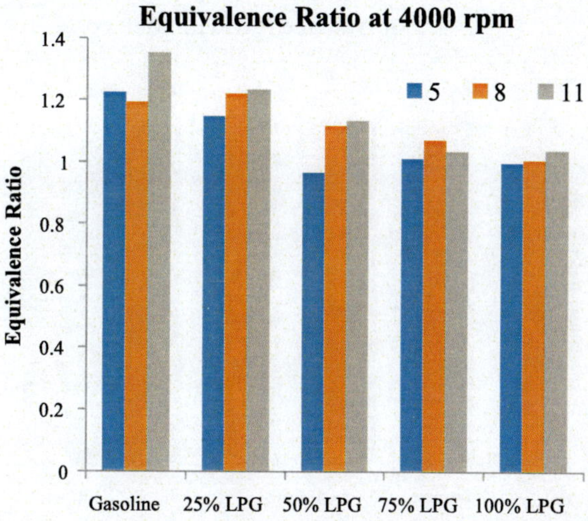

Fig. 17 Comparison of φ w.r.t static ignition timing at 4000 rpm full load condition

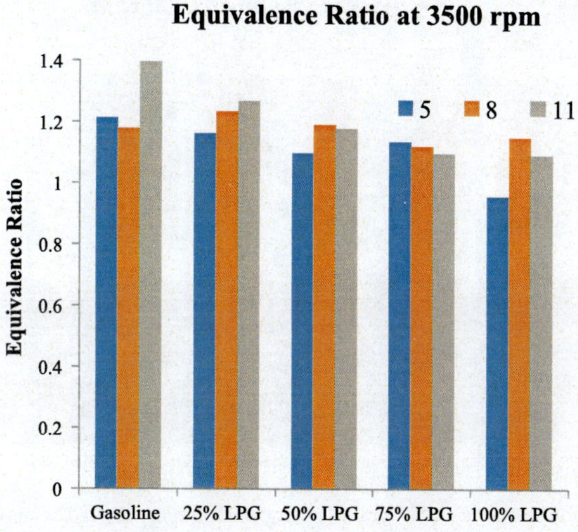

Fig. 18 Comparison of φ w.r.t static ignition timing at 3500 rpm full load condition

leaner, stoichiometric or richer mixture. As the engine speed increases, the equivalence ratio will try to remain in same value. But as static ignition timing advanced to 8 deg. bTDC, engine started to work near to stoichiometric region for gasoline for all engine speed conditions. So advancing the static ignition timing resulted in the better combustion properties for gasoline. Still advancing the static ignition timing resulted in the richer mixture in the combustion chamber for gasoline. As the percentage of

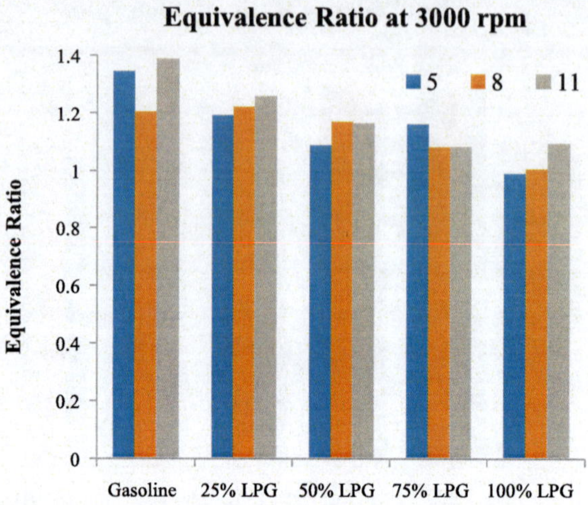

Fig. 19 Comparison of φ w.r.t static ignition timing at 3000 rpm full load condition

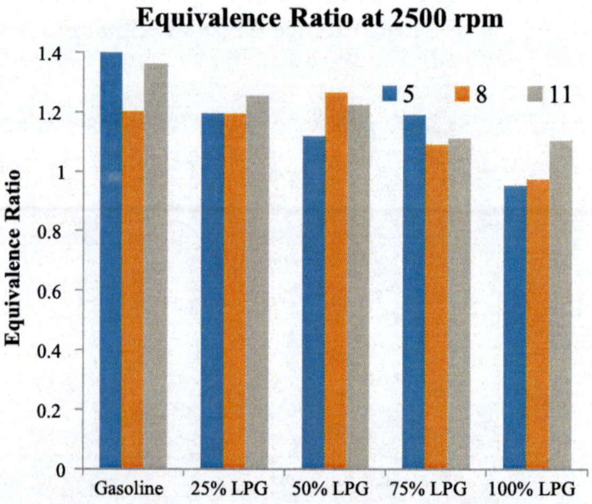

Fig. 20 Comparison of φ w.r.t static ignition timing at 2500 rpm full load condition

LPG increases in the fuel blend, fuel–air ratios within the combustion chamber are coming closer to stoichiometric region which results in the better combustion properties. But by advancing the static ignition timing, there is no much variations in the equivalence ratio for the LPG fuel blend. At lower speed for 100% LPG, advancing static timing to 11 deg. bTDC resulted in slightly richer region because of more fuel consumption.

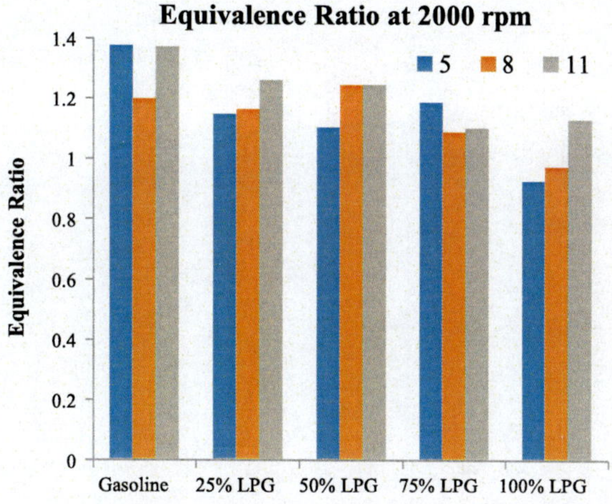

Fig. 21 Comparison of φ w.r.t static ignition timing at 2000 rpm full load condition

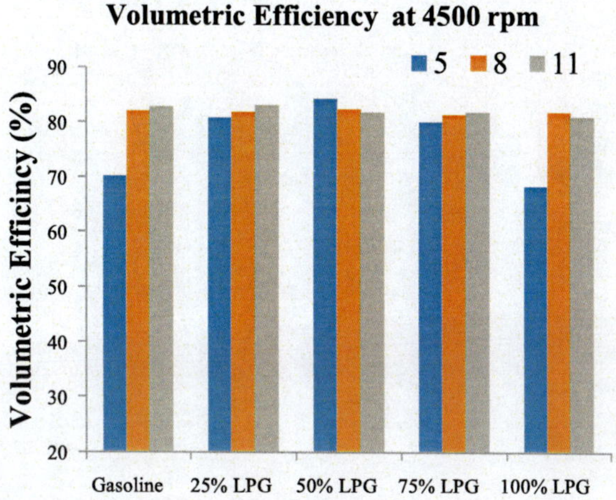

Fig. 22 Comparison of Vol. Eff. w.r.t static ignition timing at 4500 rpm full load condition

3.1.4 Volumetric Efficiency

The induction system includes the air-filter, injector, and throttle plate in spark ignition engine. This system in SI engine will restrict the amount of inducting air into the engine. So volumetric efficiency will indicate the air breathing capacity or suction air into the engine cylinder. Figures 23, 24, 25, 26, 27, and 28 compare volumetric efficiency with respect to static ignition timing for different speeds and fuel blends at full load condition. From the results, it is revealed that for gasoline and 100% LPG as the static ignition timing is advanced, volumetric efficiency is going to increase

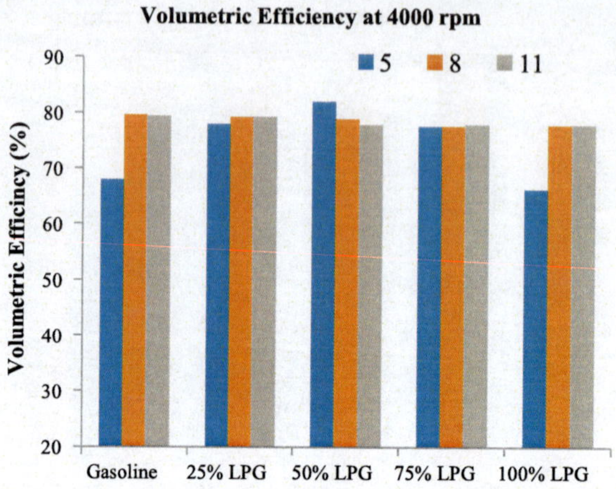

Fig. 23 Comparison of Vol. Eff. w.r.t static ignition timing at 4000 rpm full load condition

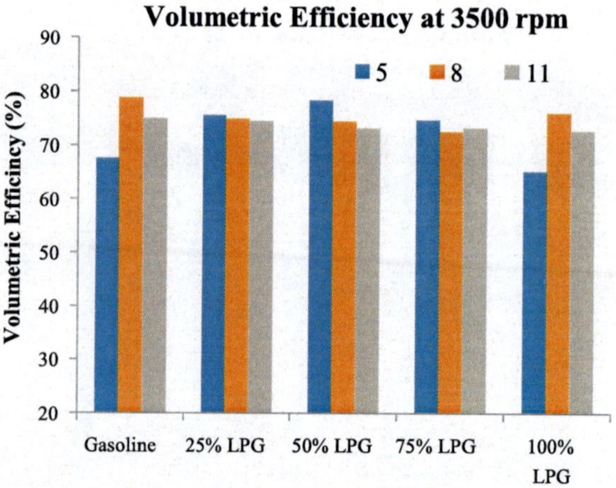

Fig. 24 Comparison of Vol. Eff. w.r.t static ignition timing at 3500 rpm full load condition

for all fuel blend conditions. Since the equivalence ratio is going to leaner side with advancing the static ignition timing for gasoline and 100% LPG fuel, volumetric efficiency will increase. For gasoline, maximum volumetric efficiency found to be 82.73% at 11 deg. bTDC, whereas at 8 deg. bTDC, it is 82.03% and 5 deg. bTDC 70.20%. Volumetric efficiency values for 100% LPG are 68.26%, 81.76%, and 80.92% at 5, 8, and 11 deg. bTDC, respectively.

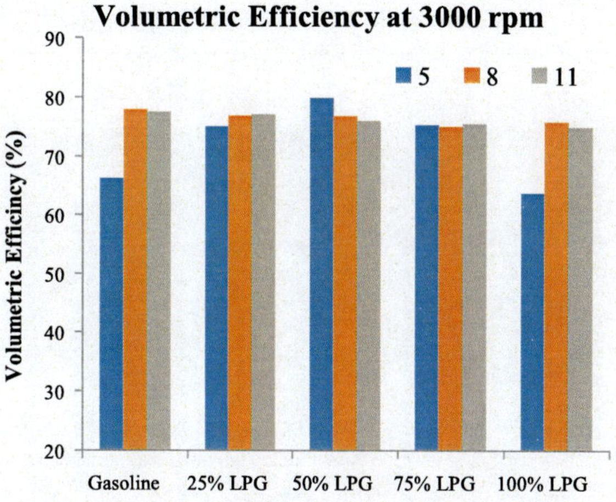

Fig. 25 Comparison of Vol. Eff. w.r.t static ignition timing at 3000 rpm full load condition

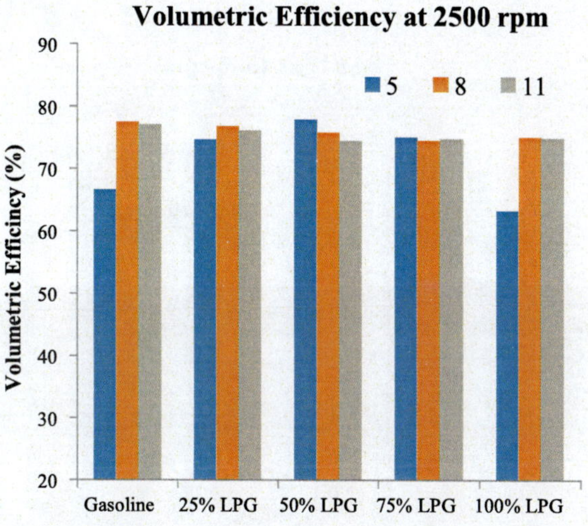

Fig. 26 Comparison of Vol. Eff. w.r.t static ignition timing at 2000 rpm full load condition

3.1.5 Brake Specific Energy Consumption (BSEC)

BSEC will indicate to produce one kilowatt of power in one hour how much energy is consumed. This comparison will give the fuel economy for an engine where different calorific values of fuels are being used. Figures 29, 30, 31, 32, 33, and 34 compare the BSEC with respect to static ignition timing for different speeds and fuel blends at full load condition. For the advanced static ignition timing of 8 deg. bTDC, the energy consumed to produce will be less or gasoline for all engine speed condition.

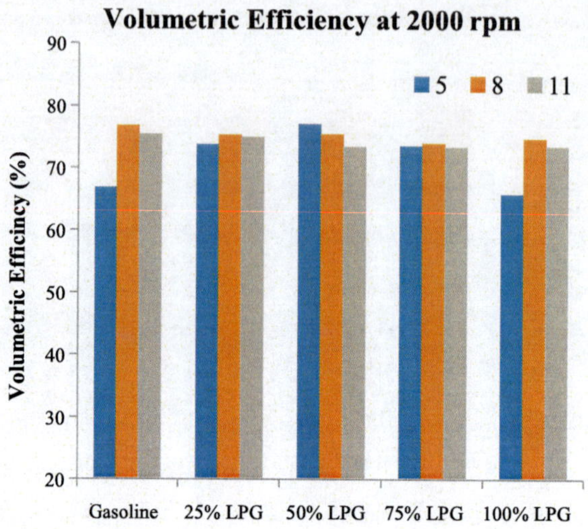

Fig. 27 Comparison of Vol. Eff. w.r.t static ignition timing at 2000 rpm full load condition

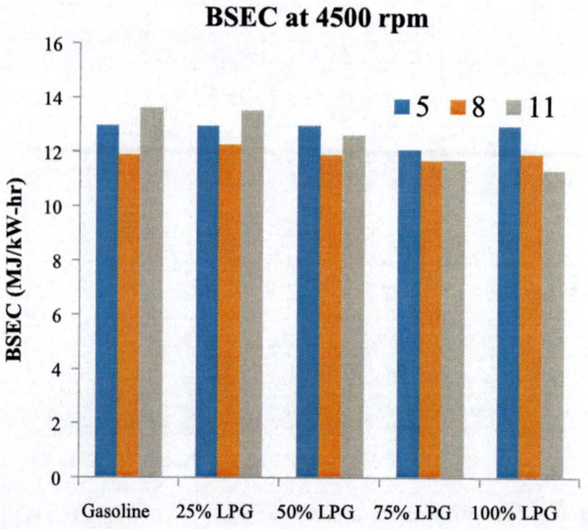

Fig. 28 Comparison of BSEC w.r.t static ignition timing at 4500 rpm full load condition

Further increase in the timing will lead to increase in the energy consumption and intern fuel economy. This reduction in the energy consumption is because gasoline will work near to stoichiometric region and fuel consumption also reduced so energy consumed also reduced. As the percentage of LPG increased, advancing the timing will result in reduction in the energy consumption and it is minimum for 11 deg. bTDC. 100% LPG gives the minimum energy consumption at 11 deg. bTDC. Since LPG is having higher calorific value than gasoline which results in the lower fuel

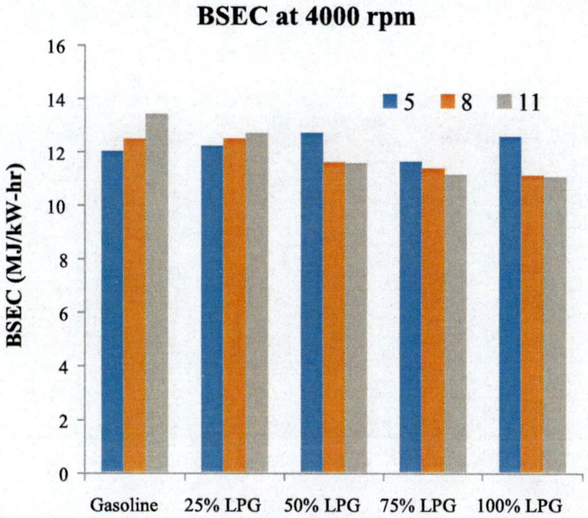

Fig. 29 Comparison of BSEC w.r.t static ignition timing at 4000 rpm full load condition

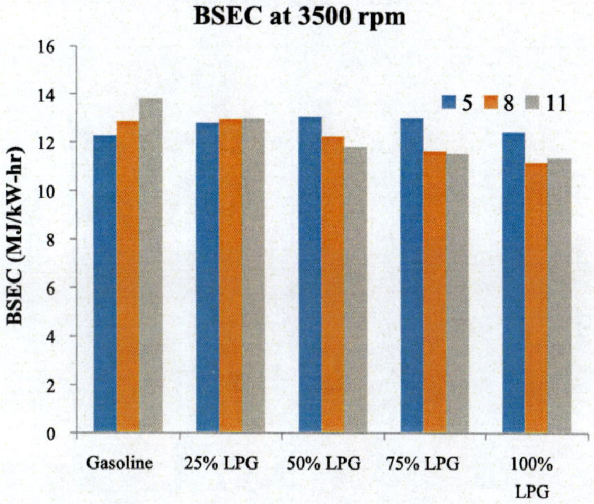

Fig. 30 Comparison of BSEC w.r.t static ignition timing at 3500 rpm full load condition

consumption, hence energy consumed for LPG will be minimum. Average decrease in the BSEC value for gasoline is 8.18% and −5.02% at 8 and 11 deg. bTDC, respectively, when compared to 5 deg. bTDC at 4500 rpm. Similarly, for 100% LPG is 7.84% and 12.55% at 8 and 11 deg. bTDC, respectively, when compared to 5 deg. bTDC at 4500 rpm.

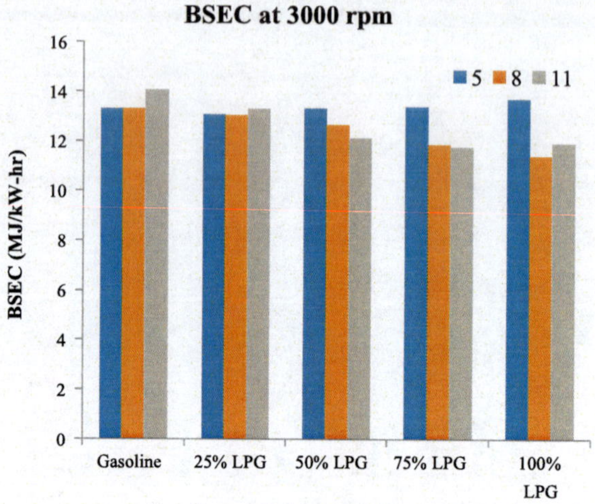

Fig. 31 Comparison of BSEC w.r.t static ignition timing at 3000 rpm full load condition

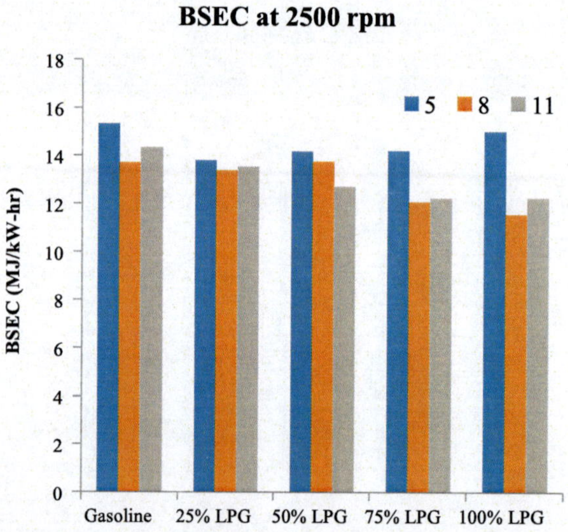

Fig. 32 Comparison of BSEC w.r.t static ignition timing at 2500 rpm full load condition

3.2 Emission Characteristics

3.2.1 CO Emissions

The comparison of CO emissions with respect to static ignition timing for full load condition at different speeds and fuel blend operations are shown in Fig. 35, 36, 37,

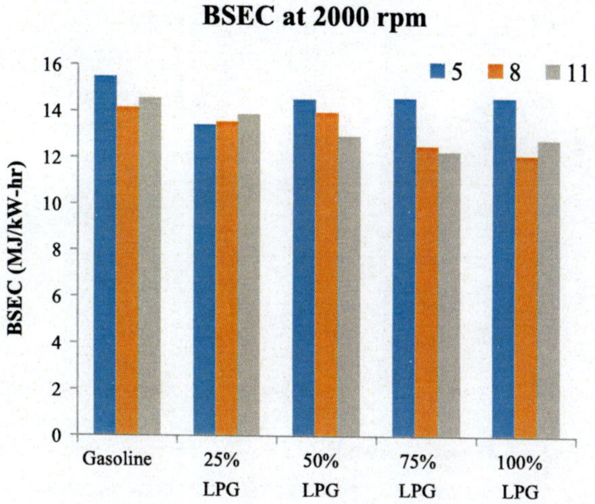

Fig. 33 Comparison of BSEC w.r.t static ignition timing at 2000 rpm full load condition

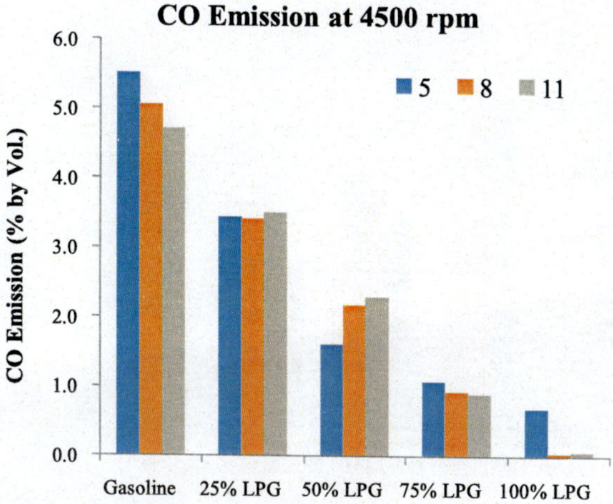

Fig. 34 Comparison of CO w.r.t static ignition timing at 4500 rpm full load condition

38, 39, and 40. CO emission is mainly influenced by the oxygen availability in the combustion zone. Hence, CO emission is expected to be continuing to decrease as the air–fuel mixture becomes leaner. Also CO emission requires higher exhaust temperature and longer residence time to oxidize. At full load condition as the percentage of LPG increases CO emission going to be reduced since LPG is working near to stoichiometric region and higher flame propagation speed of LPG [14]. As advancing the static ignition timing, not remarkable changes found in CO emission. At higher speed condition, all the fuel blend shown decreasing characteristic for CO emission.

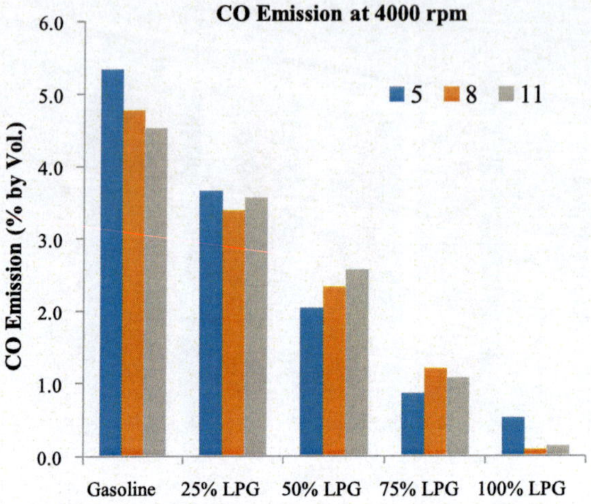

Fig. 35 Comparison of CO w.r.t static ignition timing at 4000 rpm full load condition

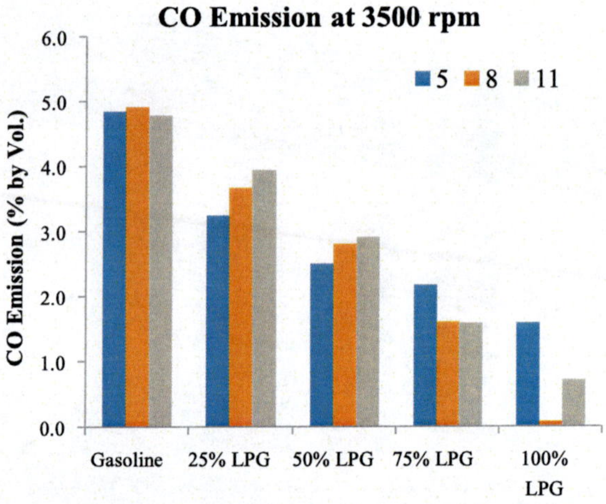

Fig. 36 Comparison of CO w.r.t static ignition timing at 3500 rpm full load condition

But as the speed decreased, exhaust gas temperature and residence time reduce, and hence CO emission going to slightly increase for all fuel blend except for 100% LPG till 8 deg. bTDC then again reduces. Very low CO emission is obtained for 100% LPG. Average decrease in the CO emission value at 4500 rpm for gasoline is 8.34% and 14.7% at 8 and 11 deg. bTDC, respectively, when compared to 5 deg. bTDC. Average decrease in the CO emission value at 4500 rpm for 100% LPG is 90% and 88% at 8 and 11 deg. bTDC, respectively, when compared to 5 deg. bTDC.

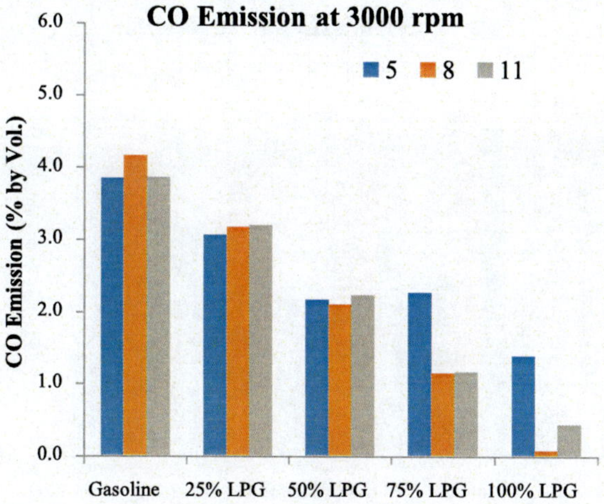

Fig. 37 Comparison of CO w.r.t static ignition timing at 3000 rpm full load condition

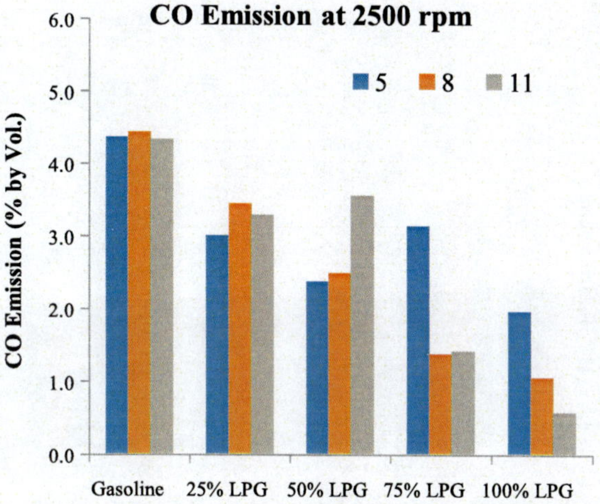

Fig. 38 Comparison of CO w.r.t static ignition timing at 2500 rpm full load condition

3.2.2 Hydro Carbon (HC) Emission

Figure 41, 42, 43, 44, 45, and 46 show the comparison of HC emissions on static ignition timing for different fuel blends and speeds at full load condition. HC emissions are mainly influenced by the combustion quality and availability of oxygen. To achieve minimum HC emission, best parameter is to decide is equivalence ratio. Whenever equivalence ratio is 0.9, then HC emission is going to be minimum as static ignition timing advanced HC emission is increased for all speed conditions

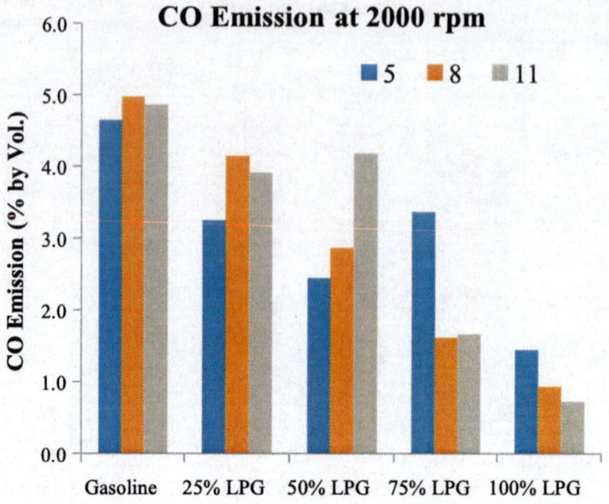

Fig. 39 Comparison of CO w.r.t static ignition timing at 2000 rpm full load condition

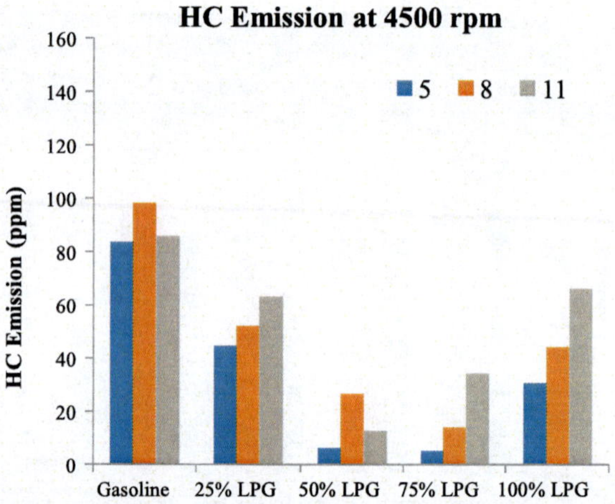

Fig. 40 Comparison of HC w.r.t static ignition timing at 4500 rpm full load condition

with gasoline and 100% LPG fuel. This is mainly because of two reasons. First, as static ignition timing is advanced, in-cylinder pressure will rise which results in greater mass of hydrocarbons trapped in the crevice volume. Second, due to advance in the ignition timing exhaust gas, temperature will reduce hence trapped hydrocarbon in the crevices volume will have less oxidation which results in higher HC emissions. In this way, HC emission is increasing for advanced static ignition timing. But at higher speed, HC emission will decrease for 100% LPG drastically because of higher flame propagation speed for LPG will lead to enhanced combustion of LPG fuel and hence higher combustion temperature is obtained [15].

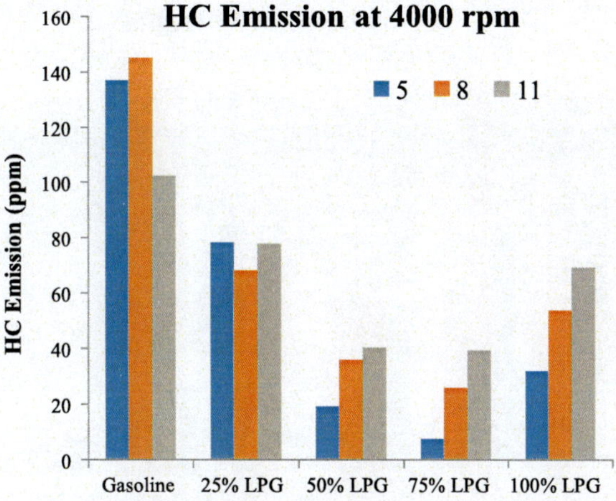

Fig. 41 Comparison of HC w.r.t static ignition timing at 4000 rpm full load condition

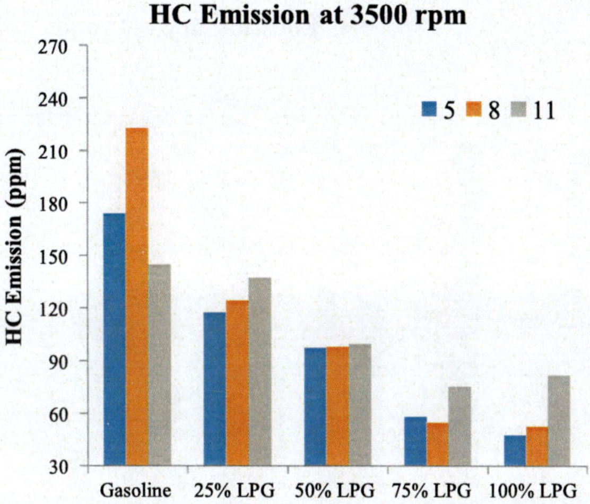

Fig. 42 Comparison of HC w.r.t static ignition timing at 3500 rpm full load condition

3.2.3 NO_x Emission

Figures 47, 48, 49, 50, and 51 show the comparison of NO_x versus static ignition timing for different engine speeds and fuel blends at full load conditions. NO_x concentration in the exhaust is mainly depending on the flame temperature and availability of oxygen. Equivalence ratio is one of the parameter is influencing the NO_x emission formation. Also NO_x emission is mainly depending upon in-cylinder temperature and NO_x emission exponentially proportional to the in-cylinder temperature [16]. During full load condition, as the static ignition timing is advanced, NO_x concentrations will

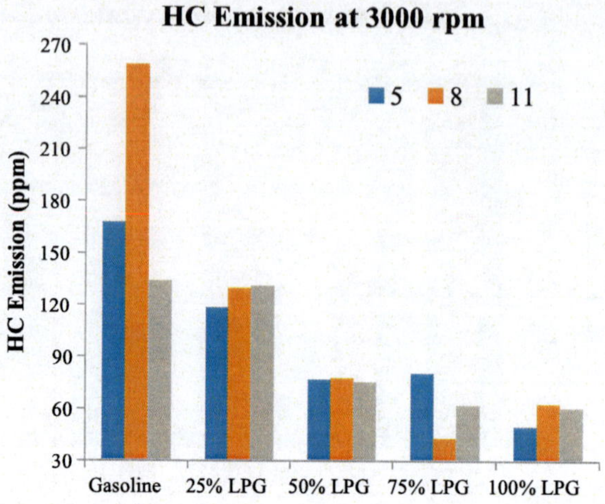

Fig. 43 Comparison of HC w.r.t static ignition timing at 3000 rpm full load condition

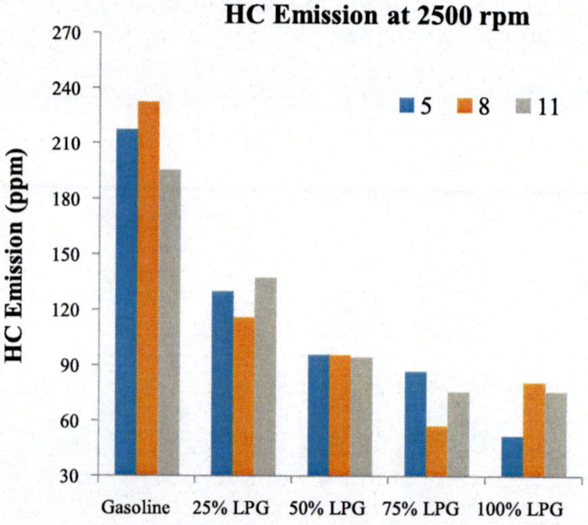

Fig. 44 Comparison of HC w.r.t static ignition timing at 2500 rpm full load condition

also increase since equivalence ratio is near to stoichiometric region. NO_x emission is maximum for 100% LPG fuel at all speed and load conditions. As the timing is advanced, the in-cylinder temperature rises and better combustion will take place in the cylinder hence NO_x concentration will increase. Also as the percentage of LPG increases, NO_x concentration will increase exponentially for all speed conditions. At full load, 4500 rpm increase in NO_x concentration for 100% LPG is 71% and 107% at 8 and 11 deg. bTDC, respectively, when compared to 5 deg. bTDC.

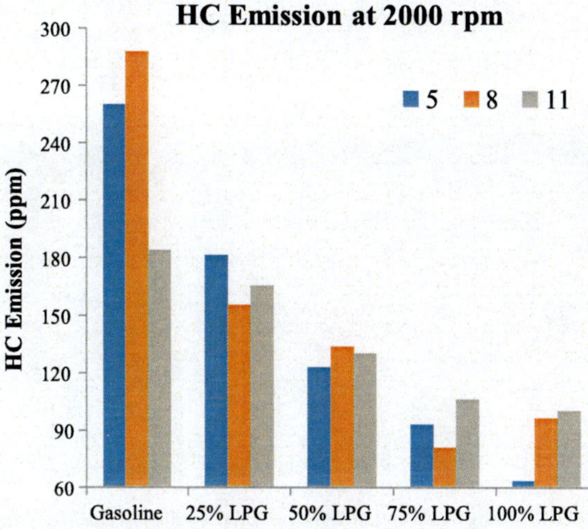

Fig. 45 Comparison of HC w.r.t static ignition timing at 2000 rpm full load condition

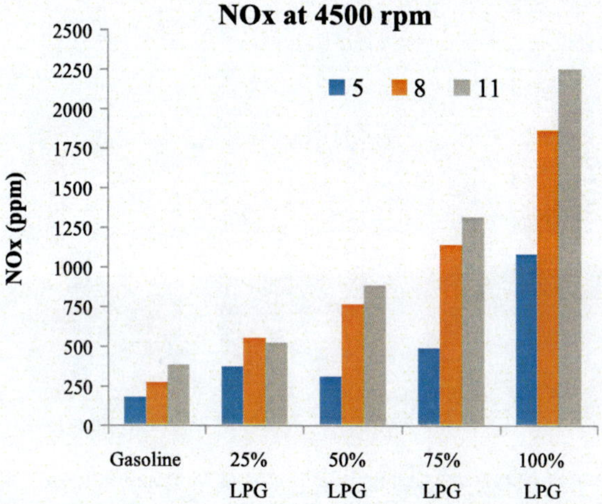

Fig. 46 Comparison of NOx w.r.t static ignition timing at 4500 rpm full load condition

4 Conclusion

The results of the experiments with advancing static ignition timing on engine performance and emission characteristics can be summarized as follows:

- Results have shown that advancing the static ignition timing will increase the BP by 12% at 11 deg. bTDC & 7% at 8 deg. bTDC for gasoline. Whereas for 100% LPG increased in BP is 5% at 11 deg. bTDC & 2% at 8 deg. bTDC.

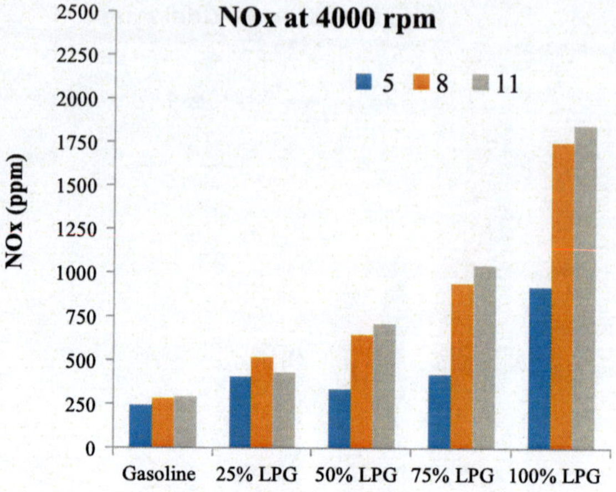

Fig. 47 Comparison of NOx w.r.t static ignition timing at 4000 rpm full load condition

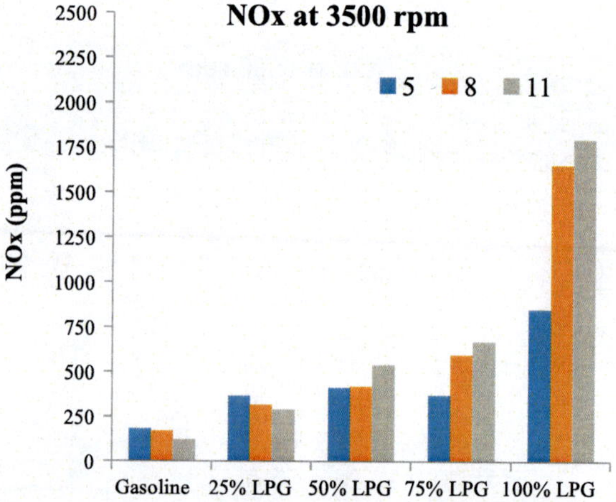

Fig. 48 Comparison of NOx w.r.t static ignition timing at 3500 rpm full load condition

- BTE also increased for both gasoline and LPG when advancing static ignition timing. Also advancing the ignition timing will engine will work leaner side hence reduction in the fuel consumption.
- As the ignition timing advanced, volumetric efficiency will increase for all fuel conditions.
- CO emission will drastically reduce when static ignition timing advanced to 8 deg. bTDC after that not significant reduction in CO emission. 100% LPG showing major reduction in CO emission is obtained with advanced timing.
- But advancing the static ignition timing resulted in increased HC emission for all fuel blends.

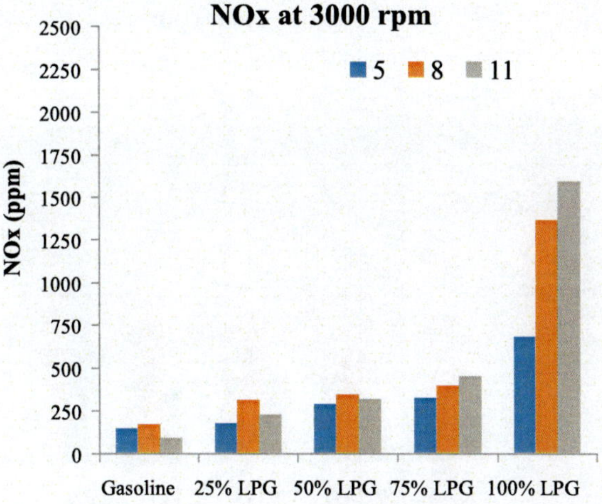

Fig. 49 Comparison of NO_x w.r.t static ignition timing at 3000 rpm full load condition

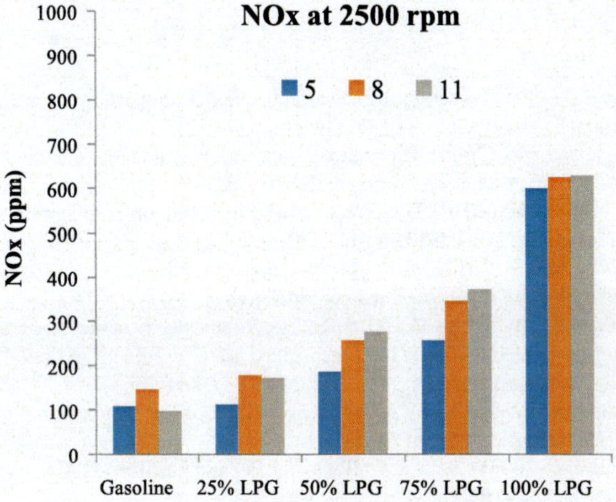

Fig. 50 Comparison of NO_x w.r.t static ignition timing at 2500 rpm full load condition

- NO_x emission also increases with advancing the static ignition timing for all fuel blends because of increase in the in-cylinder temperature.

Finally, after varying the advanced static ignition timing of 8 deg. bTDC with 100% LPG resulted in better performance and emission characteristics, hence these conditions are optimized. In the future work, suitable NO_x reduction technology with enhancement methods to the performance characteristics for optimized condition has been used.

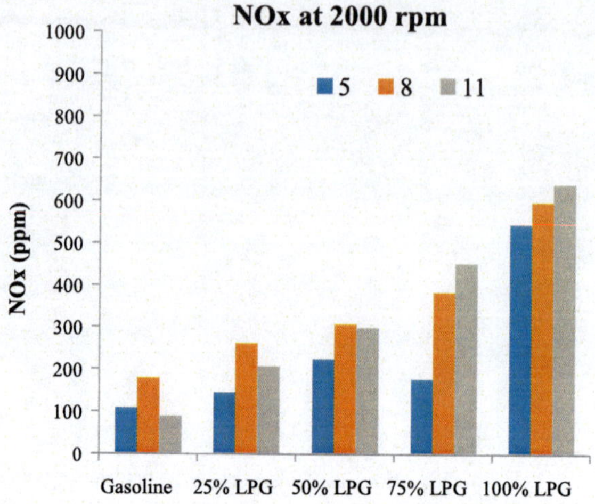

Fig. 51 Comparison of NOx w.r.t static ignition timing at 2000 rpm full load condition

References

1. Hakan B, Orhan D (2005) Investigating the effects of LPG on spark ignition engine combustion and performance. J Energy Convers Manage 46:2317–2333
2. Hakan O, Soylemez MS (2005) Thermal balance of LPG fuelled, four stroke SI engine with water addition. J Energy Convers Manage 47:570–581
3. Shankar KS, Mohanan P (2010) The effect of LPG injection on combustion, performance and emission characteristics of a MPFI engine. J Middle Eur Constr Des Cars (03):VIII
4. Shankar KS, Mohanan P (2011) MPFI gasoline engine combustion, performance and emission characteristics with LPG injection. Int J Energy Environ 2(4):761–770
5. Patil B, Nayak V, Padmanabha M (2014) Emission and performance enhancement of multi-cylinder SI engine fuelled with LPG and vaporized water methanol induction. SAE Technical Paper 2014-01-2764. https://doi.org/10.4271/2014-01-2764
6. Philip P, Shengmin G, Martin H (2004) Performance of an evaporator for a LPG powered vehicle. J Appl Thermal Eng 24:1179–1194
7. Ceviz MA, Yuksel F (2006) Cyclic variations on LPG and gasoline-fuelled lean burn SI engine. Renew Energy 31:1950–1960
8. Erkus B, Karamangil MI, Surman A (2015) Enhancing the heavy load performance of a gasoline engine converted for LPG use by modifying the ignition timings. Appl Thermal Eng 85:188–194
9. Alasfour FN (1998) NOx emission from a spark ignition engine using 30% iso-butanol-gasoline blend: Part 2-ignition timing. Appl Thermal Eng 18(8):609–618
10. Lawankar SM, Dhamande LP, Khandare SS (2012) Experimental study of effect of ignition timing and compression ratio on NOx emission of LPG fuelled engine. Int J Sci Eng Res 3(10):1–4
11. Turkoz N, Erkus B, Karamangil IM, Surmen A, Arslanoglu N (2014) Experimental Investigation of the effect of E85 on engine performance & emission under various ignition timings. Fuel 115:826–832
12. Kline SJ, McClintock FA (1953) Describing uncertainties in single-sample experiments. Mech Eng 75(1):3–8
13. Gumus M (2011) Effects of volumetric efficiency on the performance & emissions characteristics of a dual fueled (gasoline & LPG) spark ignition engine. J Fuel Process Technol 92(10):1862–1867

14. Ceviz MA, Asok Sen K, Alp Kuleri K (2012) Engine performance, exhaust emissions, and cyclic variations in a lean-burn SI engine fueled by gasoline hydrogen blends. Appl Thermal Eng 36:314–324
15. Heywood JB (1998) Internal combustion engine fundamentals. McGraw Hill Book Company
16. Murillo S, Miguez JL, Porteiro J, Lopez G, Granada E, Moran JC (2005) LPG: pollutant emission and performance enhancement for spark-ignition four strokes outboard engines. J Appl Thermal Eng 25:1882–1893

Optimization of Process Parameters of Paddy Straw Gasification System Using Taguchi Methodology

Mohit Sharma and Rajneesh Kaushal

1 Introduction

In recent times, considerable changes in environmental conditions are observed due to global warming resulting in diminution of fossil fuels. Therefore, all around the world researchers start paying attention on some other energy alternatives in order to fulfil the escalating demand of energy. Keeping in view, biomass proves to be a promising energy source because of its carbon neutral feature [1–3]. Biomass is defined as biological matter that comes from animal or plants waste [4]. The share of bioenergy is approx. 25% in world's energy consumption & out of total 2% is served by the bioenergy as depicted in Fig. 1.

Biomass contains small amount of sulphur therefore on combustion the fraction of SO_2 produced is very small in spite of its low calorific value [5, 6]. It is suggested to use biomass in either gaseous or liquid state. The concept of gasification was first introduced in 1959 by Thomas Shirley. In gasifier, there are four processes are taking place namely: drying, pyrolysis, oxidation and reduction [7]. Gasification is a process of extracting the energy from wide range of biomass such as plants and animals waste, food waste, industrial waste, etc., by undergoing a series of processes in a device called gasifier.

M. Sharma (✉)
National Institute of Technology, Kurukshetra, Haryana 136119, India
e-mail: er1986.sharma@outlook.com

R. Kaushal
Department of Mechanical Engineering, National Institute of Technology Kurukshetra, Kurukshetra 136119, India
e-mail: rajneesh@nitkkr.ac.in

A. K. Gupta et al. (eds.), *Advances in IC Engines and Combustion Technology*, Lecture Notes in Mechanical Engineering,
https://doi.org/10.1007/978-981-15-5996-9_18

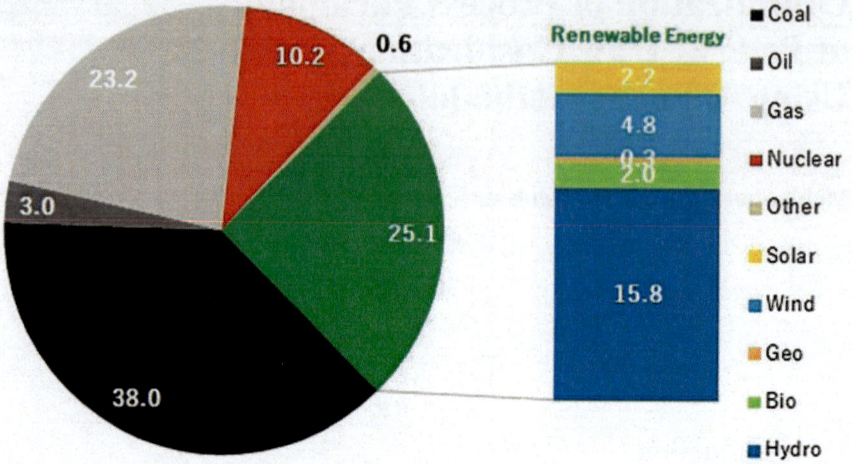

Fig. 1 Global energy consumption (2019). *Source* Statistical review of world energy (2019)

2 Taguchi Method

Taguchi is a technique proposed by Taguchi for the purpose of optimizing the process variables [8]. Earlier, Fisher developed methods for design of experiment but implementation of those methods seems to be difficult in case system involved large number of input parameters [9, 10]. This method follows a three step process, namely system design, parameter and tolerance. In Taguchi method, a numbers of experiments are minimized by employing orthogonal arrays over the entire range of process space [11, 12]. In this method, the experimental outcomes are changed into signal-to-noise ratio (S/N). Signal-to-noise ratio is a measure of quality features that are different from the required value. For the analysis, (S/N) ratios are categorized, as lower is better, higher is better & nominal is better. It is observed that higher value of (S/N) ratio implies superior quality characteristic. Therefore, level of process parameter with highest value of (S/N) ratio is considered as optimum level. Furthermore, in this study, ANOVA is conducted to check the significance of input parameters. Finally, in order to justify the optimum input parameters, affirmative experiment is performed.

3 Description of Experimental Work

In the present research work, paddy straw gasification has been done by using a downdraft gasifier to generate a combustible gas known as "producer gas" as shown in Fig. 1. The gasification unit is equipped with a reactor, charcoal filter, venturi, fine filter & fabric filter as depicted in Fig. 2. To start the process, gasifier is first loaded with paddy straw approx. 48 kg & 15 kg charcoal. Blower is turned on for creating

Fig. 2 Gasification system

a vaccum. As a result, air starts flowing inside the gasifier from top. Anemometer predicts the flow rate of air. In this study, flow rate of air and equivalence ratio are varied from 0.11 to 0.19 m/s and 0.227–0.295, respectively. Once system achieved stable conditions, high temperature gas from reactor flows towards charcoal filter for elimination of impurities. Venturi is employed for cooling of producer gas. After venturi gas is allowed to pass through fine filter & fabric filter which entraps the solid particulates. The composition of producer gas is predicted by gas analyser (Applied Techno System ATS206A). K-type thermocouples are installed at different locations in order to predict the temperature.

4 Input Parameters and Their Levels Selection

In this work, gasification of rice husk was performed on a downdraft gasifier. The preliminary input parameters are air flow velocity 0.11 m/sec, equivalence ratio 0.227. The selection of input parameters depends on its effect on gasification process. The

Table 1 Gasification Input parameters with their corresponding levels

Levels	Input parameters (units)	
	Air velocity (m/s)	Equivalence ratio
1	0.11	0.227
2	0.14	0.253
3	0.19	0.295

Table 2 Description of L_9 Orthogonal array for experimental design

Experiment No.	Input parameters level	
	Air velocity (m/s)	Equivalence ratio
1	1	1
2	1	2
3	1	3
4	2	1
5	2	2
6	2	3
7	3	1
8	3	2
9	3	3

realistic design for input parameters was obtained by changing the air flow velocity from 0.11 to 0.19 m/sec and equivalence ratio 0.227–0.295 as shown in Table 1. In the gasification process design, three levels of input variables were opted as depicted in Table 2.

5 Examination of Input Variables via Orthogonal Array

The minimum number of experiments needs to be conducted for obtaining the optimum value of input variables is calculated by using orthogonal array. The outcomes are analysed via. ANOVA. Further, the experimental settings corresponding to gasification input parameter for which fraction of H_2 & CH_4 is higher are achieved and confirmed.

The selection of orthogonal array for predicting the optimum conditions depends on the value of degree of freedom. After recognizing suitable degree of freedom, orthogonal array needs to be selected by keeping in view the desired experimental outcomes. In this research work, L9-orthogonal array was selected with total eight

Table 3 Experimental outcomes for prediction of H_2 and CH_4 content

Run	Parameters		Responses	
	Air velocity (m/s)	Equivalence ratio	H_2 (%)	CH_4 (%)
1	0.11	0.227	13.67	1.81
2	0.11	0.253	12.93	2.50
3	0.11	0.295	12.52	2.92
4	0.14	0.227	11.16	3.17
5	0.14	0.253	11.32	3.39
6	0.14	0.295	11.38	3.74
7	0.19	0.227	11.52	3.98
8	0.19	0.253	11.74	4.01
9	0.19	0.295	11.93	4.04

degree of freedom. The experimental design obtained by using L9-orthogonal array is depicted in Table 3.

6 ANOVA of Output Responses

ANOVA helps in predicting the significance of input parameters on the experimental outputs. In the present research work, input parameters are subjected to 95% confidence level. Under this condition, the input parameters for which P value (< 0.05) is measured has been considered as significant. Further, large R^2 value ($> 70\%$) decides the precision of experimental data. The response for means is highlighted in Table 4 and in Figs. 3 and 4.

The ANOVA results for H_2 and CH_4 are depicted in Tables 5 and 6, respectively. As P value of air velocity for both H_2 and CH_4 is calculated as 0.017 and 0.005, respectively, therefore air velocity is considered as noteworthy variable. Further, from ANOVA, the observed R^2 value is greater than 70%, i.e. 87.15% and 93.88%, respectively, which justifies the experimental data correctness (Fig. 5).

Table 4 H_2 and CH_4 Response Table for Means (larger is better)

Level	H_2		CH_4	
	Air velocity (m/s)	Equivalence ratio	Air velocity (m/s)	Equivalence ratio
1	13.04	12.12	2.41	2.98
2	11.29	12.00	3.43	3.30
3	11.73	11.94	4.01	3.56
Delta	1.75	0.17	1.60	0.58
Rank	1	2	1	2

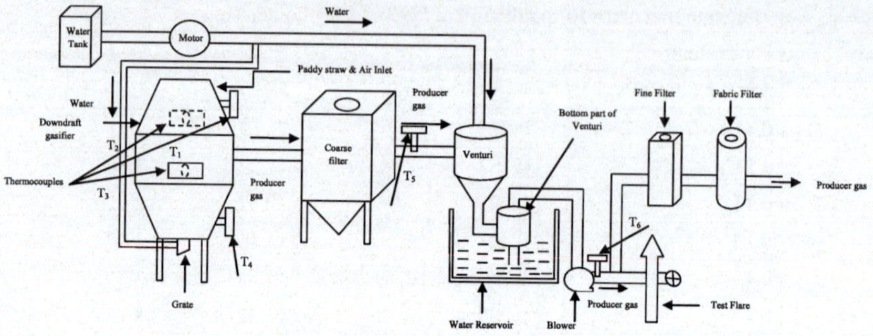

Fig. 3 Schematic of gasification system for paddy straw

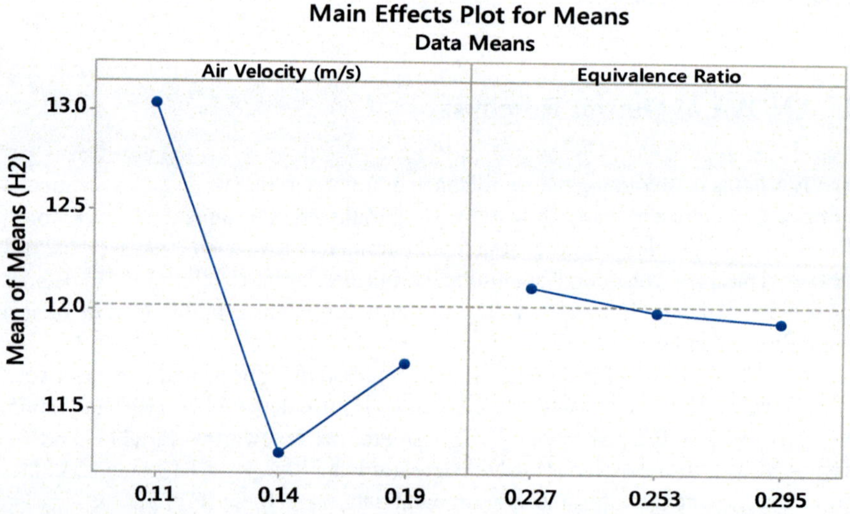

Fig. 4 Mean graph for H_2

Table 5 ANOVA for H_2 percentage

Source	DF	SS (Adj)	MS (Adj)	F-Value	P-Value	Contribution (%)	Remark
Air velocity	2	4.98682	2.49341	13.44	0.017	86.33	Significant
Equivalence ratio	2	0.04729	0.02364	0.13	0.884	0.81	–
Error	4	0.74218	0.18554	–	–	12.84	–
Total	8	5.77629	–	–	–	–	

$R^2 = 87.15\%$ R^2 (Adj) $= 74.30\%$

Table 6 ANOVA for CH_4 percentage

Source	DF	SS (Adj)	MS (Adj)	F-Value	P-Value	Contribution (%)	Remark
Air velocity	2	3.9398	1.96988	27.21	0.005	83.20	Significant
Equivalence ratio	2	0.5057	0.25284	3.49	0.133	10.68	–
Error	4	0.2896	0.07239	–	–	6.11	–
Total	8	4.7350	–	–	–	–	

$R^2 = 93.88\%$ R^2 (Adj) $= 87.77\%$

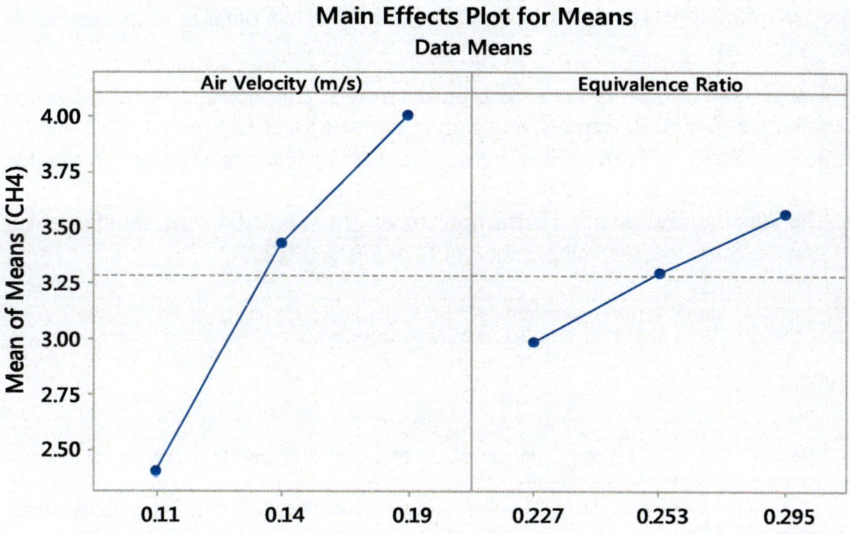

Fig. 5 Mean graph for CH_4

7 Regression Modelling

The value of output responses ($H_{2\ \%}$ and $CH_{4\ \%}$) was determined by building a regression model. The regression equation for H_2 and CH_4 has the effect of different input parameters (air velocity and equivalence ratio) as depicted by Eqs. (1) and (2).

$$H_2 = 14.67 - 13.80 * \text{Air Velocity} - 2.43 * \text{Equivalence Ratio} \quad (1)$$

$$CH_4 = -1.67 + 19.14 * \text{Air Velocity} + 8.32 * \text{Equivalence Ratio} \quad (2)$$

The R^2 value for H_2 and CH_4 was determined by performing statistical analysis with the help of regression equation on MINITAB. In the present research work, the R^2 value of H_2 and CH_4 is 87.15% and 93.88%, respectively.

8 Conclusion

In this research, Taguchi technique has been used for optimizing the fraction of H_2 and CH_4 obtained during the gasification of paddy straw. Air flow velocity and equivalence ratio were selected as input variables. The outcomes of the present study are as follows:

- The air flow velocity is observed as the most vital factor that effects the generation of both H_2 and CH_4, respectively, during rice straw gasification.
- The optimum value of air flow velocity and equivalence ratio for which fraction of H_2 is maximum, i.e. 13.67% are 0.11 m/s and 0.227, respectively.
- The maximum value of CH_4 fraction comes out to be 4.04 corresponding to air flow velocity and equivalence ratio of 0.19 m/s and 0.295.

References

1. McKendry P (2002) Energy production from biomass (part 1): overview of biomass. Bioresource Technol 83(1):37–46
2. Li X et al (2004) Biomass gasification in a circulating fluidized bed. Biomass Bioenergy 26(2):171–193
3. Li S et al (2004) Fast pyrolysis of biomass in free-fall reactor for hydrogen-rich gas. Fuel Process Technol 85(8–10):1201–1211
4. Sharma M, Kaushal R (2018) Advances and challenges in the generation of bio-based fuels using gasifiers: a comprehensive review. Int J Ambient Energy, 1–19
5. Seyboth K et al (2008) Recognising the potential for renewable energy heating and cooling. Energy Policy 36(7):2460–2463
6. Khan A et al (2009) Biomass combustion in fluidized bed boilers: Potential problems and remedies. Fuel Process Technol 90(1):21–50

7. Sansaniwal S et al (2017) Recent advances in the development of biomass gasification technology: a comprehensive review. Renew Sustain Energy Rev 72:363–384
8. Taguchi G (1986) Introduction to quality engineering: designing quality into products and processes
9. Montgomery D, Runger G, Hubele N (2017) 2012, Design and analysis of experiments. John Wiley & Sons
10. Fisher RA (2006) Statistical methods for research workers. Genesis Publishing Pvt Ltd
11. Sindhu D, Thakur L, Chandna P (2019) Parameter optimization of rotary ultrasonic machining on quartz glass using Response Surface Methodology (RSM). Silicon, pp 1–15
12. Sindhu D, Thakur L, Chandna P (2018) Multi-objective optimization of rotary ultrasonic machining parameters for quartz glass using Taguchi-Grey relational analysis (GRA). Silicon, pp 1–12

Investigation on Lean-Burn Spark-Ignition Engine with Methanol/Ethanol—Gasoline Blends

Suresh Devunuri, E. Porpatham, and Zhen Wu

1 Introduction

To meet the energy demand, growing global concern for air quality and energy security made the automakers to use the different cleaner energy resources. It also indicates to know about the various technical solutions to present novel stare toward energy deliver and ecological safety [1, 2]. Considering this cause, the usage and study of cost-effective and renewable alternative fuels like gaseous fuels (Hydrogen LPG, CNG) and liquid fuels (propane, ethanol, and methanol) were needed. Among alternatives, alcohols are favorable fuels since alcohol in particular ethanol and methanol have potential to decline the harmful emissions and boost the performance [3]. Methanol and ethanol have octane number; need more heat for vaporization, contain high oxygen percentage, and have high laminar flame speed. Gasoline and alcohol properties are presented in Table 1. Currently, small percentages of alcohol-gasoline mixtures are using in spark-ignition engines with no changes. Spark-ignition engines run with alcohols (99.9%); however, it prerequisites few modifications in the intake system of engine [4, 5].

S. Devunuri · E. Porpatham (✉)
Automotive Research Centre, School of Mechanical Engineering, Vellore Institute of Technology, Vellore 632 014, India
e-mail: porpatham.e@vit.ac.in

S. Devunuri
e-mail: suresh_d@vnrvjiet.in

S. Devunuri
Department of Automobile Engineering, VNR Vignana Jyothi Institute of Engineering and Technology, Hyderabad, India

Z. Wu
School of Chemical Engineering and Technology, Xi'an Jiaotong University, Xi'an, People's Republic of China

A. K. Gupta et al. (eds.), *Advances in IC Engines and Combustion Technology*, Lecture Notes in Mechanical Engineering,
https://doi.org/10.1007/978-981-15-5996-9_19

Table 1 Gasoline and alcohol properties [6–9]

Property	Gasoline	Ethanol	Methanol
Specific gravity	0.738	0.794	0.796
Oxygen percentage	0	34.9	50
Stoichiometric air-fuel ratio ($g.g^{-1}$)	14.7:1	9:1	6:1
Calorific value ($kJ.kg^{-1}$)	43,450	26,760	19,660
Self-ignition temperature (K)	644	738	698
Vaporization heat ($J.g^{-1}$)	311	905	1109
Octane number	85–94	109	111
Flammability values vol.%	1.3–7.5	4.4–19.1	6.4–36
Flame velocity ($m.\ s^{-1}$)	0.33	0.39	0.52

Relatively a bunch of investigations were conducted to know the performance parameters of alcohol-gasoline blends at stoichiometric combustion operation. Gravalos et al. [10] studied on stationary engine with different alcohol—gasoline mixtures as fuels varying from 10 to 30%. Test results displayed that carbon monoxide and hydrocarbon emissions from the alcohol-gasoline blends are less compare to gasoline. Zaharin [11] conducted a set of experiments on four-cylinder SI engine with different blends of alcohol-gasoline mixtures. The results stipulated that brake specific energy consumption for 15% methanol blend high relative to other blends. Tayler et al. [12] conducted a series of experiments using low and high alcohol mole mass—gasoline mixtures with different oxygen contents. The test results displayed that 40% and 75% reduction in carbon monoxide and hydrocarbon emissions were observed with alcohol-gasoline blend limited with 5% oxygen. Palmer [13] performed a study on how ethanol blends will influence the performance of the gasoline engine. The outcomes indicated that 10% ethanol-gasoline blend by volume rises the power output by 5% and 5% increment in octane number for each 10% ethanol blending with gasoline. Hüseyin et al. [6] studied on SI engine to know the effect of compression ratio and ethanol blend with gasoline at stoichiometric condition with different RPM of engine. The test results displayed that output torque raises with compression ratio and ethanol 40% and 60% blends with gasoline reduces the CO and HC levels by 11% and 10.8% correspondingly at a speed of 2000 rpm. Al Hasan [14] studied the performance characteristics of ethanol-gasoline blends varying from 0 to 25% with an increment of 2.5%. The results depict 20% ethanol-gasoline displayed better performance and emission results. Liu et al. [15] conducted the tests to know the effect of methanol-gasoline mixtures on three-cylinder port fuel SI engine. The experimental outcomes show that thermal efficiency and brake power increases while the CO and HC emissions were decreased. Alexandru et al. [16] performed the tests on scooter engine with methanol-gasoline mixtures. The results show that a decrease in torque and power up to 10% blend while the emissions were also reduced by the same amount 10%. Abu Zaid et al. [7] performed research on SI engine with different methanol-gasoline mixtures. The results displayed that

15% methanol-gasoline blend enhances performance compared to other test blends. Tiegang et al. [8] performed the study to know the influence of methanol-gasoline mixtures on combustion and emission behavior of spark-ignition engine during start conditions. The outcomes displayed that engine combustion efficiency is increased and development of flame period and burning rate shortened with methanol addition. The emission results indicated that there was a drop in HC and CO emissions.

From the previous works, several researchers have conducted experiments with alcohol-gasoline blends to know the alcohol effects on the performance behavior of SI engine at stoichiometric and rich air-fuel mixtures. The objective of the present work is to know the effect of the alcohol (ethanol/methanol) mixtures to gasoline on SI engine performance characteristics in lean conditions.

2 Materials and Methods

In the current study, alcohol-gasoline mixtures were arranged on the basis of volume. Methanol and Ethanol with 10% concentration of each separately blended to 90% gasoline to form binary blends called M10 and E10, further ethanol and methanol with concentration of 5% of each were blended to 90% gasoline to form ternary mixture; designated as E5M5.

Tests were conducted on four-stroke, 661 cc SI engine at a ratio of compression 10.5 and power of 4.4 kW at 1500 rpm. The engine details are labeled in Table 2.

Basically, the engine was modified from the Kirloskar TAF1 diesel engine by changing the compression ratio, replacing the fuel injection system with carburetor and providing the ignition system spark plug. The diesel engine was made favorable for SI mode with fuel as gasoline. The setup is displayed in Fig. 1.

The spark timing might be changed by 1^0 in the range of crank angle from 280° to 360° (0° to 80° before TDC) with ECU, based on proximity sensor input. The ratio of compression was altered by adjusting the clearance volume and clearance height simultaneously, and it was set at 10.5. In between crankcase and cylinder head, standard thickness shim plates were provided to get necessary clearance.

The intake system had a provision to place the injector at a position that fuel spray with the wall was minimum. An injection circuit with microcontroller ATMEGA 8A

Table 2 Details of engine

Category	Four-stroke, SI Engine
Version	Kirloskar, TAF1
Cylinder numbers	1
Bore × Stroke	87.5 × 110 mm
Cubic capacity	661.5
CR ratio	10.5:1
Power output	4.4 kW at 1500 rpm

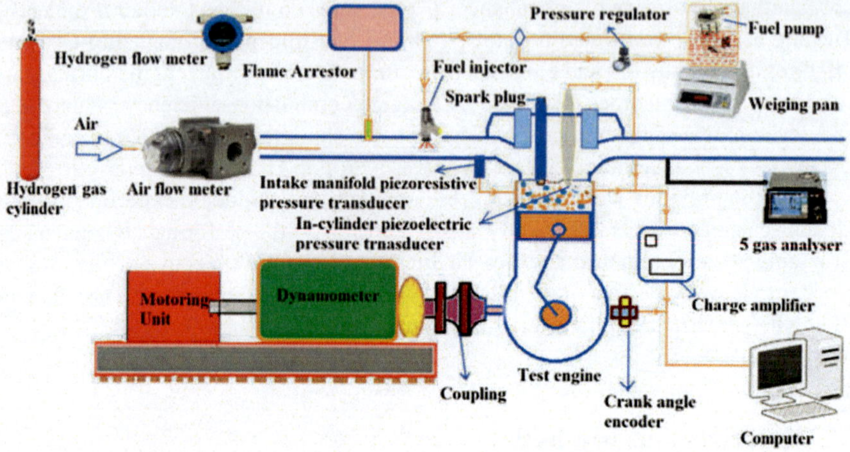

Fig. 1 Outline of setup

was prepared to produce the pulses for the appropriate amount of time to boost the injector based on the crank angle encoder input. The pulse width able to restricted up to 1 ms. The injector is capable to produce the pressure up to 3 bar with a supply voltage of 10–12 V and it was feed by the fuel pump.

The engine was combined to eddy current dynamometer, it was conditioned to run at constant speed mode to conduct tests at 1500 rpm. The throttle position sensor was linked to vary the throttle position. A five gas analyzer is used to analyze the exhaust emissions.

2.1 *Experimental Procedure*

Tests were performed at 100% throttle open condition with the help of ECU and throttle position sensor. The fuel pump was kept in the fuel tank and it is maintained at a pressure of 3 bar with the ball valve placed in the return line as displayed in Fig. 1. Flow rate of air is considered with positive displacement kind airflow meter. The fuel flow rate was calculated with a weight pan on basis of mass. The flow rates of air and fuel were useful while calculating the equivalence ratio.

The air-fuel ratio was varied from lean to rich by controlling the quantity of fuel fed to the injector. The unstable operation of the engine limits the lowest equivalence ratio. Maximum Brake Torque (MBT) spark timing was maintained at all loads. All parameters were measured after the engine reaches a steady state. Torque, speed of the engine, MBT timing, flow rates of air and fuel, pulse width modulation, temperatures of intake and exhaust and levels of exhaust emissions HC, CO, CO_2 and NO were collected at all air-fuel ratios.

3 Results

The test results were collected at 100% throttle position and 1500 rpm. The load is applied depending on the fuel inlet from lean to rich side. The emission and performance behavior was compared for all test fuels.

3.1 Analysis of Performance

Figure 2 depicts the brake power versus equivalence ratio for the test fuels. The brake power was enhanced by 5%, 9%, and 2.5% for M10 and E10 and E5M5 blends, respectively, as compared to gasoline at 0.69 equivalence ratio which is highest brake thermal efficiency position for pure gasoline. The rise in brake power with alcohol-gasoline mixtures is due to the alcohols cooling effect as the latent heat of alcohol is significantly more than pure gasoline. Hence, mixing alcohol with pure gasoline results in a cool and dense charge allowed into the engine. In addition, the high speed of alcohol flame significantly improves the rate of combustion and output brake power.

It is noted that Lean Misfire Limit (LML) was extended about 4% and 7.5% and 2% with M10 and E10 and E5M5 blends correspondingly compare with pure gasoline indicating lesser brake specific energy consumption (BSEC) as exposed in Fig. 3. Flame speed and flammability limits relatively high for alcohols compare with the gasoline so, when they were mixed with gasoline allows the engine to operate in a lean zone without any misfire and also extends the lean limit.

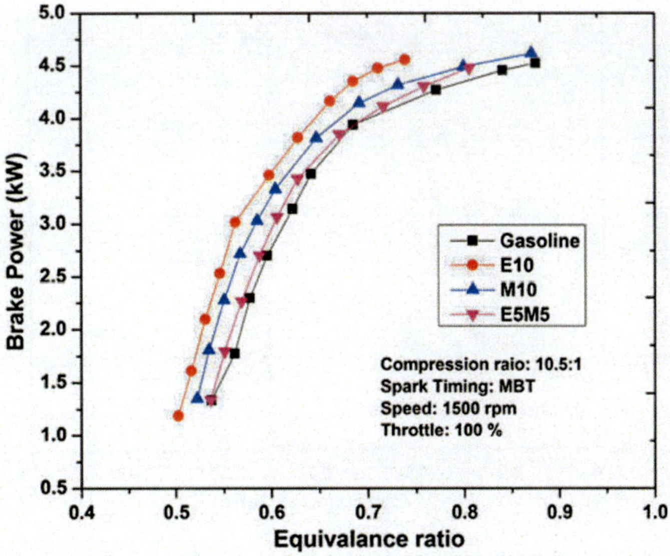

Fig. 2 Brake pressure versus equivalence ratio

Figure 4 depicts the brake thermal efficiency versus equivalence ratio for gasoline and alcohol mixtures. The brake thermal efficiency was improved by 2.5%, 7.5% and 1.5% for M10 and E10 and E5M5 blends, respectively, compared with pure

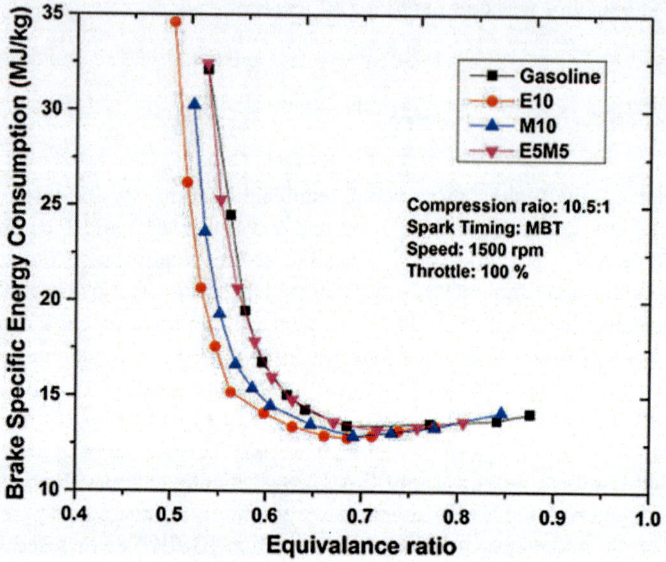

Fig. 3 Brake specific energy consumption versus equivalence ratio

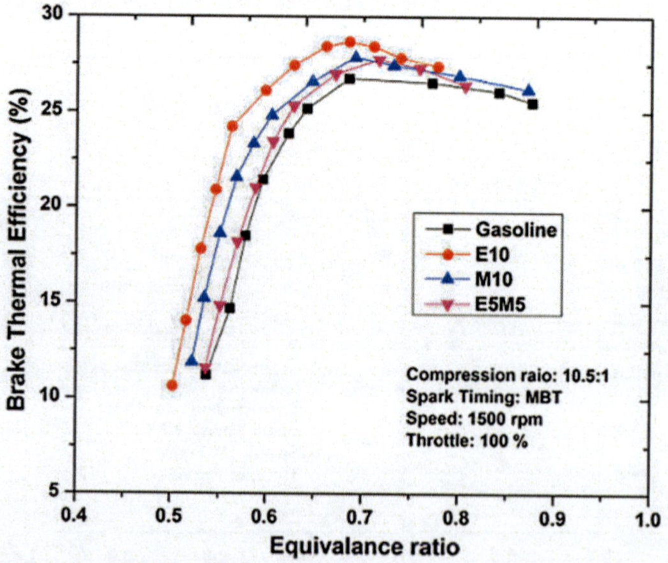

Fig. 4 Brake thermal efficiency versus equivalence ratio

gasoline at an equivalence ratio 0.69. The increment in brake thermal efficiency with alcohol blends is owing to its high latent heat vaporization relative to gasoline. The high value of the latent heat of vaporization the alcohols rise the fuel vapor, which improves the homogeneity of the fuel-air mixture and causes the efficient combustion and minimizes the loss of heat to walls of cylinder.

3.2 Emissions

Figure 5 depicts the hydrocarbon emissions with equivalence ratio for pure gasoline and alcohol blends. As the equivalence ratio approaches lower limit, a rise in hydrocarbon emissions for all test fuels owing to misfire. At the lean misfire limit of gasoline, HC emissions were reduced by 65% 75% and 60% with M10 and E10 and E5M5 blends, respectively. Further, as the mixture strength varies from lean to rich, it is observed that reduction in hydrocarbon emission levels for all test fuel blends compare with pure gasoline owing to leaning effect due to the mixing of alcohol to pure gasoline. It is also noticed that the appearance of peak hydrocarbon emission level moved toward the leaner side.

Figure 6 shows carbon monoxide emissions versus equivalence ratio for pure gasoline and alcohol blends. As all the tests are performed under conditions of lean, the carbon monoxide emissions are low with all fuels. It can be observed that there is no significant reduction in CO emission levels with the alcohol-gasoline blends because it is dependent on equivalence ratio. On the other hand, the carbon dioxide

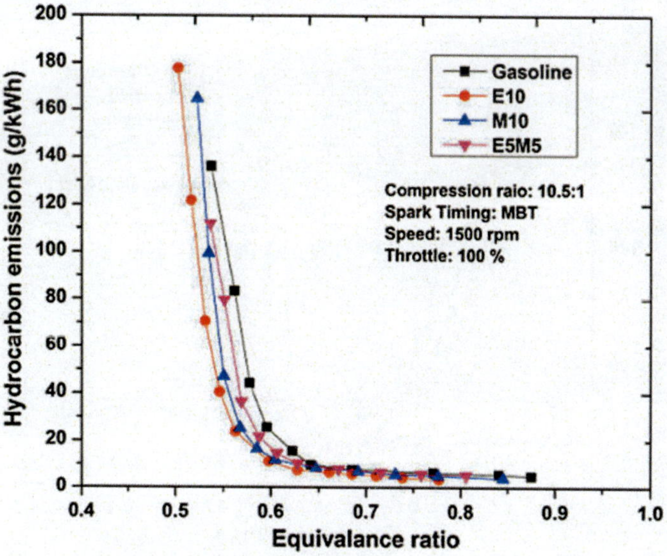

Fig. 5 Hydrocarbon emissions versus equivalence ratio

emissions as shown in Fig. 7, within the stable limit of combustion that is 0.7 equivalence ratio for the alcohol-gasoline mixtures are high compare to pure gasoline owing to complete combustion.

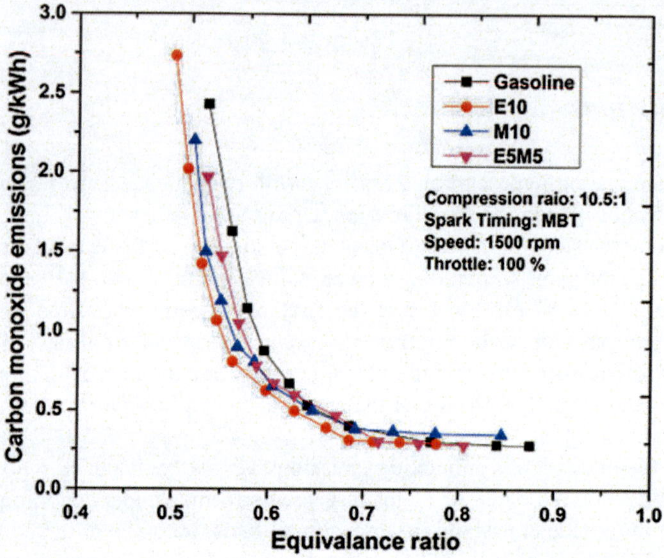

Fig. 6 Carbon monoxide emissions versus equivalence ratio

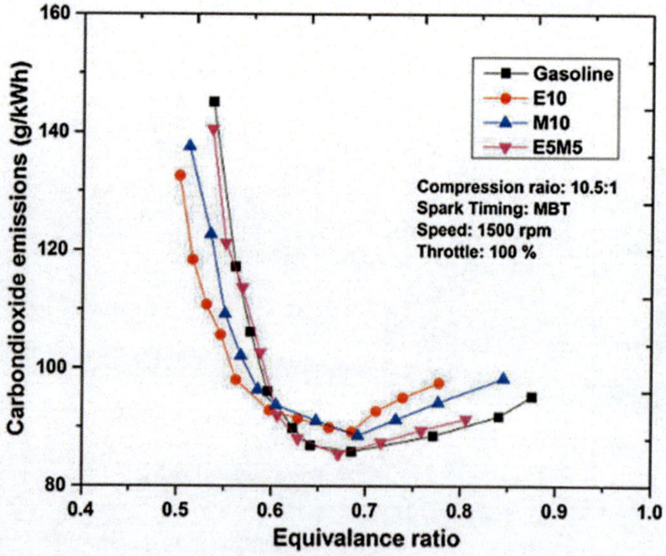

Fig. 7 Carbon dioxide emissions versus equivalence ratio

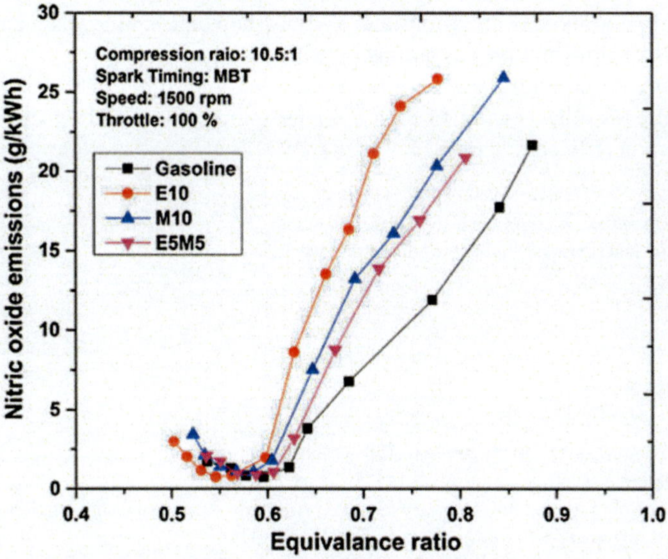

Fig. 8 Nitric oxide emission versus equivalence ratio

At high load conditions, in-cylinder temperatures are high which results in high nitric oxide emission for all fuels as observed in Fig. 8. On the other hand, as the air-fuel mixture becomes lean, a sudden drop in nitric oxide emissions for test fuels owing to in-cylinder temperature drop as a result of excess air present in the fuel mixture. Brake thermal efficiency point at 0.69 equivalence ratio, for pure gasoline, M10 and E10 blends the nitric oxide emission level is 7 g/kWh, 14 g/kWh and 17 g/kWh respectively

4 Conclusions

The performance behavior of alcohol-gasoline mixtures compared with pure gasoline. From the experiments, the following conclusions were observed.

- There was an enhancement of 5%, 9%, and 2.5% and 2.5% brake power output and 5%, 7.5% and 1.5% brake thermal efficiency with M10, E10, and E5M5 blends compare to pure gasoline. The maximum brake thermal efficiency enhanced and also the brake specific energy consumption reduced for all alcohol-gasoline blends.
- Lean limit was extended from 0.55 equivalence ratio using pure gasoline to 0.52 and 0.50 and 0.53 equivalence ratios with M10, E10, and E5M5 blend, respectively.

- The carbon monoxide emissions and hydrocarbon emission levels were less for alcohol-gasoline mixtures as compared with gasoline.
- The nitric oxide emissions were high for alcohol-gasoline mixtures compared with pure gasoline, owing to high cylinder temperatures and enhancement of combustion efficiency.

Acknowledgements Authors wish to express thanks to Vellore Institute of Technology, Vellore for the facility and financial support for doing this work.

References

1. Adem U, Semiha U (2015) A comparative analysis study of alternative energy sources for automobiles. J Hydrogen Energy 40:11178–11188
2. Salvi BL, Subramanian KA, Panwar AL (2013) Alternative fuels for transportation vehicles: a technical review. Renew Sustain Energy Rev 25:404–419
3. Muharrem E, Ahmet NO, Mustafa C, Ali T (2010) Impact of alcohol-gasoline fuel blends on performance and emission characteristics of an SI engine. Fuel 89:2713–2720
4. Sehmus A, Hakan FO, Cengiz O, Yasin V (2013) Exhaust emissions of methanol and ethanol-unleaded gasoline blend in a spark-ignition engine. Thermal Sci 17:291–297
5. Ahmet NO, Mustafa C (2011) Performance and combustion characteristics of alcohol-gasoline blends at wide-open throttle. Energy 36:2747–2752
6. Hüseyin SY, Tolga T, Can C, Inar, Melih O (2006) Effect of ethanol-gasoline blends on engine performance and exhaust emissions in different compression ratios. Appl Thermal Eng 26:2272–2278
7. Abu-Zaid M, Badran O, Yamin J (2004) Effect of methanol addition on the performance of spark-ignition engines. Energy Fuels 18:312–315
8. Tiegang H, We Y, Liu S, Longbao Z (2007) Improvement of spark-ignition (SI) engine combustion and emission during cold start, fueled with methanol/gasoline blends. Energy Fuels 21:171–175
9. Devunuri S, Porpatham E (2019) Investigations on effects of ethanol blending on performance and combustion characteristics of gasoline fuelled lean burn SI engine
10. Gravalos I, Moshou D, Gialamas Th, Xyradakis P, Kateris D, Tsiropoulos Z (2013) Emission characteristics of spark ignition engine operating on lower-higher molecular mass alcohol blended gasoline fuels. Renew Energy 50:27–32
11. Zaharin MSM, Abdullah NR, Imran DM, Hazim S, Asiah AR, Beriache M (2017) Experimental study on the effects of methanol and ethanol on gasoline engine performance and exhaust emissions. J Mech Eng 2(2):27–44
12. Taylor AB, Moran DP, Bell AJ, Hodgson NG, Myburgh IS, Botha JJ (1996) Gasoline/alcohol blends: exhaust emissions, performance and burn-rate in a multi-valve production engine. SAE Paper, 961988
13. Palmer FH (1986) Vehicle performance of gasoline containing oxygenates. International conference on petroleum-based fuels and automotive applications. In: Proceedings of international mechanical engineering conference, pp 36–46
14. Al-Hasan M (2001) Effect of ethanol-unleaded gasoline blend on engine performance and emission. Energy Convers Manage 44:1547–1561

15. Liu S, Eddy R, Cuty C, Hu T, Wei Y (2006) Study of spark ignition engine fueled with methanol/gasoline fuel blends. Appl Thermal Eng 27:1904–1910
16. Dima A, Dumitru I, Tutunea D (2017) Evaluation of performance and emission characteristics of methanol blend (gasohol) in a naturally aspirated gasoline engine. 2017 IOP conference series: material science engineering, 252 012 086

CFD Simulation Studies on Model Can-Type Combustor with Syngas and Enhanced Hydrogen Fuel Combustion

R. Ramkumar, Manoj Mannari, and A. T. Sriram

1 Introduction

The world's energy requirement has dramatically increased over few decades due to increase in world's population. Because of increase in demand for energy and declining of fossil fuels availability and along with this increase in the pollution and global warming, leading researchers and engineers are having a task of arriving renewable energy sources, improving overall efficiency of the existing power plants and arriving alternative fuels. Many renewable energy sources such as hydro, solar, wind, geothermal and bio-fuel are being considered by various researches and industries. However, various factors like the practical implementation of technologies, the initial investment, operational and maintenance cost, energy availability as compared with fossil fuels, uncertainties in continuous production rates due to environmental changes, transportation and storage requirements make these renewable energies less attractive as such now. Many techniques have been implemented for the overall improvement in the performance of the power plants used in transport vehicles as well as stationary plants. In the case of coal-fired thermal power plants, the operating pressure is increased to build super critical boilers for the enhancement of overall efficiency. At the same time, technologies such as combined cycle have evolved to use gas turbines and steam turbines in a combined way to improve overall efficiency. It is expected that the fossil fuel will remain the dominating source for power generation for at least next five decades. As an alternate to directly firing fossil fuel, there are

R. Ramkumar · M. Mannari · A. T. Sriram (✉)
Ramaiah University of Applied Sciences, Bangalore, India
e-mail: atsriram@gmail.com

R. Ramkumar
e-mail: ramkumar.vayal@gmail.com

M. Mannari
e-mail: mannari.manoj@gmail.com

A. K. Gupta et al. (eds.), *Advances in IC Engines and Combustion Technology*, Lecture Notes in Mechanical Engineering,
https://doi.org/10.1007/978-981-15-5996-9_20

advantages of generating syngas, consisting of hydrogen and carbon monoxide, and effectively utilizing it. The best example is integrated gasification combined cycle (IGCC). Here, syngas is generated from coal or natural gas, and it is fired in gas turbines. The exhaust heat of gas turbine is further used to run steam turbine. The syngas may also be derived from biomass and solid waste as alternative fuel. The main advantage of IGCC that there is not only improvement in efficiency but also possible reduction in NO_x and CO_2 emissions.

The overall design of gas turbine combustor and alternate fuels for gas turbines are discussed by Lefebvre and Ballal [1]. It is generally noticed that minor modifications are needed to existing fossil fuel-fired combustor to operate it on syngas. The composition of syngas makes considerable changes in performance and emission level. The experiments with counterflow diffusion flame configuration [2] indicated that the NO_x emissions are higher due to higher temperature. Though the presence of methane decreases the peak temperature, the prompt NO_x made to higher value of NO_x level. The diluents N_2, H_2O and CO_2 added in the oxidizer stream helped to reduce the NO_x level. In an experiment for lean premixed combustion of hydrogen–syngas/methane fuel mixture with equal amount of H_2 and CO [3] was shown that the CO and NO_x emissions were less as compared with pure methane fuel for the given flame temperature. As a practical combustor has both fuel-rich primary zone and other dilution zones, it is better to identify the temperature and species distribution with the help of computations.

The chemical kinetics plays important role in the prediction of temperature and species distribution. A four-step global reaction mechanism was proposed by Abou-Taouk [4] with the help of Siemens Industrial Turbomachinery (SIT) combustor. The effect of variation in syngas was also studied on a can combustor [5] for the emission level for given power rating. It is also noted that research on enhanced hydrogen content combustion for gas turbine applications was initiated in the recent past [6] to reduce CO_2 in exhaust gas. It is important to study the flow features in a combustor with variation in fuel. The combustor aerodynamics is responsible for flame holding, mixing and emission reduction within combustor and temperature distribution. Before considering prediction of emissions, combustor aerodynamics is considered in this study. The objective of the present work is to identify the flow characteristics and temperature distribution of a model can-type combustor for methane, syngas and enhanced level of hydrogen content fuel. The effect of flow rate is considered in this study.

2 Geometry, Mesh, Boundary Conditions and Cases for Simulation

The can-type combustor considered here is three-fourth scale model of Rolls-Royce Tay gas turbine engine. The geometrical model of the can-type combustor is shown in Fig. 1. The combustor can be divided into three parts: a hemispherical head which

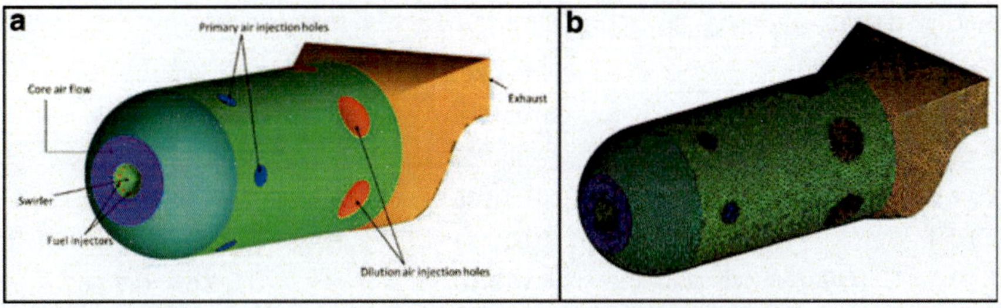

Fig. 1 Geometrical scaled-down model of Rolls-Royce Tay gas turbine engine can-type combustor. **a** Geometry **b** Grid

includes core airflow inlet from the compressor and fuel injection holes, a cylindrical portion with circumferentially placed primary and secondary cooling air holes and an exhaust portion where the geometrical transformation from cylindrical to rectangular section takes place. The total length of the combustor is 210 mm and the diameter of the mid cylindrical portion is 75 mm. Totally there are six primary air injection holes each of 10 mm diameter located at 50 mm from inlet and six secondary dilution air holes of 20 mm diameter each located at 80 mm from the primary air injection holes.

The combustor is discretized with hexahedral elements with a total count of 1.6 million. Air enters the core flow air inlet with the total mass flow rate of 0.0069 kg/s; since the swirler is not modeled separately, both axial and angular components are assigned in the boundary condition. The fuel is injected through injection holes with a total mass flow rate of 0.00176 kg/s. Air enters through primary holes with a total mass flow rate of 0.0163 kg/s and through secondary dilution holes with a total mass flow rate of 0.0768 kg/s. Turbulent intensity of 5% is considered at all the inlets and a temperature of 315 K. The hot exhaust gases exit through the rectangular nozzle with relative pressure of zero atmosphere.

Totally seven cases are considered in this study. The first case corresponds to combustion of methane fuel. The Case-2 to Case-4 is of syngas with the mole fraction of 10% of CH_4, 22.5% of CO and 67.5% of H_2. The Case-5 to Case-7 is hydrogen-enhanced fuel. The air mass flow rate is fixed in all the cases and fuel flow rate is varied for syngas and enhanced hydrogen cases. All the cases considered for the simulation are shown in Table 1.

3 Numerical Methods

All the simulations were carried out in CFX software, and the governing equations solved are mass, momentum, energy and species transport. Apart from this, to account for turbulence, k-ω model is used. It consists of partial differential equation for turbulent kinetic energy 'k' and another partial differential equation for turbulent frequency 'ω'. It is one of the widely used turbulence models for wall-bounded flows

Table 1 Various cases for simulation

Cases	Fuel	Mass flow rate (kg/s)	Reactants
Case-1	Methane	0.001760	CH_4 100%
Case-2	Syngas	0.001760	CH_4 10%, CO 22.5% H_2 67.5%
Case-3		0.001936	
Case-4		0.001584	
Case-5	Hydrogen-enhanced fuel	0.001760	CH_4 2.0%, CO 3.0% H_2 95.0%
Case-6		0.001936	
Case-7		0.001584	

in the industry because of its simplicity and accuracy for most of the flows. In general, Eddy Dissipation Model (EDM) is considered first before using EDM/finite-rate chemistry model or any other models for turbulent combustion.

EDM: This model is used in a situation where the turbulent mixing is important. The model is valid where turbulent timescales are large than the chemical timescales; that is, the Damkohler number is very large (Da $\gg$ 1). Since the air enters with swirl component into the combustor and fuel–air mixing happens due to turbulent mixing, the reactions are of mixing limited. For the RANS models, the reaction rate (net production rate) '$R_{i,r}$' is calculated for EDM through the following relation for any reaction 'r' of 'i-th' species

$$R_{i,r} = \vartheta'_{i,r} M_{w,i} A\rho \frac{\in}{k} \min_R \left(\frac{Y_R}{\vartheta'_{r,R} M_{w,R}} \right) \quad (1)$$

$$R_{i,r} = \vartheta'_{i,r} M_{w,i} AB\rho \frac{\in}{k} \frac{\sum_P Y_P}{\sum_j^N \vartheta''_{j,r} M_{w,j}} \quad (2)$$

where A and B are empirical constants with default values of 4 and 0.5.

In the case of EDM and finite-rate chemistry model, the reaction rates are estimated from both EDM and finite-rate chemistry model and the minimum of two is selected. The reaction kinetics for various fuels are listed in Table 2. As a first step, EDM model is used in the simulations. In-build NO_x models are also considered in the simulations. Since the flow features are of primary interest to this study, major flow features are presented in this paper.

For all the reactions, the pre-exponential factor has the units (m^3/mol-sec) ^ (n − 1) where 'n' is the order of the reaction, and for methane combustion, it has units of (1/s).

Table 2 Description of reactions, reaction constants and reaction rates for different fuels

Fuel	Reaction	Pre-exponential factor (A)	Activation energy (cal/mole)	β	Reaction rate (RR)
Syngas	$CH_4 + 0.5O_2 \rightarrow CO + 2H_2$	9.660840e12	35,000	0	$AT^{\beta}e^{\frac{-E_a}{RT}}[CH_4]^{0.37}[O_2]^{1.3991}$
	$H_2 + 0.5O_2 \rightarrow H_2O$	1.428428e18	40,445	−1.1682	$AT^{\beta}e^{\frac{-E_a}{RT}}[H_2]^{0.5}[O_2]^{1.1}$
	$CO + 0.5O_2 \leftrightarrow CO_2$	5.531351e14	40,735	0	$f(\varphi)AT^{\beta}e^{\frac{-E_a}{RT}}[CO_2]^{1}[O_2]^{0.5}$
	$CO_2 + H_2O \leftrightarrow CO_2 + H_2$	1.791791e12	20,845	0	$AT^{\beta}e^{\frac{-E_a}{RT}}[CO]^{1}[H_2O]^{1}$
Methane	$CH_4 + 1.5O_2 \rightarrow CO + 2H_2O$	1.50e7.0	30,000	0	$AT^{\beta}e^{\frac{-E_a}{RT}}[CH_4]^{-0.3}[O_2]^{1.3}$
	$CO + 0.5O_2 \rightarrow CO_2$	1.00e14.6	40,000	0	$AT^{\beta}e^{\frac{-E_a}{RT}}[CO]^{1}[O_2]^{0.25}$
Propane	$C_3H_8 + 3.5O_2 \rightarrow 3CO + 4H_2O$	1.00e12.0	30,000	0	$AT^{\beta}e^{\frac{-E_a}{RT}}[C_3H_8]^{0.1}[O_2]^{1.65}$
	$CO + 0.5O_2 \rightarrow CO_2$	1.00e14.6	40,000	0	$AT^{\beta}e^{\frac{-E_a}{RT}}[CO]^{1}[O_2]^{0.25}$

4 Results and Discussion

The general flow feature of the model can-type combustor is well known from the earlier experimental [7] and numerical studies [8–10]. The formation of upstream recirculation zone due to opposed jet configuration of primary jets is the essential flow feature of the combustor. In the earlier study, the mass flow distribution is adjusted to achieve the flow feature with EDM approach [9, 10]. In the present study, simulations are carried out to identify the temperature values in the recirculation zone and the strength of recirculation by plotting the axial velocity.

4.1 Effect of Syngas

The temperature distribution in the vertical mid-plane for methane and syngas combustion is shown in Fig. 2. The maximum temperature is around 2300 K for methane fuel and it is around 2490 K for syngas fuel. The temperature distribution in the core region of combustor is significantly different.

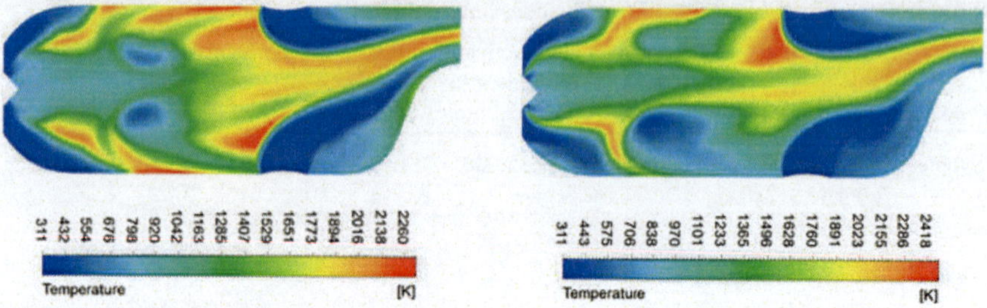

(a) Methane combustion (b) Syngas combustion

Fig. 2 Temperature distribution in vertical mid-plane for **a** Case-1 and **b** Case-2

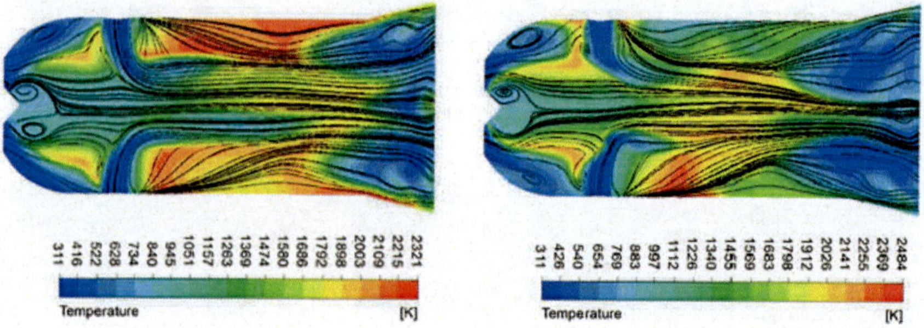

(a) Methane combustion (b) Syngas combustion

Fig. 3 Temperature distribution in horizontal mid-plane for **a** Case-1 and **b** Case-2

The temperature distribution in the horizontal mid-plane is shown in Fig. 3 for methane and syngas combustion along with superimposing streamlines. The qualitative difference between the two cases (Case-1 and Case-2) is less. Essentially, the flow is like jet in crossflow type with little higher penetration of the jet.

In order to identify the difference in temperature, line plots are made on horizontal mid-plane at an axial distance of 20 mm, 50 mm, 80 mm and 130 mm. It is noted that the overall temperatures are higher for syngas fuel considered in this study. At the axial location of $x = 20$ mm, the maximum temperature is noticed slightly away from the core region. At $x = 50$ mm, the temperature near the outer region is low. It is due to the presence of primary air holes. The temperature rises slightly away from the wall due to blockage of primary air jet against axial flow of fuel–air mixture. At the center, the temperature is slightly lower because of mixing of cold air. Figure 4c shows the temperature distribution at $x = 80$ mm, and it is 30 mm away from primary jet location. There is a jet in crossflow feature, where the streamlines move away from the wall. It is shown in Fig. 3. Basically, the axial flow of mixture is blocked near the primary air jet. It moves with horse-shoe vortex. Hence, reactions are higher in this region. Wall temperatures are relatively high in these regions. The temperature distribution is not symmetrically evolved. There is low level of unsteadiness in the

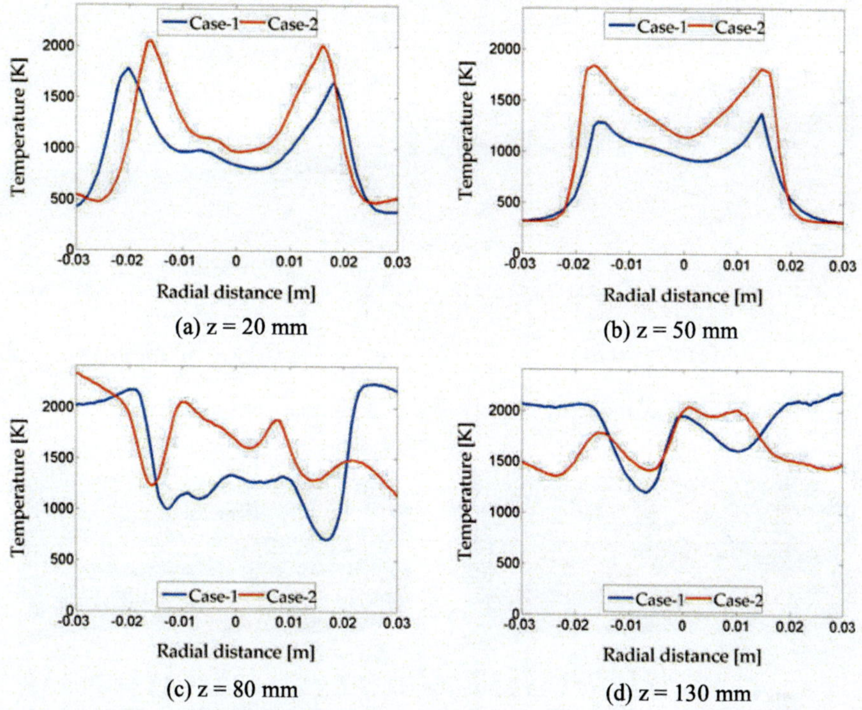

Fig. 4 Comparison of temperatures in horizontal mid-plane at axial location. Case-1 corresponds to methane and Case-2 corresponds to syngas

flow field. However, this is not changing the major flow characteristics of streamlines. As further moving away in the axial direction, at x = 130 mm, near-wall temperatures are reduced and there is little influence of secondary air. In general, the major flow is of jet in crossflow type and about to meet the opposed jet type configuration.

The axial velocity variation for the methane and syngas is compared in Fig. 5. At $x = 20$ mm, the reverse flow is slightly intense for methane fuel. It is also further seen in Fig. 5b that the axial velocity is about 58 m/s for syngas fuel as compared with maximum axial velocity about 20 m/s for the case of methane fuel.

4.2 *Effect of Mass Flow Rate of Syngas*

The mass flow rate of fuel is increased by 10% for Case-3 and 10% reduced for Case-4. The streamlines superimposed with temperature distribution for Case-3 and Case-4 are shown in Fig. 6. There is a slight difference in both the cases. The axial velocity at $x = 20$ mm and $x = 50$ mm for Case-2 to Case-4 is shown in Fig. 7. It is noticed that there is a slight shift for the case of lower mass flow rate at $x = 20$ mm.

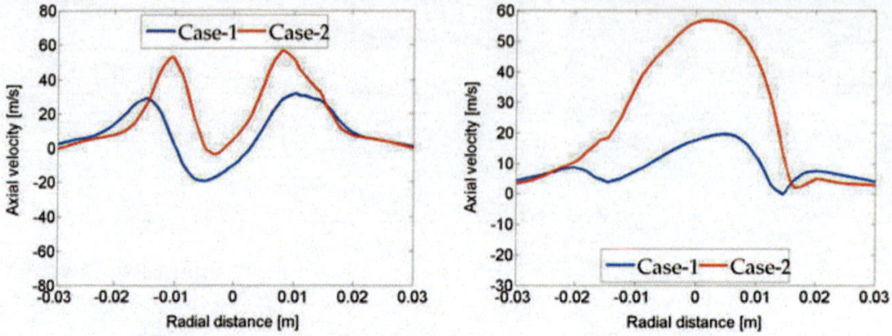

(a) z = 20 mm (b) z = 50 mm

Fig. 5 Comparison of axial velocity at axial locations for methane and syngas

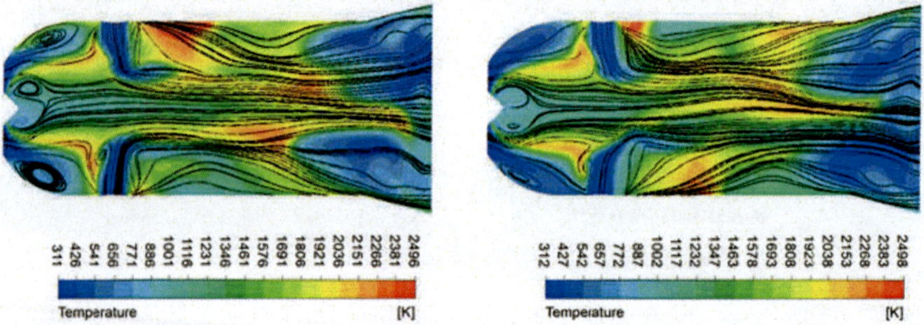

(a) Syngas with 10% more fuel flow rate (b) Syngas with10% less fuel flow rate

Fig. 6 Temperature distribution in horizontal mid-plane **a** for Case-3 and **b** Case-4

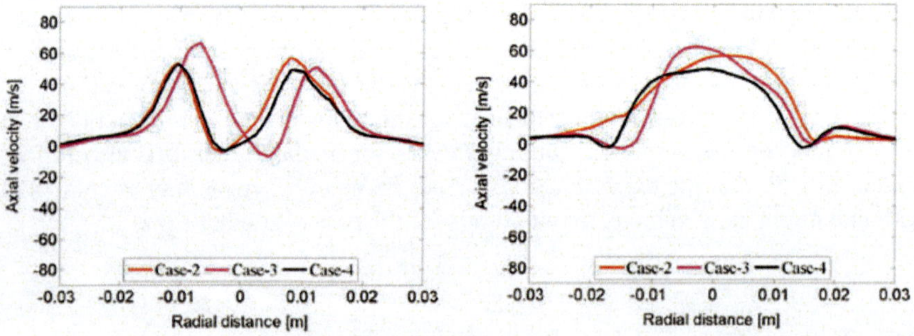

(a) z = 20 mm (b) z = 50 mm

Fig. 7 Comparison of axial velocity at axial locations

However, at the primary injection location, $x = 50$ mm, the overall trend seems to be consistent for all the three cases.

4.3 *Effect of Enhanced H2 Fuel*

The temperature distribution for enhanced hydrogen fuel case (Case-5) in the horizontal and vertical mid-plane is shown in Fig. 8. As compared with methane and syngas combustion (Figs. 2 and 3), the enhanced hydrogen fuel combustion shows stronger upstream recirculation. The flow characteristics change from jet in crossflow to opposed jet configuration, as clearly seen from streamline pattern in horizontal mid-plane. It indicates that energy content and fuel properties change the combustor aerodynamics. Once the flow characteristics change, the temperature distribution also changes. The opposed jet configuration is favorable for higher temperature at the front portion of spherical head. The high temperature zone in the core of syngas combustion describes that the combustor volume is effectively utilized (Fig. 9).

The temperature distribution along the line $z = 20$ mm and $z = 50$ mm is compared for methane and enhanced hydrogen fuel cases. It clearly indicates that the wall temperature reaches higher value for hydrogen fuel. The combustor aerodynamics allows to have intense reaction in the primary zone area.

The velocity distribution for Case-1 and Case-5 is compared in Fig. 10. The high value of negative velocity of about 60 m/s at $z = 20$ mm for Case-5 is a typical characteristic of opposed jet configuration and it helps for flame holding.

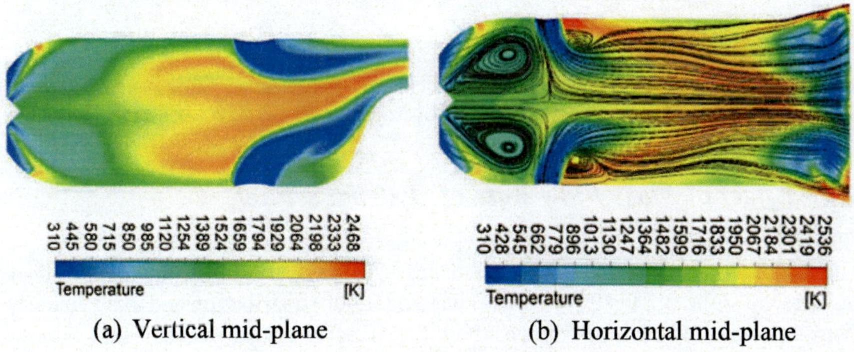

Fig. 8 Temperature distribution for the case of enhanced hydrogen fuel (Case-5)

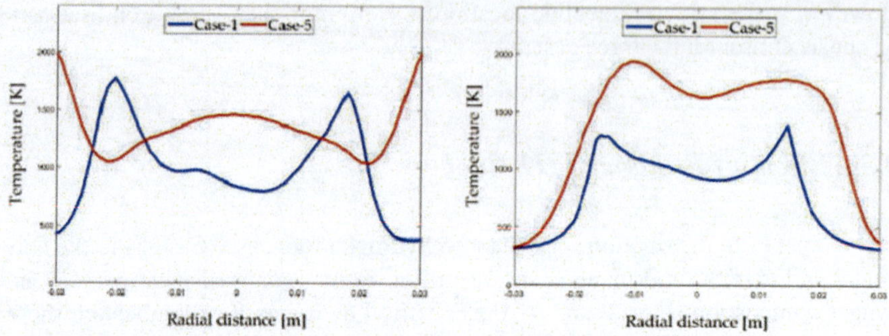

(a) z = 20 mm (b) z = 50 mm

Fig. 9 Comparison of temperatures in horizontal mid-plane at axial location. Case-1 corresponds to methane and Case-5 corresponds to enhanced hydrogen fuel

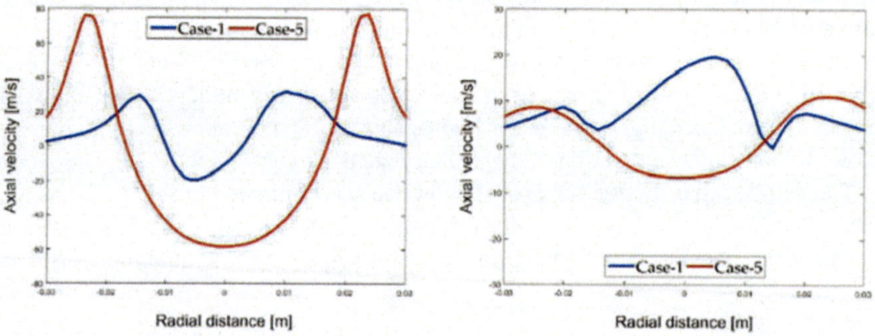

(a) z = 20 mm (b) z = 50 mm

Fig. 10 Comparison of axial velocity in horizontal mid-plane at axial location. Case-1 corresponds to methane and Case-5 corresponds to enhanced hydrogen fuel

4.4 Effect of Mass Flow Rate of Hydrogen Fuel

The fuel flow rate is varied in Case-5 to Case-7. It is observed that the overall flow features are similar in the three cases. The temperature distribution and axial velocity distribution at $x = 50$ mm are shown in Fig. 11. It clearly indicates that the temperature decreases for lower fuel flow rate. Also, the strength of reverse flow decreases with decreasing fuel flow rate.

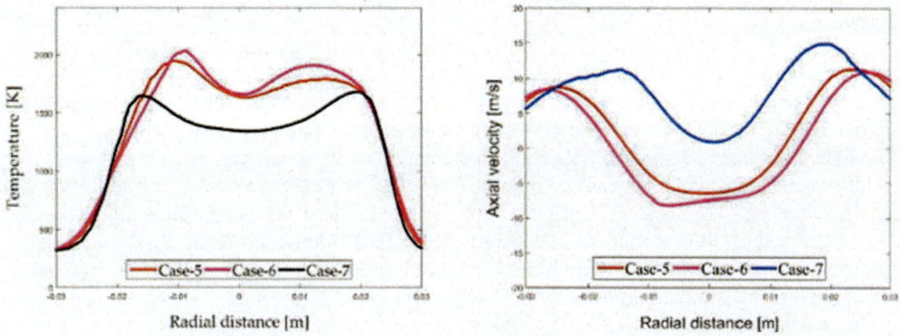

(a) Temperature (b) Axial velocity

Fig. 11 Comparison of temperature and axial velocity in horizontal mid-plane at axial location $x =$ 50 mm

5 Conclusions

The trend in thermal power plant is to use combined gas and steam turbines to enhance the efficiency. Coal and biomass gasification is one such way of providing syngas to gas turbines. It is also thought of excess power from renewable energy to produce hydrogen fuel. Design modification for existing natural gas-fired gas turbine engines with syngas and hydrogen fuel is of current interest. A well-known three-fourth scale model of Rolls-Royce can-type combustor is considered. Numerical simulations are carried out with ANSYS CFX software. Totally seven cases are considered in this study with methane, syngas and enhanced hydrogen fuel to simulate the flow features.

It is observed that the selected syngas and methane fuel combustion has shown similar flow characteristics with slightly higher value of temperatures. The primary jet flow in the horizontal mid-plane is of jet in crossflow type. In such case, the strength of upstream recirculation is lower and the temperature in the front portion of the combustor is relatively low. The hydrogen-enhanced fuel has shown significant changes in flow structure as compared with methane and selected syngas fuel. The recirculation in the primary zone is more and the streamlines show opposed jet configuration. The wall temperature at the initial portion of the combustor is high as compared with other fuels.

Acknowledgments The authors acknowledge publication committee of Ramaiah University of Applied Sciences for permitting to publish this paper.

References

1. Lefebvre AH, Ballal DR (2013) Gas turbine combustion: alternative fuels and emissions, 3rd edn. CRC Press, Taylor and Francis Group, Boca Raton (UAS, First Indian Reprint)
2. Giles DE, Som S, Aggarwal SK (2006) NOx emission characteristics of counterflow syngas diffusion flames with airstream dilution. Fuel 85(12–13):1729–1742
3. Alavandi SK, Agrawal AK (2008) Experimental study of combustion of hydrogen-syngas/methane fuel mixtures in a porous burner". Int J Hydrogen Energy 33(4):1407–1415
4. Abou-Taouk A, Sigfrid IR, Whiddon R, Eriksson LE (2012) A four-step global reaction mechanism for CFD simulations of flexi-fuel burner for gas turbines. In ichmt digital library online. Begel House Inc
5. Ghenai C (2010) Combustion of syngas fuel in gas turbine can combustor. In: Advances in mechanical engineering, Article ID 342357
6. Bancalari E, Chan P, Diakunchak IS (2007) advanced hydrogen gas turbine development program. ASME. Turbo expo: power for land, sea, and air, vol 2, Turbo Expo 2007
7. Bicen AF, Tse DGN, Whitelaw JH (1990) Combustion characteristics of a model can-type combustor. Comb Flame 80:111–125
8. di Mare F, Jones WP, Menzies KR (2004) Large eddy simulation of a model gas turbine combustor. Comb Flame 137:278–294
9. Sriram AT (2013) Numerical study of swirl of radial jets of a model can-type combustor. In: 23rd National conference on I. C. engine and combustion, SVNIT, Surat, India, 13–16 Dec 2013
10. Sriram AT, Shankapal SR (2018) Flow characteristics of a model can-type combustor. In: National conference on aerospace technologies, NCAT-2018, Hindustan Institute of Technology and Science, Chennai. 27th–28th Sept 2018

Experimental Investigation on Performance and Emission Characteristics of Lime Treated and Preheated Biogas

A. Murugesan, A. Avinash, and D. Subramaniam

1 Introduction

Biogas is a renewable energy source that can be utilized in both compression ignition (CI) and spark ignition (SI) engines [1, 2]. Biogas is a mixture of different gases produced by the breakdown of organic wastes (agricultural waste, manure, and municipal waste and plant wastes) in the absence of oxygen (anaerobic digestion) with anaerobic bacteria by the process of fermentation. It consists of 60% methane, 29% carbon dioxide and small amounts of hydrogen sulfide (H_2S) [3]. In fact, mixing of conventional diesel with biogas at various rates helps in achieving lesser fuel consumption where diesel is to be used as pilot fuel because the auto-ignition temperature of diesel in CI engine is 553 K, whereas for biogas it is two times higher [4]. Thus, biogas needs liquid fuel to start the combustion. Today, 85% of diesel can be substituted by using biogas [5]. It is also found that biogas replacement can be 90% by mass [6]. The characteristics of CI engine operated on biogas–diesel in dual fuel mode are listed in Table 1.

A CI engine can be modified to operate in dual fuel mode by controlling the pilot fuel supply and installing a simple gas mixer at inlet manifold [8]. The extracted biogas when admitted into the diesel engine at airline using venturi gas mixer causes corrosion in combustion chamber and provides low efficiency and high emissions because of the presence of carbon dioxide in biogas [9]. This is because a part of

A. Murugesan (✉)
Department of Mechanical Engineering, K.S.Rangasamy College of Technology, Tiruchengode, Tamilnadu 637215, India
e-mail: murugesanhema2000@yahoo.co.in

A. Avinash
GMR Institute of Technology, Andhra Pradesh, India

D. Subramaniam
Mahendra Engineering College, Tamil Nadu, India

A. K. Gupta et al. (eds.), *Advances in IC Engines and Combustion Technology*, Lecture Notes in Mechanical Engineering,
https://doi.org/10.1007/978-981-15-5996-9_21

Table 1 Characteristics of CI engine operated on dual fuel mode [7]

Pilot fuel	Gaseous fuel	Engine used	Engine characteristics		
			Combustion	Performance	Emission
Diesel	Biogas	Lister–Petter, single-cylinder, four-stroke, NA, air-cooled, diesel engine	Lower peak cylinder pressure Longer ignition delay	Higher BTE	Reduced HC, CO, CO_2, NO_x and smoke emission
Diesel	Biogas	Single-cylinder, four-stroke, water-cooled, diesel engine	–	Higher power output Higher BTE at low load Higher SFC	Higher HC and CO emission
Diesel	Biogas	Kirloskar TAF 1, single-cylinder, four-stroke, NA, air-cooled, CI engine	–	Higher BSEC Lower EGT	Higher HC, CO emission Lower NO, smoke emission
Diesel	Biogas	Kirloskar TAF 1, single-cylinder, four-stroke, NA, air-cooled, CI engine	Higher peak cylinder pressure Higher heat release rate Longer ignition delay Increased combustion duration	Higher BSFC Lower BTE Lower VE Lower EGT	Higher HC, CO emission Lower NO_x, CO_2, smoke emission

biogas contains impurities and moisture which lowers the calorific value of biogas. In order to overcome these problems, many technologies have emerged but they are chemical processes that are more expensive. So, the extracted biogas can be treated with lime which is natural and makes it convenient for use. As reported by Wadud et al. [9], lime treatment of biogas can significantly reduce the emissions during combustion.

The reaction takes place during lime treatment is, $Ca(OH)_2$ (aq) + CO_2 $CaCO_3$ (s) + H_2O [9].

In addition, heat treatment carried out on the lime treated biogas changes its methane concentration and removes moisture which helps in further increase in thermal efficiency at higher compression ratio as reported by Karellas et al. [5]; Wadud et al. [10]; and Violeta et al. [11] (Table 1). In this work, biogas is made to pass through the inlet manifold of the engine with air. Overall, this study deals with the comparative analysis of performance and emission characteristics of lime treated and preheated biogas with the raw biogas at various compression ratios to attain a better result.

2 Engine Experimental Setup

In this work, a Kirloskar single-cylinder (3.7 kW), four-stroke, naturally aspirated, water-cooled variable compression ratio (VCR) diesel engine (12:1 to 19:1) was used to evaluate performance and emission characteristics of diesel with raw biogas and diesel with lime treated–preheated biogas at VCR. The photographic view of experimental setup is shown in Fig. 1.

To vary the compression ratio, a tilting cylinder block setup is used without stopping the engine and without altering the combustion chamber geometry. An eddy current dynamometer is connected to engine crankshaft for loading. The venturi gas mixer is used to mix the raw biogas and treated biogas separately with atmospheric air, and the flow rate is controlled by a gate valve. The fuel was injected into the engine from the diesel tank which is placed in the panel board. The burette setup was used to find the time taken for 50 CC fuel consumption. An AVL 444 DI gas analyzer and AVL 437 standard smoke meter were used to measure exhaust gas emissions. All the experiments were conducted at rated engine speed (1500 rpm). The temperature of the biogas was maintained at 100 °C temperature.

Fig. 1 Photographic view of experimental setup

3 Result and Discussion

3.1 Pilot Fuel Replacement (PFR)

The variation of PFR with respect to load for raw and treated biogas is shown in Fig. 2a, b. At all CR, the PFR increases with an increase in load because the in-cylinder combustion temperature is high at high loads than at low loads which help the biogas to combust better. At maximum load (100%) with CR 18, the PFR is found

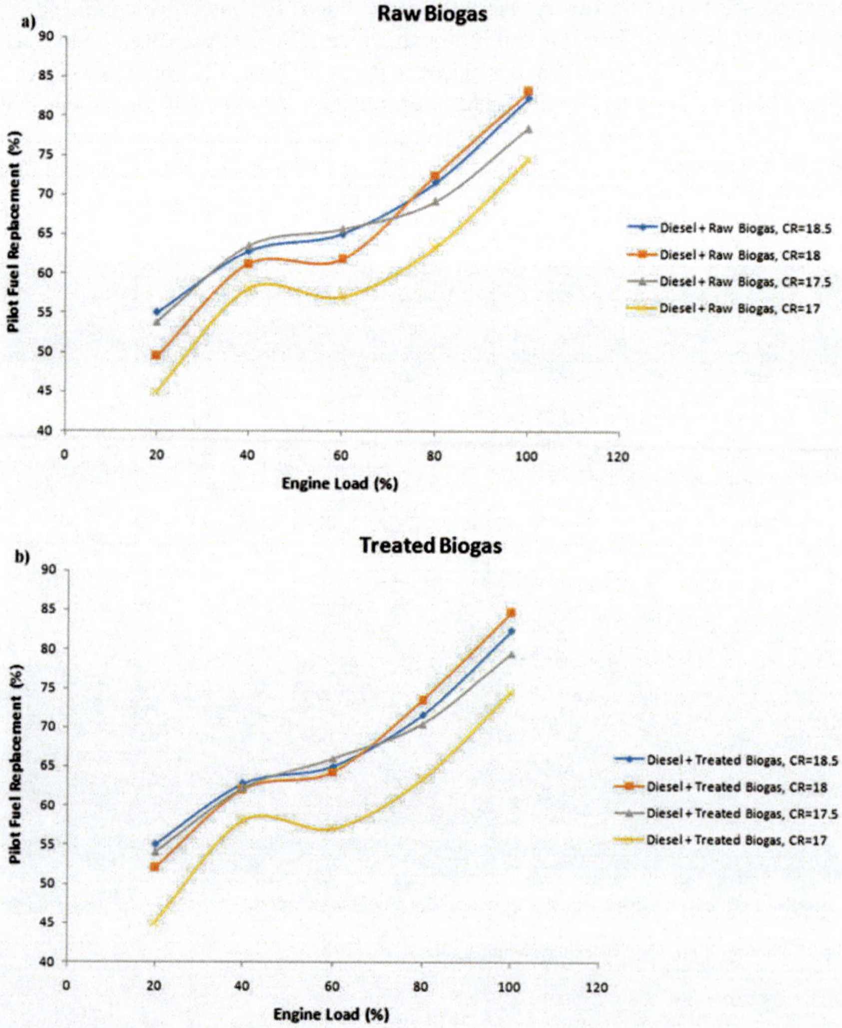

Fig. 2 **a** Pilot fuel replacement of raw biogas, **b** pilot fuel replacement of treated biogas

to be 83% for raw biogas, and for treated biogas, it is found to be 84.7%. Thus, there is a 1.7% increase in PFR for treated biogas in comparison with raw biogas.

3.2 Brake Thermal Efficiency (BTE)

Some factors like low calorific valve, low flame propagation speed, combusted residual, low combustion temperature, higher fuel flow rate and negative compression work due to high biogas mixture lead to lower thermal efficiency of biogas in dual fuel mode when compared to diesel mode [10]. On burning of lime treated–preheated biogas, the brake thermal efficiency increases as CR increases. It is found that at full load, the BTE of diesel is 27.96%, whereas for raw biogas (Fig. 3a) at CR 17, 17.5, 18 and 18.5 the calculated BTE is 17%, 18.5%, 20% and 19%, respectively. Also, for treated biogas (Fig. 3b), the corresponding BTE is 17.5%, 19%, 20.5% and 19% at CR 17, 17.5, 18 and 18.5, respectively. This shows that there is a slight increase in BTE for treated biogas than raw biogas due to the changes in concentration of methane and removal of moisture as reported by Wadud et al. [10]. Though the biogas produces less BTE than diesel, it is a renewable source.

3.3 Carbon Monoxide (CO) Emission

Incomplete combustion in an engine leads to the formation of carbon monoxide due to oxygen insufficiency. This gas is too toxic which affects humans and also the environment. The variation of CO with respect to engine load for raw and treated biogas is shown in Fig. 4a, b. This study on CO emission analysis depicts that on higher CR the emission of CO decreases during dual fuel mode because there is better combustion at higher CR. But when compared to dual fuel operation, lesser CO emission was observed in diesel mode. This is due to the flammability limits of biogas than diesel [6].

At higher CR, the temperature of the compression stroke is higher which leads to complete combustion. Lower CR produces high emission of CO than higher CR. It is possible to obtain a minimum amount of CO emission of average 3% by burning of lime treated and preheated biogas at variable CR from 17 to 18.5 in dual fuel mode than raw biogas. Comparing the various CRs, the treated gas with a moderate CR of 18 provides better results in terms of emission than the others.

3.4 Carbon Dioxide (CO_2) Emission

This study shows that the emission of carbon dioxide is more in dual fuel mode than diesel mode because of larger volume of CO_2 present in biogas [8]. The combustion of

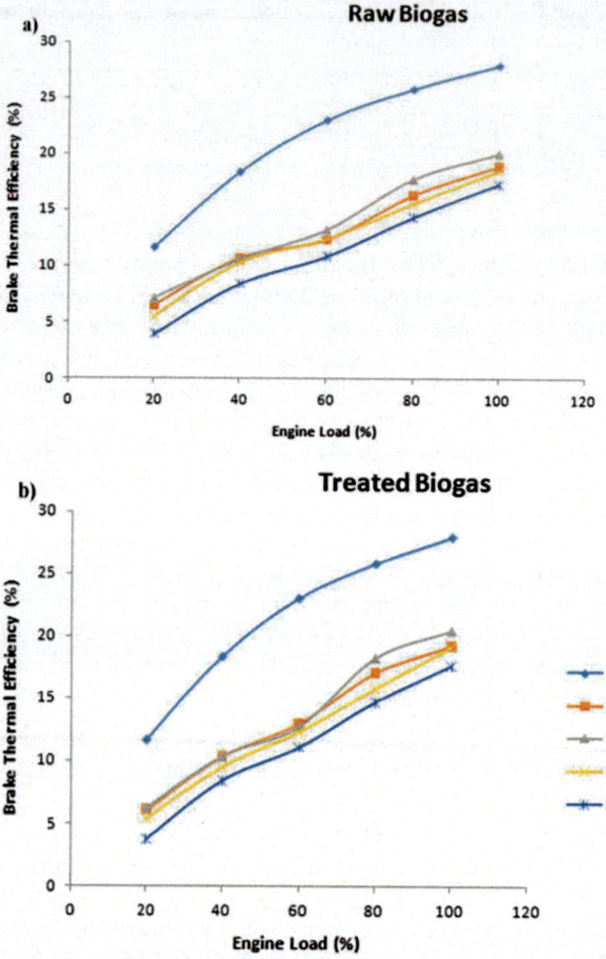

Fig. 3 **a** Brake thermal efficiency of raw biogas, **b** brake thermal efficiency of treated biogas

hydrocarbon fuel results in the production of CO_2 and water at ideal condition. When the quantity of inducted fuel increases, the CO_2 emission also increases at both modes of operation. The variation of CO_2 with respect to load for raw and treated biogas is shown in Fig. 5a, b. At higher compression ratio, the CO_2 emission increases. This is because the clearance volume decreases with increase in CR which in turn increases the temperature and pressure inside the cylinder at the end of the compression stroke, thus resulting in better combustion of air and fuel. Simultaneously, the CO_2 emission increases at higher CR and decreases at lower CR in dual fuel mode. From the experiments, it was possible to obtain 1% increase in amount of CO_2 emission by burning of lime treated and preheated biogas at variable CR from 17 to 18 in dual fuel mode than raw biogas. The results attained from the analysis are plotted in the

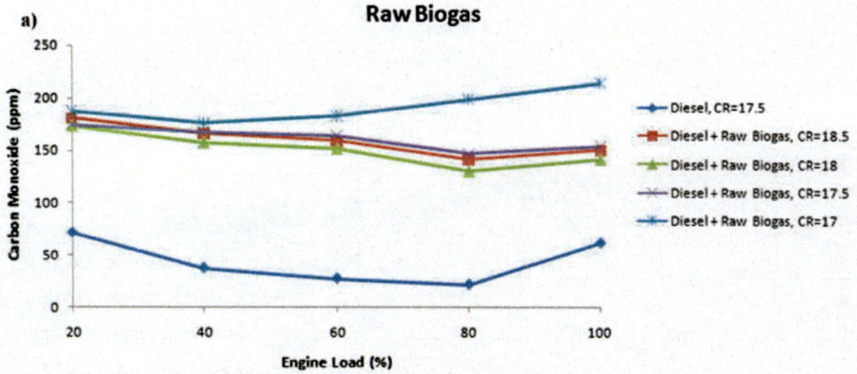

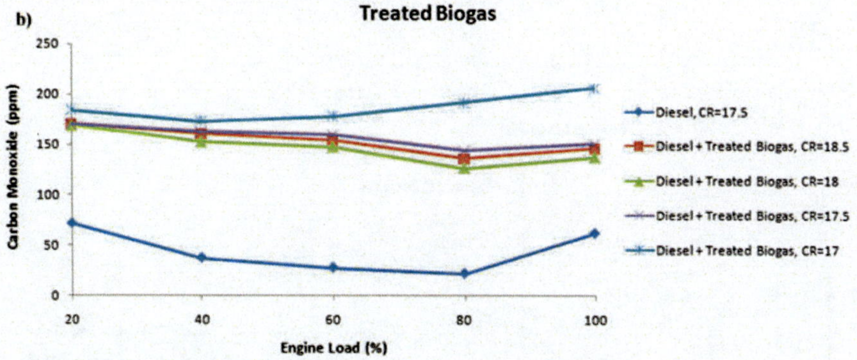

Fig. 4 **a** CO emission of raw biogas, **b** CO emission of treated biogas

form of graph, which deduces that the treated biogas with a compression ratio of 18 provides a moderate emission of CO_2 than the other ratios.

3.5 Hydrocarbon (HC) Emission

The incomplete combustion results in emission of unburnt hydrocarbons which can adversely affect the environment. The variation of HC emission with respect to load for raw and treated biogas is shown in Fig. 6a, b. The emission analysis of the raw and treated biogas (Fig. 6a, b) shows that the HC emission significantly reduces with increases in CR as seen in CO emission analysis. This is due to the lower flammability limits of biogas than diesel. It is possible to obtain a minimum amount of hydrocarbon emission of average 2.5% by burning of lime treated and preheated biogas at variable CR from 17 to 18. Similar observation was observed in the previous study conducted by Yoon et al. [8].

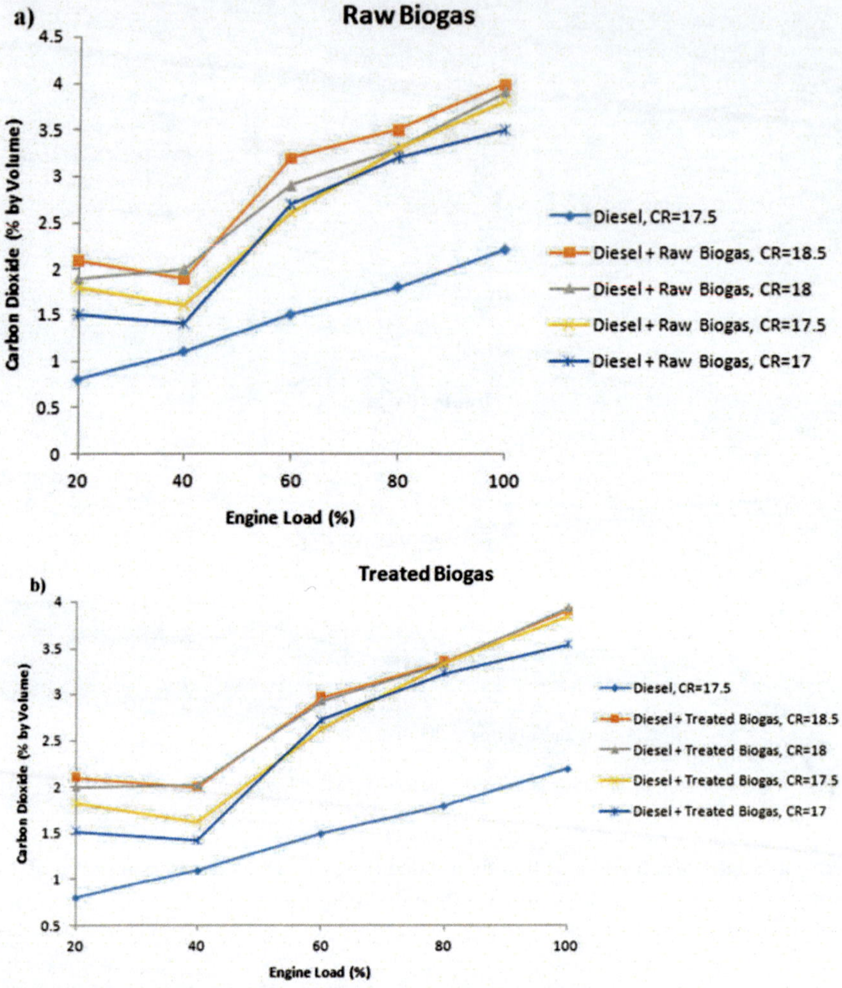

Fig. 5 a CO_2 emission of raw biogas, b CO_2 emission of treated biogas

3.6 *Oxides of Nitrogen (NO_x)*

NO_x is a toxic gas formed through oxidation of nitrogen in the intake air at high in-cylinder combustion. The composition of NO_x is NO and small amount of NO_2. When load increases, the amount of fuel also increases which results in increase in temperature inside the combustion chamber. Thus, the NO_x formation depends on the temperature of the combustion chamber [12]. On the other hand, the presence of residual gases (especially CO_2) in biogas reduces the in-cylinder combustion temperature. The variation of NO_x with respect to load for raw and treated biogas is shown in Fig. 7a, b. From this study, the NO*x* emission increases with load and

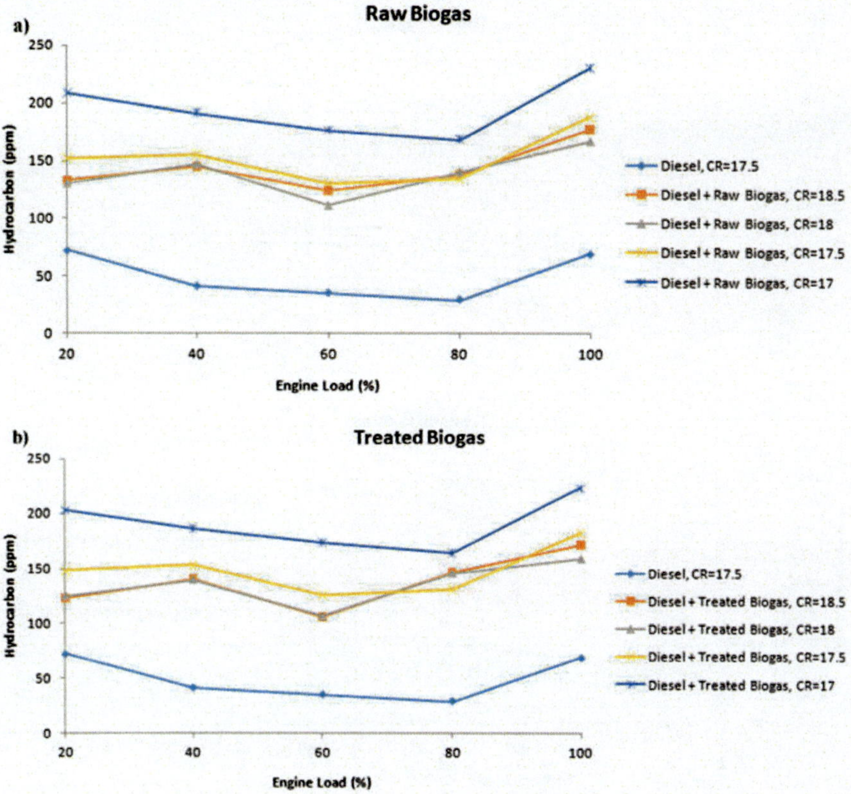

Fig. 6 **a** HC emission of raw biogas, **b** HC emission of treated biogas

fuel, and it is lower for CR 17 and higher for CR 18.5. A moderate NO*x* emission is obtained on a CR of 18 than the other compression ratios. It is possible to obtain a 1% increase amount of NO_x emission of average by burning of lime treated and preheated biogas at variable CR from 17 to 18. This increase in NO_x can be reduced by exhaust gas recirculation (EGR) method.

3.7 *Smoke Emission*

Figure 8a, b shows the variation of smoke in HSU with the load for diesel, biodiesel and lime treated biogas. For all test fuels, the smoke increases with the increase in load. The higher smoke emission indicates that the combustion is incomplete. The higher smoke emission is primarily due to the shorter delay period. The smoke emission of treated and raw biogas decreased on an average up to 10% when compared

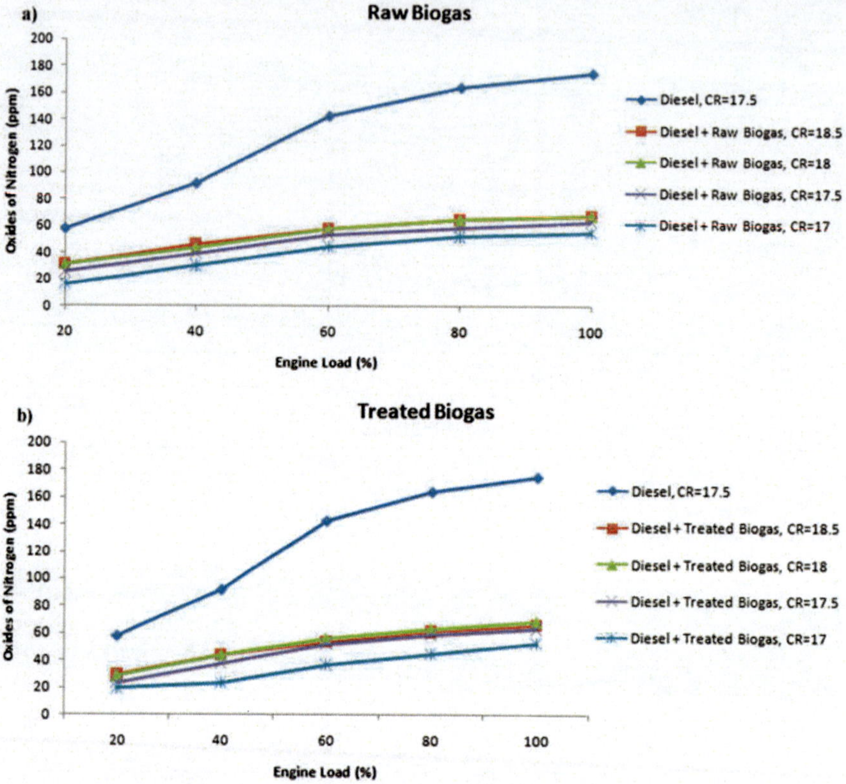

Fig. 7 **a** NO_x emission of raw biogas, **b** NO_x emission of treated biogas

to neat diesel. This is because of the increase in delay period which enhanced the fuel–air mixing, thus leading to reduction in smoke emissions.

4 Conclusion

This experiment was conducted to recognize the effect of lime treated and preheated biogas on the dual fuel engine characteristics at variable compression ratios. Single-cylinder, four-stroke, water-cooled, DI and variable compression ratio diesel engine at standard IT 23° BTDC was used to conduct this experiment. A moderate performance and emission are obtained in the treated biogas at compression ratio of 18 than the higher values of CR. Higher values of CR show lower emission. When we look at emission part, there is increase in NO*x* by 1% than raw biogas. There is reduction in hydrocarbon and carbon monoxide by 2.5% and 3% at CR 18 in comparison with raw biogas. From the result, it can be concluded that the emission characteristics of a biogas-run dual fuel mode are depend on concentration of methane, impurities,

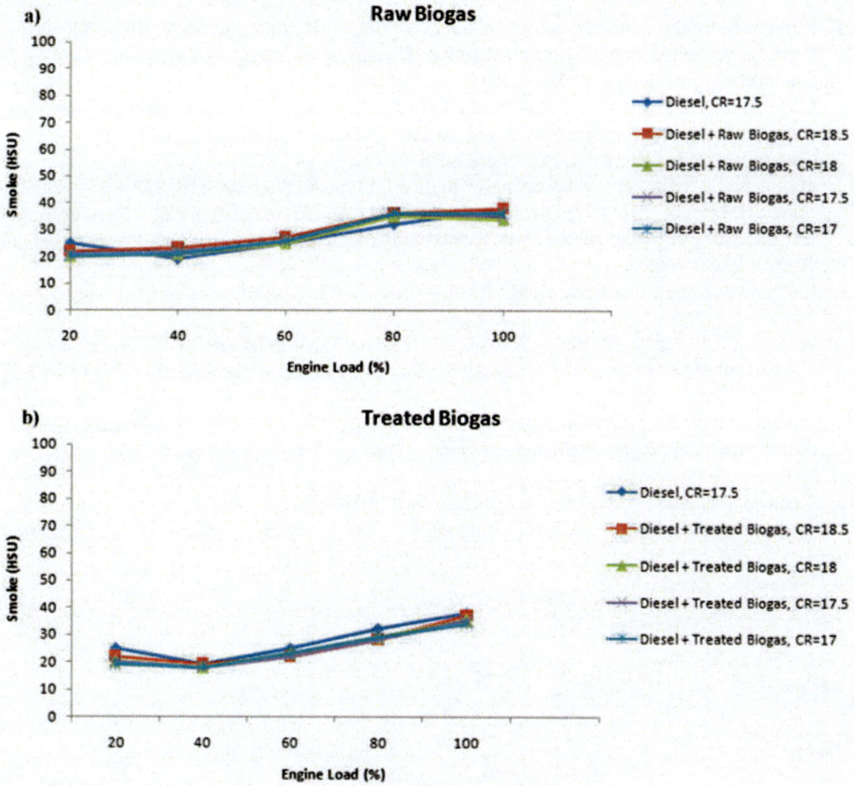

Fig. 8 **a** Smoke emission of raw biogas, **b** smoke emission of treated biogas

viscosity and moisture in biogas and value of CR of the diesel engine. Thus, the results depict that the use of treated biogas with a CR of 18 provides better results in terms of emission aspects. Also, it is possible to improve the performance of dual fuel mode by using lime treatment and preheated biogas instead of using raw biogas.

References

1. Gajendra Babu MK, Subramniam KA (2013) Alternative transportation fuels: utilisation in combustion engines. CRC press. ISSN: 13: 978-1-4398-7282-6 2013
2. Chhatwal GR, Mehra MC (1989) Environmental air pollution and its control. Anmol Publications, New Delhi, India, p 284
3. Nutt WJ, Manz DL (2008) A new energy security paradigm for the twenty first century technological forecasting and social change 75:1247–1259
4. Murphy JD, Mccarthy K (2005) The optimal production of biogas for use as a transport fuel in Ireland. Renew Energy 30:2111–2127

5. Karellas S, Boukis I, Kontopoulos, Chandra GR, Vijay VK, Subbarao PMV, Khura TK (2011) Performance evaluation of a constant speed IC engines on biogas, methane enriched biogas and biogas. Appl Energy 88:3969–3977
6. Galal MG, Abdel MM, Aal MA, Kady EL (2002) comparative study between diesel and dual-diesel engines: performance and emissions. Combust SciTechnol 174:241–256
7. Chinmay D, Saroj Kumar A, Dhirendra Nath T, Ambica M (2020) A review on performance of biogas and hydrogen on diesel engine in dual fuel mode. Fuel 260:116–337
8. Yoon SH, Lee CS (2011) Experimental investigations on the combustion and exhaust emission characteristics of biogas-biodiesel dual-fuel combustion in a CI engine. Fuel Process Technol 92:992–1000
9. Packyam S, Angelselva S, Raja kumar (2013) Flow Anal Biogas Burner Domestic Appl 3:656–668
10. Wadud AMA, Zaman T, Fazlay Rabbee M, Rajibur Rahman M (2013) Renewable energy: an ideal solution of energy crisis and economic development in Bangladesh. Glob J Res Eng 13:1–11
11. Violeta M, Egle S, Saugirdaspukalskas AR, Ricardas V (2013) Performance and emission characteristics of biogas used in diesel engine operation. Energy Conservation Manage 75:224–233
12. Agrawal Ak, Singh SK, Sinha S, Shukla MK (2004) Effect of EGR on the exhaust gas temperature and exhaust opacity in compression ignition engines. Sadhana 29(3):275–284

An Experimental Investigation on DI-CI Engine Characteristics Fueled with Green Synthesized Nanoparticle Doped with Biodiesel Blends

A. Murugesan, R. Prakash, and A. Kumaravel

1 Introduction

In India, the transportation sector acting a dominant role in monetary growth [1]. For a heavy vehicle, diesel engine is well suitable for huge amount of torque is required for long period of time and that probably used for the power generations and as well as buses, trucks, ships, and locomotives [2]. The cost of living keeps on rising because fossil fuels have more demand than there supply, which leads to energy disaster [3]. Due to the limited fossil fuel resources, biodiesel is been concentrated [4]. The benefits of using biodiesel are eco-friendly which environmental aspects will not be affected, market availability, renewable energy source, compare to gasoline the cost of biodiesel is low [5]. Using various vegetable seeds like edible and non-edible, the oil is been extracted. Biodiesel production from the oils of non-edible seeds is sustainable alternative which is recommended by many researchers who are best in that field. In order to find the best alternative fuel, several non-edible crops are recommended which comprise of neem, jatropha, pungai, mahua, neem, pignut, kusum, cottonseed, rubber seed, and flaxseed [6, 7]. As per the study, the chemical composition of neem oil is comparatively same as other non-edible seeds, so, green neem oil is chosen for the alternative fuel of diesel engines [8].

Neem oil is gained from seeds of Azadirachta Indica (Neem tree) is a deciduous and evergreen tree that typically develops in tropical and semi-tropical areas [9]. The

A. Murugesan (✉) · R. Prakash · A. Kumaravel
Department of Mechanical Engineering, K.S.Rangasamy College of Technology, Tiruchengode, Tamilnadu 637215, India
e-mail: murugesanhema2000@yahoo.co.in

R. Prakash
e-mail: velprakashr@gmail.com

A. Kumaravel
e-mail: kumaraveliitm@yahoo.com

A. K. Gupta et al. (eds.), *Advances in IC Engines and Combustion Technology*, Lecture Notes in Mechanical Engineering,
https://doi.org/10.1007/978-981-15-5996-9_22

meal versus fuel clash will not be arising because the ester from neem oil acts like a diesel engine alternative eco-friendly fuel [10]. This oil is easily available in Indian open market [11]. Basically, neem oil has high viscosity used as biodiesel in diesel engines and arising complications such as low fuel atomization, incomplete combustion, engine fouling, carbon deposit forming, and lubrication oil contaminating. The methods like direct use and blending, Transesterification, Pyrolysis (cracking), and Microemulsion are been followed for lessening the viscosity of neem oil. Almost all biodiesel production industries have followed transesterification method.

The disadvantage of using biodiesel in diesel engines has slightly greater density, a slight decrease in fuel-consumption, lower pour point, cloud point and increase oxides of nitrogen emission. The above-listed limitations are averted by several methods like modified fuel, hybrid fuel, and fuel additives which result leads to advancement in engine performance and reduced adverse emissions [12]. To promote superior fuel properties, improve efficiency of combustion, and lessen adverse emissions with the inclusion of green synthesized silver nanoadditives to biodiesel [13].

Silver act as the best-commercialized nanoparticle with the production rate per year is around four hundred tons of AgNPs and is predicted to rise in upcoming years. AgNPs play a dominant role in the high sensitivity field of biosensors, optical studies, anti-bacterial agents and catalysts activities [14–16]. There are several approach for synthesizing the AgNPs such as chemical reduction, ion sputtering, sol-gel, etc., [17–20]. This involves some difficulties with the requirements of high energy and perilous chemicals. So, another solution that demands which should be eco-friendly and also cost-efficient, by green syntheses become the main practice and the above demonstrates their potential.

Approaches for extracting nanoparticles by naturally occurring processes like plant extract, biodegradable polymers, microorganisms, etc. for reductants and capping agents act as alluring in nanotechnology [21, 22].

By comparing all other methods, the neem leaf extract green synthesized AgNPs are simple, single-step, cost-efficient, eco-friendly and relatively reproducible that offer better performance than others [23]. Conversion of silver ions into nanoparticles, which focuses on the application of a green approach to the synthesis of silver nanoparticles, as an alternate conventional method, is commonly available in India [24]. Production of AgNPs using neem leaf extract at low concentration without adding adverse chemicals/physical application methods. The performance and emission characteristics of a diesel engine fueled with various nano additives biodiesel blends are represented in Table 1. Table 1 depicts the performance and emission attributes of a diesel engine powered with different nano additives biodiesel blends.

To optimize the way in which silver nanoparticles are synthesized, the concentration of metal ions and leaf extracts has been evaluated, and this method is simple, easy to perform and stable.

However, no research has been done on neem leaf extract green synthesized silver nanoparticles used in diesel-bio diesel blends (B20) as a DI-CI engine test fuel, based on the above literature and the authors' knowledge. The present work was to prepare silver nanoparticles from neem leaf extracts doped with B20 and examine the performance, emission, and combustion attributes of the DI-CI engine.

Table 1 Performance and emission attributes of a diesel engine powered with different nano additives biodiesel blends

S. No.	Biodiesel feedstock	Nano additive	Parameters of engine	Engine performance and combustion characteristics	Emission characteristics	Ref.
1.	Neem Oil (25%)	NiO	Single cylinder, 4 stroke, Water-cooled, VCR DI-CI Engine, 4.8 kW, 1500 rpm, CR 17.5	Increased BTE by 6.3%, Reduced BSFC by 7.86%	Reduced HC Reduced CO Reduced CO_2 Higher NO_X	[25]
2.	Jatropha oil	Al_2O_3	Single cylinder, 4 stroke, Air-cooled, 4.4 kW, 1500 rpm, CR 17.5	Reduced BSFC and Reduced Ignition Delay	Reduced HC Reduced CO Reduced CO_2 Reduced NO_X Reduced Smoke	[26]
3.	Nerium Olender oil	Cerium oxide (CeO_2)	Single cylinder, 4 stroke, DI CI Engine, Water cooled, 5.2 kW, 1500 rpm, No of nozzle holes 3, Ignition pressure 200 bar	Increased BTE, Reduced BSFC	Reduced HC Reduced CO Reduced Smoke Reduced NO_X	[27]
4.	Diesel	Al_2O_3 and CuO	Single cylinder, 4 stroke, 5.1 kW, 3600 rpm, CR 20.3	Increased power by 3.28%, Increased Torque by 1%, Reduced BSFC by 0.5%(CuO) and Reduced BSFC by 1.2% (Al_2O_3)	Reduced (CO 11%, HC 13%, NO_x6%)—Al_2O_3 Reduced (CO 5%, HC 8%, NO_x2%)—CuO	[28]
5.	Mahua Oil	Al_2O_3	Four stroke, Single cylinder, CRDI Diesel engine, Water cooled, 3.7 kW, 1500 rpm, CR 17.5	Increased BTE, Reduced BSFC, Higher Heat release rate & Cylinder peak pressure	Lowered CO by 48% Lowered HC by 26.04% Increased NO_X by 4.8% Lowered Smoke	[29]
6.	Mahua Oil	Copper oxide (CuO)	Single cylinder, 4 stroke, Water cooled, DI–CI Engine, 3.7 kW, 1500 rpm	Increased BTE by 2.19%	Lowered CO by 33%, Lowered HC by 5.33% Lowered Smoke by 12.5% Higher NO_X by 3.2%	[30]

(continued)

Table 1 (continued)

S. No.	Biodiesel feedstock	Nano additive	Parameters of engine	Engine performance and combustion characteristics	Emission characteristics	Ref.
7.	Pongamia Oil	Copper oxide (CuO)	Single cylinder, 4 stroke, 3.7 kW, 1500 rpm CR 12.1–20.1	Increased BTE by 4.01%, Reduced BSFC by 1.0%	Lowered CO by 29%, Lowered HC by 7.9% Lowered Smoke by 12.8% Lowered NO_X by 9.8%	[31]

2 Preparation of Test Fuels

2.1 Preparation of Biodiesel

In this work, the production of biodiesel using green neem oil as a feedstock. The biodiesel was produced through the transesterification process. The reaction catalyst concentration (KOH) of 0.75 wt/wt% of oil, ethanol to oil molar ratio as 9:1525 rpm of mixing speed with the temperature of 60°C is maintained for 65 min. At this optimized process parameters, monoesters (biodiesel) capitulate of $93 \pm 0.24\%$ was attained. The optimum production of biodiesel process parameters was chosen based on our previous research publications [32].

2.2 Preparation of Plant Extract

Extract of fresh green neem leaf was preferred to make AgNPs from the basics of easily available and environmentally friendly. Raw green neem leaves were collected from K.S.R Educational Institutional campus, Tiruchengode. The collected fresh neem leaves were cleaned with distilled water, then thinly cut into small species and again cleaned with distilled water. 25 g of the fresh green leaves was added with 100 ml of deionized water and allowed to boiling process for one hour in a water bath. After the boiling process, the mixture was filtered to acquire the aqueous leaf extract.

2.3 Procedure for Green Synthesis of Silver Nanoparticles

The neem leaf extract interacted with 20 mg of $AgNO_3$ and 100 ml of distilled water. Conical flask is utilized for the production of AgNPs by green synthesis method. Dilute Ammonium Hydroxide (NH_4OH) was utilized to keep up the pH of the response blend in the scope of 8. The flask was kept for 4 h in a rotating shaker at 80 rpm to accomplish a homogenous reaction and attained silver nanoparticles in dry form. The soup containing AgNPs was centrifuged at 7000 rpm for 15 min. A similar procedure was used multiple times to better partition of free substances from the metal nanoparticles. The sanitized particles were dried utilizing a hot air oven up to a temperature of 700 °C. The silver nanoparticles are obtained after the reaction between neem soup and the $AgNO_3$ solution.

2.4 *Steps Involved for Green Synthesis of Silver Nanoparticles*

Figure 1 represents the pictorial view for the steps involved in preparation of green synthesized silver nanoparticles from neem leaf extract. Step 1 involves the collection of fresh neem leaves and then washed with distilled water to remove any unwanted particles. Step 2 indicates the weight the collected leaves to prepare neem leaf soup. Step 3 showing the collected leaves into small species. Take 100 ml of distilled water and add 25 g of collected neem leaves in a conical flask and set the temperature to 90 °C. After some time, the water will turn into greenish color it indicates the extracts of neem leaf shown in step 4. Followed by take 20 mg of silver nitrate in 100 ml of deionized water maintained the temperature at 60 °C and also cover with aluminum foil to avoid photodegradation of silver this process is indicated in step 5. Step 6 shows the measuring of pH value of silver nitrate in the scope of 8. Add leaf extract to the silver nitrate solution very slowly until its change color. The color change indicates the formation of silver nanoparticles solution. In step 7, using centrifuge apparatus maintaining the speed at 7000 rpm for 15 min to separate the free entities

Fig. 1 Steps involved for preparation of green synthesized nanoparticles

present in the silver nanoparticles solution. After that using hot air oven equipment to dry the silver nanoparticles in power form which shows in step 8 and 9.

2.5 Fuel Preparation

As part of this work, test fuels are diesel, diesel-bio diesel blends (B20) doped with and without green synthesized silver nanoparticles are used in DI-CI engine without engine modifications.

Ultrasonicator is commonly used in nanotechnology to disperse nanoparticles evenly in diesel-bio diesel blends. In a beaker, one liter of diesel-bio diesel blends is taken and then 1 g of green synthesis silver nanoparticles is added. After the addition of silver nanoparticles in diesel-bio-diesel blends, it is allowed into an ultrasonic device to produce uniform dispersion for the period of 30 min. Since the nanoparticles are excess in solution, this should be well shaken before use.

2.6 Fuel Properties

The properties of various nano additive biodiesel blends are compared and presented in Table 2.

Table 2 Properties of various nanoadditive biodiesel blends

Properties	Methyl Ester biodiesel (20%) with 25 ppm TiO_2	Jatropha biodiesel (100%) with 50 ppm Al_2O_3	Jatropha biodiesel (100%) with 50 ppm CeO_2	Mahua Methyl ester biodiesel (20%) with 50 ppm CeO_2	Neat diesel with 50 ppm CuO
Kinematic viscosity at 40 °C (cSt)	3.13	4.11	3.99	4.5	3.5
Calorific value (MJ/kg)	43.67	37.27	37.39	43	–
Density (kg/m^3)	845.7	877	880	861	834.1
Flash point (°C)	82	173.6	174.8	71	66
Fire Point (°C)	–	–	–	75	–
Cetane Number	52	–	–		54.5
References	[33]	[34]	[34]	[35]	[28]

Table 3 Properties of test fuel samples

Properties	Diesel	B20 Neem oil	B20GS20 (B20 + 20 ppm of AgNPs)	ASTM standards of bio diesel
Kinematic viscosity at 40 °C (cSt)	4.2	4.5	4.6	1.9–6.0
Calorific value (MJ/kg)	43.4	41.5	41.8	>40
Density (kg/m^3)	830	850	843	875–900
Flash point (°C)	56	61	58	>130
Fire Point (°C)	60	72	66	>65

The test fuel properties of diesel, diesel-bio diesel blends (B20), and diesel-bio diesel blends dopped with green synthesized nanoparticles (B20GS20) are compared and presented in Table 3. After the process of transesterification, the green neem oil properties have been enhanced slightly. However, when green synthesized nanoparticles are doped with diesel-bio diesel blends, all the properties are considerably increased similar to diesel. The test fuel properties in the table display that all are within the biodiesel ASTM standards range. The addition of silver nanoparticles (AgNPs) in B20 blends reduces the viscosity and density significantly.

3 Experimental Setup and Procedure

The pictorial representation of engine setup is shown in Fig. 2. The test conducted on a naturally aspirated, Kirloskar AV1, 1-cylinder, 4-stroke, water-cooled, DI-CI engine with 3.5 kW maximum power and 1500 rpm constant speed. Eddy current dynamometer was used to vary from 0, 20,40,60,80 and 100% engine load, respectively. Combustion analysis carried out with the help of cityzen made piezoelectric transducer, air-cooled pressure sensor mounted on an engine head and high-speed data acquisition system. DI-CI engine operated with the standard injection pressure and time of 200 bar and 23° BTDC, respectively, also to run the engine for about 20 min without any load to achieve the steady-state condition. To analyze the adverse exhaust emissions and smoke levels are measured using instruments like AVL-444 Di-gas analyzer and the AVL 437 standard smoke meter, respectively. The engine was tested at each load condition with diesel, B20, and B20GS20.

Fig. 2 A photographic view of experimental engine setup

4 Result and Discussions

4.1 Brake Thermal Efficiency

Brake thermal efficiency (BTE) refers to the conversion of fuel energy into mechanical output, which is a function of the engine parameter. For all the tested fuels, increasing the BTE with respect to load increases.

Figure 3 represents the variations of BTE with different engine load conditions of diesel, B20, and B20GS20.

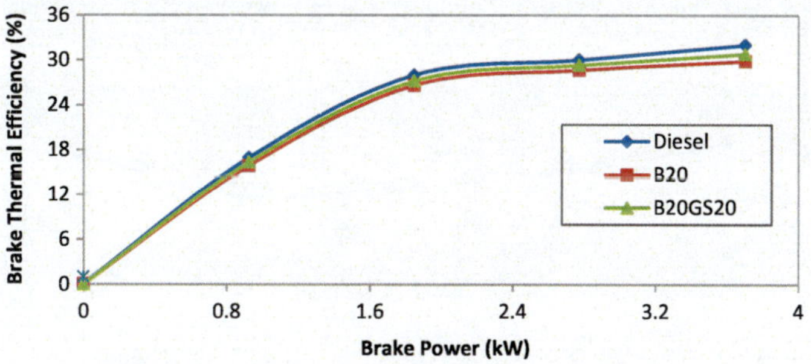

Fig. 3 Variation of brake power (kW) versus brake thermal efficiency (%)

For all load conditions, BTE of B20GS20 is slightly lower in comparison with diesel. At maximized condition of load, BTE of diesel, B20, and B20GS20 are 32%, 29.8%, and 30.8%, respectively.

BTE decreases for B20 compared to diesel because of low spray characteristics, lower heating value and higher viscosity and poor spray properties, which results in incomplete combustion [36].

Biodiesel doped with green synthesized nanoparticle leads to increase the heat content value, reducing in viscosity and density and increase the volatility of fuel at maximized load condition. It enhances the secondary atomization of B20GS20 fuel to induce the complete combustion, which results in increase the BTE compared with B20 [37].

4.2 *Carbon Monoxide Emission (CO)*

Figure 4 represents the variations of CO emissions with various engine load conditions of diesel, B20, and B20GS20. In the combustion process, formation of CO due to inadequate oxygen and time lacking in the combustion chamber [38].

At maximized condition of load, CO emission of diesel, B20, and B20GS20 are 0.115, 0.13 and 0.126%, respectively. Due to more viscosity and slighter increasing in specific gravity of B20 fuel result in incomplete combustion leads to greater CO emissions in comparison with diesel.

CO emission of B20GS20 is slightly reduced compared with diesel and B20, because the green synthesized AgNPs act as good oxygen donating catalyst and have a greater ratio of surface to volume, which gives large quantity of air react with fuel, combined with shortening of the ignition delay it leads to CO emissions would be lower [13].

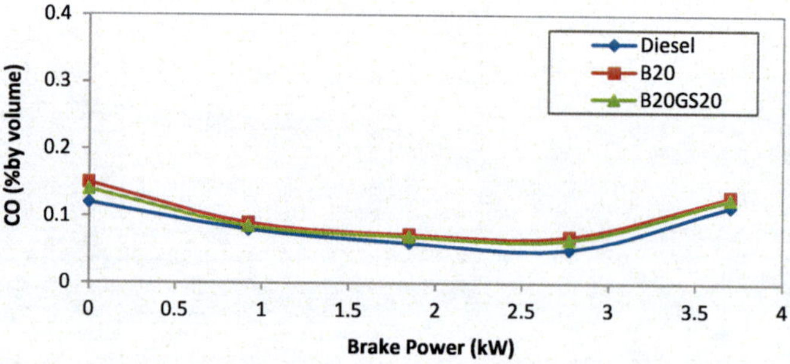

Fig. 4 Variation of brake power (kW) versus carbon monoxide (% by Volume)

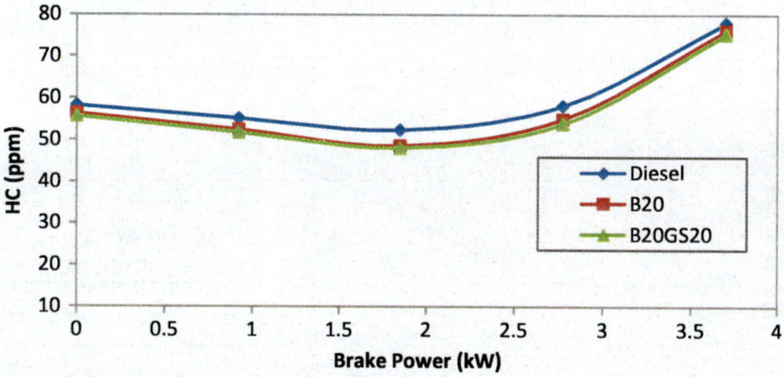

Fig. 5 Variation of brake power (kW) versus hydrocarbon (ppm)

4.3 Hydrocarbon (HC)

Figure 5 represents the variations of HC emissions with different engine load conditions of Diesel, B20, and B20GS20.

Hydrocarbon emissions of the diesel, B20, and B20GS20 are lower and then increases at medium and higher load, respectively. Mainly, due to unreacted hydrocarbon fuel is the source for producing HC emission and also, due to more amount of fuel is pumped into the combustion chamber at maximum loads, the reaction is usually less oxygen.

The maximized load condition for hydrocarbon emission of diesel is 77.9 ppm, whereas B20 is 77.1 ppm and B20GS20 is 75.2 ppm. In this case, HC emission of B20 similar to diesel because of biodiesel cetane number is higher than diesel.

Hydrocarbon emissions for the B20 and B20GS20 are lower than neat diesel by 0.9 and 2.7%, respectively. This is because the oxygen in the biodiesel leads to the proper mixing of the air-fuel ratio and an increase in oxidation properties to reduce the HC emission in comparison with diesel.

Nanoparticles doped fuel has oxygen boosting catalytic nature and a high ratio of surface to volume, it enhances the heat transfer rate leads to reduce the HC emission in comparison with diesel and B20, respectively [39].

4.4 Oxides of Nitrogen (NO_x)

Figure 6 represents the variations of oxides of nitrogen emissions with different engine load conditions of Diesel, B20, and B20GS20.

It perhaps clearly shown that NO_x emission increases with increasing load. The formation of NO_x depends upon the in-cylinder temperature, oxygen concentration in the cylinder, and engine technology. When increasing the engine load, the quantity of fuel supply increases and then increases combustion chamber temperature.

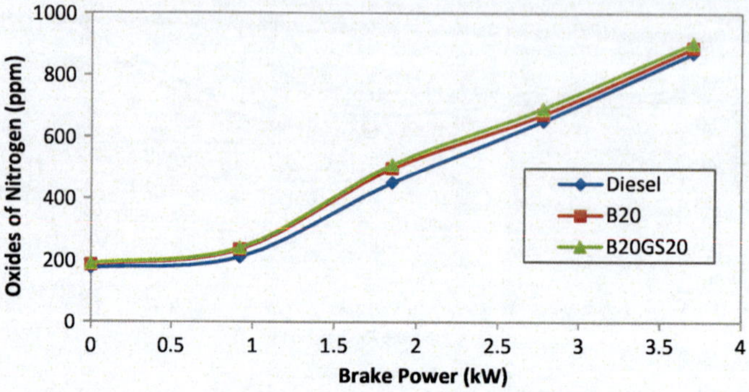

Fig. 6 Variation of brake power (kW) versus oxides of nitrogen (ppm)

At 100% load condition, emission of NO_x for diesel is 870 ppm whereas B20 is 888 ppm and B20GS20 is 903 ppm, respectively.

The B20 has slightly high viscosity which gives poor atomization along with faster-premixed combustion result in increasing the temperature of in-cylinder and which tends to higher NO_x formation than neat diesel. [40]

The NO_x emissions for B20GS20 are higher than B20 because nano additives being an oxygen boosting catalyst supports the fuel complete combustion and increasing the temperature of combustion chamber, results in increasing NO_x.

4.5 Smoke

Figure 7 represents the variations of smoke level with different engine load conditions of Diesel, B20, and B20GS20.

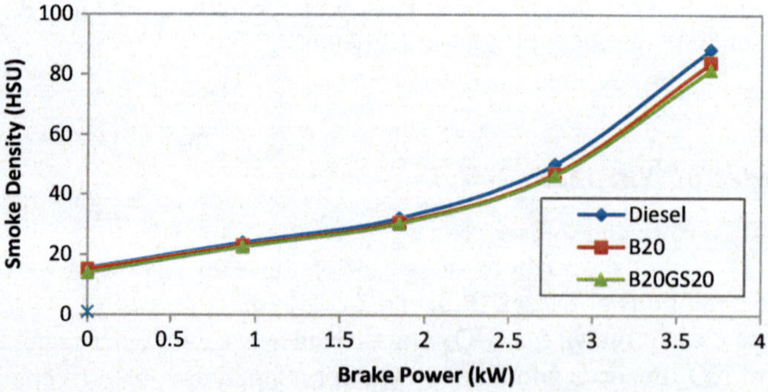

Fig. 7 Variation of brake power (kW) versus smoke (HSU)

For all test fuels, the smoke level increases simultaneously with load increases. Smoke levels are mainly caused by high viscosity and low instability, resulting in formation of poor mixing. From the graph, at maximized load condition neat diesel, B20, and B20GS20 emits smoke level are 88.3, 84.2, and 81.8 HSU, respectively.

The oxygen present in the B20 molecules decreases the rate of combustion leads to increased ignition delay of gas-phase emissions in comparison with diesel.

The smoke emissions of B20 doped with green synthesized nanoparticles are lower than B20 and diesel, because of reduced ignition delay, perfect mixture formation due to the high ratio of surface-to-volume, rapid evaporation rate, and enhanced ignition properties of the green synthesized silver nanoparticles [41].

5 Combustion Analysis

5.1 *Pressure and Crank Angle*

At maximized load condition, the variations of cylinder pressure with respect to the crank angle of diesel, B20, and B20GS20 represent in Fig. 8.

The diesel engine's maximum pressure depends on the combustion rate during the initial combustion and also quantity of fuel involved in the phase of uncontrolled combustion. The rise of maximum pressure rate will fall at high load in all cases

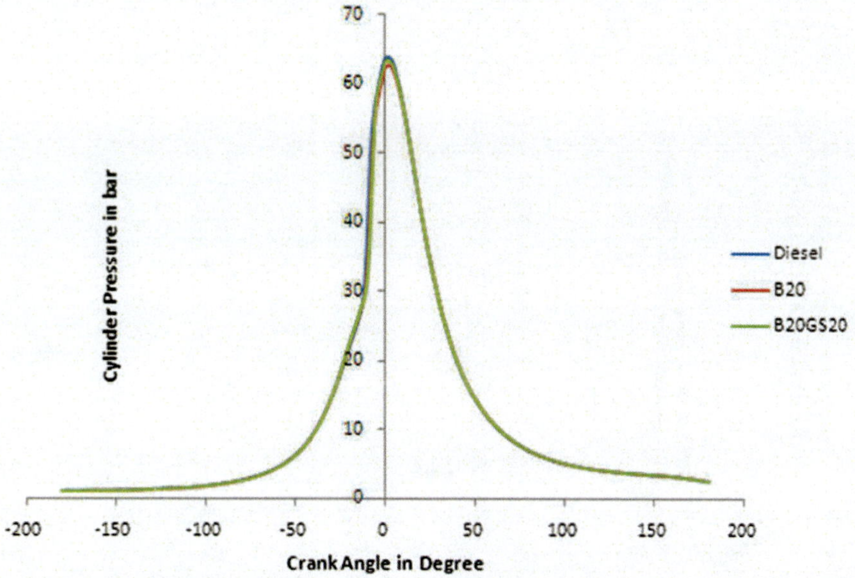

Fig. 8 Variation of crank angle (degree) versus cylinder pressure (bar)

because of cutback in ignition delay. The highest value of peak pressure is noticed with diesel followed by B20GS20 and B20, respectively. Diesel having high ignition delay and lower volatility compared with B20 and B20GS20 fuels. The maximum pressure of diesel, B20, and B20GS20 are 63.77, 62.682, and 63.243 bar, respectively.

For the diesel-bio diesel blend fuels, the maximum pressure slightly reduced because of cutback in delay period, and increase in cetane number of the biodiesel blend.

It clearly shows peak pressure of B20GS20 which is closer to diesel and higher than B20. The reason behind that is diesel biodiesel blends doped with silver nano additives have high calorific value compared to B20.

Another reason is that B20GS20 are having higher surface areas combined with catalytic effect of nanoparticle leads to better mixing it enhances the advanced combustion process than B20.

5.2 *Rate of Heat Release*

At maximized load condition, the variations of heat release rate with respect to the crank angle of diesel, B20, and B20GS20 represent in Fig. 9.

High-speed data acquisition is used to collect data on cylinder pressure for 50 consecutive engine cycles and finally averaged. The premixed combustion phase has been found to be dominant in the case of diesel. But for B20 and B20GS20 fuel diffusion combustion is dominant.

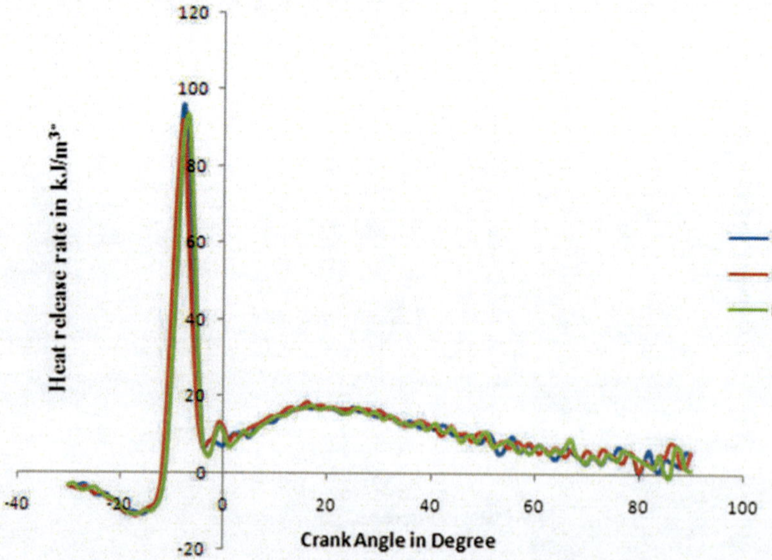

Fig. 9 Variation of crank angle (degree) versus heat release (kJ/m^3°)

The high premixed combustion phase is dependable for the diesel fuel steep peak pressure rise. The heat release highest value is noticed with diesel followed by B20GS20 and B20, respectively. Diesel having high ignition delay and lower volatility compared with B20 and B20GS20. A maximum heat release rate for diesel, B20, and B20GS20 are 95.741, 91.336, and 93.066 kJ/m$^{3°}$, respectively.

Heat release is slightly lower for B20 due to the slightly increased viscosity compared to diesel and B20GS20. The heat release rate of B20GS20 is closer to diesel and higher than B20. This is due to the nano additive doped fuels have higher thermal conductivity, it produced better combustion faster fuel burning rate.

6 Conclusion

Based on the test fuel investigations results, the subsequent conclusions were made.

- By the process of transesterification, biodiesel is extracted from neem oil, using ethanol as a solvent and KOH as a catalyst.
- Silver nanoparticles are produced using green synthesized neem leaf extract at low concentration without adding adverse chemicals/ physical application methods.
- The properties of test fuel are tested and compared with ASTM standards.
- BTE of B20GS20 fuel blend is enhanced by 1% contrasted with B20 while it compared with neat diesel reduced by 2.2%.
- B20 doped with green synthesized nano additives fuel having a higher the surface-to-volume ratio, oxygen boosting catalytic nature, it enhances the BTE by 1%, significantly reduce the adverse emissions such as HC, CO, and smoke level by 2.46, 3.07, and 2.97%, respectively, and increases NO_x by 1.66% compared to B20.
- In further work, NO*x* emissions will be reduced using hot and cold electronically controlled external re-breathing engine technique (Exhaust Gas recirculation).
- The nanoparticles present in engine exhaust emissions may additionally penetrate deep interior the human cell wall, it creates cellular damage by interacting with phosphorus- and sulfur-containing compounds. We also recommend in further studies, need to investigate the health risk issues associated with the use of silver nanoparticles doped in biodiesel blends.
- B20GS20 combustion characteristics like peak pressure rise and heat release rate are slightly enhanced by 1 and 1.9% compared to B20.
- From the experimental investigation proved that B20GS20 doped test fuel is the better blend ratio than B20 for DI-CI engine.

Acknowledgements The authors genuinely thank AICTE, New Delhi, India, for the budgetary help with completing this work under RPS (AICTE FILE NO: 8-29/RIFD/RPS/POLICY-1/2016-2017. Dated 02nd August 2017).

References

1. Haiter Lenin A, Ravi R, Thyagarajan K (2013) Performance characteristics of a diesel engine using mahua bio diesel as alternate fuel. https://doi.org/10.5829/idosi.ijee.2013.04.02.10
2. Ramreddy M (2014) Performance analysis and emission testing on diesel engine by using bio diesel as mahua oil. Int J Ethics Eng Manage Educ 1(4). ISSN: 2348-4748
3. Nandi S (2010) Performance of C.I engine by using bio diesel–mahua oil. Am J Eng Res (AJER) 02(10):22–47, e-ISSN: 2320-0847, p-ISSN:2320-0936
4. Naramsetty S, Ranjith Kumar V, Hanumantha Rao YV (2013) Performance and emission characteristics of diesel engine fuelled with bio diesel of mahua oil & diesel blends with additives. Int J Eng Res Appl (IJERA) 3(4):936–941. ISSN: 2248-9622
5. Demirbas A (2007) Progress and recent trends in biofuels. Progress Energy Combustion Sci 33:1–18
6. Ashraful AM, Masjuki HH, Kalam MA, Rizwanul Fattah IM, Imtenan S, Shahir SA, Mobarak HM (2014) Production and comparison of fuel properties, engine performance, and emission characteristics of bio diesel from various non-edible vegetable oils. Energy Conversion Manage 80:202–228
7. Hossain AK, Davies PA (2010) Plant oils as fuels for compression ignition engines: a technical review and life-cycle analysis. Renew Energy 35:1–13
8. Heroor, Shruthi H, Bharadwaj SD, Rahul (2013) Performance and emission characteristics of CI-engine using mahua (*Madhuca longifolia*) bio diesel under varying operating conditions. Int J Appl Eng Res 8(17):2005–2010. ISSN: 0973-4562
9. Mulimani H, Hebbal OD, Navindgi MC (2012) Extraction of bio diesel from vegetable oils and their comparisons. Int J Adv Sci Res Technol 2(2). ISSN: 2249-9954
10. Nurun Nabi M, Shamim Akhter MD, Zaglul Shahadat MD (2006) Improvement of engine emissions with conventional diesel fuel and diesel–bio diesel blends. Bio-resource Technol 97:372–378
11. Sukumar Puhana N, Vedaramana N, Sankaranarayanana G, Boppana B, Bharat Rama V (2005) Performance and emission study of Mahua oil (Madhucaindica oil) ethyl ester in a 4-stroke natural aspirated direct injection diesel engine. Renew Energy 30:1269–1278
12. Kadarohman A, Hernani FK, Astuti RM (2010) A potential study on clove oil, eugenol and eugenyl acetate as diesel fuel bio-additives and their performance on one cylinder engine. Transport 25(1):66–76
13. Senthilraja S, Karthikeyan M, Gangadevi R (2010) Nano-Micro Letters 2(4):306–310
14. Larue C, Castillo-Michel H, Sobanska S, Cecillon L, Bureau S, Barthes V (2014) Foliar exposure of the crop Lactuca sativa to silver nano particles. Evidence for internalization and changes in Ag speciation. J Hazard Mater 264:98–106
15. Ghadir A, El-Chaghaby, Abeer, Ahmad AF (2011) Biosynthesis of silver nano particles using pistacialentiscus leaves extract and investigation of their antimicrobial effect. Oriental J Chem 27:929–936
16. Veerasamy R, Xin TZ, Gunasagaran S, Xiang TFW, Yang EFC, Jeyakumar N, Dhanaraj SA (2011) Biosynthesis of silver nano particles using mangosteen leaf extract and evaluation of their antimicrobial activities. J Saudi Chem Soc 15:113–120
17. Bindhu MR, Umadevi M (2015) Antibacterial and catalytic activities of green synthesized silver nanoparticles. Spectrochimica Acta Part A: Mol Biomolecular Spectrosc 135:373–378
18. Pourmortazavi SM, Taghdiri M, Makari V, Rahimi-Nasrabadi M (2015) Procedure optimization for green synthesis of silver nano particles by aqueous extract of *Eucalyptus oleosa.* Spectrochimica Acta Part A: Mol Biomolecular Spectrosc 136:1249–1254
19. Padalia H, Moteriya P, Chanda S (2014) Green synthesis of silver nano particles from marigold flower and its synergistic antimicrobial potential. Arab J Chem. https://doi.org/10.1016/j.arabjc.2014.11.015
20. RathiSre PR, Reka M, Poovazhagi R, Arul Kumar M, Murugesan K (2015) Antibacterial and cytotoxic effect of biologically synthesized silver nano particles using aqueous root extract of Erythrinaindica lam. Spectrochimica Acta Part A: Mol Biomolecular Spectrosc 135:1137–1144

21. Ahmed S, Ahmad M, Swami BL, Ikram S (2015) Plants extract mediated synthesis of silver nano particles for antimicrobial applications: a green expertise. J Adv Res. https://doi.org/10.1016/j.jare.2015.02.007
22. Kharissova OV, Dias HVR, Kharisov BI, Perez BO, Perez VMJ (2013) The greener synthesis of nano particles. Trends Biotechnol 31:240–248
23. Mittal J, Batra A, Singh A, Sharma MM (2014) Phytofabrication of nano particles through plant as nanofactories. Adv Natural Sci Nanosci Nanotechnol. https://doi.org/10.1088/2043-6262/5/4/043002
24. Omoja VU, Anaga AO, Obidike IR, Ihedioha TE, Umeakuana PU, Mhomga LI, Asuzu IU, Anika SM (2011) The effects of combination of methanolic leaf extract of *Azadirachta indica* and diminazenediaceturate in the treatment of experimental Trypanosomabruceibrucei infection in rats. Asian Pacific J Tropical Med 4:337–341
25. Srinidhi C, Madhusudhan A, Channapattana SV (2019) Effect of NiO nanoparticles on performance and emission characteristics at various injection timings using biodiesel-diesel blends. Fuel 235:185–193. https://doi.org/10.1016/j.fuel.2018.07.067
26. Venu H, Madhavan V (2016) Effect of Al_2O_3 nanoparticles in biodiesel-diesel-ethanol blends at various injection strategies: performance, combustion and emission characteristics. Fuel 186:176–189. https://doi.org/10.1016/j.fuel.2016.08.046
27. Dhinesh B, Annamalai M (2018) A study on performance, combustion and emission behaviour of diesel engine powered by novel nano nerium oleander biofuel. J Cleaner Prod. https://doi.org/10.1016/j.jclepro.2018.06.002
28. Gumus S, Ozcan H, Ozbey M, Topaloglu B (2016) Aluminum oxide and copper oxide nanodiesel fuel properties and usage in a compression ignition engine. Fuel 163:80–87. https://doi.org/10.1016/j.fuel.2015.09.048
29. Syed Aalam C, Saravanan CG (2015) Effects of nano metal oxide blended Mahua biodiesel on CRDI diesel engine. Ain Shams Eng J. https://doi.org/10.1016/j.asej.2015.09.013
30. Chandrasekaran V, Arthanarisamy M, Nachiappan P, Dhanakotti S, Moorthy B (2016) The role of nano additives for biodiesel and diesel blended transportation fuels. Transp Res Part D 46:145–156. https://doi.org/10.1016/j.trd.2016.03.015
31. Perumal V, Ilangkumaran M (2018) The influence of copper oxide nano particle added pongamia methyl ester biodiesel on the performance, combustion and emission of a diesel engine. Fuel 232:791–802. https://doi.org/10.1016/j.fuel.2018.04.129
32. Avinash A, Murugesan A (2018) Prediction capabilities of mathematical models in producing a renewable fuel from waste cooking oil for sustainable energy and clean environment. Fuel 216:322–329
33. Lawrence KR, Tharigonda H, Alluru G. An experimental investigations on combined effect of EGR and SCR for diesel engine fuelled with the addition of TiO_2 nanoparticle to M. Elangi methyl ester blend. Austr J Mech Eng. ISSN Print: 1448-4846 and ISSN Online: 2204-2253
34. Hossain AK, Hussain A (2019) Impact of nanoadditives on the performance and combustion characteristics of neat jatropha biodiesel. Energies 12,921. https://doi.org/10.3390/en12050921
35. Chiranjeeva Rao S, Saravanakumar A, Chandra Sekhar G (2019) Influence of nano added MME blends on CI engine based on doe concept. Int J Mech Eng Technol (IJMET). 12,921, vol 8, pp 880–868. ISSN Print: 0976-6340 and ISSN Online: 0976-6359
36. Nurun Nabi MD, Mustafizur Rahman MD, Shamim Akhter MD (2009) Biodiesel from cotton seed oil and its effect on engine performance and exhaust emissions. Appl Thermal Eng 29:2265–2270
37. Ajin C, Sanjeevan, Sajith V (2013) Diesel engine emission reduction using catalytic nano particles: an experimental investigation, Volume 2013, Article ID 589382
38. Keskin A, Guru M, Altiparmak D (2010) The investigation of performance and emissions characteristics of tall oil bio diesel with a co-based additive, Energy sour. Part-A Recov Util Environ Eff 31:1899–1907
39. Varathavijayan M, Ponvel Murugan S, Mohanamurugan (2013) Experimental investigation of DI diesel engine using bio diesel blends with different injection pressures. Int J Eng Sci Res Technol 2(11). ISSN: 2277-9655

40. Pradeep V, Sharma RP (2007) Use of hot EGR for NO*x* control in a compression ignition engine fuelled with bio diesel from jatropha oil. Int J Renew Energy 32:1136–1154
41. Shadangi KP, Mohanty K (2014) Comparison of yield and fuel properties of thermal and catalytic Jatropha seed pyrolytic oil. Fuel 117:372–380

Thermodynamic Modelling and Experimental Investigation on a CI Engine Operated with Oxy-Hydrogen Gas as the Secondary Fuel

P. V. Manu, T. R. Navaneeth Kishan, and S. Jayaraj

1 Introduction

The major objectives of present-day engine research include substitution for fossil fuel consumption and reduction of exhaust emissions. Due to the issues of global warming and climate change, nowadays, engines with low emissions (or zero emission) are gaining much importance [1]. For an existing engine, in order to reduce the emissions and improve the specific fuel consumption, operating the engine in a dual-fuel mode is found to be a good alternative. No major engine modifications (such as design changes) are involved to implement this technique. Gaseous fuels such as hydrogen, natural gas and LPG can be used as secondary fuel in dual-fuel engines. It is a promising approach for reducing the dependence on petroleum-based conventional fuels and improving the performance of the engine. An overall comparison of the properties of different kinds of secondary fuels available in market in relation to diesel is given in Table 1.

In the present study, a compression ignition (CI) engine is operated in the dual-fuel mode with diesel and oxy-hydrogen (HHO) gas as the primary and secondary fuels, respectively. Oxy-hydrogen gas is generated using an on-board dry cell electrolyser (DCE), which is operated with a DC power source. Though it was experimentally proven that mean effective pressure and heat release rates are higher, a suitable theoretical model for prediction is yet to be developed. A two-zone analysis is carried

P. V. Manu (✉) · T. R. Navaneeth Kishan · S. Jayaraj
Mechanical Engineering Department, National Institute of Technology Calicut, Calicut, Kerala 673601, India
e-mail: pvmanu@nitc.ac.in

T. R. Navaneeth Kishan
e-mail: navaneeth762000@gmail.com

S. Jayaraj
e-mail: sjayaraj@nitc.ac.in

A. K. Gupta et al. (eds.), *Advances in IC Engines and Combustion Technology*, Lecture Notes in Mechanical Engineering,
https://doi.org/10.1007/978-981-15-5996-9_23

Table 1 Comparison of properties of secondary fuels against diesel

Properties	Diesel	Hydrogen	Natural gas	LPG
Calorific value (MJ/kg)	43.25	141.9	47.13	46.1
Relative density	0.83	0.069	0.128	0.51
Stoichiometric A/F ratio (mass basis)	14.7:1	34:1	17.2:1	15.7:1
Flammability limits in air (%)	0.6–7	4–75	5–15	1.8–8.4
Ignition temperature (°C)	250	858	540	400

out to increase the accuracy of the prediction of combustion characteristics in dual-fuel combustion. The modelling results obtained are validated by comparing with the experimental results. The HHO gas generation process from water is similar to any electrolysis process and involves an anode, cathode, electrolyte solution and the DC power source. In standard conditions, a theoretical potential difference of 1.23 V is needed to perform electrolysis to split the water molecule to produce oxy-hydrogen gas [2]. This voltage is according to the Gibbs free energy requirement for the process at the standard condition of 298 K and 1 bar. There are mainly two types of design features for electrolysers, namely wet cell and dry cell electrolysers. In wet cell electrolysers, the electrodes are immersed in the electrolyte liquid in a container. The cell does not have any electrolyte circulation, since the electrodes are immersed in the electrolyte and are placed sufficiently apart. As the set-up is not very compact, it faces problems in mobile applications. Due to the plates dipped in the electrolyte, current loss is experienced because of the electrodes being exposed. Further, wet cell electrolyser tends to get overheated during its operation, which is considered to be a major limitation of these types of electrolytic cells for practical usage [3].

In a dry cell electrolyser system, the reservoir and cell are apart. Only a small volume of electrolyte is exposed to the electrode plates. Hence, current requirement is less for dry cell electrolyser due to the volumetric size of the electrolyte within the closed chamber. The major advantage of the dry cell electrolyser is that it is very compact in design and hence can be used for the on-board generation of HHO gas in mobile applications. Two alkaline electrolysis cell configurations are possible (namely the mono-polar and the bipolar). In mono-polar configuration, alternate electrodes are parallelly connected to the opposite terminals of the DC power. High electrical currents at low voltages, causing large ohmic losses, are the major drawback of this configuration. Only electrodes at the two ends are connected to the DC power supply in the bipolar configuration. All the other electrodes are connected by the electrolyte solution, and all the adjacent electrodes form individual cells. These cells form a series connection, hence same current flows through each individual cell, and they have the advantage of reduced ohmic losses [4]. Oxy-hydrogen gas can be easily used in CI engines operating in a dual-fuel mode replacing the primary fuel (diesel). In the dual-fuel mode of operation, the HHO gas is inducted into the combustion chamber through the intake manifold along with fresh air and is compressed. Then, the charge undergoes combustion with a spray of diesel fuel, which is injected towards the end of compression stroke [5].

2 Cell Optimization

Taguchi's design of experiments (DoE) makes use of a special design of orthogonal array to examine the quality characteristics through a minimal number of experiments [6]. The experimental results based on the orthogonal array are then transformed into signal-to-noise (S/N) ratios to evaluate the performance characteristics. In the present study for analysing the DCE performance, a L18 orthogonal array is formulated to conduct experiments with varying combinations of dry cell electrolyser parameters (such as type of electrode plate surface, space between the electrode plates, concentration of electrolyte, type of electrolyte and input voltage applied to the electrolyser). The selected dry cell electrolyser parameters and their level values are given in Table 2.

The results of the experiments carried out based on Taguchi's DoE were optimized and analysed using grey relational analysis (GRA) and analysis of variance (ANOVA). GRA is used in general to determine the optimum condition of various input parameters to obtain the best quality characteristics [7]. However, the data to be used in GRA must be pre-processed into quantitative indices for normalizing raw data. Original sequence is converted into a decimal sequence between 0.00 and 1.00 by pre-processing (for comparison purpose). A grey relational coefficient (GRC) is calculated to express the relationship between the ideal and actual normalized experimental results. Normally, the average of the GRC is taken as the grey relational grade (GRG), which is calculated by taking the average of the entire calculated grey relational coefficient for all the three responses. From the determined grey relational grade, the optimum condition is derived by considering the average value of grey grades corresponding to each level value of chosen input parameters. The optimum parameter levels identified for dry cell electrolyser in the present work are rough electrode surface with 3 mm spacing between the electrode plates, considering KOH as the electrolyte, having 30% electrolyte concentration with an input voltage of 12 V.

Table 2 Control parameters and its levels

Parameter/level	Level 1	Level 2	Level 3
Type of surface	Rough	Smooth	–
Spacing (mm)	2.5	2.75	3.0
Concentration (%)	25	27.5	30
Type of electrolyte	KOH	NaOH	Na_2CO_3
Voltage (V)	8	10	12

3 Experimental Methodology

This section describes the materials required and the experimental procedures adopted for the various experimental investigations conducted. The schematic arrangement of the fabricated DCE is shown in Fig. 1. The set-up consists of a dry cell, bubbler, silica gel, DC power supply and a diesel engine. Flame arrestor is provided for safety. The gas is fed into the engine through the inlet manifold. A single-cylinder, four-stroke water-cooled diesel engine with data acquisition system for capturing the in-cylinder pressure for every degree of crank rotation is used for the present experiments (specifications are given in Table 3).

The schematic diagram of experimental set-up is shown in Fig. 2. By varying the current, flow rate of HHO was varied. At different flow rates of HHO and at different loads, various readings of engine were taken. The pollutant emissions were measured by using gas analyser.

The experiments which are performed during the present research work include: (i) DCE comparison study, (ii) long-run test of DCE, (iii) engine characteristics in the dual-fuel mode and (iv) lubricating oil and piston deposit studies. The performance of the fabricated dry cell electrolyser (Cell 3) was compared against two

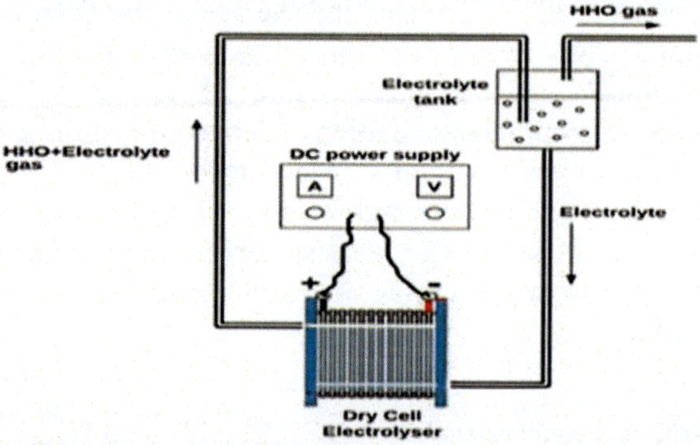

Fig. 1 Schematic arrangement of DCE

Table 3 Engine specifications

Manufacturer	Kirloskar Engines Ltd.
Engine type	Four-stroke, single-cylinder, CI engine
Rated power	3.68 kW at 1500 rpm
Bore and stroke	80 mm and 110 mm
BHP of engine	5 hp
Swept volume	553 cc
Compression ratio	16.5:1
Injection mode	Direct injection
Cooling system	Water cooling
Dynamometer	Rope brake dynamometer

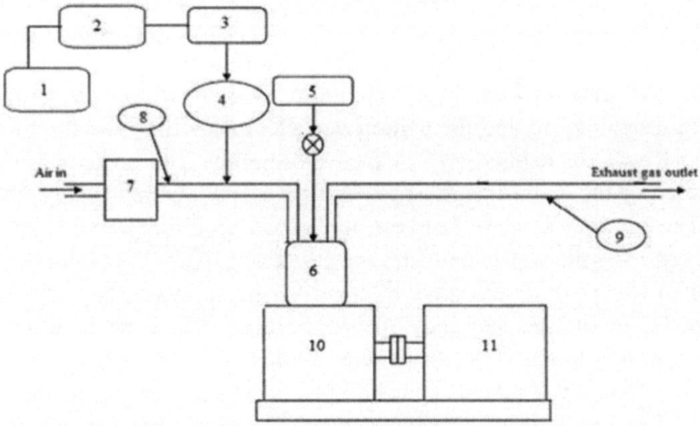

Fig. 2 Schematic diagram of the CI engine test rig with DCE. 1. DC power supply; 2. DCE; 3. bubbler; 4. fire arrestor; 5. diesel tank; 6. cylinder head; 7. air box; 8. inlet manifold; 9. exhaust manifold; 10. CI engine; and 11. eddy current dynamometer

commercially available dry cell electrolysers (Cell 1 and Cell 2). Comparison was done on the aspects of HHO generation, the temperature generated in the electrodes and electrolyte, and current consumed with respect to the applied voltage. Voltage applied to the cells was chosen as the independent variable, and four voltage values are selected for comparing the electrolytic cells. After the cell had reached a steady working condition, the required values were recorded. The DCE needs to be run for a long duration during its application in the automobiles. Therefore, it is necessary to know if the performance of the dry cell degrades with respect to time. For this purpose, a long-run test of the dry cell was conducted. A constant voltage of 12 V and current set to 10A were given as input to the cell. Then, the cell was continuously run for 100 h. At each 25 h interval, HHO gas evolution, the temperature of the electrode, the temperature of the electrolyte and amount of electrolyte consumed were recorded by adopting similar methods as for the comparative study. Initially, the engine was operated with diesel as the sole fuel (at different load conditions). Fuel consumption, exhaust emissions, smoke opacity and in-cylinder pressure readings with respect to crank angle were recorded corresponding to the different load conditions.

Thermocouples were used for measuring temperatures of cooling water inlet, outlet and exhaust. Pressure transducers were inserted in the cylinder, and sensors were provided to measure the crank angle. With the help of a data acquisition system, the readings, which are averaged for 50 cycles, were imported to a personal computer. The software of the PC-based engine set-up automatically calculates all the combustion parameters. HHO gas was generated by water electrolysis using a DCE, which was fabricated with the DoE optimized results. Engine was then modified to admit the HHO gas by inserting an inlet pipe of 6-mm diameter through the intake manifold, near to the inlet valve. HHO gas generated at a steady flow rate of 0.65 lpm was then passed through a bubbler, absorber and flame arrestor before it was introduced to the engine. Engine was operated at a constant speed of 1500 rpm in dual-fuel mode with diesel and HHO gas as primary and secondary fuels, respectively. All the tests were conducted under steady-state condition. Engine emissions were measured using a five-gas analyser and smoke metre. It is vital to study how the dual-fuel operation affects the lubricating oil and the carbon deposits inside the cylinder. Initially, the cylinder head was dismantled and cleaned thoroughly. The oil used is 20W50 as recommended by the manufacturer. The engine was first operated using diesel as the sole fuel. The engine was started and run at full load condition for 60 h. At each 20 h interval, the oil sample was taken from the crankcase. After the 60 h operation, the cylinder head was again dismantled. Piston deposits and valve deposits were carefully scraped from the surfaces and collected securely. The deposits were weighed, and also a thermogravimetric analysis was conducted. Various tests performed on the lubricant oil samples include four ball test, acid and base value analysis, total oxidation analysis and viscosity measurements. The same procedure was repeated for the dual-fuel operation of the engine.

4 Combustion Modelling

In the present study, dual-fuel combustion is considered to happen inside the engine cylinder. Diesel is the primary fuel, and the HHO gas is the secondary fuel. Dual-fuel combustion can be modelled by using the Wiebe function, which is an analytical function based on the normal distribution of a random variable. The Wiebe function is suitable to predict the burn fraction and burn rate in IC engines operating with different combustion systems and fuels. However, for dual-fuel combustion systems (where there is a significant effect in premixed combustion), the accuracy of Wiebe function was observed to be lower. So, a double Wiebe function was employed for modelling such dual-fuel combustion systems [8].

Analysis is performed using the standard thermodynamic equations. The rate of heat release is an important tool for analysing various combustion parameters like chemical energy release, energy split, burn duration, etc. These properties help in understanding of engine performance. One of the well-established methods of modelling the heat release is by employing Wiebe function to model the mass fraction burned. A double Wiebe function is also used to model the burning characteristics of the two fuels in the present work. The two Wiebe functions are then superimposed to obtain the premixed combustion and diffused combustion phases [9]. Annand's heat transfer model, which considers both convection and radiation effects, is employed here to estimate the heat transfer to the cylinder walls [10]. In the two-zone analysis, the components inside the engine cylinder are considered as two separate thermodynamic systems, namely burned and unburned region. These two regions are separated by a flame front of negligible thickness. A two-zone model can represent the combustion processes inside the cylinder in a more accurate manner than that predicted by the single-zone model. MATLAB® is used to solve the set of equations, and Euler's method is applied for the differential equations involved. Solution prediction is done within the duration between the points of inlet valve closing and exhaust valve opening.

5 Results and Discussion

Experimental and modelling studies were performed during this research work, and the results obtained are presented in the form of tables and graphs. Also, modelling results obtained are validated by comparing with the experimental results. It was evident from the cell comparison studies that Cell 3 was found to perform better on the basis of all performance parameters. The reason may be due to its design and fabrication based on the optimized conditions. Based on the observations and long-run test of DCE results, it can be concluded that fabricated DCE operation is stable and is suitable for long duration operation, such as in automobile applications. Figure 3 indicates that there is an increase in brake thermal efficiency corresponding to dual-fuel combustion, compared to diesel combustion. The faster rate of combustion in dual-fuel operation, due to the higher flame velocity of HHO gas, increases the mean effective pressure inside the engine cylinder. The presence of HHO gas in the combustion chamber improves the combustion of the primary fuel (diesel) due to its diffusive nature and higher flame velocity. Marginal increase in the brake power and hence higher brake thermal efficiency in dual-fuel mode is observed due to this.

Reduction in fuel consumption is due to the overall lean mixture formed in the cylinder in dual-fuel operation and improved lean burning characteristics of HHO gas. The favourable combustion properties of HHO gas, namely short quenching distance, higher flame velocity and wide flammability limit, also constitute for near-complete combustion during dual-fuel operation resulting in reduced specific fuel consumption as depicted in Fig. 4.

Fig. 3 Variation of BTE with BP at 1500 rpm

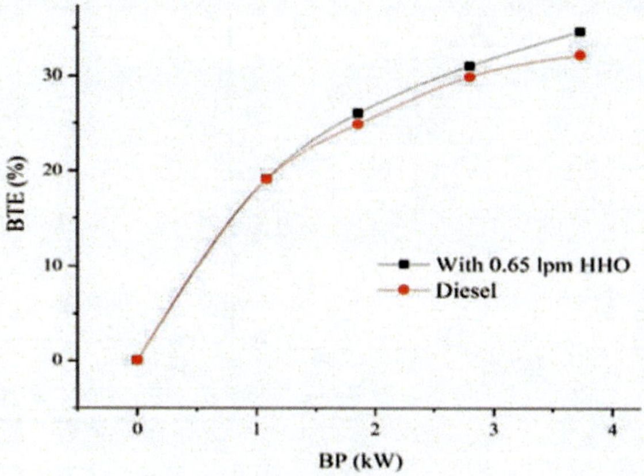

Fig. 4 Variation of SFC with BP at 1500 rpm

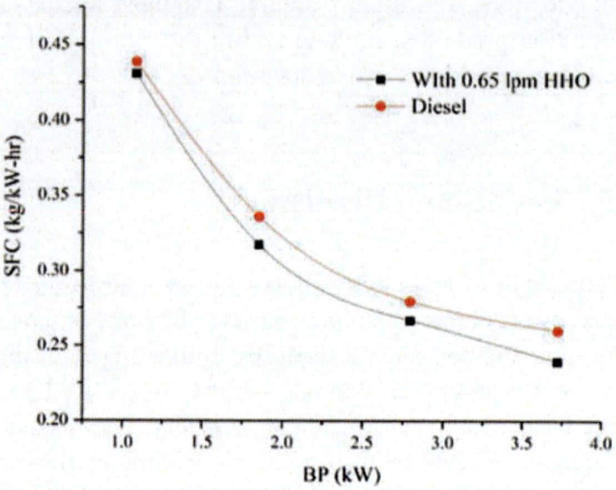

Dual-fuel engines using LPG or CNG as secondary fuel have shown poor thermal efficiency at intermediate loads due to incomplete combustion. In comparison, oxy-hydrogen gas gave better performance at part load operations owing to its highly diffusive nature and higher flame velocity.

Based on the readings obtained from emission gas analyser, it is found that got reduced considerably upon the induction of HHO gas to the engine. The better burning characteristics of HHO gas help in near-complete combustion of the primary fuel (diesel) and thereby reduce the harmful emissions. Soot formation occurs mainly due to insufficient amount of oxygen in the cylinder. The excess oxygen in the HHO gas during dual-fuel operation helps the fuel to burn more efficiently. Therefore, the opacity of smoke is found to reduce in dual-fuel combustion (Fig. 5). As per the results obtained, it could be concluded that there was not much variation in the properties of two lubrication oil samples. Hence, this study implies that there is no

Fig. 5 Variation of smoke opacity with load at 1500 rpm

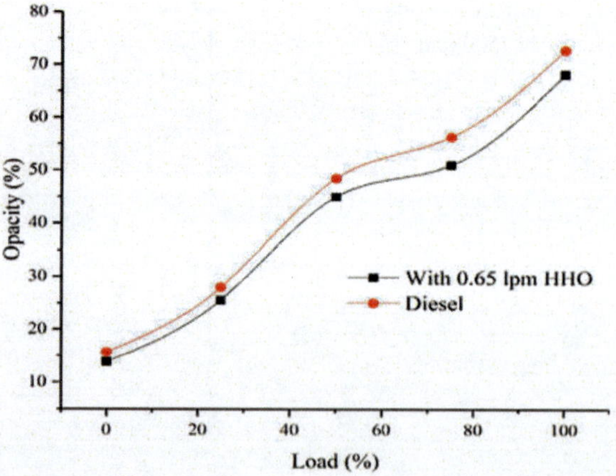

harmful effect on the properties of lubrication oil, when HHO gas is admitted to the engine. Piston and cylinder head deposits were found to be less in the case of dual-fuel combustion. This was evident from the thermogravimetric analysis (TGA) plot as depicted in Fig. 6. This indicates that more stable products were formed since dual-fuel combustion was more efficient and near to being complete.

Figure 7 shows the general variation of predicted in-cylinder pressure and experimentally obtained pressure with respect to crank angle with the two-zone model analysis. It is clear that compared to the single-zone model prediction, the peak value of the pressure is more accurate in the two-zone model result. This accuracy can be attributed to the use of double Wiebe function to determine the heat release. The error obtained in the peak value of heat release rate plot (Fig. 8) also got reduced when double Wiebe function was used. The use of burned zone temperature in the heat release gives the result with a reasonable accuracy.

A typical two-zone analysis result is shown in Fig. 9. It is assumed that there is no heat transfer between the two zones and hence the obtained behaviour clearly indicates that the rise in temperature of unburned zone is only due to the pressure

Fig. 6 TGA comparison of deposits

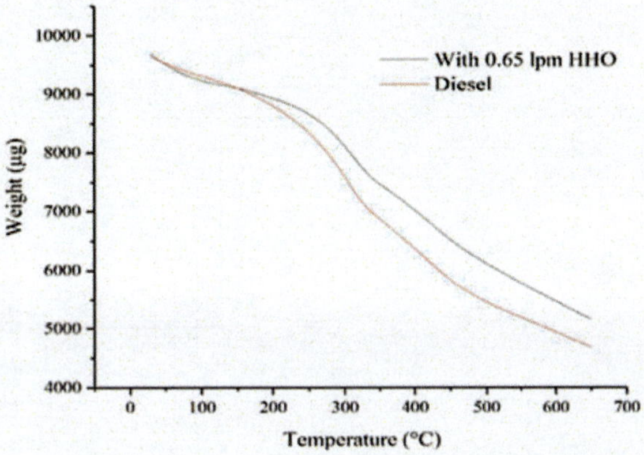

Fig. 7 Variation of in-cylinder pressure with crank angle

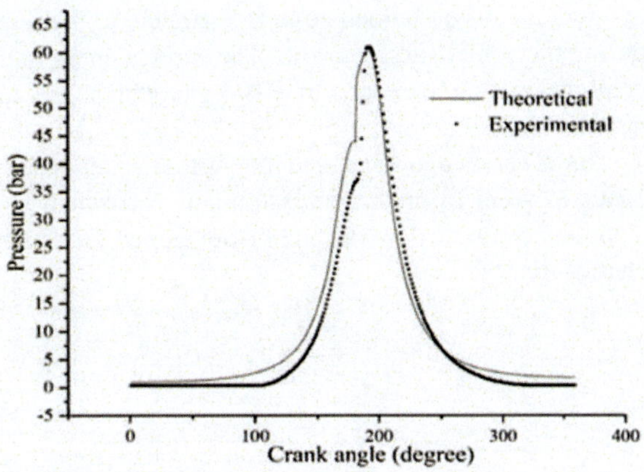

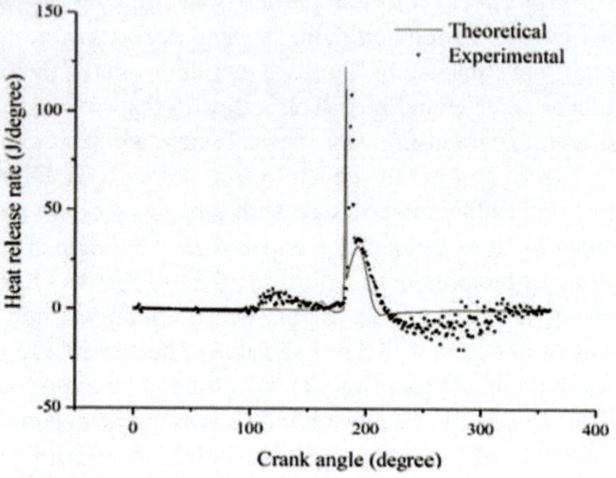

Fig. 8 Variation of heat release rate with crank angle

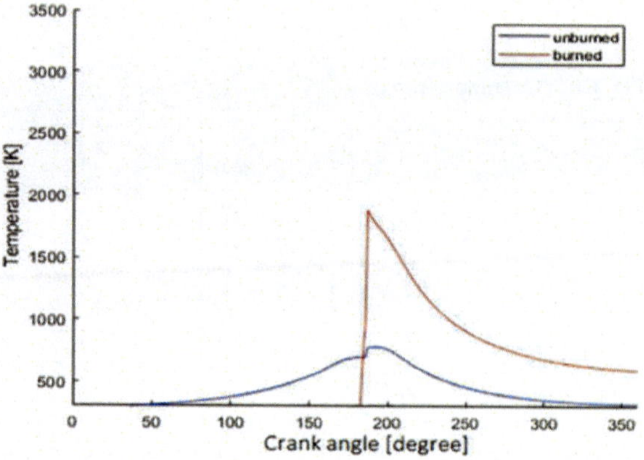

Fig. 9 Zone-wise temperature distribution

rise in the cylinder due to compression. However, the temperature rise of the burned zone is due to the heat released during the process of combustion. Sudden rise in the temperature of the burned zone is attributed to the huge amount of energy released during the combustion reaction. The sudden energy release is due to the favourable combustion characteristics of HHO gas which aids to enhance the combustion of diesel, as well.

The theoretical model can be further extended by increasing the number of zones in order to predict emission and performance characteristics. Furthermore, a phenomenological model can be developed for accurate prediction of combustion characteristics.

6 Conclusions

Based on the optimized conditions, using DoE technique, the DCE was designed, fabricated and tested. Engine characteristic tests, lubricant oil and piston deposit studies were conducted in dual-fuel mode. Combustion modelling results are compared against the experimental results and validated. The following conclusions are drawn from the investigation.

- The energy conversion efficiency of the fabricated DCE was found to be 85%, and its gas generation efficiency is found to be 67%. From comparison study, it was clear that the overall performance of fabricated cell (Cell 3) is better in all aspects, compared to two other commercially available cells (Cell 1 and Cell 2).
- On the basis of the long-run test conducted for 100 h, the results justify the suitability of using DCE to produce required quantity of HHO gas as secondary fuel in automobile applications for longer duration. However, much larger duration studies are essential before making any final conclusions with respect to this.
- A maximum increase of 7.6% in BTE and a maximum reduction of 7.8% in SFC are observed in dual-fuel operation when compared to diesel operation.
- A maximum reduction of 20%, 10% and 30% is observed, respectively, for CO, HC and NO_x emissions in the dual-fuel operation with HHO gas as the secondary fuel.
- The quality of the lubricating oil is not much affected when HHO gas was used as secondary fuel in dual-fuel operation.
- From the piston–cylinder deposit studies, it can be concluded that the combustion is more efficient in dual-fuel operation. But more detailed investigations may be required for arriving at any final conclusions regarding the deposit study.
- The developed model predicted reasonably good results for both the premixed combustion stage and the diffused combustion stage. Such a good agreement is due to the result of employing double Wiebe function with different constants in the premixed combustion stage and the diffused combustion stage.
- Annand's model can be used as a good tool for the evaluation of wall heat transfer in CI engines also. It was found that the predicted results using Annand's model were in good agreement with the experimental data. Hence, it can be concluded that the developed thermodynamic model for dual-fuel operation can successfully predict the variation of in-cylinder pressure (at different crank angles). The instantaneous and cumulative heat release rates with respect to crank angle were also captured (with reasonable accuracy).

References

1. Brayek M, Jemni MA, Kantchev G, Abid MS (2016) Effect of hydrogen—oxygen mixture addition on exhaust emissions and performance of a spark ignition engine. Arab J Sci Eng 1:1–8
2. Basile A, Lulianelli A (2014) Advances in hydrogen production. Woodhead Publishing Series in Energy, UK, Storage and Distribution
3. Egan DR, De Len CP, Wood RJ, Jones RL, Stokes KR, Walsh FC (2013) Developments in electrode materials and electrolytes for aluminium–air batteries. J Power Sources 236:293–310
4. Vogt H, Balzer RJ (2005) The bubble coverage of gas-evolving electrodes in stagnant electrolytes. Electrochim Acta 50:2073–2079
5. Ramadhas AS (2011) Alternative fuels for transportation. CRC Press, New York
6. Ross TJ (2004) Fuzzy logic with engineering applications. Wiley, USA
7. Senthilkumar N, Ganapathy T, Tamizharasan T (2014) Optimization of machining and geometrical parameters in turning process using taguchi method. Australian J Mech Eng 12(2):233–246
8. Yasar H, Soyhan HS, Walmsley H, Head B, Sorusbay C (2008) Double-wiebe function: an approach for single-zone HCCI engine modeling. Appl Therm Eng 28(11–12):1284–1290
9. Bonatesta F, Waters B, Shayler PJ (2010) Burn angles and form factors for Wiebe function fits to mass fraction burned curves of a spark ignition engine with variable valve timing. Int J Engine Res 11(2):177–186
10. Annand WJD (1963) Heat transfer in the cylinders of reciprocating internal combustion engines. Proc Inst Mech Eng 177(1):973–999

Potential Use of Low-Rank High-Ash Indian Coals Through Gasification Route

Taha Y. Poonawala, Parth D. Shah, and Salim A. Channiwala

1 Introduction

India being one of the most dynamic economic powers in the world seeks an ever-growing energy demand to sustain its developing economy. To stay abreast with the flexible global energy system comprising a mix of conventional and non-conventional energy sources, it needs to make sure that the challenges of energy poverty and energy security are met by sustainable energy technologies. In 2018, India's share in world's share of total primary energy supply (TPES) was 5.8% with a growth rate of 7.9% w.r.t. previous year. Out of this share, coal accounted for 55.9%. This indicates that coal has been a dominating source of energy supply when it comes to meeting the demands in developing countries like India [1].

The majority of coal produced in India belongs to the non-coking coal category, and over the past decade, country has witnessed an exponential growth in their imports, in 2017–18 being 20.24% of the net non-coking coal demand. Indian non-coking coal grades are categorized based on their useful heating value, UHV = 8900–138*(ASH + MOISTURE) kcal/kg [2]. Figure 1 shows this grade-wise non-coking coal production in India. It is interesting to note that the trends in production of all coal grades have declined over time while there is an increase in production of 'F' and 'E' grade coal. It can be seen from the UHV formula that the decrease in UHV is due to the increase in ash and moisture content. A higher moisture content can be dealt easily by drying, but high ash content is the main hindrance. Moreover, the effect of coal quality deterioration is seen as the reduction in performance of the end-use equipment [3].

T. Y. Poonawala (✉) · P. D. Shah · S. A. Channiwala
Mechanical Engineering Department, S. V. National Institute of Technology, Surat, India
e-mail: taha.poonawala@hotmail.com

A. K. Gupta et al. (eds.), *Advances in IC Engines and Combustion Technology*,
Lecture Notes in Mechanical Engineering,
https://doi.org/10.1007/978-981-15-5996-9_24

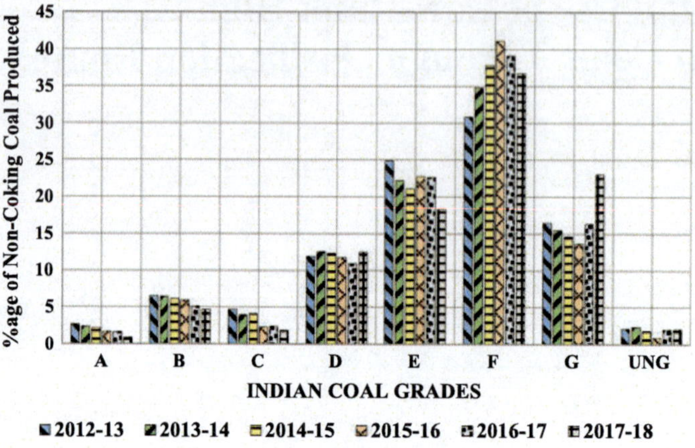

Fig. 1 India's grade-wise non-coking coal production [2]

The share of coal consumed for electricity generation in India has been decreasing as India tries to focus on the renewable sector; however, there has also been an increase in coal imports (coal consumption growth rate for 2007–17 is 5.7% [1]) which raises concerns over the use of indigenous coal production in available thermal plants. It is also foreseen that coal-based generation is likely to provide 50% of the country's energy demand in future since the abundant reserves and low cost of coal make it a preferred energy source. The challenge in future is now to adopt clean coal technologies, CCTs. The increased efficiency of CCTs results in reduced CO_2 emissions for a specified electric output. NO_x, SO_x and particulate emissions are also reduced [4]. Thus, it can be said that there is an attractive need to devise an efficient technology which can utilize the indigenous low-quality coal produce. Coal gasification is one of the CCTs; however, challenge with high ash content is the decrease in calorific value of syngas and the increase in gasifier operating temperature which determines the fusibility and slagging behaviour of ash.

The amount of work reported in the area of high-ash coal gasification is quite less than that of the effects of high-ash coals used in furnaces. Study done by Roy [5] in 1940 is the first attempt towards studying the ash behaviour of Indian coals. Various thermodynamic models with modified coefficients and parametric study for different types of fuels have been reported by Channiwala [6–9]. High-ash coal gasification modelling and experiments have been reported by Krishnudu et al. [10–13]. The authors have reported for improvement in efficiency of high-ash coal gasification under different operating parameters such as pressures higher than 1 bar and steam, O_2, CO_2, Ar/He as gasifying agents. Krishnudu et al. [10] report the earliest works on gasifying high-ash Indian coals with ash content varying from 16.3 to 34.3% wt. They have reported gas HHV of 11–12 MJ/Nm^3 when steam/oxygen and carbon/oxygen molar ratios were maintained at 7 and 4, respectively. Following them, Iyengar [3] quantified the effective decrement in the performance of the gasification process and power generation when ash content was doubled in the coal feed. This is an important analysis, as it gives an estimate on the capabilities of existing equipment to perform against the incoming fuel source degradation. An interesting result is reported by Masnadi [11] where switchgrass rich in potassium when co-gasified with low rank coal, a drastic reduction in tar generation is observed. Some of the recent works by Singh et al. [12] and Mota et. al [13] adopt advanced numerical techniques using CFD to match the experimental results; however, the prediction of CH_4 due to its highly devolatilizing nature is still an ongoing research in this field.

The present work is an adoption of a simple two-phase thermodynamic model given by Channiwala [6] modified with an additional equation to take care of sensible heat loss to ash [14]. The mathematics involved being independent of the gasifier type predicts a unique operating point of gasifier where the process efficiency is highest. Current scope of work is limited to obtaining this operating point when using air as a gasifying agent and analyse the effect of moisture content and enriched air on CGE at atmospheric pressures.

2 Thermodynamic Modelling

Dry syngas is a mixture of CO, CO_2, CH_4, H_2 and N_2; out of this, CO, CH_4 and H_2 are major energy carriers of syngas which later releases heat upon combustion for end use. It also has moisture, tar and other higher molecular aromatic compounds which are then condensed and filtered off in the downstream cooling equipment. The product gas composition thus needs to be predicted. Most basic method for such prediction is thermodynamic modelling which uses mass and energy conservation equations without involving any reaction kinetics to predict an upper limit of the process parameters [7].

In gasifier, heterogeneous and homogeneous reactions take place successively to convert the solid feedstock into gaseous fuel. The primary reaction of solid carbon in fuel with the oxidizer can be explained by heterogeneous reactions, whereas the internal reaction of the product gases can be explained by the homogenous reactions. Accordingly, in the present work, a combined two-phase thermodynamic model is used as given in [6, 7, 14]. The major chemical reactions which are assumed to be at equilibrium in this model are given in Table 1. Two different models each for

Table 1 Chemical reactions considered at equilibrium in two-phase model [6, 14]

Heterogeneous (solid–gas phase) reactions	
Boudouard reaction	$C_{(s)} + CO_2 \leftrightarrow 2CO$
Water–gas reaction	$C_{(s)} + H_2O^{(g)} \leftrightarrow CO + H_2$
Direct methanation reaction	$C_{(s)} + 2H_2 \leftrightarrow CH_4$
Homogeneous (gas–gas phase) reactions	
Water–gas shift reaction	$CO + H_2O^{(g)} \leftrightarrow CO_2 + H_2$
Methane shift reaction	$CH_4 + H_2O^{(g)} \leftrightarrow CO + 3H_2$

heterogeneous and homogenous process are developed and synchronized so as to balance the conservation equations in Eq. (1) such that the heterogeneous model shifts to homogeneous model when 100% solid carbon conversion is reached. The overall reaction stoichiometry is:

$$C_1H_xN_yO_zS_p + \text{A.ash} + \text{wH}_2\text{O}^{(l)} + a(\text{O}_2 + 3.762\text{N}_2) \rightarrow n_{\text{C}}.C_{(\text{s})} + n_{\text{CO}}.\text{CO} + n_{\text{CO}_2}.\text{CO}_2 + n_{\text{CH}_4}.\text{CH}_4 + n_{H_2}.\text{H}_2 + n_{\text{H}_2\text{O}}.\text{H}_2\text{O} + n_{\text{H}_2\text{S}}\,\text{H}_2\text{S} + n_{\text{N}_2}.\text{N}_2 + \text{A.ash} \quad (1)$$

Coal equivalent formula used in thermodynamic model is converted to single carbon atom basis, and it is given as $C_1H_xN_yO_zS_p + \text{A.ash} + \text{wH}_2\text{O}^{(l)}$. All calculations are carried out on dry basis and follow the assumptions given in [6, 7, 14]. Equivalence ratio (ER) is defined as the ratio of actual air-to-fuel ratio to the stoichiometric air-to-fuel ratio. Hence, the actual air supplied is given as $a = a_{\text{st}} * \text{ER}$. For gasification purpose, ER ranges from 0 to 1. An additional assumption is made to model the sensible heat loss to ash in the range of 400–1400 K such that no ash fusion takes place. This heat loss is given by Song [15] and shown in Eq. (2). Also, as gasification occurs in deficient air, it is assumed that H_2S is formed instead of SO_2 while N_2 remains inert due to low operating temperatures.

$$h_{\text{ash}} = 0.0002155T^2 + 0.7618T - 254.20\frac{\text{kJ}}{\text{kg}}; T\text{in}K \quad (2)$$

Five nos. of typical high-ash Indian coal samples are selected from various data sources and given in Table 2. Thermodynamic model is then solved for these samples, and the major performance parameters are calculated. The code for this model is developed in MATLAB, and the thermochemical data is used from NIST-JANAF tables [18]. The major parameters calculated are [14]:

a. **Dry Gas Yield**—amount of syngas produced per kg of coal at NTP (273 K, 1 atm). n_t is the total moles of product gas, and MW_{coal} is the molecular weight of coal

Table 2 Sample high-ash Indian coals [16, 17]

S. no	Coal sample	Dry ultimate analysis (%wt)						MC %wt	Dry C/ASH %wt/wt	LHV_{dry} kcal/kg
		C	H	N	O	S	ASH			
1	MCI	42.17	3.00	0.77	10.21	0.45	43.40	5.35	0.9715	3743
2	GEWAR	37.60	2.61	0.66	8.34	0.41	50.38	4.56	0.7465	3283
3	Sipat	36.12	2.71	0.71	6.29	0.47	53.71	15.00	0.6725	3216
4	RTC	33.01	2.36	0.60	9.15	0.26	54.62	3.78	0.6043	2797
5	Kahalgaon	30.80	3.62	0.61	8.23	0.21	56.53	18.50	0.5448	2913

$$Y_{\text{dry}} = \frac{\left(n_t - n_{\text{H}_2\text{O}}\right) \times 22.4}{\text{MW}_{\text{coal}}}; \frac{\text{Nm}^3}{\text{kg coal}} \tag{3}$$

b. **Lower Heating Value of Dry Syngas**—calculated at NTP (273 K, 1 atm)

$$\text{LHV}_{\text{dry}_{\text{gas}}} = \frac{n_{\text{CO}} \cdot (282770) + n_{\text{CH}_4} \cdot (802528) + n_{\text{H}_2} \cdot (241589)}{22.4\left(n_t - n_{\text{H}_2\text{O}}\right)}; \frac{\text{kJ}}{\text{Nm}^3} \tag{4}$$

c. **Cold Gas Efficiency (CGE)**—ratio of calorific value of syngas to calorific value of coal. It indicates the fraction of energy retained by syngas after gasification.

$$\eta_{\text{CGE}_{\text{LHV}}} = \frac{Y_{\text{dry}}\left(\text{Nm}^3/\text{kgcoal}\right) \times \text{LHV}_{\text{dry}_{\text{gas}}}\left(\text{kJ/Nm}^3\right)}{\text{LHV}_{\text{dry}_{\text{coal}}\ @273\text{K}}(\text{kJ/kgcoal})} \tag{5}$$

d. **Carbon Conversion Efficiency (CCE)**—the amount of solid carbon converted

$$\text{CCE} = (1 - n_c) * 100 \tag{6}$$

3 Results and Discussion

3.1 Model Validation

Currently, the experimental data reported for high-ash coal gasification considers steam and enriched air basis along with a pressurized reactor [3, 10–13]. The developed model is therefore validated with 22 dry-ash-free solid fuels reported by Channiwala [6]. The syngas composition predicted by the model is compared with the predictions of [6], and the average absolute error (AAE) is computed. It is found that AAE is 1.5% for all gases except CH_4 which is 6.25%. The reason for this being the amplification of small changes with change in edition of JANAF table as the conservation equations are nonlinear in nature and the high sensitivity of methane shift reaction. It is found from comparison that AAE reported by model of [6] with the experiments and AAE by current model with experiments differ by 1%. This indicates that the defined model can be used for prediction.

3.2 Model Predictions

Using air as a gasifying agent and gasification pressure as atmospheric, the coal samples given in Table 2 are fed to the model. Predictions of Kahalgaon coal w.r.t. ER variation are shown in Fig. 2. The vertical line in the figure indicates the point of 100% carbon conversion. This point is generally referred as optimum ER [6, 8] as

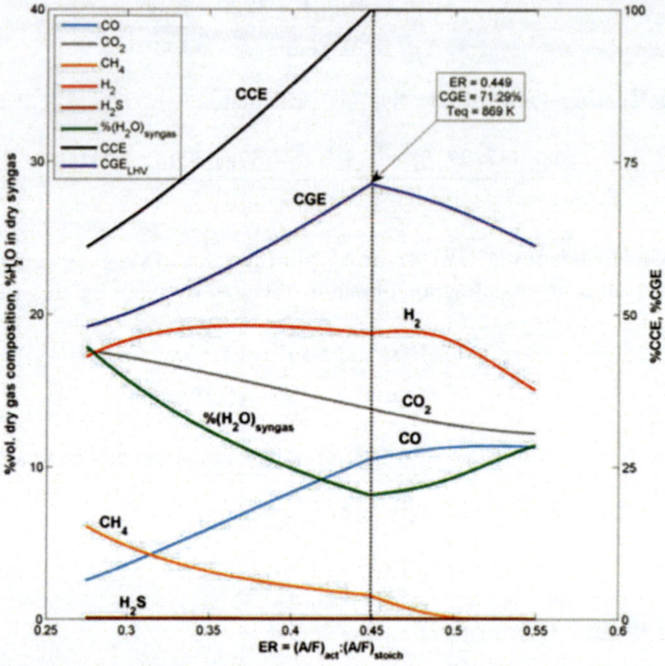

Fig. 2 Model predictions for Kahalgaon coal

the CGE and calorific value of syngas become maximum. Prediction of this ER_{opt} is important as it gives the ideal operating point for the gasifier. The experimentally obtained ER_{opt} will lie in this vicinity. This directly helps in reducing the number of experimental trials. Table 3 summarizes the results obtained for all 5 coals at ER_{opt}. It is observed that ER_{opt} depends on feedstock quality. Also, the equilibrium temperature achieved in current model is a function of the chemical equilibrium achieved amongst the participating reactions and methane shift reaction being quite sensitive; limits the model's operating temperature range to 750–1400 K. Hence, the assumption of heat loss to ash in the form of sensible heat only seems judicial. This heat loss to the ash results in reduction of equilibrium temperature which leads to high methane formation.

It is seen from Fig. 2 that CO increases up to ER_{opt} and then starts to decrease while CO_2 keeps on decreasing up to the same point and then increases. The trends in H_2O content in syngas, shown here as % MC (moisture content) in syngas, have the same trend as that of CO_2. It must be noted that the trends are not straight lines, but slightly curved as it follows from equilibrium constant. The increasing trend of water and carbon dioxide after ER_{opt} indicates that the process is now shifting towards combustion zone and at ER = 1, the products will be exactly equal to the stoichiometric products of combustion. As ER_{opt} also depends on the C, H, O of the fuel, a higher ER_{opt} always leads to N_2 dilution which causes an increase in equilibrium

Table 3 Optimum operating point predicted by model for all 5 coals

S. No.	Coal sample	Opt. ER	% vol. dry gas composition						%MC syngas	Eq. Temp (K)	Gas LHV_{dry} kJ/Nm^3	% CGE LHV basis	% heat loss to ash
			CO	CO_2	CH_4	H_2	H_2S	N_2					
1	MCI	0.429	24.89	5.57	0.21	13.31	0.12	55.91	1.85	979.10	4651.24	76.27	1.86
2	GEWAR	0.449	23.69	6.13	0.19	12.47	0.12	57.39	1.94	972.30	4404.32	75.01	2.43
3	Sipat	0.456	16.92	10.49	0.60	15.72	0.14	56.13	4.39	917.14	4046.40	72.36	2.39
4	RTC	0.465	22.06	7.44	0.22	12.15	0.09	58.04	2.28	956.51	4174.69	73.97	3.00
5	Kahalgaon	0.449	10.54	13.85	1.58	18.79	0.07	55.17	8.20	869.26	3923.25	71.29	2.48

temperatures. In case of fluidized or entrained bed gasifiers, higher bed temperatures may result in dissociation of N_2. Such endothermic reactions may hamper the auto-thermal operation of the gasifier [9]. To avoid this, secondary gasifying agents like steam, O_2, CO_2 along with a pressurized bed are used and have been reported by various researchers [3, 10–13]. The operating cost, however, increases drastically. The aim is to rather improvise the efficiency with least inputs, be it the raw feed quality or the gasifying agent. This is where the adoptability of the process and its thermodynamic investigation becomes attractive.

3.3 Trends in Cold Gas Efficiency

Effect of Moisture Content. Moisture content during gasification favours water–gas shift reaction and hence generates CO_2 by consuming CO and liberating H_2. Increase in moisture content hampers the auto-thermal operation of gasifier. This is because the latent heat required for moisture evaporation may supersede the heat generated by the exothermic reactions. Auto-thermal limit of 65% moisture content is reported by Gautam [9] beyond which self-sustaining gasification is not possible.

Moisture content in the dry fuel is varied from 0 to 20% in steps of 5% for all the coal samples, and the optimum points are collected over each moisture content. A 3D plot of maximum CGE predicted is plotted against dry C/ASH ratio of the fuel as shown in Fig. 3a. A 3D linear surface fit is obtained using the inbuilt subroutine—curve fitting of MATLAB. A similar plot is also reported by Channiwala [6]. It is seen that maximum CGE is 76.69% for MCI coal at 0% MC while minimum CGE is 69.76% for Sipat coal at 20% MC. Practically, a highest gain in CGE cannot occur at 0% MC, since some moisture is always required to attain the shift reactions, which

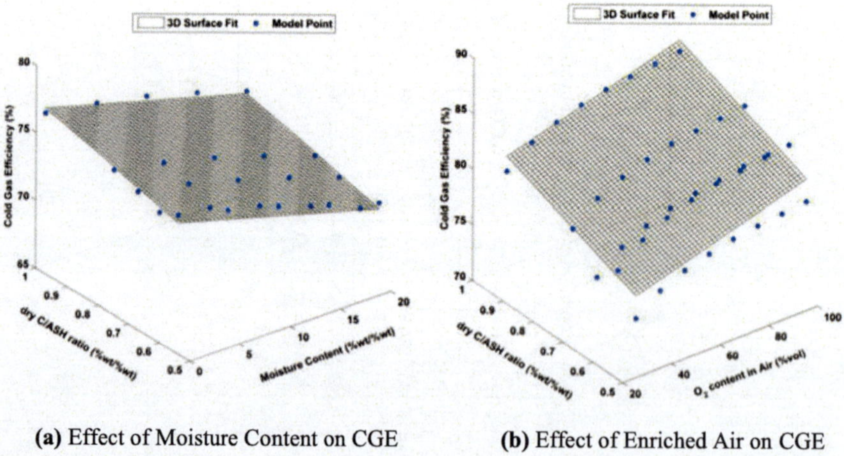

(a) Effect of Moisture Content on CGE **(b)** Effect of Enriched Air on CGE

Fig. 3 Parameters affecting cold gas efficiency (CGE)

is now balanced by the bound H_2 of the fuel. Hence, a complex equilibrium of more than one parameter is acting simultaneously to result in the most favourable outcome. It is this region where the optimum amount of MC to get a good balance of H_2 and CO in the fuel can be obtained but requires inclusion of chemical kinetics for better prediction.

Effect of Enriched Air. Various gasifying agents—steam, CO_2, O_2, He/Ar—have been identified over time which acts as a process catalyst [10–13]. Variation in O_2 content is attractive as it increases the carbon conversion rates by reducing the dilution effect of N_2. Keeping the fuel contents with moisture content, the volumetric composition of air is changed by changing the O_2/N_2 ratio in the model. It is varied from 30 to 100% in units of 10 for all coal samples, and the optimum points are collected for each such case. A 3D plot as discussed earlier is obtained to interpret the global trend in CGE variation. It is clear from Fig. 3b that with increase in O_2 content with higher carbon content coals, CGE increases. This is expected as H_2 and CO contents increase due to high depletion rates of CH_4 at higher temperatures and favourable conditions for Boudouard reaction, permitting high amount of reduction of CO_2, thereby increasing the syngas LHV. Max CGE is predicted for MCI coal as 84.2% in 100% O_2 atmosphere and minimum of 74.21% for Kahalgaon coal in 30% O_2 atmosphere. Since it is an increasing trend, it seems that a 100% O_2 atmosphere must be used to improve the process efficiency; however, there are two aspects which can give the optimum value of O_2 required for a given coal: firstly, the ash fusion temperatures must be modelled with high bed temperatures as it will consume a lot of thermal energy for slagging purpose, and secondly, the economics of the O_2 production plant used.

4 Concluding Remarks

The rising energy demand with coal as a dominant source in India has pushed itself to make use of the domestic low-rank high-ash coal produce, and gasification appears to be one of the CCTs for this. In the present work, the past coal scenario has been outlined where India has produced high amount of 'F' grade coal. Five nos. of coal samples of this category are selected whose ash content (% wt. dry basis) varies from 43.4 to 56.53% and two-phase thermodynamic model has been used to predict the upper limit of the syngas composition and CGE irrespective of the type of gasification process employed. Model predicts the ER_{opt} for each coal where CGE is maximum which is very useful in reducing the number of experimental trials. The effect of moisture content and enriched air on CGE shows that max CGE occurs with 0% MC and 100% O_2. The inclusion of chemical kinetics in this model would give better predictions and can be considered as future scope of work. The high values of CGE predicted intuitively imply that there is a potential in gasifying low-grade coals.

References

1. BP Statistical Review of World Energy, 68th edition. British Petroleum, London, England (2019). http://www.bp.com/statisticalreview
2. Provisional Coal Statistics, Government of India, Ministry of Coal, Coal Controller's Organisation, Kolkata. http://www.coalcontroller.gov.in/pages/display/20-provisional-coal-statistics
3. Iyengar RK, Haque R (1991) Gasification of high-ash Indian coals for power generation. Fuel Process Technol 27(3):247–262. https://doi.org/10.1016/0378-3820(91)90051-D
4. Gräbner M (2015) Industrial coal gasification technologies covering baseline and high-ash coal. Wiley-VCH Verlag GmbH & Co., Germany
5. Roy RKD (1940) A critical study of some Indian coal ashes. Geol Survey of India 4(3):539–548
6. Channiwala SA (1992) Biomass gasification process and technology development—some analytical and experimental investigation, Ph.D. Thesis submitted to Mechanical Engineering Dept., IIT Bombay
7. Ratnadhariya JK, Channiwala SA (2009) Three zone equilibrium and kinetic free modeling of biomass gasifier—a novel approach. Renew Energy 34(4):1050–1058. https://doi.org/10.1016/j.renene.2008.08.001
8. Jarungthammachote S, Dutta A (2007) Thermodynamic equilibrium model and second law analysis of a downdraft waste gasifier. Energy 32(9):1660–1669. https://doi.org/10.1016/j.energy.2007.01.010
9. Gautam G (2010) Parametric study of a commercial-scale biomass downdraft gasifier: experiments and equilibrium modelling, M. Tech Thesis submitted to the Graduate Faculty of Auburn University, Alabama
10. Krishnudu T, Madhusudhan B, Narayan Reddy S, Seshagiri Rao K, Vaidyeswaran R (1989) Moving bed pressure gasification of some Indian coals. Fuel Process Technol 23(3):233–256. https://doi.org/10.1016/0378-3820(89)90022-2
11. Masnadi MS, Grace JR, Bi XT, Lim CJ, Ellis N, Li YH, Watkinson AP (2015) From coal towards renewables: Catalytic/synergistic effects during steam co-gasification of switchgrass and coal in a pilot-scale bubbling fluidized bed. Renew Energy 83:918–930. https://doi.org/10.1016/j.renene.2015.05.044
12. Singh N, Raghavan V, Sundararajan T (2014) Mathematical modeling of gasification of high-ash Indian coals in moving bed gasification system. Int J Energy Res 38(6):737–754. https://doi.org/10.1002/er.3078
13. Mota R, Krishnamoorthy G, Dada O, Benson SA (2015) Hydrogen rich syngas production from oxy-steam gasification of a lignite coal—A design and optimization study. Appl Therm Eng 90(5):13–22. https://doi.org/10.1016/j.applthermaleng.2015.06.081
14. Poonawala TY (2015) Effect of operating parameters on high ash coal gasification. M. Tech seminar report submitted to Institute of Technology, Nirma University, Ahmedabad
15. Song G, Shen L, Xiao J, Chen L (2013) Estimation of specific enthalpy and exergy of biomass and coal ash. Energy Sources, Part A: Recovery, Utilization, and Environ Effects 35(9):809–816. https://doi.org/10.1080/15567036.2011.586983
16. Guoxin L (2009) A discussion about strategy of flue gas dust removal for indian coal fired boiler. In: Yan K (eds) Electrostatic precipitation. Springer, Berlin, Heidelberg. https://doi.org/10.1007/978-3-540-89251-9_101
17. Visuvasam D, Selvaraj P, Sekar S (2005) Influence of coal properties on particulate emission control in thermal power plants in india. In: 2nd international conference on clean coal combustion technologies for our future. Italy, p 44
18. Chase MW (1998) NIST-JANAF thermochemical tables, 4th edn. NIST, Maryland

Bio-methane Generation from Anaerobic Co-digestion of Eichhornia (Water Hyacinth) and Kitchen Edible Material Ravage and Waste Paper with Pond Sludge and Cow Compost by Using Chemical Pretreatment

Sonam Sandhu and Rajneesh Kaushal

1 Introduction

Worldwide exhaustion of non-renewable energy sources has prompted the quest for elective wellsprings of vitality. Biomass has the biggest potential and must be considered as the best alternative for gathering the interest and protection of future vitality (biofuel) supply in a supportable way. Biomass is very much perceived as a fuel and furthermore, positions fourth as an imperativeness resource giving generally 14% of the world's essentialness needs [2], biomass is the most significant wellspring of vitality in creating countries, giving ~35% of their vitality [2]. Biomass is utilized to meet an assortment of vitality needs as shown in Fig. 1, including producing power, warming homes, fuelling vehicles and giving procedure warmth to modern offices.

The biofuel creation from inexhaustible assets eludes especially to the ligno-celluloses biomass/materials, as this makes up most of the modest and bottomless non-nourishment materials accessible from the plants. In this manner, lignocelluloses feedstock can offer the possibility to give novel biofuel of the subsequent age [1].

Anaerobic assimilation of solidified desolate not just can treat strong waste, however, can likewise create helpful biofuel, which can be utilized for different purposes, for example, cooking, fuelling inward burning motors, etc. as shown in Fig. 1. The production of biogas has mainly four stages viz. hydrolysis, acidogenesis,

S. Sandhu (✉)
Department of Mechanical Engineering, National Institute of Technology Kurukshetra, Kurukshetra, Haryana 136119, India
e-mail: sandhusonam022@gmail.com

R. Kaushal
Department of Mechanical Engineering, National Institute of Technology Kurukshetra, Kurukshetra 136119, India

A. K. Gupta et al. (eds.), *Advances in IC Engines and Combustion Technology*, Lecture Notes in Mechanical Engineering,
https://doi.org/10.1007/978-981-15-5996-9_25

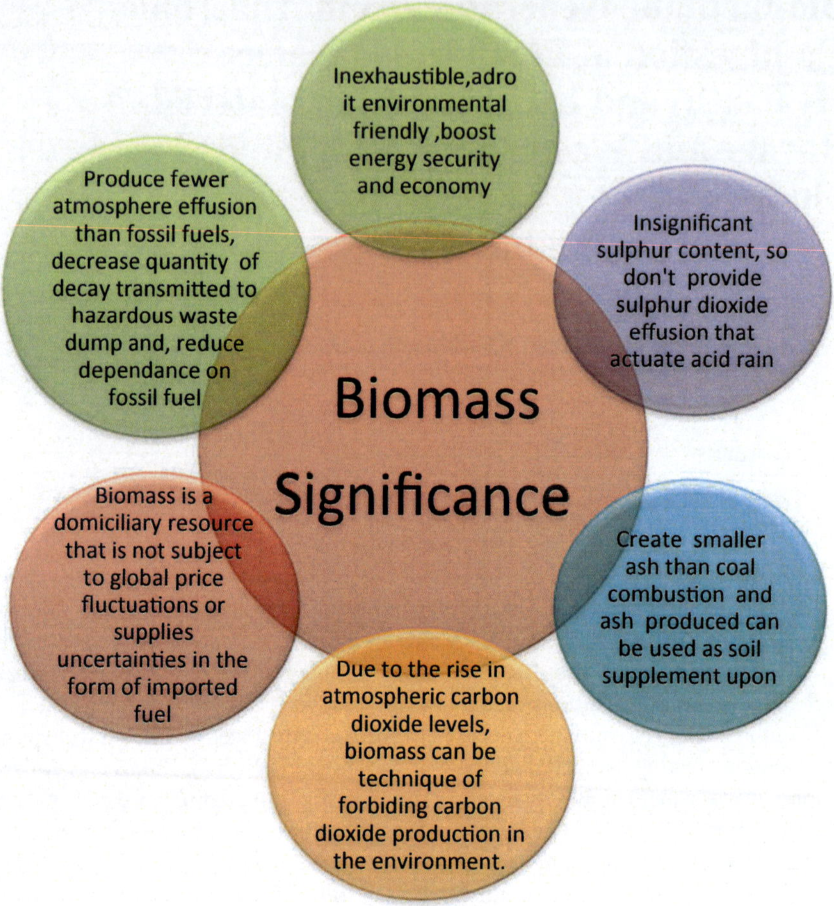

Fig. 1 Biomass significance

acetogenesis and methogenesis and is shown in Fig. 2. The hydrolysis step includes the transformation of complex organic matter into levulose; acidogenesis includes the change of levulose into natural acids; acetogenesis includes the change of these natural acids into acidic corrosive [3–5]. At last, the methanogenic stage comprises of the change of acidic corrosive into CH_4 and CO_2 [3–5]. Thus this examination was attempted to explore the plausibility of co-absorption of organic wastes like kitchen food waste, pond sludge, water hyacinth and cow dung to generate useful energy [6]. Eichhornia, pond sludge and kitchen edible food ravage, waste paper, cow compost by pretreated chemically were taken in various proportions and cluster tests were done for blend.

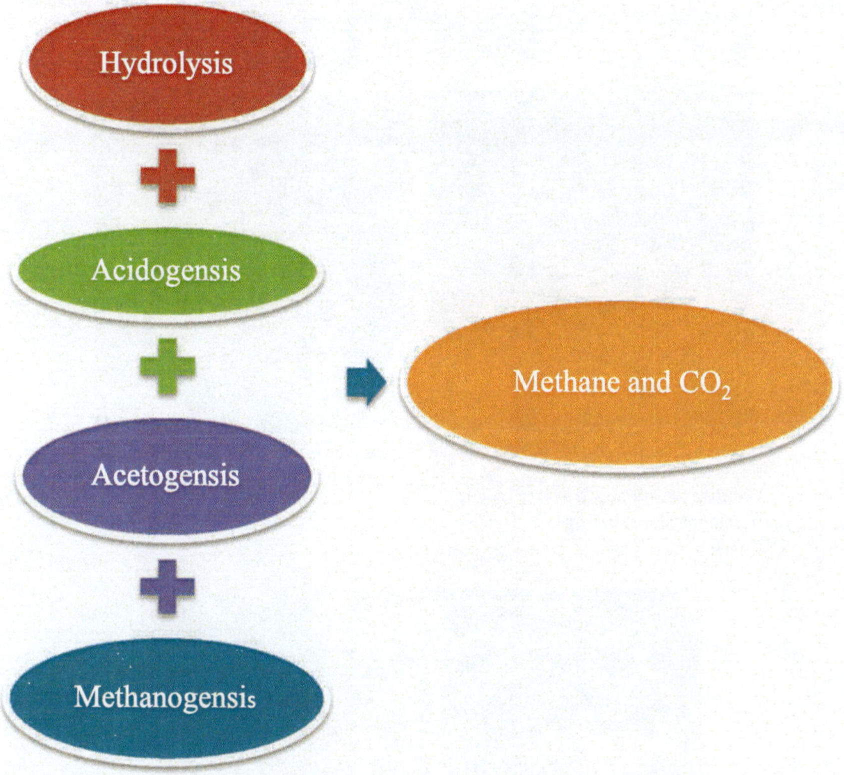

Fig. 2 Anaerobic digestion processes

2 Material

The materials used for this experiment were sludge of pond, food waste, waste paper, cow manure and water hyacinth. Wet cow manure (CM), water hyacinth and lake mud were gathered from the Dayalpur Village in the close vicinity of NIT Kurukshetra. Food waste was gathered from the Senate Hall of NIT Kurukshetra. A bio-digester of 2000 litre capacity was used to accomplish research. In continuous process mode, $CaOH_2$ and NaOH have been utilized to pre-treat the water hyacinth, waste paper, and food waste, pond sludge and cow dung. For preparation of sample, 500 g of calcium and sodium hydroxide entire of 1000 litre of ravage (combination of food waste, pond sludge, cow dung, water hyacinth, waste paper, fresh tap water) (Figs. 3 and 4). The entire biogas and its essential constituents CO_2 and CH_4 formed in bio-digester have been quantified by using biogas analyser.

The characteristics of different organic wastes used in the present study have been shown in Table 1.

Fig. 3 **a** Water hyacinth and **b** water hyacinth, food waste, waste paper, cow dung, pond sludge before pretreatment and grinding

Fig. 4 Water hyacinth and food waste, waste paper, cow dung, pond sludge after pretreatment and grinding

2.1 Experimental Test Unit and Aspect

For present research, an experimental arrangement has been facilitated in Thermal Engineering Laboratory of Mechanical Engineering Department of NIT Kurukshetra, India. The experimental test facility has a bio-digester of 2 m^3 capacity, made up of Fibre fortified plastic and has gas carrying limit of 800–900 g at normal conditions. Every day ravage bolstering limit of bio-digester is 20 kg. A PID controller has been used to maintain mesophilic state of the digester. For biogas estimation, a biogas analyser has been used. A designed electronic gauging machine of precision 0.5 g has been utilized for gauging the semiliquid mixture tests. A pH metre has been

Table 1 Characteristics of different waste [7]

Parameter	Food waste	Water hyacinth	Cow dung	Waste paper
Carbon	42.8	14.5	13.40	45
Nitrogen	2.25	–	0.666	0.6
Hydrogen	8.99	–	1.78	5.0
Oxygen	38.9	39.5 ± 6.5	11.49	45
Ash	4.81	21.1	2.68	0.1
TS	60.9	–	–	2.3
VS	58.99	–	60.91	–
Moisture content	65.36	–	69.98	7.8
Volatile matter	–	64.4	–	–

utilized for pH estimation. Hydrogen sulphide, oxygen and methane, carbon dioxide have been evaluated by the biogas analyser. A Schematic and Pictoric views of the test facility have been depicted in Fig. 5(i, ii).

3 Results and Discussion

3.1 Impact of Temperature

The pace of methane generation has been observed to be enhanced with increment in temperature for moderate temperature, i.e. 42 °C as reported [2]. However, after having maximum generation, it started decreasing in last phase and can be seen in Fig. 6. This happens since at high temperature, the survival of the microbes performing anaerobic assimilation starts affecting and, thus, the pace of gas creation altogether diminishes. Beneath 40 °C temperature, the development of the bacteria diminishes and consecutively the gas formation rate additionally decays. Consequently, the intermediate calefaction supports the most extreme generation of gas.

3.2 Impact of PH

To observe the effect of pH on the CH_4 generation rate, preliminaries were aimed at different estimations of pH of the mixture of bio-wastes used in the study at varied temperature up-to 25 °C. The cumulative methane yield with respect to HRT at different values of pH is shown in Fig. 7. The figure affirms that the greatest generation is accomplished when the blend pH is in the neutral mode and the less production rate is acquired if the pH is in the essential mode. The least creation of CH_4

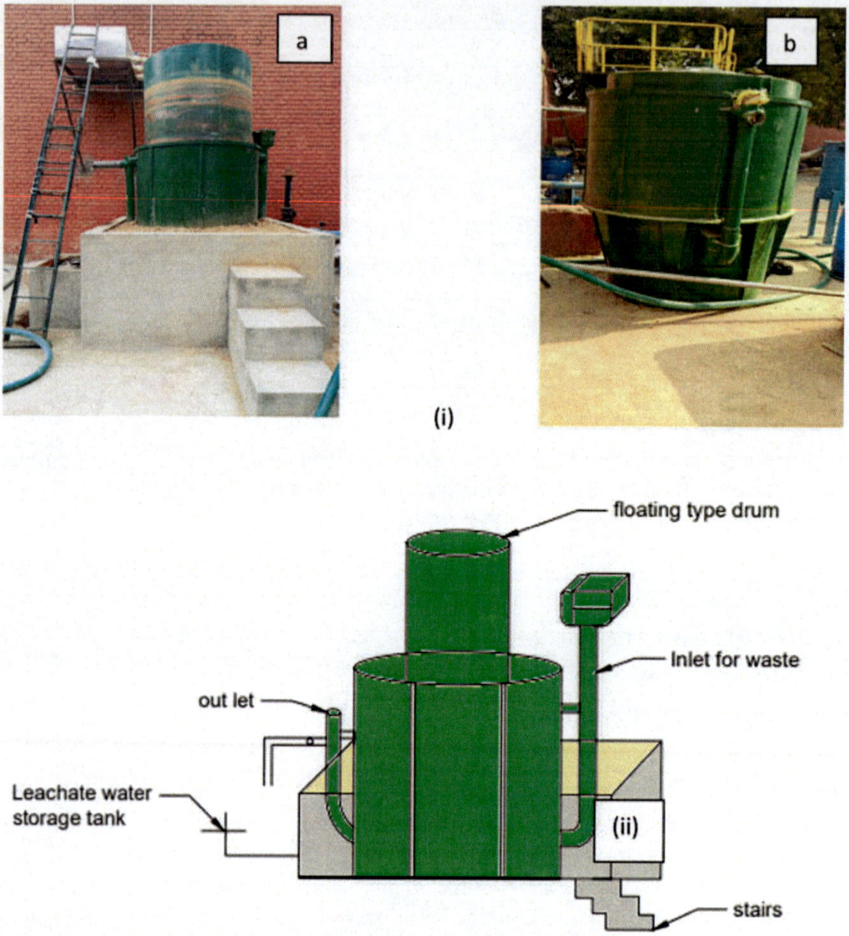

Fig. 5 (i) Pictoric view of the test facility (a) at beginning phase (b) After formation of biogas and (ii) Outline of test setup

in basic medium is accomplished by virtue of two causes: the first logic alludes to the dormancy of microbes in basic mode and the subsequent explanation is identified with the last phase (methanogenesis) of the procedure. In the final phase, acids are converted into CH_4 & CO_2. Thus, the acids are utilized for CH_4 & CO_2 generation thereby reducing the further generation of biogas. Chemical action is mediated to have a solid work by varying pH which thus impacts the cell action of the microbes. Catalysts typically stay dynamic just in a distinct pH go, yet the aggregate enzymatic movement and the natural pH have a complex function [8].

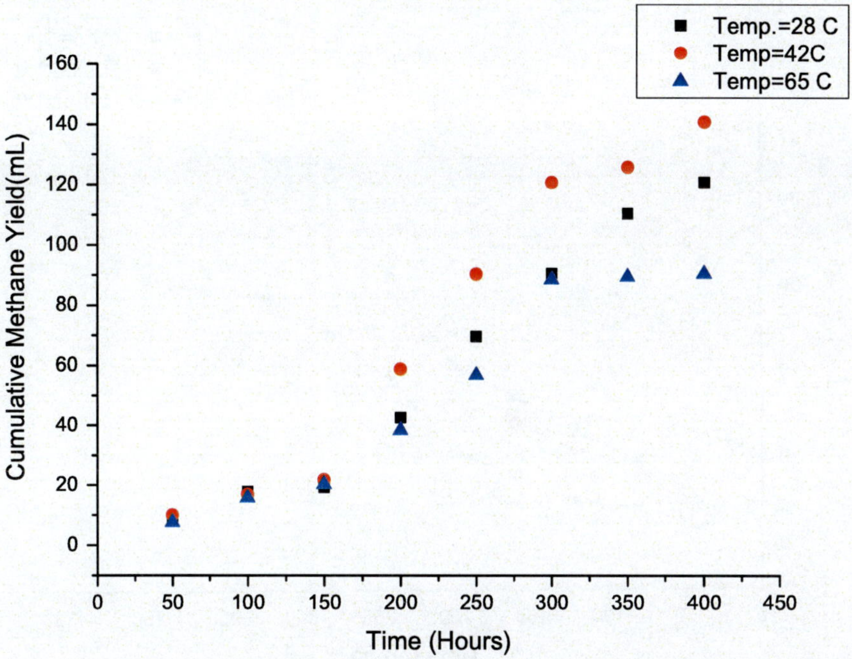

Fig. 6 Cumulative methane yield with respect to HRT at different temperatures

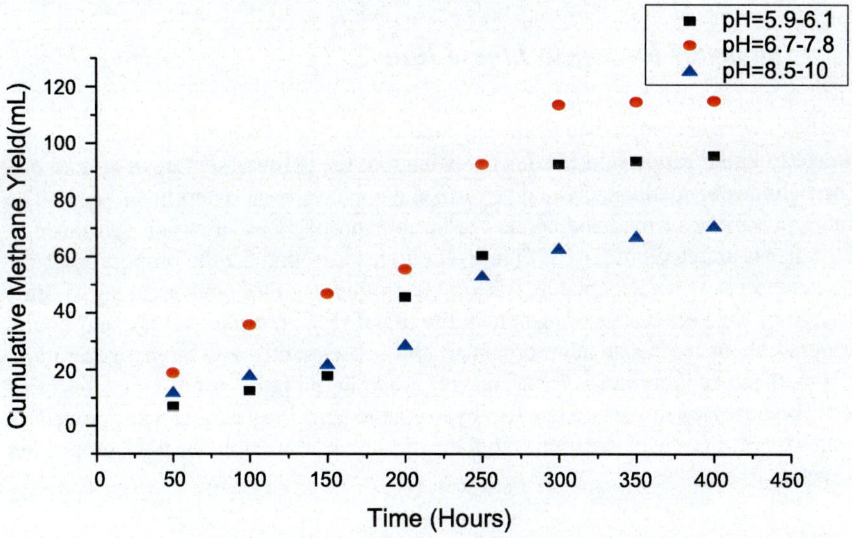

Fig. 7 Cumulative methane yield with respect to HRT at different values of pH

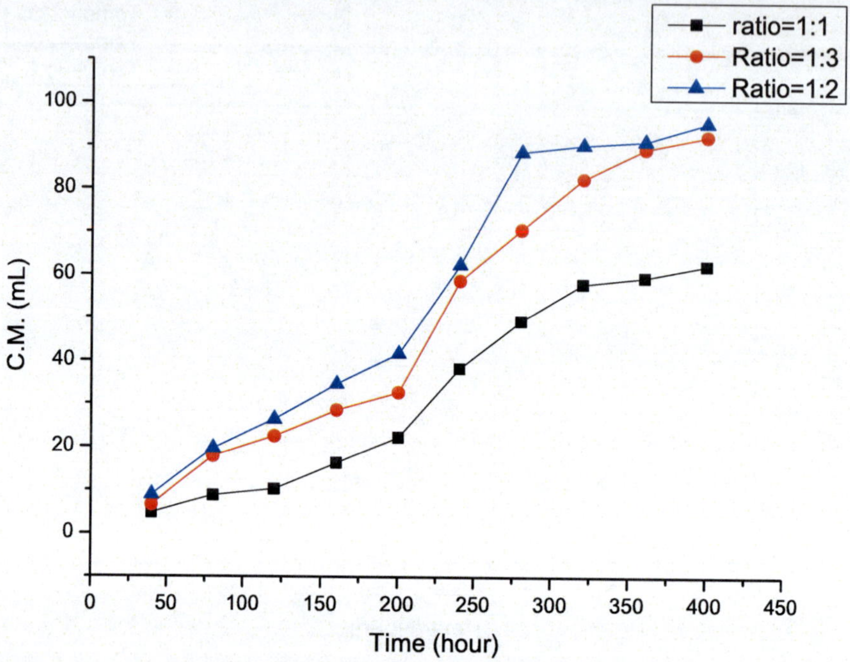

Fig. 8 Cumulative biogas production between time at different solid to waste ratio

3.3 Impact of Ravage to Liquid Ratio

Analyses were accomplished at normal conditions at various proportions of different ravage to liquid ratios, and for this investigation, the pH was kept up at around 6.2. The volumes of methane created at various days have been determined [7, 8]. The transient variety of methane content at various proportions of strong substance to the water is appeared in Fig. 8. Figure 8 demonstrates that for the proportion of 1:1 gas generation is not acceptable. It is on the grounds that for equivalent lag of time, the age of weakened acid brings down the pH of the mixture of ravage and liquid, bringing about inertia of microorganism and subsequently low biogas generation. The methane generation for 1:3 blend (ravage to liquid ratio) is not as much as that of 1:2 blends. This might be a result of over-pressurizing the slurry, and this procedure makes issue in methane creation. From the present examination, the ideal proportion is seen as 1:2.

4 Biogas Fructification from Co-assimilation of KW with PS and KW and CD

Initially, the comparison of biogas production between KW + CD (see Fig. 9) and KW + PS (see Fig. 10) demonstrated that KW with CD showed higher efficacious mixture as compared to KW and PS. While KW + PS has demonstrated the production reaches highest point before KW + CD, the production point reached by KW + CD was perceptibly higher than KW + PS. Figure 11 outlines that KW + CD delivers a higher measure of biogas contrasted with KW + PS. While it is fascinating to observe that when PS was utilized with KW, the production rate appeared to increment fundamentally simply after 140 h, though the rate for KW + CD increased dramatically after 140th hour. Additionally, the higher production acquired from KW + PS showed up appeared much abrupt than KW + CD (see Fig. 11).

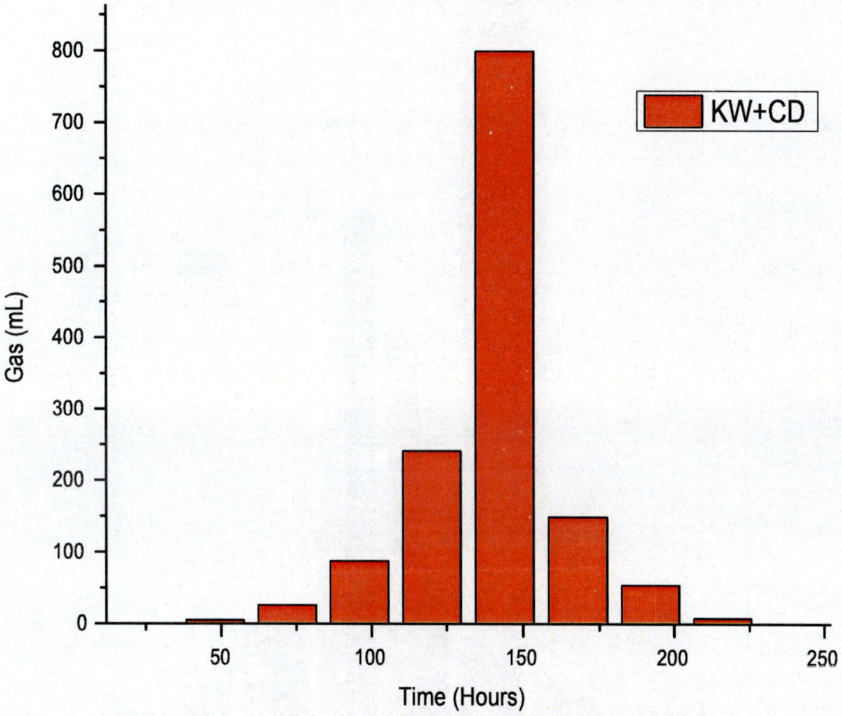

Fig. 9 Production of gas from KW + CD in 50 days at loading rating of 205 gm/L

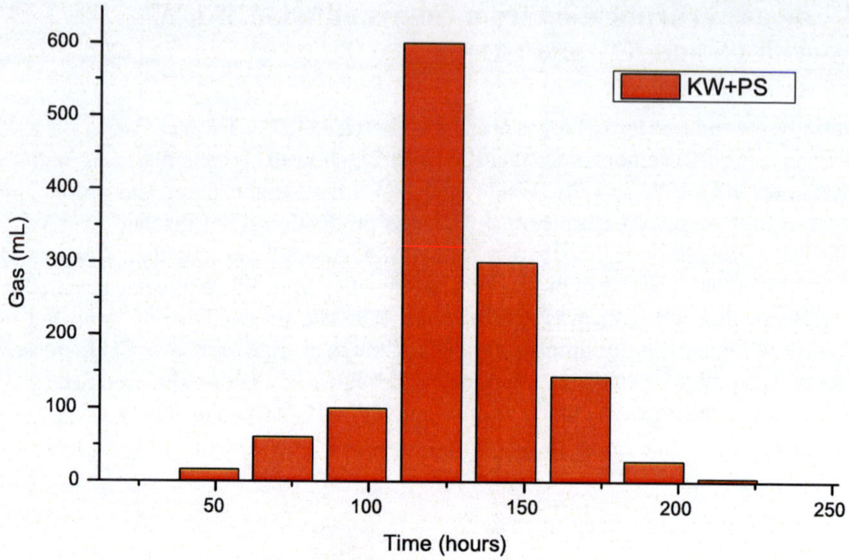

Fig. 10 Production of gas from KW + PS in 50 days at loading rating of 205 gm/L

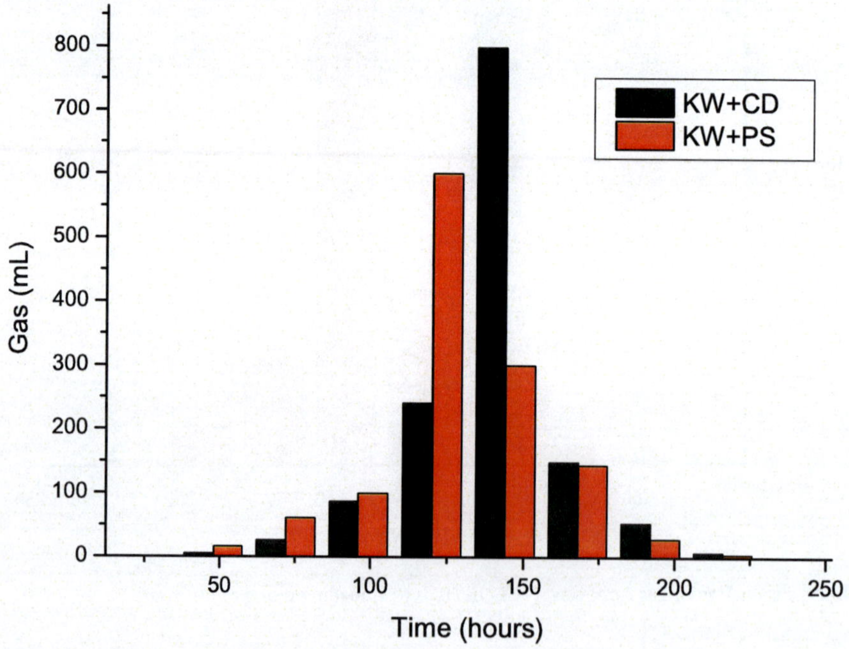

Fig. 11 Comparison of gas production from KW + CD and KW + PS in loading rating of 205 gm/L

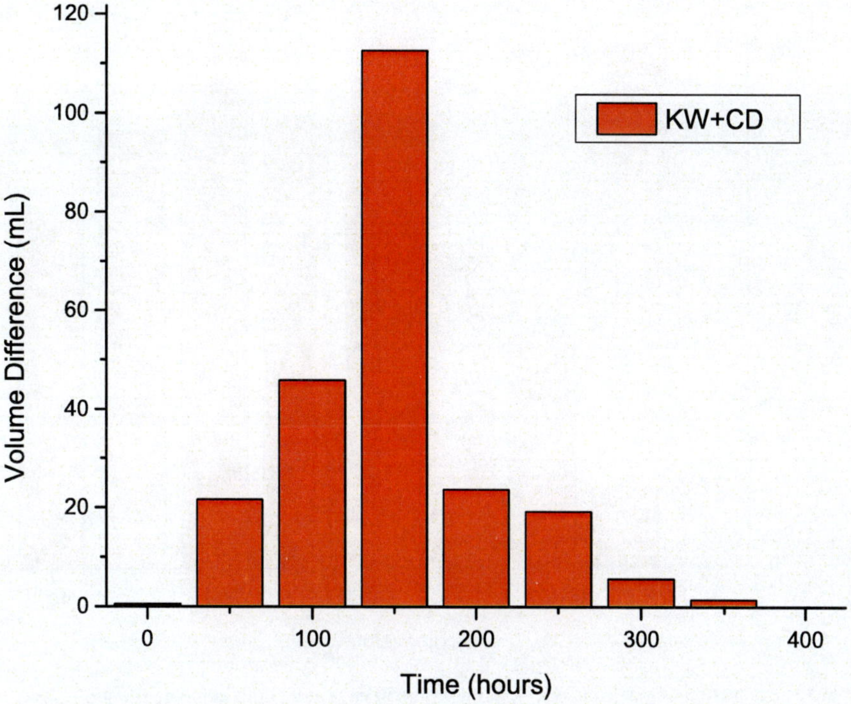

Fig. 12 Production of gas from KW + CD in 50 days at loading rating of 100 gm/L

5 Production of Biogas from KW+CD

Biogas generation from KW and CD demonstrated a dramatic increment underway after about 120 h of the investigation, though generation began falling not long after this ascent (see Fig. 12).

6 Production of Biogas from WH + CD + PS

While in different cases gas creation has been found to stop after around 240 h, gas from WH + CD alongside PS proceeded with generation even after the 374th hour (see Fig. 13).

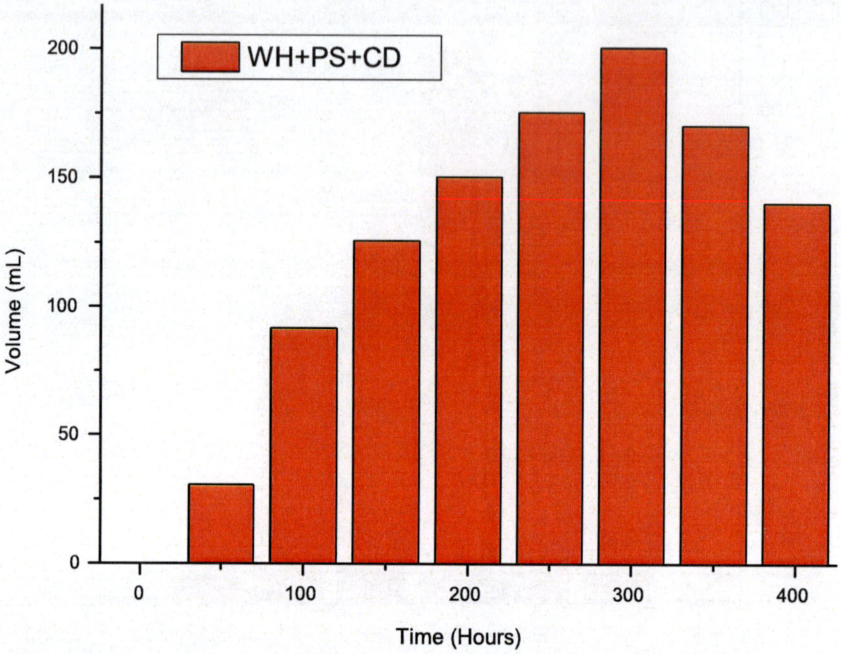

Fig. 13 Production of gas from WH + CD + PS in 50 days at loading rating of 100 gm/L

7 Comparison of Biogas Production from KW + CD & WH + CD + PS for 254 h

The enormous distinction between KW + CD and WH + CD + PS is seen because of the use of pond sludge and NaOH and Ca $(OH)_2$ with WH + CD rather than typical water. Pond sludge is effectively biodegradable since it comprises of absorbable starches and fats, contrasted with actuated ooze which comprises of composite sugars, proteins and elongated series hydrocarbons. Likewise, unadulterated water does not contain any of that. Also, Ca $(OH)_2$ and NaOH were utilized to keep up pH and start microbial activity. They help to build the generation of biogas [6] (see Fig. 14).

8 Conclusion

Dispersal of biomass and its management should be given a more serious need in lighting up our imperativeness crisis. Useful bioenergy headway could diminish progressively the waste-related issues. In general, it may be concluded that with mesophilic condition (42 °C) and co-ingestion in the conditions of OLR 100 gm/L treated with 2.0% NaOH and Ca $(OH)_2$, squander paper, cow compost and Eichhornia

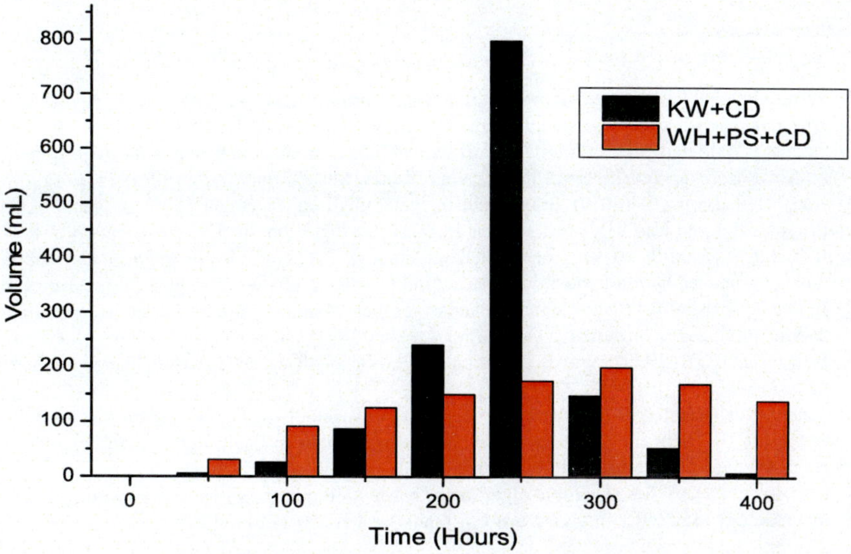

Fig. 14 Production of biogas versus time curve for KW + CD and WH + PS + CD

with lake sludge could be an essential wellspring of potential biogas for nuclear family similarly as mechanical essentialness use. Kitchen edible food ravage with CD can likewise be a sensible alternative for petroleum gas. Be that as it may, the gas creation solidness is excessively low for this situation. Further mixes of waste materials should be tried with extraordinary thought of pond sludge as a quickening agent. It is accepted that upon the use of viable mixes of devastate resources and appropriate modification in conditions, improved outcomes can be acquired. The result showed that

1. The biogas produced after 900 h of anaerobic decomposition of the above-mentioned mixture consisted of 67.8% CH_4, 13% carbon monoxide and 28.9% other gases, while the same mixture contents with volume of 335 ml, without chemical pretreatment produced 62.5% methane, 18.9% CO and 24% other gases within the same time frame.
2. Maximum methane production augmented with increased in temperature from 25 to 42 °C and additional boost in temperature reduce the rate of production.
3. Methane productions augment with increase in pH and solid to waste ratio from 6.1 to 8.5 and 1:2.

Acknowledgements The authors gratefully acknowledge the financial support provided by SERB-DST (Govt. of India) project (Grant no. EEQ/2017/000227) and infrastructural facilities provided by NIT-Kurukshetra for the present work.

References

1. Tasnim F, Iqbal SA, Chowdhury AR (2017) Biogas production from anaerobic co-digestion of cow manure with kitchen waste and Water Hyacinth. Renew Energy 109:434–439
2. Jingura RM, Matengaifa R (2009) Optimization of biogas production by anaerobic digestion for sustainable energy development in Zimbabwe. Renew Sustain Energy Rev 13(5):1116–1120
3. Kaushal R, Baitha R (2019) Biogas and methane yield enhancement using graphene oxide nanoparticles and $Ca(OH)_2$ pre-treatment in anaerobic digestion. Int J Ambient Energy 1–8
4. Baitha R, Kaushal R (2019) Experimental and numerical study of biogas, methane and carbon dioxide produced by pre-treated wheat straw and pre-digested cow dung. Int J Sustain Eng 1–8
5. Baitha R, Kaushal R (2018) Numerical and experimental study of biogas, methane and carbon dioxide produced by pre-treated slurry. Int J Ambient Energy 1–7
6. Patil JH et al (2014) Anaerobic co-digestion of water hyacinth and sheep waste. Energy Procedia 52:572–578
7. Sandhu S, Kaushal R (2019) Anaerobic digestion of vegetable, fruit and cafeteria wastes with cow dung by chemical pretreatment for biogas production in batch digester. In: Journal of physics: conference series. IOP Publishing
8. Pati A et al (2019) The anaerobic digestion of waste food materials by using cow dung: a new methodology to produce biogas. J Inst Eng (India) Ser E 100(1):111–120

Experimental Study of a Diesel Engine Using Soybean-Based Biodiesel and Diesel Blends

Arun Rathi and Rajneesh Kaushal

Nomenclature

VCRE	Variable compression ratio engine
CR	Compression ratio
BSFC	Brake-specific fuel consumption
TRCC	Toroidal re-entrant combustion chamber
IVC	Inlet valve closing
EVO	Exhaust valve opening
CFD	Computational fluid dynamics
BTE	Brake thermal efficiency
BMEP	Brake mean effective pressure
BP	Brake power

1 Introduction

The world is troubling in the current scenario with two major problems such as increasing demand for fossil fuel and the steady increase of environmental emissions. In this situation, finding the alternative of petroleum-based fuels is a better way to

A. Rathi (✉)
Department of Mechanical Engineering, National Institute of Technology Kurukshetra, Kurukshetra, Haryana 136119, India
e-mail: arun_31806112@nitkkr.ac.in

R. Kaushal
Department of Mechanical Engineering, National Institute of Technology Kurukshetra, Kurukshetra 136119, India
e-mail: rajneesh@nitkkr.ac.in

A. K. Gupta et al. (eds.), *Advances in IC Engines and Combustion Technology*, Lecture Notes in Mechanical Engineering,
https://doi.org/10.1007/978-981-15-5996-9_26

overcome this issue. Biodiesel leads its rivals among non-renewable fuel opponents to replace diesel in the transportation sector for fuel and emission control problems. Due to its carbon neutrality, renewability, biodegradability and non-toxicity, biodiesel is considered as a better alternative fuel of diesel. Biodiesel contains reactive unsaturated fatty acid esters, mostly oleic acid and linoleic acids, which during the storage process are susceptible to oxidative degradation [1]. The reactive unsaturated fatty acid esters affect engine performance and result in many problems like seal swelling deterioration etc [2]. Wang et al. [3] studied that the fundamental engine efficiency as energy output and fuel consumption are similar to that of the diesel when engine powered with edible oil-based biodiesel and its mixtures. Also, nitrogen oxide emissions are less relative to diesel fuel. Ramadhas et al. [4] experimented the diesel engine with vegetable oil-based biodiesel and found out that the engine performance and emission properties of the diesel engine are comparable to that of the diesel fuel. Thus, biodiesel may be replaced for the diesel fuel completely. Qi et al. [5] experimented that soybean oil-based biodiesel blended with diesel could be the best alternative fuel to use in the IC engines without any significant changes. But more care must be taken when blending a large quantity of biodiesel with diesel in the diesel engine to ensure the proper engine operation. Altin et al. [6] studied that, from a performance perspective, both vegetable oil-based biodiesel and its ester components are a better replacement to diesel fuel. But vegetable oil-based fuels are having some problems in usage such as atomisation, fluidity and emissions of heavy particles because of its higher viscosity, frozen in cold and storage problems. Lee et al. [7] experimented that there is very less significant variance in biodiesel and diesel ignition delays with the speed of flame propagation during the combustion. Nitrogen oxides (NO_x) emissions reduced due to less heating value of the mixture with increasing biodiesel fraction in the blend. Also, NO_x emissions affect the increase of the percentage of oxygen in the blend. Ozener et al. [8] carried out an experiment in which the IC engine is fuelled with biodiesel blends and observed that ignition delay and premixed peak get reduced significantly but with a slight increase in NO_x emissions. Qi et al. [9] experimented that soybean oil-based biodiesel performing the same stages of combustion compared to diesel, but combustion started earlier. Because of its lower heating value, the BSFC was higher for biodiesel fuels. When biodiesel is combusted in the IC engine, the average reduction in NO_x emissions is around 5%. Dhar et al. [10] studied that BSFC was comparable to mineral diesel with low quantity of Karanja oil-based biodiesel blends; however, it increases with the high volumetric percentage of biodiesel quantity in the blends. Also, it results in higher nitrogen oxides (NO_x) emissions, especially at high load condition. Çelikten et al. [11] found similar properties to those of pure diesel for rapeseed oil and soybean oils. Soybean oil-based biodiesel had higher BSFC and emission of NOx relative to normal diesel and biodiesel rapeseed. An et al. [12] investigated that the increase in the biodiesel blend ratio reduced NO emissions. Lahane et al. [13] researched that with all biodiesel diesel blends, ignition delay and pressure increase rate decreased owing to a greater biodiesel cetane number than base diesel. However, owing to the oxygen content in biodiesel, the nitrogen oxides (NO_x) emissions raised with the utility of biodiesel relative to diesel. Khan et al. [14] researched that the enhanced air movement of the TRCC motor increases the air–fuel mixing process owing to its bowl geometry, which decreases BSFC and exhaust emissions as a result of enhanced combustion. Jaichandar et al. [15] researched that compression ignition engine performance and emission features rely primarily on the method of combustion. The firing of petrol in the diesel engine relies on effective atomisation of fuel to rise the surface area for fuel to attend the enhanced mixing of air and fuel and reduced evaporation rates. Reducing the mean droplet diameter, enhance the rate of heat discharge, promotes ignition, offers great vaporisation, thus improving the effectiveness of combustion and reducing pollutant emissions. Rakopoulos et al. [16] experimented that the impact of geometry on the engine with the modified air movement design is capable of capturing the geometrical effect on the in-cylinder velocity and temperature field of the fuel firing chamber geometry and speed. Qi et al. [17] noted that BSFC rise with the rise in the biodiesel ratio in the fuel mixture due to low heating value of the vegetable oil-based biodiesel. High oxygen ratio in biodiesel improved the combustion properties and reduces the temperature of the firing chamber but raise the emissions, particularly on the elevated engine load. Ismail et al. [18] experimented the biodiesel to observe the changes in the ignition delay and combustion behaviour, which subsequently influence the procedures of formation of nitrogen oxides and smoke. Moreover, Ismail et al. [19] designed and check a decreased, zero-dimensional chemical transformation process and multidimensional motor simulations to check the engine working surroundings and used different types of biodiesel fuel. In excellent agreement with those generated using comprehensive kinetics, the ignition delay intervals and the sequential growth of significant species. The findings indicate that the mechanism proposed provides credible predictions of procedures of in-cylinder combustion and emission. Kumar [20] researched that altered piston design shows the better atomisation in the cylinder, resulting in uniform combustion at each cylinder place, increasing the BTE, BMEP and lowering the engine relative to the baseline. The exit gases temperature is higher due to improved combustion in altered systems, resulting in elevated nitrogen oxides emissions. Kegl et al. [21] researched and found out that raised NOx emissions are the result of the sophisticated injection and ignition processes and higher in-cylinder gas temperatures at the beginning of ignition.

The above literature suggests that a feasible alternative to diesel is the biodiesel in compression ignition motors. The design of air–fuel ratio, droplet-wall collision and combustion chamber set-up is a major difficulty for diesel-powered cars. Simulation of the engine using CFD can play a significant part in reducing the number of studies and provide a thorough insight into the flow, air–fuel ratios, firing and emissions of pollutants. The mathematical model can, therefore, give a better knowledge of the method of combustions and the properties that affect engine efficiency and emission rate when operating on biodiesels. Each sample of biodiesel mixed with diesel fuel consisted of 20 and 30% of biodiesel in separate proportions. The primary goal of the experiment is therefore to explore the performance of a biodiesel-assisted IC engine running under variation of chamber settings of combustion. All information is compared graphically with each other and diesel fuel.

Table 1 Fuel properties of different fuels

Source	Cetane number	Calorific value (MJ/kg)	Viscosity (40 °C) (mPa s)	Density (g/cm^3)	Oxygen content (wt%)
Soybean	47.1	39.53	3.69	0.867	14.3
Diesel	51	49.95	2.42	0.821	–

2 Soybean Oil-Based Biodiesel as a Fuel

Soybean oil-based biodiesel was regarded as a prospective alternative to diesel and utilised as a gas in the CI engine in the current inquiry. Table 1 compares the significant characteristics of diesel, soybean biodiesel. Based on the researchers' equations [22], the characteristics of the mixture fuels were determined. Soybean oil-based biodiesel shows a low heating value than diesel because of oxygen percentage in the biodiesel, whereas soybean biodiesel's dynamic viscosity and flashpoint are greater than diesel fuel. A small proportion of additional biodiesel in diesel can certainly raise the resulting mixture's flashpoint. Therefore, related to diesel, the transportation and storage of the biodiesel are much safer.

3 Engine Test Method

Each sample of biodiesel (D80B20, D70B30 and D100B0) is mixed in dissimilar proportions with pure diesel to use in the engine testing. Blends are defined in the form of 'Ba'. The 'a' value shows the proportion of the quantity of biodiesel in the blend, such as B12 is a combination of 12% biodiesel and 88% diesel. Likewise, B90 is 90% biodiesel and 10% diesel. The three distinct fuel biodiesel blends (B0, B20 and B30) are used as fuel in engine testing. All the effective characteristics that are BSFC, BTE, brake power, BMEP, engine speed, air–fuel ratio, torque and volumetric efficiency are determined in the study. The outcomes results are related to each other. All experiments were repeated at three different loads.

In this experiment, a diesel engine that has direct injection, 4-stroke, single-cylinder, MF water-cooled and 17.5 CR is used for engine tests (Fig. 1). Table 2 presents the basic engine characteristics.

4 Results

In this research, the experimental determination and comparison of effective characteristics and indicated characteristics of biodiesel samples produced from soybean oil-based biodiesel and diesel fuel were conducted. Also, the performance of the engine is tested by varying the compression ratio. The tests were carried out at

Fig. 1 Experimental test rig

Table 2 Mechanical configuration of the engine used in the experiments

Manufacturer	Accurate test equipments and engineers
Model	VCRE SI-CI
Engine type	4 stroke, single-cylinder, VCR, water-cooled, multi-fuel engine
Rated power (kW)	3.7285 @ 1800 rpm
Rated torque (N-m)	23.5 N-m @ 1500 rpm
Bore (mm)	87.50
Stroke (mm)	110.00
Length of connecting rod (mm)	232.00
Displacement volume (cc)	662.00
Fuel injecting system	Direct injection
Injection pressure (bar)	175

various loads applied to the engine. The variation of the pressure with change in the volume and with the change in the crank angle is indicated in Figs. 2 and 3, respectively. Also, the highest value of pressure is about 45 bars at 70 cc and 0° crank angle, respectively. The experimental results have been displayed graphically. The results are divided into two sections. In the first section, the composition of biodiesel in the mixture is varied to test the engine performance, and comparison among the engine performance properties is carried out to find the optimum conditions. Comparison between the BSFC, BTE, mechanical efficiency, air–fuel ratio, volumetric efficiency and engine RPM with BP is represented graphically in Figs. 4, 5, 6, 7, 8 and 9, respectively. In the second section, the value of the compression ratio is varied to test engine

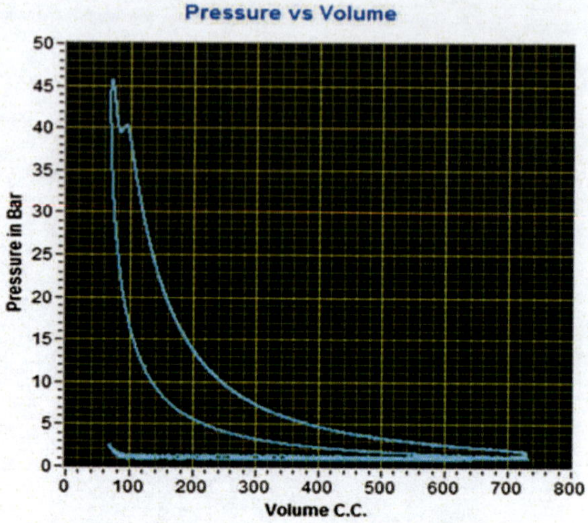

Fig. 2 Variation of the pressure with change in volume

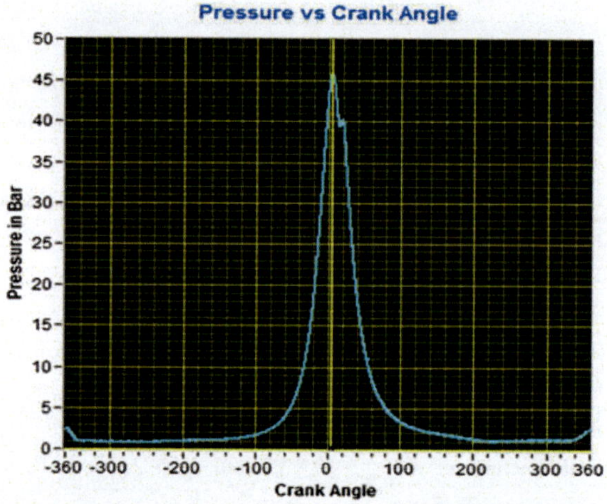

Fig. 3 Pressure variation with change in crank angle

performance. The values of compression ratio which are taken in the consideration during the experiment are 14, 16, 18 and 20. Comparison of the engine performance properties at different compression ratio is performed and indicated graphically in Figs. 10, 11, 12, 13 and 14.

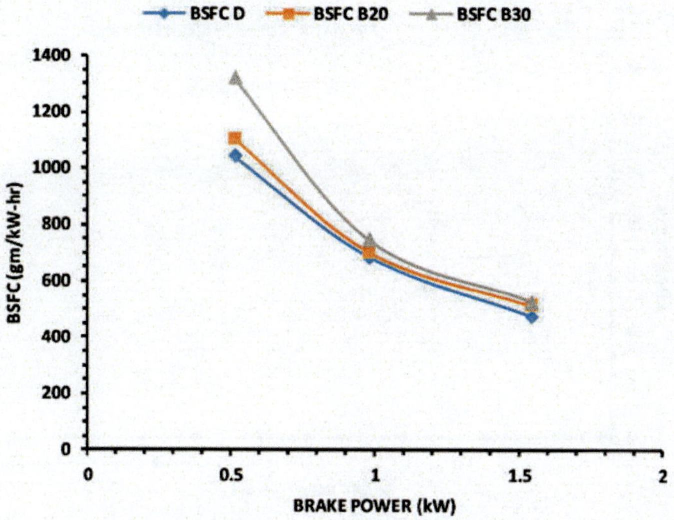

Fig. 4 BSFC (gm/kW-hr) versus BP (kW) at CR 17

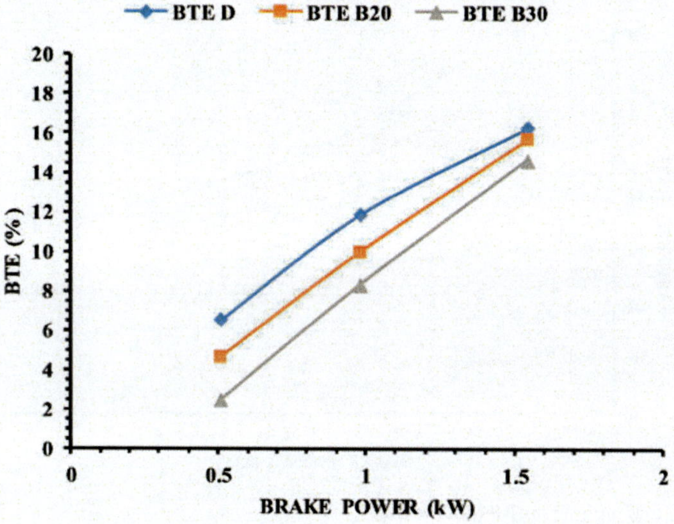

Fig. 5 BTE (%) versus BP (kW) at CR 17

4.1 Comparison of Engine Performance Characteristics for Different Fuel Blending Ratios

BSFC Versus BP. BSFC is recognised as the most important parameter to consider in evaluating the performance of the engine. It is required that an engine produce

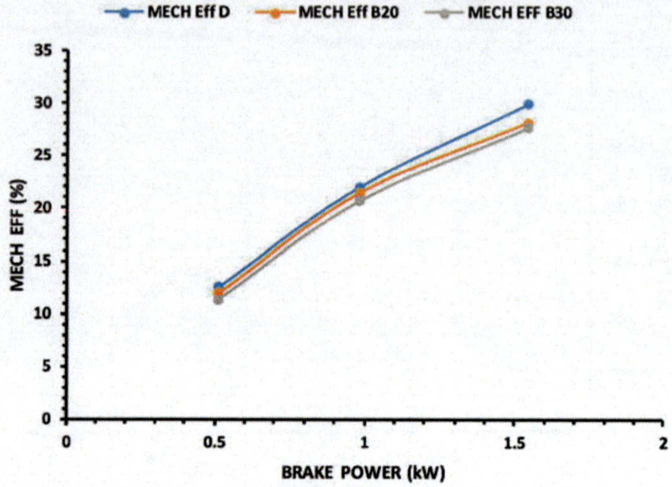

Fig. 6 Mechanical efficiency (%) versus BP (kW) at CR17

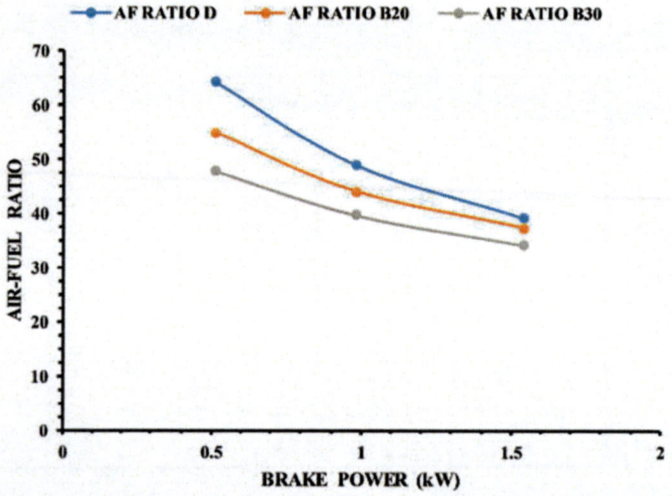

Fig. 7 Air–fuel ratio versus BP (kW) at CR17

more power on consuming less fuel, which improves the efficiency of the engine. Fuel consumption varies with change in the load or power output. Fuel consumption rises with the rise in load regardless of the constant engine speed. So, this variation is shown with the help of Fig. 2. It is found out that the BSFC is showing similar trends as that of the diesel at the constant CR. However, the fuel that has blended with 20% of biodiesel in the diesel shows better results than other blends. BSFC decreases with an increase in the brake power until 1.2 kW and then almost become constant. These

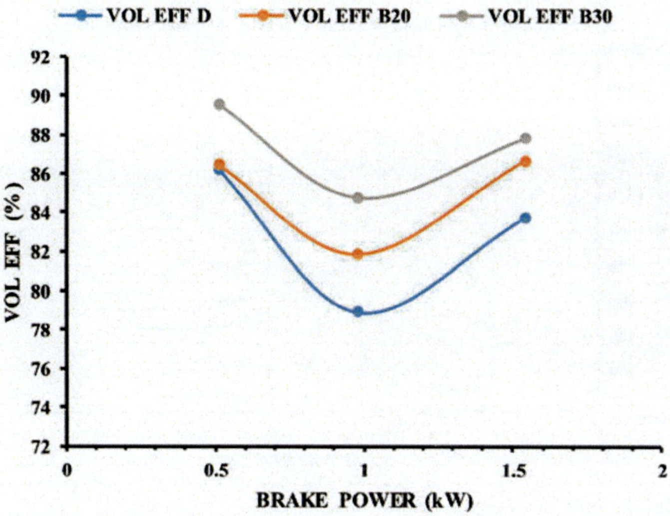

Fig. 8 Volumetric efficiency (%) versus BP (kW) at CR17

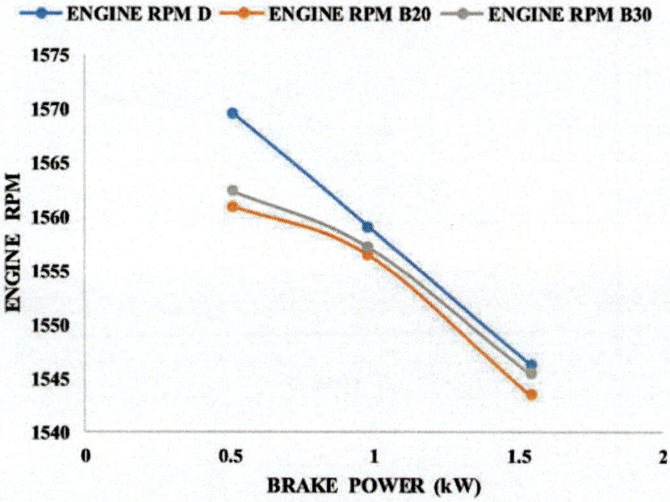

Fig. 9 Engine speed (RPM) versus BP (kW) at CR17

effects are coming in the result because of the change in the calorific value of the blended fuel. As we know, biodiesel fuels have a low calorific value related to diesel. Mixing of the biodiesel with diesel reduces the calorific value of the fuel. Moreover, the viscosity also varies in mixing, i.e. the viscosity of biodiesel is more relative to diesel. Thus, the viscosity of the blend advances with an upsurge in the composition

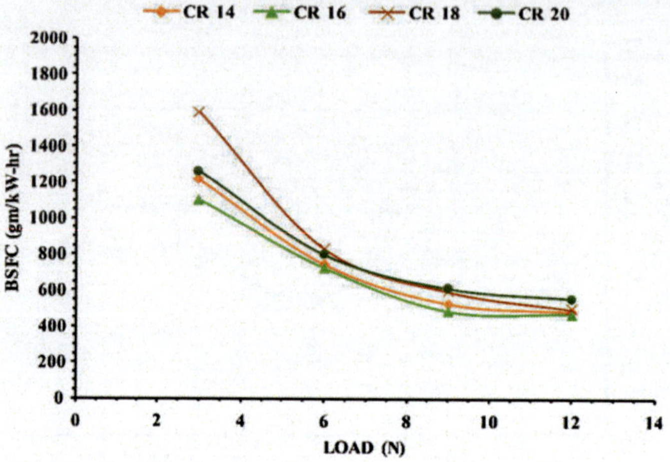

Fig. 10 BSFC (gm/kW-hr) versus load (N)

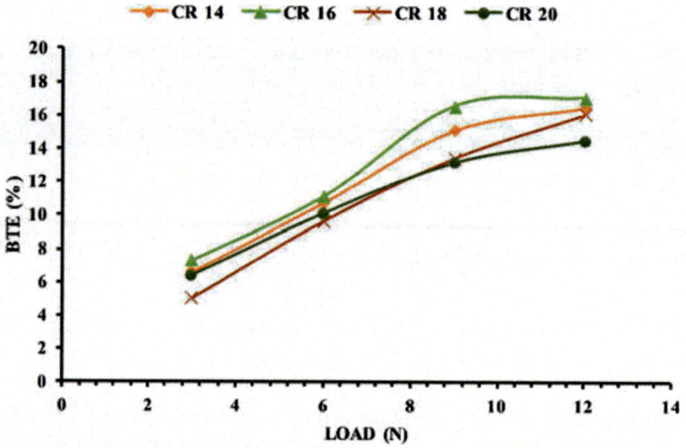

Fig. 11 BTE (%) versus load (N)

ratio of biodiesel in the fuel. The BSFC of B20 and B30 at 1 kW is 3 and 8.5% more than that of the diesel.

BTE Versus BP. BTE is considered as the most vital parameter required to evaluate the performance of the engine. BTE is explained as the fraction of the work output to the total sum of the energy produced by firing the fuel in the engine. A high value of the BTE is always desired to advance the performance of the engine. The variation of BTE for biodiesel and diesel at different BP is displayed in Fig. 3. It is found out that BTE improves with increase in the BP and it is maximum in the case of pure diesel. These variations are due to the influence on the combustion and ignition properties

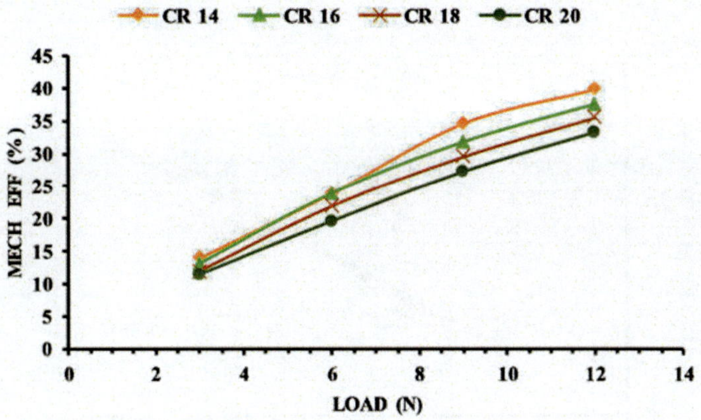

Fig. 12 Mechanical efficiency (%) versus load (N)

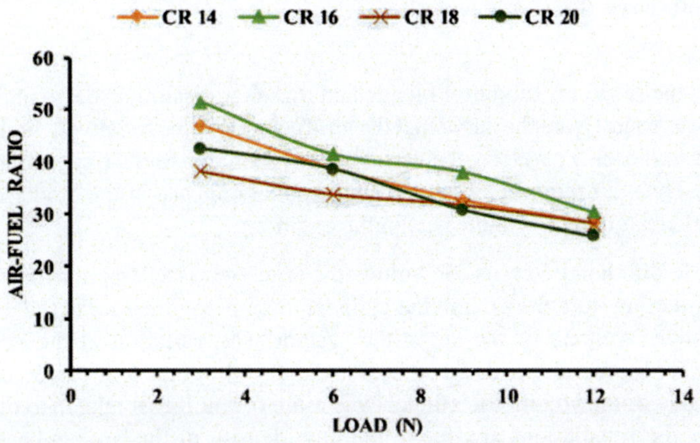

Fig. 13 Air–fuel ratio versus load (N)

of the engine. The value of the BTE was reduced by about 12 and 19% for B20 and B30 fuel mixture at 1 kW than that of the diesel, respectively.

Mechanical Efficiency Versus BP. The variation of the mechanical efficiency for diesel and biodiesel at different BP is represented in Fig. 4. It is observed form the curve that mechanical efficiency varies slightly for the biodiesel blends compared to the diesel. Mechanical efficiency is explained as the fraction of the power output to the power input. Power input remains nearly constant due to the increase in fuel consumption. Therefore, mechanical efficiency does not affect much with the change in the proportion of biodiesel in the diesel.

Air–fuel Ratio Versus BP. Air–fuel ratios were considered to check the performance of the engine due to its influence on the fuel intake in the engine. So, the variation

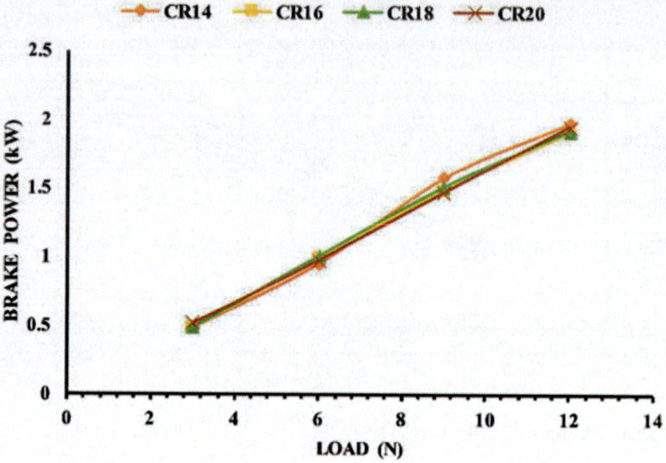

Fig. 14 Brake power (kW) versus load (N)

of the air–fuel ratio for biodiesel blends and diesel at dissimilar BP is indicated in Fig. 5. It was found from the curve that the air–fuel ratio is less relatively for biodiesel compared to diesel. Therefore, the variation was seen due to the oxygen ratio in the biodiesel. Thus, air requirement to the engine is less in case of the biodiesel due to which fuel consumption is increased in the engine.

Volumetric Efficiency Versus BP. Volumetric efficiency is the fraction of the density of the air–fuel mixture drawn into the cylinder at ambient pressure to the density of the unchanged volume of the air in the system. The variation of the volumetric efficiency for biodiesel blends and diesel is displayed in Fig. 6. It can be perceived that the value of the volumetric efficiency is more for biodiesel related to that of the diesel. So, these variations are due to the high density of the soy biodiesel. Thus, when the biodiesel is mixed with the diesel, the resultant density of the mixture is increased.

Engine RPM Versus BP. The deviation of the engine RPM for biodiesel mixtures and diesel with BP is presented in Fig. 7. Engine speed reduces with an increase in BP. As load is increasing on the engine, it reduces the engine rpm. The change in the engine rpm for biodiesel blends is negligible related to that of the diesel.

4.2 Comparison of Engine Performance Characteristics for Different Compression Ratios

BSFC Versus Load. The variation of the BSFC for different CR with the change in load is indicated in Fig. 10. It is visible that BSFC increases with increase in

CR except in the case of CR16 at load 12 N. Hence, BSFC is minimum in the case of CR16. These changes are due to the enhancement of combustion of fuel in the cylinder and reduction in heat losses. These results are obtained for 30% biodiesel composition in the mixture.

BTE Versus Load. The variation of the BTE for different CR with the change in load is indicated in Fig. 11. BTE is increasing with the increase in load because of the increase in brake power at higher load. Also, it is found out from the analysis that BTE is increasing with the increase in CR and maximum for CR16. These changes are due to enhancement in the combustion with an increase in CR and less consumption of fuel. The high peak of BTE is obtained on load 9 N and CR16, which is about 19% higher compared to the condition of the load 9 N and CR20.

Mechanical Efficiency Versus Load. The variation of the mechanical efficiency for different CR with the change in load is indicated in Fig. 12. Mechanical efficiency is increasing with the increase in load and decreasing with increase in CR. These variations are due to the increase in the value of brake power and BTE with the increase in CR.

Air–Fuel Ratio Versus Load. The variation of the air–fuel ratio for different CR with the change in load is indicated in Fig. 13. The air–fuel ratio is decreasing with increase in load due to the decrease in BSFC. Similarly, the air–fuel ratio is decreasing with increase in CR. The air–fuel ratio is maximum for CR16 due to less consumption of fuel as finding out in the comparison of BSFC with the load.

Brake Power (BP) Versus Load. The variation of the air–fuel ratio for different CR with the change in load is indicated in Fig. 14. When the compression ratio was kept steady, and the load was varied from 3 N to 12 N, brake power (BP) was increasing with an increase in the load. Similarly, when the volume percentage of biodiesel was taken 30% in the fuel mixture, and the compression ratio of the engine was varied from CR 14–20, there are no significant changes in the value of brake power (BP) at a particular value of the load. At the load 12 N, the maximum value of brake power is obtained for CR20.

5 Conclusions

The experimental study of the single-cylinder diesel engine fuelled with soybean-based biodiesel has been concluded as:

1. The soybean oil-based biodiesel was mixed with diesel in different ratios, i.e. pure diesel (BO), B20 and B30 and fuelled in the engine to test the performance of the diesel engine without any modifications in the engine.
2. BSFC was increased with a rise in the proportion of biodiesel in the fuel, but trends are similar in all the cases, i.e. diesel and biodiesel blends with the increase

in BP. The maximum value of the BSFC was found in the case of the B30 biodiesel blend.

3. BTE declined with the rise in the volume of biodiesel in the fuel. However, the trend of the curves is similar in all the cases with an increase in BP. BTE is higher for diesel relative to that of the biodiesel blends.
4. Moreover, it can be concluded that the performance of the engine is highest in the case of B30 biodiesel blend than other biodiesel blends.
5. The performance of the engine is varying with the change in CR. The performance of the engine, fuelled with biodiesel blend B30, is best at CR16.

Acknowledgements The test facility and infrastructural facility provided by NIT Kurukshetra has been gratefully acknowledged.

References

1. Pantoja SS et al (2013) Oxidative stability of biodiesels produced from vegetable oils having different degrees of unsaturation. Energy Convers Manage 74:293–298. https://doi.org/10.1016/j.enconman.2013.05.025
2. Karavalakis G, Stournas SJE, Fuels (2010) Impact of antioxidant additives on the oxidation stability of diesel/biodiesel blends. Energy Fuels 24(6):3682–3686. https://doi.org/10.1021/ef1004623
3. Wang Y et al (2006) An experimental investigation of the performance and gaseous exhaust emissions of a diesel engine using blends of a vegetable oil. Appl Therm Eng 26(14–15):1684–1691. https://doi.org/10.1016/j.applthermaleng.2005.11.013
4. Ramadhas A, Jayaraj S, Muraleedharan CJRE (2004) Use of vegetable oils as IC engine fuels—a review. Renew Energy 29(5):727–742. https://doi.org/10.1016/j.renene.2003.09.008
5. Qi DH, Lee CF (2014) Influence of soybean biodiesel content on basic properties of biodiesel-diesel blends. J Taiwan Inst Chem Eng 45(2):504–507. https://doi.org/10.1016/j.jtice.2013.06.021
6. Altın R et al (2001) The potential of using vegetable oil fuels as fuel for diesel engines. Energy Convers Manage 42(5):529–538. https://doi.org/10.1016/s0196-8904(00)00080-7
7. Lee Y, Huh KYJF (2013) Numerical study on spray and combustion characteristics of diesel and soy-based biodiesel in a CI engine. Fuel 113:537–545. https://doi.org/10.1016/j.fuel.2013.05.106
8. Özener O et al (2014) Effects of soybean biodiesel on a DI diesel engine performance, emission and combustion characteristics. Fuel 115:875–883. https://doi.org/10.1016/j.fuel.2012.10.081
9. Qi D et al (2009) Combustion and performance evaluation of a diesel engine fueled with biodiesel produced from soybean crude oil. Renew Energy 34(12):2706–2713. https://doi.org/10.1016/j.renene.2009.05.004
10. Dhar A, Agarwal AKJF (2014) Performance, emissions and combustion characteristics of Karanja biodiesel in a transportation engine. Fuel 119:70–80. https://doi.org/10.1016/j.fuel.2013.11.002
11. Celikten I, Koca A, Arslan MAJRE (2010) Comparison of performance and emissions of diesel fuel, rapeseed and soybean oil methyl esters injected at different pressures. Renew Energy 35(4):814–820. https://doi.org/10.1016/j.renene.2009.08.032
12. An H et al (2014) A numerical modeling on the emission characteristics of a diesel engine fueled by diesel and biodiesel blend fuels. Appl Energy 130:458–465. https://doi.org/10.1016/j.apenergy.2014.01.004

13. Lahane S, Subramanian KJF (2015) Effect of different percentages of biodiesel–diesel blends on injection, spray, combustion, performance, and emission characteristics of a diesel engine. Fuel 139:537–545
14. Khan S, Panua R, Bose PKJF (2018) Combined effects of piston bowl geometry and spray pattern on mixing, combustion and emissions of a diesel engine: a numerical approach. Fuel 225:203–217. https://doi.org/10.1016/j.fuel.2018.03.139
15. Jaichandar S, Annamalai KJE (2013) Combined impact of injection pressure and combustion chamber geometry on the performance of a biodiesel fueled diesel engine. Energy 55:330–339. https://doi.org/10.1016/j.energy.2013.04.019
16. Rakopoulos C et al (2010) Investigation of piston bowl geometry and speed effects in a motored HSDI diesel engine using a CFD against a quasi-dimensional model. Energy Convers Manage 51(3):470–484. https://doi.org/10.1016/j.enconman.2009.10.010
17. Qi D et al (2010) Experimental studies on the combustion characteristics and performance of a direct injection engine fueled with biodiesel/diesel blends. Energy Convers Manage 51(12):2985–2992. https://doi.org/10.1016/j.enconman.2010.06.042
18. Ismail HM et al (2013) Computational study of biodiesel–diesel fuel blends on emission characteristics for a light-duty diesel engine using OpenFOAM. Appl Energy 111:827–841. https://doi.org/10.1016/j.apenergy.2013.05.068
19. Ismail HM et al (2013) Development of a reduced biodiesel combustion kinetics mechanism for CFD modelling of a light-duty diesel engine. Fuel 106:388–400. https://doi.org/10.1016/j.fuel.2012.10.015
20. Kumar VJIJOAE (2018) Experimental investigation of piston bowl geometry effects on performance and emissions characteristics of diesel engine at variable injection pressure and timings. Appl Energy 39(7):685–693. https://doi.org/10.1080/01430750.2017.1333041
21. Kegl BJAE (2011) Influence of biodiesel on engine combustion and emission characteristics. Appl Energy 88(5):1803–1812. https://doi.org/10.1016/j.apenergy.2010.12.007
22. Şahin Z, Aksu ONJRE (2015) Experimental investigation of the effects of using low ratio n-butanol/diesel fuel blends on engine performance and exhaust emissions in a turbocharged DI diesel engine. Renew Energy 77:279–290. https://doi.org/10.1016/j.renene.2014.11.093

Performance Analysis of Soybean Oil Blended Diesel Fuelled DI Engine by Varying Compression Ratio

Pradeep Kumar Sonkar and Rajneesh Kaushal

Nomenclature

EGT	Exhaust gas temperature
MUFA	Mono unsaturated fatty acids
BSFC	Brake specific fuel consumption
ID	Ignition delay
CR	Compression ratio
BSEC	Brake specific energy consumption
SME	Soyabean methyl ester
HC	Hydrocarbon
BTE	Brake thermal efficiency
WCO	Waste cooking oil
PUFA	Polyunsaturated fatty acids
BP	Brake power
MO	Methyl oleate
VCR	Variable compression ratio

P. K. Sonkar (✉) · R. Kaushal
Department of Mechanical Engineering, National Institute of Technology Kurukshetra, Kurukshetra 136119, India
e-mail: pradeep_31806117@nitkkr.ac.in

R. Kaushal
e-mail: rajneesh@nitkkr.ac.in

A. K. Gupta et al. (eds.), *Advances in IC Engines and Combustion Technology*, Lecture Notes in Mechanical Engineering,
https://doi.org/10.1007/978-981-15-5996-9_27

1 Introduction

Energy-efficient conversions need specific properties of fuel and appropriate working conditions for engines. Researchers discovered that with minimal engine changes biodiesel meets these demands. Biodiesel made by using soybean-based vegetable oils, species of animal's fatty tissue, giving similar thermal energy as diesel. In addition, it does not have amount of sulfur, mineral and due to elevated oxygen content, it generates engine low engine emission rates. High biodiesel intensity is able to self-lubricate the engine's mechanical components. Biodiesel, actually are mono-alkyl esters that can be made from natural triglyceride found in vegetable-based oils. It can be accomplished in the attendance of a catalyst through a method of transesterification with short-chain alcohols. The standard technique for producing biodiesel is homogeneous transesterification [1, 2]. In order to maximize the output of methyl ester, various researches have been executed to upgrade transesterification variables. Each variable response was researched individually, affecting the production of biodiesel from vegetable oil by alkaline transesterification catalyzed [3, 4].

Variables influence transesterification of cottonseed oil utilizing batch mode has been enhanced using an effective surface technique with an ideal yield of 96%. The optimum condition was 55 °C, reaction timing 60 min, 0.6% KOH level, and steady mixing speed [5]. In order to predict the optimum output, statistical assessment (central concrete construction) was used. In analytical and experimental terms, the ideal biodiesel production from lard oil was 96.2 and 96%. Results were 65 °C reaction temperatures, 1.25% catalytic load, the reaction time of 40 min, and methanol/oil ratio of 6:1 with 250 rpm blending speed [6]. Using a multi-variant empirical model, method constraints were optimized for the manufacturing of biodiesel fuel from various feedstocks. Results found that methanol has the most important influence on biodiesel production, accompanied by catalyst loading. The outcome of optimized circumstances is about 98% yield [7]. Analytical analysis of reaction components and the output of ethyl ester generated from processed soybean oil is significantly influencing. The impact of ethanol/oil proportion was noted to be equal to the impact of alkalis concentration on biodiesel production. There was, however, a slight influence as the temperature of the reaction increased. In addition, yield reduces considerably due in part to soap creation at greater catalyst concentration. Results indicate that the optimum output of ethyl ester at the optimum reaction situation was 97.2% [8]. Various researchers have performed analytical and experimental research on the impact of biodiesel on engine performance and exhaust gas emission relative to diesel fuel and its blend consumption [9]. The impact of using Karanja oil manufactured biodiesel and its mixture on performance and exhaust emission parameters of natural diesel was explored. Thus, it was concluded that for distinct blends and engine loads, BTE reduced by 3–5%, CO, CO_2, UBHC, and reduced exhaust emission compare from diesel. NO_x emissions were more compared to diesel fuel [10]. The consequence Pithecellobiumdulce intermix of biodiesel "PDBD5, PDBD10, and PDBD20" on engine combustion and performance was talked about with the help of experiments. BSFC for PDBD20 was 9.565% larger as compared to diesel fuel.

Although BTE reduced by 4.34% for the unchanging mixed ratio [11]. The emissions and performance of biodiesel derived from various types of WCO utilized in a CI engine have been described. The results acquired from B10 and B5 indicate that BSFC rose by 4% and reduced BTE by 2.8% [10]. All biodiesel/diesel mixture has been found to reduce the ID period resulting in the smooth operation of the engine and a low-pressure increase [11]. The experimental analysis was conducted into the performance, combustion, and exhaust emissions of the single-cylinder DI engine fuelled by soybean-based biodiesel blends "B10, B20 and B50" at various engine speeds. The results showed a reduction in torque, where BSFC has risen [12]. Performance, combustion, and properties of the diesel-powered diesel engine, the rapeseed-based biodiesel mixtures have been studied. At the highest load state, EGT and BSEC were raising for all the biodiesel mixtures as relative to diesel fuel. BTE indicates a decrease with a growing blending fraction. Brake Thermal Efficiency was 26% smaller compared to diesel for 25% biodiesel [13]. Investigation of the performance, emission & physiochemical properties of soya bean methyl ester, along with it combined with methyl oleate in a common rail DI engine was done. Results showed a decrease in BSFC and a boost in Brake Thermal Efficiency for mixed biodiesel (S80-M50, S70-M30, and S50-M20) in regards to the SME case. Blending of MO resulted in oxidative stability, improved ignition quality (IQ), and lower temperature flow properties [14]. It was examined that the soyabean biodiesel–diesel mix may prove to be a viable alternative. It can be employed as a substitute for mineral diesel. Nonetheless, further care must be taken during the utilization of a larger percentage of biodiesel in diesel engines in order to ensure proper engine function. They also registered a slight decrement in Cetane index with a rise in the content of biodiesel, normally; the impact of the content of the biodiesel on Cetane mix index is not important [15]. It has been stated that rapeseed oil and soya bean oil methyl esters properties are close to that of diesel fuel. Soya bean methyl ester (SME) biodiesel has lower direct fuel consumption and NO_x pollution than standard diesel or even rapeseed biodiesel [16]. Studied the various volumetric percentages of diesel/biodiesel mixtures: D70B30 (70% diesel-30% biodiesel), D50B50 (50% diesel-50% biodiesel) & D30B70 (30% diesel-70% biodiesel) were prepared for powering a single-cylinder DI engine. The tests of the engine relative with diesel fuel indicate a decrease in the rate at which change of CO takes place by 33.8% for D50B50. A minute decrement in the highest cylinder pressure for a larger percentage of biodiesel mixes because of lower ignition delay and low biodiesel calorific value [17]. The BSEC rose both at a lower and a rating load. There is a slight reduction at a load of 50, 75, and 91%. BTE was lowered for biodiesel mixtures relative to diesel [18]. The investigation centered on the effect of engine emissions by powering diesel engines with natural biodiesel provided by cooking oil wastes and its blends. The experimental result obtained indicates a rise in BSFC and a reduction in BTE by the use of biodiesel as a baseline fuel relative with pure diesel [19, 20].

2 Preparation of Biodiesel

Soybean is a useful source of oil for the manufacturing of biodiesel. Biodiesel has increasingly drawn global attention due to growing environmental concerns. Biodiesel has become a prime source as a replacement fuel and is becoming a big source of renewable energy in the future. It can be simply generated from edible materials, non-edible materials, and algae materials by the process of transesterification. Developing techniques try to raise oleic and decrease the content of linolenic fatty acids in soybeans genotypes, which will improve the soybeans for meals and bait and the manufacturing of biodiesel. It generated with a large content of PUFAs has a strong tendency to oxidize, making storage hard and decreased. Even then, oils with elevated concentrations of monounsaturated fatty acids (MUFAs) are preferred to produce biodiesel because oxidation is not reduced, cold flow properties are considerable and the emission of nitrogen oxides are reduced. Besides, Soybeans oil with high concentrations of PUFAs results in low oxidative stability, leading in rancidity, a fast decrease in ideal flavor, and a reduction in storage time for produced meals products. In the paper, an effort has been made to analyze the research held on the influence of biodiesel and its mixtures utilization on the performance and exhaust emissions versus diesel, which will be helpful to low harmful gas release and better performance of the engine. The flow diagram for biodiesel production has been indicated in Fig. 1. The properties of soybean oil has been compared with the diesel in Table 1.

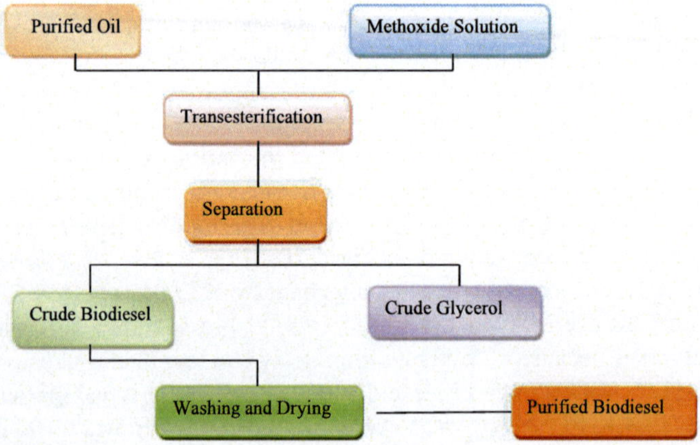

Fig. 1 Schematic of biodiesel production steps

Table 1 Comparison of soybean-based oil biodiesel with that of diesel

S. No.	Properties	Soybeans oil	Diesel
1	Kinematic viscosity (mm^2/s)	5.2	3
2	Cetane Index	50	55
3	Calorific value (kJ/kg)	39,857	45,343
4	Density ($g/cm^{3)}$	0.867	0.8508
5	Oxygen content (weight %)	14.3	38

3 Experimental Analysis

The experiment was conducted on direct injection, four-stroke, single-cylinder, VCR engine which is MF water-cooled and has 17.5 CR as shown in Fig. 2. The basic engine characteristics have been represented in Table 2. The tests were conducted by fixing RPM of the engine and by varying the load at a fixed compression ratio. Then, the tests were conducted at different loads and at different compression ratios using pure diesel. Initially, the engine was operated at the same compression ratio, i.e., CR17, by varying the composition of soybean-based biodiesel in the blends. The results obtained with soybean-based biodiesel blends at constant compression ratio were compared with the pure diesel and the comparison has been shown in Figs. 5, 6, 7, 8, 9, 10, and 11. Then, the engine was operated at different compression ratios, i.e., CR14, CR15, CR16, CR18, CR19, and CR20 and fuelled with B30 fuel. Hence, the obtained results were compared with each other, which were shown in Figs. 12, 13, 14, 15, 16 and 17.

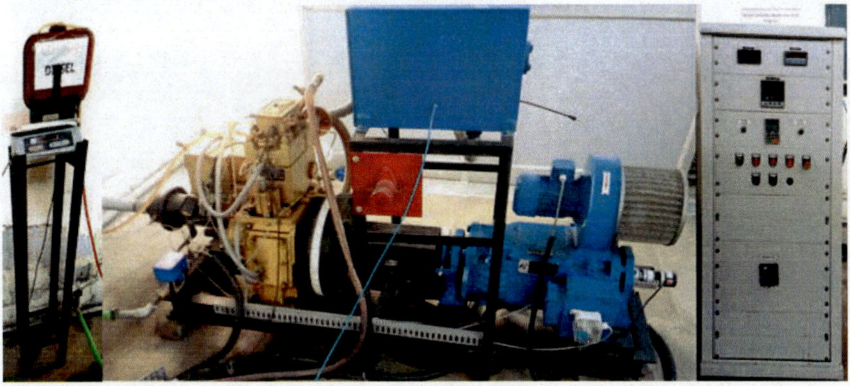

Fig. 2 Engine test rig

Table 2 IC engine test setup specifications

1	Fuel	Diesel/multi-fuel
2	Manufacture	Accurate test equipment and engineers
3	Model	VCRE.SI-CI series
4	Engine type	4 Stroke, single-cylinder, multi-fuel engine
5	Bore	87.5 mm
6	Stroke	110 mm
7	Length of connecting rod	232 mm
8	Displacement volume	662 cc
9	Cooling system	Water-cooled
10	Compression ratio	5 to 11 SI mode and 12–20 CI mode
11	Rated power	3.7285 kW
12	Rated torque	23.5 N-m@1500 rpm
13	Fuel injecting system	Direct injection

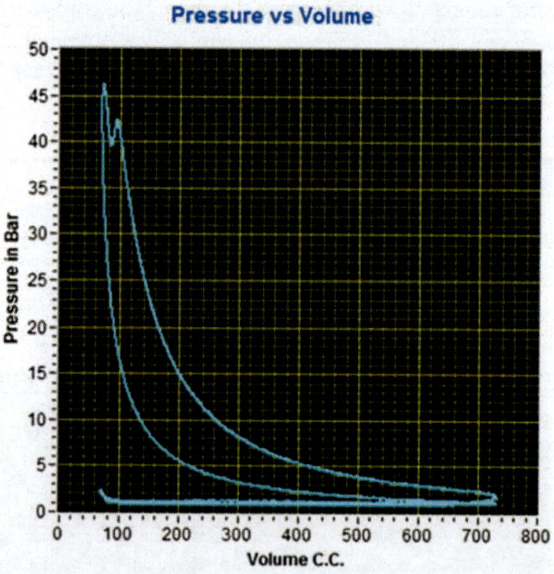

Fig. 3 Variation of pressure with volume

4 Results and Discussions

In this test, the diesel engine was fuelled with diesel–biodiesel mixture to test the engine performance. The performance of the engine depends on the type of fuel used in the engine. Firstly, the engine was tested with different fuel blending ratio.

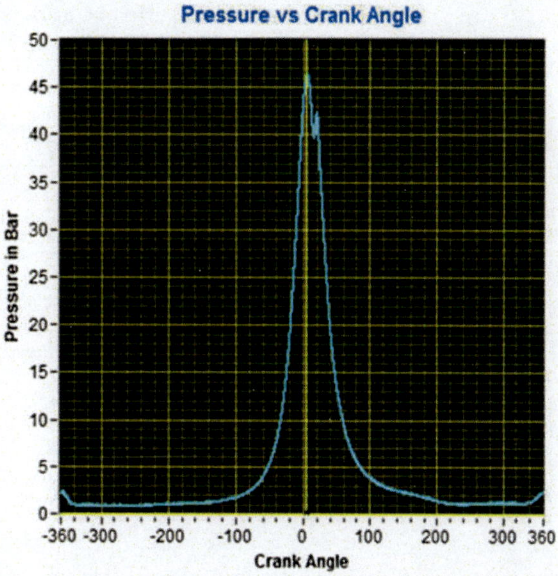

Fig. 4 Pressure variation with crank angle

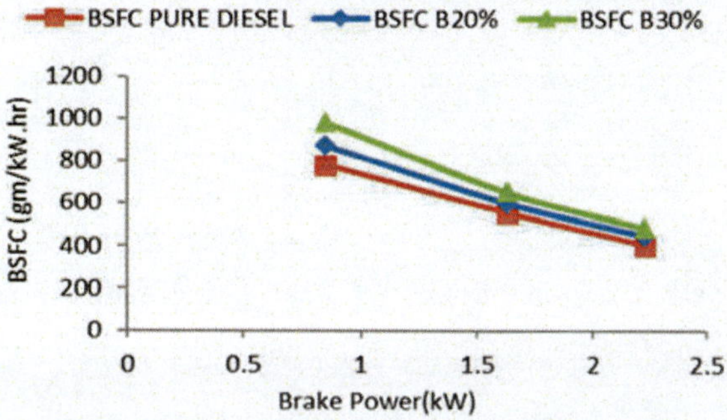

Fig. 5 Change in BSFC (g/kWh) with Brake Power (kW)

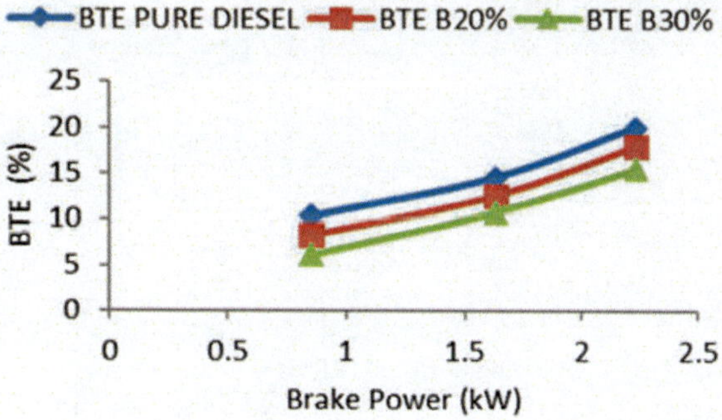

Fig. 6 Change in BTE (%) with Brake Power (kW)

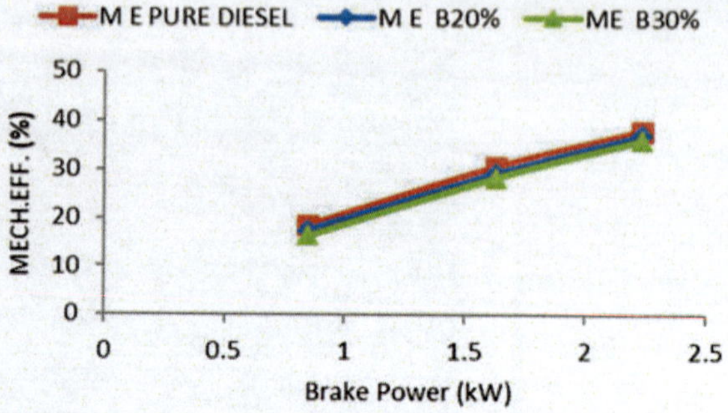

Fig. 7 Change in Mechanical Efficiency (%) with Brake Power (kW)

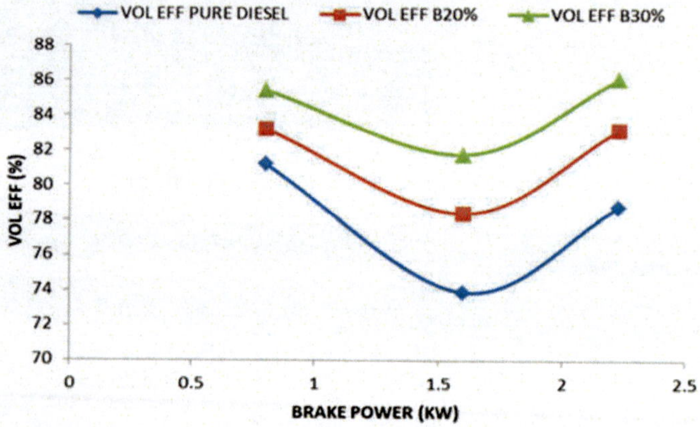

Fig. 8 Change in Volumetric Efficiency (%) with Brake Power (kW)

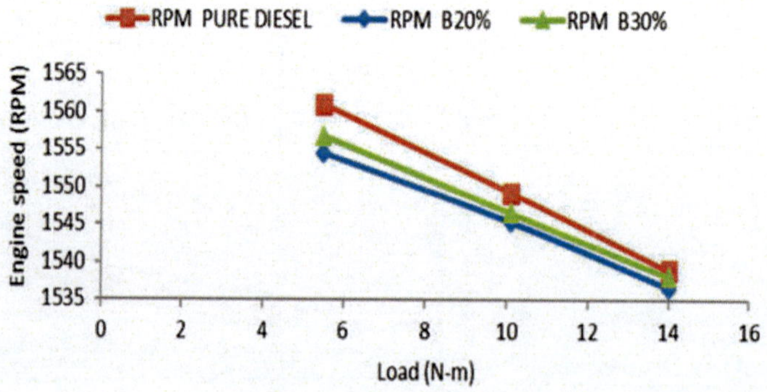

Fig. 9 Change in Engine Speed (RPM) with load (N-m)

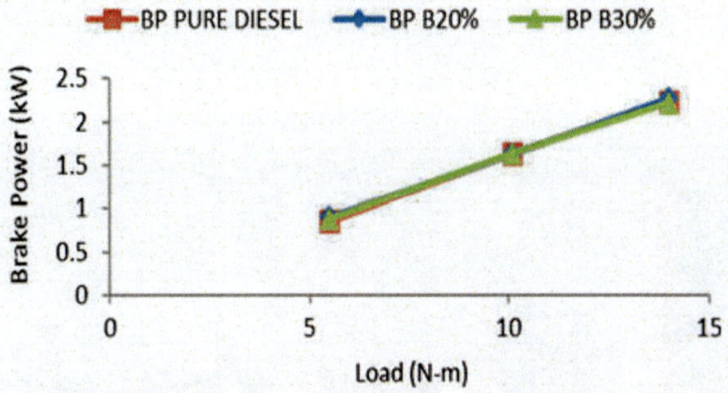

Fig. 10 Change in Brake Power (kW) with load (N-m)

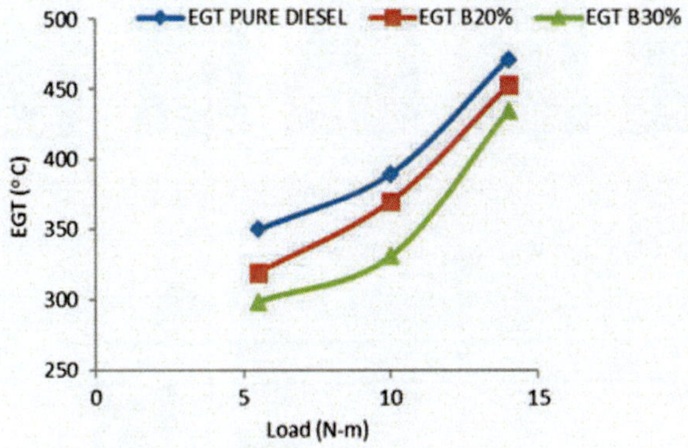

Fig. 11 Change in Exhaust Gas Temperature (°C) with load (N-m)

The soybean-based biodiesel is mixed with the diesel in different proportions, i.e., D100B0, D80B20, and D70B30. The engine is fuelled with these blends and the following properties such as BSFC, BTE, Mechanical Efficiency, Volumetric Efficiency, Brake Power (BP), Engine RPM, and Exhaust Gas Temperatures (EGT) are evaluated. Comparison of these properties is shown graphically between diesel and biodiesel blends. After this, the performance of the engine is tested on the different compression ratios, i.e., CR14, CR15, CR16, CR18, CR 19, and CR20. Also, a comparison of the mentioned properties is shown with a variation of the compression ratio of the engine.

(i) **Pressure Versus Volume and Pressure Versus Crank Angle**

The variation of pressure with volume is indicated in Fig. 3. It is clearly visible in the figure that pressure increases with a decrease in volume. Similarly, the variation of pressure with the crank angle is also evaluated and represented in Fig. 4. Combining the data obtained at five loads (5.5, 10, 14, 16.5, and 18 N-m) using D70B30 biodiesel

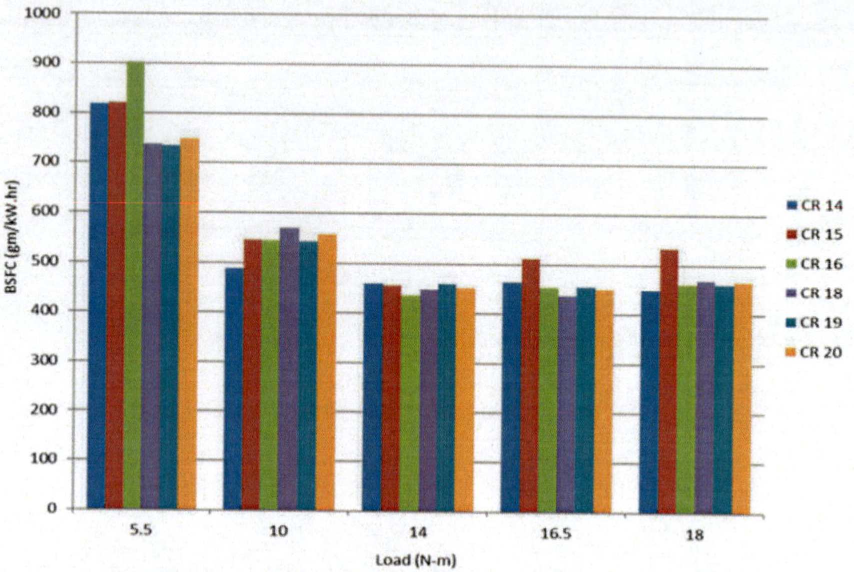

Fig. 12 Change in BSFC (g/kWh) with load (N-m)

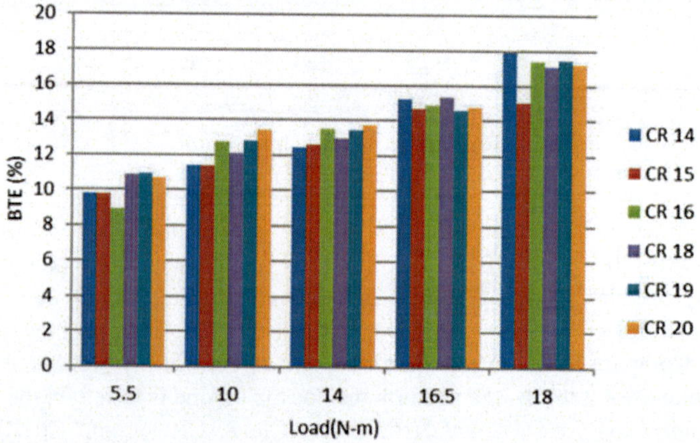

Fig. 13 Change in BTE (%) with load (N-m)

in the fuel, the graph shown in Fig. 3 is showing an average relation between pressure and volume and another graph shown in Fig. 4 is showing an average relation between pressure and crank angle. The maximum pressure is 45 bars which are found out at near about 0° of crank angle and 70 cc.

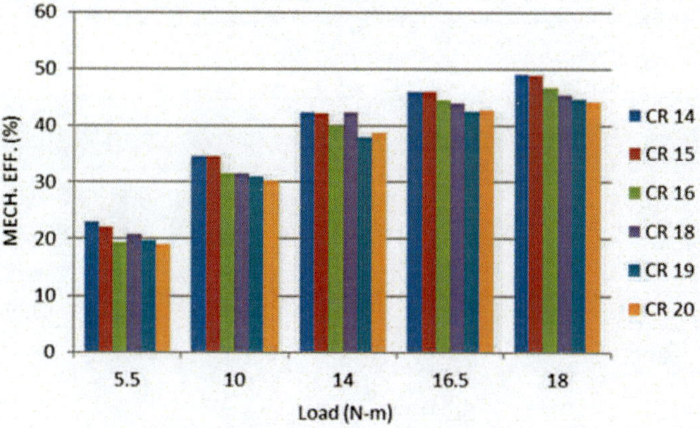

Fig. 14 Change in Mechanical Efficiency (%) with Load (N-m)

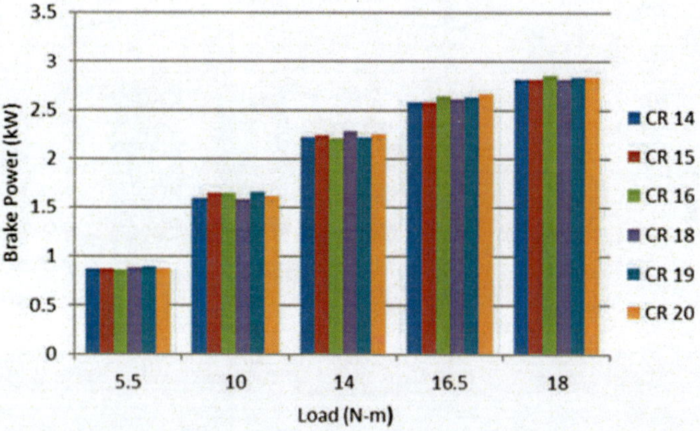

Fig. 15 Change in Brake Power (kW) with load (N-m)

4.1 Comparison of Engine Performance Characteristics at Different Fuel Blending Ratios

(ii) **Brake Specific Fuel Consumption (BSFC) Versus Brake Power**

BSFC is considered to be an essential parameter of the engine. The requirement of an engine is to produce more power even after less fuel consumption, thus improving its efficiency. The variation of fuel consumption with a change in the load or power output is such that irrespective of the constant Engine Speed, as the load increases, fuel consumption increases. The variation of the BSFC with Brake Power is shown in Fig. 5. Also, trends followed by the curve are of similar nature for different biodiesel blends at the constant CR. Fuel with 30% biodiesel has shown the highest

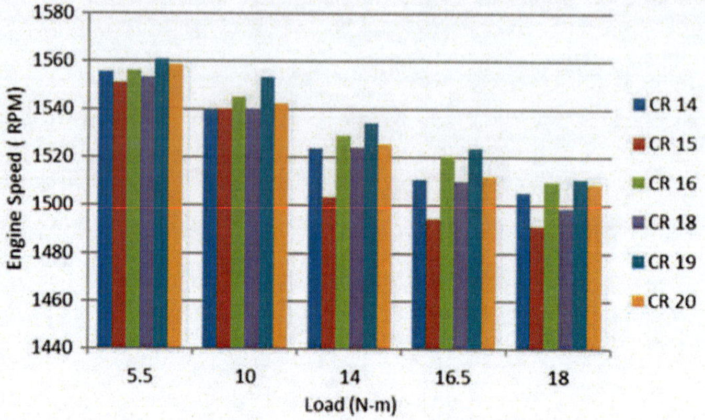

Fig. 16 Change in Engine Speed (RPM) with load (N-m)

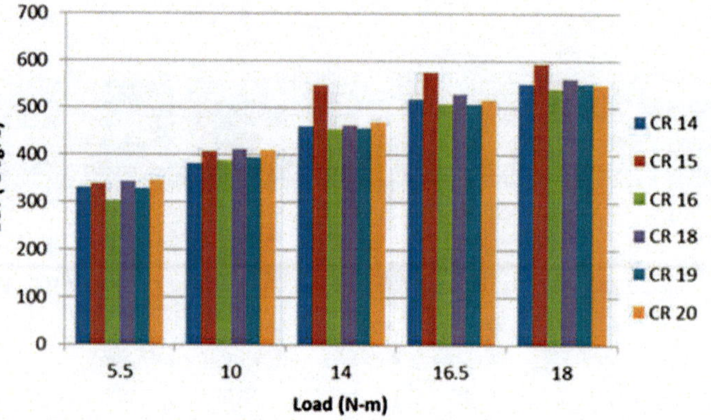

Fig. 17 Change in Exhaust Gas Temperature (°C) with load (N-m)

BSFC compared to diesel. It is clearly visible in figure that BSFC of the engine is decreasing with an increase in the Brake Power until 1.6 kW and then approximate become constant. These effects are coming in the result because of the change in the calorific value of the blended fuel. As we know biodiesel fuels have a low calorific value related to diesel. Mixing of the biodiesel with diesel reduces the calorific value of the fuel. Furthermore, the viscosity also varies in mixing, i.e., the viscosity of biodiesel is more relative to diesel. Therefore, the viscosity of the blend improves with increase in the composition ratio of biodiesel. The percentage increase in the BSFC for B20 and B30 is 6.4% and 18.52%, respectively, at the Brake Power of 1.6 kW.

(iii) **Brake Thermal Efficiency (BTE) Versus Brake Power**

BTE is considered the most vital parameter required to evaluate the performance of the engine. BTE is explained as the fraction of the work output to the total sum of the energy produced by firing the fuel in the engine. The variation of BTE for biodiesel and diesel at different BP is displayed in Fig. 5. It is found out that BTE improves with an increase in BP and it is highest in the pure diesel. The value of the BTE was reduced by about 10.71% and 22.85% for B20 and B30 fuel mixture at 1.6 kW than that of the diesel, respectively.

(iv) **Mechanical Efficiency (ME) Versus Brake Power**

Variation of the Mechanical Efficiency with Brake Power has been represented in Fig. 6. It is clearly visible in the figure that the Mechanical Efficiency of the engine is increasing with an increment in Brake Power. Also, trends followed by the curve are of similar nature for different biodiesel blends. Fuel with 30% biodiesel has shown the least Mechanical Efficiency in comparison to diesel. For B20 and B30 the percentage reduction in Mechanical Efficiency is 4.76% and 11.11%, respectively.

(v) **Volumetric Efficiency Versus Brake Power**

Volumetric Efficiency is the fraction of the density of the air-fuel mixture drawn into the cylinder at ambient pressure to the density of the unchanged volume of the air in the system. The variation of the Volumetric Efficiency for biodiesel blends and diesel is displayed in Fig. 8. So, these variations are due to the high density of the soy biodiesel. Thus, when the biodiesel is mixed with the diesel, then the resultant density of the mixtures increases.

(vi) **Engine Speed (RPM) Versus Load**

The variation of Engine Speed in the case of biodiesel mixture and diesel with load is presented in Fig. 9. It is clear that Engine Speed reduces with an increase in Engine Load. It can be seen from the graph the maximum value obtained is 1561.09 RPM for pure diesel at an Engine Load of 5.5 N-m while the minimum speed is obtained for B20 which amounts to 1536.5 RPM at an Engine Load of 14 N-m. The percentage reduction in the Engine Speed at a load, i.e., 10(N-m), has been observed approximately 0.26, and 0.22% for D80B20 and D70B30 blends.

(vii) **Brake Power (BP) Versus Load**

The deviation of the Brake Power with Loads is shown in Fig. 10. It is clearly visible in the figure that Brake Power for biodiesel blends is approximately equal to the Brake Power for diesel. Also, trends followed by the curve are of similar nature for different biodiesel blends. Fuel with 30% biodiesel has shown the highest Brake Power compared to the diesel. The Brake Power changes slightly with the variation in the biodiesel composition in the blend.

(viii) **Exhaust Gas Temperatures (EGT) Versus Load**

Variations in Exhaust Gas Temperature for various engine loads for various measured fuels have been shown in Fig. 11. The results obtained clearly point toward the fact that the EGT rises as the Engine Load rises for all blending fuel ratio. For all the biodiesel mixes, the total EGT was smaller than diesel fuel. Also, the EGT decreased by an increment in the percentage of biodiesel in fuel. Moreover, delay in ignition has a major impact on Exhaust Gas Temperature variations. Larger Exhaust Gas Temperature and delayed combustion are the result of ignition delays of longer duration. Biodiesel possesses a larger Cetane Number (CN), which results in ignition delays of smaller duration which become the major reason for reducing Exhaust Gas Temperature, however, the engine shows higher values of coolant and also of oil temperature. The average decrement in Exhaust Gas Temperature for diesel and biodiesel mixtures of D80B20 and D70B30 has been observed approximately 3.89% and 12.59%, respectively.

4.2 Comparison of Engine Performance Characteristics at Different Compression Ratios

(i) **Brake Specific Fuel Consumption (BSFC) Versus Load**

The results of the investigation showed that the consumption of Brake Specific Fuel rises when the compression ratio is reducing and decreases when the compression ratio is progressive. BSFC of esterified soybean oil was found to be lower at CR16 and CR18, and at a load of 14(N-m) and 16.5(N-m) as described in Fig. 11. Possible reason for this may be fuel density, viscosity, and heating value of the fuels. The consumption of specific energy lowers with a rise in the compression ratio. The BSFC of esterified soybean oil is 443.01 gm/kW-hr at the higher compression ratio of 20 at a load of 14(N-m), while it is 900 gm/kW-hr for a lower compression ratio of 14 at a load of 5.5(N-m). The optimum value of BSFC is found for 5.5(N-m) load in the variable compression ratio engine run with esterified soybean oil.

(ii) **Brake Thermal Efficiency (BTE) Versus Load**

The variation of Brake Thermal Efficiency for different compression ratios is indicating in Fig. 13. In the case of a higher compression ratio, the BTE has been observed to be slightly higher. While for the lower compression ratio, it has been found to be slightly lower. For a compression ratio of 14, the BTE has been found to be 19.61% at a load of 18 (N-m), which emerges to be a higher value. The result is a clear indication of the fact that there is a remarkable improvement in BTE for biodiesel in a variable compression ratio (CR) engine.

(iii) **Mechanical Efficiency (ME) Versus Load**

Observations indicate a higher Mechanical Efficiency in biodiesel with a low compression ratio and a lower Mechanical Efficiency in biodiesel with a high compression ratio at different load conditions. The maximum and minimum value of Mechanical Efficiency for biodiesel has been obtained about 45.73% and 23.19%, respectively. Mechanical Efficiency reduces with an increase in compression ratio at peak load condition.

(iv) **Brake Power (BP) Versus Load**

The values of BP for various compression ratios have been represented as shown by Fig. 15. The figures display CR14, CR15, CR16, CR18, CR19, and CR20 for biodiesel with varying operation loads. The BP rises at higher values of the compression ratio as a result of the chemical energy conversion to mechanical energy. The highest BP obtained is 2.83 kW for CR 16 at a load of 18 (N-m), while the lowest BP obtained 0.79 kW for CR 16 at a load of 5.5 (N-m).

(v) **Engine Speed (RPM) Versus Load**

The values of Engine Speed for various compression ratios have been represented by Fig. 16. The figure shows the variation in the Engine Speed with respect to load at compression ratio, i.e., CR14, CR15, CR16, CR18, CR19, and CR20. It is worth noting that as the load increases, Engine Speed decreases. The highest Engine Speed obtained is 1560 RPM for CR 19 at a load of 5.5 (N-m), while the lowest Engine Speed obtained is 1490 RPM for CR 15 at a load of 18 (N-m).

(vi) **Exhaust Gas Temperature (EGT) Versus Load**

Exhaust Gas Temperature variations have been represented in Fig. 17 for various compression ratios and for various loads. This result gives an indication that EGT decreases for CR16 in comparison to that of CR15. The highest temperature found is 594 °C for biodiesel at the CR of 15 at a load of 18 (N-m), whereas the lowest temperature obtained is 304 °C for the maximum compression ratio value of 16 at a load of 5.5 (N-m). The reason for this reduction in EGT at an increased CR is due to low calorific value possessed by biodiesel. Moreover, EGT loss depletion may be a probable reason for higher performance given by the engine.

5 Conclusion

The main conclusion drawn from the present study has been briefed as follows:

1. BSFC increased by about 11.53–17.94%, whereas BTE had an average reduction of about 15.26–36.4% for tested blends in comparison with diesel fuel.
2. From tests conducted on engine setup, it is observed that the Mechanical Efficiency decreases by about 7.14 and 24.48%, where Volumetric Efficiency increases by 5.94% in B20% and 9.11% for B30%.
3. The engine performance in the case of 30% biodiesel blend is better than other ratios of biodiesel blends in the range of 0–30%.
4. The performance of the engine for 30% biodiesel blend is best for compression ratio 18.

Acknowledgements The engine test facility provided by the Mechanical Engineering Department of NIT Kurukshetra has been gratefully acknowledged.

References

1. Shenawy EA, Elkelawy M, Bastawissi HA-E, Panchal H (20199) Comparative study of the combustion, performance, and emission characteristics of a direct injection diesel engine with a partially premixed lean charge compression ignition diesel engines. Fuel 249:277–285
2. Bastawissi HAED, Elkelawy M, Panchal H, Sadasivuni KK (2019) Optimization of the multi-carburant dose as an energy source for the application of the HCCI engine. Fuel 253:15–24
3. Chandra Sekhar S et al (2018) Biodiesel production process optimization from Pithecellobiumdulce seed oil: performance, combustion, and emission analysis on compression ignition engine fuelled with diesel/biodiesel blends. Energy Convers Manage 161:14154
4. Anwar M, Rasul MG, Ashwath N (2018) Production optimization and quality assessment of papaya (Carica papaya) biodiesel with response surface methodology. Energy Convers Manage 156:103–112
5. Onukwuli DO, Emembolu LN, Ude CN, Aliozo SO, Menkiti MC (2017) Optimization of biodiesel production from refined cotton seed oil and its characterization. Egpt J Petrol 26(1):103–110
6. Ezekannagha CB, Ude CN, Onukwuli OD (2017) Optimization of the methanolysis of lard oil in the production of biodiesel with response surface methodology. Egpt J Petrol 26(4):1001–1011
7. Miraculas GA, Bose N, Raj RE (2018) Process parameter optimization for biodiesel production from mixed feedstock using empirical model. Sustain Energy Technol Assess 28:54–59
8. Kucek KT, César-Oliveira MAF, Wilhelm HM, Ramos LP (2007) Ethanolysis of refined soybean oil assisted by sodium and potassium hydroxides. J Am Oil Chemists Soc 84(4):385–392
9. Elkelawy M et al (2018) Numerical and experimental investigation of ethyl alcohol as oxygenator on the combustion, performance, and emission characteristics of diesel/cotton seed oil blends in homogenous charge compression ignition engine. https://doi.org/10.4271/2018-01-1680
10. Chauhan BS, Kumar N, Cho HM, Lim HC (2013) A study on the performance and emission of a diesel engine fuelled with Karanja biodiesel and its blends. Energy 56:1–7
11. Sekhar SC et al (2019) Emission analysis on compression ignition engine fuelled with lower concentrations of Pithecellobiumdulce biodiesel-diesel blends. Heat TransAsian Res 48(1):254–269
12. Abed K, El Morsi A, Sayed M, El Shaib A, Gad MS (2018) Effect of waste cooking-oil biodiesel on performance and exhaust emissions of a diesel engine. Egpt J Petrol 27(4):985–989
13. Özener O, Yüksek L, Ergenç AT, Özkan MJF (2014) Effects of soybean biodiesel on a DI diesel engine performance, emission and combustion characteristics. Fuel 115:875–883
14. Hasnain SM, Sharma RJRA (2019) Evaluation of the performance and emission and spectroscopic analysis of an improved soy methyl ester. RSC Advances 9(46):26880–26893
15. Qi DH, Lee CF (2014) Influence of soybean biodiesel content on basic properties of biodiesel-diesel blends. J Taiwan Inst Chem Eng 45:504–507
16. Celikten I, Koca A, Arslan MA (2009) Comparison of performance and emissions of diesel fuel, rapeseed and soybean oil methyl ester injected at different pressures. Renew Energy 35(4):814–820
17. Elkelawy M et al (2019) Experimental studies on the biodiesel production parameters optimization of sunflower and soybean oil mixture and DI engine combustion, performance, and emission analysis fuelled with diesel/biodiesel blends 255:115791
18. Can Ö (2014) Combustion characteristics, performance and exhaust emissions of a diesel engine fuelled with a waste cooking oil biodiesel mixture. Energy Convers Manage 87:676–686
19. Lahane S, Subramanian KJF (2015) Effect of different percentages of biodiesel-diesel blends on injection, spray, combustion, performance, and emission characteristics of a diesel engine. Fuel 139:537–545
20. Cheung C, Man X, Fong K, Tsang OK (2015) Effect of waste cooking oil biodiesel on the emissions of a diesel engine. Energy Procedia 66:93–96

Internal Combustion Engine and Emission

Design and Development of Wave Rotor Technology for an Automotive Diesel Engine

J. Thangaraja, Datar Shantanu, Mayuresh Bhosale, and Razi Nalim

1 Introduction

The concept of wave rotor was first developed in the last century which works on pressure waves. The wave pressure devices have some common characteristic elements: a rotor with longitudinally shaped narrow channels within. The ports machined into the end fixed plates control the working fluids flow from the manifold pipes towards the pressure wave channels and vice versa. The energy transfer is achieved through the direct interaction between fluids by using the non-steady pressure waves. Wave rotor performs both compression and expansion in the same component [1] allowing compressed air to engine inlet.

During World War II, in 1940, Claude Seippel of the Brown Boveri & Co., trying to apply the Lebre principle to a heat pump, acknowledged that the pressure waves can efficiently transmit energy [2]. The rotor had 30 channels shaped within, rotating with 6000 rpm. Two ports on each side of the rotor allowed the passing of gas and air. The pressure ratio was initially 3:1 with a global efficiency of 69%, as reported in the tests performed in 1941–1943. According to Seippel's patents, a power boost

J. Thangaraja · D. Shantanu · M. Bhosale (✉)
School of Mechanical Engineering, Vellore Institute of Technology, Vellore, Tamil Nadu 632014, India
e-mail: mayureshb7@gmail.com

J. Thangaraja
e-mail: thangaraja.j@vit.ac.in

D. Shantanu
e-mail: shantanudatar17@gmail.com

R. Nalim
Mechanical Engineering, Indiana University—Purdue University Indianapolis, Indianapolis, IN 46202, USA
e-mail: mnalim@iupui.edu

A. K. Gupta et al. (eds.), *Advances in IC Engines and Combustion Technology*, Lecture Notes in Mechanical Engineering,
https://doi.org/10.1007/978-981-15-5996-9_28

of 80 and a 25% increase in performance were expected. The device was replaced by a heat exchanger which yielded a somewhat higher overall thermal efficiency.

In the mid-60s, Rolls-Royce company in the UK, by cooperating with Brown Boveri & Co. and with Berchtold of ETH Zurich Spalding of Imperial College London as consultants, focused on developing a wave rotor as a topping stage for a small helicopter engine, based on a reverse-flow configuration rotor incorporated into a single turbine engine [3].

In 1971, the first Comprex prototype was installed on a truck engine by the company Valmet Tractors from Finland, while around the same year Mercedes-Benz started testing the Comprex for their passenger diesel cars. Diesel engines' supercharging using pressure wave rotors became of big interest starting with 1974 when car manufacturers like Opel, Ford, Mercedes-Benz, Volvo, Peugeot and Ferrari focused their attention on Comprex supercharging. The first success that became a limited edition was the 2.3 litre supercharged engine on the Opel Senator model. The first extended application of Comprex on vehicles was in 1987, when ABB (the new company formed by merging ASEA with Brown Bovery & Co.) sold the Comprex activity to Mazda. The new Mazda 626 Capella model [4] was equipped with a 2.0 litre engine supercharged with Comprex. Mazda is considered one of the most successful commercial applications of PW technology.

2 Design and Working of Wave Rotor

Figure 1 is an exploded view of the assembly. It consists of four ports, namely rotor, air inlet, air outlet, exhaust inlet and exhaust outlet. Figure 2 is a SOLIDWORKS model of rotor. The geometry of twenty-five channels in the rotor is shown. These

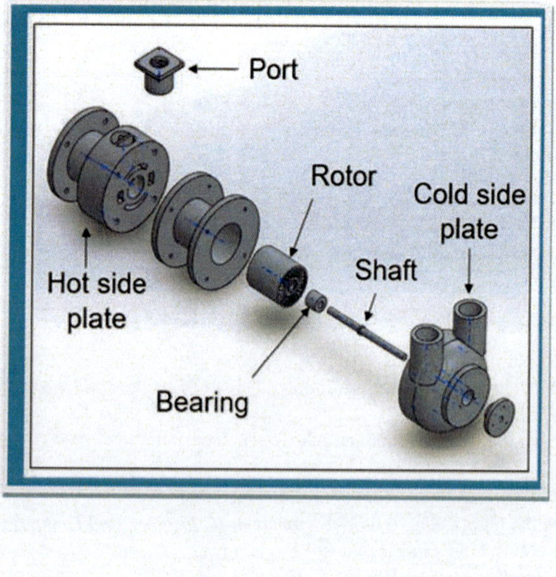

Fig. 1 Exploded view of PWS geometry

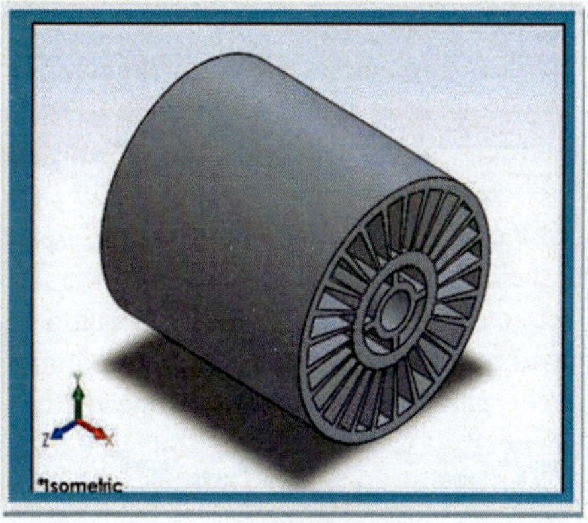

Fig. 2 CAD model of rotor

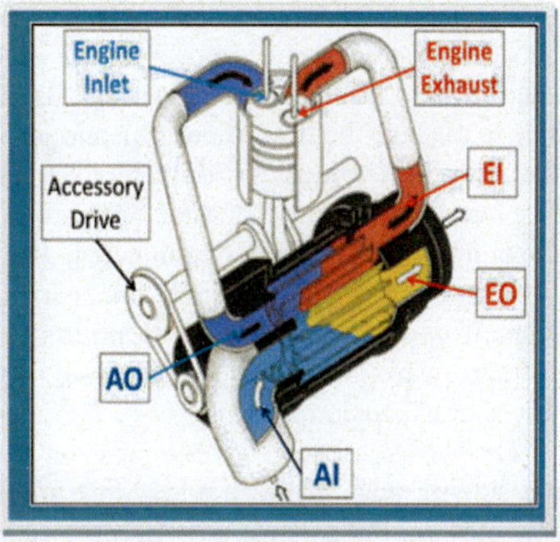

Fig. 3 Schematic of a wave rotor [5]

channels are the locations where the pressure exchange will take place between the exhaust gases and inlet air (Fig. 3).

The calculation of size of the wave rotor required for air boosting is the most critical part while designing the wave rotor. The mass flow rate of the air entering the engine is given by Eq. 1.

$$M = (N * D * V)/T \tag{1}$$

where

N number of channels in wave rotor
D density of air
V volume of wave rotor
T time required for each revolution cycle.

The volume of the wave rotor and the RPM of the engine are iterated to obtain the mass flow rate of the air at engine inlet.

A wave rotor is a device which uses the exhaust gas of a combustion process to compress the incoming charge of air, much like a turbocharger. However, while a turbocharger separates the intake and exhaust charge into separate compressor and turbine sections, the wave rotor brings them both into contact and uses the momentum of the exhaust charge to pressurize the intake charge. To prevent mixing of the two charges, the gases are brought in contact in very small tubes called channels. The gases remain in contact for a very short amount of time and then the pressurized intake charge is removed via one of the ports to be delivered to the combustion chamber.

The device consists of a rotor enclosed in a casing and two side plates, and rotor is keyed to the shaft which is supported by bearings. The bearings are press fit into side plates which support the load through the shaft.

The casing ensures no escape of gases enabling proper exchange of pressure between gases and thus conserves the efficiency of wave rotor. In rotor, pressure equalization of the intake and exhaust gas takes place and two manifolds, one on each side to manage the flow of gases in and out of the rotor. To achieve optimal function, the gases must be introduced and removed from the wave rotor in a controlled manner. The manifolds ensure that the intake and exhaust charges are timed perfectly so that the intake charge is not contaminated, and maximum pressure equalization occurs. Each port consists of an angled runner so that the angle of incidence of the gas when it enters or evacuates the rotor is minimum [6]. The wave rotor relies on well-timed pressure waves to achieve compression. The pressure waves can be either shock waves or expansion fans.

The wave rotor achieves compression by equalizing the pressure between the high pressure exhaust charge and low pressure intake charge. The exhaust gas evacuates the rotor via the exhaust out (EO) port and this causes a low pressure wave to be generated on the right as it travels left towards the intake manifold. The air in (AI) port opens as soon as the low pressure wave reaches the end wall. This causes atmospheric air to be sucked into the AI port. The low pressure wave reflects back as a high pressure wave as the reflection is against an open end. It then reflects back and forth as alternating high and low pressure waves between the two open ports AI and EO [7]. Thus, the high pressurized gas exits through AO and is sent to the combustion chamber.

3 Methodology

3.1 Schematic Layouts

Figure 4 shows the schematic layout of the experimental set-up. The air inlet is the inlet for air to be sucked into the system. The air passes through air flow meter. From

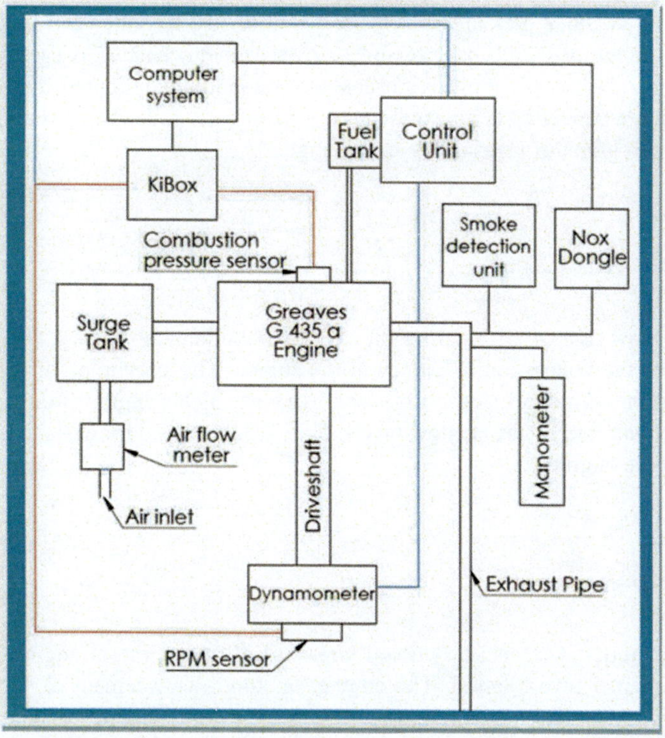

Fig. 4 Schematic layout of set-up

the air flow meter, the air further passes through a surge tank to the engine air inlet. Inside the engine, pressure sensors are installed in the cylinder head to measure the pressure in the combustion chamber. A KiBox is connected to the engine which is used to decode the data stored by the pressure sensors. This KiBox is connected to a computer system which obtains and stores the data of the pressure sensors. This data is converted to an excel file by using a MATLAB code. The dynamometer is connected to the engine through a driveshaft which sets the load on the engine. This can be adjusted manually. A smoke detection unit and NO_x dongle is connected to the exhaust manifold of the engine to measure the amount of smoke and NO_x, respectively, in the exhaust of the engine. A manometer tube is connected to the exhaust pipe to measure the pressure of the exhaust gases.

3.2 Installation of Pressure Sensors

The pressure at different regions of the engine is needed as input in the Converge software in the form of boundary conditions to get accurate simulation results. The combustion chamber pressure sensor used was of make Kistler and type 6124-A which has range of 0–300 bar pressure. An instrument, KiBox, is used to decode the dataset stored by these pressure sensors. The readings procured by this KiBox instrument are converted to an excel file by a MATLAB code. This excel file can be used to plot a graph of pressure vs. crank angle.

3.3 Mass Flow Rate

The mass flow rate of air entering the combustion chamber plays a major role in determining the volumetric efficiency of the engine. The amount of air entering the engine was measured by using a surge tank. Air entering the engine through the surge tank was monitored by the air flow meter. Time was recorded for every 0.2 m^3 of air sucked by the engine.

3.4 Testing

A single-cylinder, 4-stroke, air-cooled Greaves G435 A 3 diesel engine was used for experimental investigation. The engine has total displacement of 435 cc with compression ratio of 19:1. Maximum torque produced by engine is 20Nm at RPM range of 2200–2400. Engine testing and performance analysis along with emission testing was conducted to obtain preliminary values of combustion chamber pressure, volumetric efficiency, brake thermal efficiency, smoke emission value, NO_x, CO and CO_2 value. The set-up was made and Greaves diesel engine was tested for constant RPM 1700 and six different load conditions ranging from 0, 2, 4, 6, 8 and 10 Nm.

3.5 Exhaust Gas Analysis

With engine performance enhancement by boosting of inlet air the goal was also to analyse the smoke characteristics which included NO_x, CO and CO_2. The values of all emission testing were recorded for varying loads. The NO_x and smoke testing was conducted by using smoke detecting machine and NO_x dongle. A probe which draws a part of exhaust smoke analyses the NO_x content.

With NO_x emission values it was also important to get values of smoke particles emitted during combustion.

3.6 Exhaust Pressure Measurement

The exhaust pressure at the exhaust manifold is one of the boundary conditions required for the quasi 1-D code. The accurate measurement of this pressure was thus necessary. Due to the unavailability of a pressure sensor, a U-tube manometer was used for measurement of pressure.

3.7 Converge

Before integrating the engine with a wave rotor, it was necessary to predict the engine performance at the elevated pressure provided by the wave rotor. Converge software was used to simulate the combustion in engine Figure 5. The Converge software can produce pressure vs. crank angle graph for a specific engine when all the engine parameters are entered during the case set-up process. Converge software was used to simulate two different case set-ups. First simulation was done for naturally aspirated engine condition. The engine was run with atmospheric air as inlet. The pressure in the cylinder of the engine was measured with the installed pressure sensors. The same boundary conditions as in Table 1 were input in the Converge software and simulation was run which was validated by the experimental readings. The second simulation was done for boosted inlet air. Wave rotor when integrated with engine boosts the inlet air up to 2 bar [8].

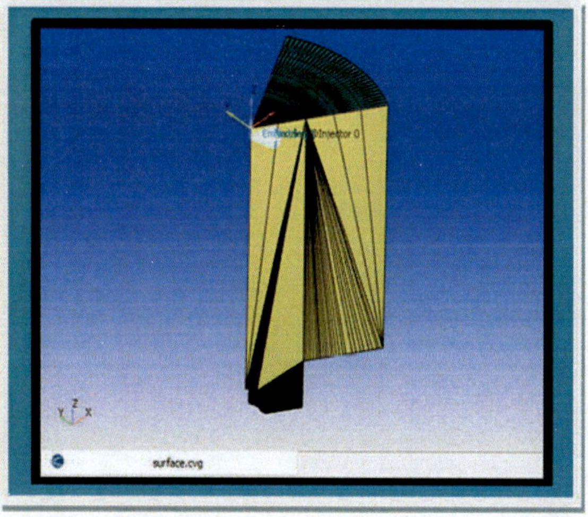

Fig. 5 Combustion chamber (converge)

Table 1 Engine specifications

Bore	86 mm
Stroke	75 mm
Connecting rod length	119 mm
Fuel	Diesel 2 (standard diesel)
Engine speed	1700 rpm

4 Results and Conclusions

The experimental results of volumetric efficiency versus load (Nm) are depicted in Fig. 6 stating about maximum volumetric efficiency of 81% approximately for naturally aspirated condition.

The comparison of experimental and simulation result is shown in Fig. 7. The simulation result of Converge software for an air inlet pressure of 1 bar (condition is for naturally aspirated engine) which is atmospheric pressure and the experimental result for the same pressure is depicted. The error between peak pressures of these two graphs is about 2.46% according to RMS error calculation which is acceptable, indicating that the simulation case set-up is accurate and can be used to predict the engine performance at boosted pressure. The raised curve depicts the simulation result for boosted pressure which is 2 bar. The maximum pressure achieved at boosted pressure is 129.25 bar which is approximately double the pressure at naturally aspirated condition. As there is significant rise in combustion pressure, NO_x

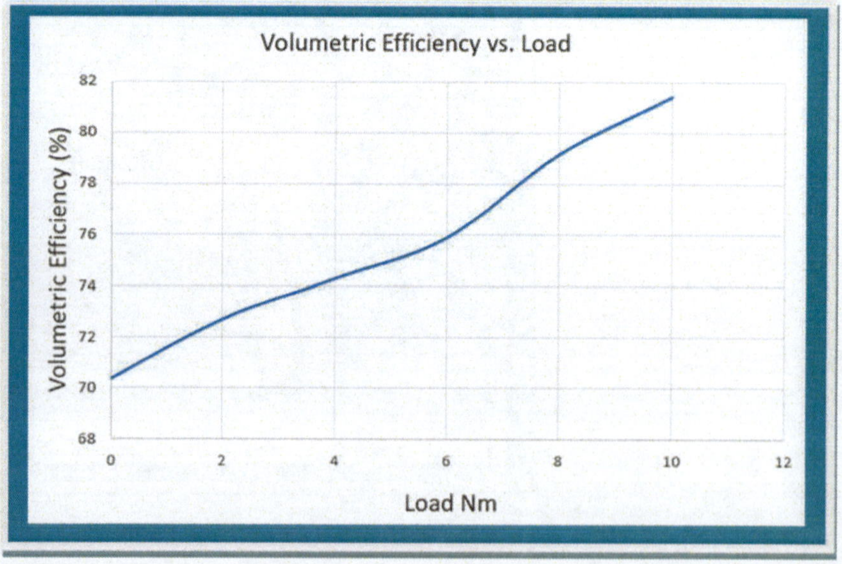

Fig. 6 Volumetric efficiency versus load

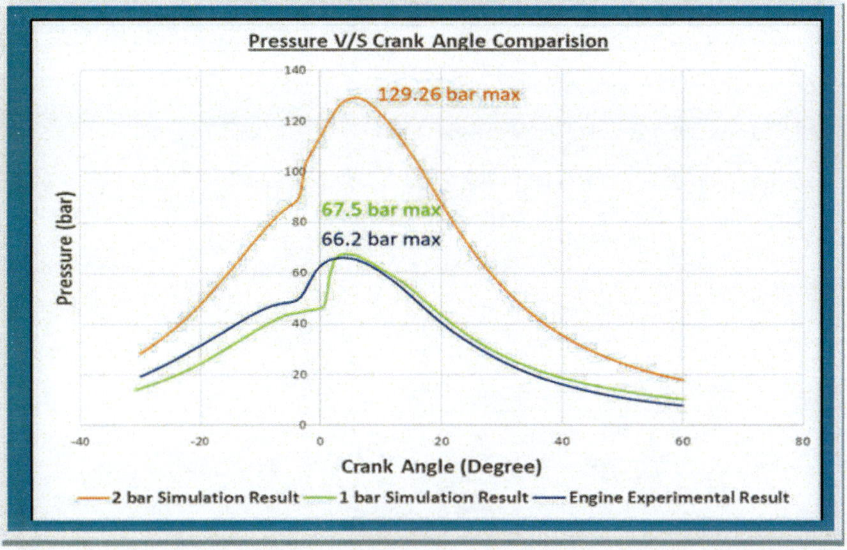

Fig. 7 Comparison of experimental and simulation result

values have increased with increase in boosted pressure by wave rotor as shown in Fig. 8. Similarly, CO values have been decreased significantly with increased boosted pressure allowing this technology to be more environment friendly Figure 9.

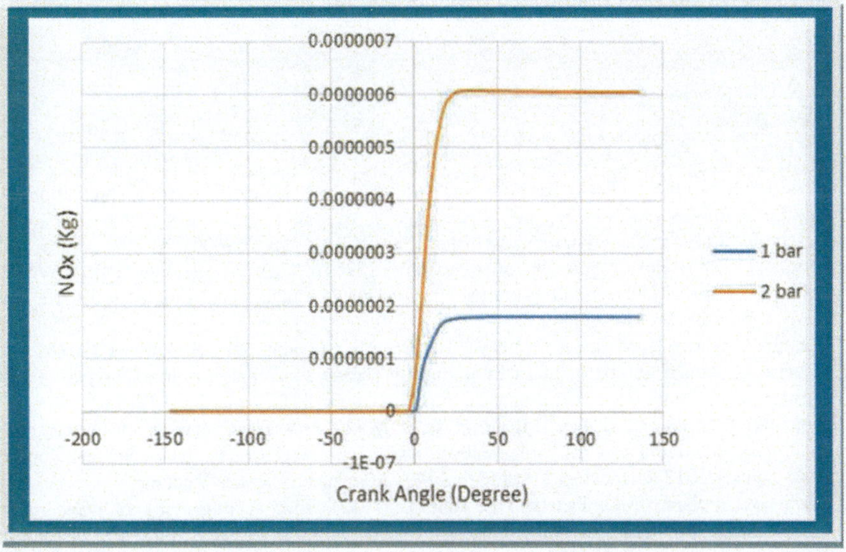

Fig. 8 Comparison of NO_x emission

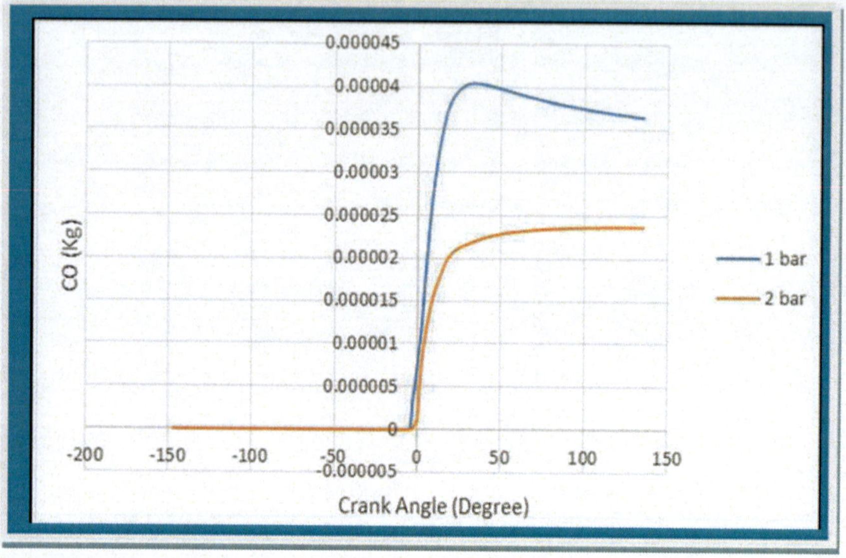

Fig. 9 Comparison of CO emission

All the graphs depict increase in performance of the engine without hampering the effects on environment and delivering more boosted pressure than a conventional turbocharger without any turbo lag. The use of wave rotor technology further on a single-cylinder diesel engine will enable us in improving downsizing of engine with same or increased power output.

References

1. Akbari P, Nalim R, Mueller N (2006) A review of wave rotor technology and its applications. J Eng Gas Turb Power 128
2. Berchtold M (1951) The comprex diesel supercharger. SAE Paper 590001, 67
3. Moritz R (1985) Rolls-royce stuy of wave rotors (1965–1970). Proceedings ONR/NAVAIR wave rotor research and technology workshop, naval postgraduate school, monterey, CA, Report NPS-67-85- 008, pp 116–124
4. Zehnder G, Mayer A, Mathews L (19989) The free running comprex®. SAE Paper 890452
5. George AC, Chiru A (2013) Internal combustion engine supercharging: turbocharger versus pressure wave compressor. Performance comparison, June 2013, Open Eng 4(2):110–118
6. Jagannath RR, Bane SPM, Feyz ME, Razi Nalim M (2017) Assessment of incidence loss and shaft work production for wave rotor combustor with non-axial channels, AIAA SciTech Forum 9–13 January 2017, Grapevine, Texas 55th AIAA Aerospace Sciences Meeting
7. Smith BD, Polanka MD, Paxson DE, John L (2012) Hoke scaling study of wave rotor turbo normalization of an internal combustion engine. In: 48th AIAA/ASME/SAE/ASEE joint propulsion conference & Exhibit AIAA 2012-3837
8. Costiuc, I, Chiru, A (2017) Evolution of the pressure wave supercharger concept. In: IOP conference series: materials science and engineering, vol 252, p 012081

Study of Emission Characteristics of a CI Engine Fueled with Water Diesel Emulsion Fuel

Bhupendra S. Patil and Rajesh C. Iyer

1 Introduction

With the rise in oil prices and depletion of petroleum reserves encourage researchers to focus on the alternate fuel products for automotive transportation purpose. Hybrid and electric vehicles are emerged as the promising solution for that. Also, researchers are trying to make current technology more efficient and economic. Emission of different hazardous automobile exhaust contents including unburnt hydrocarbons (HC), oxides of carbon (CO and CO_2), nitrogen (NO_x), and sulfur (SO_x), etc. is found responsible for global warming which is observed as one of the key problem from worldwide environment point of view. Government of India is going to apply BS-VI directly from BS-IV instead of going for BS-V by April 1, 2020 which will be very challenging task for the scientists [1]. There are many ways available in previously published literature to control emissions of different hazardous pollutants. There are two significant ways of treatment inside cylinder and outside cylinder. In the current study, treatment inside the cylinder is utilized. Water diesel emulsion (WDE) technique is one of the promising methods to fulfill objective of the current study [2]. From the literature review, it is observed that WDE is promising fuel improvement technique which is highly appreciated by researchers. WDE is widely used to get expected engine performance and exhaust emission behavior of CI engine with and without change in existing CI engine [3, 4].

The objective of current study is to research emission characteristics (CO, CO_2, HC, NO_x, and smoke opacity) of the CI engine fueled with WDE fuel having different

B. S. Patil (✉) · R. C. Iyer
Gujarat Technological University, Ahmedabad, Gujarat 382424, India
e-mail: bhupendrapatil.mech@gmail.com

R. C. Iyer
e-mail: rajesh_c_iyer@yahoo.com

A. K. Gupta et al. (eds.), *Advances in IC Engines and Combustion Technology*, Lecture Notes in Mechanical Engineering,
https://doi.org/10.1007/978-981-15-5996-9_29

concentration of water (0, 5, and 10%) and under varying loading conditions at a constant 1500 RPM.

2 Materials and Methods

In this work, diesel purchased from local petrol pump (BPCL, India) and double-distilled water are utilized as continuous phase and dispersed phase of the fuel emulsion, respectively. Mixed surfactant system (Span80 and Tween80) was used to prepare WDE having an hydrophilic lipophilic balance (HLB) value of 9 at 5% concentration which forms most stable WDE [5]. Surfactants used were manufactured by Thomas Baker (Chemicals) Pvt. Ltd. Mechanical stirrer was utilized for preparation of emulsion at a constant speed of 2500 RPM. Water was added drop by drop in a conventional diesel constantly for about 20 min to get stable WDE (Table 1).

Figure 1 shows experimental setup which is utilized in current work. The speci-

Table 1 Properties of fuel samples [6, 7]

Fuel	Density (kg/m^3)	Calorific value (MJ/kg)	Viscosity (cSt)	Flash point (°C)
Diesel	823.42	42.024	3.05	62
5% WDE	831.85	39.812	3.29	68
10% WDE	840.70	37.718	3.58	73
EN 590:2009	820–845	NA	2–4.5	Above 55

Fig. 1 Photographic view of engine test setup

Table 2 Engine specifications

Parameter	Specifications
Manufacturer	Kirloskar
Type of engine	Computerized, 4 stroke, single cylinder, water cooled
Bore × Stroke	87.5 (mm) × 110 (mm)
Compression ratio	17.5
Power output (Kw/HP)	3.5/5 @ 1500 RPM
Swept volume (CC)	661
Dynamometer	Water-cooled eddy current with different loads

Table 3 Specifications of gas analyzer (AVL-444) and smoke meter (AVL-437)

Measured quality	Range	Resolution	Accuracy volume
CO (%)	0–10	0.01	±0.1
CO_2 (%)	0–20	0.1	±0.3
O_2 (%)	0–22	0.01	±0.2
HC	0–20,000 PPM	1 PPM	±10 PPM
NO	0–5000 PPM	1 PPM	±20 PPM
Smoke intensity	0–100% opacity	0.1%	±1% of full scale

fications of engine are listed in Table 2. The engine setup is equipped with an eddy current dynamometer coupled by a rigid coupling. A strain gauge is connected with a water-cooled dynamometer for load measurement (Table 3).

3 Results and Discussion

Figure 2 shows the formation of CO emission at different loading conditions for diesel and WDE fuel. The CO emission is found higher with 10% WDE specifically than conventional diesel and 5% WDE. It is because of higher absorption of latent heat is carried out by particles of water which leads to reduction in combustion temperature, moisture content, and the high amount of OH radicals from the water that promotes oxidation of carbon to CO. With the application of WDE, reduction in combustion chamber temperature is observed which resulted in high CO emission in exhaust gas [8]. Also, more emission of CO is observed with increment of water content in diesel emulsion. This is because of reduction in peak flame temperature in combustion. Here, presence of water is also responsible for microexplosion which leads to complete combustion of emulsified fuel [6].

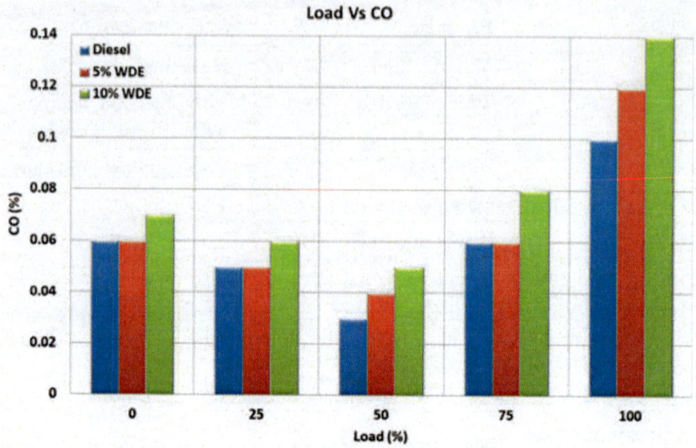

Fig. 2 Load versus CO

Figure 3 shows the formation of CO_2 emission at different loading conditions for diesel and WDE fuel. With increase in percentage of water content in emulsified fuel, increase in CO_2 emission is observed. At low to medium loading conditions, WDE shows lower CO_2 emission and it increases at high load conditions as shown in Fig. 3. It could be because of microexplosion behavior at high load condition and also water particles absorb more latent heat at low load condition [9]. At high load condition, the WDE encourages high level of CO_2 emission as compared to conventional diesel. Here, a secondary atomization phenomenon leads to complete combustion of emulsified fuel which results in high CO_2 emission.

The emission of unburnt hydrocarbon (HC) from the combustion of all considered fuel cases under all load conditions is presented in Fig. 4. Presence of water

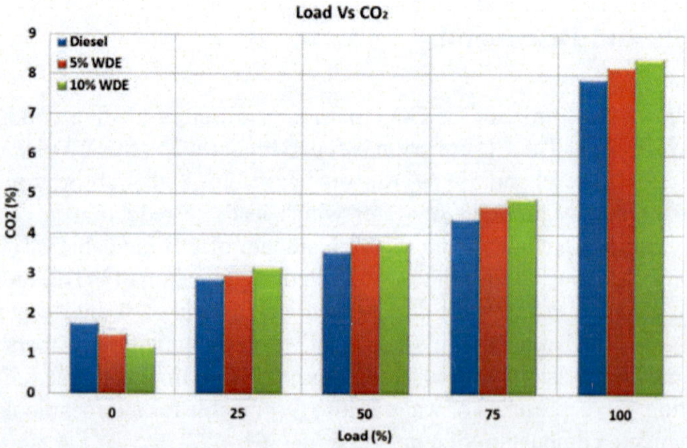

Fig. 3 Load versus CO_2

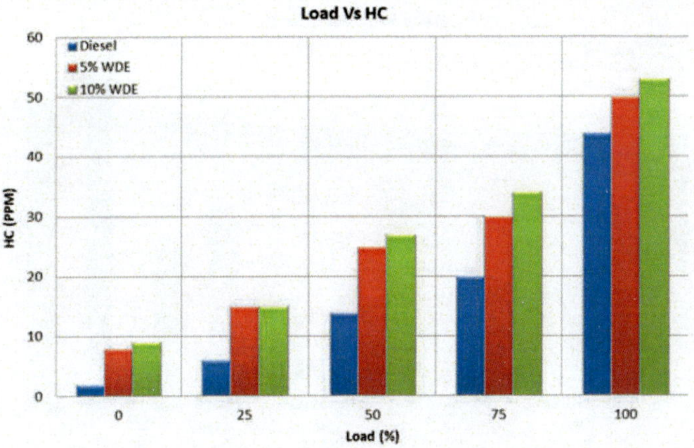

Fig. 4 Load versus HC

content leads to reduction in combustion temperature at low load conditions which resulted in more HC emission. Further, at full load condition, with increase in water content, emission HC decreases. The emission unburnt hydrocarbon in WDE fuel is found slightly higher than conventional diesel because of moisture content and low combustion temperature.

The formation of NO_x in the CI engine is generally more than SI engine due to availability oxygen and high combustion temperature. Figure 5 shows the emission of NO_x for all engine load conditions using conventional diesel, 5% WDE and 10% WDE. From the experimental outcomes, it is evident that as the proportion of water in the emulsion rises, the temperature of the exhaust gas decreases as the heat is absorbed by the extra water, which eventually reduces the NO_x as it is generated due

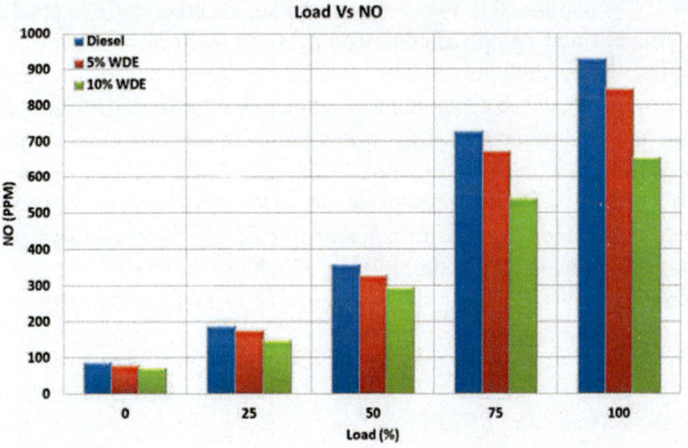

Fig. 5 Load versus NO_x

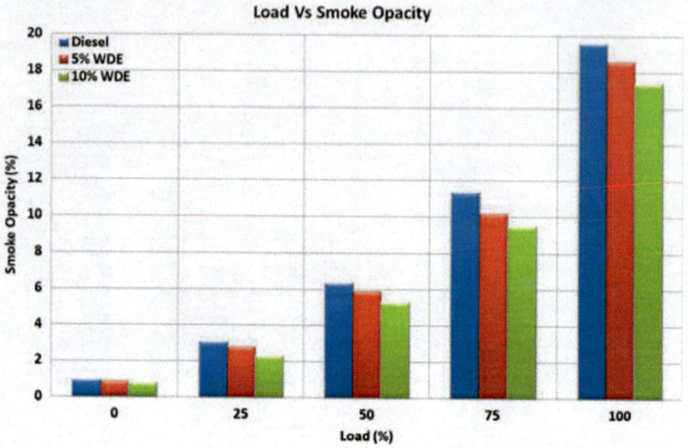

Fig. 6 Load versus smoke opacity

to reaction between nitrogen and oxygen at higher temperature. Very low emission of NO_x is observed with 10% WDE as compared to conventional diesel and 5% WDE case. This is because of the high latent heat of absorption of water which leads to decrease local adiabatic flame temperature which ultimately reduces overall combustion temperature. For all load conditions, increase in emission of NO_x is observed with combustion of all the considered cases of fuels. It could be due to high peak flame temperature. Also, with the increase in water content, results show linear reduction in emission of NO_x for individual load condition [10].

Figure 6 shows the smoke opacity of all considered fuel cases under all load conditions. Here, partial combustion of hydrocarbon is main reason of presence of carbon atom in the fuel. Here, it plays a major role in the formation of smoke at the engine exhaust [9]. From the experimental outcome, WDE found promising to reduce smoke opacity as compared to conventional diesel. The reduction in smoke opacity stipulates efficient and complete combustion of WDE. It is because of secondary atomization of diesel content in WDE with microexplosion. Also, with the rise in water content, results exhibit increment in smoke opacity for individual load condition due to fuel accumulation. Also, water diesel emulsion shows about 10% of average reduction as compared to conventional diesel for all load conditions. The depletion of smoke opacity is observed because of affluent oxygen concentration than conventional diesel. However, the microexplosion during the combustion of WDE is found responsible for better air–fuel mixing which results in efficient and complete combustion [6].

4 Conclusion

Form the experimental study, following conclusions are drawn

- The water diesel emulsion (WDE) produced by blending conventional diesel and water content (5 and 10%) can be directly used as a fuel in existing CI engine without any modifications.
- Reduction in NO_x emission is observed with the increase in water concentration in WDE fuel. WDE with 10% water content exhibits about 30% reduction in emission of NO_x at full load condition than conventional diesel fuel.
- About 10% reduction in smoke opacity is observed with WDE at all load conditions as compared to conventional diesel fuel.
- At all load conditions, WDE produces more HC and CO due to low combustion temperature.
- The WDE promotes high level of CO_2 particularly at high loading conditions as compared to conventional diesel. This is because of the secondary atomization which leads to complete combustion of fuel which results in more CO_2 emission.

Appendix

5% WDE	WDE fuel with 5% water content
10% WDE	WDE fuel with 10% water content
cSt	Centistokes
kg/m^3	Kilogram per cubic meter
MJ/kg	Mega Joule per kilogram
CC	Cubic centimeter
EN 590:2009	European standard of automotive fuel requirements.

References

1. Amit A, Sharma D, Lal Soni S, Sharma PK, Sharma S (2019) A comprehensive review on water-emulsified diesel fuel: chemistry, engine performance and exhaust emissions. Environ Sci Pollution Res ISSN (Online) pp 1614–7499
2. Iyer AA, Rane IP, Upasani KS, Bhosale YP, Gawande SH (2017) Experimental study on the effect of water injection in an internal combustion engine. Int Rev Mech Eng 11(6):379–386
3. Ravikumar TS, Basar PD, Attfield MJ (2001) Emulsified diesel—an Immediate and effective solution for diesel exhaust emission reduction, pp 259–264
4. Lombaert K, Le Moyne L, Guibert P, de Maleissye JT, Amouroux J (2004) Benefits of water-fuel emulsion on automotive diesel exhaust emission. Thermo and fluid dynamics processes in diesel engines. Springer, Berlin, Heidelberg
5. Patil H, Gadhave A, Mane S, Waghmare J (2015) Analyzing the stability of the water-in-diesel fuel emulsion. J Dispers Sci Technol 36(9):1221–1227
6. Sharma A, Kumar N, Vibhanshu V. Deep A (2018) Emission studies on a VCR engine using stable diesel water emulsion. no. February 2017, 2013. Vellaiyan and C. M. A. Partheeban.: emission analysis of diesel engine fueled with soybean biodiesel and its water blends. Energy Sources, Part A Recover. Util Environ Eff 40(16):1956–1965

7. Vellaiyan S, Amirthagadeswaran K (2019) Influence of water-in-diesel emulsion fuel and compression ratio on combustion, performance and emission characteristics of diesel engine. J Sustain Energy Eng 3(3):238–253. Retrieved 8 Nov. 2019, ISSN (Online) 2164–6295
8. Hemant Kumar MK, Shukla VK (2018) Water-in-diesel nano emulsion fuels for diesel engine: combustion properties and emission characteristics. prospects of alternative transportation fuels. energy, environment and sustainability. Springer, Singapore
9. Vellaiyan S, Amirthadeswaran KS (2016) Influence of water-in-diesel emulsion fuel and compression ratio on combustion, performance and emission characteristics of diesel engine. J Sustainable Energy Eng 3(3)
10. Singh N, Bharj RS (2014) An experimental investigation of emulsified fuel in a slow speed diesel engine for performance and emission. Int J Ind Eng Technol 6(1):1–9. ISSN 0974–3146

Comparison of Concurrent Reduction of Smoke and NO_x Emission Techniques

M. Bharathiraja, Ragupathy Karu, P. Arjunraj, and P. D. Jeyakumar

1 Introduction

1.1 A Subsection Sample

Diesel engine exhaust contributes to the creation of ozone, acid rain and weather change. Cancer, nose, eye irritation and lungs are the consequences owing to disclosure to diesel dissipate. Irritation in the lungs, which may exaggerate chronic respiratory signs and raise the frequency or strength of asthma and other health difficulties, is also produced by diesel exhaust disclosure [1]. Indian economy is diesel-based one, where the implementation of Bharath Stage VI (BS-VI) on 2020 will impose a technological change in the country. The simultaneous reductions of NO_x and smoke will be possible with the premixed and homogeneously mixed diesel combustion where achieving and controlling the combustion timing is critical. In the intake air of a CI engine, a liquid or gaseous fuel is injected by the fumigation set-up. This fuel contributed partly to the main fuel due to burn. In order to adopt fumigation in diesel engine, the important engine changes alter air intake system, fumigation tank and microcontroller control.

A variety of liquid alternative fuels are found effective in diesel engines using fumigation technique, which include propane, ethanol-water mixture, methanol, ethanol, gasoline, pine oil and methyl ethyl ketone. Gasoline fumes (GF) are tested in indirect injection diesel engines with turbo charging [2]. NO_x concentration decreases at the stages of 5–10% and maximum lessening is 20–8% GF at 2500

M. Bharathiraja (✉) · R. Karu
Bannari Amman Institute of Technology, Sathyamangalam, Tamil Nadu, India
e-mail: bharathiraja@bitsathy.ac.in

P. Arjunraj · P. D. Jeyakumar
B.S. Abdur Rahman Crescent Institute of Science and Technology, Vandalur, Chennai, Tamil Nadu, India

A. K. Gupta et al. (eds.), *Advances in IC Engines and Combustion Technology*,
Lecture Notes in Mechanical Engineering,
https://doi.org/10.1007/978-981-15-5996-9_30

Table 1 Specification of the test engine

No. of cylinder	1
No. of strokes	4
Fuel	Diesel
Power	3.5 kW
Speed	1500 RPM
CR range	18:1
Injection point variation	23° BTDC

RPM. The things of GF using carburettor on NO_x emission are planned [3] with a four-cylinder, water-cooled and indirect injection automotive diesel engine. Results display that NO_x decreases roughly 4.20%, 2.50% and 9.65% for 1/1, 3/4 and 1/2 fuel delivery ratio (FDR), respectively. An exhaust gas after treatment system is constructed and investigated in a stationary diesel engine to reduce smoke and odours to an acceptable level under varying engine running conditions [4]. The impact of an aluminium oxide nanoparticle blended diesel fuel upon the combustion, performance and emission characteristics of a diesel engine is also investigated [5]. The effect of aluminium oxide nanoparticles blended with diesel fuel for combustion, performance and exhaust emission characteristics of a diesel engine is investigated.

This paper aims to compare the performance and emission characteristics of a single-cylinder diesel engine with the fumigation and exhaust gas after treatment techniques for the concurrent reduction of NO_x and smoke emissions.

2 Materials and Methods

2.1 Engine Specification

The specification of the engine and equipment is given in Table 1.

2.2 Fuel Properties

The fuel properties are tested and presented in Table 2.

2.3 Fumigation of Gasoline

A scheme for fumigation takes in a fuel (gaseous or liquid in nature) to the stream of intake air for a compression ignited engine. Gasoline is found as suited fumigation

Table 2 Properties of the fuel used

Particulars	D	Gasoline	50ANOP-D
Density at 15 °C (kg/m^3)	815	737	819
Flash point (°C)	56	45	45
Calorific value (kJ/kg)	42,000	36,400	42,150
Cetane number	47	5	49

fuel because of its properties like volatility and calorific value. So, in this work gasoline is selected and an ECU is developed for controlling the gasoline-fumigated injection system. In fumigation mode, injection time and injection angle can be planned at each point based on load and speed. Air temperature, throttle position, coolant temperature and trigger sensor are connected to ECU which controls fuel injector.

2.4 *Fuel Map Calculation*

The gasoline fumigation is done for three different rates of 10%, 20% and 30% based on energy [6]. The performance and emission values are measured for these fumigation rates. The optimization is done using design expert software [7] and the fuel map is modified based on optimization results. The modified fuel map is used for gasoline fumigation. The fuel map for open ECU gasoline injection is used for injecting gasoline at various rates at the intake manifold.

The open time is calculated by Eq. 1.

$$\text{Open Time} = (\text{BOT} \times \text{AT} \times \text{CT} \times \text{BP} \times \text{MP}) + \text{BA} \tag{1}$$

Here, BOT is base open time in ms, AT is compensation for air temperature in %, CT is compensation for coolant temperature in %, BP is compensation for barometric pressure in %, MP is compensation for manifold absolute pressure in % and BA is compensation for battery voltage in ms.

2.5 *Nanoadditive*

The count of nanoadditives in the base fuel increases the thermo-physical properties of fuel such as high surface area to volume, thermal conductivity and mass diffusivity. In this work, aluminium oxide (Al_2O_3) nanoliquid is selected because liquid will blend more easily with diesel fuel.

Fig. 1 EGAT set-up

2.6 EGAT System

The full plan of the shower chamber and the silica gel chamber was modelled and analysed using design software. The construction is made with appropriate material. This consists of a silica gel absorption and a water spray chambers. It is completed by drenching water using nozzles through the connected pump. The silica gel adsorption chamber is recycled to absorb the smoke. Smoke gets sediment on the silica gel. The photographic view of the EGAT set-up is shown in Fig. 1.

3 Experimental Set-up and Testing Procedure

3.1 Experimental Set-up

The diameter of the existing petrol manifold is increased and is fitted with the engine. Using the ECU software is executed and gasoline fumigation is done based on the different input values. Gasoline fumigation along with aluminium oxide nanoliquid additive blended with diesel is tested in the same engine. Aluminium oxide nanoliquid is blended with diesel in the proportions of 50 ppm. The schematic of the experimental set-up is shown in Fig. 2. The EGAT arrangement is fitted to the exhaust and the readings are taken.

3.2 Testing Procedure

The engine tests were carried out based on the fuel combination, which is presented in Table 3. During testing, the engine is switched on and operated for 20 min with

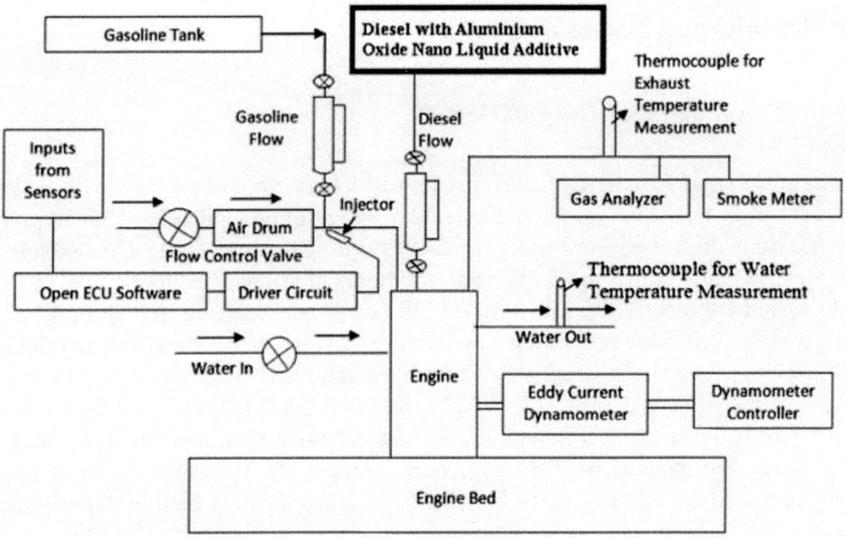

Fig. 2 Experimental set-up

Table 3 Fuel combinations

Fuel	Composition	Nomenclature
Diesel	Diesel 100%	D
Diesel and gasoline	Diesel and gasoline in open ECU	GF
Diesel, gasoline and Al_2O_3	Diesel blended with 50 ppm Al_2O_3 and gasoline	50ANOP – D +GF
Diesel with EGAT	Diesel 100% and EGAT is fitted at the exhaust	EGAT

diesel. The dynamometer is set for the load value and the readings are noted. The fuel consumption of diesel and gasoline is noted in gasoline fumigation. AVL five gas analyser and AVL smoke meter are used to measure the emissions from the engine. After taking the gasoline fumigation, diesel is mixed with 50 ppm aluminium oxide nanoadditive. Again the engine is operated and readings arc taken. Finally, the fumigation set-up is removed from the engine and the engine is fitted with EGAT set-up. Diesel is used as fuel and the engine is operated and readings are taken.

4 Results and Discussion

4.1 Performance Characteristics

Brake Thermal Efficiency (BTE) on Power. The variations of BTE with brake power for diesel, EGAT, gasoline fumigation and gasoline fumigation with 50 ppm Al_2O_3 nanoadditive are shown in Fig. 3. Since the EGAT set-up is used in the exhaust, it does not much affect the BTE. Gasoline fumigation gives reduction in BTE of 11.2% and 1.9% for 0.875 kW and 1.75 kW powers and increase in BTE of 0.15% and 9.31% at 2.625 kW and 3.5 kW powers, respectively, when compared with EGAT. Similarly, gasoline fumigation with Al_2O_3 nanoadditive gives an increase in BTE of 5.8%, 2.9%, 4.8% and 3.6% at 0.875 kW, 1.75 kW, 2.625 kW and 3.5 kW powers, respectively, when compared with EGAT. Nanoadditive aids in better combustion, and hence, the increase in BTE is obtained [2, 4, 8, 9]. The boosted rate of heat release due to the burning of the homogeneous air/fumigation fuel mixture should help to increase the BTE for fumigation.

Exhaust Gas Temperature (EGT) on Power. The difference of exhaust gas temperature on brake power is shown in Fig. 4. A decrease in exhaust gas temperature is obtained when the EGAT set-up is used. The percentage of reduce in EGT for EGAT is 14.04, 16.11, 7.54, 11.44 and 27.39 for 0 kW, 0.875 kW, 1.75 kW, 2.625 kW and 3.5 kW powers, respectively, when compared with diesel. Fumigation gives an increase in exhaust temperature of 1.9%, 5.3%, 2%, 2.4% and 26.4% at 0 kW, 0.875 kW, 1.75 kW, 2.625 kW and 3.5 kW powers, respectively, when compared with EGAT. Similarly, gasoline fumigation with 50 ppm Al_2O_3 nanoadditive gives further increase in exhaust temperatures of 10.5%, 14.2%, 6.6%, 9.2 and 36.1% at 0 kW, 0.875 kW, 1.75 kW, 2.625 kW and 3.5 kW powers, respectively, when compared

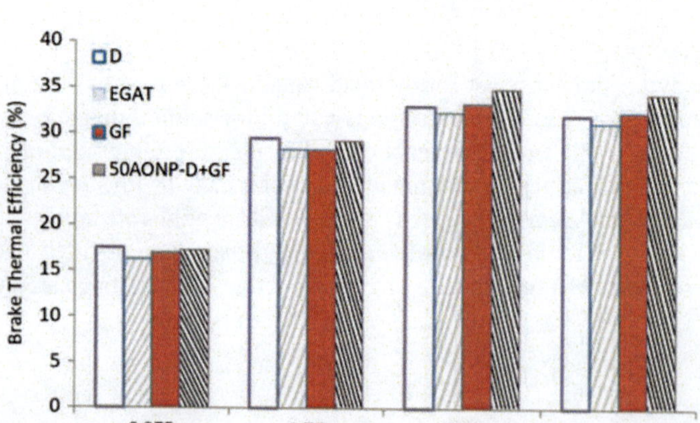

Fig. 3 BTE with brake power

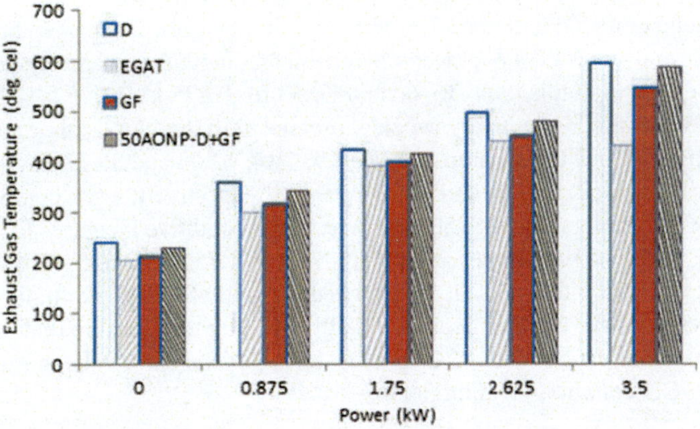

Fig. 4 EGT with brake power

with EGAT. There are decreases in the exhaust gas temperature in EGAT and the main reason behind this is that water spraying would take away a good part of the heat from the exhaust gas. Owing to complete combustion, the exhaust temperature is growing in gasoline fumigation with nanoadditive.

4.2 Emission Characteristics

CO Emissions on Power. The difference of carbon monoxide with power is shown in Fig. 5. From the figure, except 0 kW load, low CO emissions are observed for

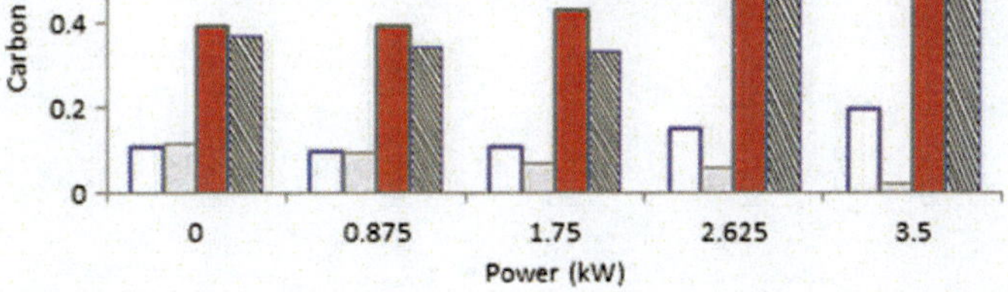

Fig. 5 CO with brake power

EGAT when compared with diesel at other remaining loads. The percentage of hike in CO emission is 9 for 0 kW power when compared with diesel and the percentage reduce in CO emission is 0, 36.36, 60 and 86.53 for 0.875 kW, 1.75 kW, 2.625 kW and 3.5 kW powers, respectively, when compared with diesel. Gasoline fumigation gives an increase in CO emission of 3.26, 3.9, 6.14, 12.03 and 35.77 times at 0 kW, 0.875 kW, 1.75 kW, 2.625 kW and 3.5 kW powers, respectively, when compared with EGAT. Further fumigation with 50 ppm Al_2O_3, nanoadditive increases CO by 3.08, 3.46, 4.76, 7.97 and 19.67 times at 0 kW, 0.875 kW, 1.75 kW, 2.625 kW and 3.5 kW powers, respectively, than EGAT. The hurried burning of vaporized fumigation fuel and following charge cooling, reduction in the cylinder temperature that might lead to incomplete oxidation of the CO to CO_2 during expansion stroke, ensuing in an increase of CO emission for fumigation.

UBHC Emissions on Power. The change of UBHC on power for diesel, EGAT, gasoline fumigation and gasoline fumigation with 50 ppm Al_2O_3 nanoadditive is shown in Fig. 6. From the figure, high UBHC in short loads and short UBHC at great loads for EGAT are observed when compared with diesel [2, 4, 8, 9]. The percentage of increase in UBHC is 50 and 40 for 0 kW and 0.875 kW powers, respectively, when compared with diesel and the percentage of decrease in UBHC is 6, 32 and 42 for 1.75 kW, 2.625 kW and 3.5 kW powers respectively when compared with diesel. Gasoline fumigation gives an increase in UBHC of 2.00, 4.08, 4.98, 7.62 and 6.92 times at 0 kW, 0.875 kW, 1.75 kW, 2.625 kW and 3.5 kW powers, respectively, when compared with EGAT. Further fumigation with 50 ppm Al_2O_3 nanoadditive increases UBHC, 1.30, 2.66, 3.23, 4.95 and 4.49 times at 0 kW, 0.875 kW, 1.75 kW, 2.625 kW and 3.5 kW powers, respectively, than EGAT. Meanwhile, fumigation fuel has cooling outcome on combustion processes, result reduced combustion temperature might not be able to ignite the unburned fumigated fuel through the expansion stroke which leads to the rise in HC emission.

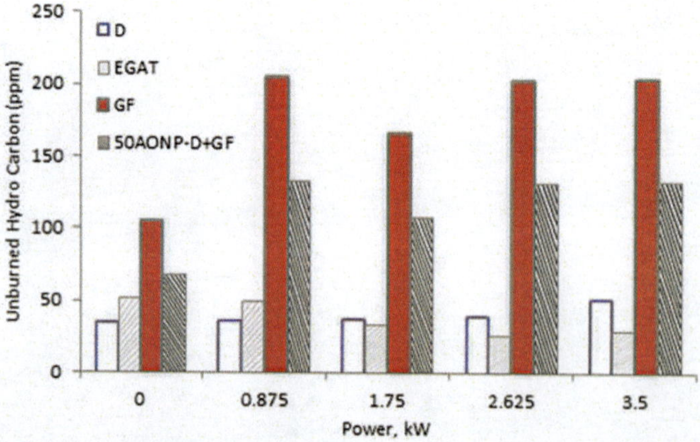

Fig. 6 UBHC with brake power

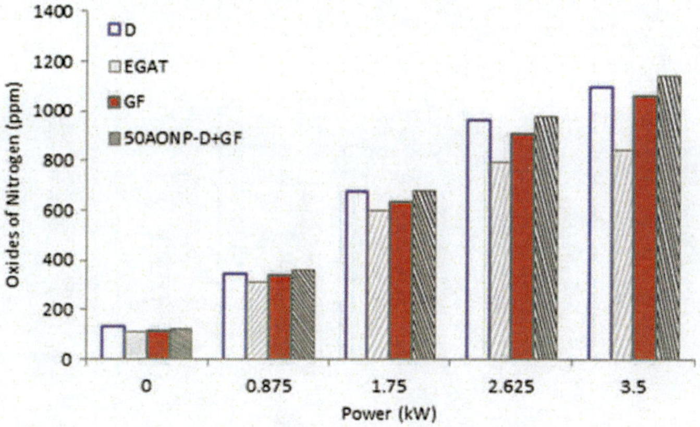

Fig. 7 NO_x with brake power

NO_x emissions on Power. The change of NO_x on brake power for diesel, EGAT, gasoline fumigation and gasoline fumigation with 50 ppm Al_2O_3 nanoadditive is shown in Fig. 7. The percentage of decrease in NO_x values is 1.5, 8.9, 10.7, 17.1 and 22.8 for 0 kW, 0.875 kW, 1.75 kW, 2.625 kW and 3.5 kW powers, respectively, when compared with diesel for EGAT. Fumigation also reduces the NO_x 0.98, 0.92, 0.95, 0.88 and 0.8 times lower for 0 kW, 0.875 kW, 1.75 kW, 2.625 kW and 3.5 kW powers, respectively, than EGAT. Further, fumigation with 50 ppm Al_2O_3 nanoadditive decreases smoke 0.91, 0.85, 0.88, 0.81 and 0.74 times for 0 kW, 0.875 kW, 1.75 kW, 2.625 kW and 3.5 kW powers, respectively, than EGAT. Fumigation fuel has a tall latent heat of vaporization; hence, the least amount of heat is out during combustion process which decreases the combustion temperature, leading to the lessening in NO_x formation, specifically under the lean conditions at lower engine loads.

Smoke Emissions on Power. The variation of smoke on brake power for diesel, EGAT, gasoline fumigation and gasoline fumigation with 50 ppm Al_2O_3 nanoadditive is shown in Fig. 8. The percentage of decrease in smoke values is 91.13, 91.87, 91.87, 93.87 and 93.01 for 0 kW, 0.875 kW, 1.75 kW, 2.625 kW and 3.5 kW powers, respectively, when compared with diesel. Fumigation also reduces the smoke 9.9, 10.49, 10.41, 13.82 and 10.95 times higher for 0 kW, 0.875 kW, 1.75 kW, 2.625 kW and 3.5 kW powers, respectively, than EGAT. Further, fumigation with 50 ppm Al_2O_3 nanoadditive also decreases smoke 9.28, 9.18, 7.80, 8.64 and 5.48 times higher for 0 kW, 0.875 kW, 1.75 kW, 2.625 kW and 3.5 kW powers, respectively, than EGAT. Less diesel fuel is burnt during combustion and combust organized with the fumigation fuel/air mixture which aids to burn faster and with higher accessibility of oxygen, leading to a reduction in PM emission.

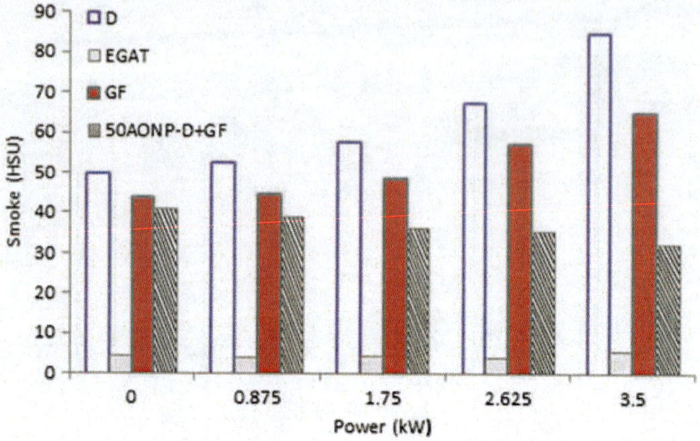

Fig. 8 Smoke with brake power

5 Conclusions

From the comparison of gasoline fumigation and exhaust gas after treatment with water scrubbing methods of simultaneous reduction of NO_x and smoke, the following conclusions are obtained. There is no much difference in BTE values when the EGAT is used in a diesel engine and gasoline fumigation shows a noticeable change in BTE with/without additive. Exhaust gas temperature is reduced with EGAT and it is further reduced when fumigation is used. CO and UBHC emissions show an increase when EGAT is used and it is further increased with fumigation. CO_2 emission shows a slight decrease with EGAT and it is increased with fumigation. NO_x emission gets reduced with EGAT set-up when compared with gasoline fumigation and gasoline fumigation with Al_2O_3 nanoadditive. Smoke emission is greatly reduced by EGAT set-up when compared with gasoline fumigation and gasoline fumigation with Al_2O_3 nanoadditive.

Acknowledgements The authors accept the testing and infrastructural facilities provided by Bannari Amman Institute of Technology, Sathyamangalam and B.S. Abdur Rahman Crescent Institute of Science and Technology, Vandalur during the entire work.

References

1. Ali R (2011) Effect of diesel emissions on human health: a review. Int J Appl Eng Res ISSN 6(11):1333–1342
2. Sahin Z, Durgun O, Bayram C (2012) Experimental investigation of gasoline fumigation in a turbocharged IDI diesel engine. Fuel 95:113–121
3. Sahin Z, Durgun O (2013) Improving of diesel combustion-pollution-fuel economy and performance by gasoline fumigation. Energy Convers Manag 76:620–633

4. Roy MM, Parvez R, Sarker RI (2011) Exhaust odor and smoke reduction of stationary DI diesel engines to acceptable level by water-scrubbing and air-dilution system. Appl Energy 88:2391–2399
5. Madhan Raj N, Gajendiran M, Pitchandi K, Nallusamy N (2016) Investigation on aluminium oxide nanoparticles blended diesel fuel combustion, performance and emission characteristics of a diesel engine. J Chem Pharm Res 8(3):246–257
6. Bharathiraja M, Venkatachalam R, Murugesan A, Tiruvenkadam N (2016) Experimental investigation of noval alcohol fumigation on single cylinder constant speed diesel engine. Int J Ambient Energy. https://doi.org/10.1080/01430750.2016.1222950
7. Bharathiraja M, Venkatachalam R, Prasanna N (2017) Experimental and mathematical investigation on performance and emission characteristics of ethanol and methanol fumigation in a single cylinder diesel engine. Asian J Res Soc Sci Humanit 7(1):632–652
8. Durgun O, Ahin Z (2009) Theoretical investigation of heat balance in direct injection (DI) diesel engines for neat diesel fuel and gasoline fumigation. Energy Convers Manage 50(1):43–51
9. Sahin Z, Durgun O, Bayram C (2008) Experimental investigation of gasoline fumigation in a single cylinder direct injection (DI) diesel engine. Energy 33(8):1298–1310

Performance and Emission Characteristics of an Engine Generator for Different Fuels

Saurav Sagar, N. K. Singh, and N. S. Maurya

1 Introduction

The 'little genie' engine generator (genset) is simple in design and controlled manually. This is kept stationery on the ground, and there is no need to bolt it with the foundation. Such engines are noisy, pollute and generate vibrations. They generate significant amount of smokes and gases whose analysis is needed. The increasing use of the above 'genset' in domestic, agriculture and manufacturing units is due to its small size, lightweight and easy to handle it. The author has not found any literature about the effects of conventional fuels (kerosene, petrol and diesel) and alternative fuels (biogas and biodiesel) in the engine used for this genset. It is an important issue to understand the nature of the above engine characteristics for different fuels. Also, there is a need to find the best alternative fuel for these engines which do not necessitate significant changes to their design.

Many researchers [1–9] have used different biofuels for estimation of diesel engine performance by minimizing harmful emissions. Hosseini and Ahmadi [10] observed that the presence of hydrogen in the combustion chamber of a diesel engine delays the ignition process which can reduce the above hazardous emissions. The performance of hydrogen-rich biogas in the spark ignition engine is studied by Zhang et al. [11] and Boulahlib et al. [12], and found the improved results. Becerra-Ruiz et al. [13] used green hydrogen and bioethanol fuels in internal combustion engines to reduce emissions. The utilization of liquid biofuels in gas turbines is reviewed by Ibrahim et al. [14]. The impact of biodiesel B100 on a small agricultural diesel engine is

S. Sagar · N. K. Singh (✉)
Department of Mechanical Engineering, National Institute of Technology Patna, Patna, India
e-mail: nilambersingh@nitp.ac.in

N. S. Maurya
Department of Civil Engineering, National Institute of Technology Patna, Patna, India

A. K. Gupta et al. (eds.), *Advances in IC Engines and Combustion Technology*, Lecture Notes in Mechanical Engineering,
https://doi.org/10.1007/978-981-15-5996-9_31

observed by Suthisripok and Semsamran [15]. The different authors [16–20] have used various blends with biofuels to reduce the injurious discharges of the internal combustion engines. The present study describes the performance and emission characteristics of the little genie engine generator which has single-cylinder, four-stroke spark ignition (SI) engine.

2 Experimental Technique

The 'genset' is operated in single-fuel operation mode for different fuels at several engine speeds and loads. Specifications of genset are frequency: 50 Hz, single phase, AC rated output: 1.5 KVA, voltage: 220 V, DC output: 12 V/10A, capacity: for petrol 0.8 l and for diesel 5.5 l (Fig. 1).

The liquid and gaseous fuels can be used in the genset with appropriate connection of intake manifold and fuel tanks or gas converter kit. The fuel consumption, output voltage, current and engine speed are measured. For every 50 ml of each fuel consumption (petrol, kerosene and diesel), it is connected with different loads (using bulbs to change voltage). After measuring different engine parameters, the same experiments are repeated for 50 g of LPG fuel. The exhaust emissions are measured with a gas analyzer. The gas analyzer consists of an exhaust probe, a modular diagnostic system (MDS) and a monitor. One end of the exhaust probe is inserted in the exhaust manifold of the engine, whereas the other end is connected with MDS which analyzes the sample of exhaust emissions and the results are displayed on the monitor.

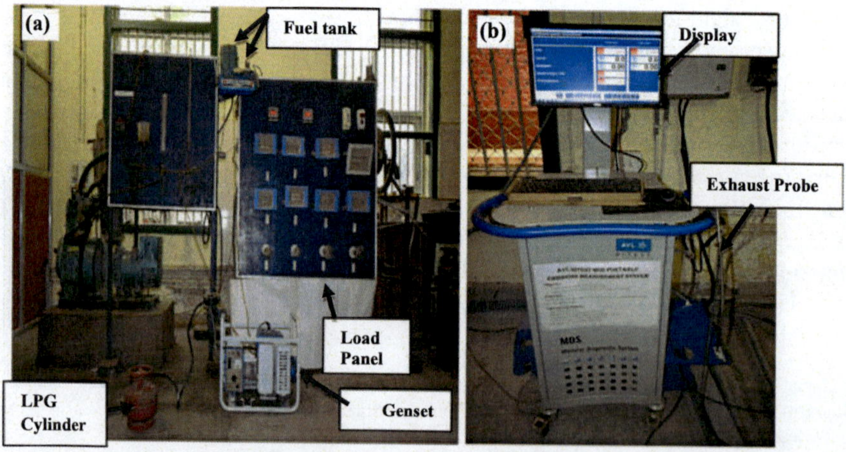

Fig. 1 Measurement of **a** performance parameters and **b** emission parameters

3 Results and Discussion

3.1 Performance Characteristics

The four different fuels (kerosene, petrol, diesel and LPG) are used to run the engine on loads 0, 250, 500, 1000 and 1500 W. The quantity of 50 ml of kerosene, petrol or diesel and 50 gm of LPG are considered at every load to determine the different parameters presented in Table 1. The LPG cylinder is directly connected to the engine through gas kit, and the reduction in weight of the cylinder is used to calculate the time required to consume 50 gm of LPG.

The brake-specific fuel consumption (BSFC) versus engine load for different fuels (kerosene, petrol, diesel and LPG) is compared in Fig. 2a. The specific fuel consumption decreases nonlinearly with increasing engine load for all fuels. The LPG engine has lowest brake-specific fuel consumption for all engine loads (0–1500 W). In Fig. 2b, the thermal efficiency versus engine load is compared for all

Table 1 Performance of engine generator with different fuels

Fuels	Load (W)	Quantity of fuel	Voltage (Volt)	Current (Ampere)	Time (s)	Engine speed (RPM)	Brake-specific fuel consumption (g/kWh)	Thermal efficiency (%)
Kerosene (ml)	0	50	223	0	251	3043	–	–
	250	50	220	2.2	234	3000	2492.35	3.36
	500	50	218	4	193	2982	1510.8	5.54
	1000	50	208	7.7	148	2913	985.13	8.49
	1500	50	192	10.9	110	2830	883.63	9.47
Petrol (ml)	0	50	222	0	249	3058	–	–
	250	50	221	2.1	207	3020	2503.29	3.14
	500	50	216	4	170	2977	1524.02	5.16
	1000	50	208	7.8	128	2917	1012	7.76
	1500	50	194	11	99	2848	872.35	9.01
Diesel (ml)	0	50	222	0	266	3058	–	–
	250	50	224	2.2	225	3038	2662.3	2.97
	500	50	219	4	211	2995	1419.48	5.57
	1000	50	209	7.8	154	2930	972.47	8.13
	1500	50	194	11	134	2862	745.05	10.62
LPG (gm)	0	50	219	0	501	3015	–	–
	250	50	218	2.2	436	2995	1651.24	4.73
	500	50	217	4	391	2990	920.16	8.48
	1000	50	209	7.8	340	2911	529.2	14.76
	1500	50	188	10.8	229	2808	504	15.49

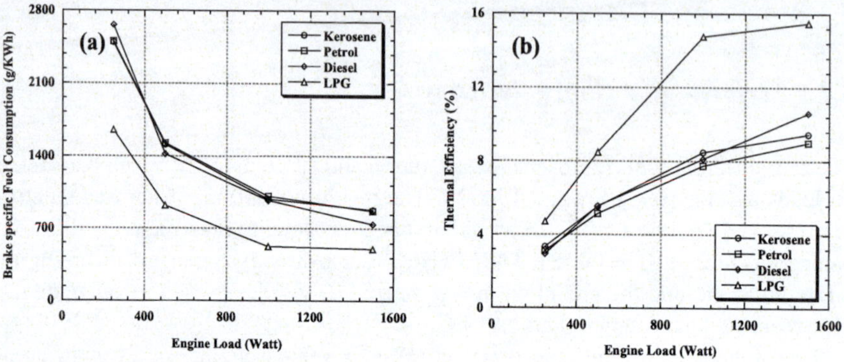

Fig. 2 Comparison for all fuels: **a** brake-specific fuel consumption (g/kWh) versus engine load and **b** thermal efficiency versus engine load

fuels. Thermal efficiency of the engine is maximum with LPG fuel compared to other fuels (kerosene, petrol and diesel) for all engine loads (0–1500 W). The engine shows almost similar thermal efficiency with kerosene, petrol and diesel fuels; however at higher engine loads (≥1200 W), diesel engine has more efficiency.

3.2 Emission Characteristics

The harmful discharge elements from the engine for above fuels are measured with the help of gas analyzer and are presented in Table 2. One end of the exhaust probe is inserted in the exhaust manifold of the engine, whereas the other end is connected with the modular diagnostic system for analyzing exhaust emissions. The emission of carbon monoxide (CO) is maximum for petrol at all engine loads except at no load where kerosene has more CO emission compared to the other fuels and the minimum emission of this gas taking place from LPG. Diesel has less CO emission than kerosene for all engine loads. For 0–250 W loads, the exhaust of diesel has higher volume percentage of O_2 as compared to the exhausts of other fuels. The content of NO is minimum at lower loads (<400 W), intermediate loads (400 W < load < 1100 W) and higher loads (>1100 W) for LPG, kerosene and petrol, respectively. The comparison of engine exhausts versus engine loads is shown in Fig. 3.

Table 2 Emissions from engine generator using different fuels

Fuels	Parameter	Actual values				
		0 W	250 W	500 W	1000 W	1500 W
Kerosene	CO_2 (vol.%)	5.82	5.97	5.29	5.04	4.70
	CO (vol.%)	3.75	3.31	5.80	8.03	9.12
	O_2 (vol.%)	8.72	8.67	7.95	5.99	5.40
	HC (ppm)	131	109	236	206	317
	NO (ppm)	55	76	36	21	37
Petrol	CO_2 (vol.%)	5.24	5.06	4.49	4.85	4.99
	CO (vol.%)	3.19	5.43	6.35	8.93	9.83
	O_2 (vol.%)	9.81	8.53	8.52	5.91	4.90
	HC (ppm)	188	126	219	265	227
	NO (ppm)	60	42	48	28	5
Diesel	CO_2 (vol.%)	4.83	3.02	6.01	6.07	6.42
	CO (vol.%)	1.86	1.76	4.20	6.24	7.11
	O_2 (vol.%)	11.78	14.63	7.94	6.14	4.80
	HC (ppm)	172	147	320	392	442
	NO (ppm)	58	37	106	90	108
LPG	CO_2 (vol.%)	5.86	6.17	6.71	5.84	9.01
	CO (vol.%)	0.11	0.48	1.60	4.3	0.19
	O_2 (vol.%)	11.60	9.27	7.46	6.63	6.66
	HC (ppm)	74	34	67	137	7
	NO (ppm)	19	34	60	77	114

4 Conclusions

The analysis of the above engine characteristics reveals that:

- On increasing engine load, the brake-specific fuel consumption decreases and the thermal efficiency of the engine increases.
- The emissions of carbon monoxide (CO) and hydrocarbon (HC) are minimum in LPG engines, whereas those emissions are maximum in petrol and diesel engines, respectively, for engine loads (0–1500 W).
- The emissions of carbon dioxide (CO_2) and oxygen (O_2) gases are maximum in LPG engines, whereas those emissions are minimum in petrol and kerosene engines, respectively, for engine loads (0–1500 W).
- The contents of nitrogen oxide (NO) are minimum at lower loads (<400 W) for LPG, at intermediate loads (400 W < load < 1100 W) for kerosene and at higher loads (>1100 W) for petrol.
- The LPG is found as the best fuel compared to the other fuels (kerosene, petrol and diesel).

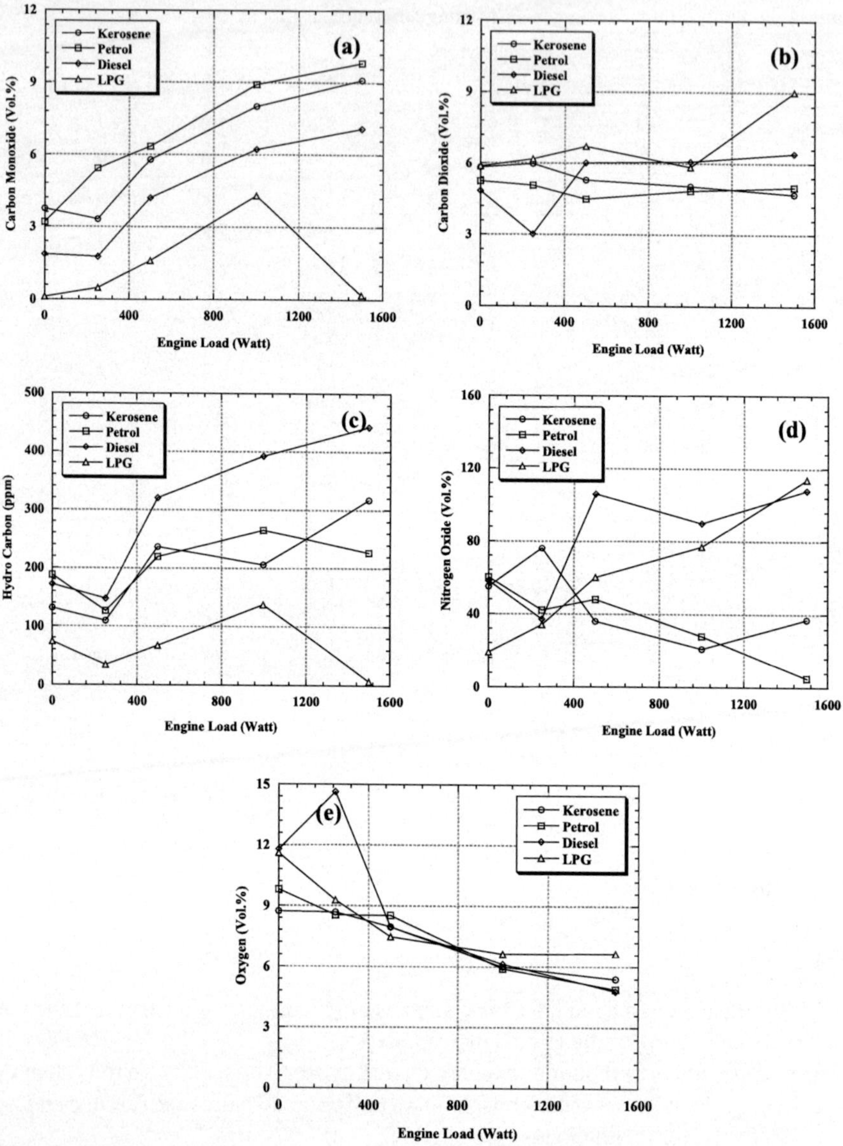

Fig. 3 Comparison of engine exhausts versus engine loads for all fuels

References

1. Chandra R, Vijay VK, Subbarao PMV, Khura TK (2011) Performance evaluation of a constant speed IC engine on CNG, methane enriched biogas and biogas. Appl Energy 88:3969–3977
2. Yoon SH, Lee CS (2011) Experimental investigation on the combustion and exhaust emission characteristics of biogas-biodiesel dual fuel combustion in a CI engine. Fuel Process Technol 92:992–1000

3. Thangaraj S, Govindan N (2018) Evaluating combustion, performance and emission characteristics of diesel engine using karanja oil methyal ester biodiesel blends enriched with HHO gas. Int J Hydrogen Energy 43:6443–6455
4. Kalsi SS, Subramanian KA (2017) Effect of simulated biogas on performance, combustion and emissions characteristics of a bio-diesel fueled diesel engine. Renew Energy 106:78–90
5. Ambarita H (2017) Performance and emission characteristics of a small diesel engine run in dual-fuel (diesel-biogas) mode. Case Studies in Thermal Eng 10:179–191
6. Monirul IM, Masjuki HH, Kalam MA, Mosarof MH, Zulkifli NWM, Teoh YH, How HG (2016) Assessment of performance, emission and combustion characteristics of palm, jatropha and Calophyllum inophyllum biodiesel blends. Fuel 181:985–995
7. Mofijur M, Rasul MG, Hyde J, Azad AK, Mamat R, Bhuiya MM (2016) Role of biofuel and their binary (diesel–biodiesel) and ternary (ethanol–biodiesel–diesel) blends on internal combustion engines emission reduction. Renew Sustain Energy Rev 53:265–278
8. Chang YC, Lee WJ, Lin SL, Wang LC (2013) Green energy: water-containing acetone–butanol–ethanol diesel blends fueled in diesel engines. Appl Energy 109:182–191
9. Chang YC, Lee WJ, Wu TS, Wu CY, Chen SJ (2014) Use of water containing acetone–butanol–ethanol for NO_x-PM (nitrogen oxide-particulate matter) trade-off in the diesel engine fueled with biodiesel. Energy 64:678–687
10. Hosseini SM, Ahmadi R (2017) Performance emissions characteristics in the combustion of co-fuel diesel –hydrogen in a heavy duty engine. Appl Energy 205:911–925
11. Zhang Y, Zhu M, Zhang Z, Chan YL, Zhang D (2018) Combustion and emission characteristics of simulated biogas from two phase anaerobic digestion in a spark ignition engine. Appl Therm Eng 129:927–933
12. Boulahlib MS, Medaerts F, Boukhalfa MA (2018) Experimental study of combustion performances and emissions of a spark ignition cogeneration engine operating in lean conditions using different fuels. Int J Hydrogen Energy 43:3586–3596
13. Becerra-Ruiz JD, Gonzalez-Huerta RG, Gracida J, Amaro-Reyes A, Macias-Bobadilla G (2019) Using green-hydrogen and bioethanol fuels in internal combustion engines to reduce emissions. Int J Hydrogen Energy 44(24):12324–12332
14. Ibrahim I, Enagi KA, Al-attab Z, Zainal A (2018) Liquid biofuels utilization for gas turbines: a review. Renew Sustain Energy Rev 90:43–55
15. Suthisripok T, Semsamran P (2018) The impact of biodiesel B100 on a small agricultural diesel engine. Tribol Int 128:397–409
16. Krishna SM, Salam PA, Tongroon M, Chollacoop N (2019) Performance and emission assessment of optimally blended biodiesel-diesel-ethanol in diesel engine generator. Appl Therm Eng 155:525–533
17. Li Y, Tang W, Chen Y, Liu J, Lee CF (2019) Potential of acetone-butanol-ethanol (ABE) as a biofuel. Fuel 242:673–686
18. Patel C, Tiwari N, Agarwal AK (2019) Experimental investigations of Soyabean and Rapeseed SVO and biodieselson engine noise, vibrations, and engine characteristics. Fuel 238:86–97
19. Jia B, Wu D, Smallbone A, Lawrence C, Roskilly AP (2018) Design, modelling and validation of a linear Joule Engine generator designed for renewable energy sources. Energy Convers Manag 165:25–34
20. Aydın H (2016) Scrutinizing the combustion, performance and emissions of safflower biodiesel–kerosene fueled diesel engine used as power source for a generator. Energy Convers Manag 117:400–409

Carburetion and Port Fuel Injection Metering Strategies for Natural Gas Spark-Ignited Engine

Jim Alexander and E. Porpatham

1 Introduction

The internal combustion engines have placed significant role in several automotive applications and other prime mover equipment, for which greater demand was always present towards the use of conventional petroleum-based fuels. With the evolving technology features, stringent emission norms are formulated on a global scale having the overall objective to minimize the emission effluents. The alternative fuels offer potential benefits having similar properties of conventional fuels that improves engine thermal efficiency, reduces exhaust emission and forms mutual balance of energy resources [1]. The gaseous alternative fuels are known for its compatible features with spark ignition (SI) engines such as wider flame limits and higher octane rating able to operate at higher compression ratio, and have reduced carbon content, hence reduced emissions [2–4]. Among various gaseous alternative fuels, CNG a promising fuel comprises mostly methane is recognized as dedicated fuel in several vehicles. It has higher octane rating that enables higher compression ratio operations benefitting towards higher thermal efficiency and power [5]. The higher hydrogen-to-carbon ratio results in reduced greenhouse gas emission [6]. Ever since the origin of SI engine technologies, there has been consistent strive towards advancing better fuel metering mechanism in order to attain the determined air–fuel mixture. Most of the existing SI engine vehicles have retrofitted CNG system that are broadly two categories of fuel metering devices such as carburetion and the advanced throttle body injection (TBI) system. The construction of the carburetion is one of the oldest metering devices that functions based on differential vacuum which prepares the necessary air–fuel ratio by directly drawing the gas from the vapourizer sending into the intake manifold [7].

J. Alexander · E. Porpatham (✉)
Automotive Research Centre, School of Mechanical Engineering, Vellore Institute of Technology, Vellore 632014, India
e-mail: porpatham.e@vit.ac.in

A. K. Gupta et al. (eds.), *Advances in IC Engines and Combustion Technology*, Lecture Notes in Mechanical Engineering,
https://doi.org/10.1007/978-981-15-5996-9_32

This is also similar to TBI; TBI resembles the design of a carburetor but differs by precise ECU control that meters flow of gas injected in the stream of air in the intake manifold, provided with a nozzle of macro diameter size.

Optimizing an injector and associated components specifically for gaseous fuelled vehicle is a necessary priority towards better thermal efficiency and reduced emission. This is governed primarily by factors such as the rate at which the mixing of air and fuel occurs, nozzle diameter and intake chamber design. Literature states that during the process of preparing the air-fuel mixture conventional mixing system face difficulties with gaseous fuelled engines limiting to meet the emission norms [8]. Lower volumetric efficiencies are some adversaries while using gaseous fuel which displaces the intake air and also has no vapourization effects unlike liquid fuels [2]. Hence knowing from this, the mixture formation depends on the distance between the fuel input and the intake port. Therefore, PFI with specially designed nozzle for CNG offers better engine performance compared to retro systems. Retrofitted CNG vehicles with the conventional design of intake manifold have straight runner causes resistance to airflow [9]. This can be overcome by specially designed intake plenum with surge profile, and positioning the PFI closer enough to intake port maximizes the intake airflow. Most of the CNG PFI system is injected at high pressure than gasoline, which has advantage towards minimizing pumping work loss [10].

The power enhancement of naturally aspirated natural gas engines shows increment with higher compression ratio, where the higher octane rating of CNG overcomes knocking [11] and moreover the density of the air increases resulting in increased power output [5], which is not feasible with gasoline fuel. Experiments were conducted with gasoline and natural gas at conditions with and without supercharging, increasing compression ratio along with supercharging resulting in 50% increase towards torque when fuelled with natural gas [12]. There were reductions in HC and CO emissions with the use of natural gas compared to gasoline where the amount of fuel present in per air–fuel mixture is higher [9]. In this, research experiments were conducted for naturally aspirated conditions under carburetion and PFI using CNG as fuel at full throttle condition having higher compression ratio of 11.5:1, the performance, emission and combustion characteristics we determined and comparisons were made stating the beneficial outcomes of PFI system.

2 Experimental Set-up

A naturally aspirated twin-cylinder, CNG-fuelled, liquid-cooled 702 cubic capacity variable speed engine was experimented with carburetion and PFI system under full throttle condition having 11.5:1 compression ratio. The factory default specification states a maximum brake power of 15.5 kW at 3400 rpm and maximum torque of 49 Nm at 2200 rpm, at its highest compression ratio. The engine was fuelled with CNG fuel by regulating the storage tank pressure from 150 bar to 0.9 bar for carburetion and 5 bar injection pressure for PFI. The properties of CNG fuel are shown in Table 1.

Table 1 Properties of the CNG fuel [13]

Properties	CNG
Lower heating value (MJ/kg)	50
Density (kg/m^3)	0.79
Flame velocity (cm/s)	34
Stoichiometric air–fuel ratio (kg/kg)	17.3
Flammability limits (vol.% in air)	5–15
Research octane number	120
Auto-ignition temperature (°C)	540

The fuel flow was controlled using a fine flow control valve for carburetion setting the stoichiometric conditions at all loads.

The flow rates of air and fuel were determined using analog FMG airflow metre and digital FOX flow metre. Figure 1 shows the layout of the experimental set-up. The engine was provided with a dedicated Woodward make ECU that takes control towards all sensors and actuators, and the graphic user interface settings are the throttle angle, injection width and ignition timing for fine-tuning to maximum brake torque (MBT) maintaining stoichiometric mode. The engine load was provided by Dynalec make eddy current dynamometer, and the speed was varied from low to high. A Kistler make piezoelectric pressure sensor was used to monitor the pressure signals from the combustion chamber, and KiBox amplifier was used to acquire the pressure data for the respective combustion cycles. The engine emissions were measured using a Horiba 5-gas analyser.

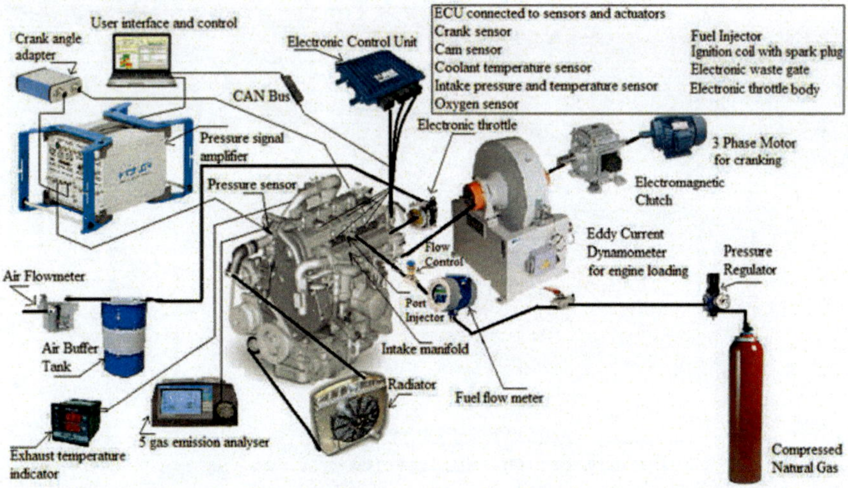

Fig. 1 Layout of the experimental set-up

3 Result and Discussion

3.1 Performance

The experiments were conducted, and comparisons were drawn between carburetion and PFI system for the performance parameters such as brake power, torque, brake thermal efficiency and specific fuel consumption.

Figure 2 shows the brake power and torque with respect to engine speed, where the CNG PFI system has attained maximum torque of 45.2 Nm at 1800 rpm and brake power of 13.6 kW at 3400 rpm compared to carburetion. The PFI responds to full throttle load demands preparing the air–fuel mixture precisely metered and injected in the intake port which determines the power output compared to conventional carburetion.

Figure 3 denotes the brake thermal efficiency and specific fuel consumption with respect to engine speed, and the PFI has the maximum brake thermal efficiency of 27.2% at 1800 rpm when compared to carburetion. The PFI system comprises higher intake pressure working environment which reduces pumping work. Moreover, the higher compression ratio increases air density for which the PFI metres and sprays the pressurized fuel into the intake port stream of air instantly, whereas carburetion is of lower pressure where the fuel entering the intake manifold causes displacement of air and draws more fuel in the mixture. The specific fuel consumption showed an increasing trend with engine speed and lesser fuel consumption was observed for PFI due to reduced throttling losses.

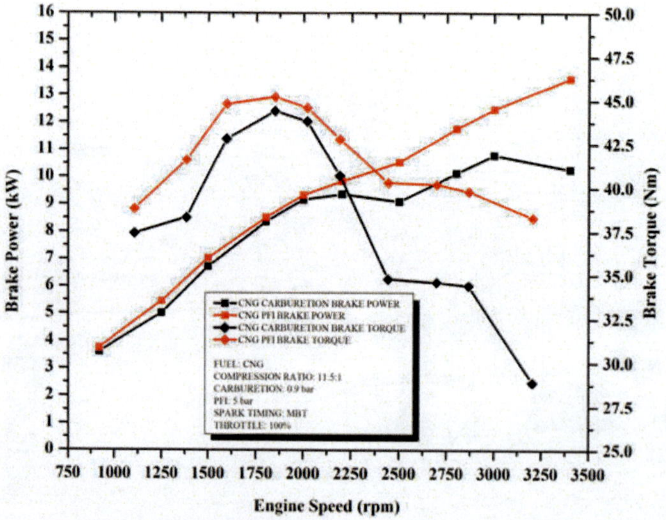

Fig. 2 Brake power and torque with engine speed

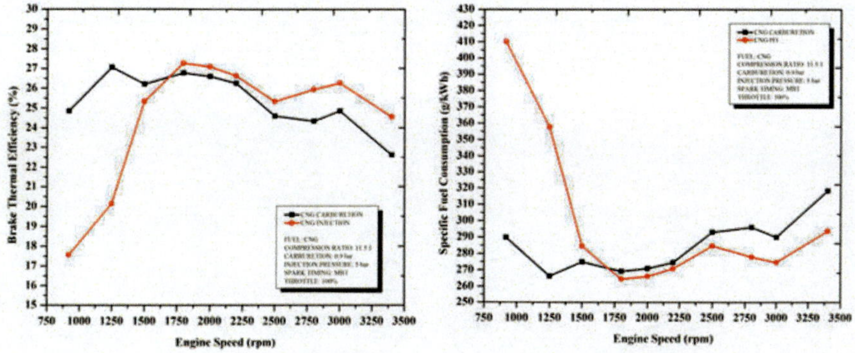

Fig. 3 Brake thermal efficiency and specific fuel consumption with engine speed

3.2 *Emission*

This section describes the emissions such as HC, CO, NO and MBT spark timing setting at stoichiometric mode and its comparison between CNG PFI and carburetion.

Figure 4 shows the HC and CO emissions with engine speed, and the PFI resulted in lower emission levels in both the cases compared to carburetion. This is due to the higher injection pressure of CNG PFI which reduces pumping losses and in turn raises the volumetric efficiency, and with higher compression ratio, the increase in air density results in reduced emissions. In carburetion, the CNG is less dense due to lower working pressure and this results in improper mixture formation causing emission. Figure 5 shows the variation of nitric oxide emission and spark timing with engine speed, and the proper mixture formation of the PFI results in complete combustion that raises the combustion temperature in the presence of

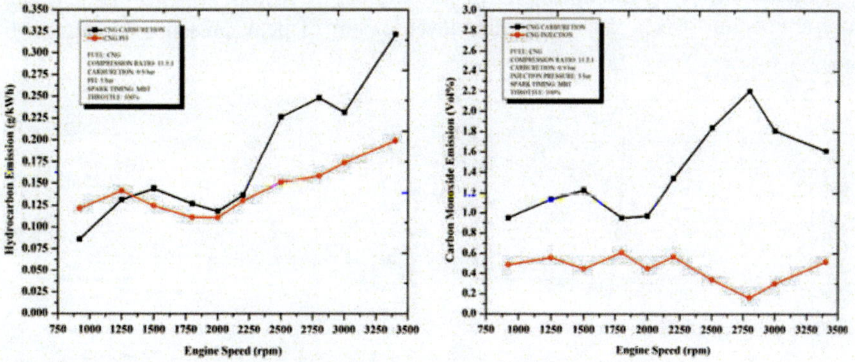

Fig. 4 HC and CO emissions with engine speed

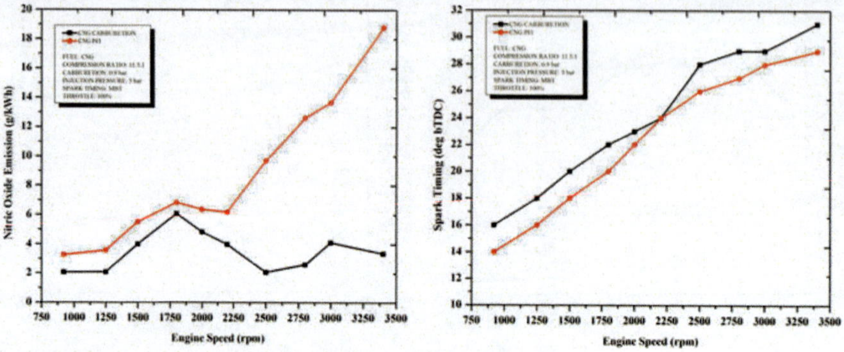

Fig. 5 NO emission and spark timing with engine speed

oxygen (increased air density) causing NO emission. The spark timing was generally advanced with increasing engine speeds, and the PFI system requires retard ignition timing before top dead centre since it exhibits faster combustion duration than carburetion.

3.3 *Combustion*

This section describes the combustion parameters such as in-cylinder pressure, heat release rate and cyclic variations between CNG PFI and carburetion.

Figure 6 shows the in-cylinder pressure and heat release rate with engine speed, PFI resulted in a maximum peak pressure of 47.9 bar, higher compression resulted in an increase in thermal efficiency for PFI system, and along with this the raise in volumetric efficiency due to higher injection pressure results in faster flame travel and temperature rise, causing earlier development of peak pressure compared to

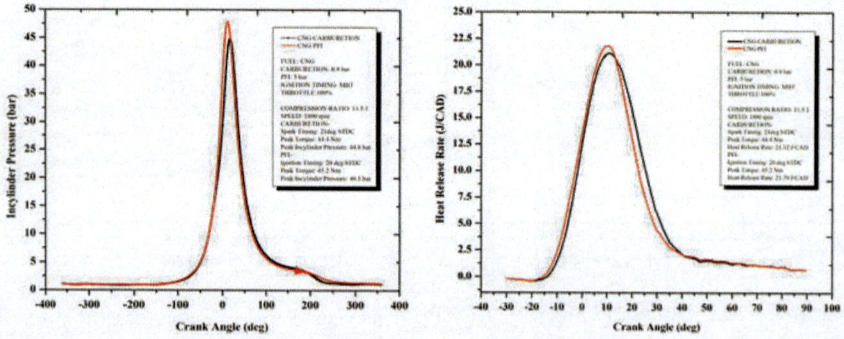

Fig. 6 In-cylinder pressure and heat release rate with engine speed

carburetion. The PFI system has higher heat release rate having maximum work transfer minimizing heat loss to cylinder walls compared to carburetion.

Figure 7 shows the cumulative heat release rate with engine speed, for which the PFI system all through has shown earlier combustion process compared to carburetion.

Figure 8 depicts the COV of IMEP for a typical 100 combustion cycle numbers at the peak torque speed of 1800 rpm; on a whole, the pattern looks alike as they are gaseous fuel operating at full throttle conditions; however, PFI has comparatively lesser variations compared to carburetion.

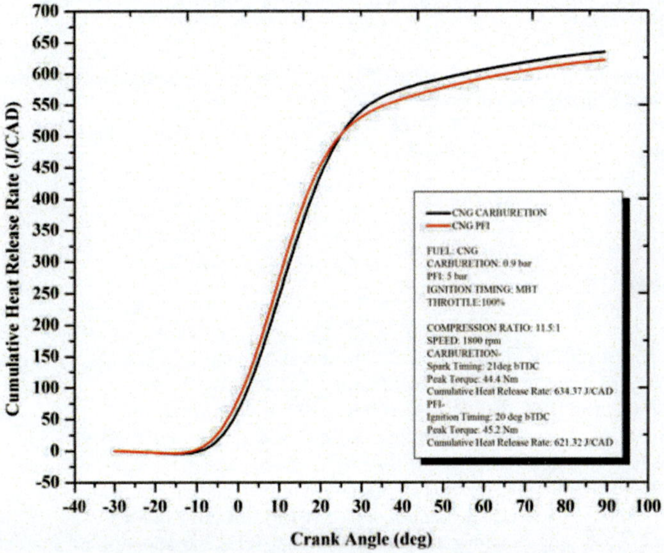

Fig. 7 Cumulative heat release rate with engine speed

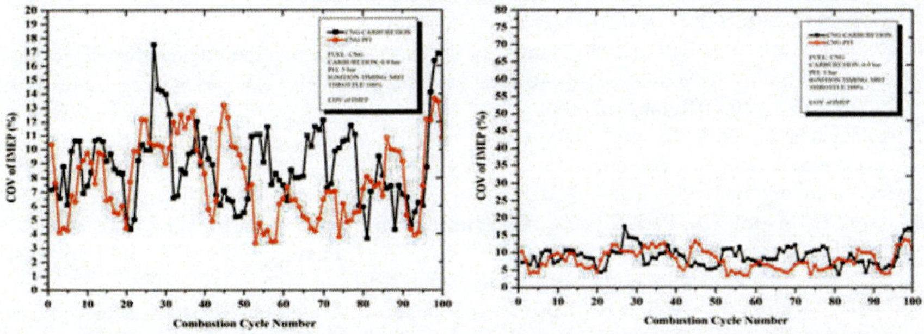

Fig. 8 Coefficient of variation of indicated mean effective pressure (COV of IMEP) with combustion cycle at speed 1800 rpm

4 Conclusion

Experiments were conducted with CNG carburetion and PFI at full throttle conditions under compression ratio of 11.5:1 on a twin-cylinder liquid-cooled SI engine:

- The PFI has increase in torque by 1.8% and power output by 32.04% when compared with carburetion.
- The PFI resulted in an increase in brake thermal efficiency by 1.87% having reduced fuel consumption by 1.77%.
- The HC and CO emissions dropped, while NO increased for PFI.
- PFI also had the peak pressure owing superior combustion.

Acknowledgements The authors wish to thank the Science and Engineering Research Board (SERB), Department of Science and Technology (DST), project No: EMR/2016/004138, for their valuable funding, and Vellore Institute of Technology (VIT), for their support to carry out the research work.

References

1. Bae C, Kim J (2017) Alternative fuels for internal combustion engines. Proc Combust Inst 36:3389–3413. https://doi.org/10.1016/j.proci.2016.09.009
2. Ran Z, Hariharan D, Lawler B, Mamalis S (2019) Experimental study of lean spark ignition combustion using gasoline, ethanol, natural gas, and syngas. Fuel 235:530–537. https://doi.org/10.1016/j.fuel.2018.08.054
3. Pham PX, Vo DQ, Jazar RN (2017) Development of fuel metering techniques for spark ignition engines. Fuel 206:701–715. https://doi.org/10.1016/j.fuel.2017.06.043
4. Linzenich A, Arning K, Bongartz D (2019) What fuels the adoption of alternative fuels? Examining preferences of German car drivers for fuel innovations. Appl Energy 249:222–236. https://doi.org/10.1016/j.apenergy.2019.04.041
5. Korakianitis T, Namasivayam AM, Crookes RJ (2011) Natural-gas fueled spark-ignition (SI) and compression-ignition (CI) engine performance and emissions. Prog Energy Combust Sci 37:89–112. https://doi.org/10.1016/j.pecs.2010.04.002
6. Sodre R, Helena P, Za B (2009) Journal of natural gas science and engineering evaluating carbon emissions reduction by use of natural gas as engine fuel. J Nat Gas Sci Eng 1:216–220. https://doi.org/10.1016/j.jngse.2009.11.002
7. Lee Y, Kim C, Oh S, Kang K (2005) Effects of injection timing on mixture distribution in a liquid-phase LPG injection engine for a heavy-duty vehicle. JSME Int J, Ser B 47:410–415. https://doi.org/10.1299/jsmeb.47.410
8. Kwak H, Myung CL, Park S (2007) Experimental investigation on the time resolved THC emission characteristics of liquid phase LPG injection (LPLi) engine during cold start. Fuel 86:1475–1482. https://doi.org/10.1016/j.fuel.2006.11.023
9. Mello P, Pelliza G, Catalun R (2006) Evaluation of the maximum horsepower of vehicles converted for use with natural gas fuel. Fuel 85:2180–2186. https://doi.org/10.1016/j.fuel.2006.04.002
10. Huang J, Crookes RJ (1998) Assessment of simulated biogas as a fuel for the spark ignition engine. Fuel 77:1793–1801. https://doi.org/10.1016/s0016-2361(98)00114-8

11. Thurnheer T, Soltic P, Eggenschwiler PD (2009) S.I engine fueled with gasoline, methane and methane/ hydrogen blends: heat release and loss analysis. Int J Hydrogen Energy 34:2494–2503. https://doi.org/10.1016/j.ijhydene.2008.12.048
12. Bach C, Lämmle C, Bill R, Soltic P (2004) Clean engine vehicle a natural gas driven Euro-4/ SULEV with 30% CO_2-Emissions. SAE technical paper 2004-01-0645. https://doi.org/10.4271/2004-01-0645
13. Porpatham E, Ramesh A, Nagalingam B (2012) Effect of compression ratio on the performance and combustion of a biogas fuelled spark ignition engine. Fuel 95:247–256. https://doi.org/10.1016/j.fuel.2011.10.059

Evaluation of Performance and Emission Characteristics on Diesel Engine Fueled by Diesel–Algae Biodiesel Blend with Ignition Enhancing Additives

A. Gurusamy, A. A. Muhammad Irfan, E. R. Sivakumar, and P. Purushothaman

Abbreviations

GHG	Greenhouse gas
BTE	Brake thermal efficiency
BSEC	Brake-specific energy consumption
KOH	Potassium hydroxide
UBHC	Unburned hydrocarbon
NO_x	Oxides of nitrogen
CO_2	Carbon dioxide
CO	Carbon monoxide
PD	Pure diesel
IC engine	Internal combustion engine

A. Gurusamy (✉)
Department of Automobile Engineering, Pace Institute of Technology and Sciences, Ongole, India
e-mail: agsmeice@gmail.com

A. A. Muhammad Irfan
Department of Mechanical Engineering, Mohamed Sathak A.J. College of Engineering, Chennai, India

E. R. Sivakumar
Department of Mechanical Engineering, Pace Institute of Technology and Sciences, Ongole, India

P. Purushothaman
Department of Mechanical Engineering, Agni College of Technology, Chennai, India

A. K. Gupta et al. (eds.), *Advances in IC Engines and Combustion Technology*, Lecture Notes in Mechanical Engineering,
https://doi.org/10.1007/978-981-15-5996-9_33

1 Introduction

Significance of biofuels on the reduction of emission in IC engine is reviewed by Mofijur et al. [1], and they state that emission from on-road vehicle is basically accountable for health hazards and greenhouse gas (GHG) emission. The addition of biofuel (biodiesel and bioethanol) in IC engine is a feasible solution to reduce such emissions more than 80%. The use of biofuel in IC engine vehicle is mandatory for some developed countries; as an example, the USA needs to use 25% fermentation alcohol among 2020, Brazil has an intention to execute B20 among 2020 and India planned to use B10 throughout the nation among 2017.

The most acknowledged definition for third-generation biofuels is fuel that would be derived from algal biomass, which has an extremely distinctive growth yield as contrasted and traditional lignocellulosic biomass. The alga fuel is that the biofuel of the future will considerably decrease the necessity of fossil fuels and will facilitate to scale back the total quantity of harmful gases accountable for environmental issues [2]. Algae or microalgae produce the carbon content products and result in accumulation of lipid, as well as triacylglycerols (TAGs) using solar energy after that the biomass will be transformed into bio-alcohol and biodiesel. In comparison with the other feedstock, the microalgae grow up rapidly, are effortless to cultivate, need less attention and can exercise wastewater as nutrient for algae production. The further benefit of algae is a potential capture of GHG like CO_2 from atmosphere, which is produced by the power plants, driven by fossil fuels like coal-based thermal power plant [3]. Rastogi et al. [4] reviewed the progress of technological aspects, and the current trends going on the algae cultivation and process on production of biodiesel form the algae biomass. They reported that it is technically viable to cultivate algae in large scale for various usages, but it is a cost-effective concern with current technology.

1.1 *Potential of Algae Biodiesel in IC Engine*

Biodiesel derived from algae is the foremost economical choice in terms of using agriculture lands because it is not requiring terrestrial areas for cultivation of biomass and keeps away from competition with food crops. The benefits of biofuels produced from algae or microalgae over other source of alternative fuels are large quantities of neutral lipids/oil, grow at high rates, higher Oil yield per area, can be cultivated in saline/brackish water/coastal seawater on non-arable land and culture vessels (photo bioreactors), produce value-added co-products (biopolymers, proteins, polysaccharides, pigments, animal feed and fertilizer) [5]. Patel et al. [6] assess the performance and emission characteristics of diesel engine using algae methyl ester (Chlorella vulgaris) biodiesel blends (B10, B15 and B20) at different loads. They reported increase in BTE and NO_X (up to 36%) and reduction in BSEC, smoke, CO and HC emissions. Mwangi et al. [7, 8] conducted engine tests with biodiesel derived from

Table 1 Properties of test fuels

S. no.	Properties	Diesel	B20
1	Density @ 15 °C kg/m^3	876	890
2	Kinematic viscosity @ 40 °C in CST	2.9	4.0
3	Calorific value (MJ/kg)	44.43	40.42
4	Cetane index	48	43

the algae cultures, and they found that NO_X emission is increased for algae biodiesel. Ramon Piloto-Rodriguez et al. [9] reviewed the diesel engine performance and emission characteristics fueled with biodiesel from algae and microalgae (B5-B100). They concluded that B20 blend has higher CO_2 and NO_X emissions compared with diesel.

1.2 Biodiesel Preparation

After collection of algae sample, the algae (Chlorella vulgaris) biomass is produced by filtration and drying. Algae oil is isolated from algal biomass by utilizing Bligh and Dyer's method. During the production period of algae biodiesel, the essential quantity of catalyst 0.4% weight of KOH is suspended with methanol in the reactor later than heating the algae oil up to 343 K for the transesterification process. The reaction was carried out on 338–350 K, and the mixture was stirred at around 800 rpm for 1 h with the aid of a magnetic stirrer. After finishing the transesterification response, the blend was cooled up to working area temperature and after that exchanged to a separatory channel and separation of the glycerin and ester stages was performed by giving them a chance to remain for 8 h in the separatory pipe. The crude ester phase is washed with heated water with the proportion of 1/5 water to ester at 3 times. Subsequent to washing process, the blend held up in separatory pipe amid 30 min and by along these lines, water is isolated from algae methyl ester (AME). The deliberate physical fuel properties of diesel fuel and B20 blend are compared and are mentioned in Table 1. The drying procedure was directed by warming the biodiesel to 378 K amid 1 h until the point that bright color occurred. At long last, sifting process was done. In order to achieve good quality of biodiesel, two bunches of transesterification response were performed.

1.3 Ignition Enhancer

Cetane number is generally utilized as a parameter of indicating fuel ignition quality and period of ignition delay. Generally, the fuels have higher cetane number which have a longer chain hydrocarbons and the fuels have lower cetane number which have a the aromatics or unsaturated hydrocarbons [10]. In this study, 5% of di-ethyl

Table 2 Properties of DEE, DTBP and 2-EHN

S. no.	Properties	DEE	DTBP	2-EHN
1	Flash point (°C)	−40	1	76
2	Density (kg/m^3 @ 25 °C)	0.71	0.79	0.963
3	Autoignition temperature (°C)	160	165	215
4	Molecular weight (g/mol)	74.12	146	175.23

ether (DEE), di-tertiary butyl peroxide (DTBP) and 2-ethylhexyl nitrate (2-EHN) is blended with biodiesel. Di-ethyl ether is derived from ethanol, which has higher cetane number and oxygen content [11]. Usually, peroxides are synthesized by the response of alcohols or olefins with natural peroxide over acid catalyst while; nitrates are produced by nitration of alcohols. In spite of the fact that alky nitrates, for example, octyl nitrate, amyl nitrate and hexyl nitrate, plague the market as powerful ignition enhancers, peroxides are likewise given equivalent thought, as they are more steady than nitrates [12]. The addition of ignition enhancer with diesel and biofuel–diesel blends (biodiesel and alcohol) improved the performance and emission characteristics in IC engine [13]. Properties of di-ethyl ether (DEE), 2-ethylhexyl nitrate (2-EHN) and di-tertiary butyl peroxide (DTBP) are given in Table 2.

2 Experimental Setup

The engine utilized as a part of this investigation is a four-stroke, one-cylinder DI diesel engine, its particulars are listed in Table 3, and the schematic diagram of the engine setup along with the entire instrumentation is illustrated in Fig. 1. The emission from tail pipe was estimated with the assistance of the AVL DI gas analyzer. The smoke emanation was estimated with the assistance of the smoke meter. The

Table 3 Specification of the engine

S. no.	Details	Specification
1	Rated power and speed	5.2 kW and 1500 rpm
2	Number of cylinders	Single cylinder
3	Compression ratio	17.5:1
4	Bore and stroke	87.5 mm and 110 mm
5	Method of loading	Eddy current dynamometer
6	Dynamometer arm length	0.185 m
7	Type of injection	Mechanical pump-nozzle injection
8	Injection timing and pressure	23° before TDC and 220 bar

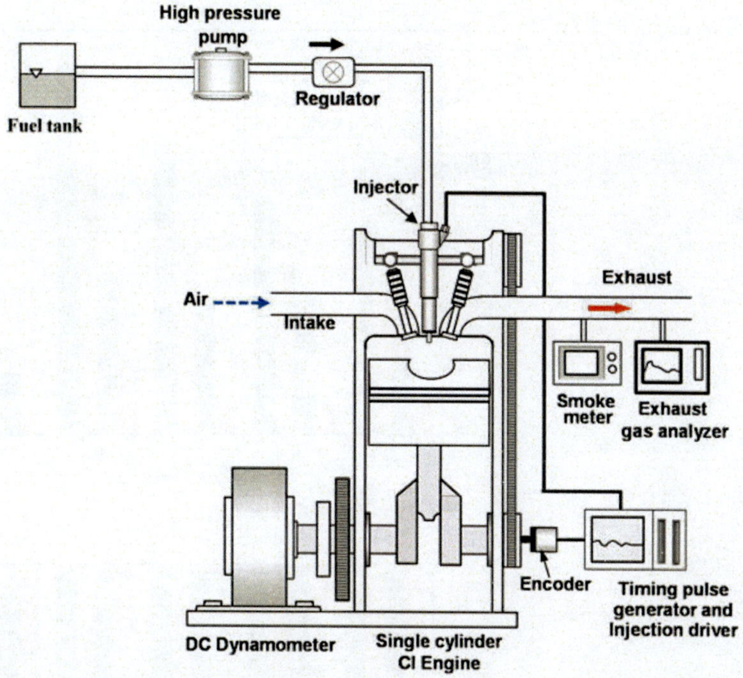

Fig. 1 Schematic diagram of the experimental setup

engine was loaded from 20% to 100% continuously in step of 20% by regulating the current provided to the vortex current dynamometer for diesel, B20 and B20 with 5% DEE, 2-EHN and DTBP. Figures 2, 3, 4, 5, 6 and 7 show the deviation of

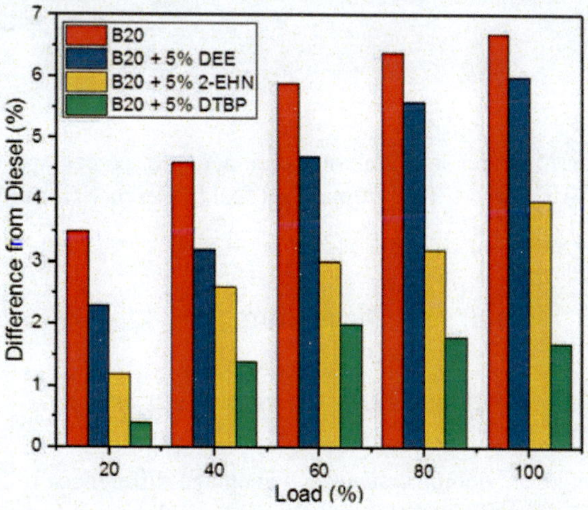

Fig. 2 Variation in BSEC for the tested fuel blends with respect to diesel

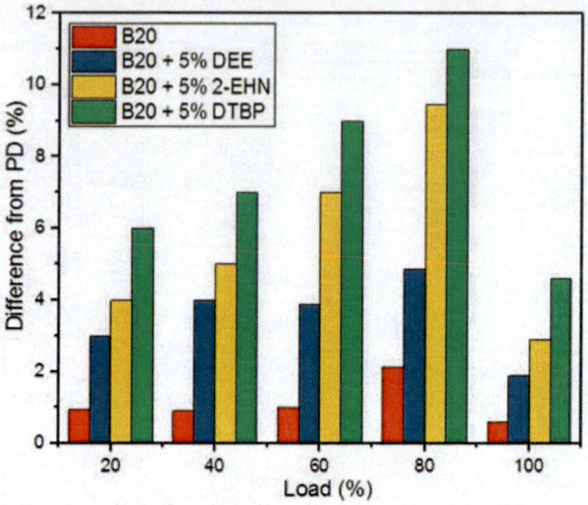

Fig. 3 Variation in BTE for the fuel blends compared to diesel

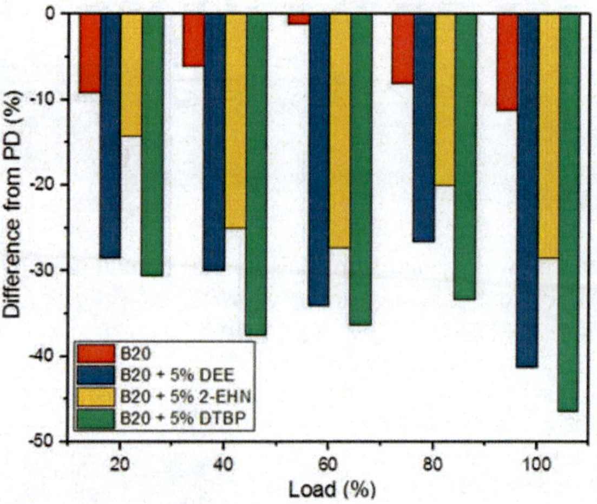

Fig. 4 Variation in CO emission for the tested fuel blends with respect to diesel

performance and emission parameters in percentage for B20, B20 with 5% DEE, 2-EHN and DTBP from diesel fuel (difference from PD).

3 Result and Discussion

3.1 Brake-Specific Energy Consumption

Figure 2 demonstrates the percentage differences in BSEC for B20 and B20 with 5% of DEE, 2-EHN and DTBP fuel blends with diesel fuel. It shows that BSEC

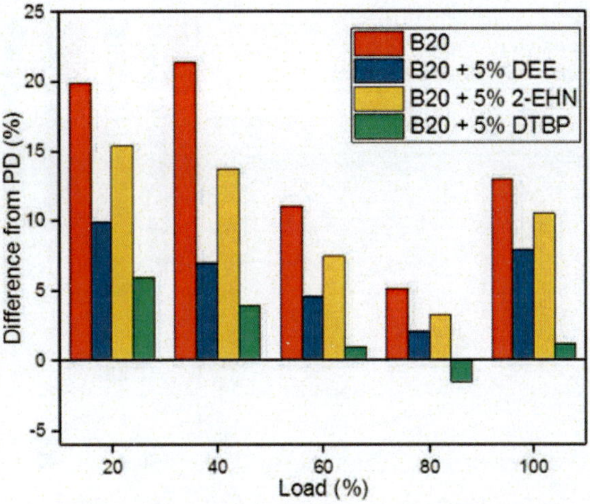

Fig. 5 Variation in NO_X emission for the tested fuel blends with respect to diesel

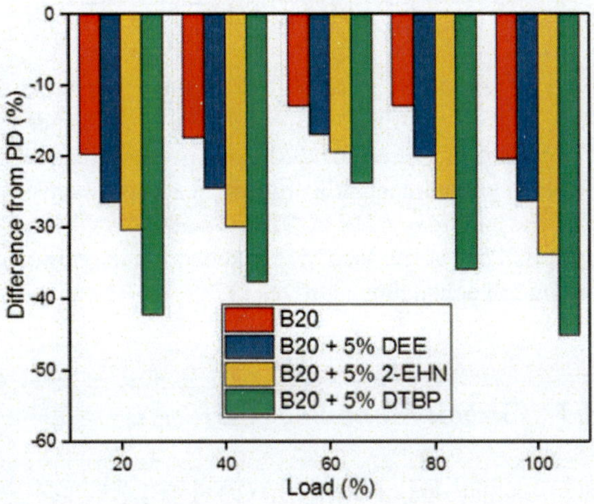

Fig. 6 Variation in smoke emission for the tested fuel blends with respect to diesel

increased by 6.7, 6, 4, 1.7% for B20 and B20 with 5% of DEE, 2-EHN and DTBP fuel blends at full load. On account of superior viscosity and lower heating value of B20 blend, engine consumes more fuel to produce same power output for B20 blend [6]. The addition of 5% of DEE, 2-EHN and DTBP in biodiesel blend (B20) blend results in increasing the cetane number of B20 blend. It leads lower BSEC [14].

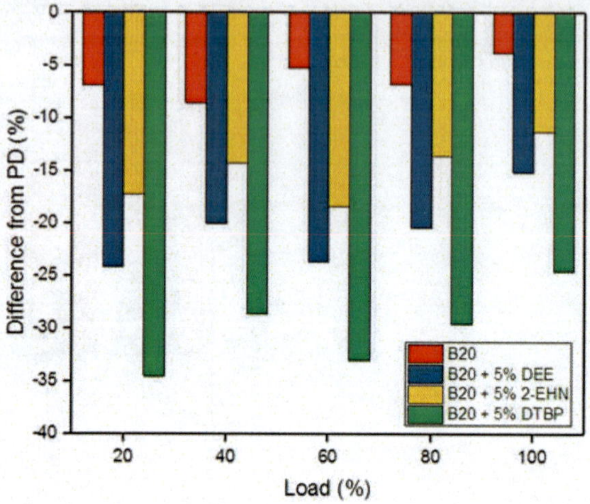

Fig. 7 Variation in UBHC emission for the tested fuel blends with respect to diesel

3.2 Brake Thermal Efficiency

Figure 3 reveals the percentage of differences in brake thermal efficiency for B20 and B20 with 5% of DEE, 2-EHN and DTBP fuel blends with baseline fuel. It illustrated that BTE is increased for B20 blend at entire load range. It might be because of air, better air entrainment, utilization of heat energy and oxygen present in algae biodiesel [6]. The addition of 5% of DEE, 2-EHN and DTBP with B20 blend increased BTE by 2, 3, 4.5% at full load, due to reduction in ignition delay and presence of oxygen in ignition enhancing additive [11].

3.3 Carbon Monoxide (CO)

Figure 4 illustrates deviation of CO emission in percentage for B20 and B20 with 5% of DEE, 2-EHN and DTBP. Compared with baseline fuel, CO emission is decreased by 11.3% for B20 fuel blend at full load. This can be justified by oxygen enrichment attributable to addition of algae biodiesel (B20) with baseline fuel [7]. With the addition of DEE, 2-EHN and DTBP in B20 blend, CO emission is decreased by 41.28, 28.57 and 46.42% at full load condition. Because of the presence of oxygen and shorter ignition delay, it leads better combustion and provides sufficient time period to convert CO into CO_2 during the period of combustion [11, 12].

3.4 Nitrogen Oxides (NO_X)

Figure 5 indicates the deviation of NO_X emission in percentage for B20 and B20 with 5% of DEE, 2-EHN and DTBP fuel blends with baseline fuel. The usage of algae biodiesel (B20 blend) in diesel engine is increased by 12% at 100% load. This might be endorsed to higher oxygen content in biodiesel, consequences increasing of residence time of the biodiesel fuel blends for the duration of period in combustion leading to a high temperature zone in the engine cylinder [8]. With the addition of ignition improver in algae biodiesel blend, NO_X emission was decreased by 7.89, 10.48 and 1.2% from 12% (B20 blend) for 5% of DEE, 2-EHN and DTBP with B20 blend at 100% load condition. This can be addressed by increasing the cetane of B20 blend with the addition of additive. It leads better combustion in premixed phase and lesser NO_X emission [13].

3.5 Smoke Emission

Figure 6 demonstrates the deviation of smoke emission in percentage for B20 and B20 with 5% of DEE, 2-EHN and DTBP fuel blends with diesel fuel. Smoke emission is decreased by 20.41% for B20 blend. Further, smoke emission is decreased by 26.27, 33.83, 45% with the addition of 5% of DEE, 2-EHN and DTBP in B20 blend. The smoke emission reduction was observed due to increased oxygen content of algae biodiesel and ignition enhancer fuel blends, which boost the process of combustion and decrease the fraction of unburned fuel in the combustion chamber [7].

3.6 Unburned Hydrocarbon Emission

The deviation of UBHC emission in percentage for B20 and B20 with 5% of DEE, 2-EHN and DTBP fuel blends with diesel fuel is shown in Fig. 7. With the addition of algae biodiesel in diesel engine, UBHC emission is decreased by 15.09, 11.32, 24.52% with the addition of 5% of DEE, 2-EHN and DTBP in B20 blend at full load. Cetane improver assists the better combustion with algae biodiesel by increasing the cetane value of by fuel blend and advances the combustion timing of biodiesel blend, which contributes to the significant reduction of UBHC emission [13].

4 Conclusion

The performance and emission characteristics of algae biodiesel fuel obtained from Chlorella vulgaris with and without the addition of ignition enhancer like DEE, 2-EHN and DTBP are compared for a diesel engine with baseline fuel as the reference. The main outcomes of this experimental study are concluded as follows: As compared to diesel, BSEC of algae biodiesel is increased at entire load range. With the addition of cetane improvers in biodiesel blend (B20 + 5% of DEE, 2-EHN and DTBP), the energy consumption is lower, but not lower than diesel fuel. The value of BSEC for B20 with 5% of DTBP is nearer to diesel fuel. Brake thermal efficiency of algae biodiesel blend (B20) is increased at entire load condition. The higher BTE is found for B20 with 5% of DTBP fuel blend. Unburned hydrocarbon, carbon monoxide and smoke emissions are found lower for algae biodiesel blend (B20) to compare with diesel fuel. Further, these emissions are lowered with the addition of ignition promoting additives. NO_X emission is higher for algae biodiesel blend to compare with diesel fuel. A significant NO_x reduction was observed for B20 with DEE, 2-EHN and DTBP blends when compared with B20 blend, but still it was higher than diesel. It may be concluded that engine can perform adequately with the B20 and 5% DTBP fuel blend over other two additives. In the future, NO_X reduction techniques will be adopted for reducing the emission for the B20 and 5% DTBP fuel blend.

References

1. Mofijur M, Rasul MG, Hyde J, Bhuyia MMK (2015) Role of biofuels on IC engines emission reduction. Energy Procedia 75:886–892. https://doi.org/10.1016/j.egypro.2015.07.211
2. Alam F, Mobin S, Chowdhury H (2015) Third generation biofuel from algae. Procedia Eng 105:763–768. https://doi.org/10.1016/j.proeng.2015.05.068
3. Maity JP, Bundschuh J, Chen CY, Bhattacharya P (2014) Microalgae for third generation biofuel production, mitigation of greenhouse gas emissions and wastewater treatment: present and future perspectives-a mini review. Energy 78:104–113. https://doi.org/10.1016/j.energy.2014.04.003
4. Rastogi RP, Pandey A, Larroche C, Madamwar D (2018) A green energy—R&D and technological perspectives for biodiesel production. Renew Sustai Energy Rev 82:2946—2969. https://doi.org/10.1016/j.rser.2017.10.038
5. Khan SA, Rashmi MZ, Hussain SP, Banerjee UC (2009) Prospects of biodiesel production from microalgae in India. Renew Sustain Energy Rev 13:2361–2372. https://doi.org/10.1016/j.rser.2009.04.005
6. Patel J, Kumar N, Deep A, Sharma A et al (2014) Evaluation of emission characteristics of blend of algae oil methyl ester with diesel in a medium capacity diesel engine. SAE Technical Paper 2014-01-1378. https://doi.org/10.4271/2014-01-1378
7. Mwangi JK, Wen-Jhy L, Liang-Ming W, Tser SW, Wei-Hsin C, Jo-Shu C (2015) Microalgae oil: algae cultivation and harvest, algae residue torrefaction and diesel engine emissions tests. Aerosol Air Qual Res 15:81–98. https://doi.org/10.4209/aaqr.2014.10.0268
8. Mwangi JK, Wen-Jhy L, Jen-Hao T, Tser SW (2015) Emission reductions of nitrogen oxides, particulate matter and polycyclic aromatic hydrocarbons by using microalgae biodiesel, butanol and water in diesel engine. Aerosol Air Qual Res 15:901–914. https://doi.org/10.4209/aaqr.2015.03.0140

9. Piloto-Rodríguez R, Sánchez-Borroto Y, Melo-Espinosa EA, Verhelst S (2017) Assessment of diesel engine performance when fueled with biodiesel from algae and microalgae: An overview. Renew Sustain Energy Rev 69:833–842. https://doi.org/10.1016/j.rser.2016.11.015
10. Biodiesel SA (2012) production and properties. Royal Soc Chem https://doi.org/10.1039/9781849734721
11. Rakopoulos DC, Rakopoulos CD, Giakoumis EG, Dimaratos AM (2012) Characteristics of performance and emissions in high-speed direct injection diesel engine fueled with diethyl ether/diesel fuel blends. Energy 43:214–224. https://doi.org/10.1016/j.energy.2012.04.039
12. Nandi MK, Jacobs DC, Liotta FJJ, Kesling HSJ (1994) Performance of a peroxide-based cetane improvement additive in different diesel fuels. SAE Trans 103:1654–1668 www.jstor.org/stable/44612452
13. Ribeiro M, Pinto AC, Quintella CM, Rocha GO, Teixeira LSG, Guarieiro LN et al (2007) The role of additives for diesel and diesel blended (Ethanol or Biodiesel) fuels: a review. Energy Fuels 21:2433–2445. https://doi.org/10.1021/ef070060r
14. Atmanli A (2016) Effects of a cetane improver on fuel properties and engine characteristics of a diesel engine fueled with the blends of diesel, hazelnut oil and higher carbon alcohol. Fuel 172:209–217. https://doi.org/10.1016/j.fuel.2016.01.013

Investigations of Combustion, Performance, and Emission Characteristics of Gasoline Engine Operated on Blends of Gasoline with Ethanol and n-Butanol

M. S. Sawant, S. P. Wategave, N. R. Banapurmath, and R. S. Hosmath

1 Introduction

Petroleum-based fuels like gasoline and diesel are becoming scarce due to diminishing resources and ever-increasing demand. Coupled with this fact, world is also confronting with environmental concerns like air pollution, global warming, etc., which has proved to be as a threat to the life on planet earth. This has encouraged researchers globally to explore new options for petroleum fuels, which are sustainable, ecological, and easily available as a plentiful.

Alcoholic fuels are being considered in this regard as these fuels are less emitting, ecological, and sustainable. Ethanol is more appealing to methanol as it can be made from lignocellulosic materials like agriculture feedstock, wastes of sugar cane, and forestry biomass in contrast to methanol which requires coal or petroleum-based materials [1, 2]. On the other hand, butanol is seeking attention of researchers globally as it also can be obtained from agricultural residue and industrial waste as feedstock with insignificant pretreatment. Many authors revealed [7–9] butanol as slightly corrosive and less susceptible to water absorption, and it can be distributed through existing pipelines used for petroleum-based fuels. As compared to ethanol, butanol has low vapor pressure along with larger flash point which avoids vapor lock and is

M. S. Sawant (✉) · S. P. Wategave · N. R. Banapurmath · R. S. Hosmath
School of Mechanical Engineering, KLE Technological University Hubli, Hubli, Karnataka, India
e-mail: m.sawant7885@gmail.com

S. P. Wategave
e-mail: sumsumit1987@gmail.com

N. R. Banapurmath
e-mail: nrbanapurmath@gmail.com

R. S. Hosmath
e-mail: rshosmath@bvb.edu

A. K. Gupta et al. (eds.), *Advances in IC Engines and Combustion Technology*, Lecture Notes in Mechanical Engineering,
https://doi.org/10.1007/978-981-15-5996-9_34

Table 1 Properties of gasoline, ethanol, and butanol [10]

Properties	Fuel		
	Gasoline	Ethanol	Butanol
Density, 20 °C (kg/m^3)	730–784	810	810
Lower calorific value (MJ/kg)	43.5–44	26.9	33.1
Octane number			
RON	91–99	130	96
MON	81–89	56	78
Cetane number	–	8	18
Air–fuel ratio for complete combustion	14.6	9	11.2
Heat of evaporation (kJ/kg)	300	920	570
Kinematic viscosity at 20 °C (cSt)	0.4–0.8	1.52	3.64
Self-ignition Temperature (°C)	415	423	390

safe [7, 8]. Lower heat of evaporation reduces cold starting problem as compared to ethanol and methanol [10]. Air-fuel ratio for 100% combustion and lower calorific value (LCV) of butanol are closer to gasoline, and it can be combined with gasoline in major proportions as compared to ethanol [10]. Following table depicts property comparison of gasoline, ethanol, and butanol [10] (Table 1).

1.1 Recent Engine Studies on Ethanol–Gasoline Blend

Li et al. [2] performed tests for blends of ethanol with gasoline (E10, E30, E60), and they found that ignition delay for ethanol is shorter than gasoline and % decrease of ignition delay increases with higher blending. In the same study, combustion rate was also seen less than pure gasoline and exhibited further decrease with increase of blending ratio. Performance and emissions study revealed, decrease of BTE for ethanol by roughly 2%, whereas BSFC was increased by around 15%. Emissions of CO were increased; however, UHC and NO_x emissions were lower.

Chansauria and Mandloi [3] reviewed effect of blends of ethanol (E5, E10) with gasoline on spark ignition engine and found increase of BTE and attributed this increase to more oxygen rate and increased octane rating of the blends than that of gasoline. Heat release rate and cylinder gas pressure quantities were also better for blends according to this review. They reported increased volumetric efficiency for blends because of increased evaporative cooling brought by the blends. They also mentioned reduced emission of CO, HC, and NO_x and increased discharge of CO_2 with increase of blending ratio as compared to gasoline.

Hasan et al. [4] observed impact of altering combustion chamber shape of SI engine run with blends of gasoline with ethanol. They varied compression ratios and found CO, HC, and NO_x emissions remarkably reduced, and CO_2 was more at high compression ratio. BSFC was increased because of ethanol's higher volumetric content at different engine conditions. Increase of compression ratio increased BMEP by 12% on an average for 10% and 20% ethanol blend. They reported slight improvement in BTE when ethanol blend was used.

To use ethanol easily in new technology of EDI + GPI, i.e., ethanol's direct injection and Gasoline's port injection, Huang and Hong [5] heated ethanol prior injection to counteract higher emissions of HC and CO arising due to lower vapor pressure and higher LHV of ethanol. They studied consequences of ethanol heating on combustion and emissions of an engine, which is fitted with EDI + GPI technology. They examined drop in CO and HC emissions by 40 and 51% in EDI only operation at spark timing of 15° BTDC and ethanol's injection at 90 °C; however, NO_x emissions were slightly increased at this condition but were less than only GPI condition. IMEP and combustion speed were slightly lowered at this condition.

Thakur et al. [6] rigorously reviewed performance parameters of ethanol-gasoline mix fuel in a gasoline engine. They concluded that ethanol blends in smaller proportions exhibited enhancement in brake power, BTE. For high ethanol amount, i.e., for E100 operation, BSFC increased by 56%. Elphaskhany [7] reported maximum power output for ethanol blending with increase of CO_2 emissions and decrease of HC and CO.

1.2 Recent Engine Studies on Butanol–Gasoline Blend

Elfasakhany [11] experimented on gasoline engine with blends of n-butanol in proportions of 10, 7 and 3% and varying engine speed between 2400 and 3400 r/min. He found emissions of CO and UHC increased with increasing speed up to a moderate speed and then reduce with further increase in speed. Emissions for blends are less than neat gasoline operation by 36% for CO_2, 13% for CO, and 13% for UHC. No significant emission differences were seen between n-butanol at proportions less than 10%. Performance of 10% blend were seen lower than neat gasoline by 5.6, 2.5, 6.6, 8.3, and 3.5% for exhaust gas temperatures, engine torque, brake power, in-cylinder pressure, and volumetric efficiency, respectively.

Mittal et al. [12] performed experiments on partially (head and valves area) coated LHR engine operated on blends of n-butanol (10 and 15%) with gasoline. They reported increase of BTE and decrease of SFC for coated engine due to rise in combustion temperature as a result of coating. Subsequently NO_x emissions for blends in coated engine increased as compared to uncoated head because of higher gas temperatures arising because coated engine would have produced adiabatic conditions. CO emissions were less for coated engine in comparison to uncoated and HC emissions reduced significantly with increase in addition of n-butanol for both the engines.

Singh et al. [13] carried out study on technical practicability of butanol–gasoline blends for medium duty gasoline engine. They combusted butanol in proportions of 5, 10, 20, 50, and 75% with gasoline. BTE with blends was achieved less than for neat gasoline, and it was because of higher auto-ignition temperature and larger enthalpy of vaporization of butanol compared to gasoline which attributed to fall in combustion efficiency. They reported lower combustion duration and heat release rate for butanol blends as compared to gasoline, and the reason for this is higher heat of vaporization and higher self-ignition temperature of butanol. However, very small variation of in-cylinder pressure was seen. Emissions of HC, CO, and NO_x were less than gasoline.

Yusri et al. [14] analyzed influence of 2-butyl alcohol blended with gasoline in the proportions of 5,10, and 15% on performance of SI engine at varying speeds of 1000–4000 rpm and at 50% WOT. They concluded increase of brake power and BMEP with increase of 2-butyl alcohol content, but as compared to pure gasoline, it is less by 12.3, 9.2, and 4.3% for 5, 10, and 15% 2-butyl alcohol.

Elfasakhany [15] also investigated iso-butanol blends with gasoline on single cylinder gasoline engine for performance and emissions. He conducted tests varying speed from 2600 to 3400 rpm for each blend with vol.% of 3, 7, and 10% of iso-butanol. Analysis showed that for all blends at lower speeds, emissions of CO_2, CO, and HC were less than gasoline; however, at moderate speed of 2900 rpm, CO and HC emission came in the same order of magnitude with gasoline. At high speed (3400 rpm), CO and HC emissions exceeded slightly than gasoline. For all speed range, level of emission of CO_2 was lower for blends than gasoline. This study revealed that brake power, torque, volumetric efficiency, exhaust gas temperature, and in-cylinder pressure for blended fuels were lesser than that of neat gasoline by an average of 3.4%, 0.8%, 3.1%, 4.4%, and 6%, respectively.

In an experiment conducted by Heorhi [10], 10% blend of butanol was seen to give no effect on engine power, torque, and SFC but increasing percentage of butanol, decreases above parameters, and it was suggested to alter ignition advance angle to check the effect of higher % of n-butanol in gasoline.

Vast experimental studies have been done on ethanol, and in many countries, the use of ethanol up to 10% in gasoline is realized. However, increased quantities of ethanol blending have addressed some issues, and research is directed toward butanol which is expected to be blended in more percentage. Some studies give comparative analysis of ethanol–gasoline and butanol–gasoline blends for SI engine. Most of the studies show that ethanol is superior to butanol slightly. In this study, efforts are made to compare combustion, performance, and emission behavior of ethanol–gasoline and butanol–gasoline blend in single cylinder engine at varying speeds.

2 Experimental Setup and Methodology

The following figure and table briefly explain experimental setup and specifications of engine. In this work, PE3 software with open ECU was used. Engine parameters like CR and spark timing was tuned by varying these parameters for gasoline. Piezosensor was used to plot combustion characteristics. National instruments data acquisition system was employed. AVL444N make five gas analyzer was selected to measure HC, CO, NO, O_2, and CO_2 emissions. Following table shows engine specifications (Fig. 1 and Table 2).

Initially, engine was operated on gasoline, and using open ECU engine operation was tuned for optimum performance. Further, pure ethanol 99.9% was mixed with gasoline on volume basis as 10, 20, and 30% with names E10, E20, and E30. Similarly, butanol was blended with gasoline for 10%, 20%, and 30% blends as Bu10, Bu20, and Bu30, respectively. The WOT was achieved with throttle body injection of fuel. Engine speed at WOT was varied from 1200 to 1800 rpm with a step increase of

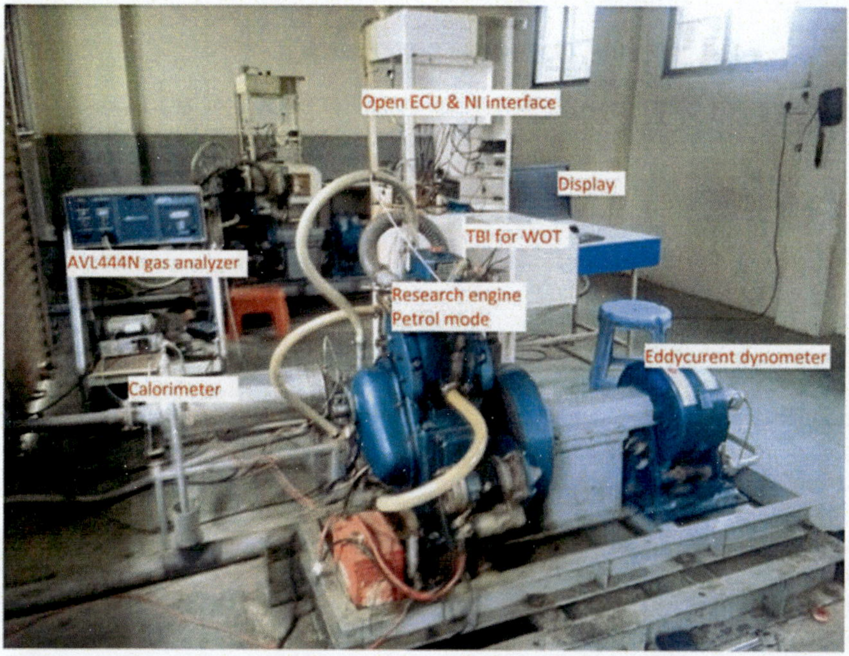

Fig. 1 Experimental set-up

Table 2 Engine specifications

Engine	Make: TV1, Kirloskar, 1 Cyl., 4St, water cooled, 3.5 kW @ 1500 RPM, speed: 1200–1800 RPM, CR:8–10
Dynamometer	Eddy current
ECU	Performance electronics, USA, PE3 model open ECU
Piezosensor	PCB, USA, up to 350 bar combustion range
Temperature sensor	RTD and thermocouples for 200–1200 °C
Interface/software	National Instrumentation for interface and engine specifications for analysis

200 rpm. The combustion, performance, and emission characteristics were studied in detail for gasoline blends with ethanol and n-butanol.

3 Results and Discussions

Results are presented in three major sections viz performance, emission, and combustion characteristics.

3.1 Performance Characteristics

Here, clearly revealed that gasoline fuel shows highest brake power at all speeds compared to alcohol blends. This may be due to higher calorific value. Brake power increased with speed and was obtained maximum at 1600 rpm and then reduced with further increase for all fuels under test. Bu20 and Bu30 were seen superior to E20 and E30 with a percentage decrease in brake power of 5.4% over gasoline at 1600 rpm, whereas decrease of 6.1% and 11.8% was seen for E20 and E30, respectively. It should be noted that higher blends of alcohol fuel limits combustion and hence performance of single cylinder, 4S, gasoline engine rated for 4.5 kW @ 1800 RPM. The said engine was not run beyond 30% blending for both ethanol and butanol with gasoline. This may be due to lean mixture (increased lambda than 1), more oxygen content, and reduced CV compared to pure gasoline.

Figure 2 reveals that E30 and Bu30 have poor brake specific fuel consumption among all test fuels. Maximum percentage increase of 15.38% and 10.25% in BSFC was occurred over gasoline for E30 and Bu30, respectively, at 1800 rpm. For all fuels, BSFC varies in direct proportion to speed except for E20 and E30 where minimum BSFC was obtained at 1400 rpm. BSFC values for E10, E20, and Bu10 were observed in very close relation to gasoline. In region of higher speed (between 1600 and 1800), BSFC of lower blend alcoholic fuels viz Bu10, E20, and E10 tried to come down below respective values of BSFC for gasoline. Roughly 2.5% of decrease was attained for E20 at 1600 rpm and Bu10 at 1800 rpm. This might have risen its fuel efficiency and largely offset the less energy content of alcohol fuels.

Higher brake thermal efficiency was attained at minimum speed of 1200 rpm and decreased with increased speed for all fuels. In medium and high (1400–1600 rpm) speed range, all ethanol blends have shown best performance with approximately 3.5% of improvement in BTE over gasoline, but at low speed, they tended to reduce BTE with an average decline of 4.4%. Perfectly reverse trend to ethanol was noticed for Bu10 with nearly 4% of improvement at low and high speed. This shows a wide scope for blending conventional gasoline with alcoholic fuels like butanol where high octane number of alcohol fuels can be used to advantage. High octane ratings

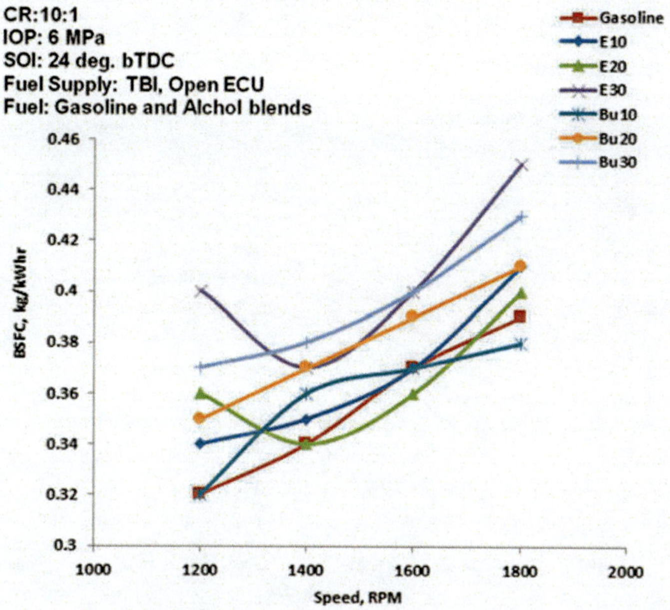

Fig. 2 Speed versus brake specific fuel consumption

and oxygen content of alcohol fuels may have bettered brake thermal efficiency by 4% as compared to gasoline fuel (Figs. 3 and 4).

3.2 *Emission Characteristics*

HC emissions were noted higher at low speeds (1200–1400 rpm). For nearly all blends, HC emissions were higher than gasoline. Actually, HC emissions would have less for oxygenated fuels like alcohol. However, cause for this reversed nature could be less in-cylinder temperature which slowed combustion along with less uniform mixture due to high latent heat of alcoholic fuels. About 50% rise of HC was observed at low speed. CO emissions were noted less at higher speed and arose with speed declination. At higher speeds (1800 and 1600 rpm), CO emissions for gasoline were more than blended fuels by about 40%, which were risen by almost 45% over gasoline at low speeds. Generally, CO levels remained more than gasoline. This may be attributed to less in-cylinder temperature and insufficient mixing (Figs. 5 and 6).

NO_x emissions were found less at speed of 1200. However, NO_x emissions were drastically increased at 1600 rpm. This may be due to slightly higher temperature arising because of complete combustion. This is also compensated by lower HC, CO emissions at 1600 rpm. CO_2 emissions remained at an average value of 2–5%

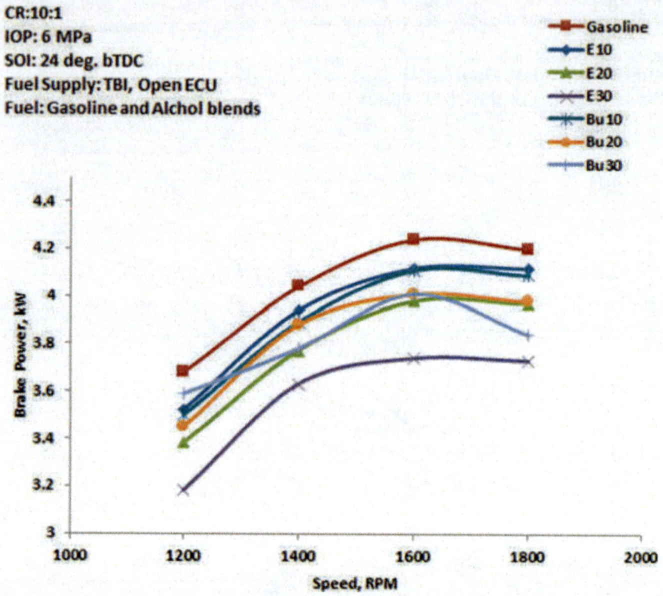

Fig. 3 Speed versus brake power

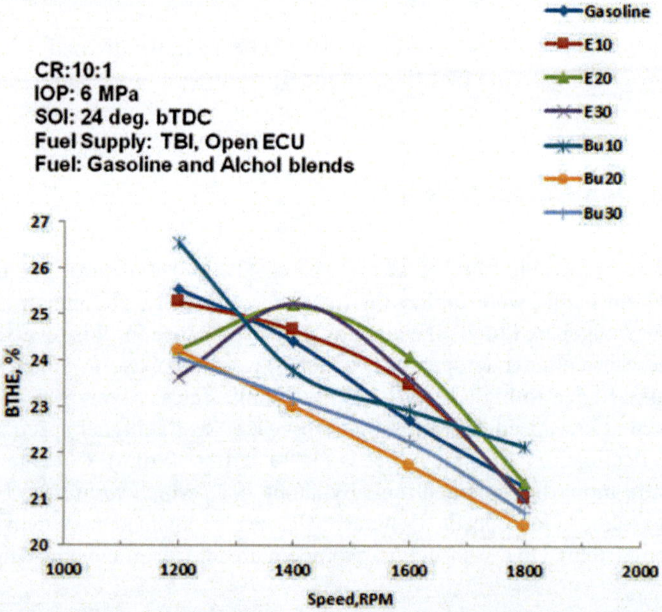

Fig. 4 Speed versus brake thermal efficiency

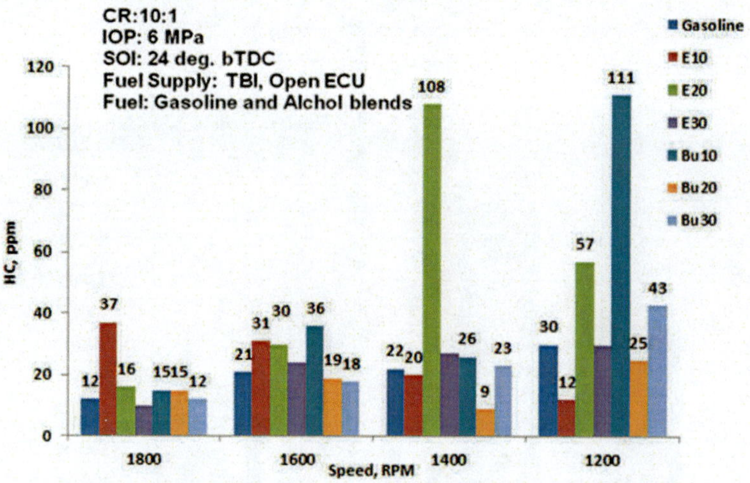

Fig. 5 Speed versus HC emissions

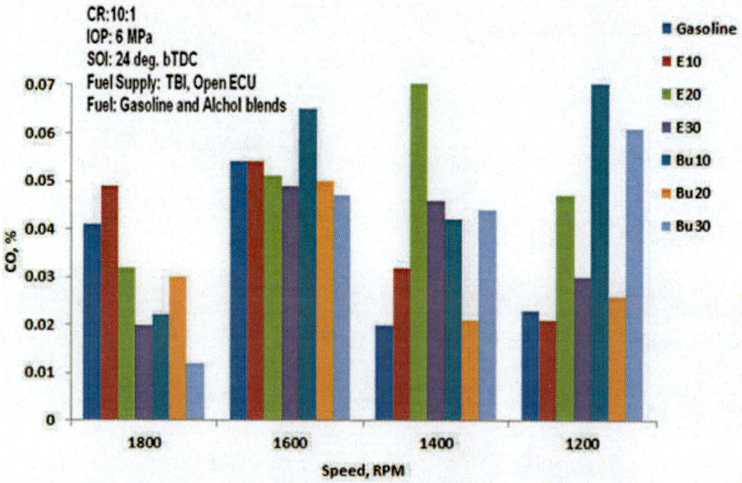

Fig. 6 Speed versus CO emissions

besides lower CO_2 at 1200 rpm. At lower speeds, CO_2 levels reduced, and at higher speeds, it remained more except for top speed of 1800 rpm. In all for gasoline, CO_2 levels were high than blended fuels. Overall, it was seen that Bu10 can be realized as a subsidiary fuel for ethanol in coming future (Fig. 7, 8, 9 and 10).

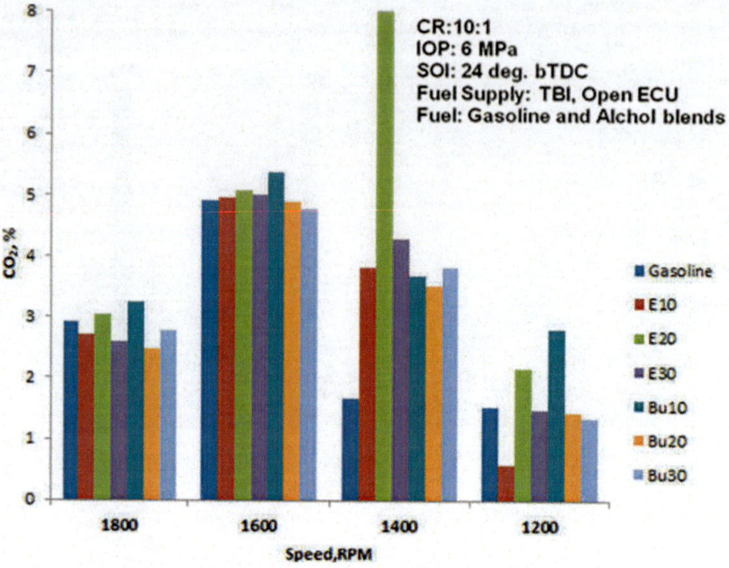

Fig. 7 Speed versus CO_2 emissions

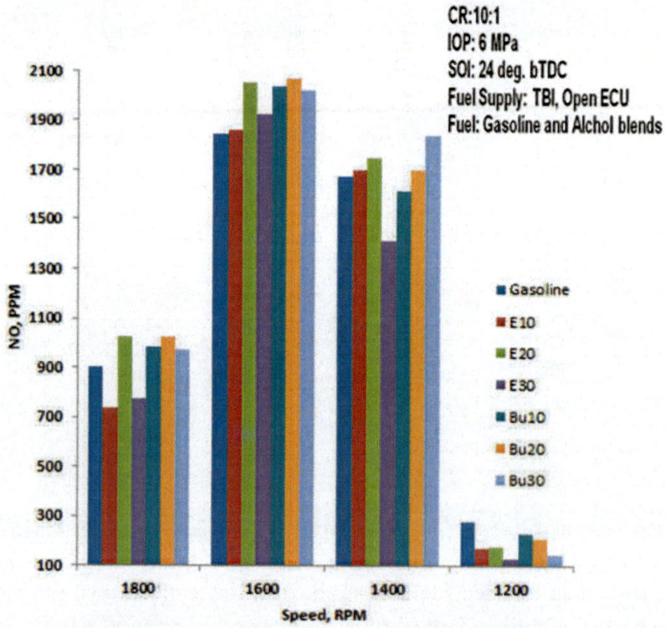

Fig. 8 Speed versus NO_x

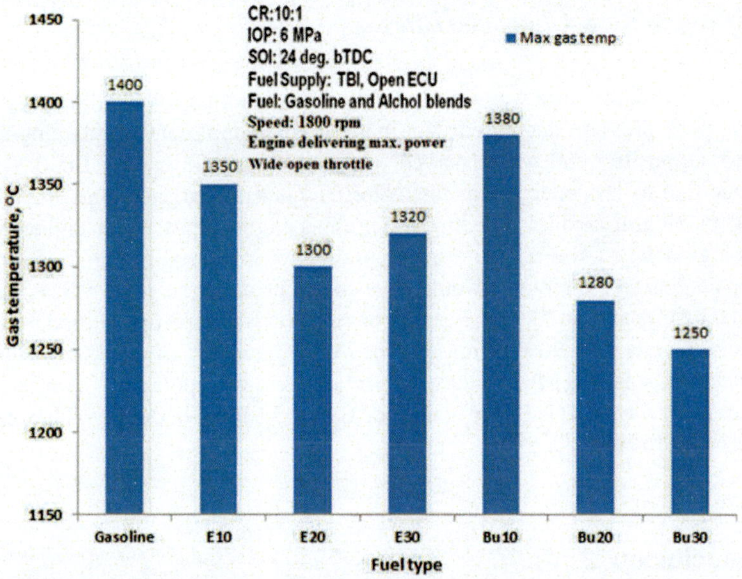

Fig. 9 Maximum gas temperature at top speed for fuels under study

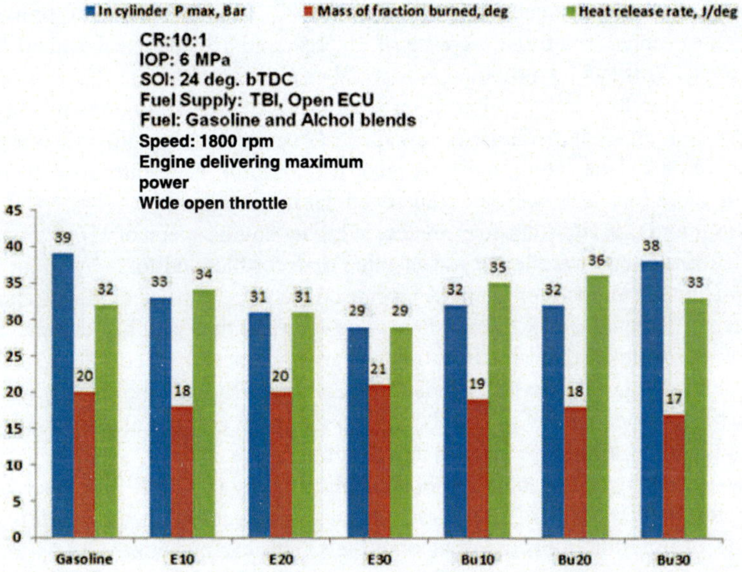

Fig. 10 Max. cyl. pressure, mass of fraction burned, and heat release rate at 1800 rpm for fuels under study

3.3 Combustion Characteristics

In-cylinder peak pressure was seen maximum for gasoline followed by Bu30. For ethanol, peak pressure decreased with increase in ethanol content attaining 29 bar for E30. Opposite trend was observed for butanol with 38 bar for Bu30. Reason might be due to less energy content (calorific value) of ethanol. This might have limited maximum ethanol blending for present engine. Further, at higher blends due to high octane number, flame speed might have increased and along with extra oxygen content in the fuel this would have contributed for the greater mass fraction burned for Bu30 and Bu20. This is seen from more mass of fraction burned for Bu30, and Bu20 blends with 90% of mass burned in 17° and 18° of crank angle. In line with peak pressure and MFB values, heat release rate was more for butanol than for ethanol. HRR for Bu10 is 35 J/deg and for E10 is 34 J/deg both of which exceeded HRR of gasoline (32 J/deg).

4 Conclusion

Bu 30 blend was seen superior than ethanol in case of peak pressure and MFB. Improved HRR was noticed for low blends. BSFC for higher blending proportion was seen to increase with an increase of 15.38% and 10.25% for E30 and Bu30, respectively. This can be attributed to less CV of alcohols.

Alcohol blending in low proportions was observed to give improved BTE than gasoline particularly at moderate-to-low speed. An increase of roughly 4% was found in BTE. High octane rating, more oxygen in fuel could have improved BTE, but blending more than 20% was not seen beneficial in this regard.

It was found that HC emissions are less at higher engine speed of 1800 rpm for all fuels. HC emissions were lower for gasoline than for blended fuels by about 50%. Low in-cylinder temperatures might be caused for this. Similar kind of trend was showed by CO emissions. CO emissions were 40% less than gasoline at high speeds which were first leveled up and then rose to nearly 45% at low speeds. Reason behind high CO levels may be less in-cylinder temperatures. High alcohol blends like E30 and Bu30 were operated with less CO due to more oxygenated alcohol fuels. CO_2 emissions for gasoline were more than alcohol blends.

It can be concluded that, blending of alcoholic fuels by 10% to 20% vol. in gasoline gives better BTE and combustion parameters. However, at low speeds, slightly more HC and CO emission limits use of alcohol fuels. This can be bettered by optimizing engine parameters for trade-off among emission, combustion, and performance.

Finally, it can be concluded that like ethanol, blending of n-butanol in lower proportions (5–10% vol.) with gasoline is reasonable in present engines with minor tuning of engine operating parameters. This is due to almost same performance and emissions characteristic of n-butanol (Bu10) to present-day E10 fuel. Author finds further experimental studies of deep insight into combustion behavior of n-butanol with higher blending proportions.

References

1. Vinod Babu M, Madhu Murthy K, Amba Prasad Rao G (2017) Butanol and pentanol: the promising biofuels for CI engines—a review. Renew Sustain Energy Rev 78:1068–1088
2. Li Y, Gong J, Deng Y, Yuan W, Jun F, Zhang B (2017) Experimental comparative study on combustion, performance and emissions characteristics of methanol, ethanol and butanol in a spark ignition engine. Appl Therm Eng 115:53–63
3. Chansuria P, Mandoi RK (2018) Effects of ethanol blends on performance of spark ignition engine-a review. Mater Today: Proc 5:4066–4077
4. Hasan A, Al-Rawashdeh H et al (2018) Impact of changing combustion chamber geometry on emissions and combustion characteristics of a single cylinder SI (Spark Ignition) engine fueled with ethanol/gasoline blends. Fuel 231:197–203
5. Huang Y, Hong G (2016) Investigation of the effect of heated ethanol fuel on combustion and emissions of an ethanol direct injection plus gasoline port injection (EDI + GDI) engine. Energy Conver Manag 123:338–347
6. Thakur AK et al (2017) Progress in performance analysis of ethanol-gasoline blends on SI engine. Renew Sustain Energy Rev 69:324–340
7. Elfasakhany A (2017) Investigation on performance and pollutant emissions of spark ignition engines fueled with n-butanol, isobutanol, ethanol, methanol and acetone-gasoline blends: a comparative study. Renew Sustain Energy Rev 71:404–413
8. Patil R et al (2019) Current challenges and advances in butanol production. Chapter 8 Sustainable Bioenergy
9. Saraswat M, Chauhan NR (2019) Comparative assessment of butanol and algae oil as alternate fuel for SI engines. Eng Sci Technol an Int J, In press
10. Heorhi K et al (2017) Operation of a spark-ignition engine on mixtures of petrol and n-butanol. Procedia Eng 187:588–598
11. Elfasakhany A (2014) Experimental study on emissions and performance of an internal combustion engine fueled with gasoline and gasoline/n-butanol blends. Energy Convers Manag 88:277–283
12. Mittal N et al (2013) Study of performance and emissions characteristics of a partially coated LHR SI engine blended with gasoline and n-butanol. Alexandria Eng J 52:285–293
13. Singh SB et al (2015) Technical feasibility study of butanol-gasoline blends for powering medium-duty transportation spark ignition engine. Renew Energy 76:706–716
14. Yusri IM et al (2017) Investigation of influences of secondary butyl-alcohol blends on performance and cycle-to-cycle variations in a spark-ignition engine. Energy Procedia 110:310–315
15. Elfasakhany A (2015) Experimental investigation on SI engine using gasoline and a hybrid iso-butanol/gasoline fuel. Energy Convers Manag 95:398–405

Effect of Operating Parameters on the Performance and Emission of a Diesel Engine Fuelled with Diesel–Methanol Blend

M. R. Sumanlal, Vineeth Satheesh, Vishal Dilip, J. Navaneeth, and M. V. Yadhukrishnan

Abbreviations

M15	15% methanol blended in diesel
P	Injection pressure
CR	Compression ratio
BTE	Brake thermal energy
BSEC	Brake specific energy consumption
EGT	Exhaust gas temperature
HC	Hydrocarbon
CO	Carbon monoxide
NO_x	Nitrous oxide
IT	Injection timing

1 Introduction

Methanol, which is a highly oxygenated fuel, has the potential to reduce crude oil imports in India and become the best alternative fuel for the Indian scenario. Methanol can also be used to produce fuel blends with petroleum derived fuels. It has high efficiency and less emissions of nitrogen oxides, sulphur oxides and particulate matter [1]. It can also be converted into another clean alternative fuel called dimethyl ether.

M. R. Sumanlal (✉) · V. Satheesh · V. Dilip · J. Navaneeth · M. V. Yadhukrishnan
Department of Mechanical Engineering, Federal Institute of Science and Technology, Angamaly, Kerala, India
e-mail: sumanlalmr@gmail.com

A. K. Gupta et al. (eds.), *Advances in IC Engines and Combustion Technology*, Lecture Notes in Mechanical Engineering,
https://doi.org/10.1007/978-981-15-5996-9_35

In the world, India being the fifth-largest country with coal reserves, must tap its potential and produce methanol/DME using coal. By using a recent technology, methanol can be easily produced through coal, which can help reduce crude oil imports.

As of now, even methanol has several drawbacks. Methanol when blended with diesel shows an increase in NO_x production than from neat diesel fuel. EGT also increases on methanol addition which is a major factor for NO_x production. The efficiency of methanol in performance and emission can now be improved by engine modifications/varying engine parameters [2]. Different engine parameters have a great impact on the combustion characteristics, engine performance and emission. Parameters like CR, P and IT can be altered, and their effect on an optimum methanol blended diesel fuel can be studied. Present engine parameters are provided for best suiting of diesel for good fuel efficiency and performance. The optimum performance and efficiency on methanol can be brought by varying these parameters under different conditions.

2 Experimental Methods and Apparatus

The experiment was conducted at Apex innovations, MIDC, Kupwad, Maharashtra. The experimental setup is shown in Fig. 1. A Kirloskar SV1 model single cylinder direct injection diesel engine was utilised in order to carry out the analysis. This is a water cooled four stroke diesel engine. The engine was also a variable compression ratio engine. It also has provisions for adjusting the injection timing. Injection pressure can be varied by using different fuel injectors with the engine. A water cooled eddy current dynamometer, namely the AG 250 model, was connected directly with the engine that produces 3.5 kW power at the rated speed which is 1500 rpm. In order to determine the amount of smoke emitted from the engine, a smoke analyser, namely AVL 437C, was used. A gas analyser, namely the AVL Di Gas 444 model, was utilised in order to record the exhaust emissions.

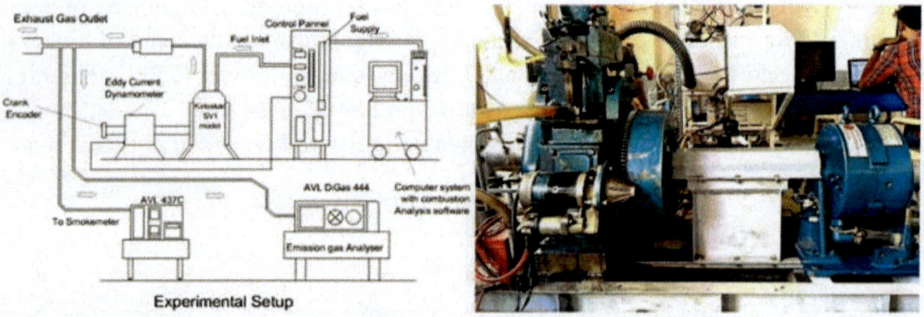

Fig. 1 Experimental setup

Interfacing of fuel flow, air flow, temperature, speed and load measurements with the computer was also done by providing the required provisions. Sensors were placed in various parts of the setup including inside the cylinders to measure the temperature. Various performance and emission characteristics were studied and analysed through this setup. Online performance and combustion evaluations were done by using ‘Engine soft’, an engine performance analysis software package.

The fuel was obtained by blending 15% of methanol in diesel. Remi model RQ5 + stirrer was used to prepare the fuel blend [3].

The tests were carried out at combinations of three compression ratios—16, 17 and 18; three injection pressure—210, 225 and 250; and three injection timing—21BTDC, 23BTDC, and 25BTDC, i.e., there were 27 different operating conditions in which the tests were conducted.

3 Result and Discussion

Performance:-

1. *Brake thermal efficiency*

BTE at different conditions is illustrated above in Fig. 2. As the compression ratio increases, BTE increases. As the compression ratio decreases, ignition delay increases and results in improper combustion with M15. That is, ignition delay along with combustion duration decreases at elevated compression ratio and load. The high combustion temperature due to the elevated compression ratio and load improves the reaction characteristics of oxygen present in the methanol [4]. Because of this, brake thermal efficiency increases.

As the injection pressure increases, as above in figures, the brake thermal efficiency increases due to the improved combustion occurring at this high pressure. The high pressure improves the spray characteristics and atomisation, and it also enhances the mixing of the fuel with air which contributes to the enhanced combustion. If the pressure is too high, fuel droplet diameter will be too fine. This causes some of the fuel to escape from the chamber which affects the efficiency. As a result, when IP increases from 225 to 250 bar, BTE shows only a slight increase. It can also

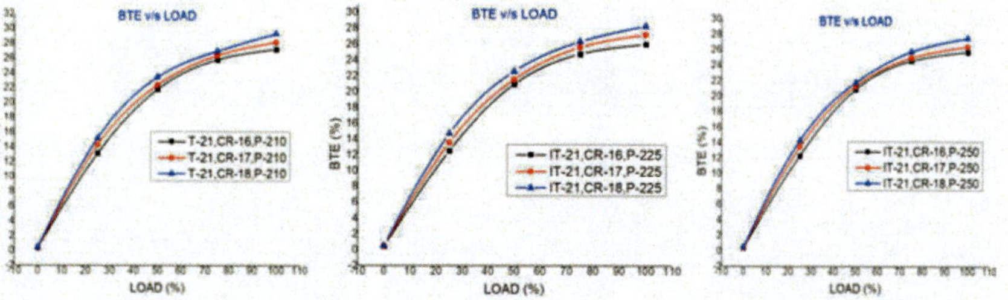

Fig. 2 Variation of BTE with compression ratios at different injection pressures (at 21 BTDC)

be seen that more fuel droplets strike the combustion chamber wall at high injection pressure due to enhanced fuel jet velocity. As a result, these fuel droplets are cooled which causes incomplete combustion [5]. This also contributes to the slight increase in efficiency at high pressures as seen in Figs. 3 and 4.

2. *Brake Specific Energy Consumption*

The variation in BSEC can be seen in the above graphs. From Figs. 5, 6 and 7, it can be seen that as the compression ratio increases, BSEC decreases for all injection pressures and timings. This is mainly due to the increase in BMEP and temperature caused by the increase in compression ratio which increases the effective power.

BSEC also decreased at elevated *P*. This could be because of better atomisation of M15 spray which enhanced combustion. On retarding the fuel injection timing, BSEC increased. The results of different conditions are clearly visible in the graphs

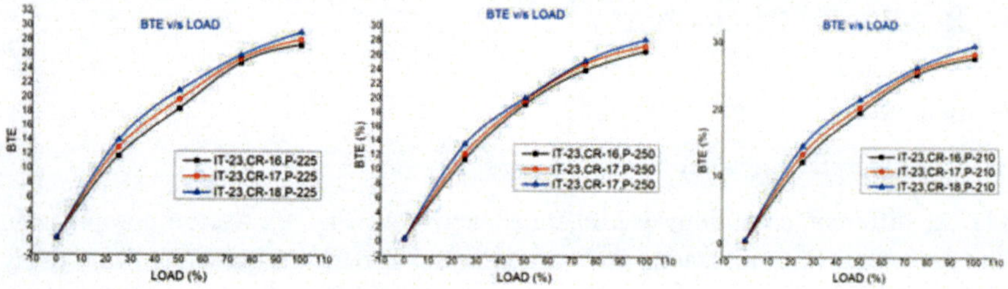

Fig. 3 Variation of BTE with compression ratios at different injection pressures (at 23 BTDC)

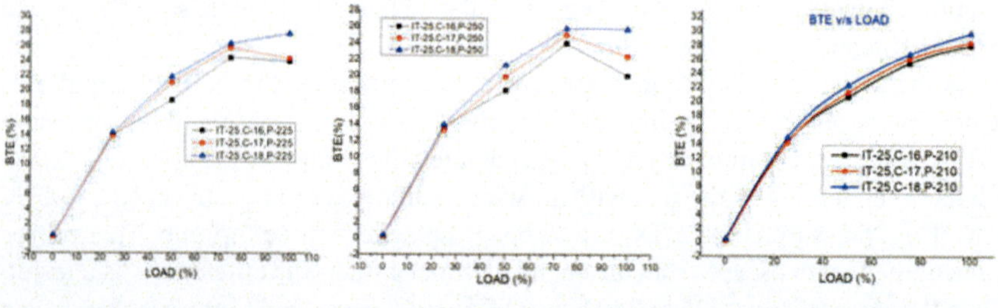

Fig. 4 Variation of BTE with compression ratios at different injection pressures (at 25 BTDC)

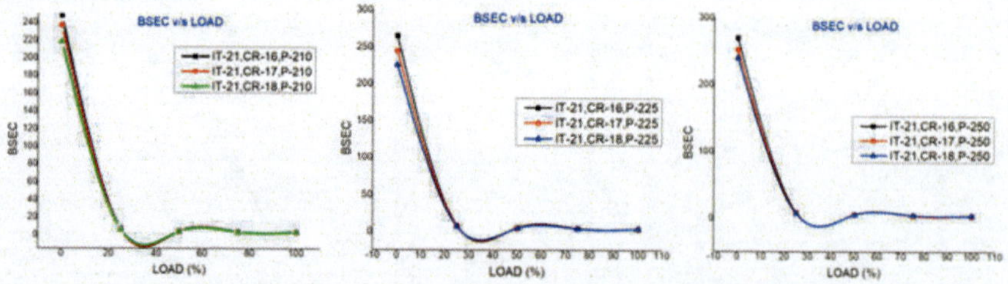

Fig. 5 Variation of BSEC with compression ratios at different injection pressures (at 21 BTDC)

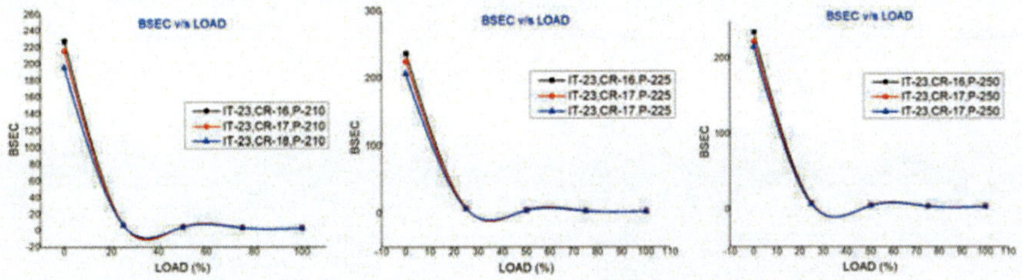

Fig. 6 Variation of BSEC with compression ratios at different injection pressures (at 23 BTDC)

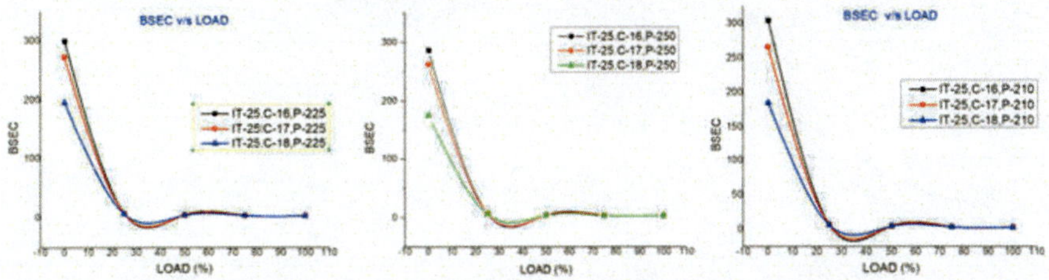

Fig. 7 Variation of BSEC with compression ratios at different injection pressures (at 25 BTDC)

3. *Exhaust Gas Temperature*

It was observed from the above graphs that the EGT increases at higher compression ratio due to enhanced compression. From Fig. 8, it was established that EGT of the blends were modest when compared with that of diesel. This happens because under high compression ratio, the combustion process shifts a bit to the prior stroke of the cycle [6]. This leads to better utilisation of the fuel energy to produce power and contributes to the overall decrease in EGT.

The exhaust gas temperature increases for M15 by advancing the injection timing. The reason is that with advancing IT, a larger part of the fuel is burning in the first phase of the combustion process which increases the peak temperature.

When the *P* is boosted, the total pressure goes increasing, and this result in high temperature in the chamber. Thus, exhaust gas temperature increases at elevated *P*. The results of different conditions are shown in Figs. 8, 9 and 10.

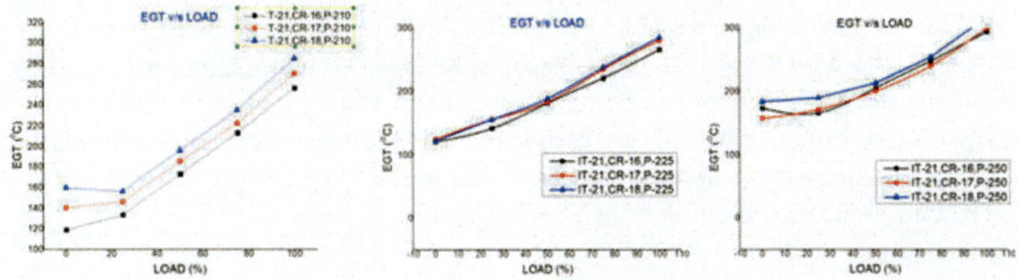

Fig. 8 Variation of EGT with compression ratios at different injection pressures (at 21 BTDC)

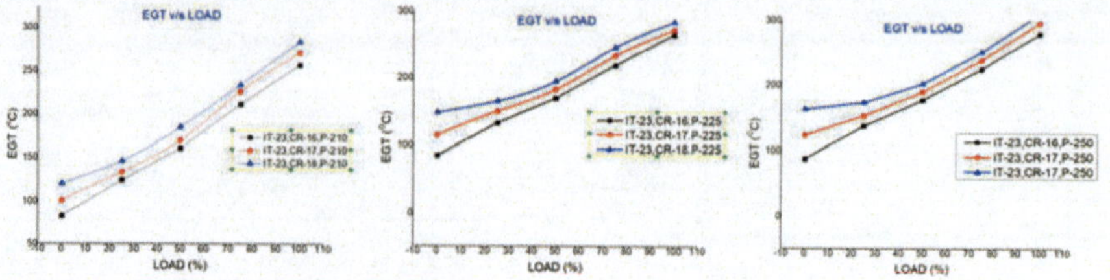

Fig. 9 Variation of EGT with compression ratios at different injection pressures (at 23 BTDC)

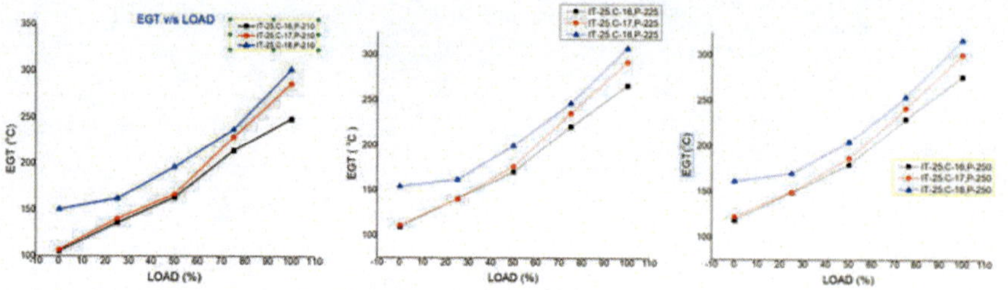

Fig. 10 Variation of EGT with compression ratios at different injection pressures (at 25 BTDC)

Emission:-

1. *Co and HC*

Generally, as seen in Fig. 9 as the compression ratio increases, there will be an elevated temperature at the end of compression phase, and chemical kinetics changes due to the availability of more oxygen in methanol. This decreases the carbon monoxide emissions. This is due to the fact that the elevated combustion temperature produced by the high compression ratio and high loads enhances the affinity of carbon with oxygen, and the effect of oxygen atom in methanol oxidation reaction enhances which causes the conversion of carbon monoxide to carbon dioxide.

It is seen from Fig. 12 that the increase in the *P* reduced the CO emission and decrease in the *P* increased the CO emission. It may be due to the elevated *P*, which contributes to an efficient fuel–air blending and due to the better combustion of the tiny fuel droplets. These factors contribute to decrease in emission of carbon monoxide.

It can be seen from Figs. 11, 12 and 13 that the carbon monoxide emission decreases with advancing IT. This is because the elevated temperature produced by advancing the IT contributes to better oxidation reactions joining the molecules of oxygen and carbon. Because of this, carbon monoxide emissions decrease. Similarly, carbon monoxide emission becomes high when the IT is retarded.

HC emission trend is similar to CO emissions.

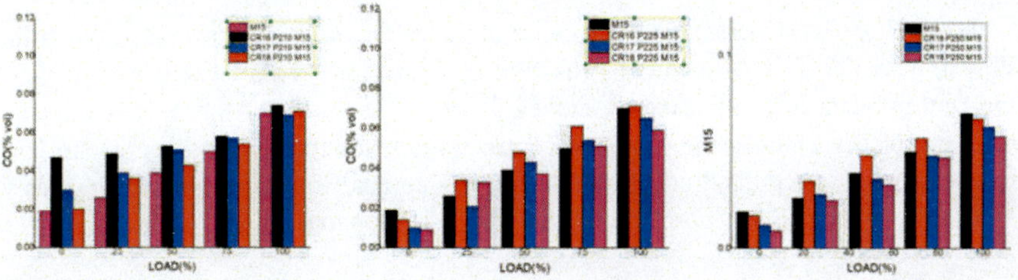

Fig. 11 Variation of CO emission with compression ratio at different injection pressures (at 21 BTDC)

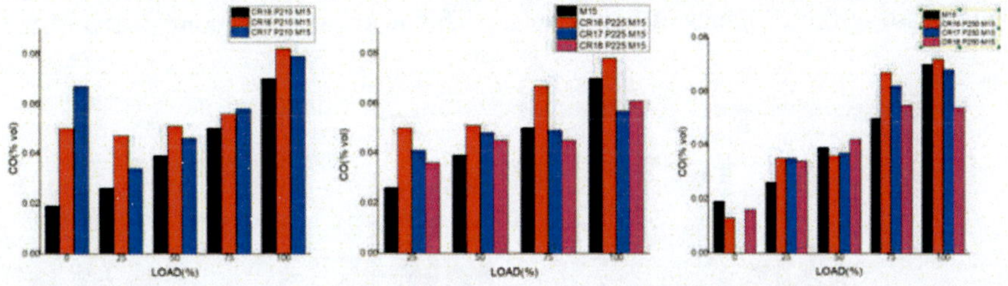

Fig. 12 Variation CO emissions with compression ratios at different injection pressures (at 23 BTDC)

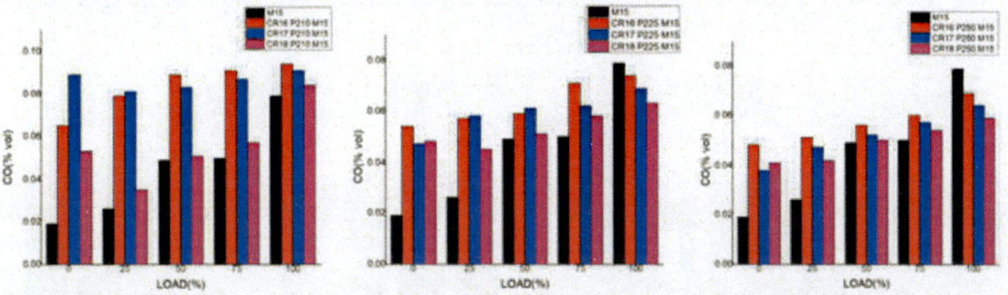

Fig. 13 Variation of CO emissions with compression ratios at different injection pressures (at 25 BTDC)

2. NO_x

The NO_x emission increases as the compression ratio increases and can be seen in the above figures. This is due to the rapid burning of M15 due to its high volatility at high compression ratio. This increases the EGT which increases the nitrogen oxide emissions.

As the *P* is increased, fuel droplet size decreases which leads to rapid combustion of the M15 fuel spray due to the high surface area-to-mass ratio of fuel spray. As a result, large amount of heat is released quickly which results in an increase in the in-cylinder temperature which directly affects the nitrogen oxide emissions causing it to increase.

From Figs. 14, 15 and 16, it can be seen that NO_x emission decreased when the IT was retarded due to a lower peak cylinder pressure. This is because much of the fuel is burnt after the piston reaches the top,. i.e., most of the heat is released after the piston reaches the top dead centre. This leads to lower peak pressure and temperature inside the cylinder which causes a decrease in NO_x emission. Similarly, NO_x emission increases when the IT was advanced due to a higher cylinder pressure.

3. *Smoke*

At elevated CR, the smoke decreases at higher load because of the better combustion and can be seen in Figs. 17, 18 and 19. From the above figures, it can be seen that smoke capacity reduces with increasing *P*. This is due to the fact that, higher *P* causes a decrease in the diameter of fuel droplets. This leads to better combustion of the

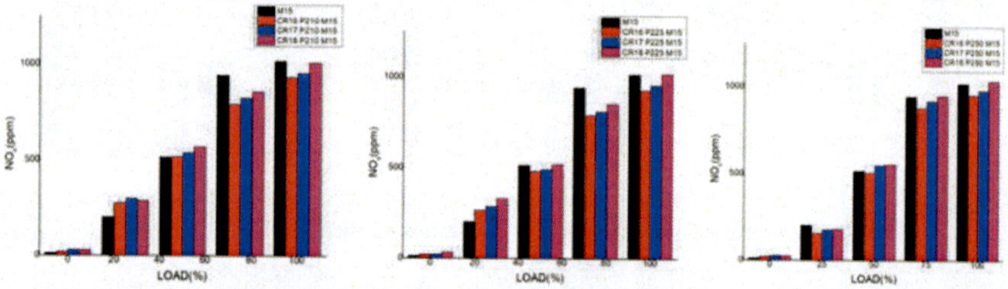

Fig. 14 Variation of NO_x with compression ratios at different injection pressures (at 21 BTDC)

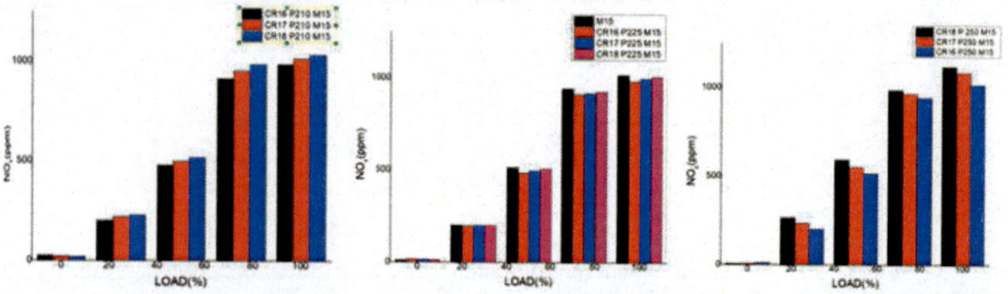

Fig. 15 Variation of NO_x with compression ratios at different injection pressures (at 23 BTDC)

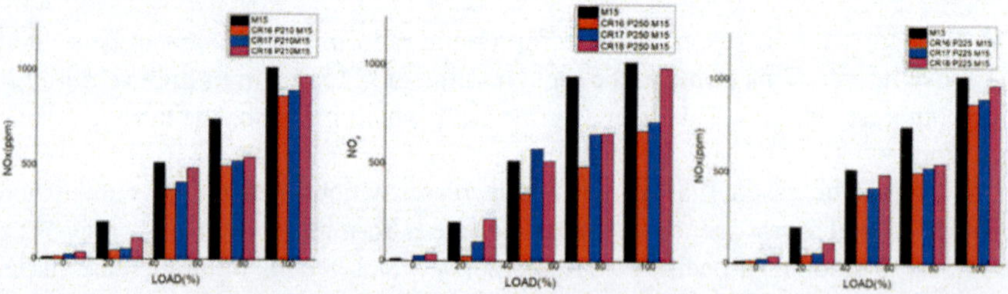

Fig. 16 Variation of NO_x with compression ratios at different injection pressures (at 25 BTDC)

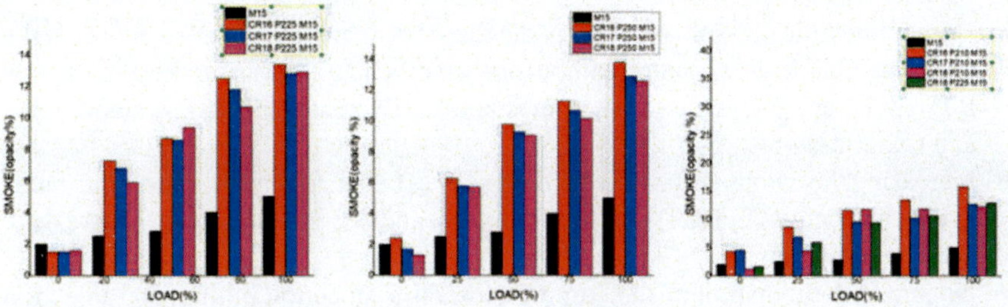

Fig. 17 Variation of smoke with compression ratios at different injection pressures (at 21 BTDC)

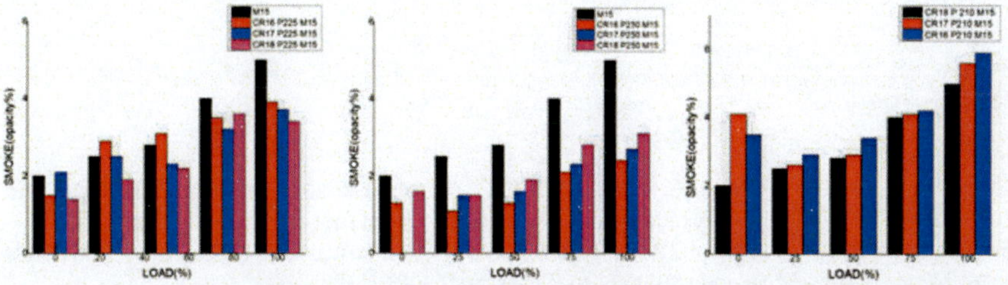

Fig. 18 Variation of smoke with compression ratios at different injection pressures (at 23 BTDC)

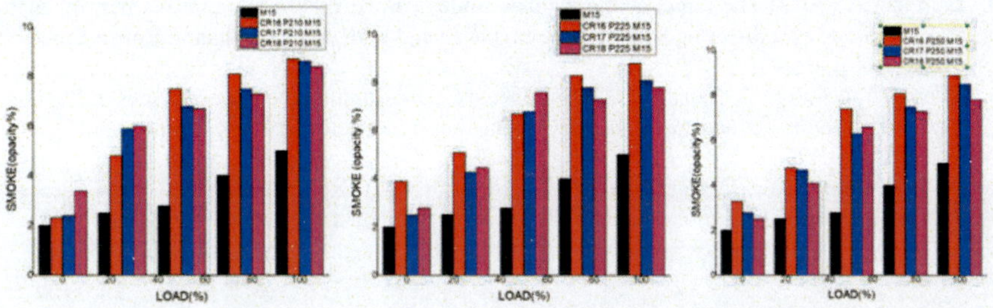

Fig. 19 Variation of smoke with compression ratios at different injection pressures (at 25 BTDC)

fuel due to an increase in surface area-to-mass ratio and due to the enhanced mixing of the fuel with air.

4 Conclusion

- Compression ratio and injection pressure are the most significant factors for emission characteristics.
- Ignition timing does not have much significant effect of the performance.

- Higher injection pressure means that the fuel and air are mixed very well inside the cylinder. This leads to better combustion, and hence, increasing the P increases BTE, EGT, and NO_x emission and substantially decreases the BSEC, smoke, HC, and CO emissions.
- Increasing the compression ratio leads to overall better combustion and contributes to an enhanced BTE, EGT, and NO_x emission and decrease in BSEC, smoke, HC, and CO emissions.
- No single combination of the compression ratio, injection timing, and injection pressure could be used for the overall improvement of performance and emission characteristics simultaneously.

References

1. Lu H, Yao A, Yao C, Chen C, Wang C (2019) An investigation on the characteristics of and influence factors for NO_2 formation in diesel/methanol dual fuel engine. Fuel 235:617–626
2. Guo Z, Li T, Dong J, Chen R, PeijunXue XW (2011) Combustion and emission characteristics of blends of diesel fuel and methanol-to-diesel. Fuel 90(3):1305–1308
3. Murat C, Huseyin K, Eyub C, Ozgur S (2011) An experimental investigation on effects of methanol blended diesel fuels to engine performance and emissions of a diesel engine. Sci Res Essays 6(15):3189–3199
4. Jamrozik A (2017) The effect of the alcohol content in the fuel mixture on the performance and emissions of a direct injection diesel engine fueled with diesel-methanol. Enegy Convers Manage 148:461–476
5. Chen C, Yao A, Yao C, Wang B, Lu H, Jun F et al (2019) Study of the characteristics of PM and the correlation of soot and smoke opacity on the diesel methanol dual fuel engine. Appl Therm Eng 148:391–403
6. Wang B, Chen C, Yao C, Dou Z, Wei H, Liu M et al (2017) Experimental study of the effect of engine parameters on ultrafine particle in diesel/methanol dual fuel engine. Fuel 192:45–52

Fuel Injection Pressure and Combustion Chamber Geometry Effects on the Performance and Emission Characteristics of Diesel Fueled CI Engine with EGR

V. D. Tamilarasan and T. Ramesh Kumar

1 Introduction

Diesel locomotives are the main source of power for the most of the industrial and domestic applications owing to their greater thermal efficiency, higher brake power output, reliability, low fuel use and durability. An innovative variety of injection systems have been designed to get higher injection pressure. In the present scenario, the source of conventional hydrocarbon fuel has been depleting with an increase in its usage as there is an accelerating demand for energy.

The mixing of air and fuel is a vital factor in combustion. The decrease in fuel penetration for the available air in the chamber causes incomplete fuel burning. Therefore, increase in fuel injection pressure and modification of combustion chamber geometry would improve the fuel penetration into air present in the combustion chamber which results in complete atomization of air and fuel [1, 2]. The performance and emission are indicatively influenced by injector type, injection pressure and chamber design in a diesel engine. The performance mostly depends on characteristics of the fuel spray which would be achieved by proportionately high pressure. In olden day's diesel engine, the fuel is injected at pressure of 250 bar but, nowadays the common rail direct injection system employs very high injection pressure at 2000 bar. In this common rail direct injection system, the fuel rail pressure keeps up the injection pressure of the fuel [3, 4].

Very small sized drops are produced by decreasing the nozzle hole diameter of injector, which leads to reduction of time taken for atomization. However, this would thin out the spray penetration due to low spray momentum of fuel. [5]. Wang et al.

V. D. Tamilarasan (✉) · T. Ramesh Kumar
Department of Mechanical Engineering, Bannari Amman Institute of Technology, Tamilnadu, India
e-mail: tamil.maritime@gmail.com

A. K. Gupta et al. (eds.), *Advances in IC Engines and Combustion Technology*, Lecture Notes in Mechanical Engineering,
https://doi.org/10.1007/978-981-15-5996-9_36

investigated the effect of injector holes numbers and different injector nozzle diameter on the structure of flame and soot formation of single cylinder compression ignition (CI) engine; the result was reduced atomization time and rapid fuel distribution throughout the combustion chamber [6]. Kuleshov et al. explored the multi-zone thermodynamic model for calculating the performance and combustion parameters of closed cycle direct injection (DI) diesel engine [7]. These studies explain the performance parameters and emission characteristics of single cylinder CI engine using multiple injection method with multi-zone thermodynamic model software. Diesel-RK software is a simulation software of diesel engine which simulates the IC engine combustion by changing the parameters like modifying the piston bowl geometry, injection spray angle, injection timing, injection location, injection pressure, valve timing, turbocharge parameters, EGR, etc. [8–14]. Exhaust gas recirculation (EGR) plays a significant role in reducing the oxides of nitrogen (NOx) of diesel engines by decreasing the flame temperature along with oxygen concentration of the air and fuel mixture in the combustion chamber [15].

From the points of literature review, it is understood that conventional diesel engine receives fuel at the pressure of somewhat up to 250 bar. However, recent advanced fuel injection system injects the fuel at a pressure of 2000 to 3000 bar that would result in reduction of soot formations [6]. The present investigation intends to improve the air efficiency in the combustion chamber. In order to achieve so, three different swirl combustion chambers have been developed and experimentally investigated to study the performance and emission characteristics with respect to injection pressure when the EGR is employed (Fig. 1).

2 Experimental Apparatus and Procedure

2.1 Experimental Setup

The experiment carried out with a single cylinder, four stroke, CRDI, variable compression ratio (VCR) diesel engine which is coupled with eddy current dynamometer. Table 1 indicates the test engine technical specifications and Fig. 2 represents the schematic layout of the experimental setup. Additional setups for measuring injection pressure, in-cylinder pressure and crank angle of the engine were included. Injection pressure and in-cylinder pressure were measured using piezoelectric sensor. The sensor was attached near the injector and head of the engine, respectively. The setup consists of fuel flow measurement devices for measuring the mass flow rate of air and fuel, respectively. Rotameters were implemented to measure the cooling water flow to the engine. Also, online measurement device was used to study the degree of exhaust gas temperature, engine cooling water intake and outlet temperature and engine load. Data acquisition was employed to protocol the signal from the measuring devices to a computer and to display the in-cylinder pressure (ICP), heat release rate (HRR) and cylinder temperature with respect to

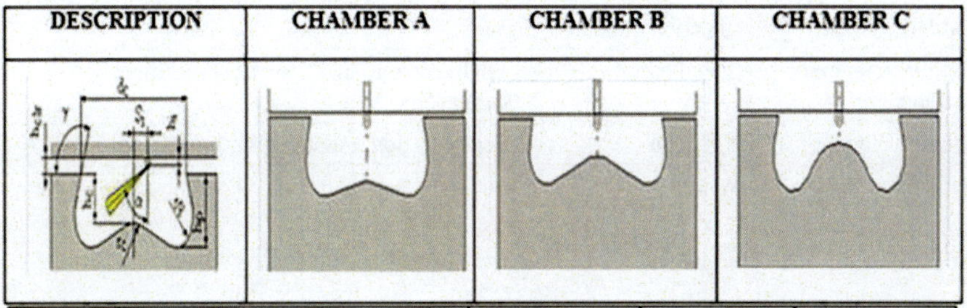

Description	Chamber-A	Chamber B	Chamber-C
External diameter, dc,[mm]	50.2	57.2	53.2
In-centre piston bowl depth, hc, [mm]	14	8	6
Radius of sphere in centre of piston bowl, rc, [mm]	15.59	13.59	15.59
Depth of combustion chamber in periphery, hp, [mm]	17.1	14.1	17.1
Radius of hollow chamber in periphery of bowl, rp, [mm]	8.31	7.31	8.31
Inclination angle of a bowl forming a plane of the piston crown, gamma, [deg]	105	105	105
Top-clearance at TDC, h_clr, [mm]	1.001	1.001	1.001
Swirl ratio	3.1431	2.534	2.8555

Fig. 1 Geometric parameters of the chambers A, B and C

crank angle. The sensed signals from the engine are linked to a computer through by a data acquisition system (DAQ) and the engine performance analyzing software "Engine Soft LV" which displays the P-H and P-V diagram. The setup supports the study of performance parameter of the engine that includes brake power (BP), BTE, SFC and combustion analysis.

2.2 Emission Measurements

An exhaust gas measuring instrument was used to measure and analyze the gas composition of the exhaust gas released by a compression ignition engine. The analyzer AVL444 is used to measure the concentration of nitrogen oxides (NO_X, in ppm), hydrocarbons (HC, in ppm), carbon monoxide (CO in % vol), carbon dioxide (CO_2 in % vol) and oxygen (% vol). It was also used for quantifying the opacity of the gas exhausted by the engine. It also quantifies the opacity of the smoke in

Table 1 Technical specification of the test engine

Engine operation parameters	Specifications	Engine operation parameters	Specifications
Engine model	Kirloskar	Fuel injection pressure	600–1400 bar
Cooling system	Water cooled	Fuel injection system	Direct injection (drive solenoid driven)
Number of strokes	4 Strokes	High pressure system	Common rail direct injection
Number of cylinder	Single cylinder	Electronic control unit	Model Nira i7r
Cylinder bore/stoke	87.5 mm/110 mm	Valve per cylinder	2 Valves (1 Inlet. 1 Outlet)
Swept volume	661.45 cc	Liner type/bases	Wet
Compression ratio (CR)	17.5:1	Exhaust gas recirculation (EGR)	SS, water cooled
Rated speed (rpm)	1500	Analysis software	"Engine soft" Engine performance analysis software

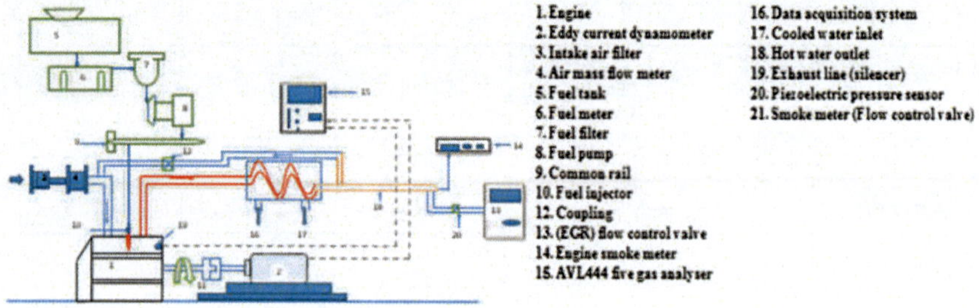

Fig. 2 Schematic diagram of the experimental test setup

Hartridge smoke unit (HSU). Figure 3 shows AVL 444 five gas analyzer and smoke meter.

2.3 *Exhaust Gas Recirculation System*

Exhaust gas recirculation is a best way to minimize the emission in diesel engines. The EGR recirculates the exhaust combusted gas about 10% to 30% into inlet manifold of the engine, during the suction stroke it gets mixed with the inlet ambient air, which in turn diminishes the combustion temperature and hence reduce the oxides of nitrogen. The recirculation percentage of EGR intends to improve the performance and reduce the emission characteristics of single cylinder CI diesel engine. It was found from Fig. 4 that connecting EGR to the inlet manifold is a better method of

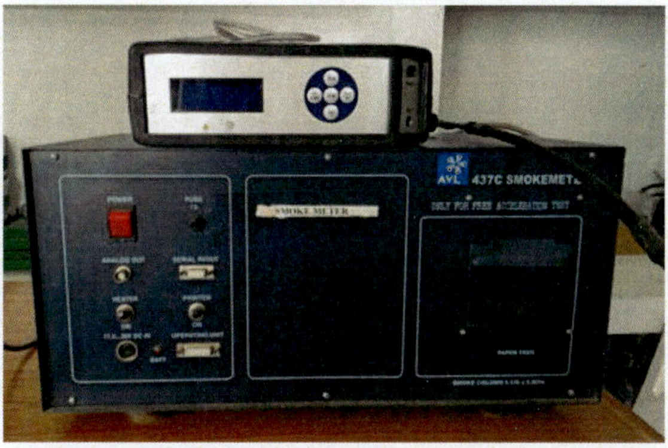

Fig. 3 Assembly of AVL 444 and smoke meter emission measurement systems

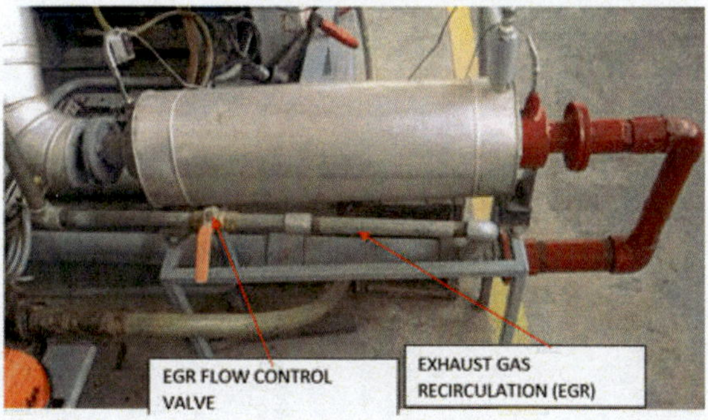

Fig. 4 Experimental exhaust gas recirculation (EGR) setup

employing EGR in a single cylinder diesel engine. About 80% of oxides of nitrogen were reduced when recycled at 15% of EGR. Based on literature review, the exhaust gas recirculation increases the reduction of combustion temperature and also reduces the NO_X. But it increases the brake specific fuel consumption.

2.4 *Experimental Procedure*

The performance and emission test for the experimental engine were carried out at various Cycle Fuel Mass (g/cycle) CFM, which was computed from the experimental TFC and it varies from 0.0111 g/cycle to 0.0289 g/cycle, from low to high load. It was used as an input fuel quantity for thermodynamic simulation software Diesel-RK. First the simulation was carried out at various CFM, by simultaneously varying the IP and EGR using diesel as fuel for CI engine. Based on the simulation results, the suitable combustion chamber was selected and the experiment was carried out in single cylinder CI diesel engine. The injection pressure was varied by using CRDI fuel injection system as shown in Fig. 5. The experiment was carried out in the base engine and suitable combustion chamber was selected from the simulation results from the

Fig. 5 Common rail fuel injection system

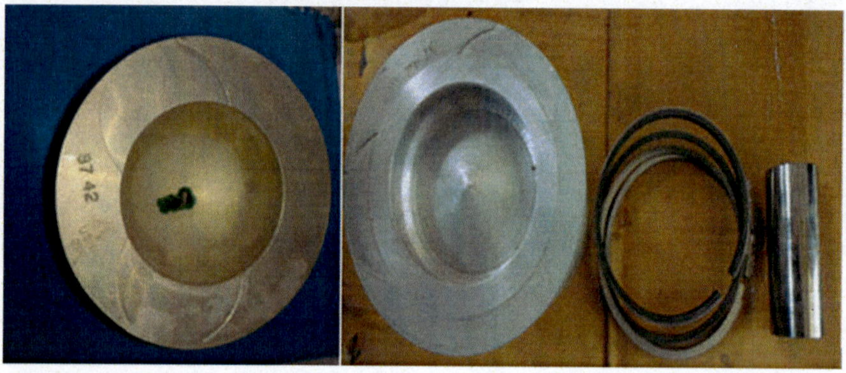

Fig. 6 Photographic view of standard and modified combustion chamber

Diesel-RK software as shown in Fig. 6. The performance and emission characteristics of the single cylinder CI diesel engine were conducted at a compression ratio of 17.5:1, at five injection pressures (600, 800, 1000, 1200 and 1400 bar) maintained at a constant speed of 1500 rpm with and without 15% EGR. The performance and emission tests were recorded from experimental setup and were assessed on the basis of SFC, BTE and emissions of particulate matter (PM) and NO_X.

3 Results and Discussion

The three modified combustion chambers were analyzed based on their performance, combustion characteristics and their emission parameters. They were also compared for SFC, NO_X, PM and ignition delay (ID). The best among the three modified combustion chambers were compared with standard diesel engine chamber for in-cylinder pressure (ICP), heat release rate (HRR), specific fuel consumption, brake thermal efficiency and oxides of nitrogen.

3.1 *Performance Analysis*

3.1.1 Specific Fuel Consumption (SFC)

Figures 7, 8, 9, 10, 11 and 12 show the variation of SFC against IP and various bowl geometry with and without 15% EGR at 0%, 50% and 100% load conditions. SFC is the ratio between fuel mass consumed by the engine to the applied brake power. Figures 7 and 8 show the comparison of SFC of three modified combustion chambers at low load condition. The SFC of chamber-A decreases with an increase in IP from 600 to 800 bars, but slightly increases with an increase in IP from 800 to 1400 bar due to the power required to increase the injection pressure. From Figs. 7, 8, 9, 10,

Fig. 7 Variation of SFC against IP and various bowl geometry without EGR (0% load)

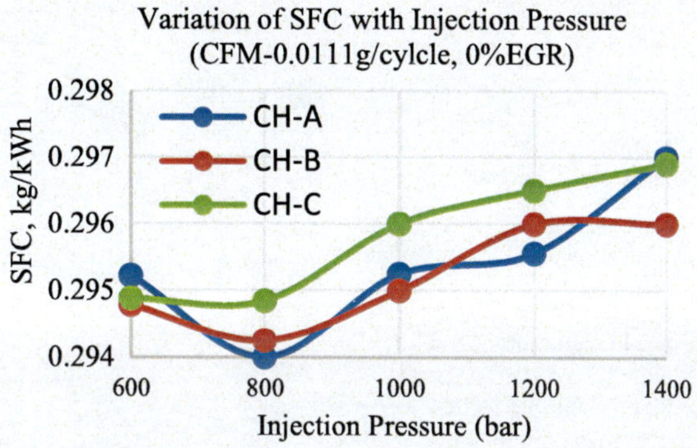

Fig. 8 Variation of SFC against IP and various bowl geometry with 15% EGR (0% load)

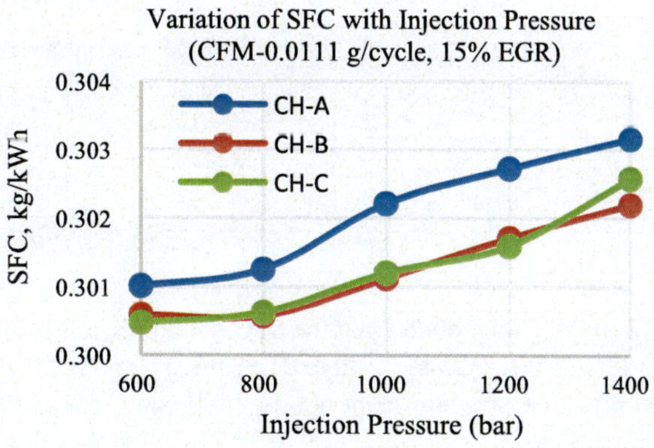

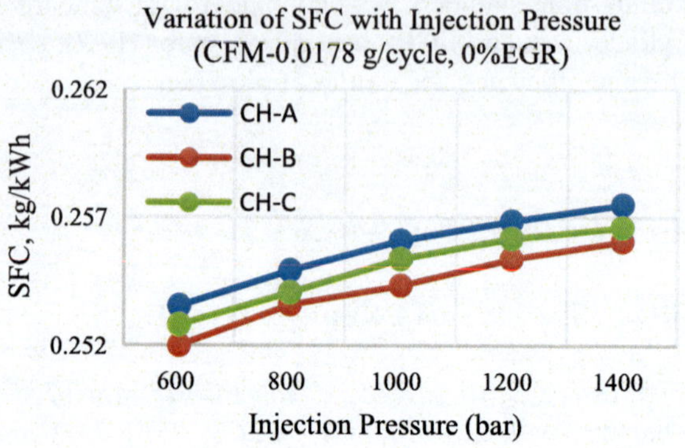

Fig. 9 Variation of SFC against IP and various bowl geometry without EGR (50% load)

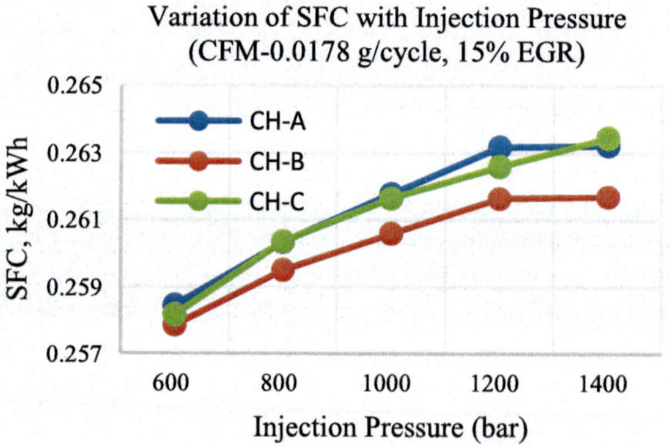

Fig. 10 Variation of SFC against IP and various bowl geometry with 15% EGR (50% load)

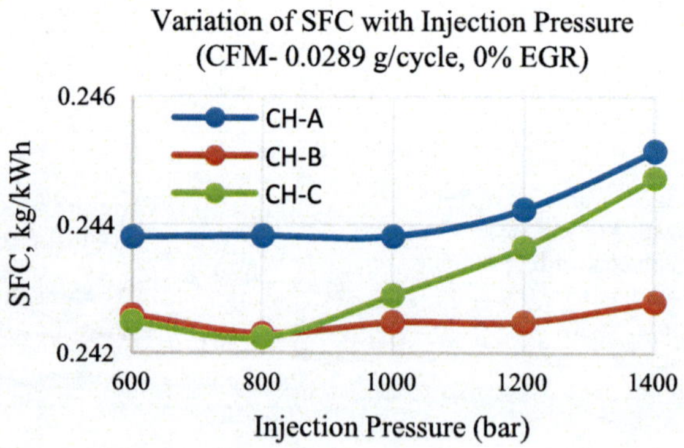

Fig. 11 Variation of SFC against IP and various bowl geometry without EGR (100% load condition)

11 and 12, it is noticed that the SFC of chamber-A is increased from 0.295 g/kWh to 0.296 g/kWh, 0.254 g/kWh to 0.258 g/kWh and 0.244 to 0.245 g/kW with increasing in injection pressure from 600 to 1400 bar at 0%, 50% and 100% load conditions, respectively. From Fig. 11, SFC of chamber-B varied from 0.243 to 0.242 g/kWh

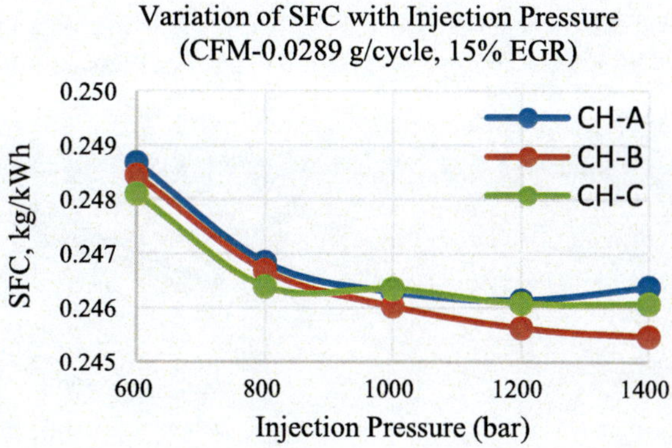

Fig. 12 Variation of SFC against IP and various bowl geometry with 15% EGR (100% load conditions)

which is slightly lower when compared to chamber-A. SFC of chamber-B is slightly lower than the chamber-C at all the conditions. The SFC of combustion chamber-A with 15% EGR mode simulation results is shown in Fig. 8, 10 and 12 and it varies from 0.301 to 0.303 g/kWh, 0.258 to 0.263 g/kWh and 0.249 to 0.246 g/kWh at different load conditions, respectively. SFC of chamber-A at 600, 800, 1000, 1200 and 1400 bar injection pressure with 15% EGR mode simulation increased at 2.1%, 1.3%, 0.8%, 0.85% and 0.84% when compared to without EGR at high load condition. The variation of SFC with IP in chamber-B is lower (about 0.4% at 0% load, 0.4 to 0.8% at 50% load, 0.4 to 0.56% at 100% load) compared to other chambers C and A. In most cases, SFC of all three chambers without EGR mode is lower when compared to with 15% EGR mode simulation. From the above observations, it can be concluded that piston bowl chamber-B is a better combustion chamber when compared to other two chambers.

3.1.2 Nitrogen Oxide (NO_X)

NO_X is one of the major pollution in CI engine which is formed by chain reaction that takes place between oxygen and nitrogen in air. This reaction is highly dependent on temperature. Figures 13, 14, 15, 16, 17 and 18 show the variation of NO_X with respect to IP and of three modified chambers in with 15% and without EGR mode at 0%, 50% and 100% load conditions. NO_X of chamber-A without EGR mode increased from 5.838 g/kWh to 7.751 g/kWh, 9.629 g/kWh to 13.674 g/kWh and 5.677 g/kWh to 10.030 g/kWh at 0%, 50% and 100% load conditions. Also, at 15% EGR mode, it increased from 0.731 to 1.52 g/kWh, 1.583 to 2.520 g/kWh and 1.684 to 3.800 g/kWh at all three different load conditions. From Fig. 18 the Cycle Fuel Mass (CFM) is as same as the value, 0.0289 g/cycle. NO_X of chamber-A at 600, 800, 1000, 1200 and 1400 bar injection with 15% mode simulation decreased by 71.2%, 54.1%, 59.1%, 57.3% and 63.6% when compared with that of 0% EGR mode. The variation of NO_X with IP of chamber-B is lower (about 2.5% at 0% load, 2–7% at 50% load, 3.9 to 10.5% at 100% load) when compared to other chambers C and A.

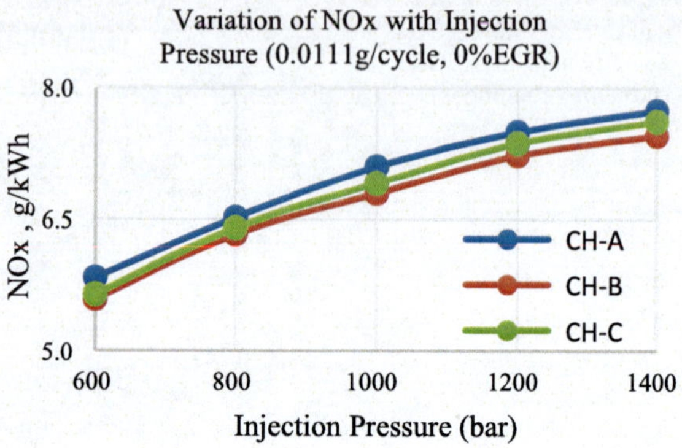

Fig. 13 Variation of NO_X against IP and various bowl geometry without EGR (0% load)

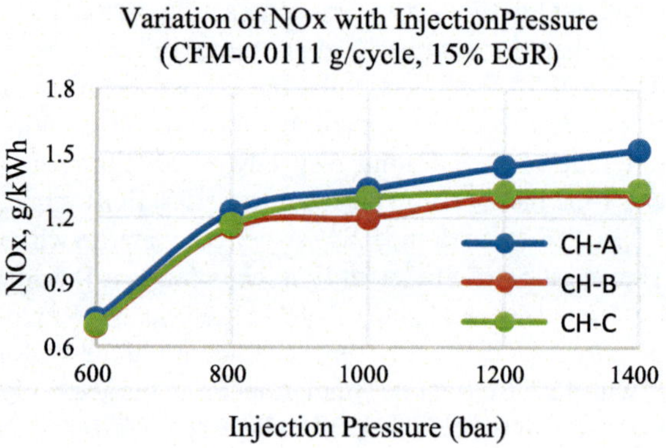

Fig. 14 Variation of NO_X against IP and various bowl geometry with 15% EGR (0% load)

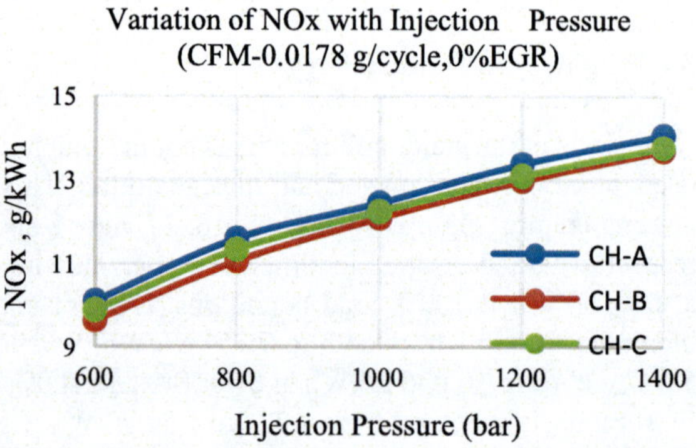

Fig. 15 Variation of NO_X against IP and various bowl geometry without EGR (50% load)

The piston bowl chamber-B is more suitable for combustion temperature than the other two chambers. It is observed that NO_X emission is relatively lesser in 15% EGR mode when compared to without EGR mode, as the oxygen content at the inlet air supply is comparatively low.

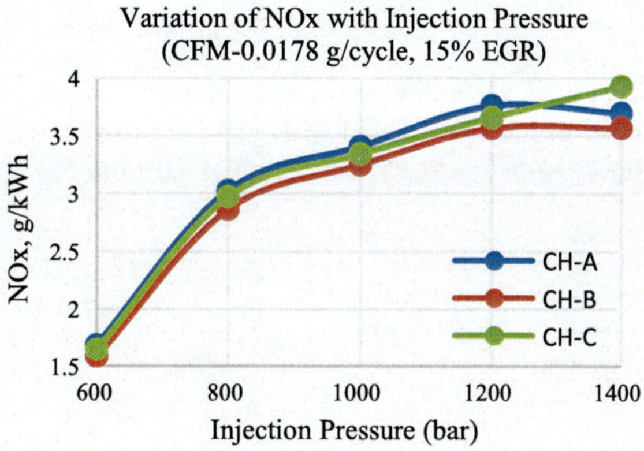

Fig. 16 Variation of NO_X against IP and various bowl geometry with 15% EGR (50% load)

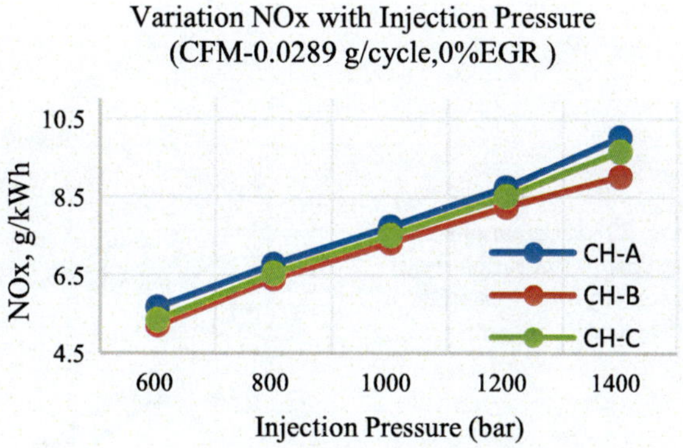

Fig. 17 Variation of NO_X against IP and various bowl geometry without EGR (100% load)

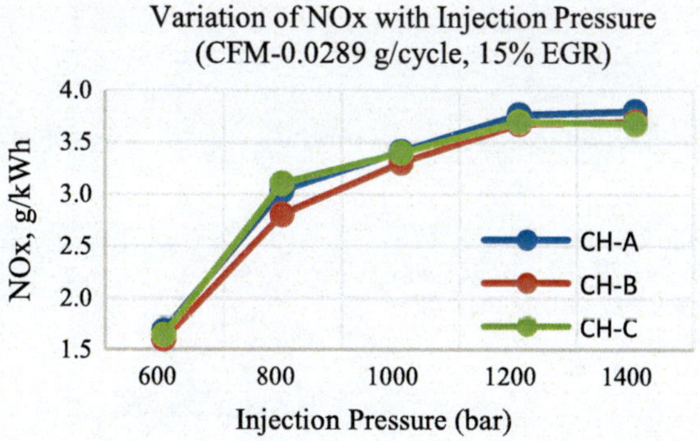

Fig. 18 Variation of NO_X against IP and various bowl geometry with 15% EGR (100% load)

3.1.3 Particulate Matter (PM)

PM emissions from diesel engines are the major contributor to the ambient air contamination problem. Particulate emission draws the attention of several researchers in recent time, as the primary reemerges in the up-gradation of fuel injection system in diesel engine injects fuel directly into the cylinder. Figures 19, 20, 21, 22, 23 and 24 show the variation of PM against injection pressure and various bowl geometry with and without 15% EGR at 0%, 50%, and 100% load conditions. Variation of PM can be observed from Fig. 19. Chamber-A decreases PM

Fig. 19 Variation of PM against IP and various bowl geometry without EGR (0% load)

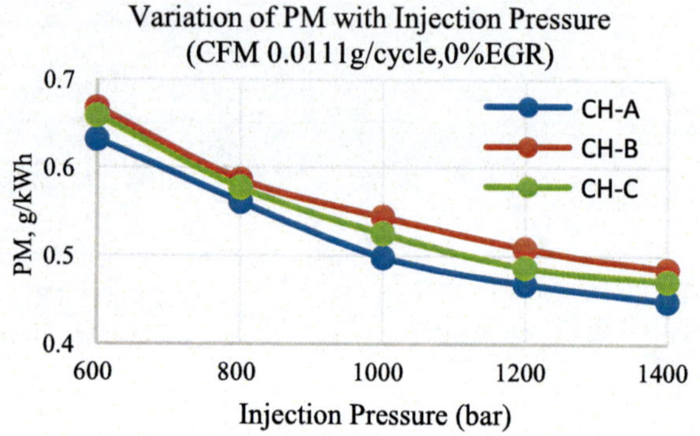

Fig. 20 Variation of PM against IP and various bowl geometry with 15% EGR (0% load)

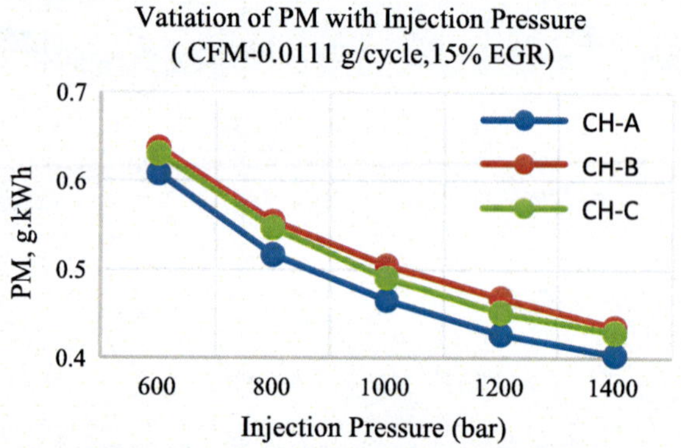

Fig. 21 Variation of PM against IP and various bowl geometry without EGR (50% load)

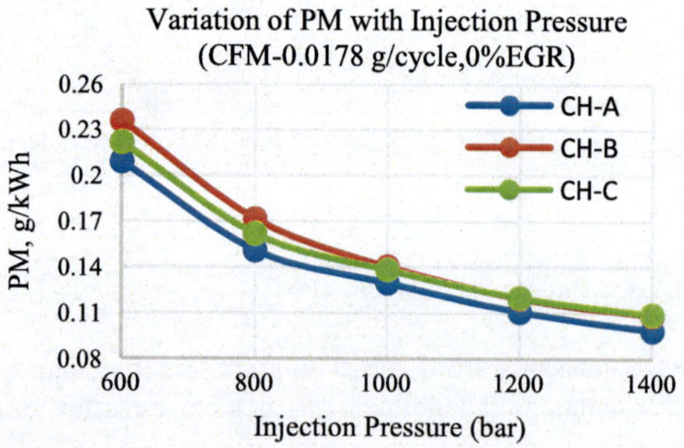

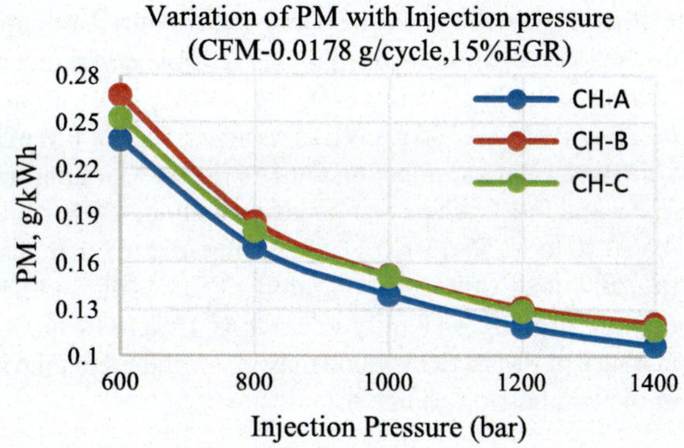

Fig. 22 Variation of PM against IP and various bowl geometry with 15% EGR (50% load)

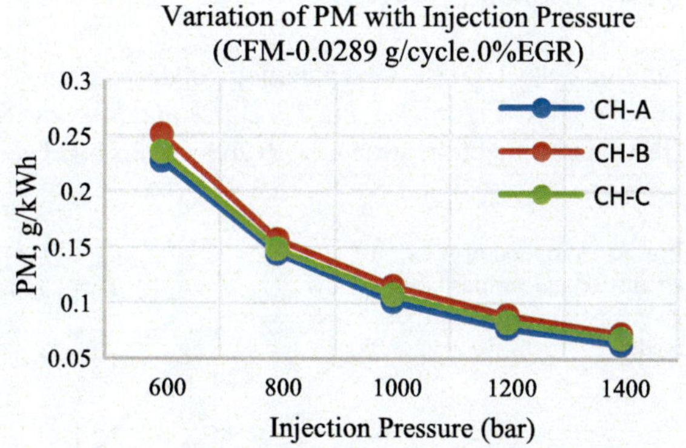

Fig. 23 Variation of PM against IP and various bowl geometry without EGR (100% load)

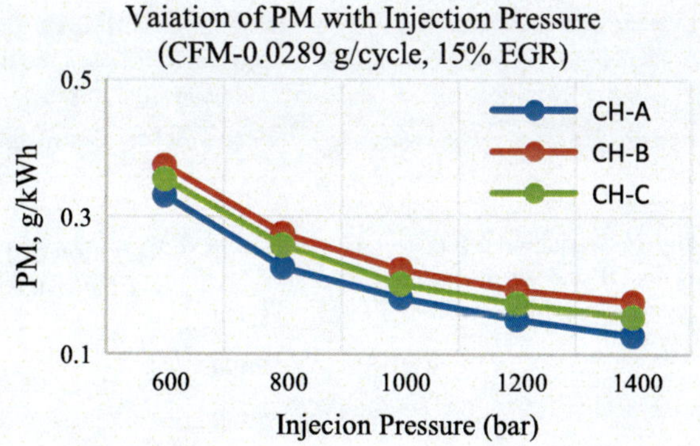

Fig. 24 Variation of PM against IP and various bowl geometry with 15% EGR (100% load)

from 0.632 g/kWh to 0.449 g/kWh with the IP increasing from 600 to 1400 bar. It is also noted that PM decreases with increase in pressure in both with and without EGR modes. Reduction in PM may be associated with higher injection pressure, that arising as the result of enhanced mixture formation of air and fuel for impoverished combustion. From Fig. 23 it is observed that PM is reduced from 0.236 at 600 bar to 0.108 g/kWh at 1400 bar. At 15% EGR simulation, results of PM vary from 0.239 g/kWh to 0.106 g/kWh as shown in Fig. 24 PM of chamber-A at 600, 800, 1000, 1200 and 1400 bar injection with 15% mode simulation increased to 14%, 11%, 7%, 7.2% and 7.4% when compared to without EGR. But chamber-A is lower (nearly 4.5%, 10 to 12.5%, 10 to 11.5%) compared to other chambers B and C at varying the 0%, 50% and 100% load conditions. The PM of modified three combustion chambers of without EGR is slightly lower than 15% EGR mode. Besides, another reason for increase of PM in EGR mode may be attributed to unavailability of oxygen contained in the combustion chamber.

3.1.4 Ignition Delay (ID)

Ignition delay (ID) is a vital parameter in determining the knocking characteristics of diesel engine. Figures 25, 26, 27, 28, 29 and 30 show the variation of Ignition delay (ID) against injection pressure (IP) and various bowl geometry with and without 15%

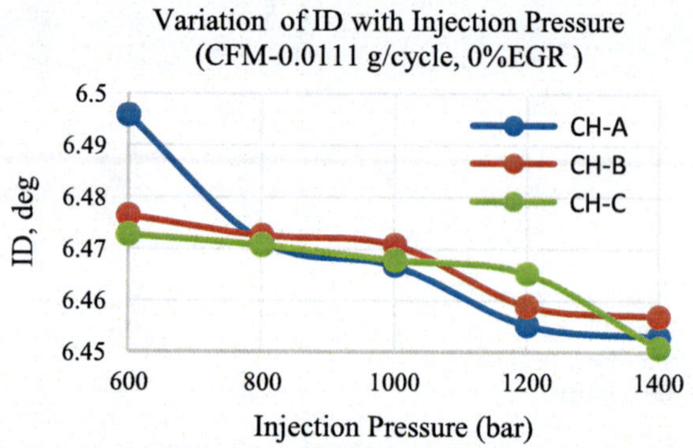

Fig. 25 Variation of ID against IP and various bowl geometry without EGR (0% load)

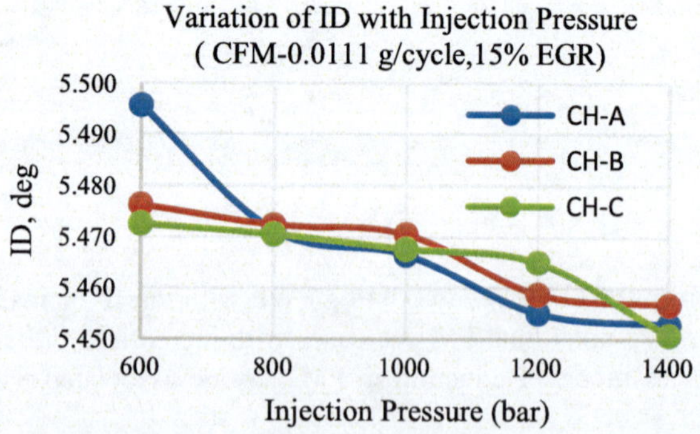

Fig. 26 Variation of ID against IP and various bowl geometry with 15% EGR (0% load)

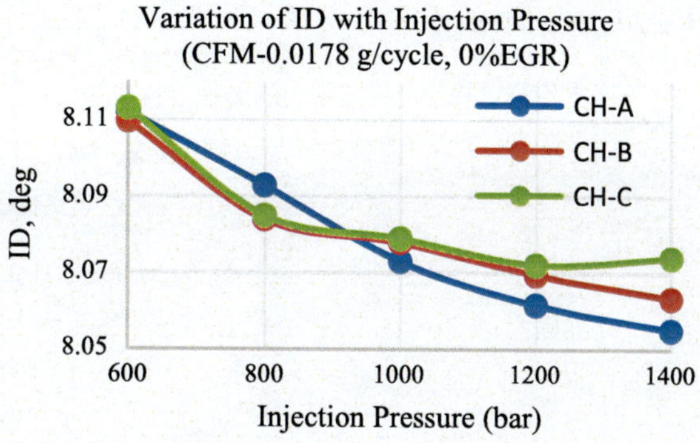

Fig. 27 Variation of ID against IP and various bowl geometry without EGR (50% load)

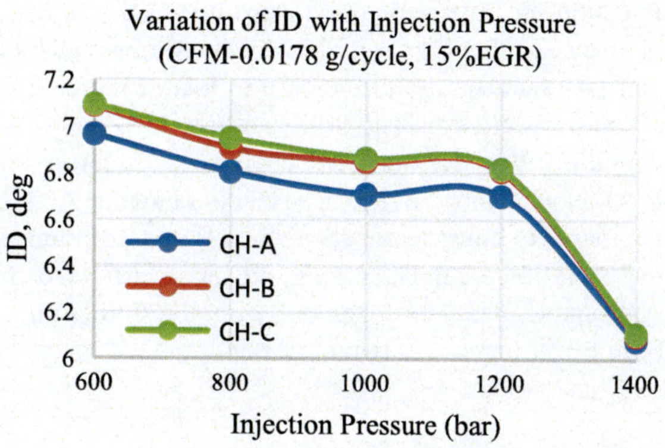

Fig. 28 Variation of ID against IP and various bowl geometry with 15% EGR (50% load)

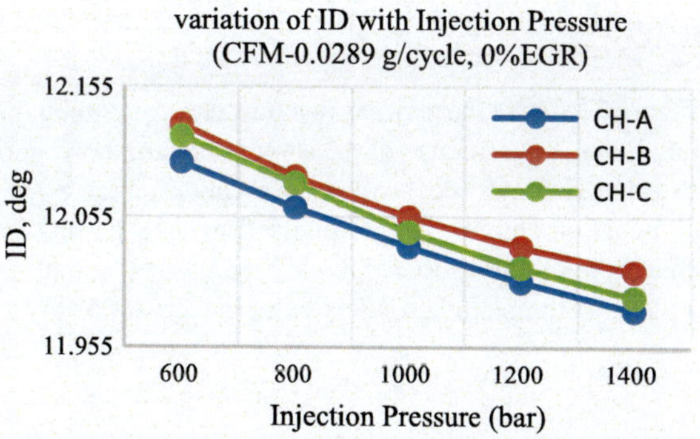

Fig. 29 Variation of ID against IP and various bowl geometry without EGR (100% load)

EGR at 0%, 50%, and 100% load conditions. Figure 25 shows the ID decreases from 6.49° to 6.45° with increase in IP from 600 to 1400 bar that means ID decreases with an increase in pressure for both with and without EGR modes. The variation of NO_X with IP, chamber-C is lower (nearly 0.3 to 1.5%) compared to other chambers B and A at low load condition. This may be due to better air fuel mixing at high

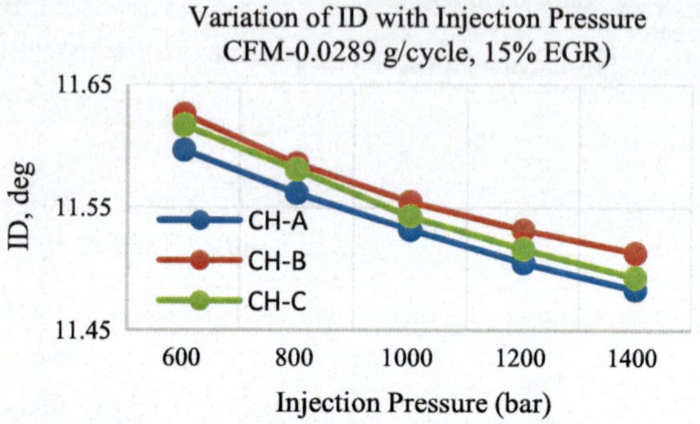

Fig. 30 Variation of ID against IP and various bowl geometry with 15% EGR (100% load)

injection pressure and temperature inlet air. It is noticed that combustion chamber geometries reducing the ignition delay decreases with increase the injection pressure and also increases with increase in load. Figures 29 and 30 show the variation of ID with injection pressure of chambers A, B and C followed same trend at all load condition. But without EGR chamber-A is lower (nearly 0.3 to 1.2%) compared to other chambers B and C. Hence, chamber-A is suitable combustion chamber compared to other chambers namely B and C. Influence of geometric characteristics and emission parameters are also been investigated and compared for the chambers A, B and C. It can be concluded from the simulation results that the chamber-A has been better than other two chambers.

4 Experimental Validation of Standard Piston (STDP) and Modified Combustion Chamber-a (CH-a)

Figure 31 shows the geometric parameters of combustion chamber of standard piston bowl. Some parameters of the standard piston bowl geometrics are discussed namely external diameter (dc) −52 mm, in-center piston bowl depth (hc) −23 mm, depth of combustion chamber in periphery (hp) −18.1 mm. From Fig. 1, piston combustion chamber-A swirl ratio = 3.14 is predicted. The inclination angle of a bowl forming a plane of the piston crown is gamma (γ) −105°

4.1 Validation of in-Cylinder Pressure with Crank Angle (CA) for Chamber-a

Variation of in-cylinder pressure with CA diagram is shown in Fig. 32; Cycle Fuel Mass (CFM) is computed from the experiment. Total Fuel Consumption (TFC) varies from 0.0111 g/cycle to 0.0289 g/cycle and used as input fuel mass/cycle in full

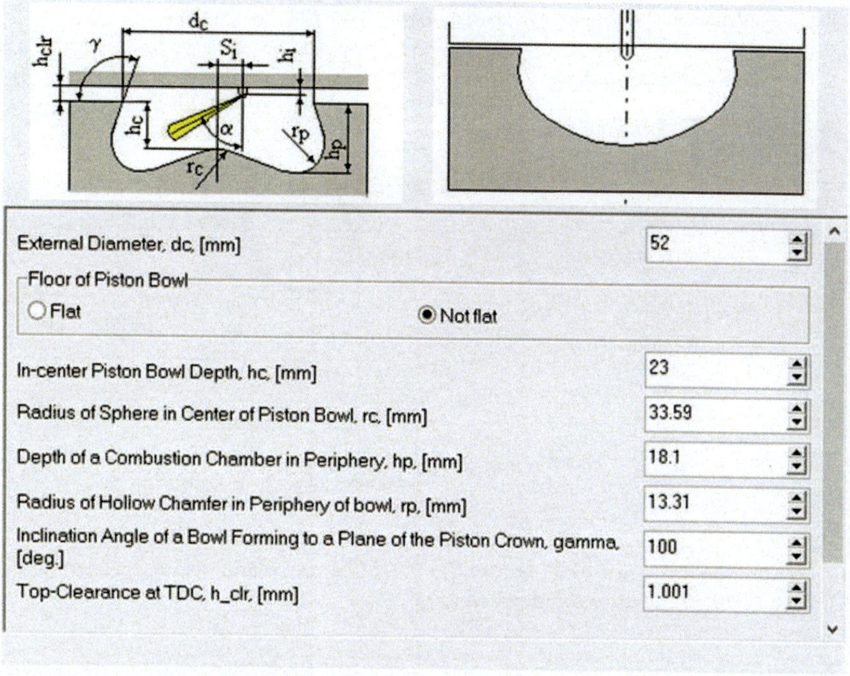

Fig. 31 Piston Geometric Parameter of Standard Piston bowl (STDP)

Fig. 32 Variation of In-cylinder pressure with CA comparison (0.5 kW)

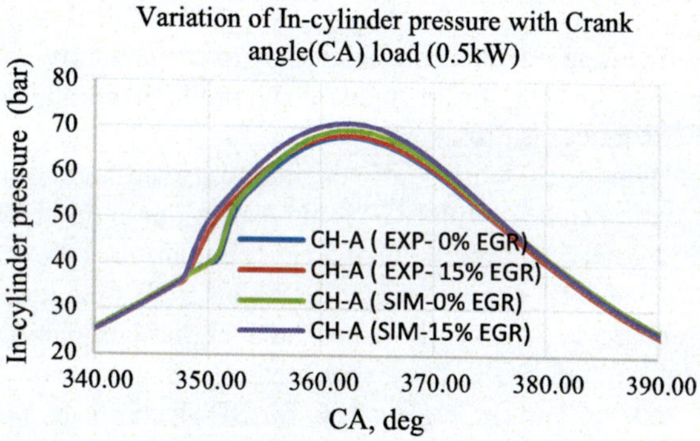

cycle simulation. Pressure variation of modified combustion chamber and standard combustion chamber follows similar form of pressure rise at all load conditions.

It is noted from Fig. 32 that simulated and experimental value of In-cylinder pressure is higher at 15% EGR condition compared to that in 0% EGR mode. This may be due to complete combustion of fuel at high inlet air temperature. Simulated result of in-cylinder pressure varied from 0.8% to 1.2% which is higher compared to experimental value. It can also be seen from Fig. 32 in 15% EGR mode the

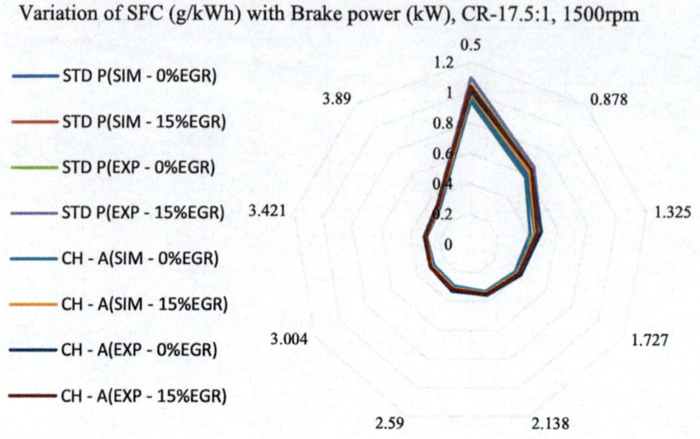

Fig. 33 Variation of SFC with BP at standard piston and modified chamber-A with and without EGR

combustion occurs at a 2 to 5° before the 0% EGR mode and also P_{max} occurs at a different crank angle after top dead center (TDC).

4.2 *Validation of SFC with BP Comparison with and Without EGR*

Simulation results are validated by experimentally namely SFC with BP as shown in Fig. 33. The SFC of standard and modified combustion chamber with and without EGR modes is discussed. The SFC for with 15% EGR is marginally higher than that of without EGR mode of standard and modified combustion chamber at low load conditions. SFC with EGR is almost similar to that of without EGR for diesel fuel at high load condition. SFC of STDP simulation is 3.6% to 10% lower compared with experimental results with and without EGR operation. SFC of experimental value of chamber-A is nearly 0.5 to 2.8% higher compared to simulation result and with and without EGR. The experimental and simulation result of STDP and CH-A is compared and the higher SFC for SDTP might be result of improper air and fuel blending, which leads to poor combustion.

4.3 *Validation of BTE with BP Comparison with and Without EGR*

Figure 34 shows the variation of brake thermal efficiency (BTE) for STDP and CH-A with respect to brake power. The BTE of STDP simulation result is higher when

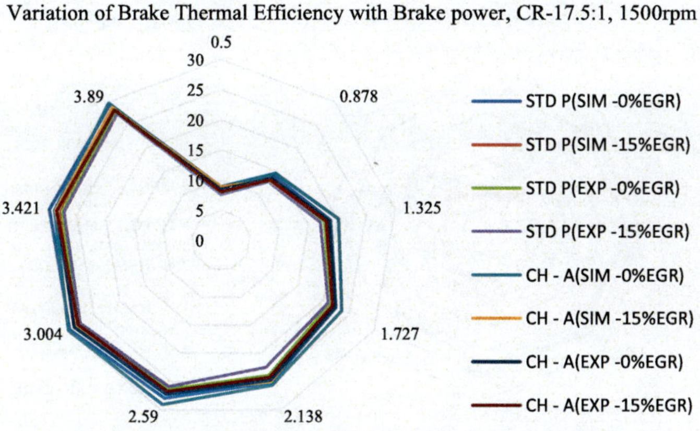

Fig. 34 Variation of SFC with BP at STDP and CH-A with and without EGR

compared to that of the experimental result at all load conditions with and without EGR. The BTE of simulation and experimental result increases for increase in BP with EGR, but at lower load it will decrease. This may be attributed to re-burning of hydrocarbon that enters during the recirculation of exhaust gases and at full load condition the BTE is nearby the same. BTE of STDP simulation is 5 to 9% higher compared experimental results with and without EGR. BTE of CH-A (EXP 15% EGR) is nearly 2 to 8% lower compared to CH-A (SIM-15%EGR). BTE of CH-A (EXP 0% EGR) is nearly 0.5% lower compared to CH-A (SIM-0%EGR). The experimental and simulation result of STDP and CH-A, the BTE for CH-A is higher compared to SDTP this may be due to of better air motion in CH-A.

4.4 Validation of NO_X with BP Comparison with and Without EGR

Figure 35 shows the variation of NO_X plotted against BP. The standard piston (STDP) simulation results are well in agreement with experimental results. The result shows that oxides of nitrogen increases with increase in the brake power due to high in-cylinder combustion temperature and also due to the excess amount of oxygen contained in combustion chamber. NO_X of CH-A (EXP 0% EGR) is nearly 1.4 to 1.7% lower compared to STDP (EXP-0%EGR). NO_X of CH-A (EXP 15% EGR) is nearly 15 to 25% lower compared to STDP (EXP-15%EGR). When the experimental and simulation result of STDP and CH-A was compared, the oxides of nitrogen for CH-A is lower to SDTP as a result of better air motion and shorter ignition delay in CH-A.

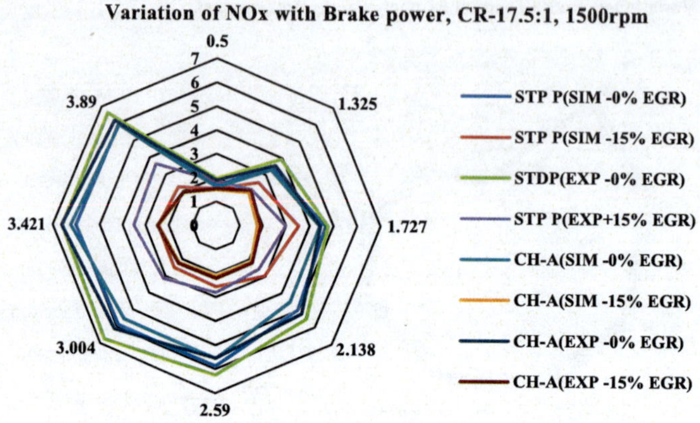

Fig. 35 Variation of NO_X with BP at STDP and CH-A with and without EGR

5 Conclusion

In this experimental investigation, the performance and emission characteristics of three modified combustion chamber are investigated. The following conclusions have been drawn from the simulation and experimental result.

- This study concludes a desirable combustion chamber among three modified chambers with respect to injection pressure 600 bar to 1400 bar, with and without EGR mode. Based on the simulation result, the SFC of chamber-B is better than the other chambers A and C at 600 bar injection pressure without EGR. NO_X of chamber-B is better than other chamber-A and C at 600 bar injection pressure with 15 EGR. PM and ID of chamber-A is better than other chambers B and C at 1400 bar injection pressure with EGR mode. Finally, it is concluded from the simulation results that the chamber-A at 1000 injection pressure with EGR mode has a better performance and emission characteristics than other combustion chambers.
- SFC of STDP simulation is 3.6% to 10% lower compared to experimental results with and without EGR operation. SFC of experimental value of CH-A is nearly 0.5–2.8% higher compared to simulation results at with and without EGR.
- BTE of CH-A (EXP 15% EGR) is nearly 2–8% lower compared to CH-A (SIM-15%EGR). BTE of CH-A (EXP 0% EGR) is nearly 4–7% lower compared to CH-A (SIM-0%EGR).
- NO_X of CH-A (EXP 0% EGR) is nearly 1.4–1.7% lower than STDP (EXP-0%EGR). NO_X of CH-A (EXP 15% EGR) is nearly 15–25% lower to STDP (EXP-15% EGR).
- The NO_X emission decreased by 22–60% in CH-A, when 15% exhaust gas recirculation was used.

- For better performance and emission reduction, the injection pressure, better bowl geometry and exhaust gas recirculation should be optimized for diesel fueled CI engine.

References

1. Bakar RA, Semin AR, Ali I (2010) Computational simulation of fuel nozzle multi holes geometry effect on diesel engine performance using GT-Power. J Appl Sci 5(2):110–116 (USA)
2. Lee BH, Song JH, Chang YJ, Jeon CH (2010) Effect of number of fuel injection holes on characteristics of combustion and emission in a diesel engine. Int J Auto Tech 11(6):783–791
3. Agarwal AK, Dhar A, Srivastava DK, Mayura RK, Singh AP (2013) Effect of fuel injection pressure on diesel particulate size and number distribution in CRDI single cylinder research engine. Fuel 107:84–89
4. Hwang J, Qi D, Jung Y, Bae C (2014) Effect of injection parameter on the combustion and emission characteristics in a common rail direct injection diesel engine fuelled with waste cooking oil biodiesel. Renew Energy 63:9–17
5. Rohit S, Singh RR, Vishwakarma SK (2013) Experimental study on effect of fuel injection nozzle hole on direct injection diesel engine. J Mech Civ Eng 7(4):67–74
6. Wang X, Huang Z, Zaung W, Kuti OA, Nishida K (2013) Effect of ultra-high injection pressure and micro hole nozzle on flame structure and soot formation of impinging diesel spray. Appl Energy 88(5):541–548
7. Rakopoulos CD, Antonopoulos KA, Rakopoulos DC, Giakoumis EG (2006) Study of combustion in a divided chamber turbocharged diesel engine by experimental heat release analysis in its chambers. Appl Thermal Eng 26(4):1611–1620
8. Kuleshov A, Grekhov L (2013) Multidimensional optimization of DI diesel engine process using multi-zone fuel spray combustion model and detailed chemistry NO_x formation model. SAE Technical Paper 2013-01-0882
9. Kuleshov AS (2009) Multi-zone DI diesel spray combustion model for thermodynamic simulation of engine with PCCI and high EGR level. SAE Int J Engines 2(1):1811–1834
10. Kuleshov AS, Kozlov AV, Mahkamov K (2010) Self-Ignition delay Prediction in PCCI direct injection diesel engines using multi-zone spray combustion model and detailed chemistry. SAE Technical Paper 1:1960
11. Kuleshov AS, Kuleshov A, Mahkamov K, Kozlov A, Fadeev Y (2014) Simulation of dual-fuel diesel combustion with multi-zone fuel spray combustion model. In: ASME 2014 internal combustion engine division fall technical conference ICEF2014-5700, pp 19–22
12. Kuleshov AS, Mahkamov K (2008) Multi-zone diesel fuel spray combustion model for the simulation of a diesel engine running on biofuel. Proc Mech Eng J Power Energy 222:309–321
13. Kuleshov AS (2006) Use of Multi-zone DI diesel spray combustion model for simulation and optimization of performance and emissions of engines with multiple injection. SAE Technical Paper, 2006-01-1385
14. Kuleshov AS (2005) Model for predicting air-fuel mixing, combustion and emissions in DI diesel engines over whole operating range. SAE Paper No. 2005-01-2119
15. Graham T, Readerb J, Hawley G, Zhenga M (2004) Diesel engine exhaust gas recirculation—a review on advanced and novel concepts. Energy Convers Manage 45:883–900

Experimental Studies on the Effect of Varying Rates of Part-Cooled EGR in High Pressure Loop on an MPFI Engine Under Variable Speed Operation

Libin P. Oommen and G. N. Kumar

1 Introduction

Combustion is the lifeline of transportation and industries and has been fuelling the progress of mankind. Researchers around the globe are focused on making combustion a more efficient and economic process following a sharp reduction in the availability of conventional fossil fuel sources. The time-to-time updation of global emission norms has also put limitations in the design and development of combustion systems. Atmospheric pollution has increased exponentially over the decades resulting in harmful implications to both the living and non-living components of the ecosystem.

Exhaust gas recirculation is a pre-treatment technique that has been widely applied to reduce the exhaust emissions. In spark-ignition engines, EGR brings about a substantial reduction in brake-specific fuel consumption and NO_x emissions [1]. The reduction in specific fuel consumption is attributed to the reduction in pumping work and a reduction in heat loss to the cylinder walls during combustion [2]. The displacement of oxygen in the intake air by EGR reduces the flame temperatures, thereby reducing the formation of thermal NO_x [3]. If the exhaust gases are directly recirculated, the average in-cylinder charge temperature is increased and the probability of occurrence of part load misfires is significantly reduced [4]. Exhaust gas traps can promote better evaporation of fuel spray, thus ensuring complete combustion [5].

The application of EGR at part loads has the potential of decreasing throttling losses but can increase the COV of IMEP, thereby reducing the stability of combustion [6, 7]. Cooled EGR is found to be effective in suppressing knock during combustion [8]. The dissociation of CO_2 to form free radicals during the recirculation is assumed

L. P. Oommen (✉) · G. N. Kumar
IC Engine Research Laboratory, National Institute of Technology Karnataka, Surathkal 575025, India
e-mail: libinpanavelil@gmail.com

A. K. Gupta et al. (eds.), *Advances in IC Engines and Combustion Technology*, Lecture Notes in Mechanical Engineering,
https://doi.org/10.1007/978-981-15-5996-9_37

to cause a marginal increase in thermal efficiency in the case of cooled EGR [9]. The reduction in laminar flame speed caused by the charge dilution prolongs the duration of combustion resulting in reduced cylinder pressures and lower end gas temperatures [10].

Fabio et al. [11] proposed that EGR improves the knock safety margin by effectively decreasing the in-cylinder peak temperatures. As lower temperatures are developed inside the engine cylinder due to the dilution caused by exhaust gases, the octane requirement is also decreased [12]. The peak pressures inside the cylinder will also be reduced at full load conditions due to the recirculation of exhaust gases [13]. Although significant reduction can be attained in NO_x emissions, the levels of HC emission increase with increased flow rates of cooled EGR [14].

2 Experimental Set-up and Methodology

The test rig consists of a Maruthi Zen made four-cylinder four-stroke MPFI engine loaded by an eddy current dynamometer. The signals incurred from various sensors are ported to the computer through NI-based USB 6210 data acquisition system. A combustion analysis software named IC Engine Soft (V 9.0) is utilized for online combustion evaluation. The specifications of the test engine are provided in Table 1.

The exhaust manifold is fitted with pipes to channel the flow of exhaust gases back to the intake manifold. The exhaust gases are partially cooled using an intercooler, and a measured mass is admitted to the inlet through an EGR valve mechanism. The remaining part of the exhaust is emitted out of the tail pipe from where its quantitative and qualitative analysis is made using an AVL-made five-gas analyser unit. The schematic representation of engine test rig is shown in Fig. 1. The EGR flow rate is calculated using the equation

$$\%\text{EGR} = \frac{\text{Volume of EGR}}{\text{total intake charge into the cylinder}} \times 100$$

Table 1 Specifications of test engine

Description	Specifications
Type	Four stroke
Cylinders	04
Bore	72 mm
Stroke	61 mm
Power	44.5 kW @ 6000 rpm
Torque	78.5 Nm @ 4500 rpm
Fuel feed	MPFI

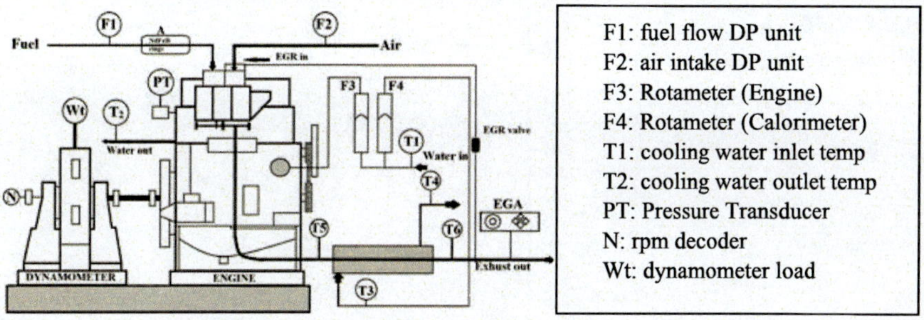

Fig. 1 Schematic diagram of experimental set-up

Experiments are conducted at four loads and four different speeds of the engine. Four different flow rates of part-cooled EGR are studied (0%, 12%, 18% and 24%) under these varying load and speed conditions to study the effect produced on the performance and emission characteristics.

3 Results and Discussion

The performance characteristics of the engine are best represented by the brake power developed, brake-specific fuel consumption and brake thermal efficiencies. The baseline operation of the engine is recorded without any addition of EGR. The flow rate of EGR is then incremented to 12%, 18% and 24%, and the characteristics are plotted against load for the four different speeds of the engine. The variation in brake power with load is shown in Figs. 2, 3, 4 and 5.

Fig. 2 Variation of BP at 3500 rpm

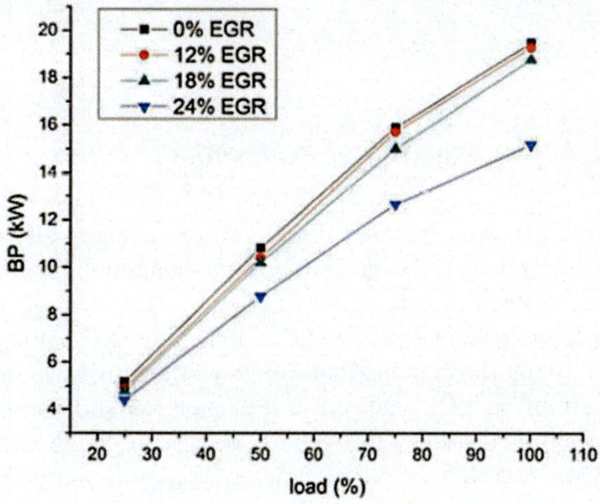

Fig. 3 Variation of BP at 3000 rpm

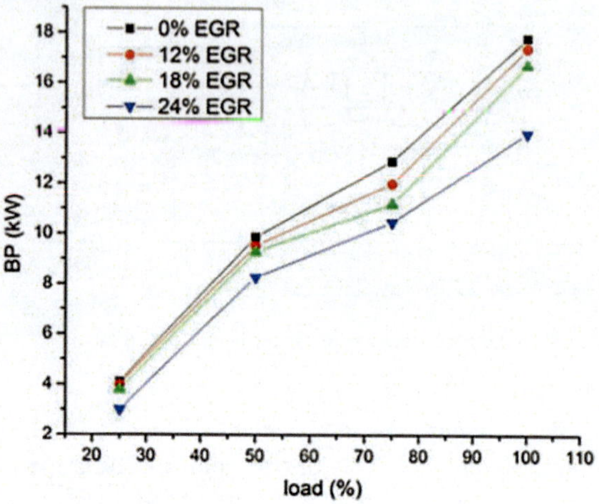

Fig. 4 Variation of BP at 2500 rpm

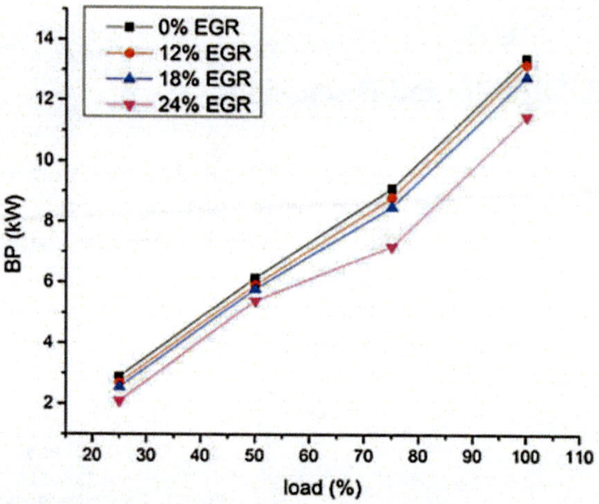

3.1 *Comparison of Performance Characteristics*

A drop in the developed brake power is noted for all four speeds and load conditions of the engine. The drop in power is minimum at lower loads and increases with increment in loading and EGR flow rates. This deterioration is attributed to the dilution effect caused by the recirculation of exhaust gases. The brake-specific fuel consumption is found to decrease with the first two stages of EGR compared to the case of no EGR. But the BSFC characteristic faces a downfall as the EGR flow rate is increased to 24%. This is because of the fact that EGR reduces the throttling losses and eliminates fuel-enriched operation up to a limit after which the combustion characteristics deteriorate. The optimum reduction in fuel consumption is obtained

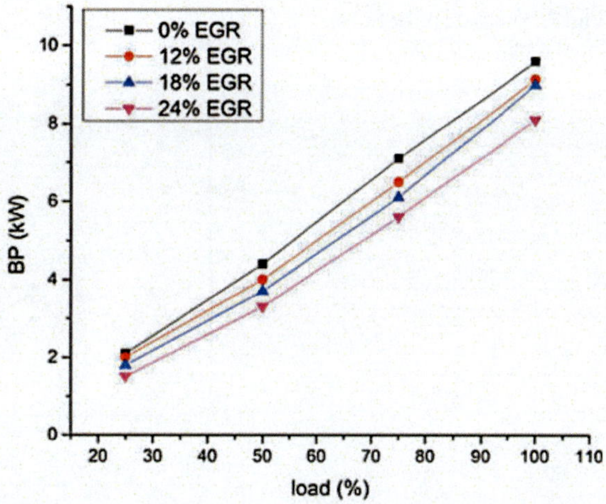

Fig. 5 Variation of BP at 3000 rpm

at 18% EGR where the heat transfer losses are minimum. The BSFC characteristics of the test engine for different flow rates of EGR are shown in Figs. 6, 7, 8 and 9.

As part-cooled EGR is employed in the engine, the laminar flame speed decreases because of the charge dilution. The flame temperature is also reduced owing to the increased heat capacity of diluents gases, thereby increasing the precombustion reaction time and reducing the octane requirement of the engine. The allocation of EGR also permits an optimized spark advancing, thus yielding better thermal efficiencies despite of the extended combustion durations [12]. The thermal efficiency curves for various EGR flow rates are illustrated in Figs. 10, 11, 12 and 13. Up to the optimum flow rate of part-cooled EGR, a slight improvement can be observed in thermal efficiencies due to improved fuel economy. It can be seen that the application of excessive amount of EGR deteriorates the performance of the engine.

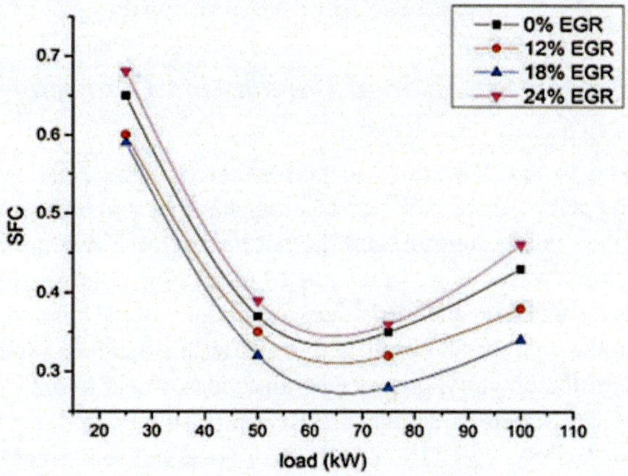

Fig. 6 Variation of BSFC at 3500 rpm

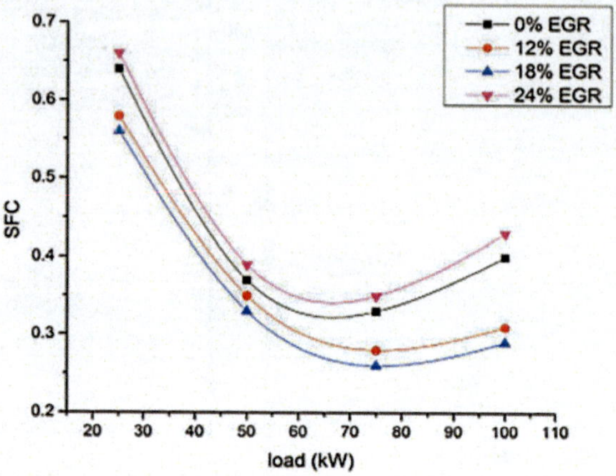

Fig. 7 Variation of BSFC at 3000 rpm

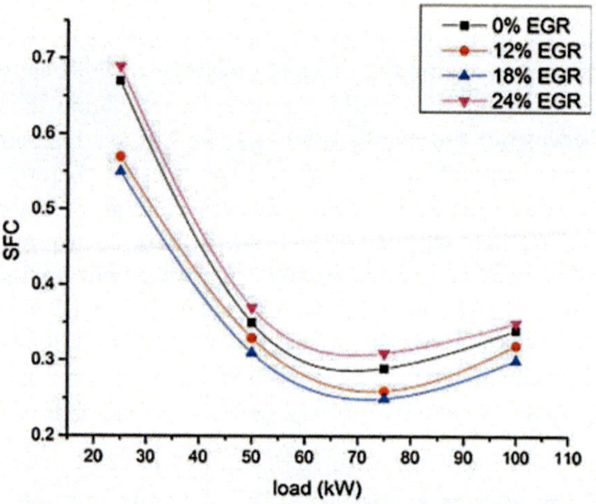

Fig. 8 Variation of BSFC at 2500 rpm

3.2 Comparison of Combustion Characteristics

The visible variations in performance characteristics occur due to the variations happening inside the combustion chamber under the tested conditions. The pressure crank angle diagram and the net heat release rate are amongst the most significant parameters that are indicative of the engine combustion. The changes in in-cylinder pressures happen mostly due to cylinder volume change, combustion and heat transfer to the chamber walls. The in-cylinder pressures are monitored using a piezoelectric transducer which phases with an accuracy near to 0.2°. Figure 14 shows the variation in developed pressures at different positions of the crank angle with varying rates of part-cooled EGR. It is clearly observed that there is a reduction in in-cylinder pressures of the engine with each stage of EGR addition. There is a shift of peak

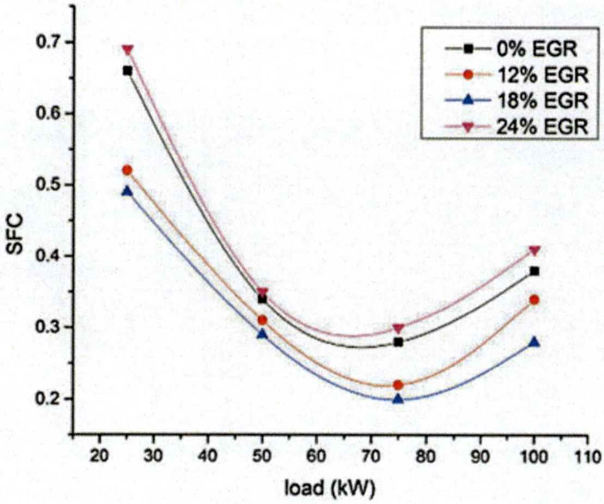

Fig. 9 Variation of BSFC at 2000 rpm

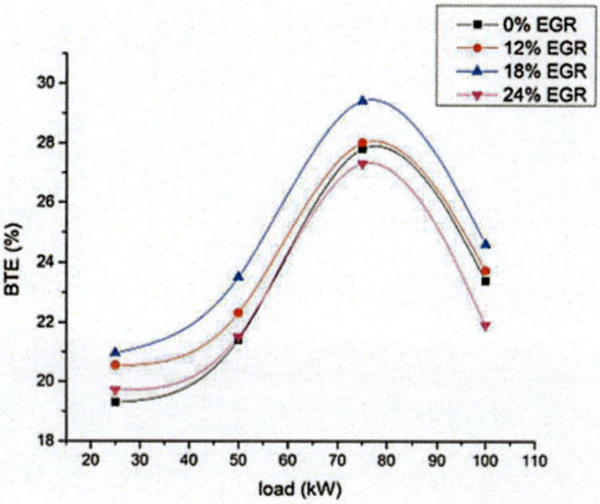

Fig. 10 Variation of BTE at 3500 rpm

points of pressure curves further away from TDC position at all stages of recirculation experimented.

The presence of burned gas in the combustion chamber due to the recirculation causes a reduction in laminar burning velocity and adiabatic flame temperature because of the reduction in heating value per unit mass of mixture [15]. Also, the combustion temperatures vary inversely with burned gas fraction due to the dilution effect. Thus, the magnitude of heat release reduces with EGR as represented in Fig. 15.

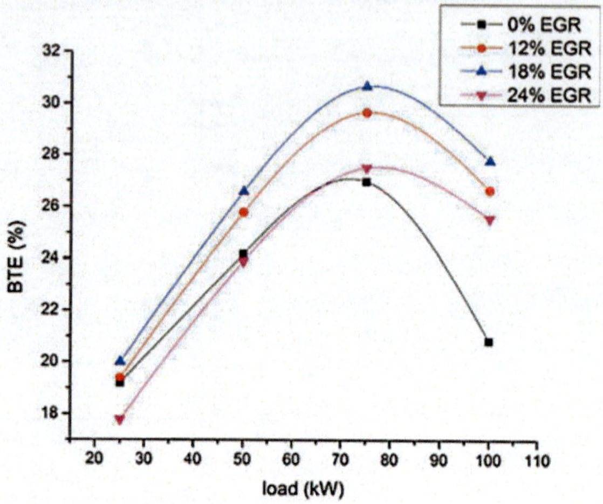

Fig. 11 Variation of BTE at 3000 rpm

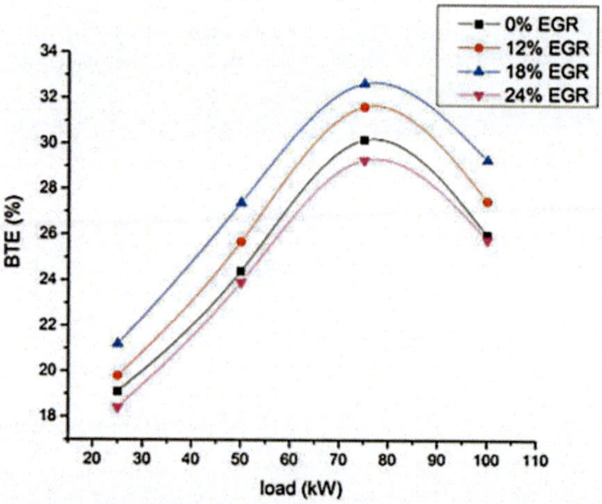

Fig. 12 Variation of BTE at 2500 rpm

3.3 Comparison of Emission Characteristics

It can be observed that the emission of carbon monoxide slightly reduces with the application of EGR up to 18% flow rate, and thereafter a sharp increase is observed because of excessive dilution of charge and prolonged combustion durations that may lead to incomplete combustion of the fuel. The emission trend of carbon monoxide under different EGR rates is represented in Figs. 16, 17, 18 and 19.

The variation of hydrocarbon emissions with varying EGR rates is illustrated from Figs. 20, 21, 22 and 23. It is observed that the emission of unburnt hydrocarbons increases with every stage of EGR addition as it is dependent on cycle temperatures.

Fig. 13 Variation of BTE at 2000 rpm

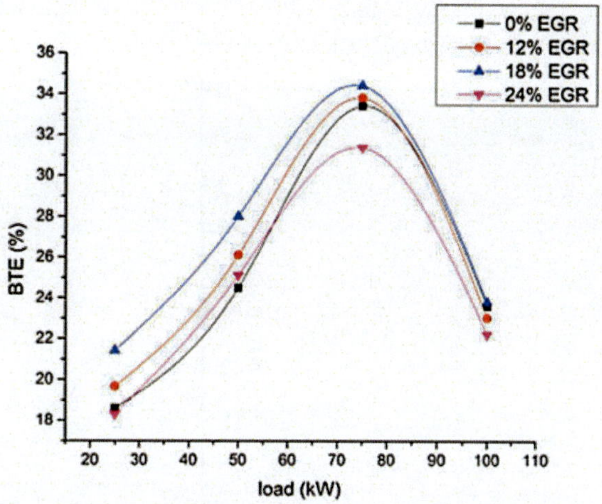

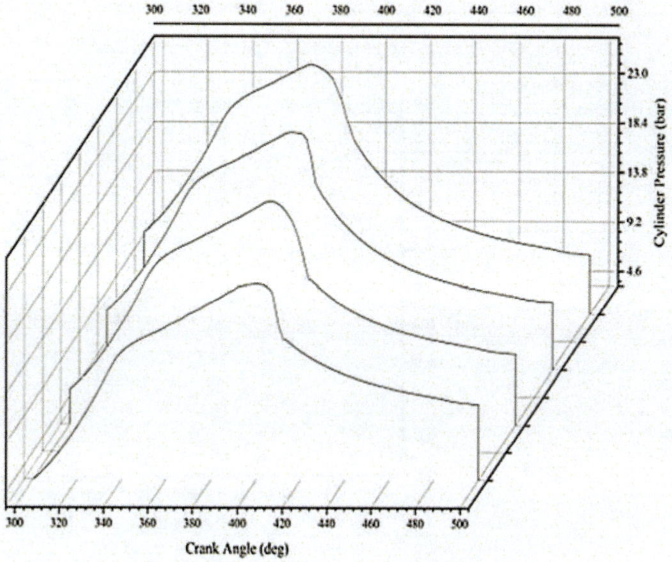

Fig. 14 Variation in in-cylinder pressures at different flow rates of EGR

A rapid rise in HC emission occurs beyond 20% EGR which indicates the occurrence of misfires and deteriorated combustion.

The most significant advantage of cooled EGR addition is the reduction it yields in NO_x emissions. The cooled exhaust gases when recirculated reduce the in-cylinder temperature which proportionally reduces the formation of thermal NO_x. As the EGR percentage increases, the availability of oxygen in the combustion chamber is reduced. The recirculated gases cause an increase in heat capacity of non-reacting

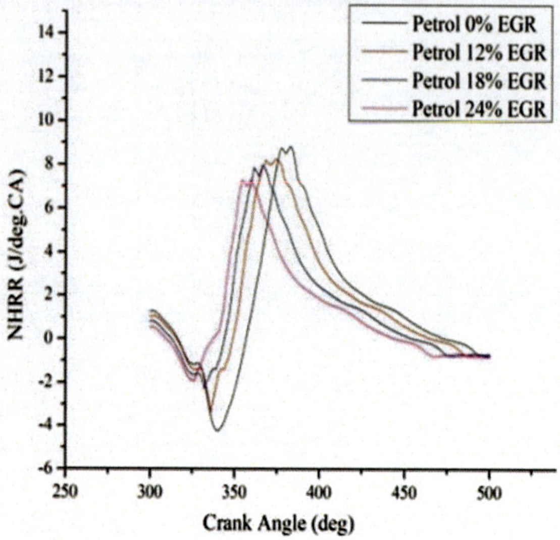

Fig. 15 Variation in NHRR at different flow rates of EGR

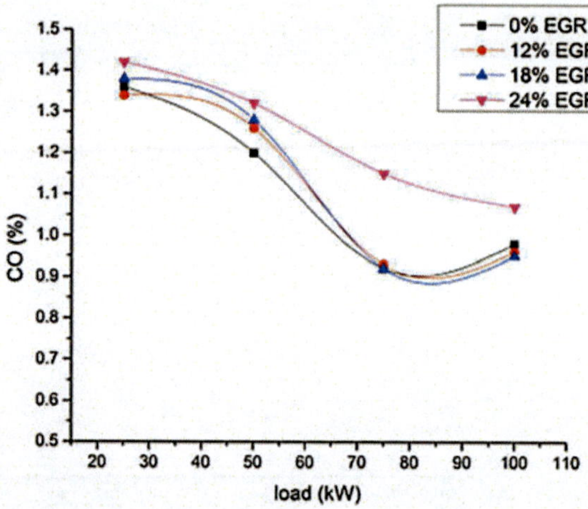

Fig. 16 Variation in CO emissions at 3500 rpm

matter which in turn results in reduced peak combustion temperatures [3]. It can be observed that a substantial reduction is obtained in the emission of NO_x for every flow rates of EGR at all engine speeds. The NO_x emission characteristics of the test engine under various EGR flow rates are shown in Figs. 24, 25, 26 and 27.

Another significant factor to be taken into consideration during exhaust gas recirculation is the effect it produces on combustion stability. The recirculated gases slow down the combustion rate and thus tend to make combustion unstable. Although it has been demonstrated in the prior works that 15–30 percentage of recirculation could be tolerated by the faster burning engines, we observed that the combustion stability

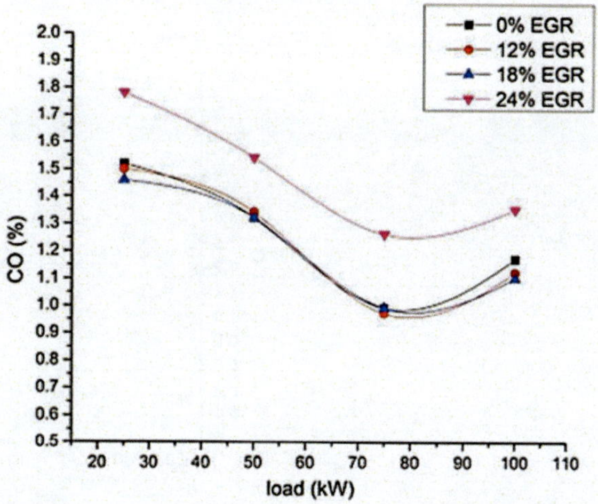

Fig. 17 Variation in CO emissions at 3000 rpm

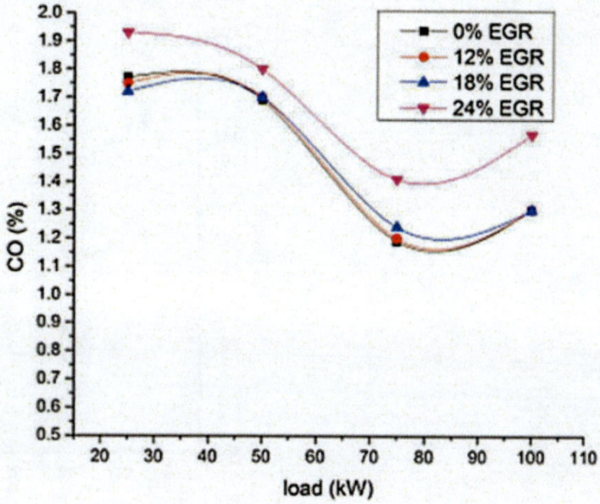

Fig. 18 Variation in CO emissions at 2500 rpm

steadily reduced in all the cases after 18% EGR. For 24% of part-cooled EGR, the coefficient of variation of IMEP was found to exceed 10% that is considered as the limiting point of combustion variability [16].

4 Conclusion

The effect of varying rates of cooled exhaust gas recirculation on the various performance combustion and emission characteristics of a multi-cylinder MPFI engine is studied under different load conditions and multiple engine speeds. Although a

Fig. 19 Variation in CO emissions at 2000 rpm

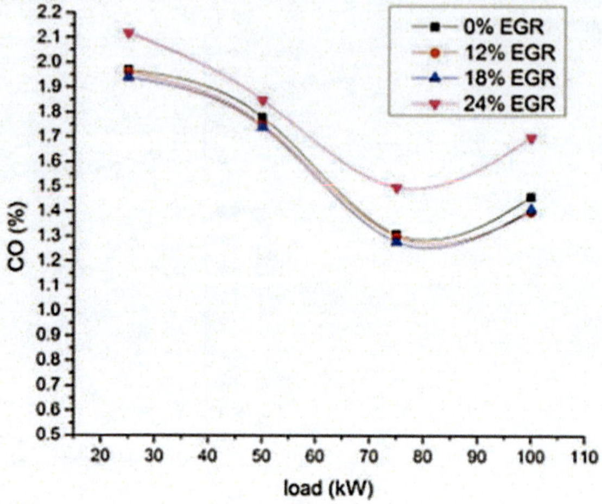

Fig. 20 Variation in HC emissions at 3500 rpm

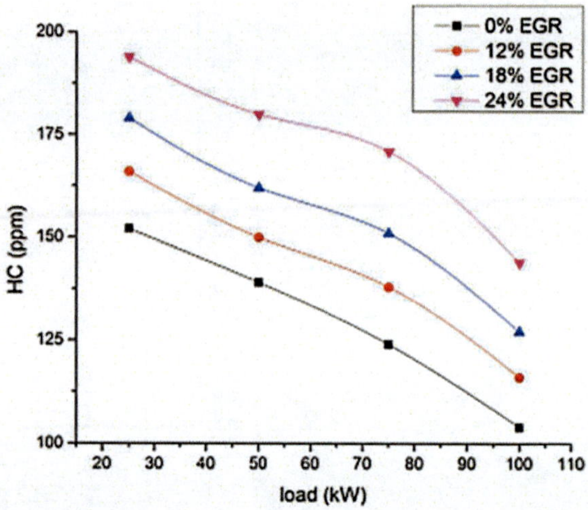

drop in obtained torque is observed with increased flow rates of recirculated gases, the thermal efficiency of the engine is found to be enhanced with increased fuel economy. The recirculated exhaust gases allow a much leaner operation, leading to fuel savings and reduced cylinder temperatures. The formation of thermal NO_x is hindered to a great extend in the expense of increased hydrocarbon emissions for all flow rates of part-cooled EGR. The optimum level of EGR addition is found to be 18% beyond which the combustion quality deteriorates because of the higher cyclic variations-induced triggering misfires. Through the recirculation of optimum flow rate of part-cooled EGR, the emission of NO_x reduced up to 8% and the fuel economy improved by 11.2%

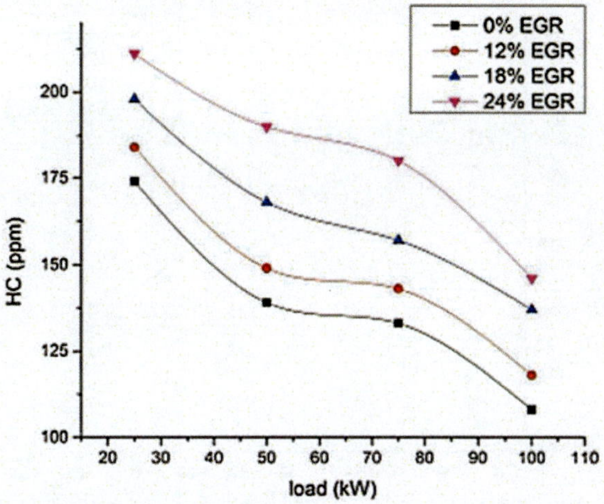

Fig. 21 Variation in HC emissions at 3000 rpm

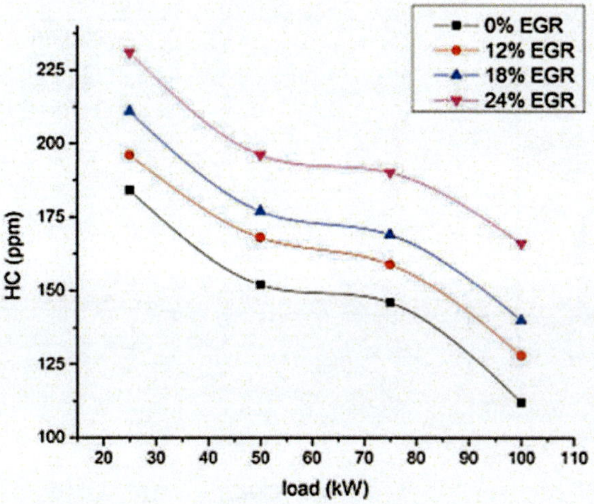

Fig. 22 Variation in HC emissions at 2500 rpm

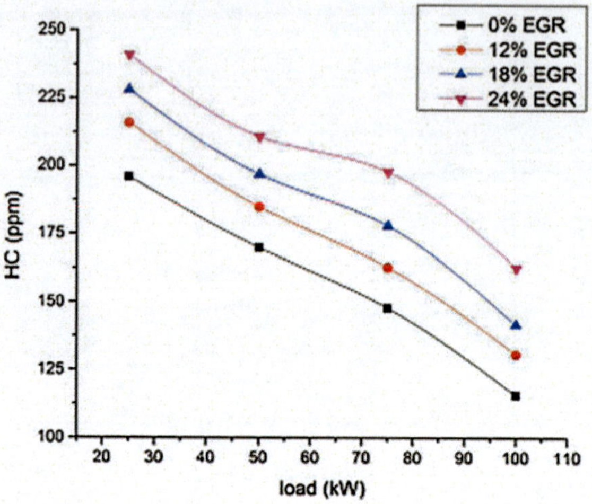

Fig. 23 Variation in HC emissions at 2000 rpm

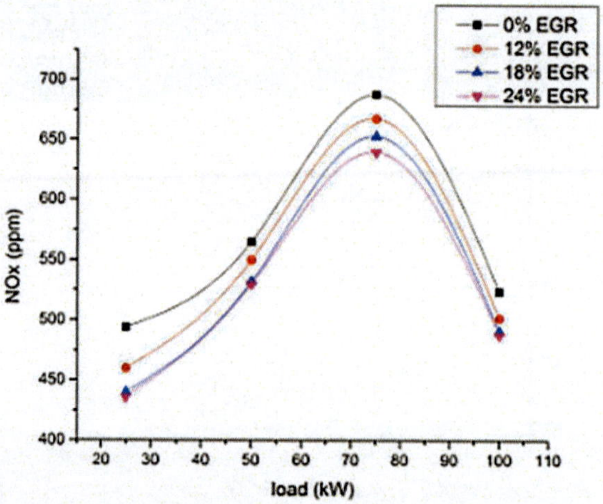

Fig. 24 Variation in NO_x emissions at 3500 rpm

Fig. 25 Variation in NO_x emissions at 3000 rpm

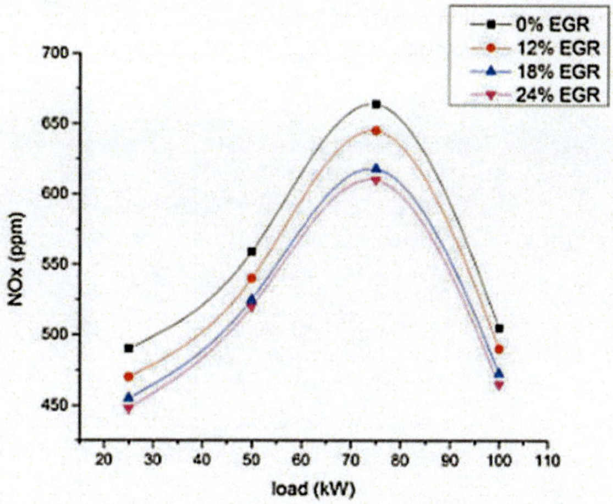

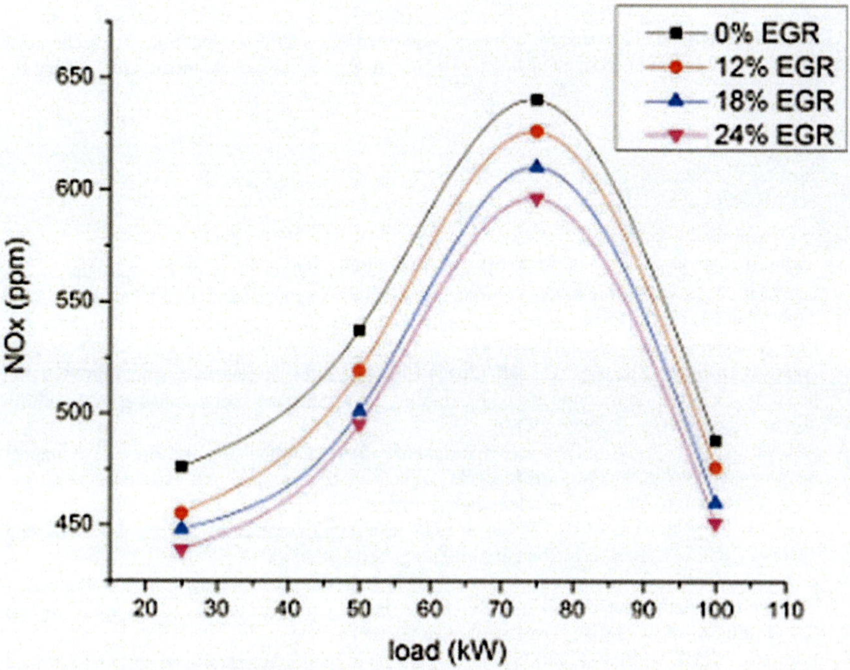

Fig. 26 Variation in NO_x emissions at 2500 rpm

Fig. 27 Variation in NO_x emissions at 2000 rpm

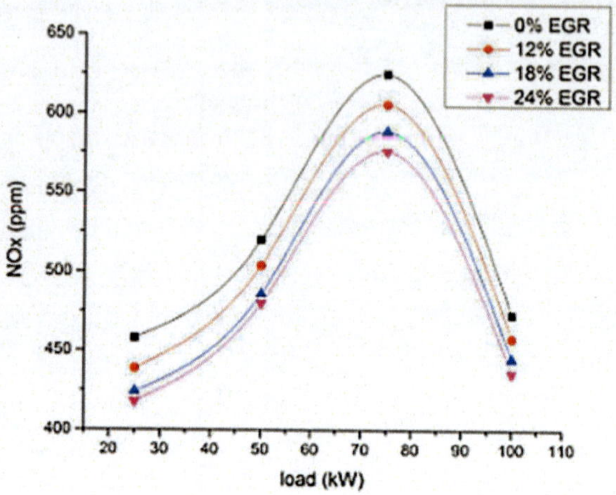

Acknowledgements This research work is supported by National Institute of Technology Karnataka, Surathkal, funded by the Ministry of Human Resources Development, Govt. of India.

References

1. Agrawal AK, Singh SK, Sinha S, Shukla MK (2004) Effect of EGR on the exhaust gas temperature and exhaust opacity in compression ignition engines. Sadhana 29(Part 3):275–284
2. Abd-Allah GH (2001) Using exhaust gas recirculation in internal combustion engines: a review. Energy Convers Manage 43:1027–1042
3. Agarwal D, Singh SK, Agarwal AK (2011) Effect of Exhaust Gas Recirculation (EGR) on performance, emissions, deposits and durability of a constant speed compression ignition engine. Appl Energy 88:2900–2907
4. Andwari AM, Pesyridis A, Esfahanian V, Said MF (2019) Combustion and emission enhancement of a spark ignition two-stroke cycle engine utilizing internal and external exhaust gas recirculation approach at low-load operation. Energy
5. Amritkar AB, Badge N (2016) Effect of EGR in internal combustion engine. Int Res J Eng Technol (IRJET) 03(03). e-ISSN: 2395 -0056. p-ISSN: 2395-0072
6. Bozic M, Vuceticl A, Kozarac D, Lulic Z (2017) Experimental investigation on influence of EGR on combustion performance in SI Engine. In: Digital proceedings of the 8th European Combustion Meeting, 18–21, Dubrovnik, Croatia-1821
7. Cairns A, Blaxill H, Irlam G (2006) Exhaust gas recirculation for improved part and full load fuel economy in a turbocharged gasoline engine. MAHLE Powertrain Ltd., 2006-01-0047
8. Chaichan MT (2016) Effect of cool and hot Exhaust Gas Recirculation (EGR) on the performance of multi-cylinder compression ignition engine fueled with blends of diesel and methanol. Al-Nahrain Univ College Eng J (NUCEJ) 91(1):76–85
9. Shojaeefard MH, Tahani M, Etghani MM, Akbari M (2013) Cooled EGR for a turbo charged SI engine to reduce knocking and fuel consumption. Int J Autom Eng 3(3)
10. Serio DD, Oliveira AD, Sodre JS (2017) Effects of EGR rate on performance and emissions of a diesel power generator fueled by B7. J Braz Soc Mech Sci Eng 39:1919–1927
11. Galloni E, Fontana G (2010) Experimental analysis of a spark-ignition engine using exhaust gas recycle at WOT operation. Appl Energy 87:2187–2193

12. Rakopoulos CD, Rakopoulos DC, Mavropoulos GC, Kosmadakis GM (2018) Investigating the EGR rate and temperature impact on diesel engine combustion and emissions under various injection timings and loads by comprehensive two-zone modelling. Energy 157
13. Park S, Cha J, Kwon J, Cho Y (2001) The effect of Exhaust Gas Recirculation (EGR) on combustion stability, engine performance and exhaust emissions in a gasoline engine. KSME Int J 15(10):1442–1450
14. Ravi K, Bhasker JP, Alexander J, Porpatham E (2017) Influence of cooled exhaust gas recirculation on performance, emissions and combustion characteristics of LPG fuelled lean burn SI engine. IOP Conf Ser Mater Sci Eng 263
15. Heywood JB (1988) Internal combustion engine fundamentals. McGraw Hill International editions, pp 836–838
16. Han S (2001) Investigation of cyclic variations of IMEP under idling operation in spark ignition engines. KSME Int J 15:81–87

Investigation of Reverse Flow Slinger Combustor with Methanol

Pooja Nema, Abhishek Dubey, and Abhijit Kushari

1 Introduction

Increasing environmental concerns due to global warming and imminent depletion of fossil fuels has motivated research toward renewable energy and efficient combustion systems in recent decades and several sustainable energy solutions have been proposed. One of the more promising alternatives to fossil fuels is methanol economy [1].

Methanol is a clean and storable liquid fuel that can be produced from variety of sources including biomass, agricultural products, and industrial waste. It can also be prepared from direct oxidation of natural gas or reduction of atmospheric carbon dioxide with hydrogen [1]. Carbon dioxide present in the exhaust products of the power plants can then be converted back to methanol, essentially resulting in zero carbon footprint [2]. The production of methanol from the atmospheric carbon dioxide presents a remarkable solution for carbon capture and reduction of CO_2 content in the atmosphere. Thus, combustion of methanol for power generation provides an environmentally friendly and sustainable solution for the increasing energy demands of the world.

Methanol is an attractive fuel for stationary land-based gas turbines because of its clean-burning nature, easy storage and transportation, and abundant availability. It can achieve improved efficiency and higher power output due to the higher mass flow rate and lower NO_x emissions due to low flame temperature [3]. Higher laminar flame speed of the methanol compared to most hydrocarbon fuels results in increased thermal efficiency [4]. However, combustion of methanol in gas turbines imposes certain technical difficulties due to its physical and chemical properties such as

P. Nema · A. Dubey (✉) · A. Kushari
Indian Institute of Technology Kanpur, Kanpur, India
e-mail: abhishek.ic1101@gmail.com

A. K. Gupta et al. (eds.), *Advances in IC Engines and Combustion Technology*,
Lecture Notes in Mechanical Engineering,
https://doi.org/10.1007/978-981-15-5996-9_38

low heating value, low lubricity, and low flash point [3]. This necessitates certain modifications in the gas turbine combustors for desired operation with methanol.

The present study employs a rotary atomizer for methanol injection in the combustion chamber. Fuel atomization and distribution are extremely crucial for the performance and operability of the gas turbine engine. Sub-optimal fuel injection can lead to non-uniform fuel–air mixtures, lower combustion efficiency, combustion instability, higher exhaust emissions, and ignition delay. Multiple studies have been conducted to study the atomization characteristics of rotary atomizers. Liquid atomization performance in different slinger geometries are dependent on operating conditions, Weber and Ohnesorge numbers, liquid–gas density ratio, and liquid–gas viscosity ratio [5]. Ignition and combustion characteristics of a gas turbine slinger combustor are function of injector rotational speed. Ignition limit and combustion efficiency increase with the increase in rotational speed. Mean droplet size decreases with increase in injector rotational speed up to a point after which it becomes constant [6].

Reverse flow slinger combustor was recently developed for potential application in aviation gas turbines [7]. It is an annular combustor with a rotary fuel injection system in which combustion air and exhaust products enter and leave the combustion chamber from the same end creating a reverse flow configuration. Such flow configuration creates a region of recirculation characterized by low axial velocity and high turbulence [8]. Flame is stabilized at this stagnation zone without the need for a swirler or a bluff body. The combustor was developed as a light weight and compact module by utilizing a rotary atomizer for fuel injection. This method of fuel injection makes the engine lighter by eliminating the need for a heavy pressure pump, which is required in conventional pressure and airblast atomizers. In the investigation of combustor performance with Jet-A [7], the results demonstrated its ability to operate at very low fuel–air ratios of up to 0.01 in a sustained manner. NO_x and CO emissions were found to be less than 10 ppm and 1500 ppm, respectively. Stable operation of combustor at very low flow rate was attributed to the preheating of the inlet air with the exhaust products similar to a counter flow heat exchanger.

The current work investigates the performance and emission characteristics of reverse flow slinger combustor with methanol at atmospheric pressure. The ability of combustor to sustain combustion at very low heat release rate or lean conditions provides motivation for evaluation of combustor characteristics with methanol as the fuel. Unique design features of the combustor such as internal preheating and the uniform fuel distribution from rotary atomizer highlight its prospects for application in land-based gas turbine engines with clean and renewable fuel, methanol.

2 Experimental Setup

2.1 Combustor Design

The design of the combustor is shown in Fig. 1. The combustor has a design air flow rate of 0.02 kg/s that corresponds to combustor inlet velocity of 40 m/s. In a gas turbine combustor, the high velocity of air after the compressor exit (~ 150 m/s) has to be reduced by diffuser before its entry in combustion chamber to effectively stabilize the flame and minimize the pressure losses [9]. Thus, typical values of combustor inlet velocity lie in the range of 30–60 m/s [9].

Hot combustion products leave the combustor through the outermost duct of inner diameter of 60 mm and length 135 mm. The fuel injector is rotated by a spindle motor capable of achieving 24,000 RPM. A variable frequency drive (VFD) is used to control and monitor the injector RPM.

Photograph of the combustor studied in the current study is shown in Fig. 2. The combustion chamber is a transparent quartz tube of inner diameter 60 mm and length 85 mm. The fuel injector at the center of the combustor is a hollow mild steel shaft of length 120 mm supported on metal-shielded bearings. Inner and outer diameters of injector are 4 mm and 12 mm, respectively, with two fuel injection holes of diameter 0.4 mm along the radial direction. Reactants and products enter and leave the combustor through a set of three concentric ducts. The stationary innermost duct of 5 mm diameter is used to transfer fuel to the rotating injector through a mechanical seal. The middle duct of inner diameter 35 mm and length 165 mm is for the air inlet.

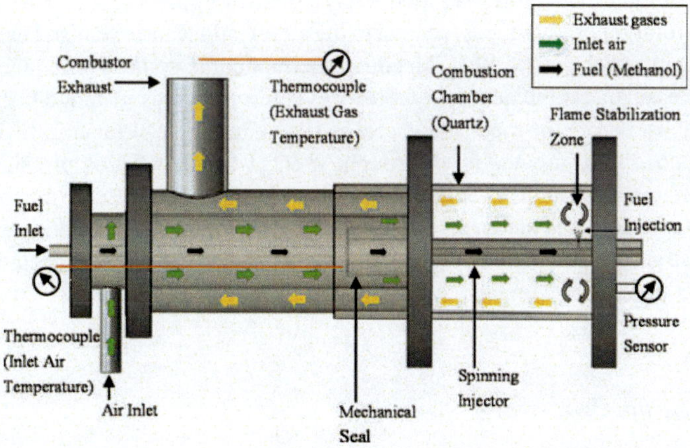

Fig. 1 Combustor design

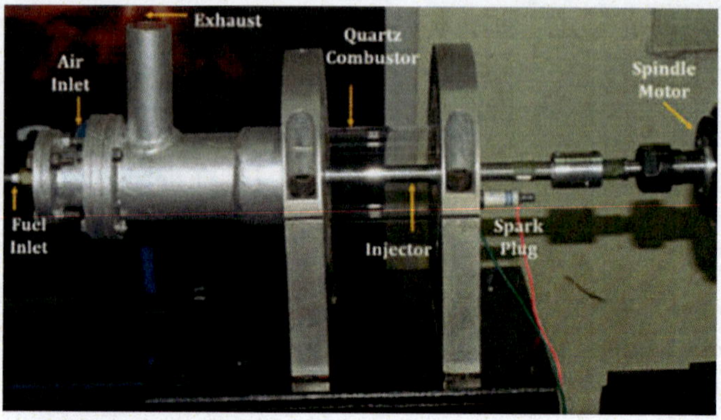

Fig. 2 Photograph of the realized combustor

2.2 Instrumentation

Combustion air is supplied through an electric compressor and is stored in a tank pressurized at 10 bar. Air is passed through moisture separator to remove water content. An air rotameter (60–600 LPM) of ± 2% accuracy is used to measure the flow rate of air. The fuel for the current study is methanol with lower calorific value of 20 MJ/kg. It is supplied through a tank pressurized at around 2 bar. As the fuel is injected by utilizing the centrifugal energy of the spinning injector, a very low supply pressure is needed for the fuel injection. The fuel flow rate is measured using a liquid rotameter of range 15–150 LPM. Two K-type thermocouples with ± 2 °C accuracy were used for temperature measurement in the air inlet duct and exhaust. Due to space constraints inside the inner most inlet duct, thermocouple could not be placed at the point where air enters the quartz combustor. Therefore, it was installed horizontally in the duct just before the mechanical seal as shown in Fig. 1. Temperature data from these thermocouples are acquired through NI-9211 DAQ. A Huba pressure sensor (1 bar), connected through NI-9205 DAQ, is used for chamber pressure measurement. Lancom 3 portable gas analyzer was used to measure the species concentration in the exhaust gases. Samples of gases from the exhaust duct were collected using a built-in gas pump and were analyzed by electrochemical sensors.

2.3 Ignition Procedure

For the study of combustor with Jet-A [7], LPG was premixed with the inlet air and the combustible mixture was ignited by spark plug. This was followed by Jet-A injection in the chamber. LPG supply was cut off once sustained combustion with liquid Jet-A was achieved. In the current study, similar procedure for methanol ignition was attempted but due to its high volatility, reliable and controlled ignition was not achieved. Therefore, a different method for methanol ignition was implemented that utilized its high volatility. Liquid methanol was injected through the spinning injector in the presence of the desired airflow and it was externally heated by a pilot flame creating a combustible mixture of vaporized methanol and air in the combustion chamber. This mixture was then ignited by the spark plug.

3 Results and Discussion

Subsequent to the studies performed with Jet-A [7], certain essential modifications related to mechanical seal and bearings were performed in the combustor to adapt it for methanol. After this, the combustor was able to achieve sustained combustion with methanol. Photograph of methanol combustion is shown in Fig. 3.

Figure 4 shows the variation of chamber pressure and temperature during combustion. A slight increase in the chamber pressure was observed during the cold flow and the combustion process. This is similar to the previous study of the combustor with Jet-A [7]. Ignition of fuel is identified by rise in exhaust gas temperature (EGT) as shown in Fig. 4. Due to restricted placement of the thermocouple in the inlet air duct, true value of the air temperature after internal preheating could not be obtained. However, an average temperature rise of 15 K was obtained in the inlet air with the thermocouple placed at the location shown in Fig. 1. Small fluctuations in the chamber pressure were observed during the sustained methanol combustion. The combustor was able to perform stable combustion with methanol at very lean fuel–air ratio (FAR) of 0.022. As combustion at lean conditions is susceptible to instabilities and flashback due to low heat release rate, sustained operation of methanol at such low fuel–air ratio highlights the importance of internal preheating and uniform flame stabilization facilitated by the combustor design. Experiments were performed at FAR of 0.022 at different injector rotational speeds (RPM).

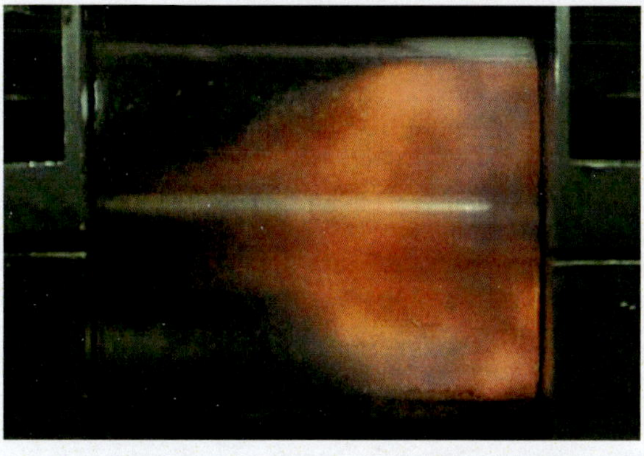

Fig. 3 Photograph of methanol combustion (FAR = 0.02, RPM = 5000)

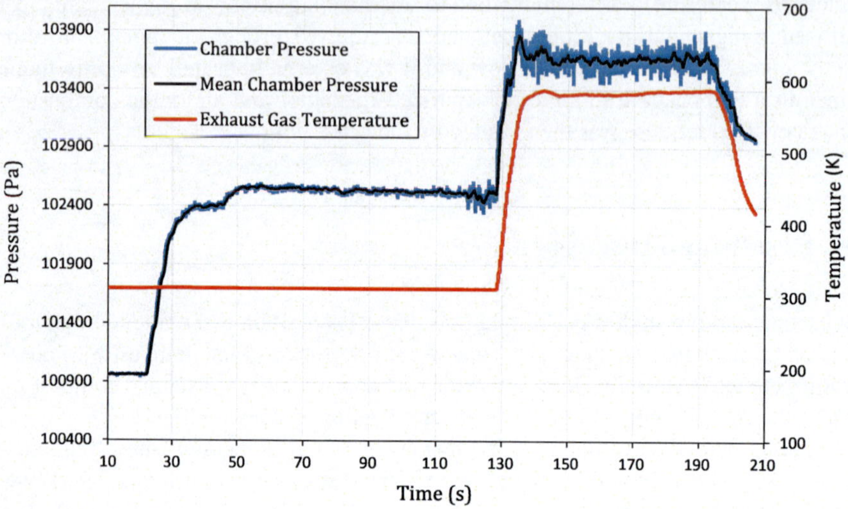

Fig. 4 Variation of parameters during the combustion process (FAR = 0.022, RPM = 5000)

Emission concentrations of NO_2, CO, CO_2, and unburnt hydrocarbons (C_xH_x) at the combustor exit are shown in Fig. 5. CO and NO_2 emissions obtained were less than 1500 ppm and 2 ppm, respectively. CO emissions are higher as the combustor is operating near the lean blowout (LBO) region. FAR of 0.022 employed in the current study corresponds to global equivalence ratio of 0.33 which is near the LBO of the combustor with methanol. Lower combustion temperature achieved due to low heat release at this condition inhibits the oxidation of CO into CO_2 resulting in higher CO concentration and hence, sudden rise in CO emissions near in the lean combustion region is considered to be an indication of LBO [8]. CO emissions obtained in the current study are higher than those obtained by [8] due to lower LBO of 0.33 as compared to 0.45. Although, both the studies employed a reverse flow configuration with flame stabilization at the stagnation zone, the lean blowout in the current study with methanol is remarkably lower than that obtained by [8]. This can be attributed to internal preheating of the inlet air by exhaust gases. NO concentrations obtained were almost close to zero. Uniform fuel–air mixture created by methanol's higher volatility reduces local hot spots that inhibit the oxidation of N_2 resulting in reduced NOx emissions. NO_x formation is also inhibited by low combustion temperature due to combustion at extremely lean conditions. Very small concentrations of UHC, less than 0.5% (by vol), were obtained which was also confirmed by insignificant soot deposition on transparent quartz.

Combustor or burner efficiency (η_b) was calculated by the known enthalpy difference at the inlet and the exit of the combustor. Exit enthalpy was calculated by estimating the specific heat at the exit by the exhaust species concentration obtained from emission analyzer. Figure 6a shows the combustor efficiency for different injector rotational speeds at FAR of 0.022. An average efficiency of 55% was achieved

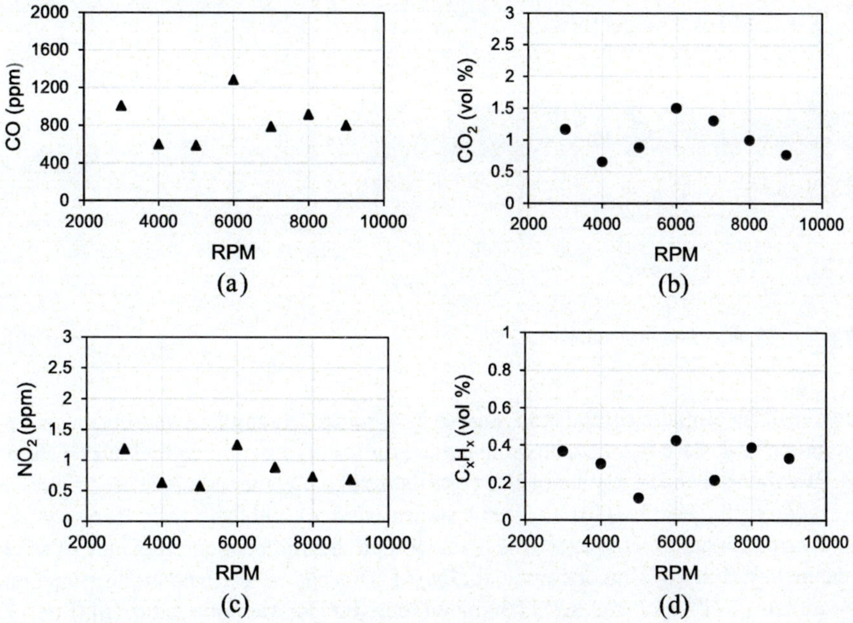

Fig. 5 Concentration of various exhaust emissions: **a** CO **b** CO_2 **c** NO_2 **d** C_xH_x

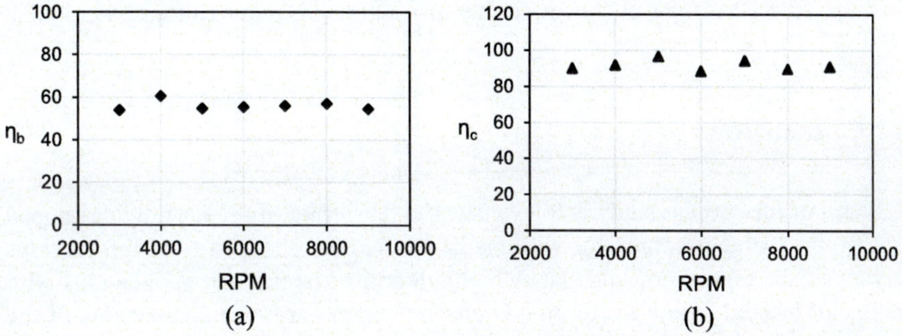

Fig. 6 **a** Burner efficiency, **b** Combustion efficiency

at different RPMs. Due to the restriction on injector rotational speed because of technical limitations, very high quality of atomization could not be achieved in the chamber, resulting in poor mixing of fuel and air. However, it is emphasized that the quality of the atomization and fuel–air mixing is expected to increase with increase in injector rotational speed in slinger combustor [6]. Burner efficiency obtained for the combustor with methanol is around 30% higher than that obtained with Jet-A [7]. This higher efficiency of methanol as compared to Jet-A is because of its lower

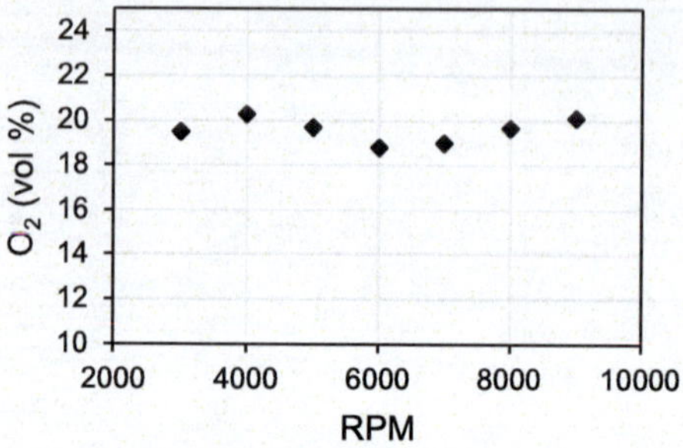

Fig. 7 O_2 concentration at combustor exit

degree of incomplete combustion. Due to its significantly higher volatility, evaporation time scales are reduced, creating a more uniformly distributed fuel–air mixture. Higher flame speed of methanol also contributes to the higher combustion efficiency by reducing the heat loss [4]. Figure 6b shows average combustion efficiency (η_C) of 92% with methanol. However, this is lower than the combustion efficiency of 99% obtained in study of slinger combustor by [6]. This can be attributed to low injector rotational speed of 11,000 RPM compared to 30,000 RPM employed in [6] (Fig. 7).

Due to narrow range of injector RPM employed in the current study, a distinct trend of the measured parameters with the injector RPM was not obtained. Thus, it can be inferred that there is no significant change in droplet SMD with the increase in this injector RPM range. However, further investigation is required for its validation.

4 Conclusion

The aim of the present study is to evaluate the feasibility of the recently developed reverse flow slinger combustor with methanol. Results obtained from the previous studies of the combustor with Jet-A highlighted its viability for applications with methanol in land-based gas turbine engines. Certain hardware components of the combustor were modified to adapt it for methanol because of its different physical properties. Eventually, sustained methanol combustion was achieved at a FAR of 0.022 with low exhaust emissions.

Preliminary performance investigation of the combustor has shown its potential for application in practical land-based gas turbines with methanol. However, for the complete feasibility assessment, it is recommended to investigate its performance at a much wider range of operating conditions and chamber pressures.

References

1. Olah GA (2005) Beyond oil and gas: the methanol economy. Angewandte Chemie Int Edn 44(18):2636–2639. https://doi.org/10.1002/anie.200462121
2. Olah GA, Goeppert A, Prakash GS (2008) Chemical recycling of carbon dioxide to methanol and dimethyl ether: from greenhouse gas to renewable, environmentally carbon neutral fuels and synthetic hydrocarbons. J Organic Chem 74(2):487–498. https://doi.org/10.1021/jo801260f
3. General Electric (2001) GE position paper: feasibility of methanol as gas turbine fuel
4. Celik MB, Özdalyan B, Alkan F (2011) The use of pure methanol as fuel at high compression ratio in a single cylinder gasoline engine. Fuel 90(4):1591–1598. https://doi.org/10.1016/j.fuel.2010.10.035
5. Dahm WJA, Patel PR, Lerg BH (2002) Visualization and fundamental analysis of liquid atomization by fuel slingers in small gas turbine engines. In: 32nd AIAA fluid dynamics conference and exhibit
6. Choi S, Lee D, Park J (2008) Ignition and combustion characteristics of the gas turbine slinger combustor. J Mech Sci Technol 22:538–544. https://doi.org/10.1007/s12206-007-1106-6
7. Dubey A, Kushari A (2017) Investigation of reverse flow slinger combustor. In: 11th international high energy materials conference & exhibits
8. Crane J, Neumeier Y, Jagoda J, Seitzman J, Zinn BT (2006) Stagnation-point reverse-flow combustor performance with liquid fuel injection. In: ASME Turbo Expo 2006: power for land, sea, and air, pp 927–936. American Society of Mechanical Engineers. https://doi.org/10.1115/gt2006-91338
9. Cohen H, Rogers GF, Saravanamuttoo HI (1996) Gas turbine theory, 4th edn. Padstow, Cornwall, UK, pp 237, 244

Experimental Analysis of Performance and Emissions of a Diesel Engine Fueled with Diesel–Water Emulsions

V. Abhinay, S. V. S. S. R. Krishna P, T. Karthikeya Sharma, and G. Ambaprasad Rao

1 Introduction

Air pollution caused by automobiles is currently a grave issue. Reduction of the polluting emissions from the vehicle engines without compromising engine performance is a challenge. Major reasons for the unwanted emissions from the engines are incomplete combustion and high peak temperatures in the engine cylinder. It should be noted that the chemical equilibrium constant for the formation of NO_x is highly dependent upon temperature. When the temperature is in the range of 2500–3000 K inside the combustion chamber of the diesel engine, a significant amount of Nitrogen atoms is generated. This leads to the formation of nitrogen oxide at very high temperatures inside the engine. Diesel–water emulsion is a capable fuel that can reduce emissions by enhancing the combustion of the fuel and reduce peak temperature inside the engine cylinder. It acts as engine coolant and needs no modification in the engine.

Four major techniques of introducing water into diesel engine have been reported [1–5]: direct injection into the engine through separate injectors, hybrid injection by the usage of single injector, fumigating the water into the engine intake air and diesel–water emulsion. The preparation of diesel–water emulsion is not easy as they are immiscible fluids and at the same time, this emulsion stability is a concern. Baskar and Senthil Kumar [6] have worked on diesel–water emulsion with oxygen enriched air and its performance characteristics. Ithnina and Yahyaa [7] along with

V. Abhinay
NIT W, Warangal, Telangana 506004, India

S. V. S. S. R. Krishna P (✉) · T. Karthikeya Sharma · G. Ambaprasad Rao
NIT AP, Tadepalligudem, AP 534101, India
e-mail: paruchuri.nitw@gmail.com

G. Ambaprasad Rao
e-mail: ambaprasadrao@gmail.com

A. K. Gupta et al. (eds.), *Advances in IC Engines and Combustion Technology*, Lecture Notes in Mechanical Engineering,
https://doi.org/10.1007/978-981-15-5996-9_39

other scholars have made a report on emulsifier-free diesel–water emulsion and its stability behavior. Griffin et al. [8] and Kothencz et al. [9] gave detailed information about HLB values of various surfactants. Rastogi et al. [10] studied experimentally Span 80 and Tween 80 mixtures for diesel–water emulsification and concluded that such surfactant mixtures can form chemically stable emulsified fuels. Seifi et al. [11] used statistical techniques to analyze diesel–water emulsifier fuels and found that fuel properties are enhanced and emission characteristics were improved when employed emulsified fuels when compared to conventional diesel fuel. Rai and Sahoo [12] studied water in diesel emulsions using mathematical techniques and predicted that net power density and efficiency of the engine effectively increase while total heat loss decreases for 5% water in diesel and 10% water in diesel fuels with an increase in compression ratio. It was further suggested by the researchers [13] that by using charged nanoparticles for preparing diesel–water emulsions, their stability can be further enhanced by using micro-emulsions which have thermodynamic stability and can form very small droplets of the order 10 nm when used for emulsification. Two surfactants with hydrophilic–lipophilic Balance (HLB) values nearly 4 and 16 are required for meeting the need. Sorbitan monoleate (Span 80) and Polysorbate 20 (Tween 20) are selected for this work. Two diesel–water emulsions are tested and it was experimentally proven that the emulsified diesel fuel not only capable of reducing NO_x emissions but it can also enhance the brake thermal efficiency of the engine.

2 Experimentation

Components used for making emulsion are diesel, distilled water, surfactants (Span 80 and Tween 20), measuring jar, burette, pipette, and mechanical agitator. Diesel, water, and surfactants were taken in required proportions and blended for 40–50 min in agitator and the obtained emulsion is checked for stability. The variations for the blends after rejecting unstable emulsions are used for experimentation. Conventional diesel fuel, diesel–water emulsions "92% Diesel, 5.5% Water, 1% Tween 20, 1.5% Span 80" and "90% Diesel 7.5% Water, 1% Tween 20, 1.5% Span 80" are used as fuels one after the other in Kirloskar made single-cylinder four-stroke direct injection diesel engine coupled with an eddy current dynamometer. The engine has 3.5 kW rated power with constant 1500 rpm. Auto exhaust multi-gas analyzer is used for measuring emissions. The engine is connected to computer to collect combustion and performance data. The engine was run at no-load, 2.46 kg, 4.92 kg, 7.4 kg, 9.8 kg, and full load (12.3 kg).

Figures 1 and 2 show diesel engine used for the experimentation (Thermal Engineering Laboratory, NIT Warangal, Telangana, India) and emulsion sample preparation setup used for preparing sample 1 and sample 2 discussed in this section. It was trial and error method that is followed in order to finalize samples that run the engine satisfactorily for at least half an hour during the testing.

Fig. 1 Diesel engine used for experiment

Fig. 2 Sample preparation setup

Table 1 Engine specifications

Name of the parameter	Value of the parameter
Name of the model	TV1- KIRLOSKAR
Engine type	Four stroke, Constant speed, Water-cooled diesel engine
Loading device	Eddy current dynamometer
Number of cylinders	1
Rated power	3.5 kW
Swept volume	661 cc
Speed	1500 rpm
Cylinder bore	87.5 mm
Stroke length	110 mm
Connecting rod length	234 mm
Compression ratio	17.5:1

Table 1 shows the detailed specifications of the engine used in this experimentation.

3 Results and Discussions

Diesel, emulsion 1 (92% Diesel, 5.5% Water, 1% Tween 20, and 1.5% Span 80), and emulsion 2 (90% Diesel 7.5% Water, 1% Tween 20, and 1.5% Span 80) were used as fuel in the experimentation and tested for emissions and performance of the diesel engine. The data gathered from IC-Engine-Soft and exhaust gas analyzer was tabulated and studied.

It is observed From Fig. 3 that the performance of the engine is not significantly reduced when water emulsions are introduced and in fact, brake thermal efficiency of the engine slightly improved when it is fueled with emulsion 1. Emulsion 2 having

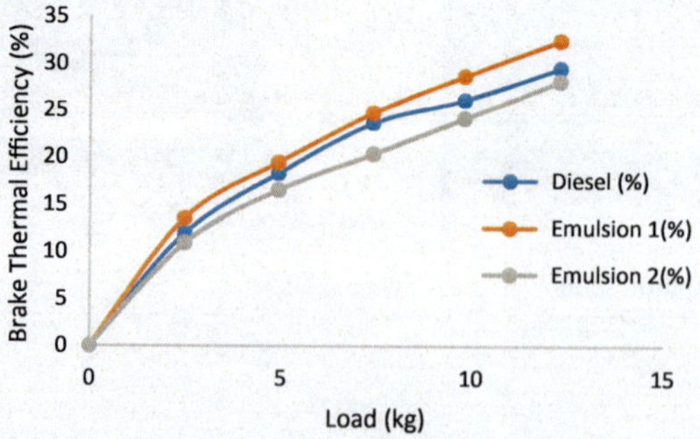

Fig. 3 Brake thermal efficiency compared among the three distinct fuels

higher water content has slightly lower brake thermal efficiency. Thermal efficiency increases with increase in load.

It is observed from Fig. 4 that the mean gas temperature decreases with increase in water content.

From Fig. 5, it has been observed that using diesel–water emulsion as fuel, a reduction in the amount of NO_x emissions was achieved. Emulsion 2 having greater water content compared to emulsion 1 has the lowest NO_x of the three distinct fuels used. NO_x emissions increased with increase in load due to the raise in in-cylinder temperature at higher loads.

From Fig. 6, it was identified that the CO emission decreases as the load on an engine increases. This is due to the more amount of oxygen taken in at higher loads yielding to oxidation of this CO to CO_2.

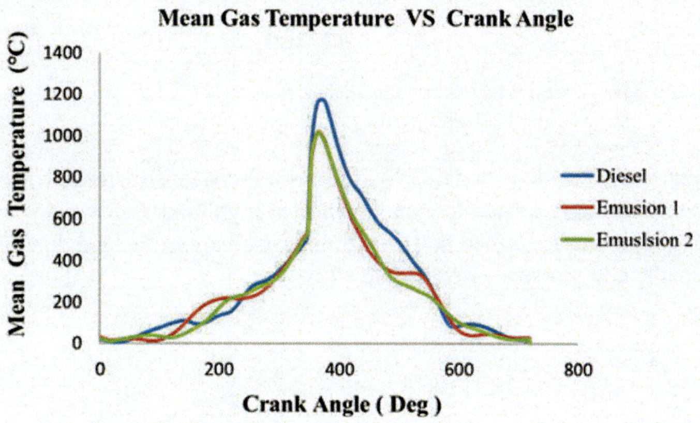

Fig. 4 Variation of mean gas temperature in engine cylinder for three distinct fuels

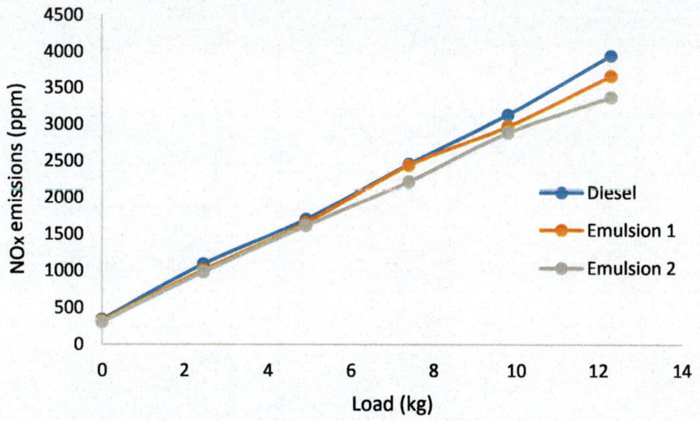

Fig. 5 NO_x emissions at various loads for three different fuels

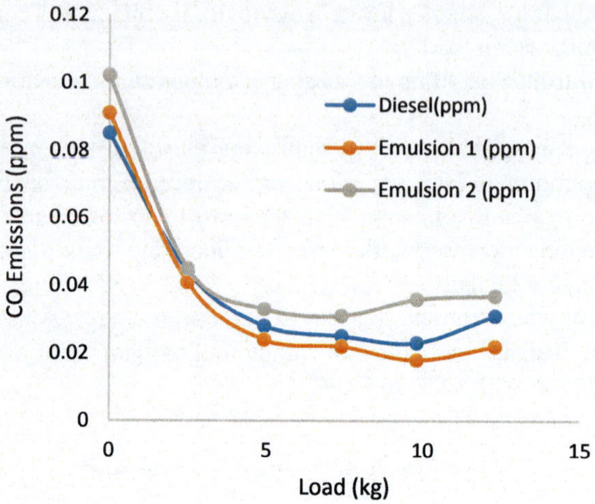

Fig. 6 CO emissions at various loads for three distinct fuels

From Fig. 7, it was identified that hydrocarbon emissions increase with increase I load, the reason being incomplete combustion at high loads.

From Fig. 8, it was identified that CO_2 emissions increase in load due to fact that at higher loads, CO is also converted into CO_2.

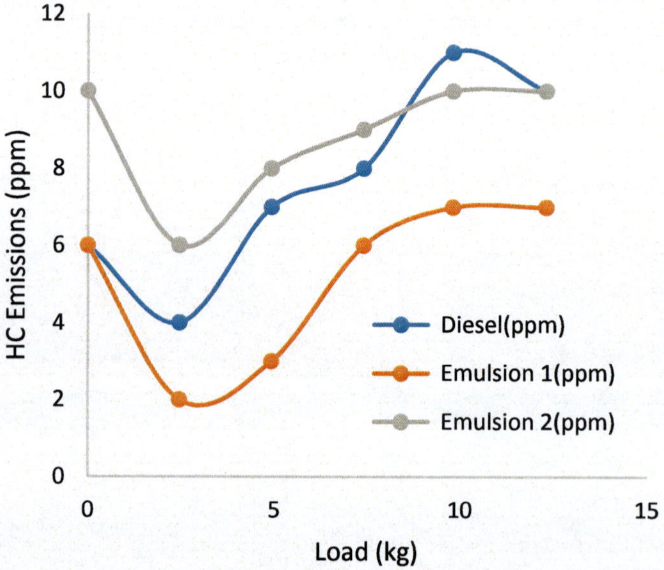

Fig. 7 HC emissions at various loads for three distinct fuels

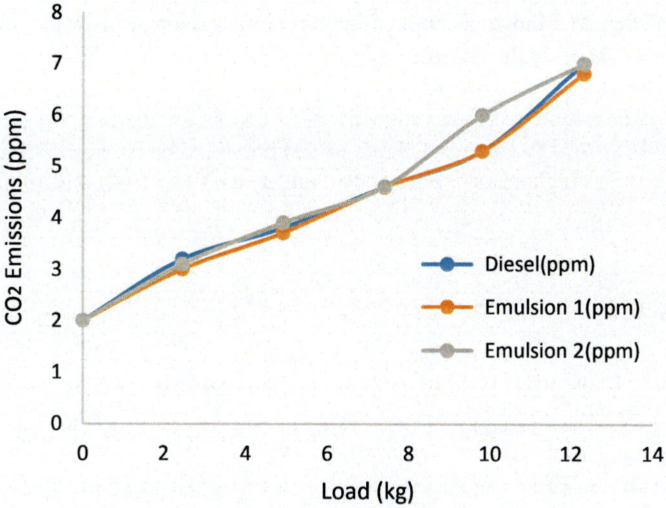

Fig. 8 CO_2 emissions at various loads for three distinct fuels

4 Conclusions

The present work used water-diesel emulsions with varying proportions of water and extensive experiments. Performance and emission characteristics are plotted. Here are the specific conclusions drawn from the study.

- The engine could run satisfactorily without any harsh noise and could take full load without difficulty.
- Nitrogen oxides emissions decreased using diesel–water emulsifier; Emulsion 1 and emulsion 2 showed a reduction in nitrogen oxides. This reduction continues as the amount of water in the emulsifier increases.
- CO emissions are lowered with increase in load on the engine for all three fuels used. By using emulsifier
- 1 (5.5% water) emissions fell more compared to that of diesel fuel.
- Hydrocarbon emissions showed a downward trend to the lowest value and a sharp increase in all three fuels used. Here, emulsifier 1 (5.5% water) showed a significant reduction in HC emissions compared to other fuels.
- CO_2 emissions showed a very close value for all three fuels with emulsifier 1 (water 5.5%) containing the largest amount of carbon dioxide and emulsifier 2 (7.5% water) with the lowest value.
- The brake thermal efficiency almost increases linearly with increase in load. Emulsion 1 (5.5% water) has the best brake thermal efficiency over the entire load range.
- The mean gas temperature inside the cylinder shows a bell-shaped trend with the peak at the center. The use of water theoretically would reduce the temperatures

inside which was shown as both emulsions recorded lower peak and average gas temperatures inside the engine.

- This reduction in mean gas temperatures is reflected in the amount of emissions as clearly the emulsions have reduced emissions and are thermally more efficient.
- For emulsifying 1 (92% diesel, 5.5% water) proved to be the best fuel with lower emulsions and higher brake efficiency coupled with low fuel consumption.

References

1. Singh RK, Murugan S (2011) Performance and emission studies in a diesel engine using bio oil-diesel blends
2. Lif A, Holmberg K (2006) Water-in-diesel emulsions and related systems. Adv Colloid Interface Sci 123:231–239
3. Badrana O, Emeishb S, Abu-Zaidc M, Abu-Rahmaa T, Al-Hasana M, Al-Ragheba M (2011) Impact of emulsified water/diesel mixture on engine performance and environment. Int J Thermal Environ Eng 3(1):1–7
4. Kannan K, Udayakumar M (2009) NOx and HC emission control using water emulsified diesel in single cylinder diesel engine. J Eng Appl Sci 4(8):59–62
5. Chadwell CJ, Dingle PJG (2008) Effect of diesel and water co-injection with real-time control on diesel engine performance and emissions. No. 2008-01-1190. SAE Technical Paper
6. Baskar P, Senthil Kumar A (2017) Experimental investigation on performance characteristics of a diesel engine using diesel-water emulsion with oxygen enriched air. Alexandria Eng J 56(1):137–146
7. Ithnin AM, Yahya WJ, Ahmad MA, Ramlan NA, Kadir HA, Sidik NAC, Koga T (2018) Emulsifier-free water-in-diesel emulsion fuel: its stability behaviour, engine performance and exhaust emission. Fuel 215:454–462
8. Griffin WC, Lynch MJ (1968) Surface active agents. Handbook of food additives, 2nd edn (Ed.: Furia, TE), 397
9. Kothencz R, Nagy R, Bartha L (2017) Determination of HLB values of some nonionic surfactants and their mixtures. Studia Universitatis Babes-Bolyai, Chemia 62
10. Preetika R, Mehta PS, Kaisare NS, Basavaraj MG (2019) Kinetic stability of surfactant stabilized water-in-diesel emulsion fuels. Fuel 236:1415–1422
11. Seifi MR, Desideri U, Ghorbani Z, Antonelli M, Frigo S, Hassan-Beygi SR, Ghobadian B (2019) Statistical evaluation of the effect of water percentage in water-diesel emulsion on the engine performance and exhaust emission parameters. Energy 180:797–806
12. Rai RK, Sahoo RR (2019) Effective power and effective power density analysis for water in diesel emulsion as fuel in diesel engine performance. Energy 180:893–902
13. Lin Q, Xu M, Cui Z, Pei X, Jiang J, Song B (2019) Structure and stabilization mechanism of diesel oil-in-water emulsions stabilized solely by either positively or negatively charged nanoparticles. Colloids Surfaces A Physicochem Eng Aspects 573:30–39

Enhancement of Hydrogen Energy Share in an Automotive Compression Ignition Engine Using EGR

Anilkumar Shere and K. A. Subramanian

1 Introduction

The diesel-fueled CI engines are generally used in heavy-duty vehicles, agricultural and power generation sectors, construction equipment, and diesel generators due to its many advantages like high performance, high fuel economy, lower fuel consumption, and low carbon emissions. However, it leads to the formation of NO_x and smoke emissions. The available technologies could not able to simultaneously reduce emissions. These problems can be addressed utilizing hydrogen in CI diesel engines under dual-fuel mode (hydrogen-diesel).

The hydrogen-fueled dual-fuel CI engines provide high thermal efficiency and combustion efficiency [1–3]. However, hydrogen-fueled diesel engines suffered the main barrier of its operation with lower hydrogen share due to knocking. The knocking phenomena lead to abnormal combustion which deteriorates the engine's performance [4]. The engine could damage with combustion with knocking.

Hydrogen fuel is a clean and green fuel as it produces water as a by-product during combustion. The hydrogen (100%) cannot be used in CI fueled engine as it is having low cetane number, but the fuel could be used through dual-fuel approach (diesel + hydrogen) in which gaseous hydrogen can be inducted into the intake system and liquid diesel can be inducted into the combustion chamber of the engine [5].

Researchers studied and reported the hydrogen utilization in CI diesel engine and examined its effect on the engine. There have been few studies based on the maximum quantity of hydrogen induction in hydrogen-diesel-fueled engines. Yang et al. reported the limitations of the hydrogen induction in six cylinders turbocharged diesel engine without exhaust gas recirculation. The results showed that with the hydrogen supply, the cylinder pressure first increased significantly and then decreased

A. Shere · K. A. Subramanian (✉)
Engines and Unconventional Fuels Laboratory, CES, IIT Delhi New Delhi, India
e-mail: subra@ces.iitd.ac.in

A. K. Gupta et al. (eds.), *Advances in IC Engines and Combustion Technology*, Lecture Notes in Mechanical Engineering,
https://doi.org/10.1007/978-981-15-5996-9_40

along with the reduction in PM emissions [6]. The single-cylinder four-stroke CI diesel engine was used to investigate the performance in the hydrogen-fueled dual-engine by Tsujimura et al. The experimental tests show that the dual-fueled engine attained higher thermal efficiency than a conventional 100% diesel-fueled engine at higher load conditions [7]. Chaichan performed the experimental investigation by adding the gaseous hydrogen in a diesel engine using hot EGR condition. The results show the reductions in NO_x, CO, and PM emissions by hydrogen substitution to EGR percentage. The results also cleared that engine noise gets reduced by substituting high level of EGR [8]. Bose et al. conducted and analyzed the experimental work on diesel engines with single-cylinder, four-stroke with a speed of 1500 rpm using hydrogen-diesel blends with an EGR system to reduce the NO_x emissions [9]. Shin et al. inspect the performance and exhaust emission characteristics of four-cylinder CI engine with EGR. The results demonstrate that hydrogen substitution increased the thermal efficiency while reduced the NO_x emission under EGR condition [10]. Sunmeet et al. investigated the experimental work on single-cylinder CI engine at 1500 rpm under CNG-diesel mode with EGR. The analysis indicates that the thermal efficiency improved slightly at with the CNG percentage share up to 37%. NO_x emission reduced drastically with EGR and the smoke emission increased significantly [11]. Chintala et al. studied and demonstrated the experimental work on single-cylinder CI engine under hydrogen-diesel mode. They reported the improvement in HES with water addition as compared to without water addition. They also reported that the NO_x emission gets reduced by 37% with water addition [12]. Pavlos et al. demonstrated the application of hydrogen fuel in a hydrogen-diesel CI engine at various operating loads. The results showed that at low load conditions, 98% of HES was achieved with a significant drop in emissions such as CO, HC, and NO_x [13].

The purpose of this paper is to show the effect of cooled EGR technique on the HES and exhaust emissions in diesel-hydrogen-fueled CI diesel engines.

2 Experimental Setup Details

The investigation was performed on a single-cylinder, four-stroke, DI automotive compression ignition engine. The dynamometer was coupled with the engine for loading purposes. The speed and torque of the engine were controlled by dyno-controller. The test rig of the test setup is presented in Fig. 1 and specifications of the test engine are listed in Table 1. An experimental test setup was developed by inducting the hydrogen fuel in the intake system with minor modification and converted it into a hydrogen-diesel mode engine. The properties of fuels are listed in Table 2. The hydrogen gas was inducted into the intake manifold whereas diesel fuel was inducted directly into the engine cylinder. The flow rate of hydrogen and diesel was recorded by using flow meters. The fuel injection timings were controlled by using Electronic Control Unit system. The flame arrestor was attached in the supply line of hydrogen fuel to avoid any type of risk due to the occurrence of backfire.

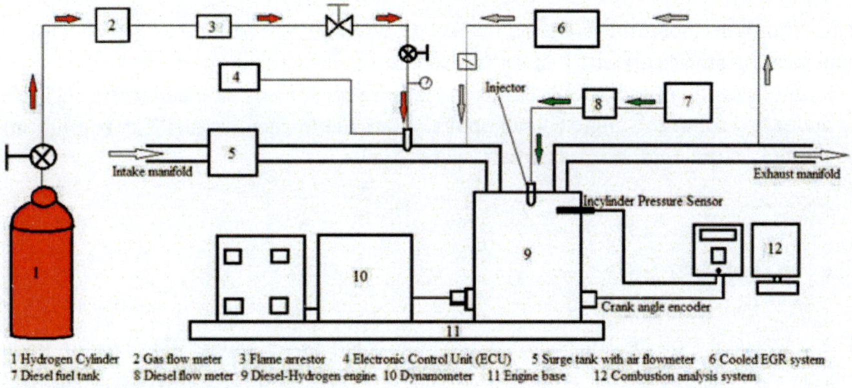

Fig. 1 Test rig of the setup

Table 1 The diesel engine's specifications

Make, model	Greaves, G435 A IV
Type of engine	Direct injection diesel engine
Cylinder	1
Strokes	4
Compression ratio	19:1
Power output (rated), hp	7.4
Speed (rated), rpm	3200
Length of connecting rod, mm	118
Displacement volume, cm^3	435
Bore and stroke, mm	86 and 75

Table 2 Properties of hydrogen and diesel fuel

Parameters	Diesel	Hydrogen
Molecular structure	$C_{16}H_{34}$	H_2
Cetane number	50	–
Calorific value, MJ/kg	43.5	120
Autoignition temperature, K	532	857
Density at 100 kPa and 15 °C, kg/m^3	850	0.082
Viscosity at 40 °C, cSt	2.87	–
Flame velocity, m/s	0.3	2.65–3.25
Flammability limits, vol%	0.7–5	4–75
Diffusivity in air, cm^2/s	0.034	0.62

The emissions including NO_x, HC, and CO were measured by using AVL Di-gas analyzer and smoke opacity meter was used for the measurement of smoke emission.

For measuring in-cylinder pressure, a pressure sensor was attached to the engine's cylinder and for crank angle (°), an optical crank angle encoder (0.1°CA resolution) was fitted on the engine's crankshaft, respectively.

2.1 Error Analysis

The uncertainty of the measured parameters such as air, diesel, H_2 rate of flow, engine's speed, torque and exhaust gas temperature, emission characteristics, gaseous HES, brake thermal efficiency, EGR percentage were calculated using Eq. (1).

$$\Delta q = \left[\left(\frac{\partial q}{\partial x_1}\Delta x_1\right)^2 + \left(\frac{\partial q}{\partial x_2}\Delta x_2\right)^2 + \cdots + \left(\frac{\partial q}{\partial x_n}\Delta x_n\right)^2\right]^{0.5} \tag{1}$$

Experimental Procedure. The single-cylinder four-stroke CI diesel engine was tested at the speed of 2400 rpm with maximum torque in three phases namely 100% diesel fuel mode, dual-fuel mode (hydrogen-diesel) and dual-fuel mode with EGR. The test engine was run under diesel-hydrogen fuel mode at the same speed and load by increasing the hydrogen percentage to the maximum limit till knocking occurred. At 2400 rpm, the maximum limit of hydrogen energy share was limited to 57% for normal and smooth engine running conditions. The hydrogen energy share (HES) can be calculated by using Eq. (2)

$$E_{H_2} = \left[\frac{(m_{H_2} \cdot CV_{H_2})}{(m_{H_2} \cdot CV_{H_2}) + (m_D \cdot CV_D)}\right] \times 100 \tag{2}$$

where m_{H_2} is rate of hydrogen flow (g/s), m_D is rate of diesel flow (g/s), CV_{H_2} is the calorific value of hydrogen gas, i.e., 120 MJ/kg and CV_D is the calorific value of diesel, i.e., 43 MJ/kg, respectively.

The experiments were performed using different EGR ratio (5, 10, 15%, and so on) at optimal hydrogen energy share. EGR percentage was calculated using Eq. (3).

$$\text{EGR\%} = \left[\frac{(m_{\text{air}}\ \text{without EGR}) - (m_{\text{air}}\ \text{with EGR})}{m_{\text{air}}\ \text{without EGR}}\right] \times 100 \tag{3}$$

where m_{air} is airflow rate (kg/s), respectively.

The efficiency (η_{th}) of the diesel-hydrogen CI engine without and with EGR can be analyzed by using Eq. (4).

$$\eta_{\text{th}} = \frac{B.P.(\text{kW})}{(m_{H_2} \cdot \text{CV}_{H_2}) + (m_D \cdot \text{CV}_D)} \times 100 \tag{4}$$

3 Results and Discussion

Figure 2 shows the impact of hydrogen addition on the brake thermal efficiency (BTE) of the engine without exhaust gas recirculation (EGR) at 2400 rpm with maximum torque. The BTE of the engine with diesel-hydrogen mode is higher than operation with 100% diesel as it increased from 29.32% to 33.20%, respectively. This could be attributed due to the combined effects of the hydrogen's higher diffusivity and availability of enough time for mixing of charge resulting in improved performance and better combustion. The maximum HES without knocking in diesel-hydrogen engine is limited up to 57%. The knock occurs at 63% hydrogen energy share.

Figures 3 and 4 show the impact of H_2 addition on CO and HC emissions in the engine without exhaust gas recirculation (EGR). Carbon-based emissions reduced to zero levels with an increased in HES in diesel-hydrogen-fueled diesel engines. This

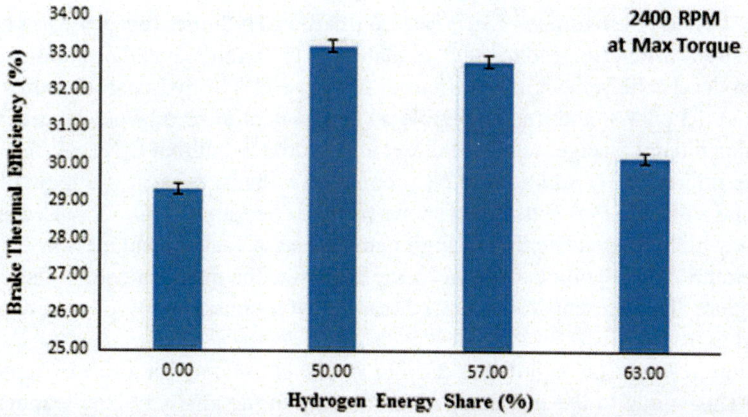

Fig. 2 Impact of H_2 addition on brake thermal efficiency without EGR

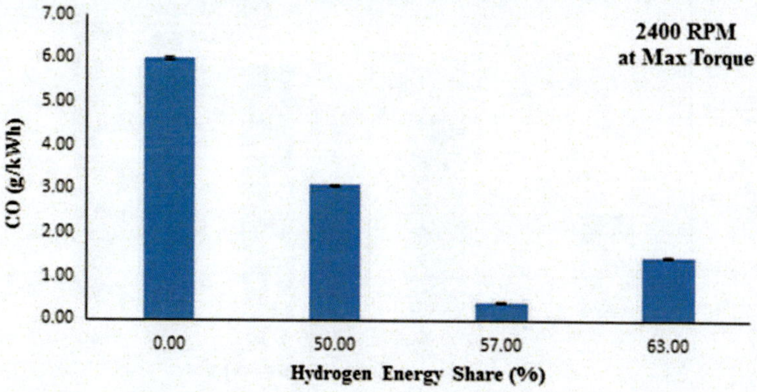

Fig. 3 Impact of H_2 addition on CO emission without EGR

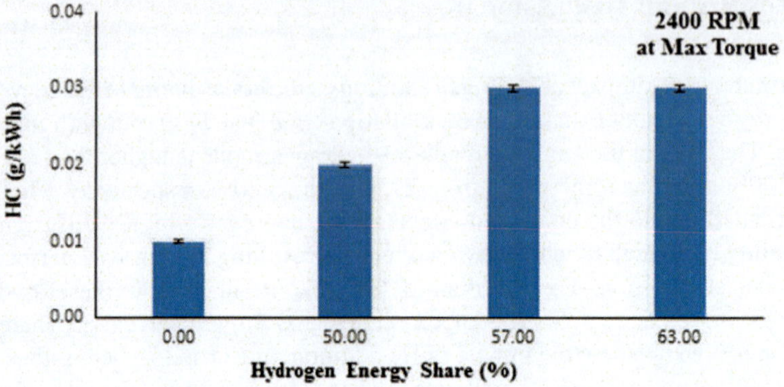

Fig. 4 Impact of H_2 addition on HC emissions without EGR

may be because the hydrogen fuel is free from carbon. Hydrogen fuel possesses higher flame speed which can lead to quick oxidation of the mixture could be another cause for zero level achievement in emissions of CO and HC. The hydrocarbon varies from 0.01 to 0.03 g/kWh and it is negligible and equivalent to zero levels. Hydrocarbon emission did not change with base diesel and hydrogen without EGR.

The impact of H_2 addition on NO_x emission without exhaust gas recirculation (EGR) is shown in Fig. 5. Its effect shows that the emission of NO_x increased significantly with hydrogen because of high flame speed of fuel and engine's in-cylinder temperature. The significant rise in NO_x emission is the main concern in diesel- H_2 CI engines. This present study showed that the NO_x emission got reduced by using cooled EGR.

Figure 6 shows the significant drop in smoke emission with a rise in supply of HES. This is due to the increment in uniformity in the air-fuel blend leading to a reduction in the smoke emission level.

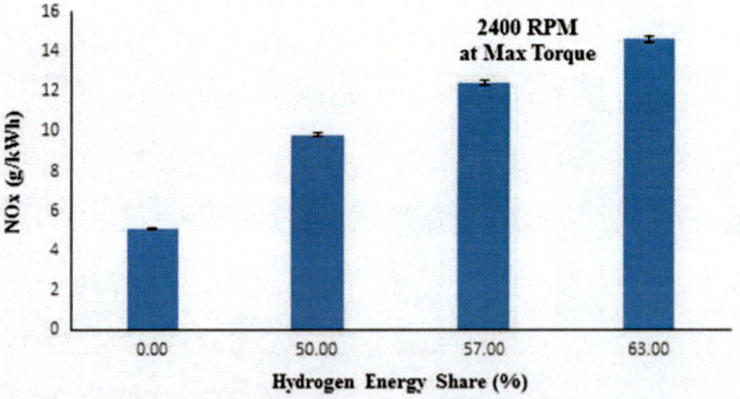

Fig. 5 Impact of H_2 addition on the NO_x emission without EGR

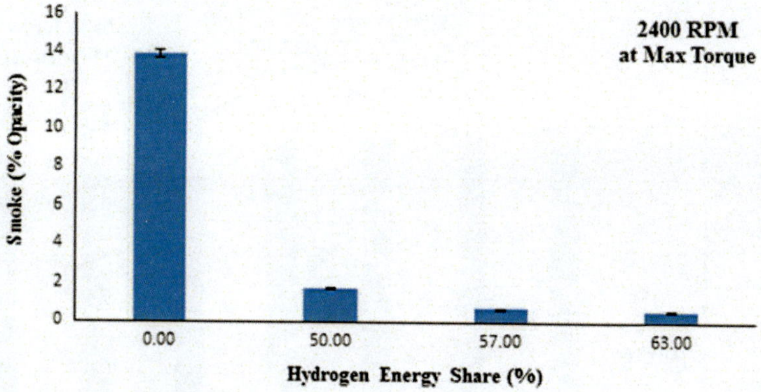

Fig. 6 Reduction in smoke emission using H_2 addition without EGR

The impact of H_2 addition on the in-cylinder pressure without exhaust gas recirculation at 2400 rpm with maximum torque is shown in Fig. 7. The complete oxidation of the mixture due to its high flame speed and cylinder temperature results in the enhancement of in-cylinder pressure significantly. Figure 7 clearly shows the knocking occurrence at 2400 rpm, when it is increased the supply of hydrogen gas at 63% and above. Hence, the optimal and maximum HES for the diesel-hydrogen engine for normal engine running condition is 57% without knocking.

Figure 8 demonstrates the variation in the percentage of HES for different EGR ratio at 2400 rpm. The EGR strategy is adopted with optimal hydrogen energy share percentage (57%) at 2400 rpm to boost the share of hydrogen energy and NO_x emission drop. Figure 8 shows that the hydrogen energy share without knocking is extended to 65% using 25% exhaust gas recirculation.

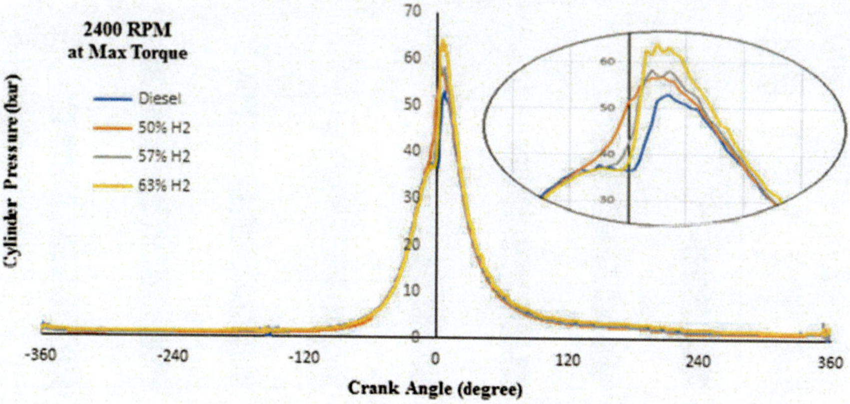

Fig. 7 Impact of H_2 addition on the cylinder pressure without EGR

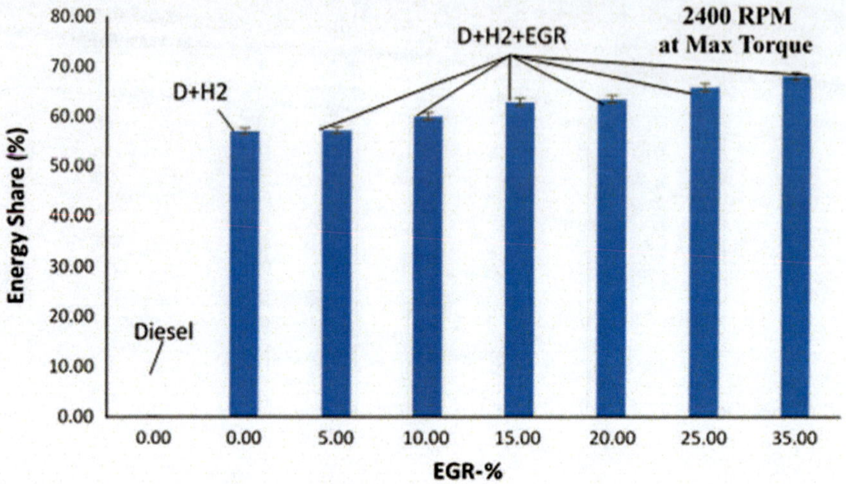

Fig. 8 Variation in the HES for different EGR percentage

Figure 9 shows the effect of EGR percentage on the BTE of the diesel-hydrogen engine at 2400 rpm. Figure 8 indicates that the HES increases as EGR percentage rises. However, from Fig. 9, it can be shown that the brake thermal efficiency of diesel-hydrogen engine enhanced from 29.32% using diesel (100%) mode to 32.79% by means of 57% HES whereas it also increases and gets maximum to 33.79% with 25% EGR percentage at 2400 rpm. The efficiency increased to some extent up to 25% EGR but it started to decrease further because of lower in-cylinder temperature, low flame velocity of air-fuel mixture and the availability of enough time in heat transfer

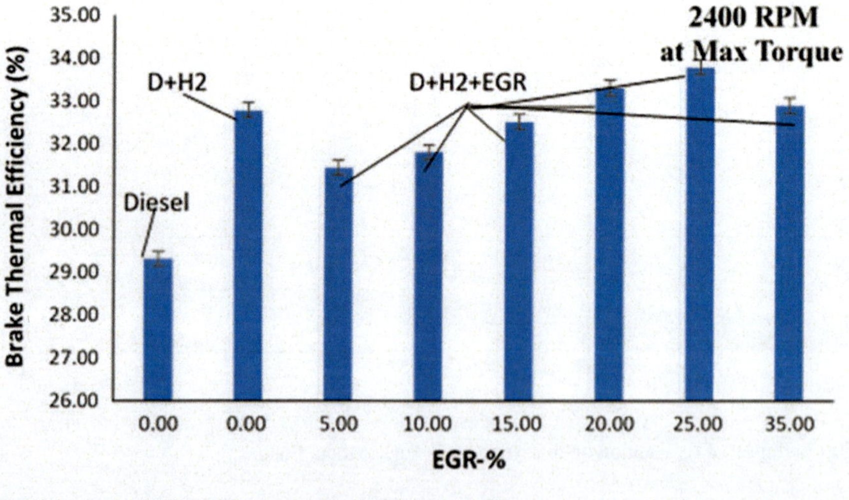

Fig. 9 Impact of EGR percentage on the BTE of the engine

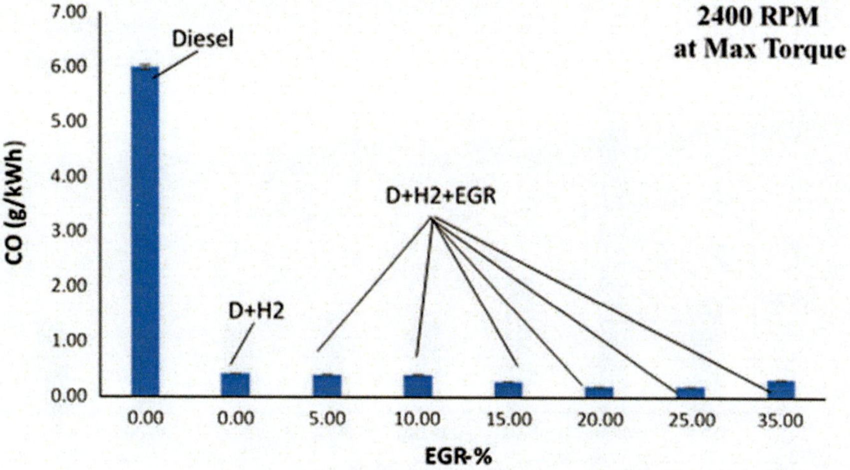

Fig. 10 Impact of EGR percentage on CO emission in hydrogen-diesel mode

to cylinder walls. Thus, the efficiency found higher with 25% EGR which is called as optimal EGR condition at 2400 rpm.

Figure 10 shows the effect of EGR percentage on carbon monoxide emission in diesel-hydrogen engine at 2400 rpm. The emission of CO decreased from 6.0 g/kWh using conventional diesel to 0.42 g/kWh with 57% optimal HES and it is further reduced to approx. zero levels with 25% EGR percentage under duel fuel mode operation. During the EGR operation in diesel-hydrogen engine, the oxygen content in the intake leads to a rise in dilution effect. The dilution effect helps in reducing the rate of combustion reaction leading to lower in-cylinder temperature and a slight rise in CO emission at higher EGR percentage. Up to 25% EGR, CO emission decreased significantly to zero level.

Figure 11 demonstrates the impact of EGR percentage on hydrocarbon emissions in diesel-hydrogen engine 2400 rpm. It can be inferred from the figure that zero level of HC emission was achieved with EGR percentage due to improvement in the rate of reaction with the recirculation of radical species of the previous cycle. HC emission varies from 0.03 g/kWh at without EGR to 0.01 g/kWh at 35% EGR.

Figure 12 indicates the impact of EGR percentage on the NO_x emission in diesel-hydrogen engine at 2400 rpm. It is conclusive from the figure that NO_x emission reduces as EGR percentage is increased. The decline in the concentration of oxygen in the mixture enhanced the dilution effect and leading to lower combustion pressure and temperature resulting in suppressing the formation of NO_x emission. NO_x emission decreased from 12.4 g/kWh at without EGR to 0.25 g/kWh at 35% EGR. Zero level of NO_x emission was achieved as the higher percentage of EGR was inducted in dual-fuel operation.

Figure 13 shows the impact of EGR percentage on smoke emission in diesel-hydrogen engine at 2400 rpm. As the EGR ratio is increased, the engine emitted high level of smoke emission. This is as of a significant drop in cylinder temperature

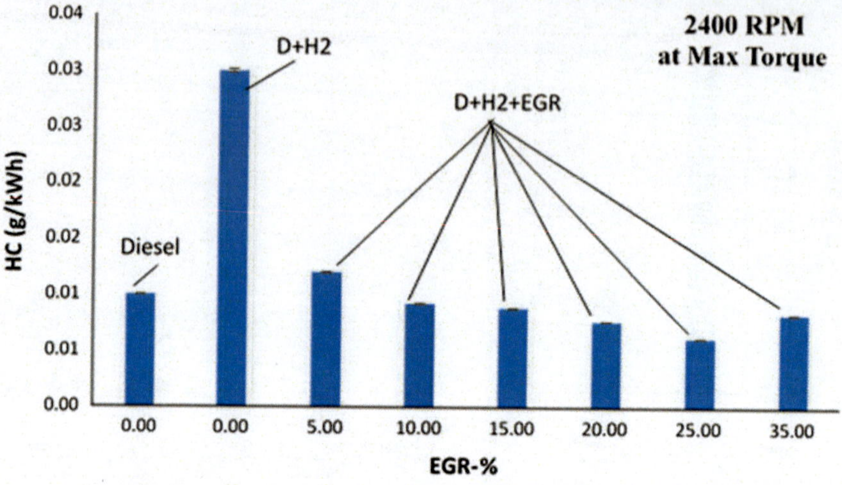

Fig. 11 Impact of EGR percentage on hydrocarbon emissions in diesel-hydrogen engine

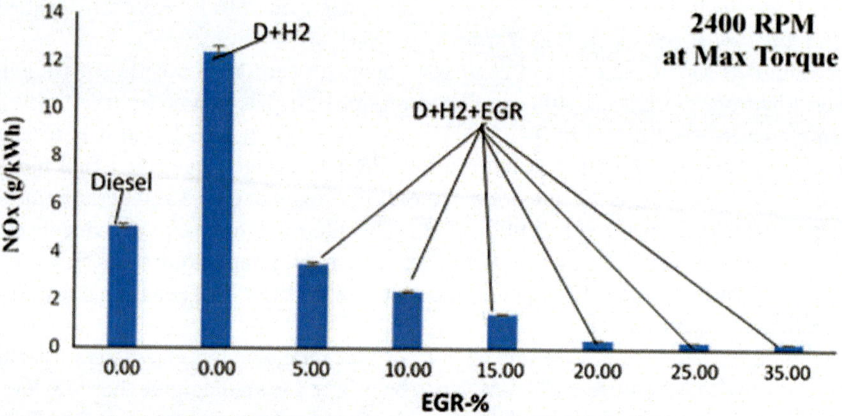

Fig. 12 Impact of EGR percentage on the NO_x emission in diesel-hydrogen engine

and pressure that could decrease the concentration of oxygen which enhances the formation of particulate matter.

Figure 14 indicates the impact of EGR percentage on cylinder pressure in diesel-hydrogen engine at 2400 rpm. The cylinder pressure decreased as EGR percentage increases. This may be because of the dilution effect and decrease in the content of oxygen which decreased the in-cylinder temperature as well. Hence, the optimal EGR percentage at 2400 rpm was found out 25%, respectively.

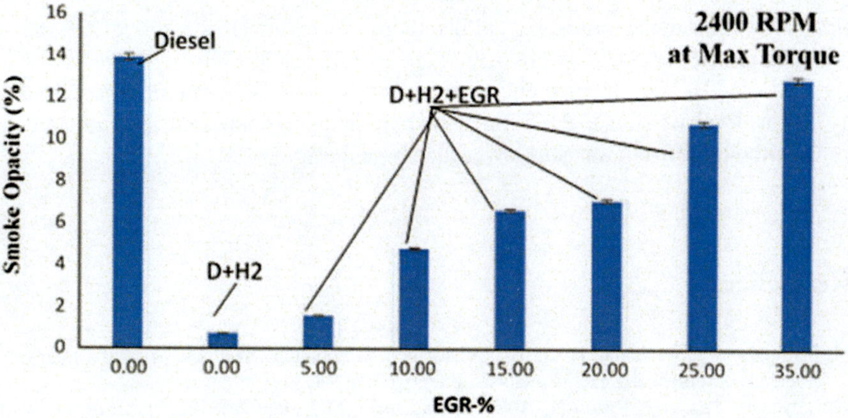

Fig. 13 Impact of EGR percentage on smoke emissions in diesel-hydrogen mode

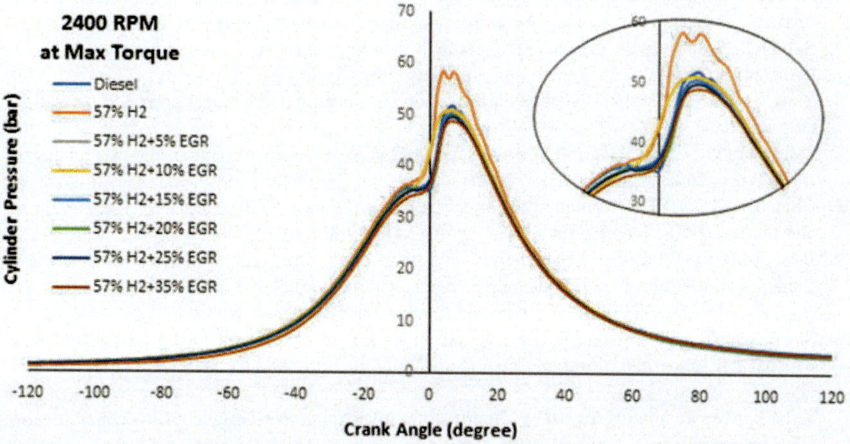

Fig. 14 Impact of EGR percentage on cylinder pressure in diesel-hydrogen mode

4 Conclusions

The influence of EGR percentage and supply of hydrogen percentage on the diesel-hydrogen engine are examined. The main conclusions are as follows:

(i) The maximum share of hydrogen energy improved from 57 to 65% by using EGR.
(ii) The brake thermal efficiency improved from 29.32% using diesel to 32.79% through hydrogen and 33.79% with hydrogen with 25% EGR. Hence, optimum EGR is chosen as 25%.

(iii) NO_x emission augmented significantly by increasing HES without EGR, but it decreased considerably almost to zero level by using EGR.
(iv) CO and HC emissions reduced considerably using EGR rate of up to 25%.
(v) The level of smoke emission with EGR slightly increased as compared to without EGR but it is still lower than base diesel.

References

1. Chintala V, Subramanian KA (2014) Assessment of maximum available work of a hydrogen fueled CI engine using exergy analysis. Energy 67:162–175
2. Fayaz H, Saidur R, Razali N, Anuar FS, Saleman AR, Islam MR (2012) An overview of hydrogen as a vehicle fuel. Renew Sustain Energy Rev 16(8):5511–5528
3. Babu MKG, Subramanian KA (2013) Alternative transportation fuels, utilisation in combustion engines. CRC Press, Taylor & Francis Group
4. Abdelaal MM, Rabee BA, Hegab AH (2013) Effect of adding oxygen to the intake air on a dual-fuel engine performance, emissions, and knock tendency. Energy 61:612–620
5. Salvi BL, Subramanian KA (2015) Sustainable development of road transportation sector using hydrogen energy system. Renew Sustain Energy Rev 51:1132–1155
6. Yang Z (2015) Effects of hydrogen addition on combustion and exhaust emissions in a diesel engine. Fuel 139:190–197
7. Tsujimura T, Suzuki Y (2017) The utilization of hydrogen in hydrogen/diesel dual fuel engine. Int J Hydrogen Energy 42:14019–14029
8. Chaichan M (2018) Performance and emission characteristics of CIE using hydrogen, biodiesel, and massive EGR. Int J Hydrogen Energy 48:5415–5435
9. Bose P, Maji D (2009) An experimental investigation on engine performance and emissions of a single cylinder diesel engine using hydrogen as inducted fuel and diesel as injected fuel with exhaust gas recirculation. Int J Hydrogen Energy 34:4847–4854
10. Shi B, Cho Y, Han D, Song S, Chun K (2011) Hydrogen effects on NO_x emissions and brake thermal efficiency in a diesel engine under low-temperature and heavy-EGR conditions. Int J Hydrogen Energy 36:6281–6291
11. Kalsi S, Subramanian KA (2017) Experimental investigations of effects of hydrogen blended CNG on performance, combustion and emissions characteristics of a biodiesel fueled reactivity-controlled compression ignition engine (RCCI). Int J Hydrogen Energy 42:4548–4560
12. Chintala V, Subramanian KA (2016) Experimental investigation of hydrogen energy share improvement in a compression ignition engine using water injection and compression ratio reduction. Energy Convers Manag 108:106–119
13. Pavlos D, Madan K, Taku T, Yasumasa S (2018) Combustion and emission characteristics of a hydrogen-diesel dual-fuel engine. Int J Hydrogen Energy 43:13605–13617

Effect of Valve Timing on Performance and Emission Characteristics of Producer Gas Fired S.I. Engine

Parth D. Shah, Taha Y. Poonawala, and Salim A. Channiwala

1 Introduction

This work presents the effect of intake and exhaust valve timing on performance and emission characteristics of producer gas fired S.I. engine. To estimate the performance and emission parameters of producer gas fired S.I. engine, heat transfer and dissociation of combustible products are being considered for actual cycle simulation with progressive combustion.

In conventional S.I. engine, charge is mixed together homogeneously in the intake and inducted through the intake valve into the cylinder where it mixes with residual gases and then compressed. Then, combustion is initiated by an electric discharge though spark plug. A turbulent flame develops after the ignition and propagates through this charge and the residual gas in the clearance volume until it approached to the walls of combustion chamber.

Combustion process strongly depends on three combustion parameters, i.e., efficiency factor (a), form factor (m), and combustion duration which are represented by Wiebe function. These parameters are the instant at which heat release rate becomes positive, duration of combustion and instantaneous crank angle of combustion [1, 2]. The Wiebe function equation is as seen in (1).

$$x_b = 1 - \exp\left[-a\{(\theta - \theta_0)/\Delta\theta_c\}^{m+1}\right] \quad (1)$$

The different values of (a), (m) [1–9] and combustion duration [10] are suggested in the literature. The different researchers have also proposed the methods to find out the efficiency parameter, slope parameters, and combustion duration of Wiebe

P. D. Shah (✉) · T. Y. Poonawala · S. A. Channiwala
Sardar Vallabhbhai National Institute of Technology, Surat 395007, India
e-mail: ds14me004@med.svnit.ac.in

A. K. Gupta et al. (eds.), *Advances in IC Engines and Combustion Technology*, Lecture Notes in Mechanical Engineering,
https://doi.org/10.1007/978-981-15-5996-9_41

function [11, 12]. For the producer gas fired engine, values of (a) and (m) have been suggested as 5.77 and 1.10, respectively [10, 13].

Heat transfer also plays vital role in the performance and emission characteristics of the engine. Therefore, it is important to model heat transfer model in simulation. For gas exchange process, Woschni's heat transfer model has been used. In compression and expansion process, heat transfer model proposed by Anand has been adopted [14].

To predict emissions, dissociation of combustion products has also been considered during the simulation.

2 Methodology

The simplest mathematical simulation of a thermal system has been carried out to estimate engine performance and emission parameters through thermodynamic modeling which is based upon the basic equations such as conservation of mass and energy and equilibrium of a set of chemical reactions. Thermodynamic models are useful in analyzing the qualitative trends as well as generating quantitative data regarding parametric influences. Different models have been used which are presented herein.

2.1 Major Processes and Assumptions

The entire simulation cycle has been divided into five processes, viz. suction, compression, combustion, expansion, and exhaust. Charge behaves as an ideal gas and mixed homogeneously. The charge gets completely burnt. Heat transfer area comprises of piston head, cylinder head, and cylinder wall surface only. Re-association of the dissociation products is neglected in exhaust process.

2.2 Selection of Engine and Fuel Data

For the simulation purpose a four-cylinder engine has been selected. Specification of this engine is as shown in Table 1 and fuel data used for simulation is as shown in Table 2.

Table 1 Engine specifications

Parameter	Value
Rated power	15 kWe
Bore * Stroke	98 * 98 mm
Speed	1500 rpm
Compression ratio	12:1
No. of cylinder	4
Connecting rod length	168 mm

Table 2 Producer gas composition [13]

Constituent	CO	CO_2	H_2	CH_4	N_2	H_2O	Σ
Composition (% volume basis)	25.6	9.5	22.5	0.3	35.7	6.4	100

2.3 *Dissociation Model*

At the completion of the compression stroke, combustion takes place, initiated by an electric spark from spark plug. During the combustion process, the chemical energy of the fuel is released in the form of heat energy, producing a temperature rise to several hundred degree Celsius and pressure soars to many tens of atmosphere. During combustion process, dissociation effect has also been considered. The analysis of dissociation model is presented below:

Analysis of Dissociation Model

Reaction stoichiometry by considering dissociation can be given as (2):

$$\begin{aligned} & n^f_{\mathrm{CO}}.\mathrm{CO} + n^f_{\mathrm{CO_2}}.\mathrm{CO_2} + n^f_{\mathrm{CH_4}}.\mathrm{CH_4} + n^f_{\mathrm{H_2}}.\mathrm{H_2} + n^f_{\mathrm{H_2O}}.\mathrm{H_2O} + n^f_{\mathrm{H_2S}}.\mathrm{H_2S} \\ & + n^f_{\mathrm{N_2}}.\mathrm{N_2} + a(\mathrm{O_2} + 3.762\mathrm{N_2}) \rightarrow n_{\mathrm{CO}}.\mathrm{CO} + n_{\mathrm{CO_2}}.\mathrm{CO_2} + n_{\mathrm{H_2}}.\mathrm{H_2} \\ & + n_{\mathrm{H_2O}}.\mathrm{H_2O} + n_{\mathrm{O_2}}.\mathrm{O_2} + n_{\mathrm{N_2}}.\mathrm{N_2} + n_{\mathrm{OH}}.\mathrm{OH} + n_{\mathrm{NO}}.\mathrm{NO} + n_{\mathrm{SO_2}}.\mathrm{SO_2} \end{aligned} \tag{2}$$

Constituent balance:

$$ttlc = n^f_{\mathrm{CO}} + n^f_{\mathrm{CO_2}} + n^f_{\mathrm{CH_4}} = n_{\mathrm{CO}} + n_{\mathrm{CO_2}} \tag{3}$$

$$ttlh = 4n^f_{\mathrm{CH_4}} + 2n^f_{\mathrm{H_2}} + 2n^f_{\mathrm{H_2O}} + 2n_{\mathrm{H_2S}} = 2n_{\mathrm{H_2}} + 2n_{\mathrm{H_2O}} + n_{\mathrm{OH}} \tag{4}$$

$$\begin{aligned} ttlo = {} & n^f_{\mathrm{CO}} + 2n^f_{\mathrm{CO_2}} + n^f_{\mathrm{H_2O}} + (2a) = n_{\mathrm{CO}} + 2n_{\mathrm{CO_2}} + n_{\mathrm{H_2O}} + 2n_{\mathrm{O_2}} \\ & + n_{\mathrm{OH}} + n_{\mathrm{NO}} + 2n_{\mathrm{SO_2}} \end{aligned} \tag{5}$$

$$ttln = 2n^f_{\mathrm{N_2}} + 3.762 * (2a) = 2n_{\mathrm{N_2}} + n_{\mathrm{NO}} \tag{6}$$

$$ttls = n^{f}_{H_2S} = n_{SO_2} \tag{7}$$

Total no. of moles,

$$n_t = n_{CO} + n_{CO_2} + n_{H_2} + n_{H_2O} + n_{O_2} + n_{N_2} + n_{OH} + n_{NO} + n_{SO_2} \tag{8}$$

Chemical equilibrium:

$$H_2O \rightleftharpoons H_2 + \frac{1}{2}O_2;\ K_1 = \frac{p_{H_2}.p_{O_2}^{1/2}}{p_{H_2O}} = \frac{n_{H_2}.n_{O_2}^{1/2}}{n_{H_2O}}.\left(\frac{P}{n_t}\right)^{\frac{1}{2}} \tag{9}$$

$$H_2O \rightleftharpoons \frac{1}{2}O_2 + OH;\ K_2 = \frac{p_{O_2}^{1/2}.p_{OH}}{p_{H_2O}} = \frac{n_{O_2}^{1/2}.n_{OH}}{n_{H_2O}}.\left(\frac{P}{n_t}\right)^{\frac{1}{2}} \tag{10}$$

$$CO_2 \rightleftharpoons CO + \frac{1}{2}O_2;\ K_3 = \frac{p_{CO}.p_{O_2}^{1/2}}{p_{CO_2}} = \frac{n_{CO}.n_{O_2}^{1/2}}{n_{CO_2}}\left(\frac{P}{n_t}\right)^{\frac{1}{2}} \tag{11}$$

$$\frac{1}{2}N_2 + \frac{1}{2}O_2 \rightleftharpoons NO;\ K_4 = \frac{p_{NO}}{p_{N_2}^{1/2}.p_{O_2}^{1/2}} = \frac{n_{NO}}{n_{N_2}^{1/2}.n_{O_2}^{1/2}} \tag{12}$$

Energy balance:

$$\sum U_R = \sum U_P$$

$$\therefore \sum n_R.[h^{\circ}_{f,R} + \Delta h_{T_R} - R^{\circ}T_R] = \sum n_P.[h^{\circ}_{f,P} + \Delta h_{T_e} - R^{\circ}T_e] \tag{13}$$

By solving (2)–(13) simultaneously and using thermochemical data from [15], composition of combustion products and equilibrium temperature has been obtained.

3 Results and Discussion

The complete modeling of each process and calculation of cycle have been coded in C language. It is worth to observe that the actual cycle analysis of laboratory scale S.I. engine does offer reasonably good efficiency 29.60% but power produced is about 50.41% less than the rated power. This is basically attributed to lower mixture calorific value at intake. The present results focus a need toward reconfiguration of engine dimensions and combustion chamber geometry and thus prove the worthiness of the analysis as it does provide design guidelines, too.

Figure 1 shows pressure versus cylinder volume curve based on overall cycle simulation results whereas Fig. 2 shows emissions chart from the onset of dissociation. It is observed from the figure that the peak value of the gases increases with

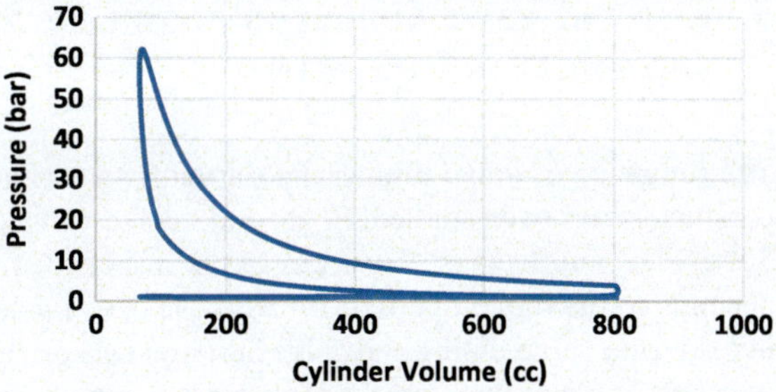

Fig. 1 Pressure versus cylinder volume

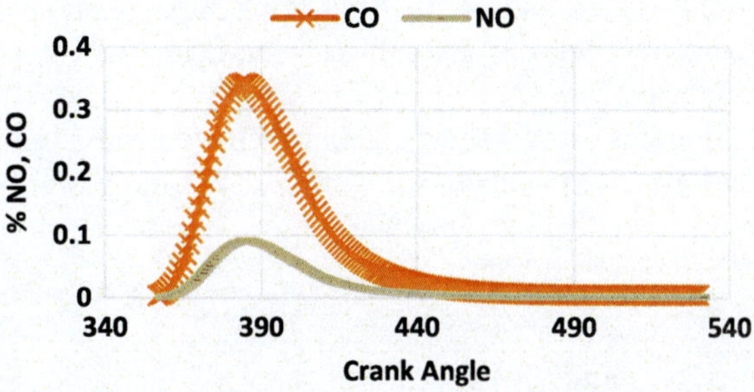

Fig. 2 Emissions at the onset of combustion

the increase in the temperature and decreases with the decrease in temperature. The maximum value of CO and NO is found to be at 385° and 384° of crank angle, respectively. This is probably because of the combustion rate and temperature reaches to its peak at this crank angles.

4 Validation of Results

The obtained results from the analysis is quantitatively validated with results obtained by the experimentally of Cummins six-cylinder natural gas G6B5.9G engine. The efficiency obtained from the experiments is 26.50% @ 35 kW electrical loading as against that 28.33% obtained by simulation at equivalence ratio obtained of 0.6568. The quantitative deviation in the power and efficiency may be due to the different composition of gas, electrical loading @ 35 kW, different engine dimensions, assumptions of valve timing and its dimensions, and duration of all processes.

5 Effect of Valve Timing on Engine Performance and Emission Parameters

The influence of valve timing and duration on engine performance and emissions parameters has been studied. In general, intake valve opening is in between 10° and 30° BTDC and exhaust valve closing is in between 10° and 35° ATDC [1, 2]. It depends on the fuel properties. Efforts have been made to carry out parametric study in respect to valve timing and valve duration because these parameters are not easily traceable in open literatures for producer gas fired engine. Hence, valve timing of intake valve and exhaust valve is varied in the step of 4° in present study from 12° to 28° BTDC and 14° to 30° ATDC, respectively. It worth to note that in all cases intake valve closing and exhaust valve opening are kept at 26° ATDC and 10° BBDC, respectively.

Figures 3 and 4 show combined effect of valve duration of both valves on engine performance parameters. Figure 3 shows the power of the cycle, and Fig. 4 shows

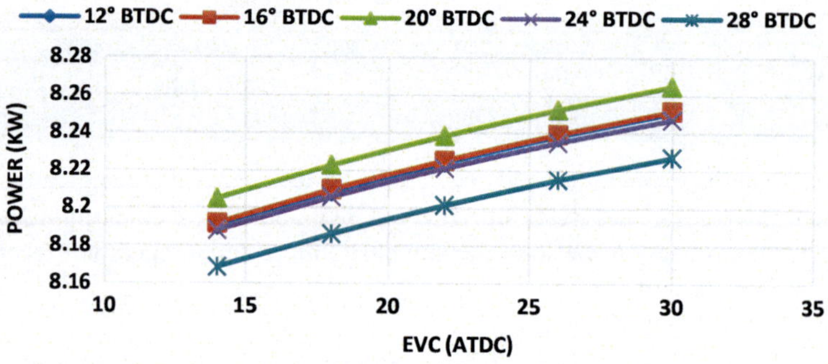

Fig. 3 Power versus intake valve timing versus exhaust valve timing

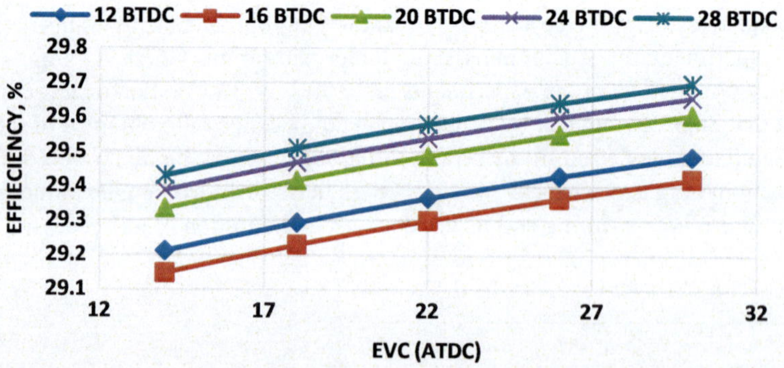

Fig. 4 Efficiency versus intake valve timing versus exhaust valve timing

the efficiency of the cycle. It shows the increasing trend as valve timing duration of the exhaust valve increases in all cases.

However, for intake valve opening as 20° BTDC, maximum power is obtained in all cases of exhaust valve timing duration. This may be because at IVO of 20° BTDC, the suction pressure might be achieving minima permitting maximum charge intake and thereby offer higher power and efficiency. Further, as exhaust process governs the scavenging, its influence on charge intake is dominant. The well-scavenged cylinder with increase in timing will certainly permit more and more charge intake resulting in better power and efficiency.

Figures 5 and 6 show the effect of valve timing on emissions of CO and NO, respectively. It is seen from the figures that no specific trend has been obtained. The emissions are usually governed by combustion temperature which in turn depends on power. For present case, power produced is only 50.41% of rated one with producer gas as fuel due to much lower mixture calorific value. Hence, neither sizable difference nor remarkable change in emissions is observed with respect to valve timing.

The study of these results shows that minimum CO emissions are obtained with IVO as 20° BTDC and EVC as 22° ATDC while minimum NO is observed to be at IVO as 20° BTDC and EVC as 26° ATDC. With respect to power, efficiency and

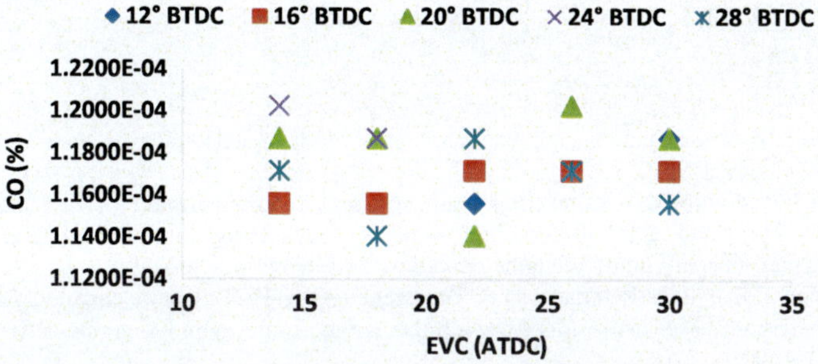

Fig. 5 CO emission versus intake valve timing versus exhaust valve timing

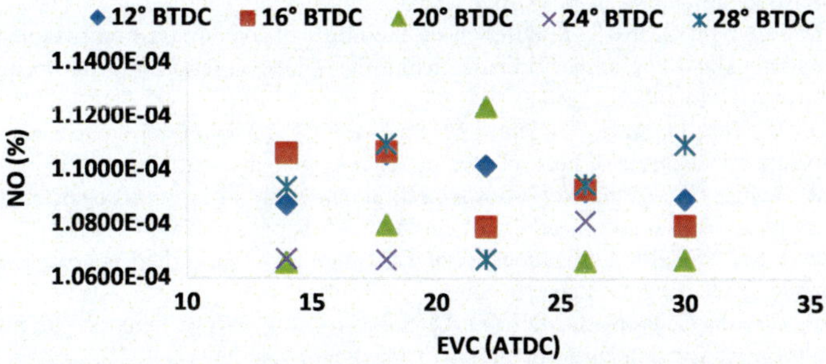

Fig. 6 NO emission versus intake valve timing versus exhaust valve timing

emissions, the most optimal combination of valve timing seems to be IVO as 20° BTDC and EVC as 30° ATDC.

6 Conclusions

- From the analysis of complete cycle, efficiency is found to be 29.60% while power is 8.27 kW, which is almost 50.41% lower than the targeted power of the engine, which shows that reconfiguration of the engine is required.
- The present analysis clearly predicts the inception of pollutant generation and the instantaneous variations of these pollutants with respect to crank angle, which is quite unique feature of this work.
- Minimum CO emission is obtained with IVO as 20° BTDC and EVC as 22° ATDC.
- Minimum NO is observed to be at IVO as 20° BTDC and EVC as 26° ATDC.
- With respect to power, efficiency, and emissions, the most optimal combination of valve timing seems to be IVO as 20° BTDC and EVC as 30° ATDC.

References

1. Heywood JB (1988) Internal combustion engine fundamentals, 1st edn. Tata McGraw Hill Book Company, USA
2. Stone R Introduction to internal combustion engines, 3rd edn. Palgrave
3. Watts PA, Haywood JB (1980) Simulation studies of the effects of turbo charging and reduced heat transfer on spark ignition engine operation. SAE paper no. 800289
4. Johansson T, Borgqvist P, Johansson B, Tunestal P (2010) HCCI heat release data for combustion simulation, based on results from a turbocharged multi cylinder engine. SAE paper no. 2010-01-1490
5. Liu Z, Karim GA (1997) Simulation of combustion process in gas-fueled diesel engines. Part A, J Power Energy 211(A2):159–169
6. Galindo J, Climent H, Pla B, Jimenez VD (2011) Correlations for Wiebe function parameters for combustion simulation in two-stroke small engines. Appl Therm Eng 31:1190–1199
7. Kim J, Bae C, Kim G (2013) Simulation on the effect of the combustion parameters on the piston dynamics and engine performance using the Wiebe function in a free piston engine. Appl Energy 107:446–455
8. Ravaglioli V, Moro D, Serra G, Marelli M, Ponti F (2011) MFB50 on-board evalution based on a zero-dimensional rate of heat release model. SAE paper no. 2011-01-1420
9. Kumar R, Reader GT, Zheng M (2004) A preliminary study of ignition consistency and heat release analysis for a common-rail diesel engine. SAE paper no. 2004-01-0932
10. Channiwala SA Modelling of processes of I.C. engine—a simplified approach. Design of Components of I.C. Engine, Lecture Notes
11. Yeliana Y, Cooney C, Worm J, Michalek D, Naber J (2008) Wiebe function parameter determination for mass fraction burn calculation in an ethanol-gasoline fuelled SI engine. J KONES Powertrain Transp 15(3):567–574
12. Abbaszadehmosayebi G, Ganippa L (2014) Characterising Wiebe equation for heat release analysis based on combustion burn factor (Ci). Fuel 119:301–307

13. Channiwala SA (2013) Zero effluent discharge coal gasification system. In: Thematic lecture, 23rd NCICEC, 13–16 Dec 2013, SVNIT, Surat
14. Anand WJD (1963) Heat transfer in the cylinders of reciprocating internal combustion engines. Proc I Mech E 177
15. Chase MW (1998) JANAF thermochemical data. NIST, 4th edn

Influence of Electronically Controlled Hot and Cold External ReBreathing System in DI-CI Diesel Engine for Reducing NO_x Emission

C. Ramesh, A. Murugesan, and G. Mylsami

1 Introduction

A human men generally breaths 10.8 m^3 of air within 24 h, and it is very essential for maintaining the quality of the air. But the diesel engine outlet pollutes the air including fine, coarse and hydrocarbons, gaseous irritants, ultra-fine particles, and PAH; these are very dangerous for health [1, 2]. Diesel engines are used in aggravated applications as the primary power source of vehicles. The intensified engine includes off-highway construction, large trucks, buses, and extract equipment. Moreover, diesel engines are increasing the high impact on the commercial vehicle market all over the world. High recommendation of the diesel engine rotates around its resilience, reliability, and fuel efficiency. High compression ratios and more oxygen attention in diesel combustions are responsible for better fuel efficiency and lower CO and HC emissions when ironically to a match with the petrol engine. The carbon monoxide occupies the residual gas in a diesel engine and is also less similar to a petrol engine. Moreover, these common components result in more nitrogen oxide emissions [3].

The diesel engine gives higher proficiency, and it can be used to lessen CO_2 emissions, which can be a principle explanation behind making a dangerous atmospheric deviation. Diesel outsources also have soot particles, nitrogen oxides (NO_x), and toxic gases. The best route to get lesser amount of the nitrogen oxide emissions of a

C. Ramesh (✉) · A. Murugesan · G. Mylsami
Department of Mechanical Engineering, K.S. Rangasamy College of Technology, Tiruchengode, Tamilnadu 637215, India
e-mail: rameshc@ksrct.ac.in

A. Murugesan
e-mail: murugesanhema2000@yahoo.co.in

G. Mylsami
e-mail: gm161966@gmail.com

A. K. Gupta et al. (eds.), *Advances in IC Engines and Combustion Technology*, Lecture Notes in Mechanical Engineering,
https://doi.org/10.1007/978-981-15-5996-9_42

Table Table 1 Description of the engine

Make	Kirloskar engine
Type of cooling	Water
Speed	1500 rpm
Bore diameter	80 mm
Brake power	3.7 kW
Stroke length	110 mm
Type	Vertical engine

diesel engine is the use of the external rebreathing system. Here, parts of the exhaust gases are changing the route into the combustion area [4].

This leads to a lowermost combustion temperature which in turn compresses the formation of NO_x. In modern turbocharged engines, it can be problematic to provide the amount of external rebreathing system that is needed to reach the emission boundaries [5].

As knowledge of the author's lot of papers is published to manage the nitrogen oxide emissions from a diesel engine using manually controlled exhaust gas recircular system, the intention of the recent research work is to focus on electronically precious control of the external rebreathing system (EGR) to study be diesel engine characteristics (Tables 1 and 2) [6, 7].

2 Methodology

The present study provides electronically controlled a simultaneously compare the different ERS systems, such as Hot external rebreathing system and cold external rebreathing system with a reed valve and a system with a pump. Both the steady-state performance and emission features of the DI-CI engine are compared. In unchanging conditions, the focus is on fuel efficiency [8]. In various conditions both their action on changed ERS-demands and the torque, the response is analyzed. Experiments were conducted on a diesel engine with (cold and hot) EGR and without EGR. To reach the intention of this study, the engine was accelerating with diesel at 10% EGR rate. The engine is also run with diesel and 0% EGR to generate basic readings. Finally, engine performance and emission features are compared to the baseline data.

3 Experimental Engine Setup

High-speed, one-cylinder, four-stroke, H_2O-cooled DI-CI diesel engine connected with high-speed DAQ system was chosen for the study along with AVL 444 di-gas analyzer and AVL 437 standard smoke meter used to measure the discharge from the

Table Table 2 Correlation of present work with other researchers works detailed in kinds of the literature

Researchers name	EGR with Fuel	BTE	CO	HC	CO_2	Smoke	NO_x
Present work	EGR with pure diesel (electronical control)	Increasing	Increasing	Increasing	Decreasing	Increasing	26% Decreasing
Deepak Agarwal et al. [8]	EGR with 20% biodiesel blend	Increasing	Increasing	Increasing	Decreasing	Decreasing	12% Decreasing
Senthil Kumar et al. [9]	EGR with pure diesel (manual control)	Increasing	Increasing	Increasing	Decreasing	Increasing	19% Decreasing
Odaka et al. [10]	EGR with pure diesel (water injection)	No change	Increasing	Increasing	No information	No change	Decreasing

Fig. 1 Experimental engine setup with an external rebreathing system

engine fired with pure diesel with and without external rebreathing system as shown in Fig. 1.

3.1 Description of the Engine

Figure 2 shows the schematic diagram of the external rebreathing setup; in this setup, the ERS driver can be connected to the Ethernet switch, and this total setup is connected with the computer. The total exhaust gas is calculated using the differential pressure sensor. It acts as a transducer, and it generates a signal as a function of the gas flow imposed. It is also measuring the ERS percentage instead of manual measurement. The 10% hot and cold ERS measurements were calculated and compared to the baseline data of the engine.

For rebreathing of the discharge gas, necessity piping was done. The quantity of the EGR was regulated by a control valve connected to the electronically controlled digital meter, which is mounted in the EGR loop along with the line diagram for the investigational EGR setup as shown in Fig. 3.

An airbox was fixed in the EGR loop to moisten the instability of the rebreathing system. An orifice was introduced into the exhaust recirculation loop, attached to

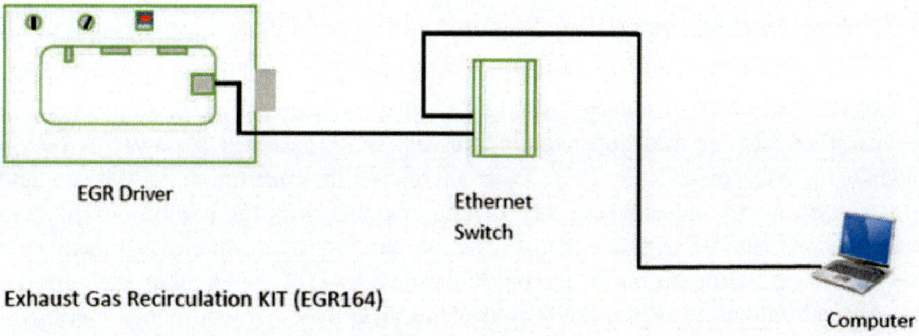

Fig. 2 Schematic diagram of external rebreathing KIT

Fig. 3 Electronically controlled external rebreathing system

the airbox in the manner to calculate the rate of flow of the rebreathing system. To measure the flow of air precisely, smooth flow equipment was invented before the inlet manifold.

The heat level of the cooling water inlet, outlet, and discharge gas was measured with type K thermocouples. An automatic room temperature compensation facility was used with a digital indicator. The temperature indicator was graduated equal to interval time. The flow of water was measured by the flow meter to know the quantity of water flowing per minute. Volumetric fuel intake measurement was done. Discharge gas emission calculations were calculated by exhaust gas analyzer.

3.2 External Rebreathing System

Controlling the NO_x discharges crudely requires a decrease in internal chamber temperatures. From the diesel engine fumes, EGR assumes a powerful job in decreasing NO_x discharges [11]. EGR includes the substitution of oxygen and nitrogen of outside air entering the burning chamber with the carbon dioxide and water vapor from the engine exhaust. The rebreathing of part of exhaust gases into the engine admission air builds the particular heat limit of the blend and lessens the oxygen centralization of the admission blend. The over two components consolidated lead to a huge decrease in NO_x emissions. EGR (%) is characterized as the mass percent of the recirculated exhaust (MEGR) in absolute admission blend (Mi) [12].

$$\text{EGR}(\%) = \text{MEGR/Mi} * 100 \quad (1)$$

Be that as it may, the utilization of EGR brings about higher fuel utilization and emanation punishments. EGR builds HC, CO, and PM discharges alongside marginally higher explicit fuel utilizations. In heavy-duty engines, the NO_x delivered during high load activity is significantly high. Applying EGR open or close full burden tends to build the smoke exhaust since the engine works close stoichiometric air-fuel proportion. This expands the particulate outflow as well as requires the derating of the engine to keep up adequate smoke levels. The utilization of EGR can likewise antagonistically influence the greasing up oil and engine toughness. It is comprehensively recognized that sulfur oxides present in exhaust gas stimulate wear [13]. Without applying EGR, vitality effective diesel engines ordinarily produce fumes that contain oxygen from around 5% at full burden to around 20% during sitting. As the intemperate fumes oxygen diminishes with increments in engine burden, the particular warmth of fumes rises as a result of the expansion in CO_2 fixation [14]. Therefore, the adequacy of NO_x decreases by EGR and additionally fluctuates with burden. Test outcomes demonstrate that high EGR rates should be connected at low burden, and low EGR rates are adequate for high load. Likewise, as burden expands, diesel engines will, in general, create more smoke due to diminished oxygen. In this manner, EGR albeit compelling to diminish NO_x further expands the smoke and PM emanations. Engine activity with diesel fuel and EGR brings about NO_x decreases without disintegrating brake explicit fuel utilization (BSFC) and discharges.

4 Experimental Test with Results

Sequences of investigations have been done on a one-cylinder four-stroke 3.7 kW brake power diesel engine filled with diesel. The experiments were conducted in the referred injection pressure and timings. The conformation results of the practical tests

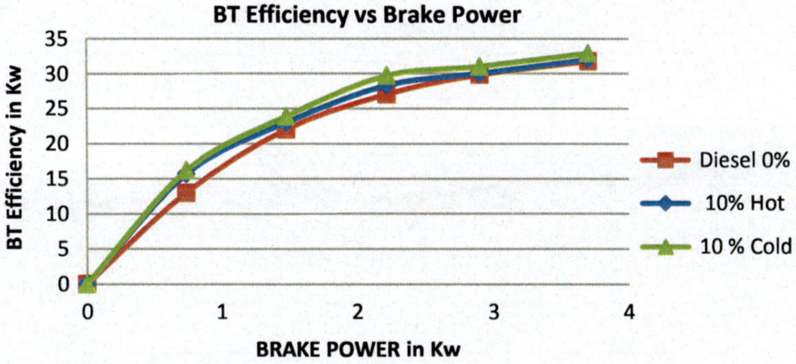

Fig. 4 BP (kW) versus BT Efficiency (kW)

are represented in the following graphs. Sequences of exhaustive engine performance are calculated using 0% EGR, 10% EGR Hot, and 10% EGR cold to evaluate the characteristics and emission values of the engine. The characteristics and discharge values are calculated for thermal efficiency, hydrocarbon, CO, CO_2, nitrogen oxide, and smoke opacity.

4.1 Brake Thermal Efficiency

Figure 4 shows the variations of BT efficiency with BP for the effects of 10% hot and cold burned gases compared with without EGR. At the maximum load, the BT efficiency of diesel is 31 kW, hot EGR is 32.7 kW, and cold EGR is 31.5 kW. The brake thermal efficiency slightly increased to 10% hot and cold EGR compared to 0% EGR. This improvement due to re-firing of HC enters into the combustion area during suction with the rebreathing exhaust gases at high load process. On other hand, higher inlet temperature of air helps to total combustion. Adding to this, somewhat higher weight than barometrical weight external rebreathing may likewise have diminished siphoning loses. The compound endeavors related to the separation of carbon monoxide to frame a free radical can likewise improve productivity.

4.2 Emission Characteristics

Figure 5 shows the differentiation of hydrocarbon emission with brake power for diesel engine with and without EGR. It's due to a shortage level of oxygen in the combustion chamber resulting is observed that hydrocarbon discharge increases with

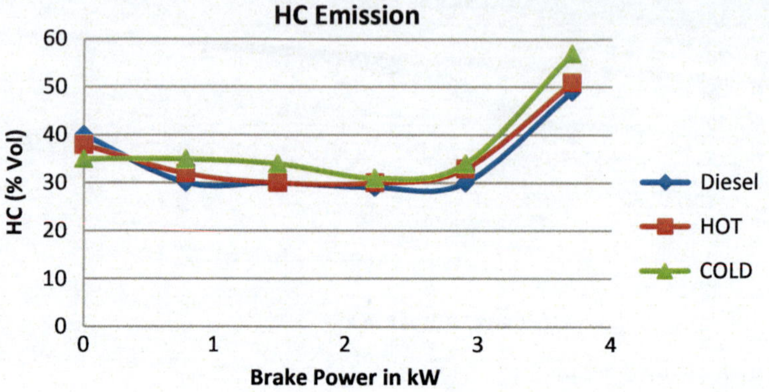

Fig. 5 BP (kW) versus HC (% Vol)

the increase of load on the unfinished combustion. At the maximum load, the hydrocarbon emission of diesel is 49% in volume, hot EGR is 51% in volume, and cold EGR is 57% in volume.

The HC emission slightly increased in 10% hot and cold EGR compared to 0% EGR. This is due to the reduced accessibility of O_2 to the heterogeneous mixture in the inner cylinder. This mixture may proper less burn resulting an increased value of HC emission.

Various levels of carbon monoxide emission against brake power are shown in Fig. 6. The carbon monoxide discharge of diesel at full load is 0.06% in volume whereas that of cold EGR is 0.07% and hot EGR is 0.08%. However, emissions are higher for the cold EGR and hot EGR compared to without EGR. This is because of a decrease in the accessibility of abundance oxygen for ignition. Oxygen focus brought about a rich air–fuel blend at various areas inside the combustion chamber. This heterogeneous blend proper less burning might have brought about higher CO emissions.

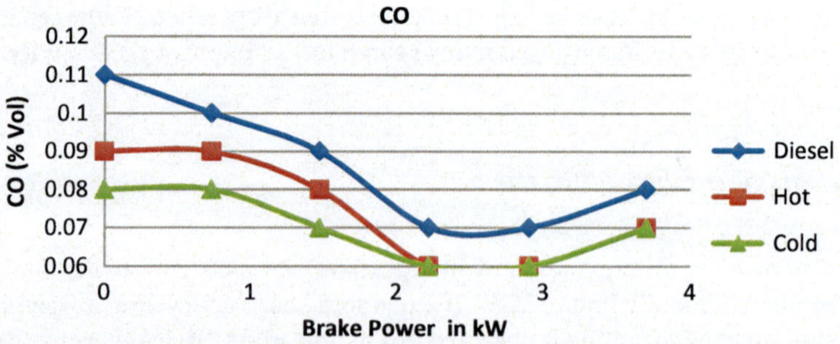

Fig. 6 BP (kW) versus CO (% Vol)

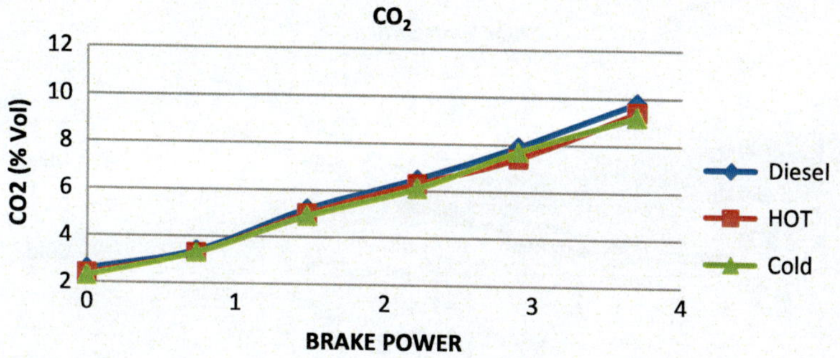

Fig. 7 BP (kW) versus CO_2 (% Vol)

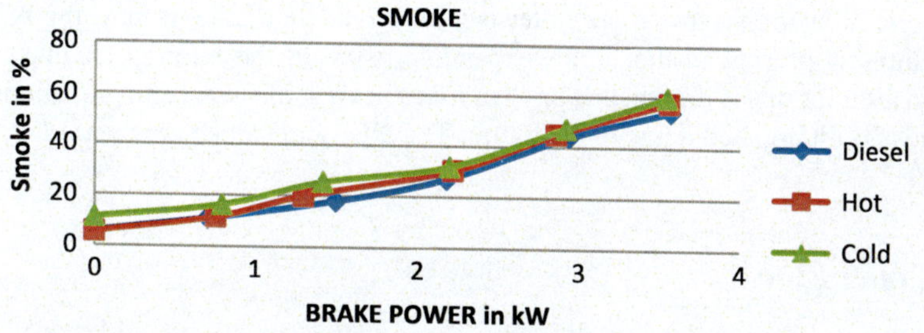

Fig. 8 BP (kW) versus Smoke (%)

The variation of CO_2 emission against BP is shown in Fig. 7. The carbon dioxide is expanding with the expansion of load conditions. The CO emission of diesel at full load is 9.8% in volume whereas that of cold EGR is 9.2% in volume and hot EGR is 9.5% in volume. This is due to reduction in the availability of O_2 leads to may not complete combustion showing in the decreasing of CO_2.

Smoke emission for diesel with and without EGR is shown in Fig. 8. The smoke increased with the increase in load conditions. The smoke emission of diesel at full load is 54.1% whereas that of cold EGR is 59.7% and hot EGR is 55.3%. This is due to external rebreathing system which reduces the availability of O_2 for complete combustion of fuel. These outcomes in a moderately inadequate ignition, prompting an expanding smoke emission.

Figure 9 shows the variation of NO_x emissions with brake power from the diesel engine. The NO_x emission expended with the expansion of load conditions. The NO_x emission of diesel at full load is 1074 ppm in volume whereas that of cold EGR is 792 ppm in volume and hot EGR is 820 ppm in volume. This is because EGR reduces the oxygen concentration in charge; consequently, the combustion pressure and temperature are decreased because of the high specific heat of EGR. Cooling of the external rebreathing system expands the charge density; consequently, it improves the volumetric productivity of the engine. Likewise, it gives extra benefits by bringing

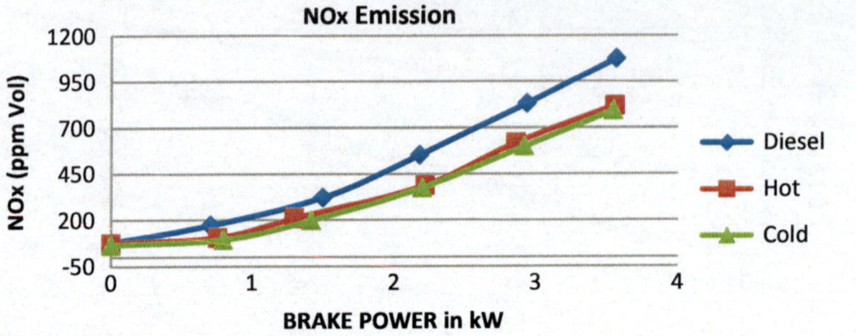

Fig. 9 BP (kW) versus NO_x (ppm Vol)

NO_x emissions down to a more noteworthy degree. Be that as it may, the buildup of dampness present in the fumes expands erosion in the burning chamber. The explanation for this is diminished oxygen focus in view of a weakening of admission charge and diminished flame temperature. The NO_x and smoke are the trade-offs.

5 Conclusion

These results taken from the electronically controlled DAQ system are very accurate compared to the manual graph values of EGR for hot and cold methods.

The engine performance characteristics at the maximum load are the BT efficiency of diesel is 31 kW, hot EGR is 32.7 kW, and cold EGR is 31.5 kW e brake thermal efficiency slightly increased in 10% hot and cold EGR compared to 0% EGR.

EGR upsurges the hydrocarbon and carbon monoxide emissions. Also, higher brake-specific fuel consumption and particulate emission are observed.

Utilizing a chilled external rebreathing system raises the density of the admission air in this way, and the measure of air entering the ignition chamber increments. Complete burning at that point is accomplished, along these lines lessening the generation particulate matter.

The nitrogen oxide emission of diesel at full load is 1074 ppm in volume whereas that of cold EGR is 792 ppm in volume and hot EGR is 820 ppm in volume. Therefore, it was concluded that the engine is operated with diesel while engaging hot and cold external rebreathing system results in nitrogen oxide decrease without trading off engine execution.

Acknowledgements The authors agree honestly "The All India Council for Technical Education" (AICTE), New Delhi, India, for the budgetary help in completing this research work under the research proposal scheme.

References

1. Mofijur M, Rasul MG, Hyde J, Azad AK, Mamat R, Bhuiya MMK (2016) Role of biofuel and their binary (diesel–biodiesel) and ternary (ethanol–biodiesel–diesel) blends on internal combustion engines emission reduction. Renew Sustain Energy Rev 53:265–278
2. Mofijur M, Rasul MG, Hyde J, Bhuyia MMK (2015) Role of biofuels on IC engine emission reduction. Energy Proc 75:886–892
3. Tabata M, Yamamoto T, Fukube T (1995) Improving NO_x and fuel economy for mixture injected si engine with EGR. SAE Trans J Engine 104(3):1221–1230 (SAE Paper No. 950684)
4. Agarwal AK (1996) Vegetable oil test fuels for diesel engines: formulation and analyses. Center for Energy Studies, Indian Institute of Technology, Delhi, India
5. Srivastava A, Ram Prasad (2000) Triglycerides-based diesel Fuels. Int J Renew Sustain Energy Rev, 111–133
6. Buyukkaya E, Cerit M (2007) Experimental study of NO_X emissions and injection timing of a low heat rejection diesel engine. Int J Thermal Sci
7. Feuge RU, Gros AT (1949) Modification of vegetable oils, VII Alkali catalyzed interesterification of peanut oil with ethanol. J Am Oil Chem Soc 26:97–108
8. Agarwal D, Sinha S, Agarwal AK (2006) Experimental investigation of control of NO_X emissions in biodiesel-fueled compression ignition engine. Renew Energy 31:2356–2369
9. Senthilkumar R, Ramadoss K, Manimaran R (2013) Experimental investigation of performance and emission characteristics by different exhaust gas recirculation methods used in a diesel engine. Double-Blind Peer-Reviewed Int Res J 13(1):15–20
10. Matsuo O, Noriyuki K, Yujiro T, Kazuysawa, Koichi Y (1991) Effects of EGR with supplemental manifold water injection to control exhaust emissions from heavy-duty diesel-powered vehicles. Soc Autom Eng. SAE paper no. 910739
11. Kanda T, Hakozaki T, Uchimoto T, Hatano J, Kitayama N, Sono H (2006) PCCI operation with fuel injection timing set close to TDC 2006. SAE Paper No. 2006-01-0920
12. Douglas R, Kee R, Carberry B (1997) Analysis of in-cylinder pressure data in two-stroke engines. SAE Paper No. 972792
13. Freedman B, Pryde EH, Mounts TL (1984) Variables affecting the yields of fatty esters from transesterification vegetable oils. J Am Oil Chem Soc 61:1638–1643
14. Freedman B, Butterfield RO, Pryde EH (1997) Transesterification kinetics of soybean oil. J Am Oil Chem Soc 74(11):1457–1463

Use of Chemical Scrubbing Method to Minimise the Pollutants

Jasbir Singh, Dipesh Popli, Harshit Nailwal, and Himanshu

1 Introduction

Pollution is one of the major crises currently faced by the world. Pollution is often the by-product of industrialisation, construction, vehicles and various other activities. It is responsible for global warming, intoxication of natural resources, death and extinction of various flora and fauna, etc. According to an estimate, in coming fifteen years, earth will face a major irreversible climate change. Air pollution is currently one of the hazardous problems of this world. It caused a domino effect that started the greenhouse effect, ozone hole and climate change, etc. Air pollution is caused by gases or particles suspended in the air. The air pollution [1] from automobiles possesses serious challenges to the environment. There are six parameters used to measure air pollution, i.e. carbon monoxide, carbon dioxide, nitrogen oxides, sulphur oxides, hydrocarbons and shoot particles. Therefore, it is necessary to control the pollution levels from the automobiles [2].

The chemical scrubbing method [3–5] can be used to clean the exhaust gases of internal combustion engines, bulky generators and other machines running on petrol. In this research paper, we are going to focus on one of the way to reduce air pollution by using a chemical scrubbing method. The main principle of our research is versatility of sodium hydroxide in reacting to various flue gases produced by various industries [6]. So, an attempt has been made to develop an economical device to reduce the air pollution caused by exhaust gases of a petrol engine. The different chemical as a scrubbing reagent has been used to clean thc different kind of exhaust

J. Singh (✉) · H. Nailwal · Himanshu
Department of Mechanical Engineering, Gurukula Kangri Vishwavidyalaya, Haridwar, Uttarakhand, India
e-mail: Jasbir.nitk@gmail.com

D. Popli
Bharat Institute of Engineering and Technology, Hyderabad, Telangana, India

A. K. Gupta et al. (eds.), *Advances in IC Engines and Combustion Technology*, Lecture Notes in Mechanical Engineering,
https://doi.org/10.1007/978-981-15-5996-9_43

gases. The quantitative analysis of aqueous salt produced after the chemical reaction has been performed.

2 Experimental Set-Up

The device is made to separate pollutants by chemical scrubbing method as shown in Fig. 1. The main body of the device is made in the form of a cuboidal shell of dimension 15 × 15 cm^2 in area and 50 cm in height. The shell is made up of stainless steel. A 15 cm diameter pipe of length 50 cm is welded to connect the connection tube and used as inlet for exhaust gases. The exhaust gases produced during the working of petrol engine have high pressure and temperature (100–150 °C). The vulcanised rubber tubes have been used as a connection tube because these tubes can endure high pressure and temperature.

Sawdust cleaner is made up of cylindrical vessel filled with sawdust which acts as a filter device to separate shoot particles, i.e. unburnt carbon particles. The sawdust acts as fine particles which collide with the shoot particles and adsorb them. The remaining pollutants in the exhaust gases react with the chemical present in the chemical chamber to clean the pollutant particles by chemical scrubbing method. The chamber is provided with a number of small holes to make the chamber permeable to gases and have capacity to hold the packing material, i.e. bituminous coal. The scrubbing material stored in the bucket is uniformly regulated over the chamber with the help of an electric motor. A rubber pipe is used to carry chemical from bucket to chemical chamber. To distribute the chemical uniformly, the pipe end at chemical chamber is pierced with number of holes. When chemical flows through the pipe, it spills over the holes provided in the chemical chamber and reaches to the parts of chamber. The discharge of chemicals is regulated with the help of electric motor.

3 Experimental Work

Based on literature survey, a number of chemicals were used to control the pollutant of exhaust gases. Various factors such as pressures, temperatures required for the reaction, total heat of enthalpy of the reaction, by-products and effects of the by-products on environment are considered while selecting the chemical. The chemicals such as $CaCO_3$, $MgCO_3$ and NaOH are taken for the study and chemical reaction of these chemicals with pollutants, i.e. NO_2, SO_2 and CO_2 are presented.

- Calcium carbonate ($CaCO_3$):

$$NO_2 + CaCO_3 \rightarrow Ca\,(NO_3)_2 + CO_2$$

$$CaCO_3 + SO_2 + O_2 \rightarrow CaSO_4 + CO_2$$

Fig. 1 Petrol engine connected to emission control device

$$CaCO_3 + CO_2 + H_2O \rightarrow CaHCO_3 \text{(very slow reaction)}$$

- Magnesium carbonate ($MgCO_3$):

$$MgCO_3 + NO_2 \rightarrow Mg\,(NO_3)_2 + CO$$

$$MgCO_3 + SO_2 + O_2 \rightarrow MgSO_4 + CO_2$$

$$MgCO_3 + H_2O\ \text{(Hot)} \rightarrow Mg_2CO_3(OH)_2 + CO_2$$

$$Mg_2CO_3(OH)_2 + CO_2 \rightarrow 2\,MgCO_3 + H_2O\ (180\,°C - 220\,°C)$$

- Sodium hydroxide (NaOH):

$$2NO_2(g) + 2NaOH\,(aq) \rightarrow NaNO_2(aq) + NaNO_3(aq) + H_2O$$

$$SO_2(g) + 2NaOH\,(aq) \rightarrow Na_2SO_3(aq) + H_2O$$

$$CO_2(g) + 2NaOH\,(aq) \rightarrow Na_2CO_3(aq) + H_2O$$

The by-products of chemical reactions of exhaust gases with $CaCO_3$ and $MgCO_3$ were harmful gases. Moreover, some of the reactions were either very slow or require high temperatures. The sodium hydroxide (NaOH) fulfilled our desired objective, i.e. no harmful by-product gases, and the reaction can take place at atmospheric temperature and pressure. Therefore, sodium hydroxide (NaOH) is used as a chemical to react with different exhaust gases in order to reduce their quantity. The exhaust gases are made to pass through a sawdust [7] cleaner with the help of connection tubes before they go into the set-up device. The function of sawdust collector is to absorb the particulate matters. An exhaust fan is provided to create forced draught in the device so that all the gases move upwards. The bituminous coal is used as wet scrubber to treat the variety of pollutant present in the exhaust gases. The bituminous coal absorbs various pollutant particles. A continuous shower of aqueous NaOH solution is sprinkled over the bituminous coal. The NaOH reacts with the oxides of carbon, nitrogen, sulphur into their respective salts and water. Most of the water gets vapourised due to the large heat produced during the reaction and rest of the water containing different salts settled in the bottom of the device. The confirmatory tests have been performed on salts to check the presence of pollutant, which are absorbed during the process. The schematic diagram presents the working of the device in Fig. 2.

4 Results and Discussion

The aqueous salts are tested in laboratory in order to confirm that device absorbs pollutants from exhaust gases. Confirmatory tests for anions are performed by using water extract when salt is soluble in water and by using sodium carbonate extract when salt is insoluble in water. In present study, salts formed after reaction with the chemical in chemical chamber are soluble in water. Therefore, water extract has been used for laboratory testing.

Test for carbonate ions ($CO_3{}^{2-}$):

The diluted sulphuric acid has been added to the salt. The CO_2 gas is evolved with brisk effervescence which turns lime water milky. If there is effervescence with the evolution of a colourless and odourless gas on adding diluted H_2SO_4 to the salt, this indicates the presence of carbonate ion. The gas turns limewater milky due to the formation of $CaCO_3$.

$$Na_2CO_3 + H_2SO_4 \rightarrow Na_2SO_4 + H_2O + CO_2$$

$$Ca(OH)_2 + CO_2 \rightarrow CaCO_3 + H_2O$$

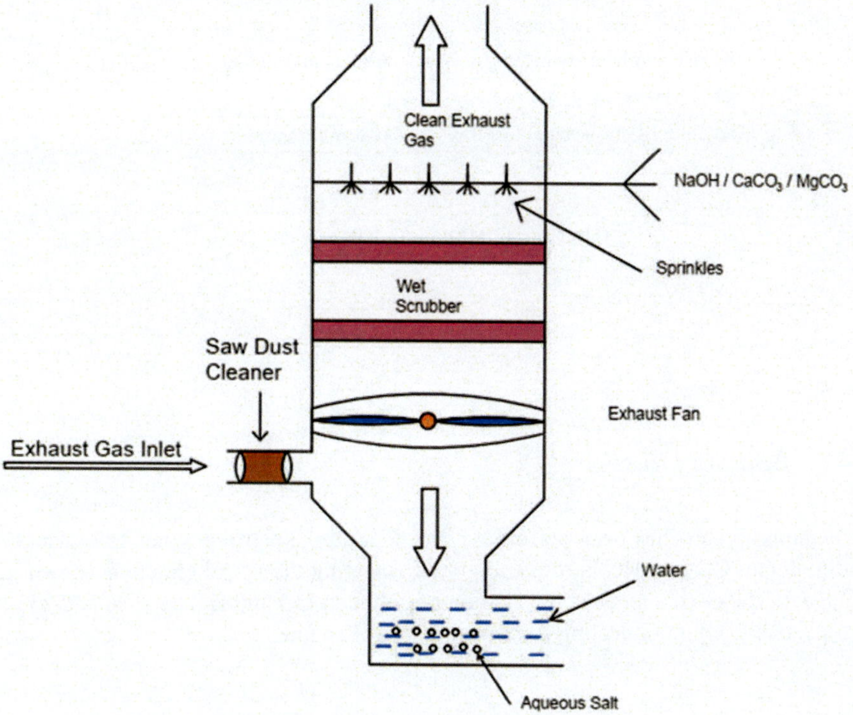

Fig. 2 Schematic diagram depicting the working of device

Test for sulphite ion (SO_3^{2-}):

On treating sulphite with warm diluted H_2SO_4, SO_2 gas is evolved which is suffocating with the smell of sulphur.

$$Na_2SO_3 + H_2SO_4 \rightarrow Na_2SO_4 + H_2O + SO_2$$

$$K_2Cr_2O_7 + H_2SO_4 + 3\ SO_2 \rightarrow K_2SO_4 + Cr_2(SO_4)_3 + H_2O$$

The gas turns potassium dichromate paper acidified with diluted H_2SO_4 into green colour. This change in colour indicates the presence of sulphate ions.

Test for nitrite ion (NO_2^-):

On treating solid nitrite with dil. H_2SO_4 and warming, reddish brown fumes of NO_2 gas are evolved. Addition of potassium iodide solution to the salt solution followed by freshly prepared starch solution and acidification with acetic acid produces blue colour. These reactions confirm the presence of nitrite ions.

$$2NaNO_2 + H_2SO_4 \rightarrow Na_2SO_4 + 2HNO_2$$

Table 1 Emission test of exhaust gases

S. no.	Gases	Without connecting device	After connecting device	Efficiency (%)
1	CO	0.71	0.53	27
2	HC	143 (n-hexane)	108 (n-hexane)	25

$$3HNO_2 \rightarrow HNO_3 + 2NO + H_2O$$

$$2NO + O_2 \rightarrow 2NO_2$$

4.1 *Emission Test*

The emission test has been performed on a four-stroke petrol engine. The quantity of pollutant CO and HC before and after connecting the petrol engine is shown in Table 1. The device reduces the amount of pollutant CO and HC by 27% and 25%, respectively, after connecting the device to petrol engine.

5 Conclusion

After performing various tests, it was observed that the device works efficiently and exhaust gases react with the scrubbing chemical which results in the reduction of the pollutant gases. The blackish smoke produced by the petrol engine was cleaned and reduces the quantity of particulate matter and the unburnt carbon particles. By using inorganic salt test and pollution test, it is evident that exhaust gases such as NO_x, SO_x, CO, CO_2, HC and PM are reduced by the device. There is scope to improve the efficiency of the device and create an automatic control system to manipulate discharge of scrubbing material according to the volume of intake of exhaust gas.

References

1. Elliott MA, Nebel GJ, Rounds FG (1955) The composition of exhaust gases from diesel, gasoline and propane powered motor coaches. J Air Poll Control Assoc 5:103–108. https://doi.org/10.1080/00966665.1955.10467686
2. Raja RAA, Priyadarshini M, Akilan J (2015) Paper on emission control techniques. In: National conference in Magna on recent trends in mechanical engineering, pp 229–237
3. Andreasen A, Mayer S (2007) Use of seawater scrubbing for SO_2 removal from marine engine exhaust gas. Energy Fuels 21:3274–3279. https://doi.org/10.1021/ef700359w

4. Diao YF, Zheng XY, He BS, Chen CH, Xu XC (2004) Experimental study on capturing CO_2 greenhouse gas by ammonia scrubbing. Energy Convers Manag 45:2283–2296. https://doi.org/10.1016/j.enconman.2003.10.011
5. He B, Zheng X, Wen Y, Tong H, Chen M, Chen C (2003) Temperature impact on SO_2 removal efficiency by ammonia gas scrubbing. Energy Convers Manag 44:2175–2188. https://doi.org/10.1016/S0196-8904(02)00230-3
6. Haridass R, Ramesh K, Rajeshkumar T, Vignesh E (2017) Performance improvement of pollution control device used in small scale foundry industry. Int J Res Sci Eng 3:118–122
7. Nwigbo MN (2019) Design, fabrication and performance evaluation of a cyclone system for saw dust emission control 5:19–27

Performance Characteristics and Emission Analysis of Nano Additives Added Mustard Oil Biodiesel in a Compression Ignition Engine with EGR

Govindaraj Elavarasan, P. Rajakrishnamoorthy, Muthu Kannan, Duraisamy Karthikeyan, and C. G. Saravanan

1 Introduction

Because of the depletion of fossil fuel based products, many researches have focused to work in the field of Renewable source of energy. Automobile sector utilizes most of the crude oil products worldwide and it is kept on increasing due to the increased number of vehicles. This tends to the increase of exhaust emissions from the Automobiles. The major emissions of Spark Ignition (SI) Engine were HC, NO_x, and CO. In Compression Ignition (CI) Engine the emissions were HC, CO, NO_x, and smoke in addition [1]. The environmental protection activist of various countries has made strict rules to reduce the toxic emissions from the Internal Combustion Engines. Each country has different testing methods for determining the automobile exhaust emission and different emission standards were implemented according to their convenience. The maximum permissible CO, HC, and NO_x emissions for passenger car

G. Elavarasan (✉) · P. Rajakrishnamoorthy · D. Karthikeyan · C. G. Saravanan
Annamalai University, Chidambaram, Tamil Nadu 608001, India
e-mail: elavarasanmit@gmail.com

P. Rajakrishnamoorthy
e-mail: prk.spm1984@gmail.com

D. Karthikeyan
e-mail: dkarthi66@yahoo.com

C. G. Saravanan
e-mail: rsdk66@yahoo.com

M. Kannan
NCR Campus, SRMIST, Ghaziabad, Uttar Pradesh 201204, India
e-mail: royalkannan11@gmail.com

A. K. Gupta et al. (eds.), *Advances in IC Engines and Combustion Technology*, Lecture Notes in Mechanical Engineering,
https://doi.org/10.1007/978-981-15-5996-9_44

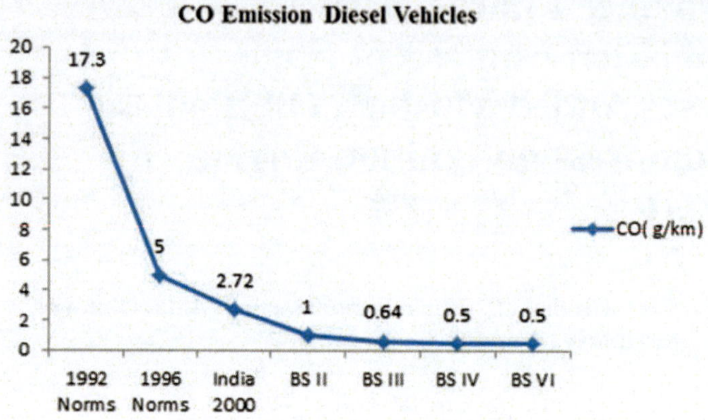

Fig. 1 Indian Emission standards for CO emission in passenger cars

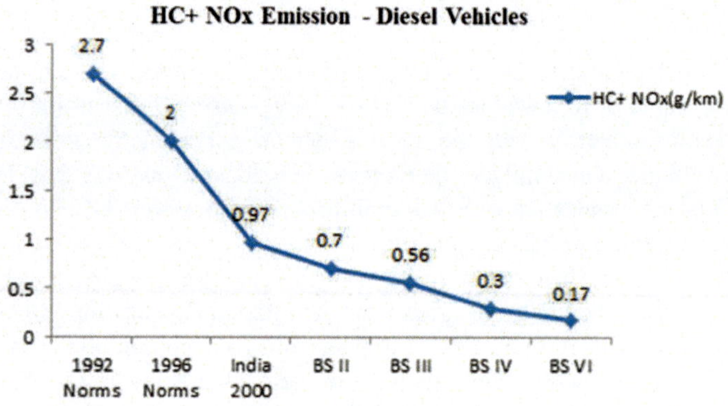

Fig. 2 Indian Emission standards for HC + NO_X emission in passenger cars

segments followed in India were shown in Figs. 1 and 2 [2]. The general vegetable oils considered for the purpose of biodiesel production were corn, cotton, soybean, mustard, Camelina, Safflower, Sunflower, Canola, Rapeseed, Jatropha, coconut and palm oil [3–5]. Mustard oil was taken as a source of biodiesel for testing. The triglyceride mustard oil was first converted into biodiesel by trans-esterification process to remove the fatty acids and the same was blended with diesel in the three volumetric proportions namely MBD5 (5% MBD and 95% diesel), MBD10 (10% MBD and 90% diesel) and MBD15 (15% of MBD and 85% of diesel) and added with 100 ppm of Al_2O_3 nanoparticles [6, 7]. The Investigation of the Al_2O_3 nanoparticles additive on the mustard oil based biodiesel was not reported. The same was tested in a constant speed diesel engine at different load conditions. Then the performance characters like Brake Thermal Efficiency (BTE), Specific Fuel Consumption (SFC), and the emissions generated by the engines CO, HC, NO_x, and smoke were calculated for every blend individually and compared with that of the diesel [8]. The results show that the MBD15 blend has less HC, CO, and smoke emissions and the high NO_x emissions than that of other test fuels [9]. In order to reduce the excess NO_x emissions formed the 10% of EGR was introduced into the combustion chamber and the results obtained show the all the emissions formed [10].

2 Preparation of Biodiesel

Mustard oil mainly contains the fatty acids like oleic acid, linoleic acid, and alpha-linolenic acid in it [11]. These fatty acids in the mustard oil should be removed by trans-esterification process [12]. In this process, the available fatty acid in the mustard oil will be converted into fatty acid methyl ester by adding the methanol in the presence of sodium hydroxide as a catalyst [13]. Initially, the catalyst 1 g of NaOH was mixed with the 1litre of distilled water. The methanol and the mustard oil were mixed in the ratio of 10:1 in a separate conical flask and added with a small amount of phenolphthalein (color indicator) [14, 15]. Then, the prepared catalyst was added to the solution drop by drop until it turns to pink. Then the solution is stirred at 600 rpm and to obtain the homogenous mixture for about two hours. Then the mixture was transferred into a separation funnel and kept undisturbed for two hours. A mass of glycerol was formed in the bottom layer of the funnel was removed with the help of the tap [16]. Then the oil was washed with distilled water and heated at 90 °C for about 30–40 min to remove the excess water content in the oil. Then the oil was blended with the diesel in the volumetric proportion of 5, 10, and 15% for the testing purpose. The prepared mustard oil biodiesel properties and the diesel oil properties were listed in Table 1. Then the blends of the biodiesel were added with 100 ppm of Al_2O_3 nanoparticles with the help of an ultrasonicator (40 kHz for about 45 min) and kept undisturbed for about 48 h [17, 18].

Table 1 Comparison of mustard oil biodiesel with that of diesel

S. No.	Properties	Mustard oil	Diesel
1	Kinematic viscosity (mm^2/s)	6.6	3
2	Flash point (°C)	105	63.6
3	Fire point (°C)	125	91
5	Pour point (°C)	−12	−6
6	Density (g/ml)	0.895	0.831
7	Pour point (°C)	−13	−6
8	Cetane index	50	55
9	Calorific value (kJ/kg)	39,857	45,343

3 Experimental Method

The constant speed engine will work at the constant rpm of 1500 rpm irrespective of the load. With the increase in load, the engine will consume more fuel in order to maintain the 1500 rpm. The photographic view of the engine was shown in Fig. 3. So the fuel consumption will increase with increase in the load. Initially, the engine was allowed to run with diesel fuel in the no-load condition. Then the load was added to the engine with the help of an eddy current dynamometer step by step and for each load, the performance and emission characters were measured.

Then for the same load conditions, the three blends (MBD5, MBD10, and MBD15) added with Al_2O_3 nanoparticles were tested and the performance and emission characteristics were measured. The specification of the engine used for the testing is shown in Table 2. The smoke emission was measured with the help of DX260 Smoke meter and the harmful emissions like HC, CO, and NO_x were

Fig. 3 Test engine

Table 2 IC engine set up specification

1	Fuel	Diesel
2	Type	1 cylinder, 4 stroke, constant speed (1500)
3	Max power	3.50 kW
4	Dynamometer	Eddy current
5	Swept volume	661.45 cc
6	Compression ratio	17.50:1
7	Cooling	Water cooled
8	Cylinder Bore	87.50 mm
9	Stroke length	110.00 mm

Table 3 Accuracy and uncertainty analysis

Parameters	Range	Accuracy/uncertainty
Smoke emission	0-100%	0.1% uncertainty
HC	0.01–4 (%)	350 ppm accuracy
NO_X	0–3000 (ppm)	8 ppm
CO	0–10,000 (ppm)	5 ppm
Brake power	0–3.5 kW	±0.25% uncertainty
Fuel flow	0–10 g/h	±0.5% uncertainty
Fuel consumption	0–10 cc	±0.1% accuracy
Engine speed	1500 rpm	±1% accuracy

measured with the help of AVL DIGAS 444 Analyzer. The accuracy and the uncertainty of the various parameters contributed to the experiment were shown in Table 3.

4 Results and Discussions

4.1 BP Versus Brake Thermal Efficiency (BTE)

BTE is the amount of energy in the fuel utilized for useful work of the engine [19]. The variations of the BTE with respect to the engine brake power generated were shown in Fig. 4. It can be concluded that with the increase in the brake power of the engine the brake thermal efficiency is increasing and also the BTE of the diesel (MBD0) was found to be highest in all the operating conditions.

The reason for the reduction of the BTE with increase in the composition of the mustard biodiesel in the diesel is to the less calorific value of biodiesel [20]. So, the calorific value of the fuel is found to be decreasing in nature with increase in the

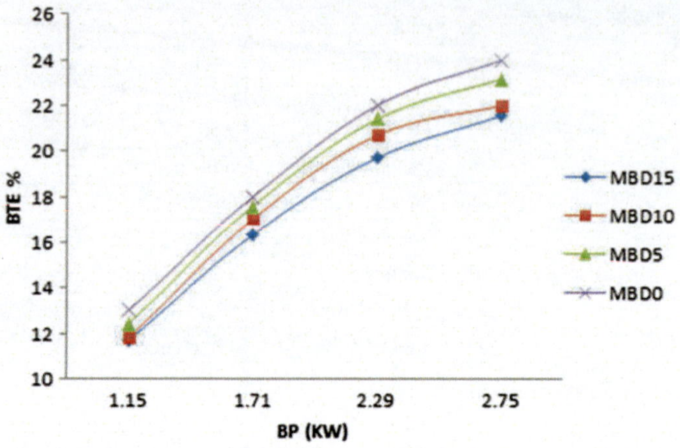

Fig. 4 BP in KW versus BTE in %

composition of the mustard oil blend in the diesel. For 2.29 BP the BTE of MBD5, MBD10, and MB15 is 2.7, 5.9, and 10.4% lesser than that of the MBD0 blend.

4.2 Brake Power Versus Specific Fuel Consumption (SFC)

As the engine needs to maintain the constant speed irrespective of the increase in load, the fuel consumption will be more with an increase in the load [21]. The mustard oil biodiesel already has less calorific value results in more fuel consumption than that of the diesel in the same load condition. Also, the viscosity of the fuel will increase with the increase in the volumetric composition of the blend which also contributes to the increase in fuel consumption [22]. The variation of the SFC with respect to the brake power of the engine was shown in Fig. 5. It can be concluded that the MBD15 blend has a maximum SFC than other test fuels. The SFC of MBD5, MBD10, and MBD15 at 1.71 kW are 4.3, 12, and 20% more than that of the MBD0 blend.

4.3 Brake Power Versus Smoke Emission

Figure 6 shows the variation of smoke emission with respect to that of the BP of the engine. From Fig. 6, it is clear that the smoke emission is reducing with the addition of blend because of more oxygen content in the mustard oil biodiesel [23]. Smoke emission is formed due to the unburned fuel molecules in the liquid state inside the combustion chamber [24]. The diesel fuel is having the maximum smoke emission and it is decreasing with an increase in the blend. As the mustard oil blend will increase the oxygen content of the fuel, the fuel will burn completely, so the

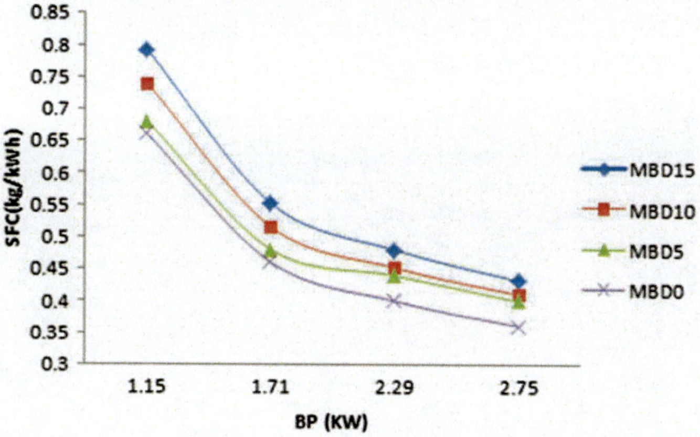

Fig. 5 BP in KW versus Specific fuel consumption in kg/kWh

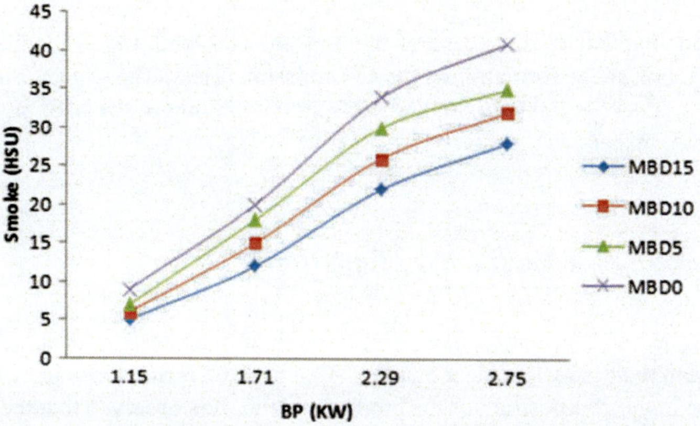

Fig. 6 BP in KW versus smoke in ppm

smoke emission of the fuel decreased with increase in the blend. The observed smoke emission of MBD0 at 2.29 kW is 11.7, 23.5, and 35.29% more than that of the MB5, MBD10, and MBD15, respectively.

4.4 *Brake Power Versus CO Emission*

The variation of the brake power with respect to that of the CO emission release was shown in Fig. 7. It is clear from the figure that the CO emissions formation is decreased with an increase in the proportion of the mustard oil biodiesel. The main reason for the formation of the CO emission is the deficiency of oxygen in the

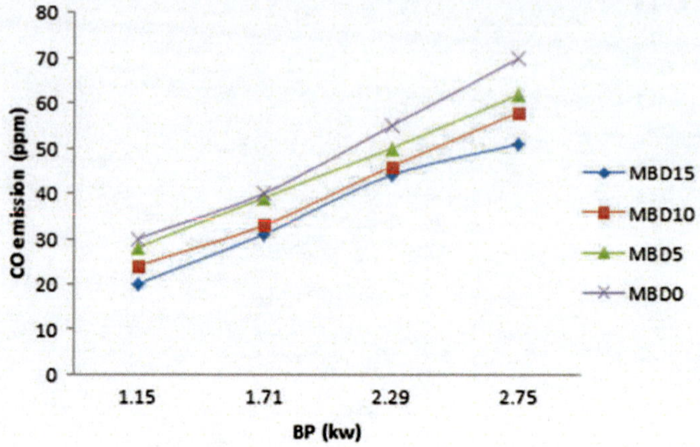

Fig. 7 BP in KW versus CO emission in ppm

combustion chamber [25]. In case of the mustard oil blend, the availability of the oxygen is more, so the formation of the CO emission is less. The CO emission of the MBD0 at 2.29 kW is 9.09, 16.36, and 20% more than that of the MBD5, MBD10, and MBD15, respectively.

4.5 Brake Power Versus NO_x Emission

Figure 8 shows the formation of the NO_x emission with respect to the BP of the engine. From the figure, it is clear that the NO_x emission formation is increased with an increase in the composition of the biodiesel blend. It is observed that the MBD15

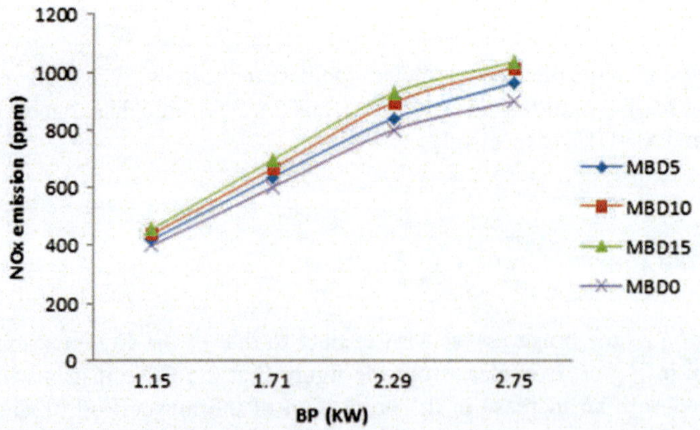

Fig. 8 BP in KW versus NO_x emission in ppm

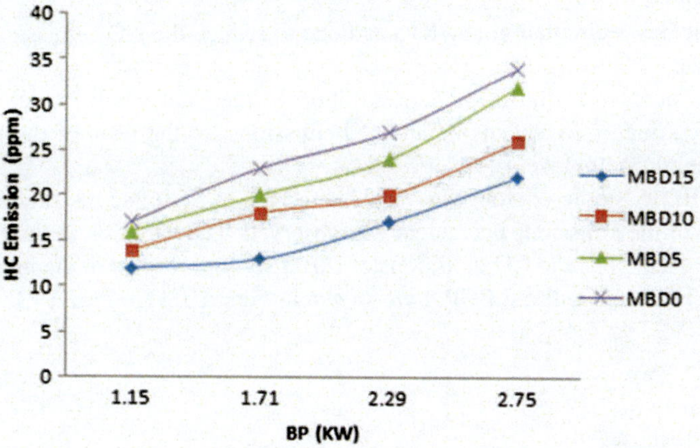

Fig. 9 BP in KW versus HC emission in ppm

has the maximum NO_x emission among all test fuels. Because of the enhanced combustion in the biodiesel blend and the high temperature inside the combustion chamber, the NO_x emission formation is increasing with an increase in the proportion of the biodiesel in the blend [14]. The NO_x emissions of the MBD0 at 2.29 kW are 5, 12, and 16% less than that of the MBD5, MBD10, and MBD15, respectively.

4.6 *Brake Power Versus HC Emission*

The main reason for the formation of the HC emission is the unburned fuel-air mixture inside the combustion chamber. But the mustard oil blend has better combustion properties than that of the diesel fuel, so the HC emission formation is less with an increase in the mustard oil blend [25]. Figure 9 shows the variation in the formation of the HC emissions with respect to that of the engine brake power. It can be seen from the figure that the HC emission is maximum in the MBD0 fuel compared to other test fuel. Thus, the biodiesel blend MBD15 has a lesser HC emission, as the formation of the HC emission will reduce with the increase in the biodiesel blend. The HC emission of the MBD0 at 2.29 kW is 11.11, 25.92, and 37.03% more than that of the MBD5, MBD10, and MBD15, respectively.

4.7 *MBD15 with EGR*

From the various emission results, it can be concluded that the MBD15 is a better choice of reducing the HC, CO, and Smoke emissions, but the NO_x emission formed is more than of the other test fuels. So, in order to reduce the NO_x emission the 10%

of Exhaust Gas recirculation (EGR) was done to reduce the NO_x emission into the engine [26].

As the increase in the EGR concentration in the intake will reduce the NO_x emission but increases the smoke and CO emissions, so the level of the EGR was kept to the minimum level [27].

The MBD15 biodiesel alone was tested with 10% of EGR and the variation in the formation of the emissions characters was observed. The variation in the formation of NO_x, smoke, HC, and CO emissions of MBD 15 with respect to the brake power with 10% EGR and without EGR were shown in Figs. 10, 11, 12, and 13.

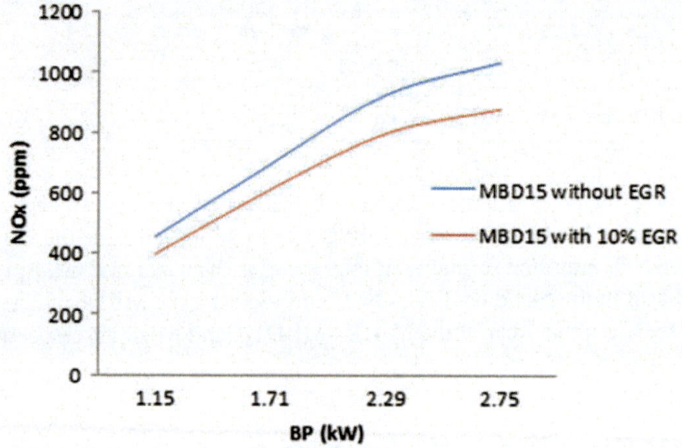

Fig. 10 NO_X emission of MBD15 with and without EGR

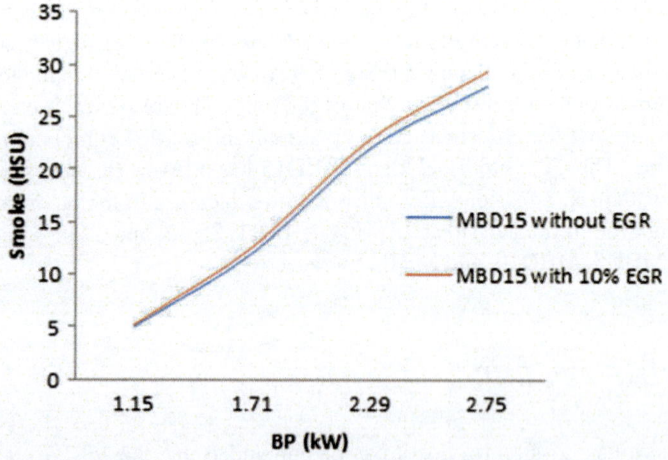

Fig. 11 Smoke emission of MBD15 with and without EGR

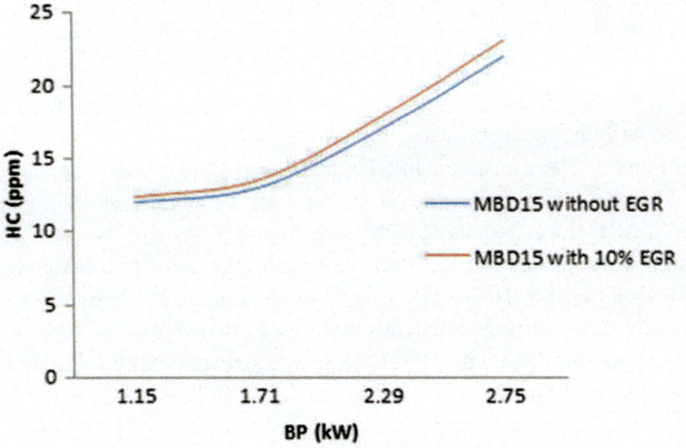

Fig. 12 HC emission of MBD15 with and without EGR

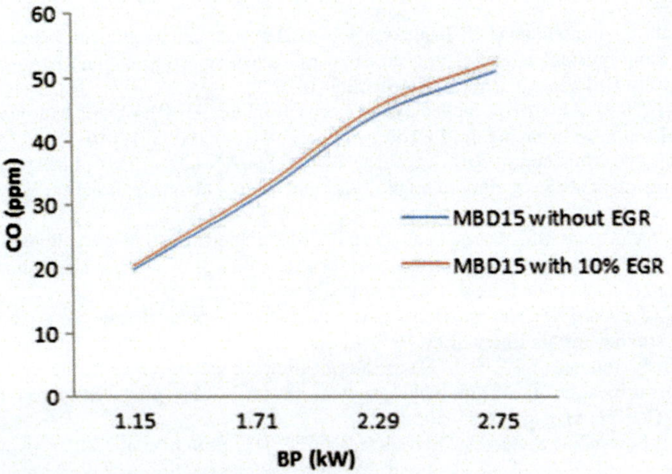

Fig. 13 CO emission of MBD15 with and without EGR

It is clear from the figures that the NO_x emission was decreased about 12%, smoke emission was increased about 5% (3.4% less than diesel), HC emission increased about 3% (3.8% less than diesel) and CO emission increased about 4% (18.2% less than that of diesel) for MBD15 with EGR when compared to MBD15 without EGR.

5 Conclusion

Thus, the mustard biodiesel was produced and it was blended with 100 ppm Aluminum Oxide nanoparticles was mixed with diesel in various ratios and tested in a diesel engine. The emissions like HC, CO, and smoke were decreased with an increase in biodiesel blend. But the NO_x emissions were increased with an increase in the blend. MBD15 is a better choice if we want to reduce the HC, smoke, and CO emission. As it produces more NO_x emission, this requires a selective catalytic reduction technique. So, 10% EGR was added to reduce the NO_x emission in the MBD15 fuel and the various emissions were measured. From the results, it can be concluded that the MBD15 with 10% EGR is the best way to reduce all the HC, CO, NO_x, and smoke emissions.

References

1. Gobinath S, Senthilkumar G, Beemkumar N (2019) Air nanobubble-enhanced combustion study using mustard biodiesel in a common rail direct injection engine. Energy Sour Part a-Recovery Util Environ Effects 41(15):1809–1816
2. Arat HT (2019) Alternative fuelled hybrid electric vehicle (AF-HEV) with hydrogen enriched internal combustion engine. Int J Hydrogen Energy 44(34):19005–19016
3. Devarajan Y, Munuswamy DB, Nagappan B, Pandian AK (2018) Performance, combustion and emission analysis of mustard oil biodiesel and octanol blends in diesel engine. Heat Mass Transf 54(6):1803–1811
4. Karthic SV, Kumar MS, Nataraj G, Pradeep P (2019) Experimental investigations on the influence of hydrogen and LPG mixtures on performance behavior of a mahua bio oil-powered dual fuel engine. Int J Green Energy 16(12):878–889
5. Dixit S, Kanakraj S, Rehman A (2012) Linseed oil as a potential resource for bio-diesel: a review. Renew Sustain Energy Rev 16(7):4415–4421
6. Fadhil AB, Abdulahad WS (2014) Transesterification of mustard (Brassica nigra) seed oil with ethanol: purification of the crude ethyl ester with activated carbon produced from de-oiled cake. Energy Convers Manag 77:495–503
7. Verma P, Stevanovic S, Zare A, Dwivedi G, Van TC, Davidson M et al. (2019) An overview of the influence of biodiesel, alcohols, and various oxygenated additives on the particulate matter emissions from diesel engines. Energies 12(10)
8. Yildirim S, Tosun E, Calik A, Uluocak I, Avsar E (2019) Artificial intelligence techniques for the vibration, noise, and emission characteristics of a hydrogen-enriched diesel engine. Energy Sour Part a-Recovery Util Environ Effects 41(18):2194–2206
9. Chintala V, Benaerjee D, Ghodke PK, Porpatham E (2019) Hydrogen rich exhaust gas recirculation (H2EGR) for performance improvement and emissions reduction of a compression ignition engine. Int J Hydrogen Energy 44(33):18545–18558
10. Jham GN, Moser BR, Shah SN, Holser RA, Dhingra OD, Vaughn SF et al. (2009) Wild Brazilian Mustard (Brassica juncea L.) seed oil methyl esters as biodiesel fuel. J Am Oil Chem Soc 86(9):917–926
11. Ganesan S, Rathinam S, Sajin JB, Yuvarajan D (2019) Performance and emission study on the effect of oxygenated additive in neat biodiesel fueled diesel engine. Energy Sour Part a-Recovery Util Environ Effects 41(16):2017–2027
12. Al-dobouni IA, Fadhil AB, Saeed IK (2016) Optimized alkali-catalyzed transesterification of wild mustard (Brassica juncea L.) seed oil. Energy Sour Part a-Recovery Util Environ Effects 38(15):2319–2325

13. Pandian AK, Munuswamy DB, Radhakrishana S, Ramakrishnan RBB, Nagappan B, Devarajan Y (2017) Influence of an oxygenated additive on emission of an engine fueled with neat biodiesel. Pet Sci 14(4):791–797
14. Sanjid A, Masjuki HH, Kalam MA, Rahman SMA, Abedin MJ, Palash SM (2013) Impact of palm, mustard, waste cooking oil and Calophyllum inophyllum biofuels on performance and emission of CI engine. Renew Sustain Energy Rev 27:664–682
15. Devarajan Y, Munuswamy DB, Radhakrishnan S, Mahalingam A, Nagappan B (2019) Experimental testing and evaluation of neat biodiesel and heptanol blends in diesel engine. J Test Eval 47(2):987–997
16. Antova GA, Angelova-Romova MI, Petkova ZY, Teneva OT, Marcheva MP (2017) Lipid composition of mustard seed oils (Sinapis alba L.). Bul Chem Commun 49:55–60
17. Yuvarajan D, Babu MD, BeemKumar N, Kishore PA (2018) Experimental investigation on the influence of titanium dioxide nanofluid on emission pattern of biodiesel in a diesel engine. Atmos Poll Res 9(1):47–52
18. Kumar N, Tomar M (2019) Influence of nanoadditives on ignition characteristics of Kusum (Schleichera oleosa) biodiesel. Int J Energy Res 43(8):3223–3236
19. Shrestha DS, Van Gerpen J, Thompson J (2008) Effectiveness of cold flow additives on various biodiesels, diesel, and their blends. Trans Asabe 51(4):1365–1370
20. Eryilmaz T, Ogut H (2011) The effect of the different Mustard oil biodiesel blending ratios on diesel engines performance. Energy Educ Sci Technol Part a-Energy Sci Res 28(1):169–180
21. Yuvarajan D, Ramanan MV (2016) Effect of magnetite ferrofluid on the performance and emissions characteristics of diesel engine using methyl esters of mustard oil. Arab J Sci Eng 41(5):2023–2030
22. Uyumaz A (2018) Combustion, performance and emission characteristics of a DI diesel engine fueled with mustard oil biodiesel fuel blends at different engine loads. Fuel 212:256–267
23. Madheshiya AK, Vedrtnam A (2018) Energy-exergy analysis of biodiesel fuels produced from waste cooking oil and mustard oil. Fuel 214:386–408
24. Issariyakul T, Dalai AK, Desai P (2011) Evaluating esters derived from mustard oil (Sinapis alba) as potential diesel additives. J Am Oil Chem Soc 88(3):391–402
25. Shahzadi I, Sadaf S, Iqbal J, Ullah I, Bhatti HN (2018) Evaluation of mustard oil for the synthesis of biodiesel: pretreatment and optimization study. Environ Prog Sustain Energy 37(5):1829–1835
26. Li YY, Wang P, Wang SQ, Liu JP, Xie YK, Li WJ (2019) Quantitative investigation of the effects of CR, EGR and spark timing strategies on performance, combustion and NO_x emissions characteristics of a heavy-duty natural gas engine fueled with 99% methane content. Fuel 255
27. Rajasekar E, Murugesan A, Subramanian R, Nedunchezhian N (2010) Review of NO_x reduction technologies in CI engines fuelled with oxygenated biomass fuels. Renew Sustain Energy Rev 14(7):2113–2121

Aerospace Flame and Combustion

Comparative Acoustic Analysis of Modified Unmanned Aerial Vehicle's Propeller

R. Vijayanandh, M. Ramesh, K. Venkatesan, G. Raj Kumar, M. Senthil Kumar, and R. Rajkumar

1 Acoustic Impact on Unmanned Aerial Vehicle

Generally, UAVs are the best alternate for the tired pilots, who has been assigned to work in dirty and dangerous environments. Pre-programmed or remote-controlled UAVs are implemented in complex applications, in which mostly pre-programmed autonomous UAVs are more suitable for complicated environments [1]. The Complex environment includes intruder detection at border, forest surveillance, border spying, and naval surveillance. UAVs are a prime invention in the aerospace industry, so the high lifetime of an UAV is a unavoidable factor [2]. Also, the fundamental work is to detect and solve the lifetime affecting issues of a UAV towards the attainment of its high lifespan with a low probability of failures [3].

Due to advanced technique, compact in size the UAVs are implemented in all the complex applications, but some of the issues in the UAV's components may collapse the entire unmanned aircraft system (UAS) usage [4]. The issues mainly comprise of propeller noise, propulsive heat, electromagnetic interference, etc. Therefore, to increase the level of implementation of UAVs in complex applications, it is mandatory to reduce the problem associated with UAV's components [5]. One of the prime detection factors of a UAV in complex regions is propeller noise, so this work deals

R. Vijayanandh (✉) · M. Ramesh · G. R. Kumar · M. S. Kumar
Department of Aeronautical Engineering, Kumaraguru College of Technology, Coimbatore, Tamil Nadu, India
e-mail: vijayanandh.raja@gmail.com

K. Venkatesan
Department of Mechanical Engineering, Alagappa Chettiar Government College of Engineering and Technology, Karaikudi, Tamil Nadu, India

R. Rajkumar
Department of Science and Humanities – Maths Division, Kumaraguru College of Technology, Coimbatore, Tamil Nadu,, India

A. K. Gupta et al. (eds.), *Advances in IC Engines and Combustion Technology*, Lecture Notes in Mechanical Engineering,
https://doi.org/10.1007/978-981-15-5996-9_45

with the numerical study of propeller noise variations and their reduction techniques [6]. It is understood that the prime contribution of the acoustic is generated due to the rotor vortices because of rotor geometrical parameters, such as twisting angle and blade thickness. The methodology proposed in this work is edges of the propeller are modified in order to reduce the blade thickness and thereby reduce the energy imbalance between the working and non-working environments [7].

2 Acoustic Analysis on the Fundamental Models

2.1 Conceptual Design

Naturally, understanding of the problem can provide requirements and its clear view, which will support the construction of the conceptual design. In conceptual design, requirement and its supportive effects directly play a vital role in the final output, which means component finalization and boundary selection are explained.

In this work, the fundamental component is propeller, which is shortlisted based on its output thrust. The output thrust force of the finalized propeller is 500 g within the dimensions of 5 inch in diameter and 4.5 inch in pitch. In most of the working cases, the maximum take-off weight is 2 kg, so the requirement of the average thrust of the propeller is around 500–750 g; hence, the short-listed propeller of this work is most suitable for multi-rotor UAV applications. Generally, the selection parameters involved in the propeller are outer boundary (diameter) and twisting angle (pitch). While dealing with the compact UAVs, the propeller must fit within compact size with a capability to provide required reasonable amount of thrust [8]. Hence, the shortlisted propeller is most suitable to act as a fundamental component of this work, which is shown in Fig. 1.

Generally, air particles are very sensitive, in which the way of dealing with the air particle will provide advantages and/or disadvantages. In order to get advantage

Fig. 1 Normal propeller

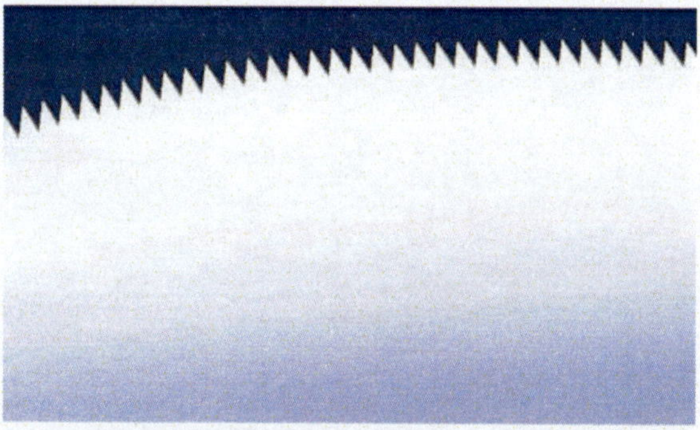

Fig. 2 Propeller with saw-tooth cut (Case-I)

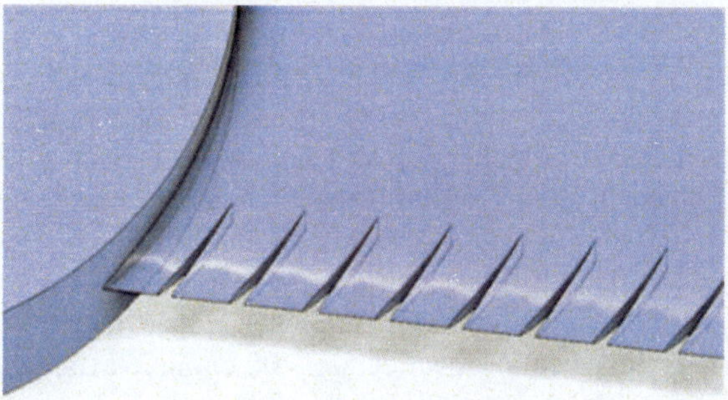

Fig. 3 Propeller with curved cut (Case-II)

output, the rotor must avoid the shocking environment in the working regions. One of the primary reasons for air disturbing elements is turbulence, which can be reduced for comfortable operation. In this work, initially, three different leading edge modifications are executed, which are saw-tooth cut, curved cut, and part-retained cut. The dimensions used in these cuts are obtained in the previous works [1, 9]. The saw-tooth cut equipped propeller is shown in Fig. 2. Figures 3 and 4 revealed the pictorial representations of modified leading edge based propellers with curved and part-retained cut, respectively.

2.2 Numerical Simulation

The complexity involved in this work is to deal with the air turbulence and its side effects in a positive manner by using numerical simulation. Generally, CFD results are approximate, but with the help of standard boundary conditions and good meshing,

Fig. 4 Propeller with part-retained cut (Case-III)

the post-processing outputs are able to change approximate into acceptable solutions. Generally, the control volume totally depends on the test model; in this case, the fundamental model is propeller, which has the dimensional data of 12.7 cm in diameter and 11.43 cm in pitch. Therefore, the reference parameter is 12.7 cm, which predominately contributes to the construction of control volume. From the previous works [10], it is understood that the control volume dimensions are min. 5 times and max. 20 times greater than the reference parameter. Hence, in this work, the control volume dimensions are calculated, which are: 400 cm in *X*-axis, 300 cm in *Y*-axis, and 300 cm in *Z*-axis. The discretized models of the various propellers with control volume are shown in Figs. 5 and 6.

The standard boundary conditions comprise of a selection of solver, selection of inlet, selection of turbulence model, and implementation of higher order in solver

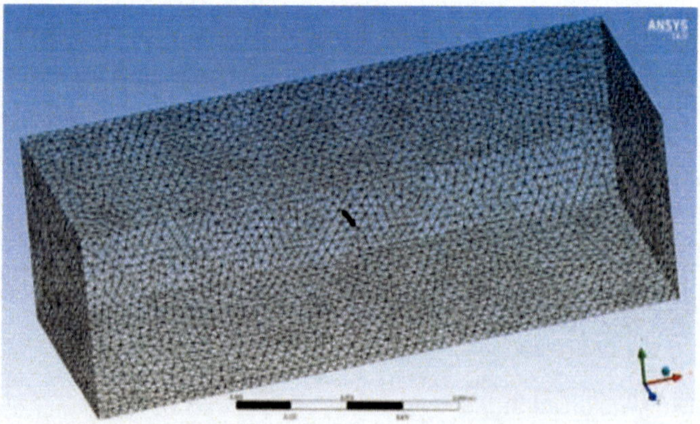

Fig. 5 A typical view of control volume

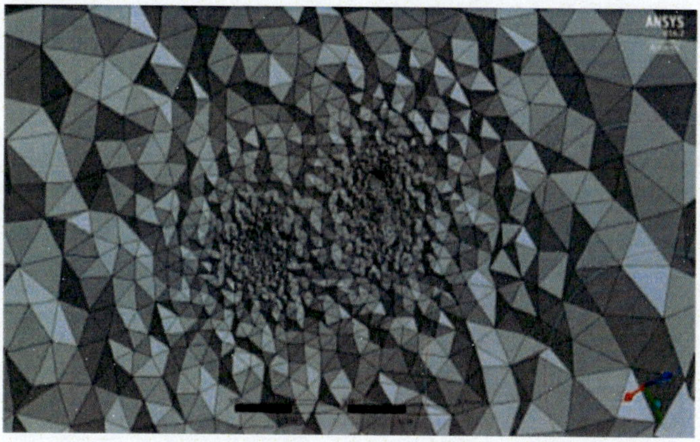

Fig. 6 Discretized model for modification

control. All these boundary conditions, fundamentally depend upon the working environment such as velocity and pressure. This work intentionally selected a velocity-based working environment. In order to attain the Vertical Take-Off and Landing (VTOL) and Horizontal Take-Off and Landing (HTOL) speed values, the literature survey is executed and working specifications of mini-UAVs are obtained, in which maximum VTOL and HTOL velocities are 15 m/s. Hence, the pressure-based solver is finalized because the flow is incompressible up to 0.3 Mach number. Also, the working velocities that are selected randomly vary from 1 to 15 m/s [working environment of existing UAV], for numerical simulation. For turbulence-based external analysis problems, the k-epsilon turbulence model has a high capability to predict the turbulence, so the k-epsilon model is implemented for this work. The design and working environments are complicated so in order to provide reliable data, the accuracy output is needful one by means of higher order implemented in the solver control step. The acoustic generations on the various propellers are estimated which are shown in Figs. 7 to 14.

Similarly, the other two cases (Case-III and Case-IV) are executed, and thereby, the comparative results are carefully noted for comparison, which are shown in Fig. 15.

3 Acoustic Analysis of High-Efficient Propellers

From the previous analysis, the saw-tooth cut propeller is induced low acoustic, so the modifications are mainly targeting on sharpen cut, in which two models are proposed: First modification is both leading edge cut based propeller and the second modification is leading and trailing edges cut based propeller.

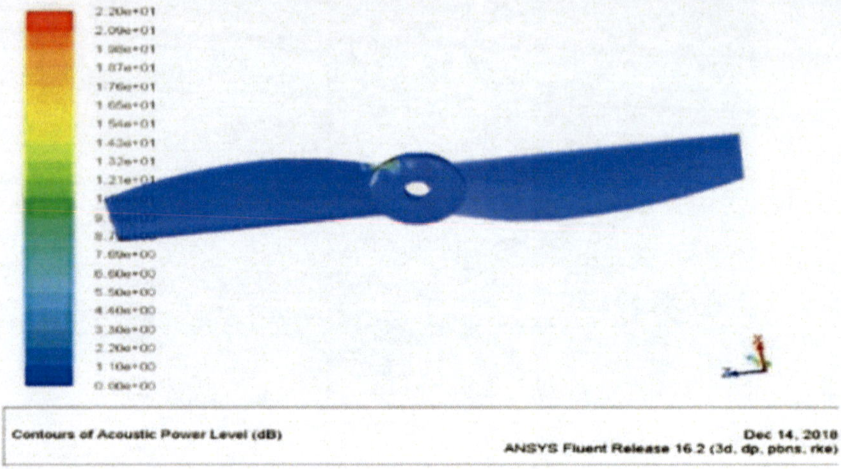

Fig. 7 HTOL—10 m/s—base

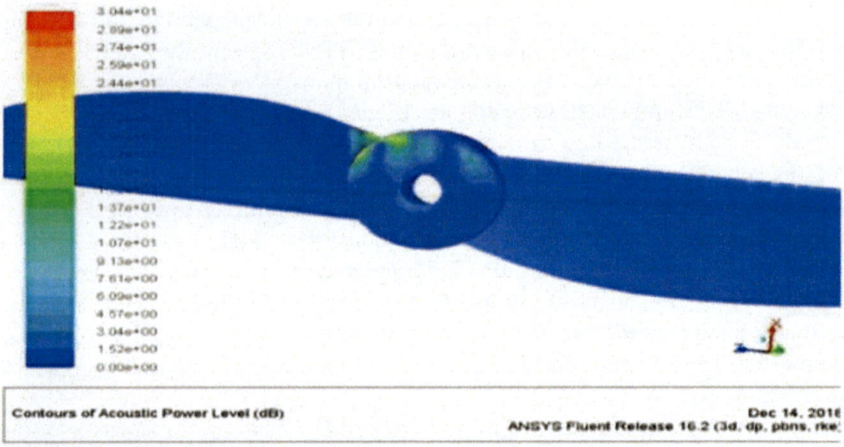

Fig. 8 HTOL—13 m/s—base

3.1 Conceptual Design

In this modification, saw-tooth cuts are plays a vital role, in which the location takes place at both sides of the leading edge of the propeller. The fundamental aim behind the finalization of location is to reduce the leading edge vortices and thereby reduce the acoustic induction. Figure 16 revealed the conceptual design of both leading edge modified propeller.

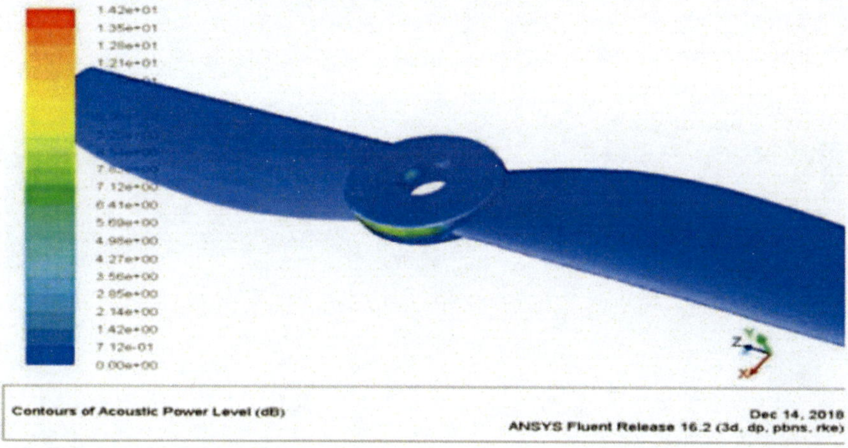

Fig. 9 VTOL—8 m/s—base

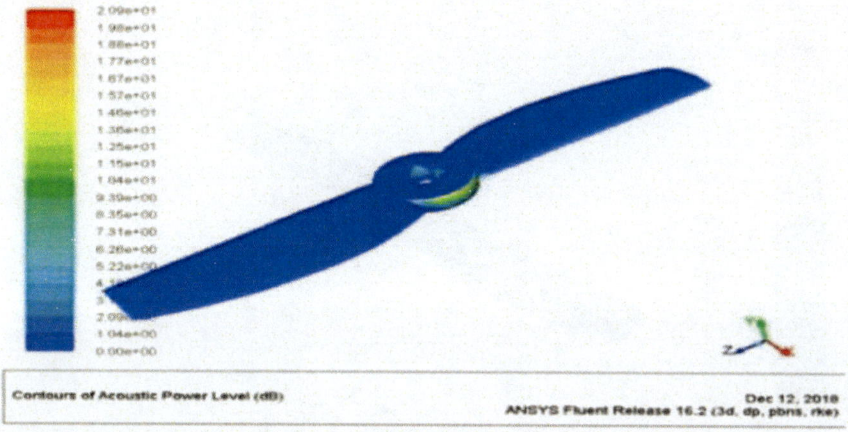

Fig. 10 VTOL—10 m/s—base

The Propeller of UAV is complex in design because of its twist. Therefore, the prediction of turbulence due to flow separation is quite high, and thereby, the estimation of acoustic generation tends to a more complicated one. So in this modification, the sharp edges are located on one side of the leading edge and another side of the trailing edge. The complete conceptual design of this modification is revealed in Fig. 17.

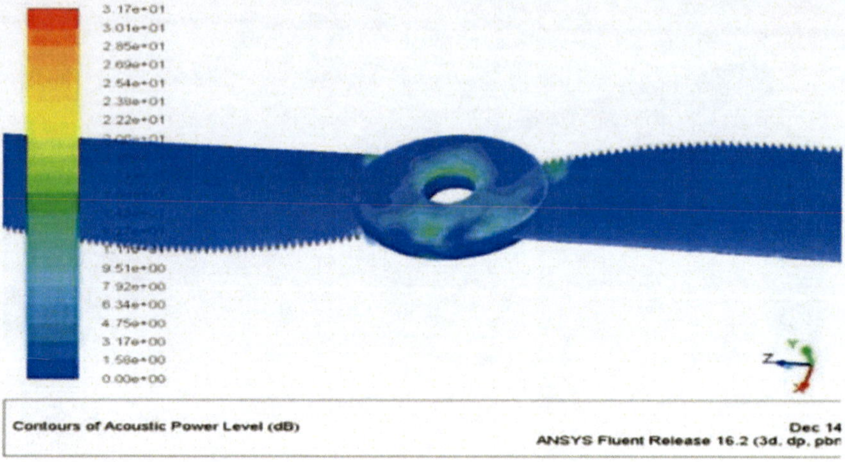

Fig. 11 HTOL—13 m/s—Case-I

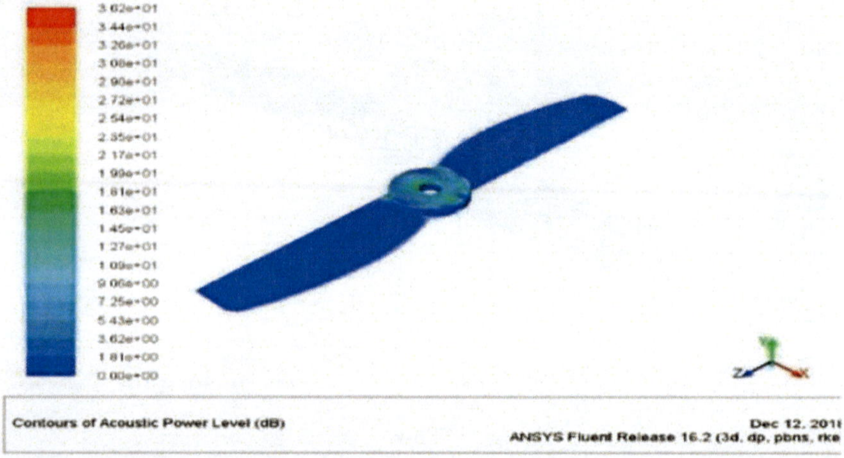

Fig. 12 HTOL—15 m/s—Case-I

3.2 *Numerical Results of Optimized Propellers*

The acoustic analyses of optimized propellers are executed with the help of ANSYS Fluent 16.2. Figures 18 and 19 showed the acoustic variations on the Case-IV for the HTOL operations. Similarly, the VTOL operations are executed, which are the primary maneuvering of compact UAVs. Figures 20 and 21 revealed the acoustic distributions of VTOL operations of Case-IV propeller [11].

The HTOL operation is secondary maneuvering in the multi-rotary UAVs, which needs to be analyzed furthermore in order to get efficient UAV's component. The

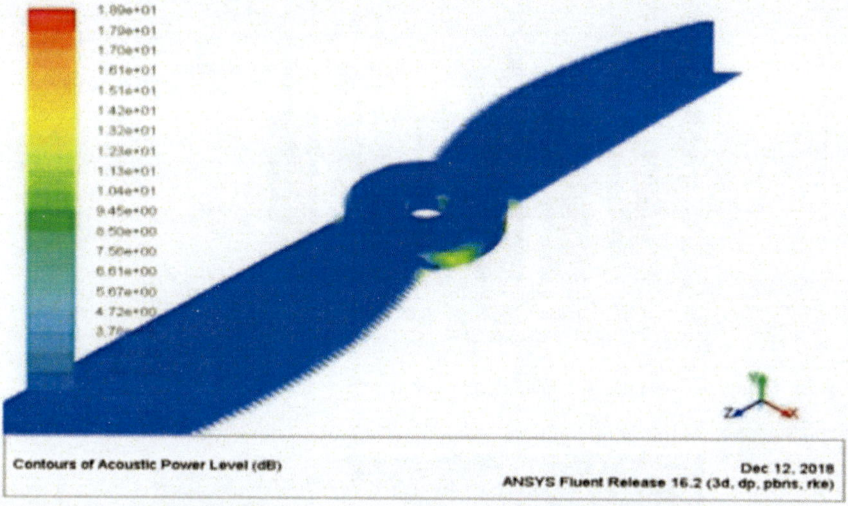

Fig. 13 VTOL—10 m/s—Case-I

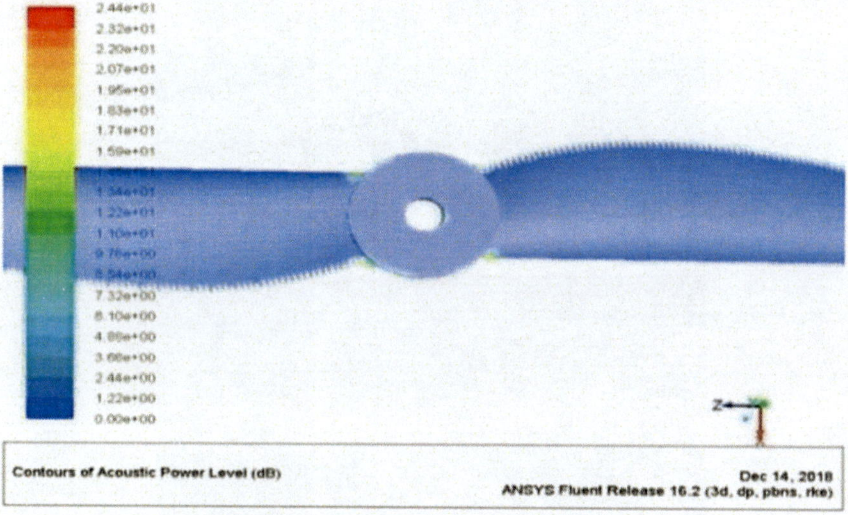

Fig. 14 VTOL—12 m/s—Case-I

different velocities of HTOL operations of Case-V are shown in Figs. 22 and 23. Figures 24 and 25 reveal the distributions of acoustics on Case-V propellers, in which VTOL cases are executed with the random velocities of 8 and 10 m/s.

The comparative analyses are examined, which is shown in Fig. 26.

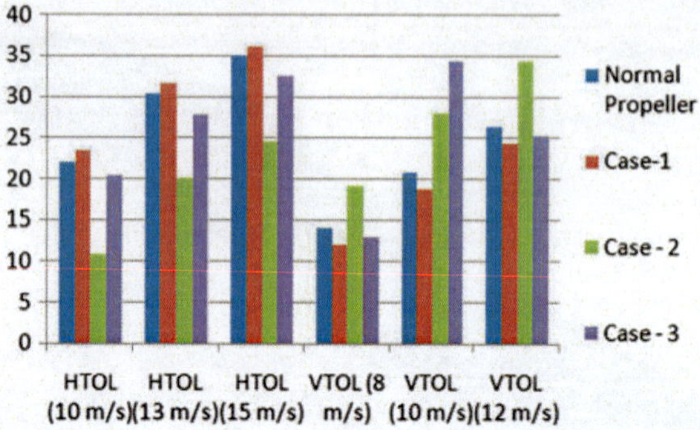

Fig. 15 Comparative values of fundamental cases

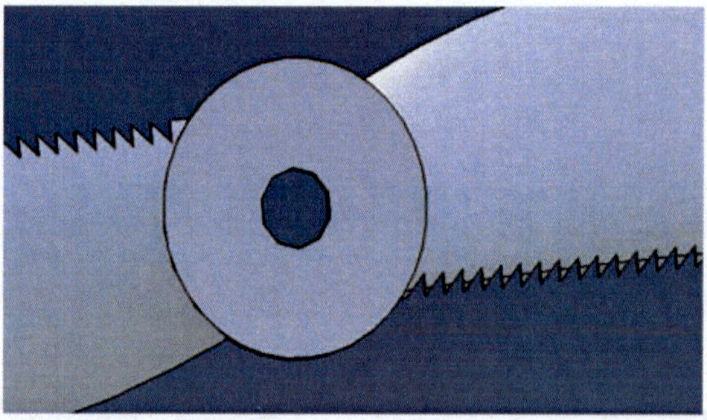

Fig. 16 Both leading edge cut (Case-IV)

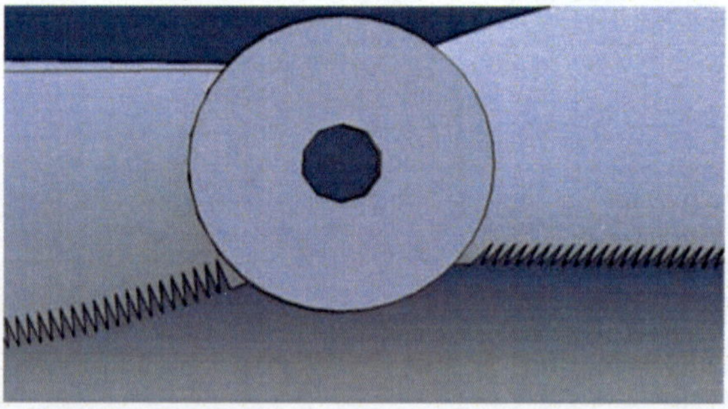

Fig. 17 One leading edge cut and one trailing edge cut (Case-V)

4 Conclusions

From the first stage of acoustic simulation of this work, it is understood that the 15% of noise induction is reduced in the implementation of the saw-tooth cut-based propeller

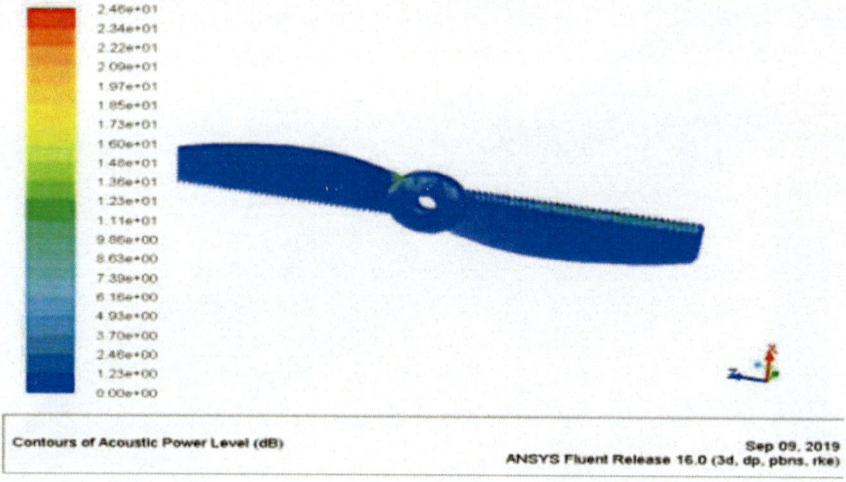

Fig. 18 HTOL—10 m/s—Case-IV

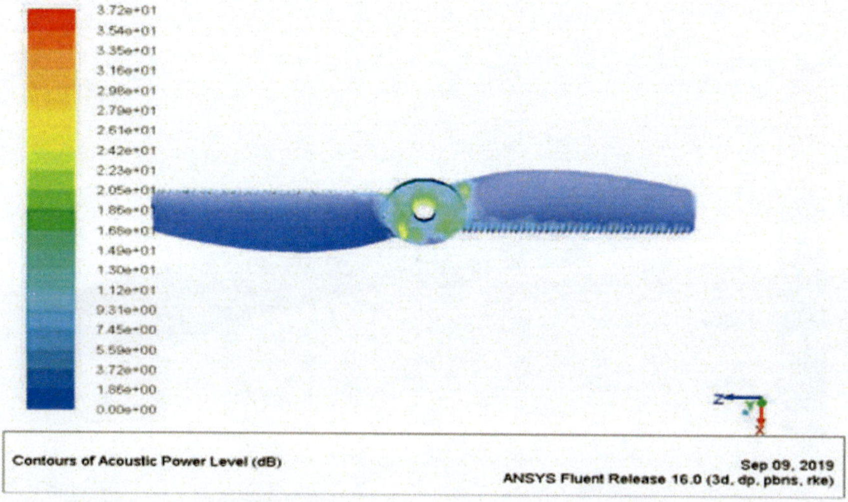

Fig. 19 HTOL—15 m/s—Case-IV

in VTOL operation, also learned that sharp edges drastically deal with the turbulence noise in a positive manner. Therefore, the second stage of numerical simulation is planned for furthermore sharp edge modifications on the propeller edges, which are successfully executed with the help of ANSYS Fluent 16.2. From both the cases, it is clearly concluded that both trailing edges propeller modified with sharp edges are more suitable to provide low noise, which is suggested to use in the quite UAVs.

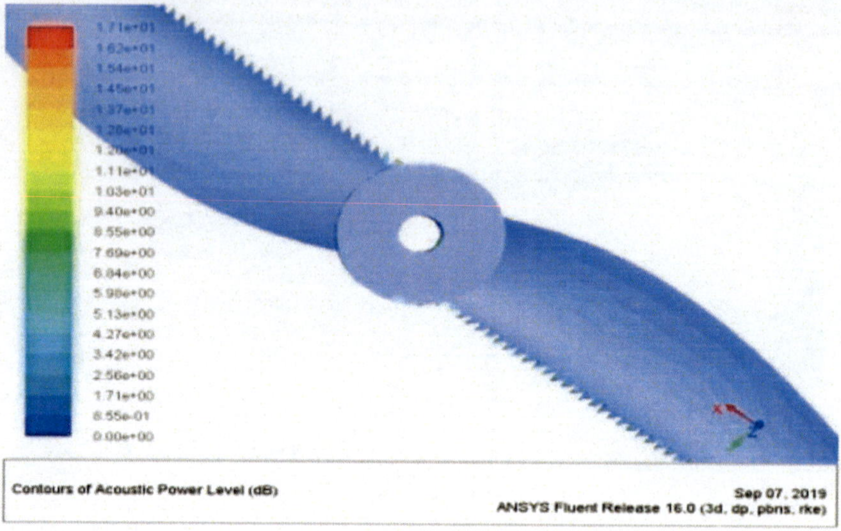

Fig. 20 VTOL—8 m/s—Case-IV

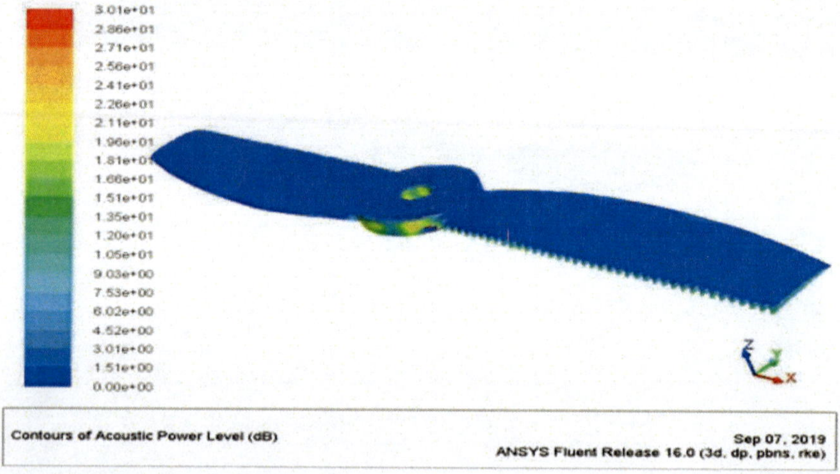

Fig. 21 VTOL—12 m/s—Case-IV

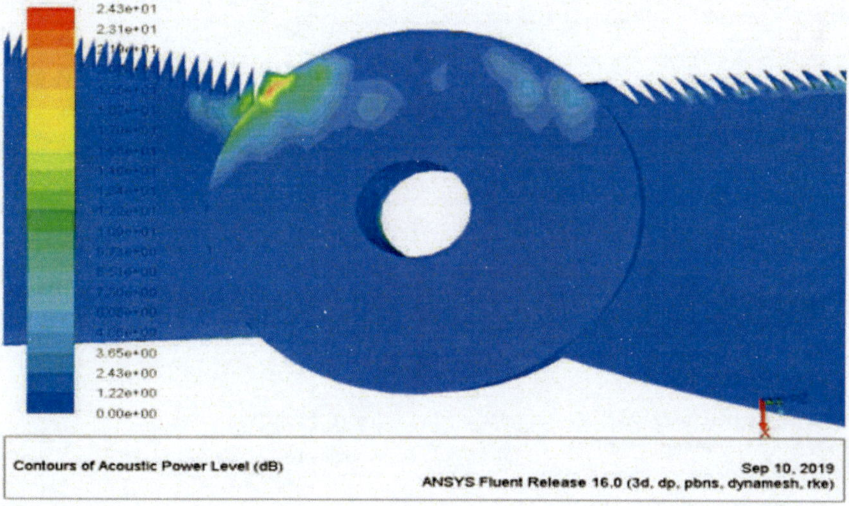

Fig. 22 HTOL—10 m/s—Case-V

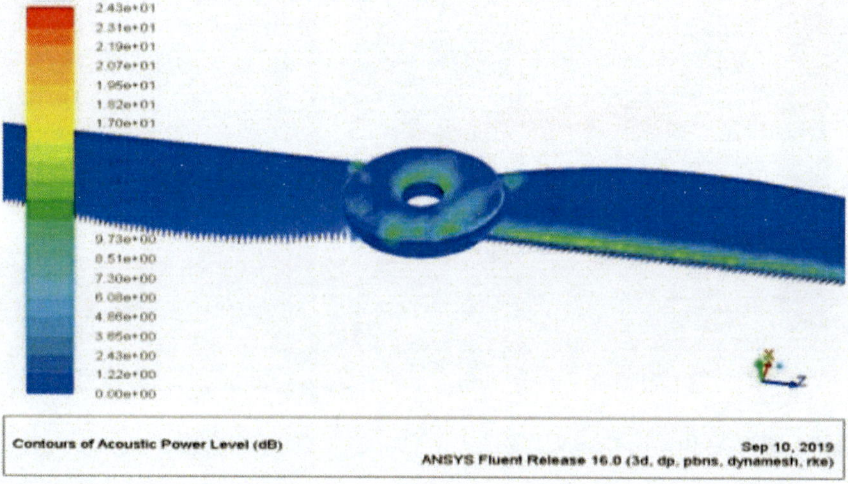

Fig. 23 HTOL—13 m/s—Case-V

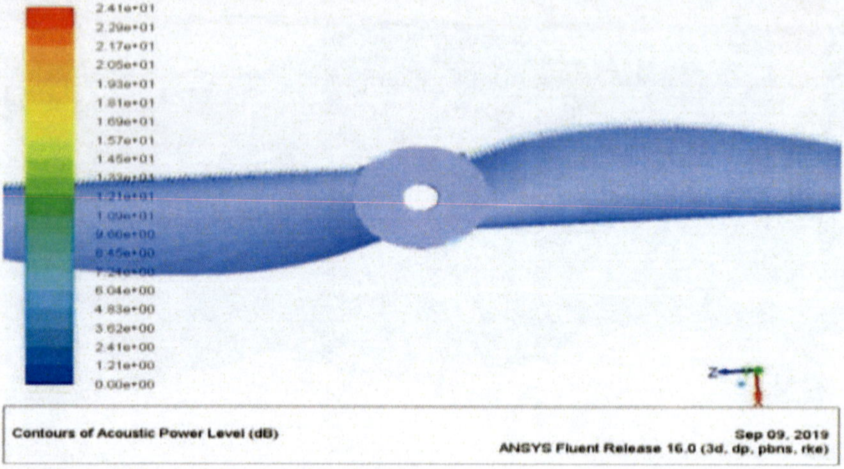

Fig. 24 VTOL—8 m/s—Case-V

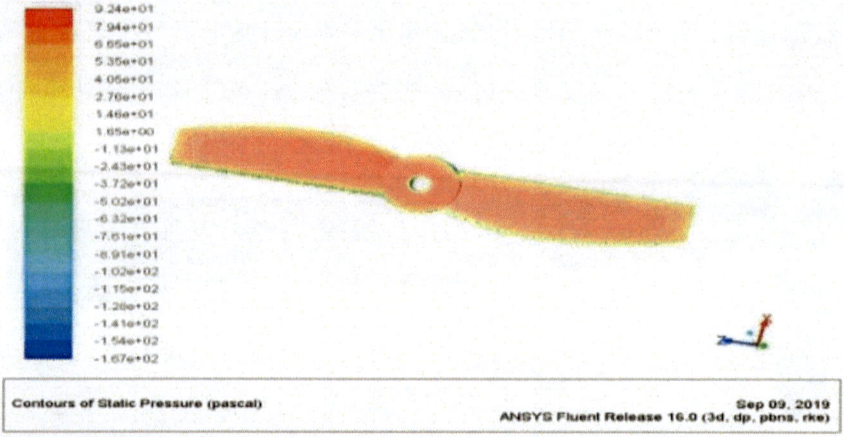

Fig. 25 VTOL—10 m/s—Case-V

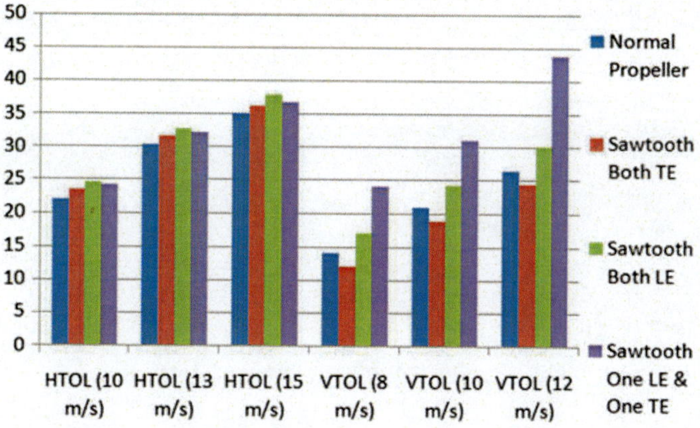

Fig. 26 Comparative values of modified cases

References

1. Vijayanandh R, Ramesh M, Raj Kumar G, Thianesh UK, Venkatesan K, Senthil Kumar M (2019) Research of noise in the unmanned aerial vehicle's propeller using CFD. Int J Eng Adv Technol (IJEAT) 8(Issue-6S):145–150. ISSN: 2249-8958
2. Vijayanandh R et al. (2018) Acoustic analysis and comparison of chevron nozzle using numerical simulation. Int J Mech Prod Eng Res Dev 8(7):1089–1103. ISSN(E): 2249-8001
3. Vijayanandh R et al. (2017) Design, fabrication and simulation of hexacopter for forest surveillance. ARPN J Eng Appl Sci 12(12):3879–3884. ISSN 1819-6608
4. Wisniewski CF, Byerley AR, Heiser WH, Liller WR (2015) Experimental evaluation of open propeller aerodynamics performance and aero-acoustic behavior. In: Proceedings of 33rd AIAA applied aerodynamics conference. Dallas American Institute of Aeronautics and Astronics Inc.
5. Vijayanandh R et al. (2017) Numerical study on structural health monitoring for unmanned aerial vehicle. J Adv Res Dyn Control Syst 9(6):1937–1958
6. Callender MN (2017) UAS propeller/rotor sound pressure level reduction through leading edge modification. J Appl Mech Eng 6:254. https://doi.org/10.4172/2168-9873.1000254
7. Vijayanandh R et al. (2017) Conceptual design and structural analysis of integrated composite micro aerial vehicle. J Adv Res Dyn Control Syst 9(14):857–881
8. Vijayanandh R et al. (2018) Design, fabrication of Tilt-Hexacopter with image processing for critical applications. Int J Pure Appl Math Special Issue, 118(9):935–945. ISSN: 1314-3395
9. Vijayanandh R (2018) Design optimization of advanced multi-rotor unmanned aircraft system using FSI, Book Title " Innovative Design, Analysis and Development Practices in Aerospace and Automotive Engineering (I-DAD 2018)", Springer Series Title—Lecture Notes in Mechanical Engineering, eBook ISBN - 978-981-13-2718-6, Chapter number 28, https://doi.org/10.1007/978-981-13-2718-6
10. Vijayanandh R et al. (2017) Energy harvesting from a gas turbine engines exhaust heat using quantum well thermoelectric materials generator. Mater Today: Proc 4:8378–8386
11. Lyu B, Azarpeyvand M, Sinayoko S (2015) A trailing-edge noise model for serrated edges. In: Proceedings 21st AIAA/CEAS Aeroacousitcs conference Dallas

Investigation of Nature of Cyclic Combustion Variations in RCCI Engine

Ajay Singh, Mohit Raj Saxena, and Rakesh Kumar Maurya

1 Introduction

Transportation sector is one of the main contributors of the global air pollution. Rapid increase in the vehicles demands for the efficient and cleaner combustion engines. The RCCI engine is a promising combustion regime which has a potential to improve the engine efficiency while simultaneously maintaining the NO_x and soot emissions which is one of the major challenges in the diesel engine (due to the trade-off between these species) [1]. In RCCI combustion, in-cylinder blending of two fuels of different reactivity (their proportions and injection timing) is used to control the equivalence ratio and charge stratification. The RCCI regime has better ignition and combustion phasing (CA_{50}) control as compared to other low temperature combustion regimes (such as HCCI and PCCI.). Combustion process of an engine has complex dynamics which is dependent on the various operating parameters such as local temperature, pressure, equivalence ratio, diesel injection timing and injection pressure. [2]. Excessive variations in the engine combustion cycle cause to poor drivability (variations in engine speed and torque), lower thermodynamic efficiency, higher emissions and may also cause to damage the engine structure. Combustion stability of an engine is generally characterized by measuring the variations in the parameters related to in-cylinder pressure, engine combustion, concentration of exhaust gases and characteristics of flame front [3]. Combustion stability is one of the major challenges in RCCI engine which limit the lower operating load boundary [4]. Combustion process in RCCI is very much sensitive to slight changes in pressure and temperature of the charge, amount of residual gases and injected mass of fuel [5]. Even though the phasing of combustion and its duration in RCCI is controlled by spatial stratification and local equivalence ratio of the two different reactivity fuels, still the variations

A. Singh · M. R. Saxena · R. K. Maurya (✉)
Indian Institute of Technology Ropar, Punjab 140001, India
e-mail: rakesh.maurya@iitrpr.ac.in

A. K. Gupta et al. (eds.), *Advances in IC Engines and Combustion Technology*,
Lecture Notes in Mechanical Engineering,
https://doi.org/10.1007/978-981-15-5996-9_46

in the engine combustion cycles are significantly higher [6]. The cyclic combustion variations are relatively higher in RCCI engine as compared to conventional compression ignition engine [7]. These variations are more sensitive to input parameters [7]. The port and directly injected fuel mass, exhaust gas recirculation (EGR) and temperature of the trapped gases are the major sources of cyclic combustion variations in the engine combustion cycle [5]. Combustion phasing in RCCI combustion is more sensitive to injected diesel mass per stroke, EGR and charge temperature [5]. Higher diesel injection pressure and retarded injection timing lead to increase the local equivalence ratio (charge stratification) which reduces the cyclic combustion variations [8]. A study reported that retarded diesel injection timing (30°–20° bTDC) leads to decrease the cyclic combustion variations with a negative effect on NO emissions [9]. A study found CNG-biodiesel RCCI engine having lesser variations in the IMEP in comparison with CNG-diesel RCCI operation [10]. In a study, the authors investigated the influence of diesel injection timing on the cyclic combustion variation of gasoline–diesel and methanol–diesel RCCI engine [4]. Their results indicate advanced diesel injection leads to increase the variations in the IMEP and THR. Too advanced diesel injection from a certain limit results in misfire condition [4]. Additionally, they have found the cyclic combustion variations in methanol–diesel RCCI to be comparatively higher. The present study experimentally investigated the effect of diesel injection timing with single and double injection strategy on the nature of cyclic combustion variations in total heat released of gasoline–diesel RCCI engine using return map technique.

2 Experimental Setup and Methodology

In this study, a single cylinder diesel engine having a displacement volume of 625 cc and a compression ratio of 18:1, with common rail direct injection, is modified to run in gasoline–diesel RCCI combustion mode. To achieve the RCCI mode of combustion, gasoline (a low reactivity fuel) is injected into the inlet manifold by using a solenoid port fuel injector during the intake stroke at −350 CAD aTDC at a constant pressure of 3 bar, while high reactivity fuel (diesel) is directly injected into the combustion chamber at an injection pressure of 500 bar. Injection timings of high reactivity fuel are controlled by a development ECU. The experimental setup is shown in Fig. 1 with further more details provided in [4]. The experiments were conducted at a constant engine speed 1500 rpm and a load of 1.5 bar BMEP for several diesel injection timings using single and double injection strategy. For double injection strategy, pilot injection is fixed at 55 CAD bTDC, while main injection is varying from 5 to 50 CAD bTDC. For double injection strategy, the fuel injected in each cycle is divided equally in the pilot and main injection. The detail about the injection strategy is available in the previous study by the authors [11]. The limiting point of main injection timing in double injection strategy has been decided as a point where there starts a rapid rise in variations in the THR and IMEP for single injection strategy. A piezoelectric transducer is installed in the cylinder head for measuring

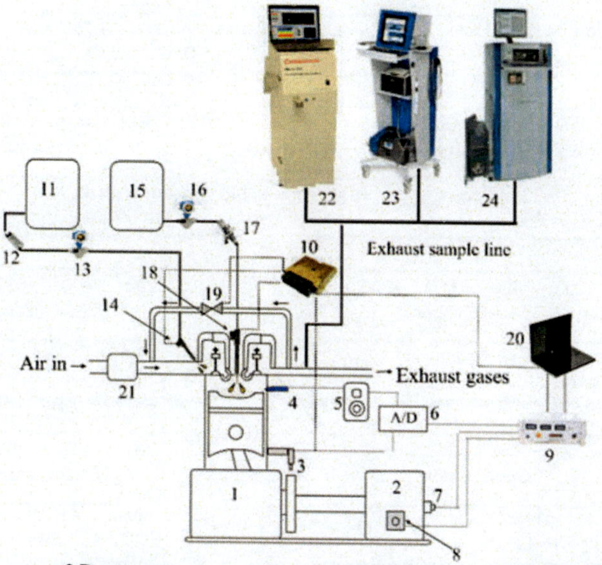

1 Diesel engine	2 Dynamometer	3 TDC position sensor	4 Piezoelectric pressure sensor
5 Charge amplifier	6 A/D converter	7 Encoder	8 Load cell
9 Dyno controller and DAQ	10 Development ECU	11 Low Reactivity (LR) Fuel Tank	12 LR Fuel Pump
13 LR Fuel Flow Meter	14 Port Fuel Injector	15 High Reactivity (HR) Fuel Tank	16 HR Fuel Flow Meter
17 HR Fuel Pump	18 HR fuel injector	19 EGR Valve	20 Computer
21 Orifice	22 DMS 500	23 Smoke opacity meter	24 THC Analyzer (FID)

Fig. 1 Schematic diagram of RCCI engine test setup

the in-cylinder pressure. A crank angle encoder of 0.1 CAD resolution assists in measuring the combustion pressure at high frequency. A digital Butterworth filter is employed to discard out the high frequency signal noise from the pressure data. Total heat released is calculated from experimentally recorded in-cylinder pressure for 1000 consecutive engine cycle to analyse the cyclic variations in RCCI combustion.

3 Results and Discussions

In this section, the analysis of cyclic combustion variations in the total heat release (THR) of 1000 engine combustion cycles of gasoline–diesel RCCI engine is presented. Total heat release is obtained by integrating the rate of heat release from start of combustion to end of combustion. Single-zone thermodynamic model [3] has been used to calculate the heat release rate. Figure 2 shows the time series of THR

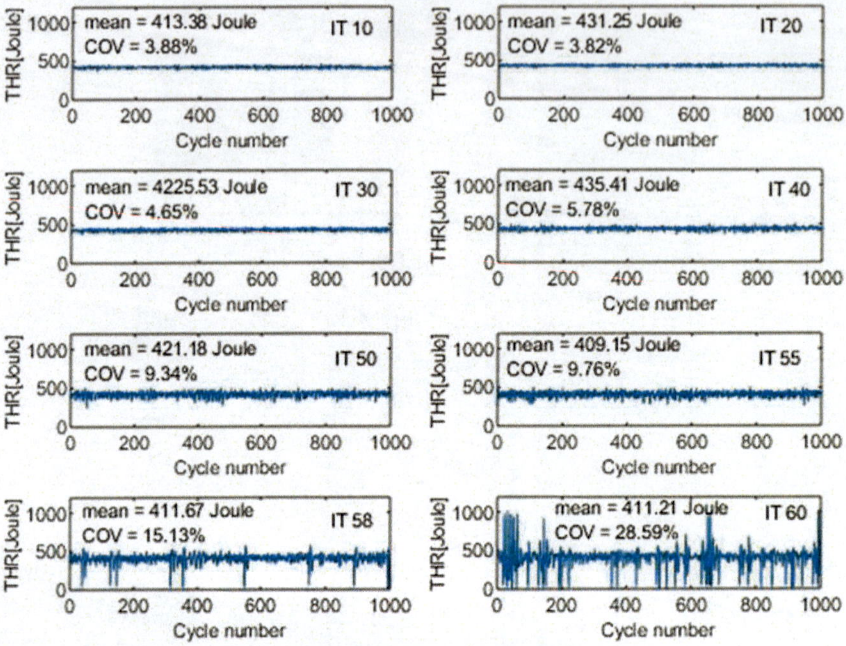

Fig. 2 Time series of THR for different injection timings for single injection operation

of 1000 consecutive engine combustion cycles of a gasoline–diesel RCCI engine for various diesel injection timings. It can be observed from Fig. 2 that mean value of THR firstly increases when injection timing advanced from 10 to 20 CAD bTDC and then decreases with further advanced injection timing. In the previous study, the authors reported that for 10 and 20 CAD diesel injection timings, the combustion (heat release rate) seems similar to conventional dual fuel combustion mode [4]. For further advanced diesel injection timings, there is enough time for diesel to mix with gasoline–air mixture, and almost, entire charge burns in premixed combustion phase. When diesel injection timing sweeps from 10 to 20 CAD bTDC, the start of combustion is advanced, correspondingly the combustion phasing is also advanced, providing larger combustion duration, and thus, total heat released increases. Also, the diffusion combustion starts before TDC in this case, and its peak is comparatively higher which leads to a higher value of THR along with higher values of peak pressure [4]. Local equivalence ratio decreases due to over mixing of charge as the diesel injection time is further advanced from 20 CAD bTDC resulting into a retarded start of combustion, and correspondingly, combustion phasing is also retarded; thus, a lower peak pressure, mean combustion temperature and total heat release are obtained. With further advanced diesel injection timing (from 20 CAD bTDC to 60 CAD bTDC), combustion phasing is retarded more in expansion stroke, and the charge temperature decreases due to counter action of high rate of volume expansion on increase in temperature due to combustion. Lower peak temperature at retarded combustion phasing increases incompleteness of combustion and thus leads to decreased combustion efficiency. Also, combustion reactions become increasingly sensitive to temperature changes with more retarded combustion phasing [6]. Together all these factors lead to an increase in cyclic combustion variations which is represented by the coefficient of variation in total heat release (COV_{THR}). It can be seen from Fig. 2 that for a single diesel injection strategy, the COV_{THR} is significantly higher and beyond the acceptable limit after 40 CAD bTDC diesel injection timing which can be put to acceptable limits by employing a pilot injection along with a main injection of high reactivity fuel [4]. To determine the nature of the combustion stability for both the diesel injection strategies (single and double), return maps technique is used.

Return maps are the simplest form of phase space reconstruction plots used to observe the probable interaction between consecutive points in a time series. These maps use a unit time delay along with an embedding dimension of two. In the previous studies, the return maps have been used to study nature of combustion instabilities for conventional spark-ignition (SI) combustion, lean mode operated SI combustion and HCCI combustion strategy [12–16]. In a random time series, the successive values are not correlated, and the return map obtained is an unstructured cloud of points gathered about a fixed point on the diagonal line. This kind of nature is observed when return maps were plotted for heat release time series [12] and gross mean effective pressure time series [13] for a port fuel injection SI engine operating with equivalence ratio close to stoichiometric ratio. Similar kind of behaviour is observed in return maps plotted for heat release time series for a HCCI engine when operated with fuels with 3, 4 octane numbers [16]. For a time series having some inherent determinism, the return plot shows some structure of points as observed in moderately and very lean operated SI engines [12, 13] and in a HCCI engine operated with fuels of octane numbers of 5, 6 and 7 [16]. In present study, time series of THR for various diesel injection timings are investigated by using return maps. THR value of a cycle 'i' is taken on the x-axis and that of cycle '$i + 1$' on y-axis to make a return plot of scatter type. In Fig. 3 for injection timing from 10 to 40 CAD bTDC, the return plots resemble a shotgun pattern similar to that obtained for high dimensional noise. Points are gathered around a fixed point which represents mean value of integrated heat release at a particular operating condition, and it can be said that there is no existence of correlations among consecutive THR values for these injection timings. Up to injection timing of 40 CAD bTDC, combustion instability is under acceptable limits, the combustion process may be affected by a large number of independent factors leading towards a normal distribution of cyclic variations according to central limit theorem [17]. Cyclic variations start increasing at a higher rate for further advanced diesel injection timing and points on return plot start to separate away from the fixed point on the diagonal line to form some structures. For direct injection timings of 50 and 55 CAD bTDC, there are several points for which the difference between THR of consecutive cycles is approximately half of mean value of THR (Fig. 3).

An engine cycle is called as a partial burn if the THR in it is less than 10% of that of a normal burn cycle, while it is said to be a misfiring one when the total amount of

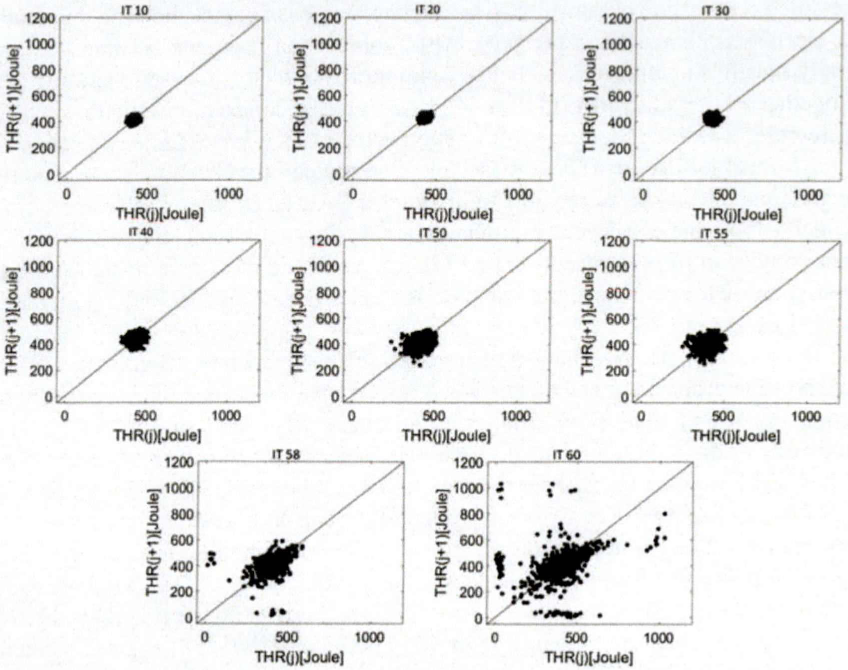

Fig. 3 Return maps of time series of THR for various diesel injection timing for single injection strategy

heat released is less than 50% of that of a well burn cycle [18]. At injection timing of 58 CAD bTDC, there exist events of misfire and partial burn due to too retarded combustion phasing, but the amount of heat released in a cycle before and after a misfired or partial burn cycle remains approximately equal to the mean value of THR. The number of misfire and partial burn events increase for further advanced injection timing along with cycles for which the amount of THR is more than the double of mean value of total heat released. Such events of very high energy exist both before and after a normal burn cycle in addition to after a misfiring cycle. It can be observed that there are times when two such strong burn cycles of very high THR and two misfired/partial burn cycles exist in succession. The last two return maps where there exist excessive cyclic variations suggest existence of low dimensional deterministic chaotic nature of combustion. At these conditions in place of large number of independent factors, a single dominant factor such as changes in local equivalence ratio may determine the nature of combustion.

The limitation of the return maps is that the skewness of data may result into misinterpretation of even un-correlated data to be correlated one. A possible solution to tackle with this problem is to plot return maps of quantile values in place of the original data [13]. Quantiles are similar to percentiles and are smaller by a factor of 100 than percentiles. A time series of 'm' points is converted into corresponding

Fig. 4 Return maps of time series of quantiles of THR for various diesel injection timing for single injection strategy

quantiles by replacing the smallest value of the time series by '1/*m*', next smallest by '2/*m*' and the largest value by unity. In this way, both rank and order of the data are kept same, while the data distribution is converted into uniform distribution between 0 and 1. A time series with random values fills the quantile return map with evenly distributed points, while with a correlated data, there will be areas of higher and lower density of points [13]. It can be observed from Fig. 4 that in return map plots for injection timing up to 40 CAD bTDC, points are distributed uniformly in the plot area, and it can be concluded that successive combustion events are un-correlated. For further advanced injection timings, the density of points decreases when moving from centre to the top left and bottom right of the plot area. At injection timing of 60 CAD bTDC, there exist strong correlations according to quantile plot. In the upper left and lower right quadrant of the plot area, the density of points is too less in comparison with the area along diagonal connecting bottom left and top right corners of the plot area. Near to the end points of this diagonal, the density of the points is comparatively higher indicating that there exist both consecutive strong burn cycles and misfiring/partial burn cycles. Also, one can observe the presence of time symmetry in the data points. Similar but weaker correlations may be observed at injection timing of 58 CAD bTDC. It is observed that the deterministic character and correlations in data comes into play because of incompleteness of combustion which is not acceptable for commercially available engines. Cyclic variations can be controlled by utilizing a pilot injection before the main injection [4]. As the cyclic variations increase significantly beyond injection timing of 55 CAD bTDC for single injection strategy, this point is chosen as a limit for pilot injection, while main injection timing sweeps from 5 CAD bTDC to 50 CAD bTDC. It would be interesting to know about the nature of cyclic variations, while the former are being reduced below the acceptable limits using double injection strategy.

It can be observed from the return plots Fig. 5 that the correlations between consecutive cycles do not exist for THR time series when main injection timing is 5 and 10 CAD bTDC. For further advanced main injection at 20 to 50 CAD bTDC in place of points gathered around a fixed point, some kind of structure about diagonal line is visible in return maps which indicates the presence of correlations among successive cyclic events. These results are also supported by density of points in quantile plots for double injection strategy (Fig. 6). It can be seen that for main injection timing of 5 and 10 CAD bTDC, the points are uniformly distributed in the plot area, while for further advanced diesel injection timing, the density of points decreases when moving towards upper left and lower right corner of the plot area which indicates the presence of correlations among successive points in time series. Thus, for double injection strategy, deterministic nature visualized in return maps for main injection timing of 20–50 CAD bTDC may be helpful for developing some controller for combustion, while for single injection strategy, the deterministic nature of combustion instability has no practical significance due to the presence of cyclic variations beyond acceptable limit.

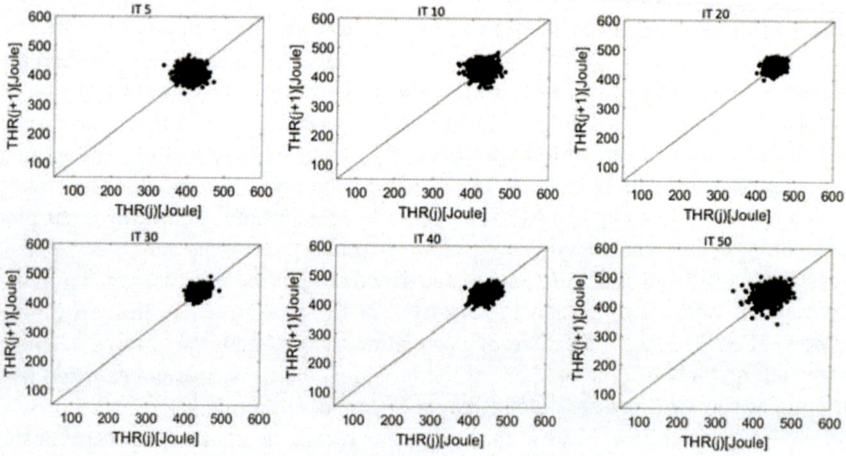

Fig. 5 Return maps of time series of THR for different diesel injection timing for double injection strategy

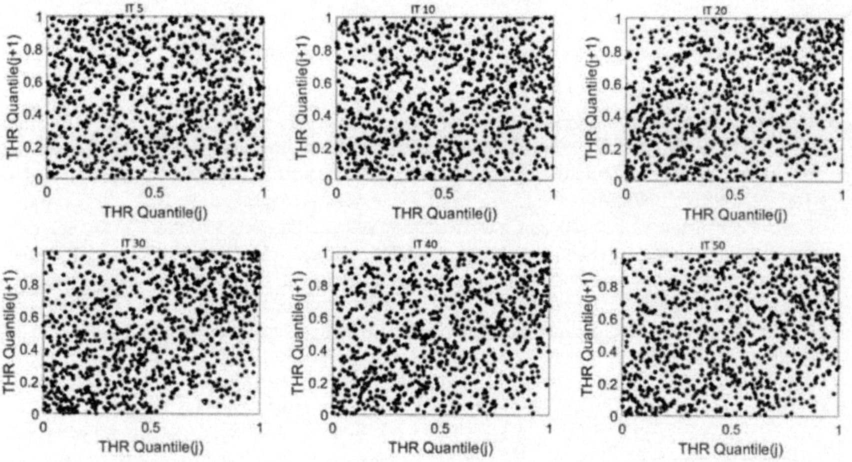

Fig. 6 Return maps of time series of quantiles of THR for different injection timings with double injection operation

4 Conclusions

It has been observed from the THR data series that when injection timing of high reactivity fuel is advanced further beyond 40 CAD bTDC in gasoline–diesel operated RCCI engine, it leads to increase in cyclic variability beyond acceptable limits. Return plots indicate the presence of randomness for retarded injection timings and correlations for advanced injection timings where cyclic variations are too high. It can be concluded that the deterministic nature and correlations in data come into existence because of incomplete combustion which is not acceptable in practical engine operation. Cyclic variations have been controlled by utilizing double injection strategy in place of single injection. Deterministic nature of cyclic variations when using a pilot injection along with main injection with main injection timing of 20–50 CAD bTDC may be useful for developing some controller in the mentioned range of injection timing.

Acknowledgements The authors gratefully acknowledge the research funding provided by DST-SERB, Government of India through Project No. ECR/2015/000177 titled "Soot Particle Number Emission Characterization and Investigation of Load Constraints in Reactivity Controlled Compression Ignition (RCCI) Engines."

References

1. Saxena MR, Maurya RK (2017) Impact of fuel premixing ratio and injection timing on reactivity controlled compression ignition engine. In: Agarwal AK, De S, Pandey A, Singh AP (eds) Combustion for power generation and transportation: technology, challenges and prospects. Springer, Heidelberg, pp 277–296. ISBN:978-981-10-3784-9. https://doi.org/10.1007/978-981-10-3785-6_13
2. Saxena MR, Maurya RK (2018) Characterization of cycle to cycle variations in conventional diesel engine using wavelets. In: Srivastava DK, Agarwal AK, Dutta A, Maurya RK (eds) Advances in internal combustion engine research. Springer, Heidelberg, pp 135–155. ISBN 978-981-10-7574-2
3. Maurya RK (2019) Reciprocating engine combustion diagnostics. Springer, Heidelberg. ISBN 978-3-030-11953-9
4. Saxena MR, Maurya RK (2018) Effect of diesel injection timing on peak pressure rise rate and combustion stability in RCCI engine (No. 2018-01-1731). SAE Technical Paper
5. Klos D, Kokjohn SL (2015) Investigation of the sources of combustion instability in low-temperature combustion engines using response surface models. Int J Engine Res 16(3):419–440
6. Maurya RK (2018) Characteristics and control of low temperature combustion engines: employing gasoline, ethanol and methanol. Springer, Heidelberg. ISBN 978-3-319-68507-6
7. Dempsey A (2013) Reactivity controlled compression ignition (RCCI) with alternative fuels (Doctoral dissertation, PhD Thesis, University of Wisconsin-Madison, Madison, Wisconsin, USA, 2013. Google Scholar)
8. Jia M, Dempsey AB, Wang H, Li Y, Reitz RD (2015) Numerical simulation of cyclic variability in reactivity-controlled compression ignition combustion with a focus on the initial temperature at intake valve closing. Int J Engine Res 16(3):441–460
9. Klos D, Janecek D, Kokjohn S (2015) Investigation of the combustion instability-NO_x tradeoff in a dual fuel reactivity controlled compression ignition (RCCI) engine. SAE Int J Engines 8(2):821–830
10. Gharehghani A, Hosseini R, Mirsalim M, Jazayeri SA et al (2015) An experimental study on reactivity controlled compression ignition engine fueled with biodiesel/natural gas. Energy 89:558–567. https://doi.org/10.1016/j.energy.2015.06.014
11. Saxena MR, Maurya RK (2018) Effect of fuel injection strategy on nano-particle emissions from RCCI engine. SAE 2018-01-1709, SAE Technical Paper
12. Daw CS, Finney CEA, Green Jr JB, Kennel MB, Thomas JF, Connolly FT (1996) A simple model for cyclic variations in a spark-ignition engine. SAE Trans, 2297–2306
13. Scholl D, Russ S (1999) Air-fuel ratio dependence of random and deterministic cyclic variability in a spark-ignited engine (No. 1999-01-3513). SAE Technical Paper
14. Wagner RM, Daw CS, Green JB (2001, March) Characterizing lean spark ignition combustion instability in terms of a low-order map. In: Proceedings of the 2nd Joint Meeting of the US Sections of the Combustion Institute
15. Daw CS, Wagner RM, Edwards KD, Green JB Jr (2007) Understanding the transition between conventional spark-ignited combustion and HCCI in a gasoline engine. Proc Combustion Inst 31(2):2887–2894
16. Ghazimirsaied A, Koch CR (2012) Controlling cyclic combustion timing variations using a symbol-statistics predictive approach in an HCCI engine. Appl Energy 92:133–146
17. Sen AK, Litak G, Finney CE, Daw CS, Wagner RM (2010) Analysis of heat release dynamics in an internal combustion engine using multifractals and wavelets. Appl Energy 87(5):1736–1743
18. Bahri B, Aziz AA, Shahbakhti M, Said MFM (2013) Understanding and detecting misfire in an HCCI engine fuelled with ethanol. Appl Energy 108:24–33

Computational Studies on Combustion Instabilities for Various Configurations of Afterburner

Srinivasa Rao Gurrala and Andavan Shaija

1 Introduction

Military Jet aircraft require minimum thrust while cruise and maximum thrust during take-off or maneuvers and/ or emergencies. Designing the jet aircraft, keeping in view of the thrust required during take-off or emergency conditions, will increase the engine size and specific fuel consumption. Hence, the best way is to have an afterburner, which only works for a specific period, for supplying the required additional thrust. However, the instabilities, like Screech and Buzz, are inevitable with the use of afterburner. Particularly the high frequency (500 to 5000 Hz) screech instabilities, travelling perpendicular to the axis of the afterburner, are detrimental to the aircraft as the same involves the catastrophic failure including the loss of human life [1]. Hence, the flow in the afterburner needs to be analyzed for the mitigation of these instabilities. Experimentations on the jet aircrafts with afterburner are costly and complex owing to the huge complex experimental setups, along with the associated sophisticated instrumentation in addition to the building and land requirements.

Hence, very few publications related to the experimental work are seen in the literature. Useller et al. [2] experimentally investigated the influence of combustion chamber length on the afterburner performance under conditions simulating altitude flight and found that the most favourable combustion chamber length is 5 feet, considering both performance and cooling requirement. Unaune and Ganesan [3] examined the afterburner for two different air-fuel ratios of 30 and 46 and predicted that the minimum air-fuel ratio of 30 is best suitable as the pressure loss is within the design limit.

S. R. Gurrala (✉)
Indian Naval Academy, Ezhimala, District Kannur, Kerala 670310, India
e-mail: raosgurrala@gmail.com

A. Shaija
National Institute of Technology, Calicut, Kerala 673601, India

A. K. Gupta et al. (eds.), *Advances in IC Engines and Combustion Technology*,
Lecture Notes in Mechanical Engineering,
https://doi.org/10.1007/978-981-15-5996-9_47

Kabiraj and Sujith [4] investigated the nonlinear transition undergone by thermoacoustic oscillations to flame blowout via intermittency and observed the transition to limit cycle oscillations from a non-oscillatory state to occur via a subcritical Hopf bifurcation to quasi-periodic oscillations. Kabiraj et al. [5] observed complex thermoacoustic oscillations experimentally in a simple laboratory combustor and found that the application of nonlinear time series analysis, reveals rich dynamical behaviour and the existence of several complex states. Nair and Sujith [6] studied the transition in dynamics from low-amplitude, aperiodic, combustion noise to high-amplitude, periodic, combustion instability in confined, combustion environments experimentally and shown that the low-amplitude, irregular pressure fluctuations acquired during stable regimes, display scale invariance and have a multifractal signature that disappears at the onset of combustion instability. Noiray et al. [7] proposed a methodology to investigate the nonlinear stability of burners by associating the flame describing function with a frequency-domain analysis of the burner acoustics and found that the calculated growth rates can be used to predict amplitudes of oscillation. Palies et al. [8] investigated the dynamics of premixed confined swirling flames by examining their response to incident velocity perturbations and found that the net result of the dynamical response and unsteady heat release rate is determined by the combined effects of the axial and induced azimuthal velocity perturbations. Palies et al. [9] experimentally analyzed thermoacoustic instabilities by making use of a nonlinear representation of flame dynamics based on the FDF and found that these data can be used to determine regions of instability, frequency shifts with respect to the acoustic Eigen frequencies.

The availability of highly configured computing machines at lower costs coupled with tested and reliable software, to carry out the simulations, enables prediction of these instabilities and design the best afterburner for absorption of maximum acoustic energy, which also saves a considerable amount of time. However, these normal computational facilities are not sufficient enough to carry out transient CFD simulations of the afterburner [10].

Hence, a novel method is suggested in this work with the objective of capturing the probable acoustic cavity modes, which may get excited during screech phenomenon leading to combustion instabilities, for the different configurations of afterburner with different area ratios (where area ratio (AR) is the ratio of aperture area to cross-sectional area of the afterburner) of 4, 5.2 and 5.9 with equivalence ratio (ER) of 0.58 and convergent-divergent (CD) nozzle angle (NA) of 6°; different nozzle angles (three convergent nozzle angles of 0°, 4° and 8° and one convergent-divergent NA of 6° with AR of 5.2 and ER of 0.58) and different equivalence ratios (0.58, 0.49 and 0.32 with AR of 5.2 and convergent-divergent NA of 6°). The mean flow characteristics of CFD analysis were interpolated, into the acoustic mesh modelled for acoustic analysis. The mean flow effect on the acoustic propagation by imposing duct modes with time variations on the mean flow characteristics enables the study of mean square pressure and transmission for the different area ratios. The best afterburner model with maximum absorption of acoustic energy thus reducing the screech instabilities is suggested.

2 Governing Equations for Computational Analysis

2.1 Equations for CFD Analysis

The equations used for CFD analysis are presented as given in ANSYS® Fluent Theory Guide [11].

Mass conservation equation:

$$\frac{\partial \rho}{\partial t} + \nabla \cdot (\rho \vec{v}) = S_m \tag{1}$$

For both the compressible and incompressible flows, Eq. (1) is applicable. In this equation, t denotes time, S_m represents any other any user-defined sources including the mass added to the continuous phase from the dispersed second phase whereas $\vec{v}$ is the velocity vector.

Momentum Conservation Equation

$$\frac{\partial}{\partial t}(\rho \vec{v}) + \nabla \cdot (\rho \vec{v} \vec{v}) = -\nabla \mathrm{p} + \nabla \cdot \left(\overline{\overline{\tau}}\right) + \rho \overrightarrow{g} + \vec{F} \tag{2}$$

In the Eq. (2) above, static pressure is denoted by p, $\overline{\overline{\tau}}$ denotes the stress tensor, $\rho \overrightarrow{g}$ represents the gravitational force, and $\vec{F}$ represents the external body forces (forces that arise from interaction with dispersed phase). The other model-dependent source terms like porous media and any other user-defined sources are also included in $\vec{F}$.

The stress tensor [4] is defined by

$$\overline{\overline{\tau}} = \mu \left[\left(\nabla \vec{v} + \nabla \vec{v}^T \right) - \frac{2}{3} \nabla \cdot \vec{v} I \right] \tag{3}$$

where molecular viscosity is denoted by μ and the effect of volume dilation is represented by I which is the unit tensor as shown in second term on the right side of the equation.

Energy Conservation Equation

$$\frac{\partial}{\partial t}(\rho E) + \nabla \cdot (\vec{v}(\rho E + p)) = -\nabla \cdot \left(k_{\text{eff}} \nabla T - \sum_j \overline{h_j j_j} + \left(\overline{\overline{\tau}}_{\text{eff}} \cdot \vec{v} \right) \right) + S_h \tag{4}$$

In the equation above, the effective conductivity k_{eff} is the sum of K the thermal conductivity and K_t the turbulent thermal conductivity whereas the diffusion flux of species J is represented by $\overline{j_j}$.

Turbulence modelling with Realizable $\mathbf{k - \varepsilon}$ model

$$\frac{\partial}{\partial t}(\rho k) + \frac{\partial}{\partial x_i}(\rho k u_j) = \frac{\partial}{\partial x_j}\left[\left(\mu + \frac{\mu_t}{\sigma_k}\right)\frac{\partial k}{\partial x_j}\right] + G_k + G_b - \rho\varepsilon - Y_M + S_k \quad (5)$$

$$\frac{\partial}{\partial t}(\rho\varepsilon) + \frac{\partial}{\partial x_i}(\rho\varepsilon u_j) = \frac{\partial}{\partial x_j}\left[\left(\mu + \frac{\mu_t}{\sigma_\varepsilon}\right)\frac{\partial \varepsilon}{\partial x_j}\right] + \rho C_1 S_\varepsilon - \rho C_2 \frac{\varepsilon^2}{K + \sqrt{\nu\varepsilon}} + C_{1\varepsilon}\frac{\varepsilon}{k}(G_k + C_{3\varepsilon}G_b) - C_{2\varepsilon}\rho\frac{\varepsilon^2}{k} + S_\varepsilon \quad (6)$$

$C_1 = \max\left[0.43, \frac{n}{n+5}\right]$, $n = S\frac{k}{\varepsilon}$, $S = \sqrt{2S_{ij}S_{ij}}$.

In the above equations, the turbulence kinetic energy generated due to the mean velocity gradients is represented by G_k and turbulence kinetic energy generated due to buoyancy is represented by G_b. $C_{1\varepsilon}$, $C_{2\varepsilon}$ and $C_{3\varepsilon}$ are constants. The contribution of the fluctuating dilatation in compressible turbulence to the overall dissipation rate is represented by Y_M. The turbulent Prandtl number for k is represented as σ_k and ε as σ_ε. S_k and S_ε are user-defined source terms.

Species Transport Equations

The local mass fraction of each species Y_i, is predicted by ANSYS® Fluent, through the solution of a convection-diffusion equation for the i - th species. The conservation equation is then given as below.

$$\frac{\partial}{\partial t}(\rho Y_i) + \nabla \cdot (\rho \vec{v} Y_i) = -\nabla \cdot \vec{J_i} + R_i + S_i \quad (7)$$

In the above equation, the net rate of production is R_i of species I by chemical reaction whereas the rate of creation S_i is the sum from the dispersed phase plus any user-defined sources.

Discrete Phase Model

The Euler–Lagrange approach is used in the Lagrangian discrete phase model. The trajectory of a discrete phase particle (or droplet or bubble) is predicted by ANSYS® Fluent by integrating the force balance on the particle. This force balance equates the particle inertia with the forces acting on the particle [5]:

$$\frac{d\vec{u}_p}{dt} = F_D(\vec{u} - \vec{u}_p) + \frac{\vec{g}(\rho_p - \rho)}{\rho_p} + \vec{F} \quad (8)$$

In the above equation, $\vec{F}$ is an additional acceleration (force/unit particle mass), whereas $F_D(\vec{u} - \vec{u}_p)$ is the drag force per unit particle mass and $F_D = \frac{18\mu C_D Re}{\rho_p d_p^2 24}$.

$\vec{u}_p$ represents the particle velocity, whereas, $\vec{u}$ is the fluid phase velocity. the molecular viscosity of the fluid is represented by μ, the density is indicated by ρ, the drag coefficient by C_D, the density of the particle is represented by ρ_p, the particle diameter is represented by d_p. The relative Reynolds number is represented by Re, which can be written as $Re \equiv \frac{\rho d_p |\vec{u}_p - \vec{u}|}{\mu}$.

2.2 *Equations for Acoustic Analysis*

The equations solved in Actran™ for acoustic analysis are presented below, which are given in Actran 16.1 User Guide [12].

The continuity equation can also be written as

$$\frac{\partial \rho}{\partial t} + \nabla \cdot (\rho v) = 0 \tag{9}$$

where ρ is density and v is velocity.

Momentum equation can be written in the form

$$\rho \frac{\partial v}{\partial t} + \rho \nabla B = \rho T \nabla s + \rho v \times (\nabla \times v) - \nabla \tau \tag{10}$$

where T is temperature, τ is viscous stress tensor, s is entropy, h is enthalpy and $B = h + \frac{v^2}{2}$ = Total enthalpy.

Energy conservation

$$\rho \frac{Dh_T}{Dt} - \frac{\partial p}{\partial t} = \nabla(\lambda \nabla T) + \nabla(v \cdot \tau) \tag{11}$$

where λ is material conductivity.

Neglecting the power dissipated by viscous forces and heat

$$\rho \frac{Dh_T}{Dt} = \frac{\partial p}{\partial t}$$

Combining the mass conservation, momentum conservation and energy conservation equations the Mohring equation for convected acoustics obtained is used for the acoustic analysis.

$$\frac{\partial}{\partial t}\left[\frac{\rho}{\rho_T^2 c^2} \frac{Db}{Dt}\right] + \nabla\left[\frac{\rho v}{\rho_T^2 c^2} \frac{Db}{Dt} - \frac{\rho}{\rho_T^2} \nabla b\right] = R \tag{12}$$

where $c^2 = \left[\frac{\partial p}{\partial \rho}\right]_{s=\text{const}}$; ρ_T = Total Density and

$$b = \text{Scaled enthalpy defined } \frac{Db}{Dt} = \rho_T \frac{DB}{Dt}$$

$$\frac{\partial}{\partial t}\left[\frac{\rho}{\rho_T^2c^2}\frac{\partial b}{\partial t}+\frac{\rho v_i}{\rho_T^2c^2}\frac{\partial b}{\partial x_i}\right]+\frac{\partial}{\partial x_i}\left[\frac{\rho v_i}{\rho_T^2c^2}\left(\frac{\partial b}{\partial t}+v_i\frac{\partial b}{\partial x_i}\right)-\frac{\partial b}{\partial x_i}\frac{\rho}{\rho_T^2}\right]=R \quad (13)$$

where

$$R=-\frac{\partial}{\partial x_i}\left[\frac{\rho}{\rho_T}(\overline{v}X\overline{\omega})_i-\frac{\partial \tau_{ij}}{\partial x_j}\right]+\frac{\partial}{\partial x_i}\left[\frac{v_i}{\rho_T}\left(\frac{\partial p}{\partial s}\right)_p\frac{\partial s}{\partial t}-\frac{\rho T}{\rho_T}\frac{\partial s}{\partial x_i}\right]$$
$$+\frac{\partial}{\partial t}\left[\frac{1}{\rho_T}\left(\frac{\partial p}{\partial s}\right)_p\frac{\partial s}{\partial t}-\frac{\rho v_i}{\rho_T^2}\frac{\partial \rho_T}{\partial x_i}\right]$$

In the present case, the propagation of acoustic perturbations within a homogeneous moving medium was considered for the study, taking into consideration, the limitations of computational facilities and software capabilities. Hence, the equation, in terms of the propagation of acoustic perturbations within a homogeneous moving medium, can be expressed in terms of acoustic pressure as follows:

$$\left(ik+\vec{M}.\nabla\right)^2+\Delta p=0 \quad (14)$$

where k is the wave number and $\vec{M}$ is the uniform mean velocity vector divided by the speed of sound. This equation is also called the convected Helmholtz equation, which is used in the present case to calculate the acoustic pressure.

3 Components of an Afterburner

The various components of the afterburner with a schematic diagram is as shown by Gurrala and Shaija [13]. The flow regions of the afterburner are shown in the 360° wireframe model as shown in Fig. 1.

4 Modelling Methodology

Eight afterburner models with ER 0.58, AR 5.2 and NA 6° (model 1); ER 0.49, AR 5.2 & NA 6° (model 2); ER 0.32, AR 5.2 and NA 6° (model 3); ER 0.58, AR 5.2 and NA 0° (model 4); ER 0.58, AR 5.2 and NA 4° (model 5); ER 0.58, AR 5.2 and NA 8° (model 6); ER 0.58, AR 4 and NA 6° (model 7) and ER 0.58, AR 5.9 and NA 6° (model 8); were analyzed using ANSYS® Fluent and steady-state reacting flow analysis CFD results were obtained. The various above-listed models, which were considered for the study, are shown in Fig. 2, for ease of appreciation. These CAD models were later imported into Actran™, by using either *.cas files

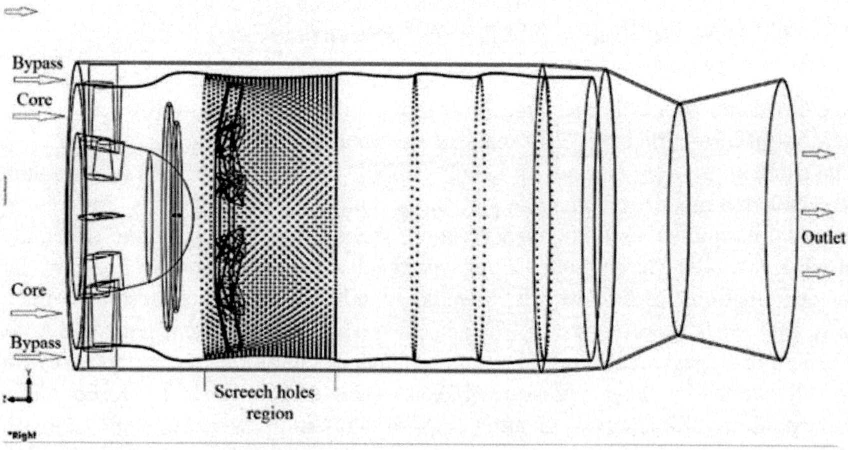

Fig. 1 360° wireframe model of afterburner

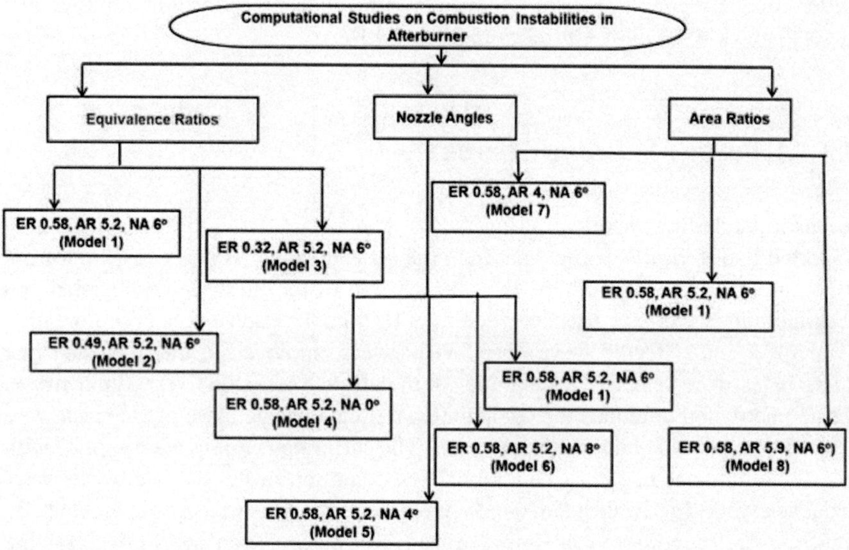

Fig. 2 Various afterburner models

or *.egf (encasgold format) files and meshing was done. The CFD results, obtained for the respective model, were then imported and interpolated using Actran™ ICFD, using*.dat files or *.egf files. The acoustic analysis for each model was then carried out. The modelling procedure was explained below.

4.1 CFD Modelling with ANSYS® Fluent

The 60° sector of afterburner models were modelled for the eight different configurations of afterburner for CFD analysis, as explained by Gurrala and Shaija [13]. The simulations were carried out using ANSYS® Fluent with SIMPLE algorithm, realizable k-ε model and energy equation for temperature calculations. For solving the multiple simultaneous chemical reactions species transport with finite-rate/ eddy dissipation model was employed along with the discrete phase model for performing the computations. Mass flow inlet boundary conditions were selected for the main inlet with mass flow rate as 11.67 kg/s, pressure as 353 kPa and temperature as 873 K. For bypass inlet, mass flow inlet boundary conditions were imposed with mass flow rate of 2.33 kg/s, pressure of 396 kPa and temperature of 373 K. For outlet and extended volume, pressure outlet boundary conditions were selected. Kerosene ($C_{12}H_{23}$) was considered as fuel which was injected through virtual fuel injection points. Further aspects of geometrical modelling, defining injection points, the procedure followed for performing the numerical calculations, specifying the inlet and outlet boundary conditions, convergence criteria, analysis and validation of results are explained in detail by Gurrala and Shaija [13].

4.2 Acoustic Modelling (Actran™)

Acoustic modelling was done using *Actran*™ software. Each model is separately imported into *Actran*™ using *.dat file. The acoustic analysis was carried out from 0 Hz to 3000 Hz. Accordingly, the average size of the mesh was calculated and the minimum mesh size was considered as 10 mm. The constrained (combination of incident and reflected wave) duct modes were employed at the inlet, and free duct modes were selected at the outlet of the afterburner. Due to the symmetry in the geometry and to reduce the computational time, the side faces of the mesh were applied with periodic boundary conditions. The perforated regions were applied with transfer admittance boundary conditions. The other portions of the afterburner were acoustically rigid and reflect the sound waves internally. Accordingly, the acoustically rigid boundary conditions were imposed at all these places. Then the CFD results were interpolated, onto the respective afterburner model in *Actran*™ by using *Actran*™ ICFD. Then acoustic analysis was carried out for the frequency range from 0 Hz to 3000 Hz.

4.3 Assumptions for Carrying Out Acoustic Simulations

Steady-state CFD mean flow parameters are simulated here to account for changes in properties of the medium due to temperature and flow. Neither transient conditions were considered for the analysis nor aeroacoustics sources were modelled. The duct mode propagation in core & bypass duct in the presence of mean flow (convected propagation) was considered with the amplitude of duct modes as 1 Pa. Since the purpose was to calculate transmission loss, the amplitude doesn't matter. Liners were represented as admittance boundary conditions and not modelled physically in the mesh for acoustic analysis. One-third model of duct was used with symmetry boundary conditions to reduce the degree of freedom as it is sufficient to capture the transmission loss and excited modes of the full duct. Any thermoacoustic instabilities and reverse effect, i.e. effect of acoustics on flow and combustion were not considered.

5 Results and Discussion

Eight afterburner models with ER 0.58, AR 5.2 and NA 6°; ER 0.49, AR 5.2 and NA 6°; ER 0.32, AR 5.2 and NA 6°; ER 0.58, AR 4 and NA 6°; ER 0.58, AR 5.9 and NA 6°; ER 0.58, AR 5.2 and NA 0°; ER 0.58, AR 5.2 and NA 4° and ER 0.58, AR 5.2 and NA 8° were computationally analyzed using ANSYS® Fluent. Later these CFD results were imported into Actran™ for acoustic analysis and the results were presented.

The Mach number contours for the afterburner with ER 0.58, AR 5.2 and NA 6° is shown in Fig. 3a. The maximum supersonic speeds of 1.3 to 1.5 Mach of the exhaust gases was observed at the exit of afterburner. The maximum Mach number was observed for the convergent-divergent nozzle angle of 6° with AR 5.2 and ER 0.58, due to the negative static pressure, at the exit of the nozzle. The velocity vectors of afterburner with ER 0.58, AR 5.2 and NA 6° are shown in Fig. 3b. The formation of eddy in the clockwise direction was observed at the top portion of V-gutter whereas the flow appears to be moving in the anticlockwise direction in the remaining portion. In the middle portion, the flow appears to be complex where the mixing takes place.

Figure 4 shows the variation of O_2 for all the eight configurations of afterburner, along the axis. Even though the mass fraction of O_2 available in all these cases at the inlet was observed to be 0.23 at the inlet, the minimum O_2 mass fraction of 0.144, was observed in the case of afterburner with ER 0.58, AR 5.2 and NA 6° at the exit.

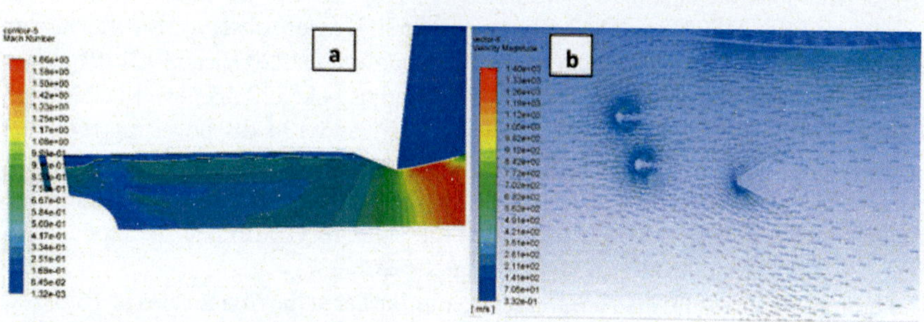

Fig. 3 Contours of **a** Mach number, **b**Velocity vectors for ER 0.58, AR 5.2 & NA 6°

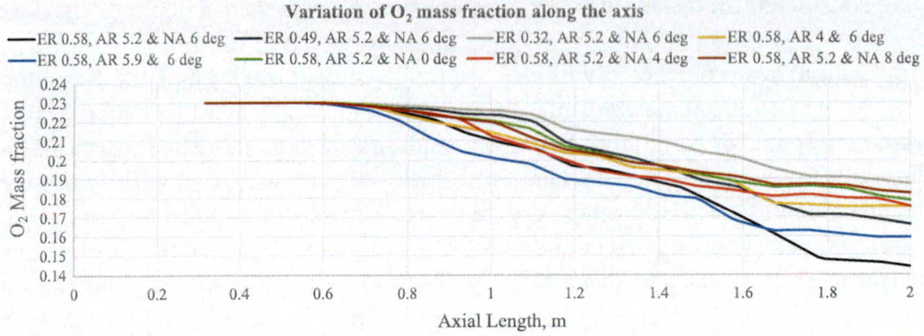

Fig. 4 Variation of O_2 along the axis for different configurations of afterburner

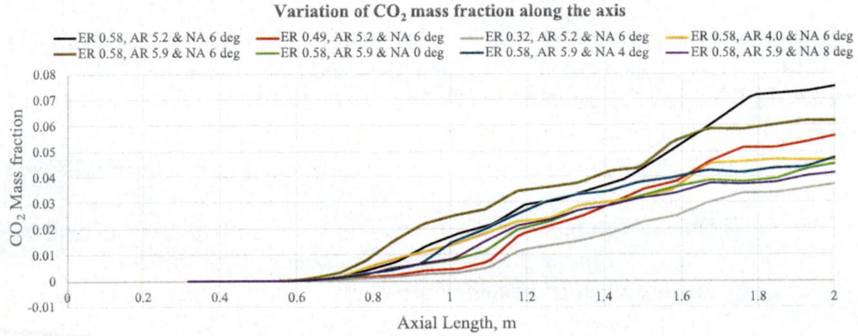

Fig. 5 Variation of CO_2 along the axis for different configurations of afterburner

This again indicates proper combustion and release of maximum heat energy due to maximum consumption of O_2 in the case of convergent-divergent nozzle angle of 6°. The variation of CO_2 for all these eight configurations of the afterburner is shown in Fig. 5. The maximum production of CO_2 was observed, for the afterburner model of ER 0.58, AR 5.2 and NA 6°, with mass fraction of 0.075 at the exit of the afterburner.

The variation of mean square pressure (dB power) with frequency is shown in Fig. 6. The peak mean square pressures (dB power) of 145 dB power at 2440 Hz, 143 dB power at 2480 Hz and 142 dB power at 2530 Hz were observed for afterburner model with ER 0.58, AR 5.2 and 4° NA; 128 dB power at 1220 Hz and 121 dB power at 1990 Hz for the model with ER 0.58, AR 4 and 6° NA; 127 dB power at 2260 Hz for the model with ER 0.58, AR 5.9 and 6° NA and 134 dB power at 1690 Hz for the model with ER 0.49, AR 5.2 and 4° NA were observed. The corresponding mean square pressures (dB power) of 15–20 dB less than the above values, at these frequencies, were observed and restricting the value to 120 dB, for the model with ER 0.58, AR 5.2 and 6° NA, at all the frequencies.

The acoustic pressure distribution in the afterburner at the frequencies of 1690 Hz, 2260 Hz and 2440 Hz is shown in Fig. 7a–c, respectively, whereas Fig. 7d shows the

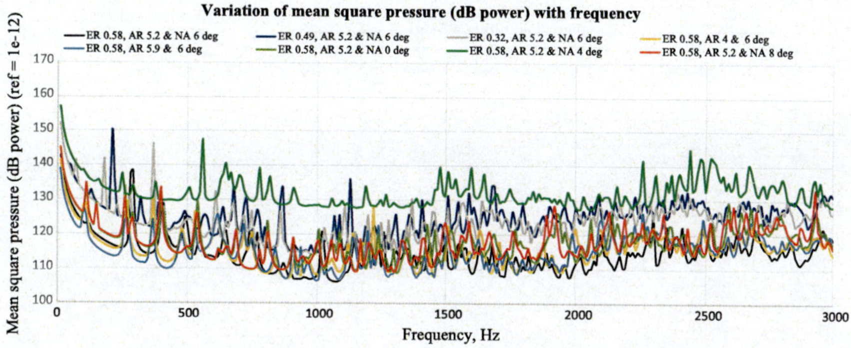

Fig. 6 Variation of mean square pressure with frequency for different configurations

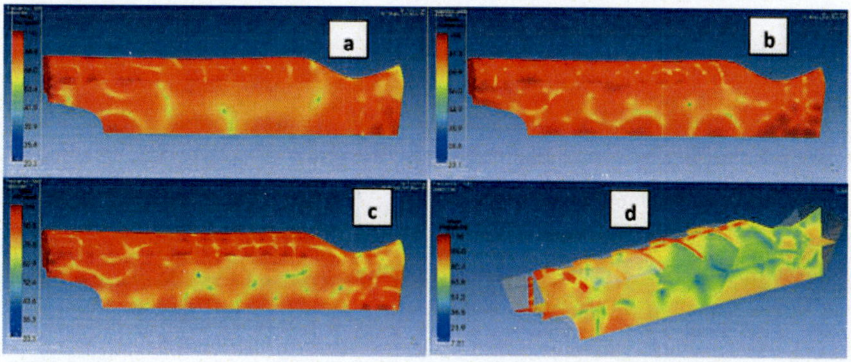

Fig. 7 Acoustic pressure **a** 1690 Hz, **b** 2260 Hz, **c** 2440 Hz, **d** 1940 Hz for ER 0.58, AR 5.2 & NA 6°

acoustic pressure distribution, at 90° cut sections of the afterburner, at the frequency of 1940 Hz.

Figure 8 shows the variation of transmission loss with frequency for the different configurations of the afterburner. The maximum transmission loss (the peak values of 75.5 dB at 530 Hz, 70.7 Hz at 990 Hz and 50.9 dB at 1940 Hz) was observed for the ER 0.58, AR 5.2 and 6° NA, indicating the maximum absorption of acoustic energy, for the specified conditions of mass flow rate. Hence, it can be concluded that the model with ER 0.58, AR 5.2 and 6° NA angle is the optimum nozzle angle for mitigation of screech, for the specified design and inlet mass flow conditions. It was observed that the transmission loss is high, throughout all the frequencies, for the afterburner with ER 0.58, AR 5.2 and 6° NA. This is attributed to the flame length, static pressure, proper combustion, formation of recirculation zone, arrangement of screech holes and design of the liner, air-fuel ratio, desired Mach number, low mean square pressure and high transmission loss.

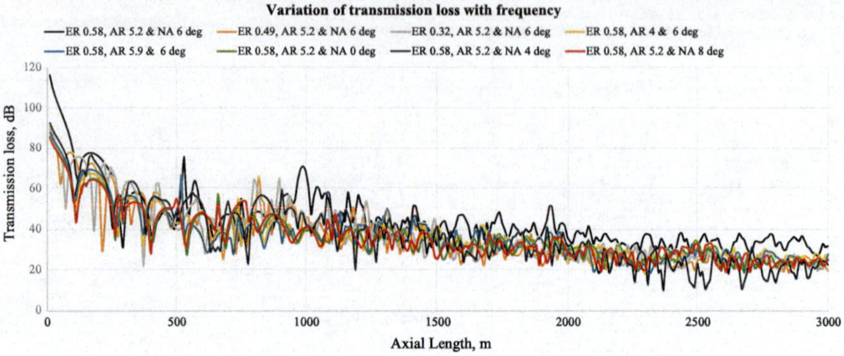

Fig. 8 Variation of transmission loss with frequency for different configurations

6 Conclusions

A novel method is introduced in this study, to capture the probable acoustic modes responsible for high mean square pressure in the afterburner, by carrying out CFD simulations followed by acoustic analysis for eight configurations of afterburner with different area ratios, different equivalence ratios and different nozzle angles. The effectiveness of convergent-divergent nozzle angle in attaining supersonic speeds is established and the best design for absorption of acoustic instabilities is suggested. The effectiveness of area ratio in reduction of screech instabilities is established and best design of screech liner is suggested. The effectiveness of equivalence in controlling the instabilities is established. It is found that the afterburner with liner, with convergent-divergent nozzle angle of 6°, area ratio of around 5, and equivalence ratio of 0.58, is predicted to be the best suitable afterburner for absorption of acoustic energy for specified design conditions. This will help in reduction of number of failures during the experiments and saves a lot of time and expenditure.

References

1. Kampa A (2007) Combustion instability screech in gas turbine afterburners. Ph.D. thesis, Department of Aerospace Engineering, Indian Institute of Science, Bangalore
2. Useller JW, Braithwaite WM, Rudy CJ (1954) influence of combustion chamber length on afterburner performance. Lewis Flight Propuslion Laboratory, USA
3. Unaune SV, Ganesan V (2005) Analysis of reacting flows in an aero-engine afterburner using computational fluid dynamics. Indian J Eng Mater Sci 11:31–37
4. Kabiraj L, Sujith RI (2012) Nonlinear self-excited thermoacoustic oscillations: intermittency and flame blowout. J Fluid Mech 713:376397
5. Kabiraj L, Saurabh A, Wahi P, Sujith RI (2012) Route to chaos for combustion instability in ducted laminar premixed flames. CHAOS 22:023129
6. Nair V, Sujith RI (2014) Multifractality in combustion noise: predicting an impending combustion instability. J Fluid Mech 747:635655

7. Noiray N, Durox D, Schuller T, Candel S (2008) A unified framework for nonlinear combustion instability analysis based on the flame describing function. J Fluid Mech 615:139–167
8. Palies P, Durox D, Schuller T, Candel S (2010) The combined dynamics of swirler and turbulent premixed swirling flames. Combust Flame 157:1698–1717
9. Palies P, Durox D, Schuller T, Candel S (2011) Nonlinear combustion instability analysis based on the flame describing function applied to turbulent premixed swirling flames. Combust Flame 158:1980–1991
10. Ebrahimi JB (2006) Overview of gas turbine augmentor design, operation and combustion oscillation. In: ILASS Americas, 19th annual conference on liquid atomization and spray systems, Toronto Canada, May 2006
11. ANSYS, Inc.: ANSYS fluent theory guide. Canonsburg, PA, U.S.A., Release 15.0, ANSYS, Inc., November 2013
12. Free Field Technologies. Actran 16.1 User Guide. 2014
13. Gurrala SR, Shaija A (2018) Computational analysis of reacting flows in afterburner. Heat Transfer Eng. https://doi.org/10.1080/01457632.2018.1546946

Experimental Study of Ignition Delay of Homogeneous Supercritical Fuel Sprays of Dieseline Blend in Constant Volume Combustion Chamber

Sanaur Rehman and Shah Shahood Alam

1 Introduction

Increasing environmental concern and stringent emissions regulations about pollutant and greenhouse gas emissions from internal combustion engines (ICE) along with depleting petroleum fuel resources clearly demands a breakthrough technology for ICE. One promising solution was proposed by some researchers [1–7] towards improving fuel conversion efficiency and significantly reducing exhaust emissions from diesel engines. The core of this proposed solution is the injection of fuel as supercritical fuel (SCF) or in SC state directly into the combustion chamber of the engine. It is found in some studies [1, 5] that 80–90% NOx and PM have been reduced in SC fuel spray injection and combustion. Thermodynamically, SC state is that state of substance when the pressure and temperature of a substance are above its critical point. Diesel fuel (DF) has enticing properties in the SC state [8, 9]. But at higher temperatures required for achieving SC state of DF, coke, gum, and tar formation occur in DF [10, 11]. This DF coking and thermal decomposition may plug the injector nozzle and hole and therefore to avoid this coking problem at elevated temperatures some anticoking agents are mixed with DF [3, 4, 10–12]. Anticoking or diluents such as EGR, CO_2, water, gasoline can be mixed with DF before injection into the combustion chamber. Gasoline is a good choice as an anticoking agent for DF because of its many useful properties [7, 12]. Dieseline (DL) blend is a mixture of DF and gasoline and was used by some researchers for SC fuel spray combustion study [12]. Therefore, in present work, DL blend has been used for SC fuel spray combustion study. ID is an important parameter in diesel engine combustion process

S. Rehman (✉) · S. S. Alam
Department of Mechanical Engineering, AMU, Aligarh, India
e-mail: sanaur.rehman@zhcet.ac.in

S. S. Alam
e-mail: sshahood2004@yahoo.co.in

A. K. Gupta et al. (eds.), *Advances in IC Engines and Combustion Technology*, Lecture Notes in Mechanical Engineering,
https://doi.org/10.1007/978-981-15-5996-9_48

as it affects the engine performance considerably [13–15]. ID depicts the quality of combustion process and affects pollutant formation in DI and IDI diesel engines. Constant volume combustion chamber (CVCC) is generally used for basic combustion studies because of its some advantages [13–16]. Hence in the present study, ID of SC fuel sprays are studied comprehensively under diesel engine-like operating conditions in CVCC as this work is missed in literature. Also, ID of SC fuel sprays is compared with the study on [17] normal spray under similar experimental conditions.

2 Experimental Setup

2.1 *Experimental Setup and Its Description*

Present experimental setup and its block diagram are shown in Fig. 1 (a, b) given below. The experimental set up is designed and fabricated to measure ignition and combustion characteristics of homogeneous SC fuel sprays in CVCC under various operating conditions. Cylindrical combustion chamber has two side plates. Combustion chamber has Bosch fuel injector on one side of plate and hot surface on the other side of plate just opposite to each other. Single-hole pintle nozzle is used for producing hollow conical fuel sprays impinging on hot surfaces. Bosch fuel injection pump is connected to the injector through high-pressure fuel line. Fuel injection pump is operated manually with the help of lever mechanism. The lever actuating the plunger of the pump is operated by hand. Pressure gauge is mounted on fuel injection pump for measuring fuel IP. Another pressure gauge is mounted on one side plate

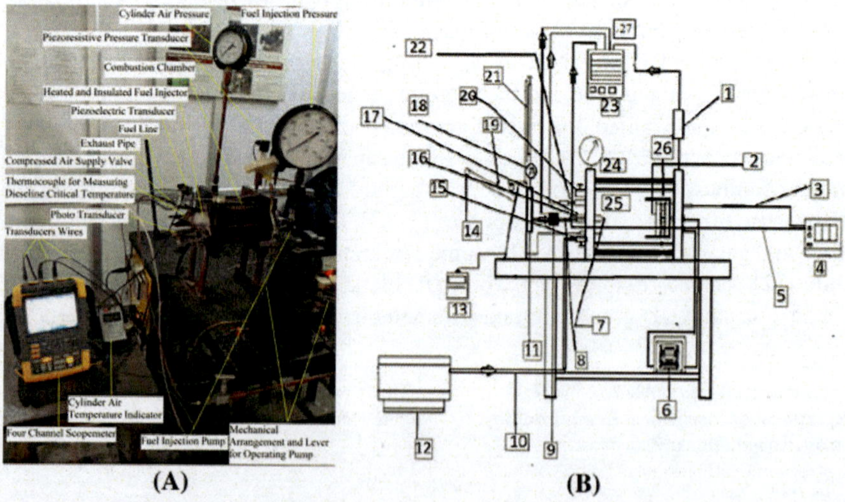

Fig. 1 **a** Experimental setup, **b** Block diagram of experimental setup

of the combustion chamber for measuring CP. Piezoresistive pressure transducer is mounted on other side plates of combustion chamber for measuring dynamic pressure rise during combustion. One K-type thermocouple is placed on hot surface to measures HST. Another thermocouple is inserted in fuel return valve of fuel injector for measuring DL blend's critical temperature (Tc). Another thermocouple inserted inside plate of the combustion chamber for measuring cylinder air or exhaust gas temperature. The details and specifications of each component are given in reference [17].

2.2 Experimental Methodology

The commercially available DF and DL50 are used for studying normal diesel spray combustion and SC fuel spray combustion, respectively. Properties of DF used in this work are given in the study [17]. DL50 is a blend having 50% gasoline and 50% DF by v/v. Experimental conditions are shown in Table 1 given. For SC fuel spray combustion study, in beginning, HST is maintained at 673 K, and then compressed air (from reciprocating air compressor) at various CP is maintained in the combustion chamber, i.e. at 20 bar, 30 bar, and 40 bar. The HST and CP are maintained inside the combustion chamber for steady-state. After maintaining the HST and CP conditions insider combustion chamber, DL50 blend temperature is monitored and maintained equal to Tc of DL50. Tc and critical pressure (Pc) of DL50 blend are given in Table 2. The Tc for DL50 blend was estimated using Kay's rule [18] and Pc for DL50 blend was estimated through Thompson's relation [19]. The DL50 blend is heated by the electrical heating of an insulated fuel injector. The thermocouple is inserted in the fuel return valve of an insulated fuel injector for measuring DL50 blend's Tc. The SC condition, i.e. Tc and Pc for DL50 blend are nearly equal to 629.3 K and 17.75 bar, respectively, as estimated. Critical points for different fuels and blend are given in Table 2. DF is assumed as n-hexadecane for estimating its critical point using Zhou correlation [21]. Gasoline is assumed as n-octane for determining its critical point

Table 1 Experimental conditions

Operating variable	Typical values
Fuel IP	100 bar, 200 bar, and 300 bar
CP	20 bar, 30 bar, and 40 bar
HST	673 K and 723 K

Table 2 Critical constants of diesel fuel, dieseline blend, and gasoline

Fuels	Tc	Pc
Diesel Fuel (DF)	714.71 K	19.23 bar
Dieseline (DL50)	629.30 K	17.75 bar
Gasoline	543.9 K [20]	25.7 bar [20]

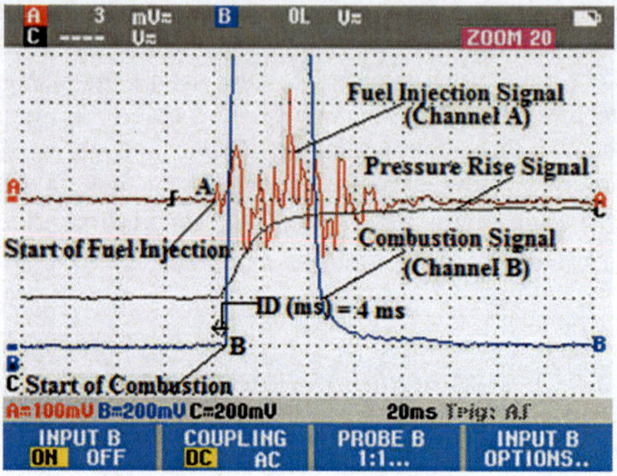

Fig. 2 Scopemeter İmage showing measurement of ID at one operating condition (HST = 673 K, CP = 20 bar and IP = 100 bar)

[20]. After heating DL50 blend inside injector above the required Tc (356.3 °C), DL50 blend is pressurized by Bosch fuel injection pump up to various IP (100 bar, 200 bar, and 300 bar) which are above DL50 blend's Pc (17.75 bar) with the help of lever mechanism and in this way SC spray condition of DL50 blend is achieved. After achieving SC spray condition, DL50 blend is injected into the compressed air inside combustion chamber through a heated injector. Injection, ignition, and combustion events of SC spray are recorded on the screen of four-channel scopemeter with the help of various transducers (sensors). ID is measured in millisecond (ms) as time interval between start of fuel injection (channel A, red colour) and start of combustion (channel B, blue colour) as shown in Fig. 2. Piezoelectric transducer mounted on fuel line detects start of fuel injection point and photo transducer detects start of combustion point and signals are sent to scopemeter as shown in Fig. 2.

After noting down readings from scopemeter, exhaust gases are expelled out from combustion chamber through exhaust needle valve. At each operating condition, five readings are taken, and mean value is calculated to minimize error in measurement. Same procedure is repeated at HST equal to 723 K. In case of conventional spray combustion study [17], heater of injector is turned off and fuel is injected at its normal temperature. Error analysis is done and it is found that the maximum percentage deviation in the measurement of ID is around 10%.

3 Results and Discussion

3.1 *Comparison of ID of SC and Normal Diesel Spray*

ID is an important characteristic of diesel and diesel-like fuels [13]. In diesel engines combustion systems, ID may be defined as time difference in ms or in crank angle (CA) between the start of fuel injection point and start of combustion point. The start of fuel injection into the combustion chamber is usually taken as the instant when the injector needle lifts off its seat (determined by the needle lift indicator) [15]. The start of the combustion point is more difficult to determine precisely [15]. The best way to detect the start of combustion point is through the change in slope of curve on the rate of heat release (ROHR) diagram, which takes place at ignition [15]. ID determined with the help of ROHR diagram is known as pressure-based ID. Flame luminosity detectors are also employed to determine the first appearance of flame in combustion chambers. ID determined with the help of flame luminosity detector is called luminous ID [16]. In the present study, photo transducer similar to flame luminosity detector is used to detect the start of ignition/combustion process, and hence luminous ID is measured in this study. In the rest of this paper, ID is always referred to as luminous ID. Figures 3 and 4 show graphs presenting the effect of fuel IP on ID at a different charge or combustion chamber conditions for both conventional diesel spray and homogeneous SC spray combustion systems. Graphs of Figs. 3 and 4 show the effect of CP on ID of normal diesel and SC sprays at constant HST of 673 K and 723 K, respectively. Also, each graph of Figs. 3 and 4 shows a comparison of ID magnitude in normal spray and SC spray combustion systems under similar operating conditions. The results presented in graphs of Figs. 3 and 4 indicate that the ID of normal diesel fuel sprays is significantly higher than ID of SC sprays at all operating or charge conditions. The significantly reduced ID in the case of SC sprays combustion as compared to normal diesel sprays combustion system is due to the attractive thermophysical properties of SCF such as density, volatility, diffusivity, surface tension, thermal conductivity and viscosity [7–9, 12]. SCF has a smaller ID and faster combustion than normal or droplet spray combustion [23]. In SC spray combustion system, DL50 blend is injected as DF-diluent mixture in SC state into compressed cylinder air. SC sprays of DL50 blend are injected as homogeneous and single-phase fluid [23–25] into compressed cylinder air. SC sprays have reduced penetration lengths and wider spray cone angles than normal diesel sprays [12]. Also, SC sprays have higher volatility due to more lighter/volatile components from gasoline, higher diffusivity, and reduced surface tension [12]. The characteristics properties of SCFs are density, diffusivity, and viscosity. The values of these properties of SCFs lie in between liquids and gases [9]. A SCF is a substance with both liquid-like and gas-like properties. SCF has higher rate of diffusion, typically an order of magnitude higher than liquid. SCF has lower viscosity than liquid by an order of magnitude [8]. The SCFs have liquid-like densities which significantly increase solubility and gas-like diffusivities and viscosities and zero surface tension which enhances mass transfer and fluid mixing. All these desirable and attractive changes in the thermophysical properties of SC fluids cause homogeneous and instantaneous mixing of SC sprays with compressed air. The mixing of SC fuel-diluent with cylinder air occurs nearly instantaneously due to very high molecular diffusion in SC fluids. This homogenous and instantaneous fuel-air mixing results in homogeneous and rapid combustion of SC sprays in combustion chamber. The SC combustion is homogeneous and nearly complete and is faster and cleaner than traditional droplet spray combustion due to the complete solubilization of fuel with the bulk oxidant [23]. Homogeneous and single-phase combustion of SC sprays also

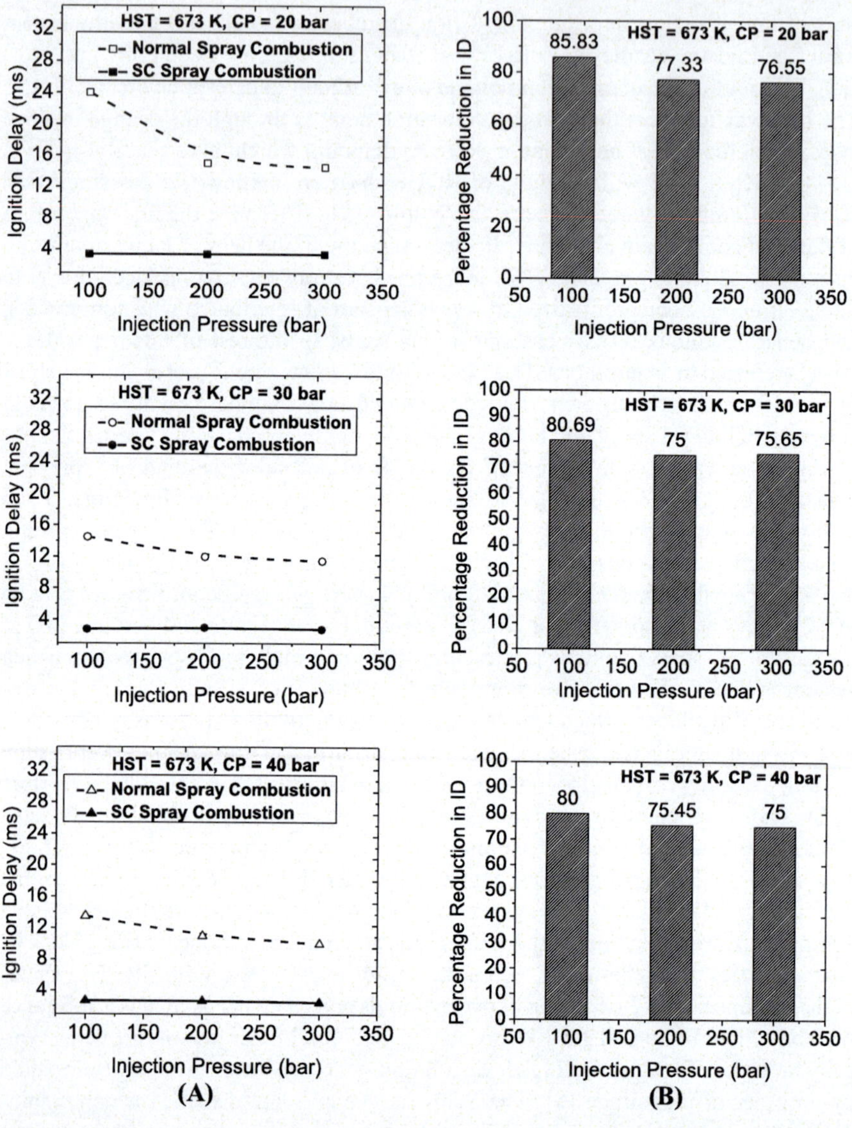

Fig. 3 **a** Variation of ID with IP at HST = 673 K and at different CP (20 bar, 30 bar, and 40 bar) in both Combustion Systems, **b** Variation of percentage reduction in ID with IP at HST = 673 K and at Various CP

results in a drastic reduction in PM formation due to the absence of droplet formation and vaporization [1, 23–25] and PM reduction is not due to reduced ID. It can be seen from Figs. 3 and 4 that percentage reduction in ID of SC sprays as compared to ID of normal diesel sprays is more than 55% in SC sprays than normal diesel sprays

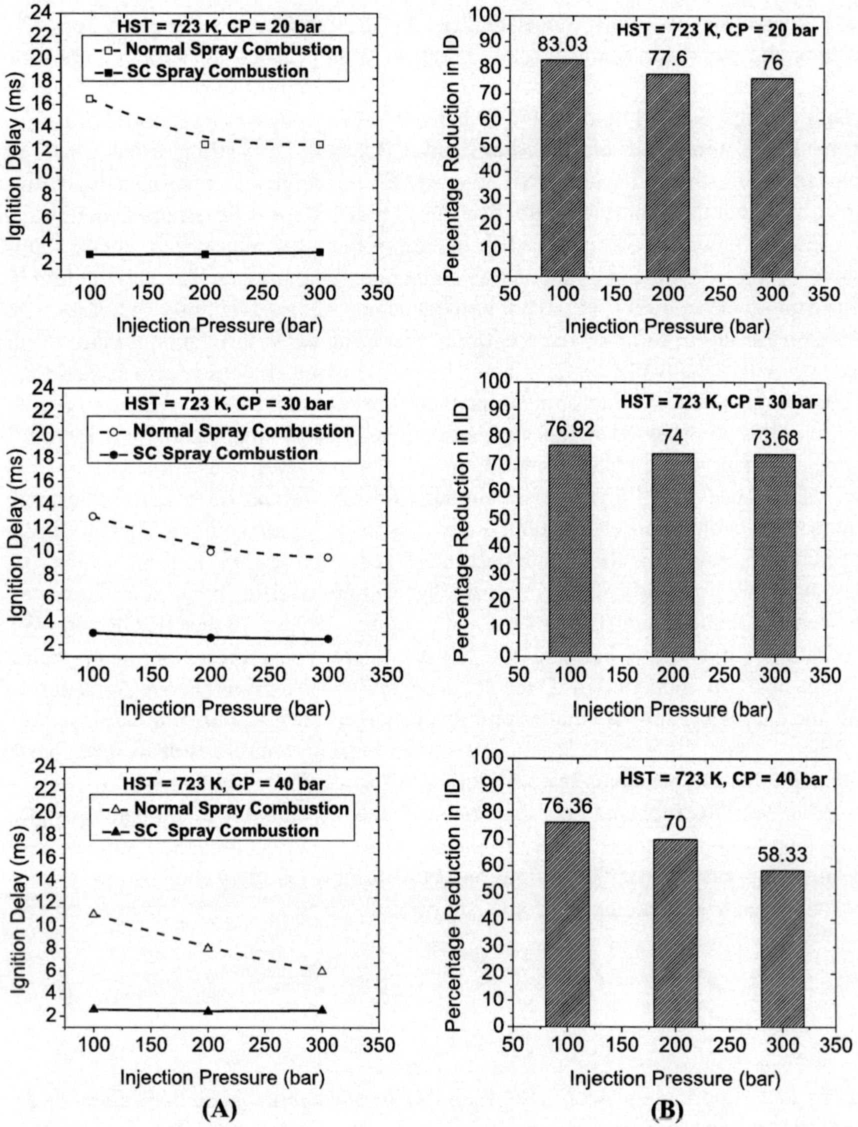

Fig. 4 **a** Variation of ID with IP at HST = 723 K and at Different CP (20 bar, 30 bar, and 40 bar) in both combustion systems, **b** Variation of percentage reduction in ID with IP at HST = 723 K and at various CP

at all operating conditions. Since smaller ID is desirable in diesel engines from efficiency and emissions point of view therefore homogeneous SC spray combustion technology is a breakthrough technology for diesel engines [23–26]. It can also be found in Figs. 3 and 4 that CP and IP have very insignificant effect on ID of homogeneous SC sprays but on the other hand, CP and IP have strong effect on ID of normal diesel sprays. In case of SC sprays of DL50 blend, hot air combustion seems to occur rather than hot surface ignition due to reduced spray penetration lengths and wider cone angles of SC sprays [12]. In case of hot air combustion of sprays, ID is mainly affected by the cylinder charge temperature and pressure [15, 22] but fuel IP has a minor effect on ID [15]. ID of homogeneous SC sprays slightly decreases with an increase in CP from 20 bar to 40 bar. Since increases in intake pressure of air increases the density of cylinder air, and hence increased air density also causes more air entrainment into SC fuel sprays, and therefore enhances mixing of SC sprays with surrounding compressed cylinder air. Also, there is a very small decrease in ID of SC sprays with increase in HST from 673 to 723 K at all CP. Increase in HST from 673 to 723 K causes small increase in cylinder air/charge temperature therefore a small increase in cylinder air/charge temperature results in an insignificant decrease in ID of SC fuel sprays. It is found that percentage reduction in ID is highest at lower IP, CP, and HST, i.e. nearly 80%, which are diesel engine starting conditions. However, percentage reduction in ID is lowest and more than 55% at higher IP, CP, and HST conditions (300 bar, 40 bar, and 450 °C) which are typical diesel engine operating conditions. Any reduction in ID results in decrease in premixed charge forms during ID and causes burning of smaller premixed charged during premixed burning stage of diesel combustion [15]. Smaller premixed burning would result in lower peak temperatures and pressures reached during this stage of diesel combustion.

Moreover, SC fuel spray combustion also takes place at lower temperatures [12, 27]. Due to both of these factors (reduced ID and lower combustion temperature) relatively less amount of NO_x formation would occur in SC spray combustion systems as compared to normal diesel droplet spray combustion.

4 Conclusions

In present experimental work, SC fuel spray combustion is studied under diesel engine-like operating conditions inside CVCC. It is found that ID of SC sprays is significantly smaller than normal diesel sprays at all operating conditions studied. At all experimental conditions, the percentage reduction in ID of SC sprays is more than 50%. At diesel engine starting conditions (100 bar IP, 40 bar CP, and 673 K HST), the highest percentage reduction in ID of SC spray occurs, i.e. 80%. At typical normal diesel engine operating conditions (300 bar IP, 40 bar CP, and 723 K HST), lowest percentage reduction in ID of SC spray occurs (nearly 58%). The best combination of experimental conditions in the present study at which maximum percentage reduction in ID occurs (nearly 85.83%) is found to be HST = 673 K, CP = 20 bar, IP = 300 bar. It is also investigated that an insignificant decrease in ID of SC sprays occurs with an increase in CP, HST, and IP. It is further found that SC spray combustion mainly occurs as homogeneous hot air combustion rather than hot surface combustion/ignition. Significant reduction in ID of SC spray as compared to normal diesel spray would cause drastic reduction in NOx and also homogeneous single-phase combustion of SC spray would result in substantial reduction in PM formation as reported in the literature. Hence, SC spray combustion is a single-phase, homogeneous, faster, and cleaner combustion technology than conventional diesel droplet spray combustion due to simultaneously dramatic reduction of major diesel engine pollutant, i.e. NOx and PM for next-generation automotive diesel engines.

Acknowledgements I would like to express my deep gratitude to Professor S. S. Alam for his generous support and guidance during the whole work. This work was supported by TEQIP-II Grant, MHRD, Govt. of India.

References

1. Ahern B, Djutrisno I, Donahue K, Haldeman C, Hynek S, Johnson K, Valbert J, Woods M, Corp SC, Taylor J, Tester J (2001) Dramatic emissions reductions with a direct injection diesel engine burning supercritical fuel. Water Mixtures. SAE Tech Pap 2001-01-3526, pp 1-8
2. Anitescu G (2008) Supercritical fluid technology applied to the production and combustion of diesel and biodiesel fuels. Ph.D. Thesis, Syracuse University
3. Tavlarides LL, Anitescu G (2009) Supercritical diesel fuel composition, combustion process, and fuel system. US Patent 7488357, issued Feb 10
4. Anitescu G, Tavlarides LL, Geana D (2009) Phase transitions and thermal behavior of fuel-diluent mixtures. Energy Fuels 23(6):3068–3077
5. Canter N (2010) Tech Beat, Tribology and lubrication technology, pp 10–11
6. Anitescu G, Lin R-H, Tavlarides LL (2009) Preparation, injection and combustion of supercritical fuels, poster P-2. Directions in Engine- Efficiency and Emissions Research (DEER) Conference; Dearborn, MI, 3 – 6 Aug 2009. http://www1.eere.energy.gov/vehiclesandfuels/pdfs/deer_2009/poster1/deer09_anitescu.pdf. Accessed April 2012
7. Anitescu G, Bruno TJ (2011) Volatility of gasoline and diesel fuel blends for supercritical fuel injection, poster P-2. Directions in Engine- Efficiency and Emissions Research (DEER) Conference, Detroit, MI, 3 – 6 Oct 2011. http://www1.eere.energy.gov/vehiclesandfuels/pdfs/deer_2011/wednesday/posters/p-02_anitescu.pdf. Accessed Apr 2012
8. Hitchen SM, Dean JR (1993) Properties of supercritical fluids. Appl Supercrit Fluids Ind Anal Springer Sci
9. http://barron.rice.edu/Courses/475/475_2013/475_projects_2013/Hizir_Draft.pdf
10. Lin R, Tavlarides LL (2013) Thermal stability and decomposition of diesel fuel under subcritical and supercritical conditions. J Supercrit Fluids 75:101–111
11. Lin R (2011) Issues on clean diesel combustion technology using supercritical fluids: thermophysical properties and thermal stability of diesel fuel, PhD Dissertation, Syracuse University, Syracuse
12. Anitescu G, Bruno T, Tavlarides L (2012) L, Dieseline for supercritical injection and combustion in compression ignition engines: Volatility, phase transitions, spray/jet structure and thermal stability. Energy Fuels 26:6247–6258 (ACS Publications)
13. Lapuerta M, Sanz-Argent J, Raine R (2014). Ignition characteristics of diesel fuel in a constant volume bomb under diesel-like conditions. Effect of the operation parameters. Energy Fuels 28:5445–5454

14. Rabl S, Davies TJ, McDougall AP, Cracknell RF (2015) Understanding the relationship between ignition delay and burn duration in a constant volume vessel at diesel engine conditions. Proc Combust Inst 35:2967–2974
15. Heywood JB (2011) Internal combustion engines, 2nd edn. McGraw Hill International, New York
16. Ghojel JI, Tran X-T (2010) Ignition characteristics of diesel-water emulsion sprays in a constant-volume vessel: effect of injection pressure and water content. Energy Fuels 24:3860–3866
17. Rehman Sanaur (2018) Hot surface ignition and combustion characteristics of sprays in constant volume combustion chamber using various sensors. Cogent Eng 5:1464879
18. Borgnakke C, Sonntag RE (2012) Fundamentals of thermodynamics. 7th edn, John Wiley & Sons
19. Thomson GH, Brobst KR, Hankinson RW (1982) An improved correlation for densities of compressed liquids and liquid mixtures. A I Ch E Journal 28:671–676
20. Poling BE, Prausnitz, JM, O'Connell JP (2001) The properties of gases and liquids, 5th edn. McGraw-Hill
21. Zhou P (1984) Correlation of the average boiling points of petroleum fractions with pseudocritical constants. Int Chem Eng 24:731–741
22. Mathur, Sharma (2008) Internal combustion engine, 8th edn. Dhanpat Rai Publication (P) ltd
23. Millikin M (2009) Use of supercritical diesel fuel for ımproved efficiency and reduced emissions. Green Car Congress, 4 Aug 2009
24. Ashley S (2009) Supercritical fuel ınjection could reduce diesel emissions. Society of Automotive Engineers International, 26 Aug 2009
25. Graham-Rowe D (2009) Supercritical fuel ınjection: a supercritical diesel engine could ıncrease efficiency and cut emissions. Technology (MIT Review), 11 Aug 2009
26. Peckham J (2009) Supercritical diesel fuel might solve problems, but needs more R&D. Diesel Fuel News, 21 Sept 2009
27. Supercritical dieseline as an optimized fuel for low temperature combustion. Green Car Congress, August 2014

Numerical Characterization of Circular and Elliptical Central Port Inverse Jet Diffusion Flame

Vishnu Hariharan and Debi Prasad Mishra

1 Introduction

Diffusion flames are commonly used due to their wider flame stability characteristics in aircraft and gas turbine applications. The requirement of cleaner combustion has sparked interest in improving the existing combustion technologies to enhance the degree of mixing and entrainment. The operating air-fuel jet velocity conditions and the burner geometry have a significant impact on mixing and entrainment characteristics.

The inverse jet diffusion flame burner configuration exhibits the advantage of both premixed and diffusion flame [1, 2]. Sze et al. [3] reported that higher peak temperature and shorter flame height was achieved for circumferential arranged fuel port Inverse Jet Diffusion Flame (CAFP IJDF) when compared to a co-axial (CoA) burner due to larger air-fuel and fuel-fuel interactions. Dong et al. [4] experimentally explored the effect of air-fuel diameter ratio and they concluded that smaller diameter central air jet exhibited blue flame with better thermal characteristics. The effect of nozzle length in CAFP IJDF was experimentally studied by Zhen et al. [5]. They reported that a shorter nozzle produced complete combustion due to faster air/fuel mixing. Recently, Hariharan and Mishra [6] experimentally observed the static flame stability aspects of a CAFP IJDF with 32 ports at a constant air jet velocity with varying fuel jet velocity and at constant fuel jet velocity with varying air jet velocity. Further, they studied thermal, emission, and flame height characteristics of this burner [7] based on a physical parameter called momentum flux ratio.

V. Hariharan (✉) · D. P. Mishra
Combustion Laboratory, Department of Aerospace Engineering, Indian Institute of Technology Kanpur, Kalyanpur, Kanpur, Uttar Pradesh 208016, India
e-mail: vishnuh@iitk.ac.in

A. K. Gupta et al. (eds.), *Advances in IC Engines and Combustion Technology*, Lecture Notes in Mechanical Engineering,
https://doi.org/10.1007/978-981-15-5996-9_49

Further enhancement of combustion performance in IJDF can be achieved through a passive mechanism by the use of the asymmetrical central jet. The asymmetrical nature induced by the elliptical jet has a favorable effect on flow turbulence characteristics. The vortex ring in a non-circular jet which propagates in a direction normal to jet exit attempt to take asymmetrical shape as the jet propagates downstream. Interestingly, during this motion, the major axis of the vortex ring rotates by 90° and in-plane and out of plane alterations repeats itself. The switching of the vortex axis was termed as vortex induction by Ho and Gutmark [8]. These inherent properties of asymmetrical jets can enhance mixing and combustion properties, which suit a wide range of applications. Kashkousha et al. [9] investigated the elliptical normal jet diffusion flame in co-flow and cross-flow cases. They concluded that the elliptical jet exhibited increased jet entrainment when compared to circular jets.

The present numerical study is based on IJDF, wherein the central air jet is surrounded by circumferentially arranged fuel port. The literature has pointed out that elliptical jet can have higher entrainment characteristics when compared to a symmetric jet. The present study was motivated towards the combination of an elliptical central jet with circumferentially arranged fuel ports to explore the possibility of enhancement of mixing and entrainment features in IJDF.

2 Computational Details

The burner geometry considered in the present work is identical to the experimental work of Hariharan and Mishra [6]. The central port has a circular orifice of 6 mm diameter air jet surrounded by 32 circular fuel ports with each of diameter 1.5 mm. The present computational study uses natural gas composed of 95% CH_4, 3% C_2H_6, 1% N_2, and 1% CO_2 as the fuel. An elliptical orifice replaces the central circular orifice with a major diameter, 8 mm, and a minor diameter, 4.5 mm to investigate asymmetric jet characteristics. Further, the center-to-center air-fuel port distance is optimized and fixed to be 13 mm for an elliptical and circular port central jet. The computational domain, with a diameter of 100 mm as shown in Fig. 1 extends to a length of 400 mm.

2.1 *Governing Equations*

The turbulent flame is considered as an ensemble of a one-dimensional steady laminar flamelet model (SLFM) in which the reactive terms are converted to mixture fraction space. The Favre averaged governing equations for mass, momentum, and energy are solved. The finite rate detailed chemistry model, GRI 3.0 mechanism [10] is used in the present computational work with 53 species and 325 reactions. Besides mass momentum and energy equations, the governing flame equation to represent reactive diffusive layer in terms of scalar dissipation rate, χ in the transformed space.

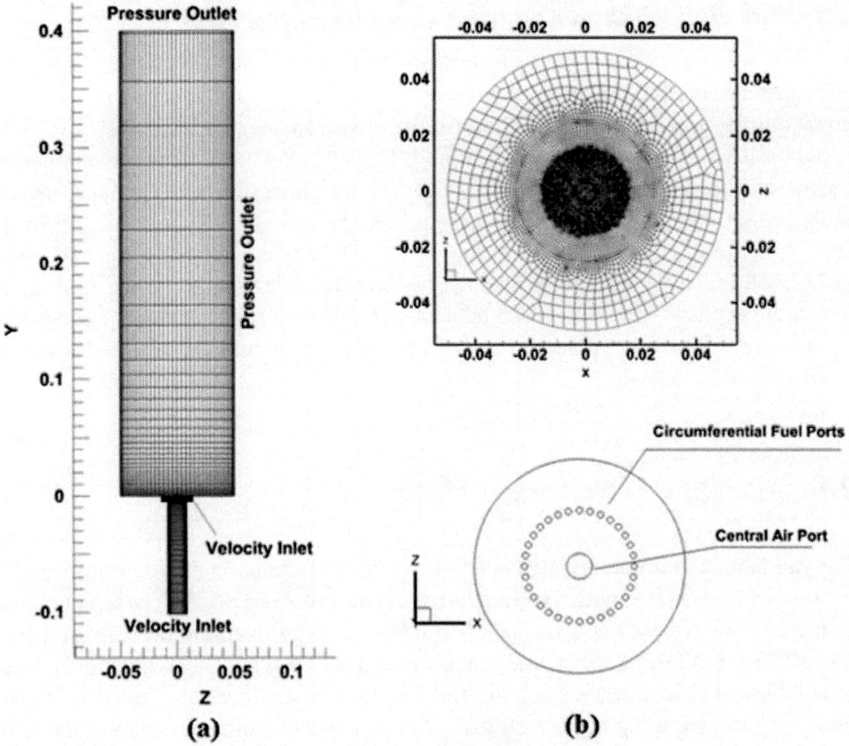

Fig. 1 Computational domain of CAFP IJDF with 32 fuel ports (**a**) Front view of burner geometry (**b**) plan view

The present computation considers non-adiabatic SLFM with mean enthalpy for calculation of flame temperature and mean density. The flamelet is generated using a β-PDF approach in terms of mean quantities of scalar $\bar{\phi}\left(\bar{z}, \bar{z}''^2, \bar{\chi}\right)$ as function of mixture fraction, variance, and scalar dissipation rate. The flamelet library is coupled with turbulent calculations using the following transport equations of $\bar{z}$ and $\bar{z}''^2$ expressed as:

$$\frac{\partial}{\partial t}(\rho\bar{z}) + \frac{\partial}{\partial x_k}\rho\widetilde{u_k}\bar{z} = \frac{\partial}{\partial x_j}\left(\frac{\mu_t}{\sigma_t}\frac{\partial\bar{z}}{\partial x_k}\right). \quad (1)$$

$$\frac{\partial}{\partial t}\left(\rho\bar{z}''^2\right) + \frac{\partial}{\partial x_k}\rho\widetilde{u_k}\bar{z}''^2 = \frac{\partial}{\partial x_j}\left(\frac{\mu_t}{\sigma_t}\frac{\partial\bar{z}''^2}{\partial x_k}\right) + C_g\mu_t\left(\frac{\partial\bar{z}''^2}{\partial x_k}^2\right) - C_d\mu_t\frac{\varepsilon}{k}\bar{z}''^2 \quad (2)$$

2.2 Solution Methodology and Computational Grid

The present computational work uses ANSYS Fluent 13.0 to solve the transport equations mentioned in Sect. 2.1. The pressure-based solver with the SIMPLE algorithm deals with pressure velocity coupling. The RNG k-ε model the flow turbulence characteristics. The radiation is modeled using a weighted sum of the grey gas model along with P1 approximation. The computational grid after grid independence study, for elliptical and circular central orifice CAFP IJDF is made of hexahedral and tetrahedral cells. The velocity inlet boundary condition is defined at air and fuel inlet. The present unconfined flame is modeled with pressure outlet boundary condition for all sides of the computational domain and remaining boundaries are defined as wall.

2.3 Validation of Numerical Model

The validation of the numerical model is carried out by comparing the thermal characteristics of CAFP IJDF with the circular central air-port IJDF [6, 7]. The temperature characteristics are obtained at a position 0.005 m away from the burner rim using a Pt-Rh13% (+)/Pt (−), with wire diameter 0.13 mm using an *XYZ* positioner. The radial thermal characteristics at a location near to burner rim ($Y = 5$ mm) are satisfactory as shown in Fig. 2. The numerical model also captures the trend of thermal characteristics at the main flame.

Besides this, maximum center-line temperature for the test case $V_a = 23.6$ m/s and $V_f = 1.3$ m/s are compared with experimental and numerical predictions. The peak center-line temperature obtained experimentally is 1743 K and numerical prediction is 1525 K which implies a variation of 15% from the experimental value. The

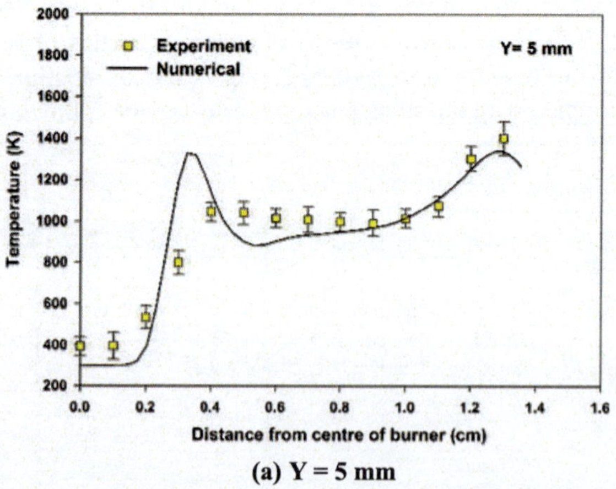

Fig. 2 Validation of numerical model with experimental values at axial locations $Y = 5$ mm along the *z*-axis

deviation of the numerical data from the experiments can be attributed to the inaccurate prediction of the flame zone width by the k-ε turbulence model. The maximum temperature within the flame zone is found to be 1928 K which is offset from the central jet axis line due to probed measurements within the flame.

3 Result and Discussions

3.1 Flame Shape

The flame shape from the numerical predictions is compared for elliptical and circular central port CAFP IJDF. The model considered for validation ($V_a = 23.6$ m/s and $V_f = 1.3$ m/s) with the experimental model is used to study the flame shape of both the configuration.

The OH mole-fraction distribution is used as a flame marker for circular and elliptical CAFP IJDF as shown in Fig. 3a–c. In the case of the elliptical central air jet, the flame shape of CAFP IJDF is visualized in major and minor planes due to the asymmetry of the jet profile. Further, regardless to the change in central port geometry, the flame shape can be characterized by base flame, main flame separated by a constricted flame neck. The difference in momentum between air and fuel jets invokes pressure differential, which entrains circumferential fuel jets towards the central air jet. The main flame which ensues from the flame neck is the region where

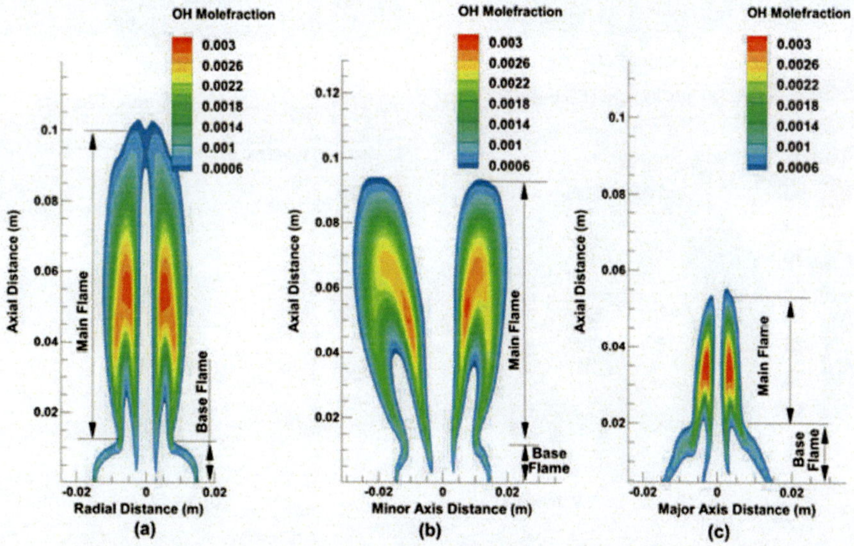

Fig. 3 OH mole-fraction distribution for CAFP IJDF (**a**) with circular port (**b**) with elliptical port on minor axis plane (**c**) elliptic port on major axis plane

the mixing is predominant. On the other that hand, the elliptical port with aspect ratio 1.8 displays variation in flame shape in major and minor axis planes owing to the asymmetric nature of the elliptical jet. The minor axis plane as shown in Fig. 3b depicts the greater spread of OH mole-fraction, which signifies that jet spread is more extensive on the minor axis plane. The distance from the burner base to the flame neck is found to vary for circular and elliptical central port CAFP IJDF. In the case of, elliptical port the location of the flame neck situated from the burner base differs by 28% on the major and minor axis as depicted in Fig. 3b and c. The flame neck is formed earlier on a minor axis plane compared to the major axis which denotes that the circumferential fuel jets meet the central air jet earlier, resulting from forming flame neck at a lesser axial distance from the burner rim on major axis plane.

The measure of the extent of mixing can be obtained from the turbulent kinetic energy contours. The turbulent kinetic energy for circular and elliptical CAFP IJDF at two different axial locations $Y = 25$ and 50 mm are compared, as shown in Fig. 4 to identify the degree of mixing. It is found that the degree of mixing at the location near

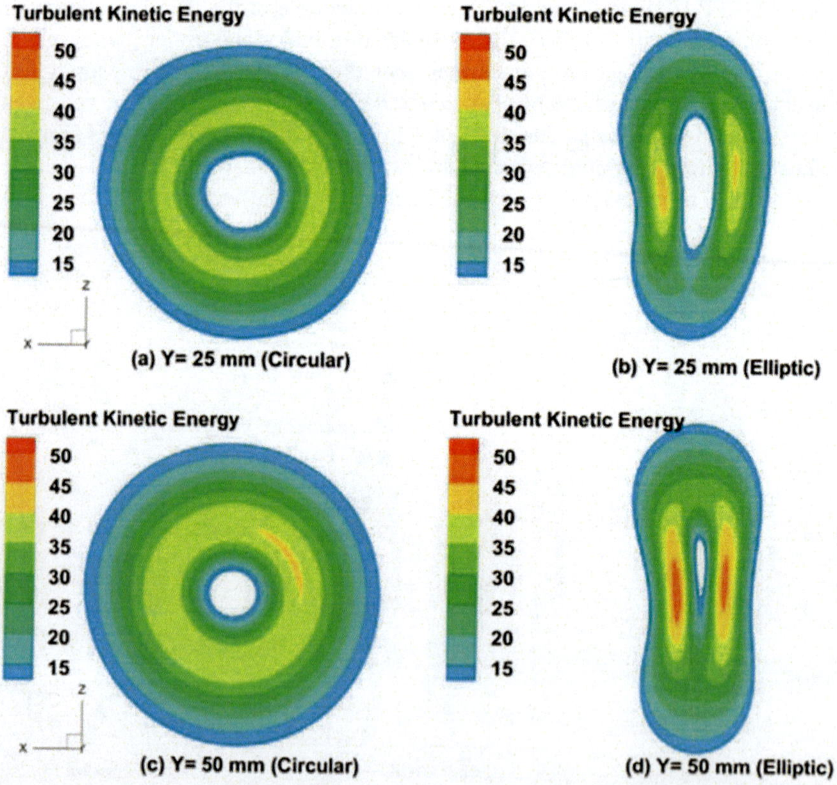

Fig. 4 Turbulent kinetic energy (m^2/s^2) contours for CAFP IJDF with a circular and elliptical central air-port at $Y = 25$ and 50 mm

to burner rim is comparable for both the configurations, whereas at a downstream location $Y = 50$ mm, the turbulent kinetic energy for elliptic configuration is found to be higher. The major axis sides are found to contribute larger turbulent kinetic energy indicating augmented mixing characteristics when compared to circular CAFP IJDF configuration.

The augmented mixing can be attributed to the inherent property of asymmetrical jets causing swapping of the major and minor axis in the elliptic jet. Figure 5 shows the axial velocity distribution at different axial locations to identify the point of axis switching in the case of elliptical CAFP IJDF. The axis switching occurs at an axial distance of 10 mm from the burner rim. As a result, the inner jet almost assumes a circular shape as shown in Fig. 5b. A higher jet spread is noted along the minor axis when compared to major axis at downstream of the jet as shown in Fig. 5c–d. The turbulent elliptical central jet can result into a vortex ring due to the viscous interaction of jet and ambient environment [8]. Due to this peculiar phenomenon of axis switching higher velocity is induced due to greater curvature. Thus, more ambient air is entrained causing greater jet spread and larger turbulent kinetic energies for elliptical jets when compared to circular jets.

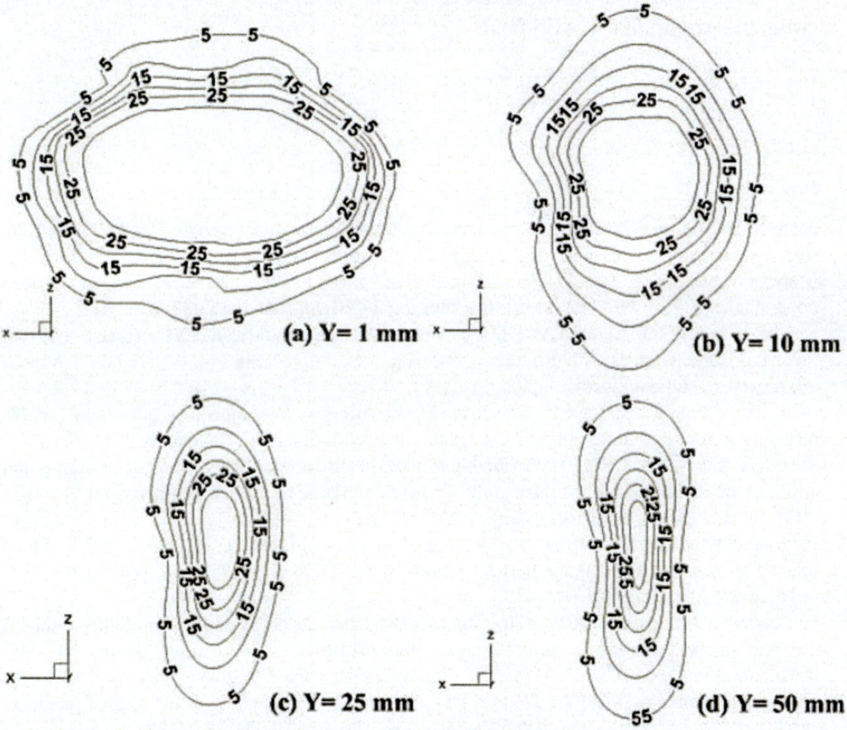

Fig. 5 Axial velocity (m/s) contours at different axial distances depicting axis switching for elliptical CAFP IJDF with aspect ratio 1.8

4 Conclusion

In the present numerical study, elliptical and circular central jet circumferentially arranged fuel port inverse jet diffusion flames (CAFP IJDF) are investigated to identify the flame shape, mixing characteristics. The following conclusions are made from this numerical investigation:

1. Asymmetric flame shape for elliptic CAFP IJDF is obtained for major and minor axis planes, which are visualized through OH mole-fraction distribution depicting the difference in the jet spread in major and minor axis planes.
2. The jet spread is more considerable for elliptical CAFP IJDF when compared with circular CAFP IJDF.
3. The flame neck is situated at a distance of 2.1d on the minor axis and 2.7d on the major axis plane, where d is the diameter of the central air jet.
4. The degree of mixing is found to be comparable at the near burner exit for circular and elliptical CAFP IJDF.
5. At the downstream of the elliptical CAFP IJDF, turbulent kinetic energy, an indicator of the extent of mixing, is found to be higher due to the inherent property of axis switching in an asymmetric jet which increases the extent of mixing compared to circular CAFP IJDF.

References

1. Mahesh S, Mishra DP (2008) Flame stability and emission characteristics of turbulent LPG IDF in a backstep burner. Fuel 87(12):2614–2619. https://doi.org/10.1016/j.fuel.2008.02.001
2. Mahesh S, Mishra DP (2015) Characterization of swirling CNG inverse jet flame in recessed coaxial burner. Fuel 161:182–192. https://doi.org/10.1016/j.fuel.2015.08.022
3. Sze LK, Cheung CS, Leung CW (2006) Appearance, temperature, and NO_x emission of two inverse diffusion flames with different port design. Combust Flame 144(1-2):237–248. https://doi.org/10.1016/j.combustflame.2005.07.008
4. Dong LL, Cheung CS, Leung CW (2011) Combustion optimization of a port-array inverse diffusion flame jet. Energy 36(5):2834–2846. https://doi.org/10.1016/j.energy.2011.02.025
5. Zhen HS, Choy YS, Leung CW, Cheung CS (2011) Effects of nozzle length on flame and emission behaviors of multi-fuel-jet inverse diffusion flame burner. Appl Energy 88(9):2917–2924. https://doi.org/10.1016/j.apenergy.2011.02.040
6. Hariharan V, Mishra DP (2019) Static flame stability of circumferentially arranged fuel port inverse jet non-premixed flame burner. Combust Sci Technol 1–27. https://doi.org/10.1080/00102202.2019.1611567
7. Hariharan V, Mishra DP (2019) Experimental characterization of circumferentially arranged fuel port inverse jet diffusion flame burner combust. Sci Technol 1–20. https://doi.org/10.1080/00102202.2019.1643847
8. Gutmark E, Schadow KC, Parr TP, Parr DM, Wilson KJ (1989) Combustion enhancement by axial vortices. J Propul Power 5(5):555–560. https://doi.org/10.2514/3.23189

9. Kashkousha OA, Kamal MM, Abdulaziz AM, Nosier MA (2012) Concentric elliptical jet diffusion flames with co-and cross-flows. Exp Therm Fluid Sci 41:177–187. https://doi.org/10.1016/j.expthermflusci.2012.03.033
10. Frenklach M, Bowman T, Smith G, Gardiner B. http://combustion.berkeley.edu/gri-mech/version30/text30.html, GRI mechanism 3.0

Performance Analysis of an LPG Cooking Stove for Improvements and Future Usability Perspective

Rohit Singh Lather

1 Introduction

A large population living in developing and under-developed does not have access to clean domestic energy, in particular, electricity and cooking gas, leading to energy poverty [1]. Many countries around the world are putting efforts to provide clean domestic energy through various schemes by either providing improved cookstove design or LPG cooking stoves [1–5]. A large population in India still uses conventional biomass-based cooking methods resulting in environmental and health issues [2, 6]. Implementation of the "Pradhan Mantri Ujjwala Yojana (PMUY)," in India, to distribute 50 million gas cylinders to below poverty line (BPL) women increased the LPG penetration in India. India is the world's second-largest LPG consumer. The cumulative annual growth rate of LPG consumers in India is 15%, presented in Fig. 1. It is expected to grow with the increasing population in the country. LPG penetration in India is presented in Fig. 2. Indian rural population is largely moving towards the use of LPG gas stoves from biomass-based domestic cooking [2]. Socio-economic factors, Pradhan Mantri Ujjwala Yojana (PMUY), education, and awareness are the motivators for increasing the use of LPG in rural India. With such a large scale of use, improvement in the fuel consumption of LPG cookstoves will have a huge impact on fuel savings and economy [7].

Kimemia and Niekerk [6] compared LPG cookstoves with kerosene, methanol, ethanol, and gel. They reported that LPG cookstoves have superior power, efficiency, lower CO emissions, fuel toxicity, stove price durability, stability, and operation. Bogagavarapu, Ray, and Ravikrishna [7] experimentally and numerically investigated the thermal efficiency of gas stove burner using LPG and PNG for inserts and radiant

R. S. Lather (✉)
Energy Conversion & Efficiency Lab, Mechanical Engineering Department, The NorthCap University, Sector 23A, Gurugram 122017, India
e-mail: rohitsinghlather@gmil.com

A. K. Gupta et al. (eds.), *Advances in IC Engines and Combustion Technology*, Lecture Notes in Mechanical Engineering,
https://doi.org/10.1007/978-981-15-5996-9_50

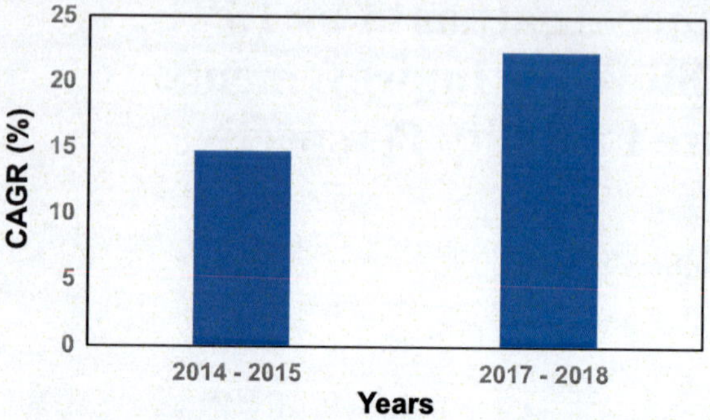

Fig. 1 LPG Consumer growth in India from the year 2014–2015 to year 2017–2018. The figure is indicative of the growing LPG consumer in India [8]

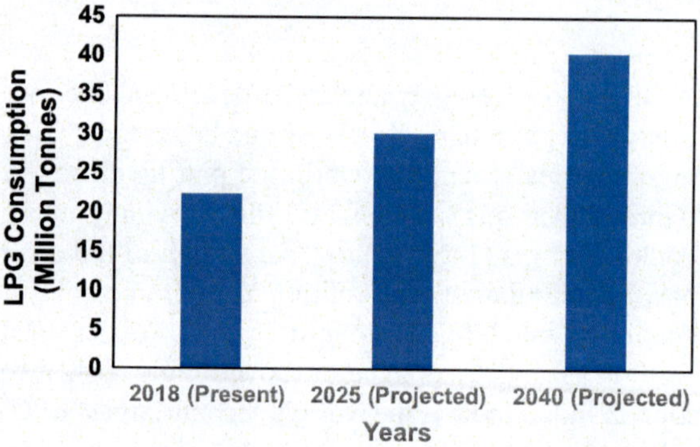

Fig. 2 LPG Consumption in India in the year 2018 (present) to year 2025 (projected) and Year 2040 (projected). The figure indicated the large-scale consumption of LPG in the time to come [8]

sheet. For LPG gas stove using inserts and radiant sheet, they were able to increase the thermal efficiency by 2.5% for LPG and 10% for PNG. Loading height was identified as one of the key parameters for improving thermal efficiency. It was also reported that an optimized loading height will give the best thermal efficiency for a gas stove. As per the Bureau of Energy Efficiency, the thermal efficiency of the gas stove should be above 68% for BEE star label [9].

LPG gas stoves are partially aerated naturally aspirated gas burners, commonly used for gaseous fuels like LPG, natural gas, etc. [10]. Many methodologies have been resented by Jones for efficiency improvement. Dahiya et al. [11] carried out an analytical analysis of the domestic LPG gas stove and analyzed the operational factors influencing the heat losses in conventional LPG burners. Some of the factors are air entrainment, utensil shape, loading height, radiation, convective heat losses, and burner design. Ways to improve the thermal efficiency by changing loading height and reducing radiation losses are also suggested by the authors.

Ko and Lin [12] investigated a gas burner using natural gas for gas composition, primary aeration, gas flow rate, gas supply pressure, and loading height on thermal efficiency and carbon monoxide emissions. They observed that lower heating value results in lower thermal efficiency and higher CO emissions, increasing the flow rate of the gas decreases the thermal efficiency, lower gas pressure reduces CO emissions and increase thermal efficiency, primary aeration did not affect thermal efficiency but reduced the CO emissions, increasing the loading height reduces emissions, however, an excessive loading height may result in lower thermal efficiency. The thermal efficiency in actual operating conditions drops due to varying operational parameters, ambient conditions, operation conditions, and heat losses.

Pantangi [13] reported that the thermal efficiencies of LPG cooking stoves in India range from 60 to 65% and the emissions levels are above the World Health Organization (WHO) standards. To improve the thermal efficiencies porous medium combustion (PMC) and porous surface combustion (PSC) were used in cooking gas stoves. Efficiencies ranging from 68 to 73% were achieved for different combinations of porous material. Gohil and Channiwala [14] experimentally investigated a commercially available gas stove for thermal efficiency and emissions. The thermal efficiency of the gas stove was 66.27% and the burner power was 1.8 kW. The emissions of carbon dioxide were 0.9 vol. % and carbon monoxide were 50 ppm.

2 Experimental Setup

Figure 3 shows the block diagram of the experimental test-setup developed for testing the cookstove as per the IS 4246:2002. A mass-produced, commercially available LPG cookstove, confirming to ISI standards was purchased for the study, from the local market. The burner of the cookstove is a double ring type and the material is brass. Figure 4 shows the 3D model of the cookstove burner used for the study. The burner is naturally aspirated; therefore, the primary air intake is dependent on the pressure of the gas supply. A commercially available 5 kg LPG gas cylinder was used for testing the gas consumption and thermal efficiency of the burner. Table 1 presents the properties of the LPG. A load cell was used to measure the weight of the LPG cylinder. The resolution of the load cell was 0.1 g and range 0–10 kgs. A rotameter was used to measure the flow rate of the LPG. The resolution of the rotameter was 0.1 lpm and the range of the rotameter is 0.1–3 lpm. Figure 3 shows the burner hood fabricated using galvanized iron, for flue gas sampling, as per IS 4246:2002 specifications. Two points were provided for flue gas sampling. The burner hood is used to prevent dilution of the flue gases by ambient air. As per IS 4246:2002, the gas supply pressure should be in the range of 35–40 mbar. A pressure regulator was installed on the LPG supply line to measure the supply pressure. The pressure regulator has a resolution of 1 mbar and an operating range of 0–40 mbar. A variable pressure regulator was installed between the gas cylinder and rotameter to regulate the gas pressure. An aluminum pan of 190 mm diameter, as per IS 4246:2002 was selected and used for the test. RTD was used for measuring the temperature of the

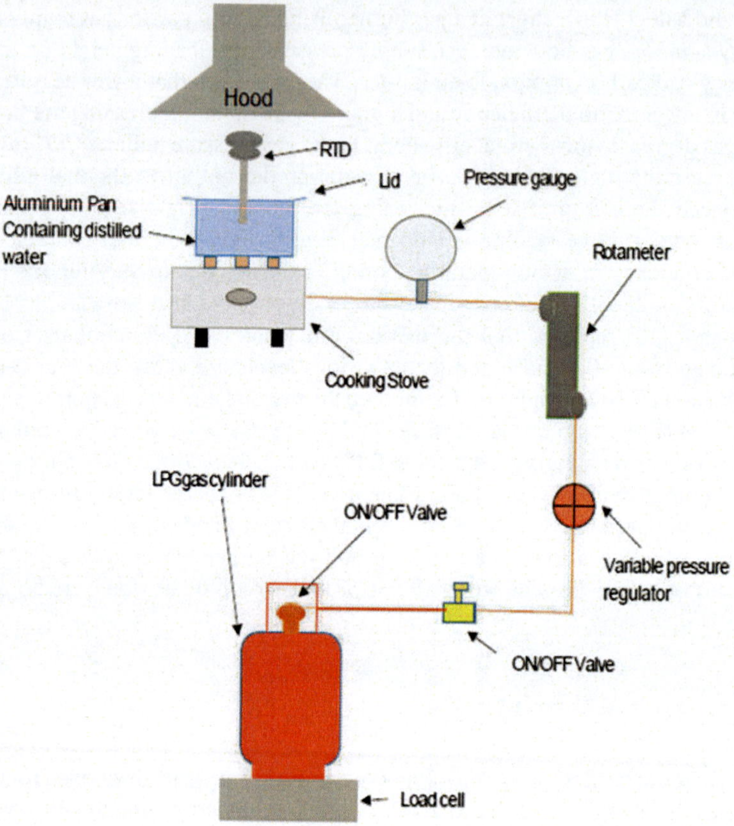

Fig. 3 Schematic of the experimental setup developed for testing the LPG cookstove as per IS 4246:2002

water in the pan. The pan was fully covered using a lid, to avoid any heat loss. All the equipment was calibrated before the tests.

The calculation of the thermal efficiency was carried out using the following equation:

$$\text{Thermal efficiency} = \frac{(m_w \times C_w + m_p \times C_p)(T_2 - T_1)}{m_f \times C_v}[10] \quad (1)$$

m_w = mass of the distilled water
C_w = specific heat capacity of water
C_p = Specific heat capacity of the aluminum pan
m_p = mass of aluminum pan along with lad and stirrer
m_f = mass of the fuel
Calorific value of the fuel (C_v) = 45,636.12 kJ/kg
Specific heat of water (C_w) = 4.1826 kJ/kg-K

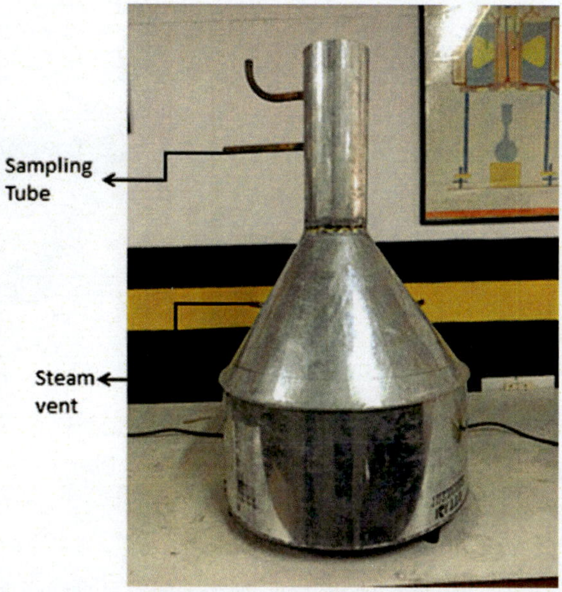

Fig. 4 The cookstove burner hood fabricated for testing and measurement of the flue gas emissions of LPG cookstove as per IS 4246:2002

Table 1 The physicochemical properties of LPG [15]

Properties	LPG
Chemical formula	C_3H_8
Energy content (MJ/m^3)	93.2
Energy content (MJ/kg)	49.58
Flame temperature (°C)	1967
Gas volume (m^3/kg)	0.540
Specific gravity	1.5219
Density @15 °C (kg/m^3)	1.899
Auto-ignition temperature	470 °C

Specific heat of Aluminum (C_p) = 0.8959 kJ/kg-K

For changing the loading height of the utensil on the LPG cookstove, the loading height mechanism was fabricated and installed on the cookstove. Figure 5 presents the loading height arrangement installed on the cookstove. A measuring scale is provided in the loading height mechanism to measure the change in the loading height in millimetres. The loading height can be varied from 0 mm (reference point, original loading height) to a maximum of 7.5 mm.

Fig. 5 The loading height mechanism installed on the cookstove for changing loading height. A scale arrangement is also provided to measure the change in height in millimetres

3 Results and Discussion

The following section presents and discusses the results from the thermal efficiency tests, flue gas emissions, loading height variation versus thermal efficiency and gas consumption, and usability evaluation of an LPG cookstove.

3.1 Thermal Efficiency

Figure 6 presents a set of five thermal efficiency tests carried out using the cookstove. Variation in the thermal efficiency was observed in all the tests. The average thermal efficiency of 60.71% was observed for the cookstove. The thermal efficiency varied from 59.12 to 63.2%. As the thermal efficiency is lower than 68% (BEE Star label rating standard), there is scope for improvement by using suitable modifications to the cookstove, one, in particular, is loading height [11, 12], and design modifications to reduce heat losses [13, 15, 16].

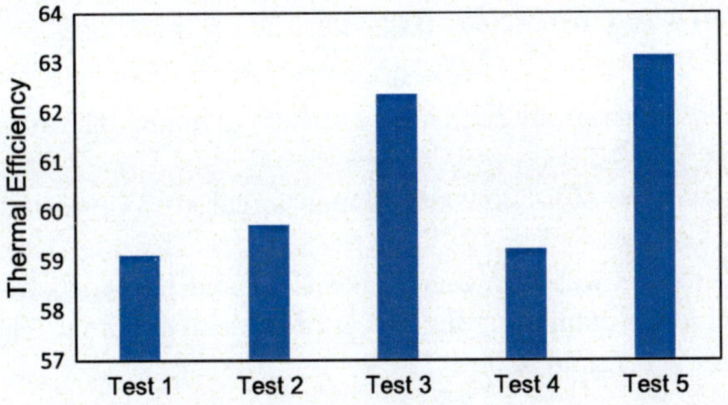

Fig. 6 The results of five thermal efficiency tests of the LPG cookstove. The variation in thermal efficiency is observed for each test with an average value of 60.71%

3.2 *Flue Gas Emissions*

Figure 7 presents the flue gas emissions measured during the tests. The flue gas emissions were measured using the AVL Di-gas analyzer. The burner hood was placed on the burner-pan setup. The probe of the analyzer was inserted in the sampling point provided in the burner hood. The flue gas emissions were noted at two conditions, as per IS 4246:2002 standard., with the hood damper fully open and hood damper fully closed. Two liters of distilled water were used for the test. The supply pressure of the gas was set to 29 mbar and the flow rate to 0.6 lpm as per IS 4246:2002 standard. Emissions test indicate the household related pollution. It signifies the air quality in the kitchen when cooking is being carried out. Lower emissions from the cookstoves are expected for meeting the WHO standards for household air quality.

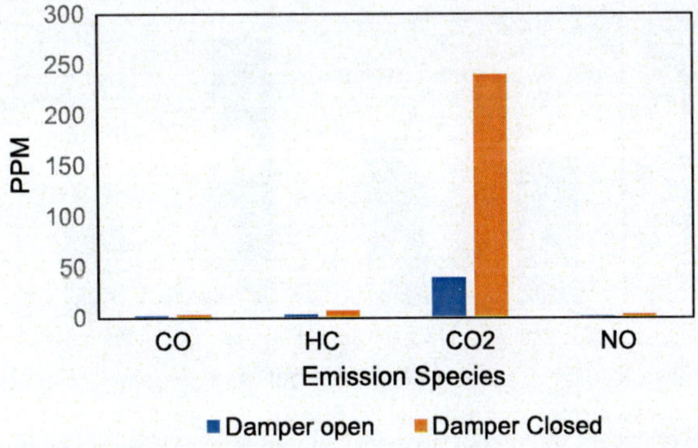

Fig. 7 The flue gas emission species, CO, HC, CO_2 and NO in part per million (ppm) measured for the cookstove

3.3 *Power of the Cookstove Burner*

The thermal load/power of the cookstove burner is given by the following equation:

$$\text{Power (kW)} = \text{Mass flow rate of fuel} \times \text{calorific value of the fuel.} \quad (2)$$

The LPG flow rate was measured using LPG and the standard LPG calorific value was selected for computing the power of cookstove burner. The power of the cookstove burner was found to be 1.12 kW.

3.4 *Variation of Thermal Efficiency with Loading Height*

Figure 8 presents the thermal efficiency with loading height varied from 0 (reference height), 2.5, 5, and 7.5 mm varying loading heights. The zero-loading height here indicated the original loading height provided by the gas stove manufacturer. By varying, loading height variation in thermal efficiency was observed. With an increase in the loading height, increased thermal efficiency was measured. After reaching a maximum, a drop in the thermal efficiency was seen. The thermal efficiency trend observed during the study is in concurrence with the literature [17]. The trend is attributed to the fact that loading height changes the shape of the flame, resulting in cooler core and wider higher temperature zone, leading to higher heat transfer and thermal efficiency [17]. The highest thermal efficiency was measured for 5 mm loading height and lowest was measured for 7.5 mm loading height.

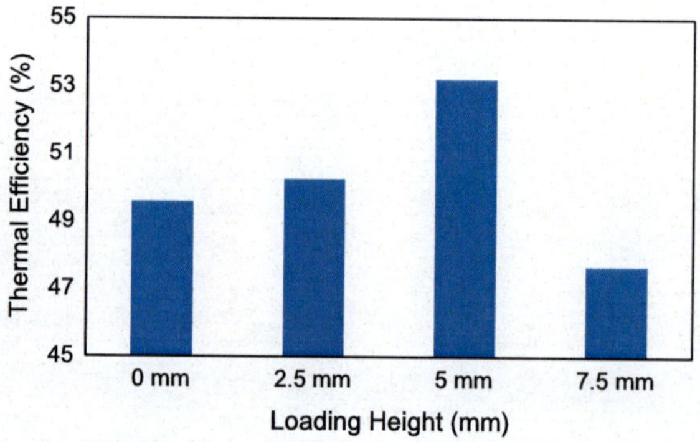

Fig. 8 The thermal efficiency of the cookstove for 0 mm (reference loading height), 2.5 mm, 5 mm, and 7.5 mm variation in loading height

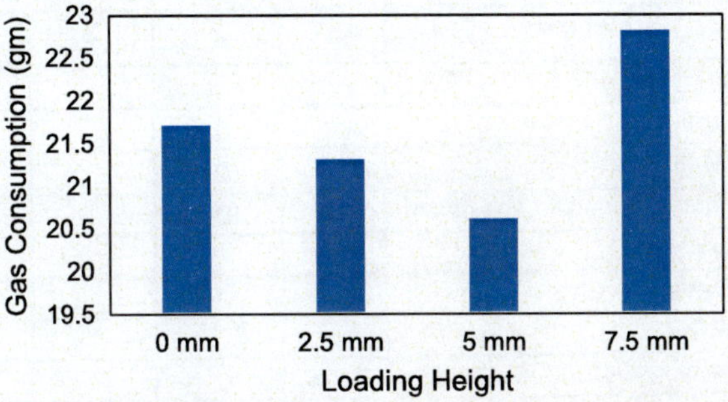

Fig. 9 The gas consumption of the cookstove for 0 mm (reference loading height), 2.5, 5, and 7.5 mm variation in loading height

3.5 Variation of Gas Consumption with Loading Height

Figure 9 represents the gas consumption trends for different loading height. The lowest gas consumption was measured for 5 mm loading height and the highest was measured for 7.5 mm height. It is noteworthy that, higher thermal efficiency is observed for higher loading higher in comparison to the original loading height provided by the gas stove manufacturer.

4 Usability Evaluation

Cookstove usability and user satisfaction have been identified as key factors for the purchase of cookstoves by consumers [18]. Gas consumption, energy-saving, and thermal efficiency happen to be an important consideration while buying a gas stove [19]. Usability and technical features were identified from Moses, Pakravan, and McCarty [18] and, Chahat [19]. Essential features that consumers look at while purchasing a cooking stove were carefully selected. Cookstove usability evaluation was done by carrying out a survey among LPG cookstove users. A survey form designed with the selected features were evaluated on a scale of 1–5, with 1 carrying the lowest weight and five, the highest weight. Figure 10 presents the usability matrix formed after the survey was completed. The key factors that have the highest weightage are ISI certification, safety, fuel-saving, time-saving, indoor air quality, maintenance, and durability.

Group	Feature	1	2	3	4	5
Usability	Time Saving				o	
	Multiple fuel		o			
	Multifunctionality		o			
	Position of the inlet connection		o			
	Body design			o		
	Maintenance				o	
	Durability				o	
	Mode of ignition		o			
	Smooth operation			o		
	Types of burners			o		
	Product finish			o		
Technical	ISI Certification					o
	Fuel Savings				o	
	Safety					o
	Indoor Air Quality				o	
	Emissions			o		
	Combustion Efficiency			o		
	Thermal Efficiency			o		
	Price			o		

Fig. 10 The usability evaluation of an LPG cookstove on various parameters selected for evaluation using a weighted average matrix

5 Conclusion

LPG is becoming a preferred energy source for household use, because of its ease of use and clean-burning nature. Its penetration is going to increase in the times to come. There is scope for improving the thermal efficiency of LPG cookstoves. Incorporating suitable mechanisms for loading height variation and reduced heat loss can help improve the thermal efficiency of the gas stove. Lower gas consumption and lower emissions are thrust areas of improvement. Lower gas stove emissions will help in improving the indoor air quality of the household and promote better health among people. Stove manufacturers should be encouraged to achieve higher star ratings as per the Bureau of Energy Efficiency (BEE). Understanding the need of the user and incorporating the desired modifications to improve the usability of the cookstove, will result in improved performance and fuel savings.

Acknowledgements The author would like to thank the management of The NorthCap University for providing the funds to carry out the research under the "VC Innovation Fund," scheme at the University. The author would like to thank Prof. Pramod Bhatia, HOD Mechanical Department, NCU, for his support. Dr. Akanksha Mathur, Assistant Professor, Mechanical Engineering Department, The NorthCap University, and Mr. Vijender Pal, Technical Assistant, Energy Conversion and Efficiency Lab, for their contribution to the project. The author is also thankful to Mr. Rishabh

Solanki, Mr. Rajat Bhola, Mr. Puneet, Mr. Shivain Solanki, Mr. Vikram Saini, and Mr. Vikrant Solanki for their contribution to the project.

References

1. Calzada J, Sanza A (2018) Universal access to clean cookstoves: evaluation of a public program in Peru. Energy policy. 118(2018):559–572
2. Sharma A, Parikh J, Singh C (2019) Transition to LPG for cooking: a case study from two states of India. Energy Sustain Dev 51(2019):63–72
3. Kimemia D, Annegarn H (2016) Domestic LPG interventions in South Africa: challenges and lessons. Energy Policy 93(2016):150–156
4. Gould FC, Urpelainen J (2018) LPG as a clean cooking fuel: adoption, use and Impact in rural India. Energy Policy 122(2018):395–408
5. Thoday K, Benjamin P, Gan M, Puzzolo E (2018) The mega conversion program from kerosene to LPG: lessons learned and recommendations for future clean cooking energy expansion. Energy for Sustain Dev. https://doi.org/10.1016/j.esd.2018.05.011
6. Kimemia D, Niekerk Van A (2017) Cookstove options for safety and health: comparative analysis of technological and usability attributes. Energy Policy 105(2017):451–457
7. Boggavarapu P, Ray B, Ravikrishna RV (2014) Thermal efficiency of LPG and PNG-fired burners: Experimental and numerical studies. Fuel 116(2014):709–715
8. PTI 2019. The economic times. https://economictimes.indiatimes.com/industry/energy/oil-gas/india-becomes-worlds-2nd-largest-lpg-consumer-after-governments-ujjawla-push/articleshow/67849012.cms
9. Bureau of energy efficiency. https://beestarlabel.com/Content/Files/Final.LPG%20schedule%20(25.02.2014).pdf
10. Jones HRN (1989) The application of combustion principles to domestic gas burner design. E. & F.N. Spon Ltd.
11. Dahiya D, Lather SR, Bhatia P (2017) Analytical analysis of domestic, partially aerated, LPG burner for identification and mitigation of heat losses. International conference on emerging trends in mechanical and industrial engineering. ISBN: 978-81-9336-12-7
12. Ko CY, Lin HT (2003) Emissions and efficiency of a domestic gas stove burning natural gases with various composition. Energy Convers Manag 44(2003):3001–3014
13. Pantangi KV (2010) Development and performance analysis of porous radiant burners for cooking applications. Indian Institute of Technology Guwahati
14. Gohil P, Channiwala AS (2011) Experimental investigation of performance of conventional LPG cooking stove. Fundam J Thermal Sci Eng 1(1):25–34
15. Sutar BK, Ravi MR, Kohli S (2016) Design of a partially aerated naturally aspirated burner for producer gas. Energy 116(2016):773–785
16. Jugjai S, Rungsimuntuchart N (2002) High efficiency heat-recirculating domestic gas burners. Exp Thermal Fluid Sci 26(2002):581–592
17. Hou SS, Ko CY (2004) Effects of heating height on flame appearance, temperature field and efficiency of an laminar jet flame used in domestic gas stoves. Energy Convers Manag 45(2204):1583–1595
18. Moses DN, Pakravan HM, MacCarty AM (2019) Development of a practical evaluation for cookstove usability. Energy Sustain Dev 48(2019):154–163
19. Suri C (2019) NIDM. https://www.nidm.net/kitchen/small-appliances/best-gas-stoves/

Numerical Analysis of Lean Premixed Micro Scale Swiss Roll Combustor

Ronak R. Shah and Digvijay B. Kulshreshtha

1 Introduction

Squeezed energy sources will have good demand for energy production in the future. Due to the rapid development of Micro electro Mechanical System (MEMS), there is a growing interest in MEMS-based micropower system development. Swiss Roll type micro combustion system is the best possibility to generate thermal energy for electricity production, which required very less space using hydrogen as a fuel in micro combustor. To arrive at the optimal condition for hydrocarbon fuels having high density using computational fluid dynamics (CFD), a study was made to know the effect of change in equivalence ratio on adiabatic flame temperature well as a pressure drop in the combustor. Present study also focuses on the profitable concept along with the technical developments of the concept.

Researchers are interested to put the efforts for developing a small movable combustion chamber for small power system. Combustion based system be preferred over electrochemical batteries for power generation due to its more energy density, more power density, small size, reducing recharging time, and enhance the working time [1–3]. Thermal and radical quenching are the main problems for flame stability in micro combustion chamber due to the higher value of surface area to volume ratio [4]. Some investigators have reported that the problem of the flame stability can be overcome by excess enthalpy combustion technique. [5, 6]. It is reported that the small scale combustion system; Catalytic combustion [7–9], Porous media

R. R. Shah (✉)
A.D. Patel Institute of Technology, Karamsad 388121, Gujarat, India
e-mail: ronak28@yahoo.com

D. B. Kulshreshtha
C.K. Pithawala College of Engineering and Technology, Surat 389001, Gujarat, India
e-mail: digvijay.kulshreshtha@ckpcet.ac.in

A. K. Gupta et al. (eds.), *Advances in IC Engines and Combustion Technology*, Lecture Notes in Mechanical Engineering,
https://doi.org/10.1007/978-981-15-5996-9_51

combustion [10–13], and use of stepped combustors [14–16] are some of the suitable approaches for avoiding the flame quenching problem. In the combustion system having hydrocarbon as a fuel is likely to give about six-time more power density over the conventional electrochemical battery [6]. Hydrogen combustion in the combustion chamber gives high combustion efficiency (over 99.95%), high heat release rate ($3.5 * 10^3$ MW/(m^3 MPa)), and low NO_x emission levels (6–25 ppm) in comparison with hydrocarbon combustion chamber [17].

Study is to be made with different parameters of combustion chamber for its productive results to achieve the positive optimal combustion conditions, the combustion system having remaining specification be design using a computational model. Computerized simulation techniques help us to find out the dimensional parameter which affecting the unit performance of the combustion system.

2 Design of Micro Combustion Chamber

The present recuperative type micro combustion chamber is developed using hydrogen for small capacity power generation. In the present micro combustion chamber shown in Fig. 1, only reaction zone is indicated. Premixed fuel enters into the micro combustor. In developed micro combustor, two separate paths for reactant and product are provided. The flow direction of the reactant and product are opposite to each other. The reactant and product wall worked like heat exchanger. The combustion process occurs at the centre of micro combustor and produces the hot

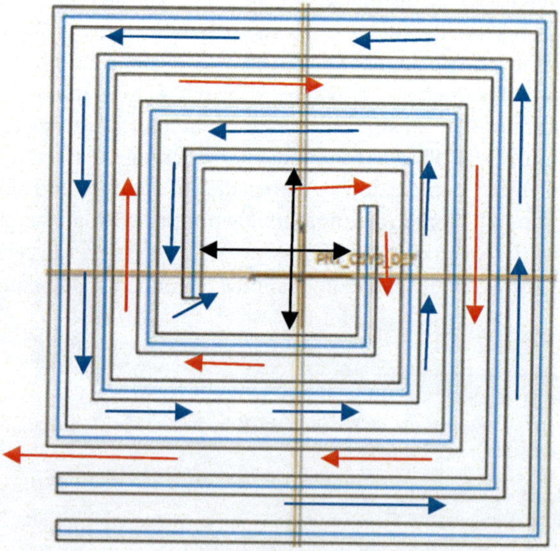

Fig. 1 Concept of combustion chamber

gases which pass through the product channel path. The premixed fuel-air enters into the combustion zone through the reactant channel path.

Both reactant and product channels are wounded about each other and separated by a channel wall. When the hot gases are generated in the combustion zone, the surface of the product channel wall absorbs the heat energy which transfers it to the surface of reactant channel. Reactant gases are preheated before entering into the combustion zone. Therefore, it requires less amount of heat energy form the source. Waste heat is utilized to preheat the air-gas mixture by allowing the heat transfer among product of combustor and reactants, similar to counter flow heat exchanger. As per primary zone volume calculations, the reaction zone volume is calculations, the reaction zone volume is calculated as 20 * 20 * 25 mm^3. The width of flow path for reactants and products is taken as 1.5 mm and 4 mm, respectively, having 1 mm thickness for channel and combustion zone wall.

3 Numerical Analysis of Micro Combustion Chamber

Computational Fluid Dynamics is an effective method for displaying liquid stream, warm exchange, and related procedures for an extensive variety of critical logical and designing issues. In the present investigation, CFD is utilized as a numerical apparatus to analyze the scientific consequences of the thermodynamic harmony model and substance energy. The conventional combustion chamber is first designed for 1 kW gas turbine engine using the Aerodynamic and Chemical Perspective. In the present case, the premix reactants (hydrogen/air) is supplied through a reactant passage. For the combustion of premixed reactants – two methodologies can be adopted in CFD. In the present study, modified O'Conaire mechanism proposed in chemical kinetic studies is selected. The present scheme consists of 19 reversible elementary reactions and three single step elementary reactions with 11 species. BVM (partially premixed) model is used for combustion simulations.

At all equivalence ratio, the flame front is confined in central reaction zone. Due to low-velocity region and recirculation in the reaction zone, flame stability is achieved. There is an increase in temperature with an increase in the equivalence ratio [18, 19]. Figure 3 indicates the variations in temperature at salient equivalence ratios for reactant temperature of 300 K. Very low temperatures in the central reaction zone of the order of 700 K is obtained at equivalence ratio of 0.132, whereas the exit temperatures are of the order of 500 K. For a higher equivalence ratio, the central reaction temperatures show sharp rise. This is due to the fact that more fuel is available for combustion. Above the equivalence ratio of 0.6, the non-uniformity of temperatures is evident. High temperatures of the order of 2000 K (Fig. 3) are witnessed near wall. The non-uniformity of temperature field suggests the operation of the present combustor below the equivalence ratio of 0.6. A similar conclusion can be derived from the expected emission analysis carried out numerically. The mass fraction of NO, H_2 is numerically studied (Fig. 2; Table 1).

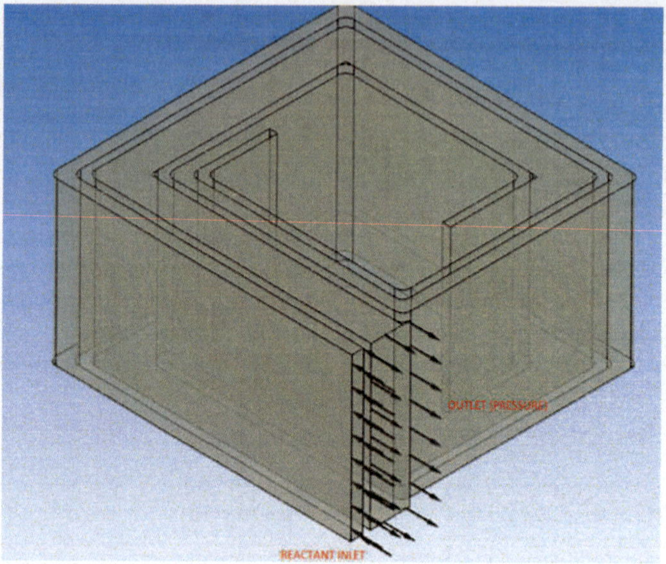

Fig. 2 CFD model with boundary conditions

Table 1 CFD model

1	Equivalence ratio	0.13–1.2 Lean to rich
2	Combustion model	BVM partially premixed
3	Reaction mechanism	Modified O'Conaire mechanism
4	Turbulence model	SST k – ω
5	Inlet boundary	Premixed fuel-air mixture
6	Outlet boundary	Pressure condition
7	Run type	Transient
8	Time step	0.001 s
9	Cycle time	5 s

The NO emissions are around 7.2 ppm at designed equivalence ratio of 0.132 and thereafter increases as equivalence ratio increases. This is expected at higher temperatures thermal NO production is more. With an increase in equivalence ratio, higher temperatures are encountered, which leads to higher NO production. The NO emissions above equivalence ratio of 0.6 show a sharp rise. The emissions are as high as 4734 ppm. This suggests that the combustor be operated below the equivalence ratio of 0.6 to reduce or for practically zero NO emissions (Fig. 4).

Figure 5 indicates the unburnt amount of hydrogen which existence of combustor at different equivalence ratios. With lower equivalence ratios, the residence time required is more for complete combustion. However, as more oxidizer is available the amount of unburnt hydrogen is lowest at designed equivalence ratio of 0.132. Due to small size of combustor, the unburnt hydrogen increases with an increase in equivalence ratio, which contradicts the results for residence time calculation. Residence time has an inverse relation to equivalence ratio till stoichiometry. However, due to the small size of the combustion chamber, the residence time effects are overcome by the amount of oxidizer available.

Figures 3, 4, and 5 shows that for equivalence ratio of 0.6 of lower value may serve as design criteria. In terms of emissions and efficient combustion, the equivalence ratio of 0.132 shows superiority compared to other values. However, the temperature levels of the order of 500 K at the exit, obtained at an equivalence ratio of 0.132, will not be feasible to develop 1 kW power from gas turbine unit. This advocates for increasing the reactant temperature. Increasing the reactant temperature may increase the exit temperature. The following section summarizes the results obtained for reactant temperature of 400 and 500 K.

Figure 6 shows the temperature distribution at 400 K reactant temperature with different equivalence ratios. The temperature trends show a similar pattern as in reactant temperatures of 300 K, but higher temperature levels are encountered. Above the equivalence ratio 0.5, the temperature distribution is not uniform. At 0.7 equivalence, the temperature near the wall is higher which suggests the operation of present combustor below equivalence ratio of 0.6.

Figure 7 shows the numerical studies of NO species at 400 reactant temperature with a difference equivalence ratio. The mass fraction of NO is higher compared to 300 K reactant temperature. This is expected as thermal NO production is more at higher temperatures. The mass fraction of NO at equivalence ratio of 0.132 is 7.1 and thereafter increases with equivalence ratio.

Ii is depicted from Fig. 8 that unburnt hydrogen at combustor exist at different equivalence ratios with reactant temperature of 400 K. With lower equivalence ratios, the mass fraction of unburned fuel is lower as more oxidizer is available for combustion. Figures 6, 7 and 8, show that for equivalence ratio of 0.6 of lower value may serve as design criteria of micro combustor. Similarly, the reactant temperature of 500 K was studied for different equivalence ratios.

Figure 9 shows the temperature distribution with inlet reactant temperature of 500 K at different equivalence ratios. The temperature trends show a similar pattern as in reactant temperatures of 300 and 400 K, but higher temperature levels are encountered. At higher reactant temperature the combustion reaction is fast and hence requires less residence time. This condition is suitable for design a micro combustor. But the increase in the reactant temperature leads to more emissions, as more thermal NO is produced. This conflicts the advantage of faster reaction rates.

Figure 10 shows the NO mass fraction with 500 K reactant temperature at different equivalence ratios. At lower equivalence ratio 0.132, the NO mass fraction is 2.5 ppm which is nearer to zero. The mass fraction of NO increases for an increase in equivalence ratio. The NO fraction is 4.5E−3 at 0.7 equivalence ratio. The emission level is higher with more than 0.6 equivalence ratio. The CFD result advocates that the micro combustor may be designed with a lower equivalence ratio.

Unburnt fuel is shown in Fig. 11 at 500 K reactant temperature with different equivalence ratios. The unburnt fuel increases as the equivalence ratio increases.

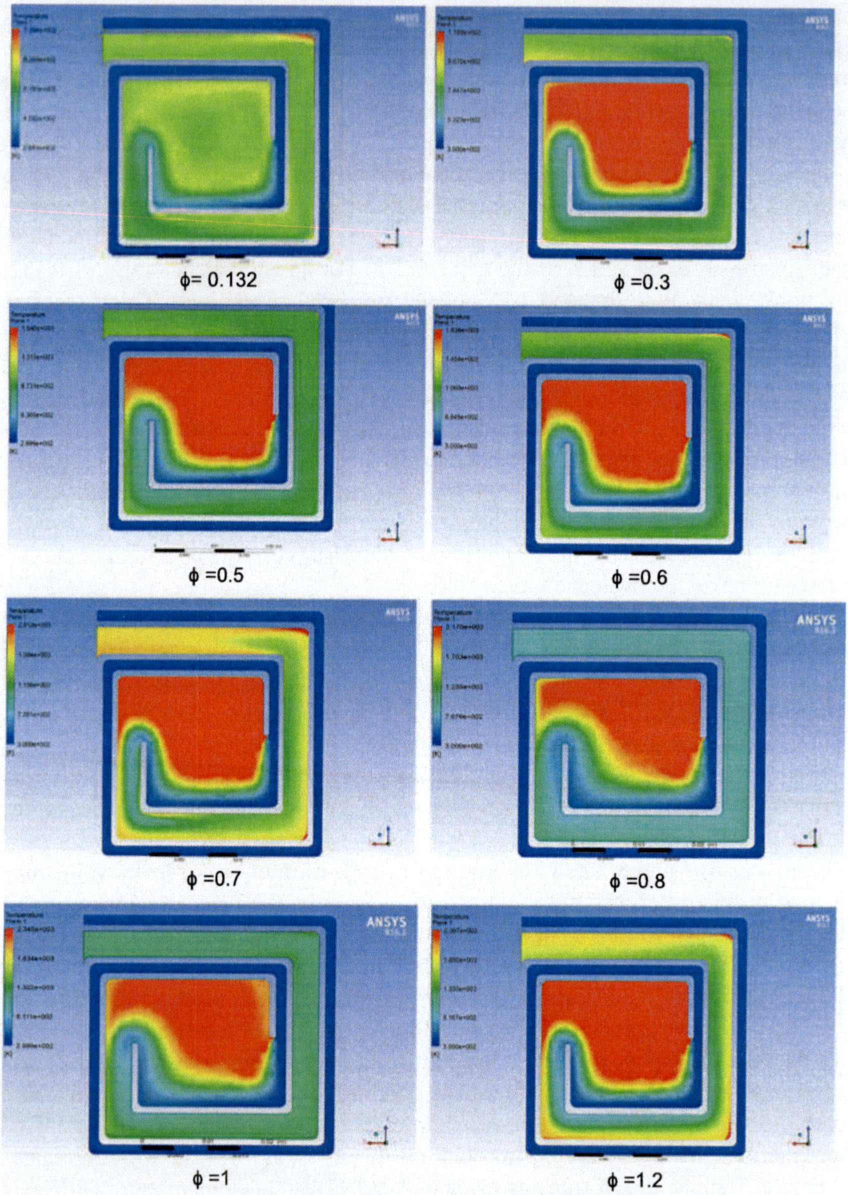

Fig. 3 Temperature variations at different equivalence ratio for reactant temperature of 300 K

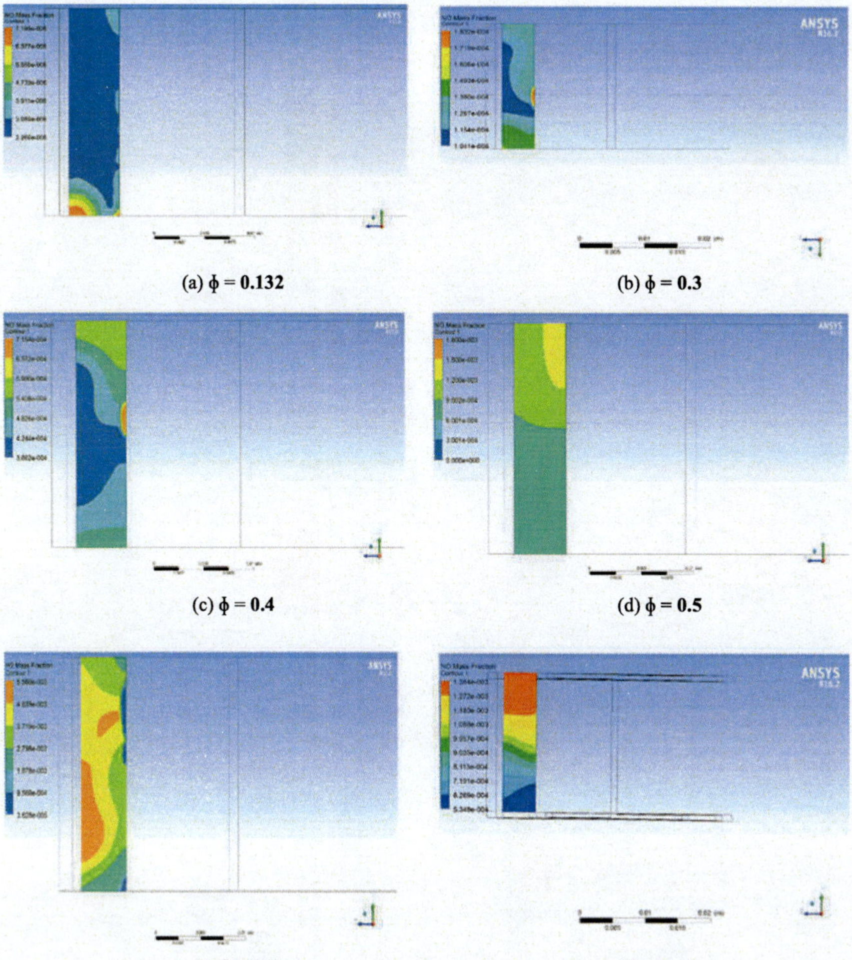

(a) $\phi = 0.132$ (b) $\phi = 0.3$

(c) $\phi = 0.4$ (d) $\phi = 0.5$

(e) $\phi = 0.7$ (f) $\phi = 0.9$

Fig. 4 NO emissions at different equivalence ratio at reactant temperature of 300 K

This may be characteristics of the fact that mass fraction of fuel increases with equivalence ratio. Therefore, less oxidizer is available at a higher equivalence ratio. Due to small size of micro combustor and with no addition of secondary or tertiary air, a part of fuel remains unburnt at higher equivalence ratios.

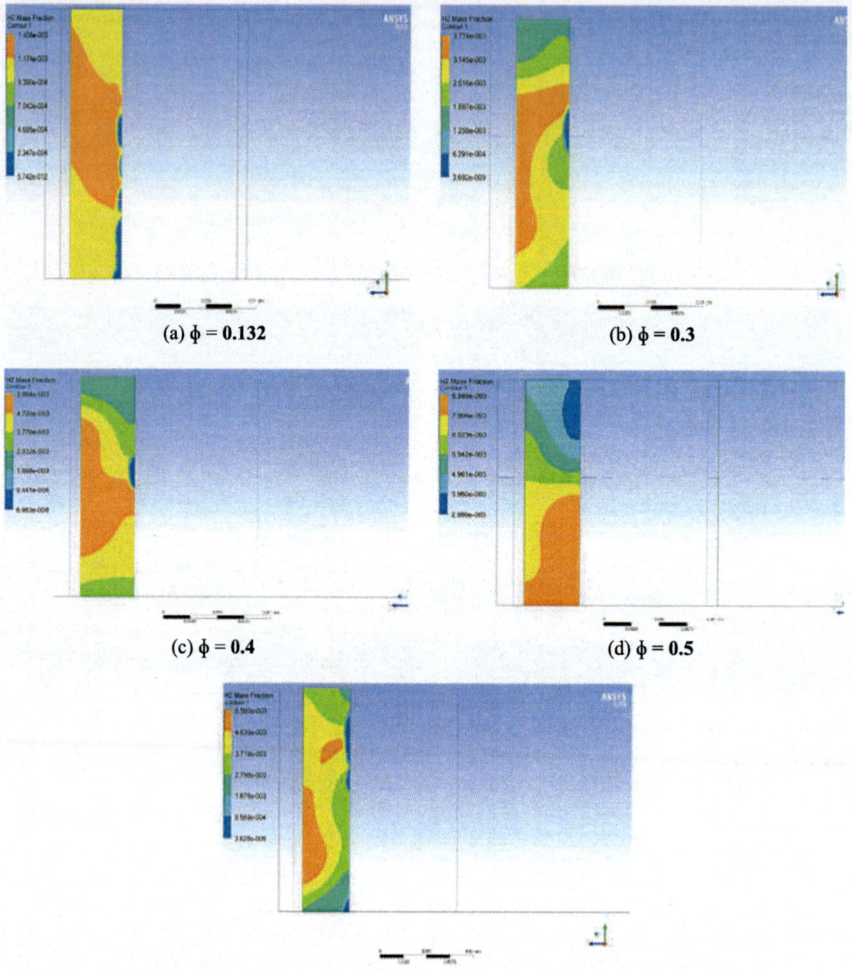

(a) $\phi = 0.132$ (b) $\phi = 0.3$

(c) $\phi = 0.4$ (d) $\phi = 0.5$

(e) $\phi = 0.7$

Fig. 5 H_2 species at 300 K reactant temperature with different equivalence ratio (ϕ)

4 Results and Discussion

Figure 12 shows the temperature with the equivalence ratio at 300, 400, and 500 K reactant temperature. It is reported that there is a rise in temperature with an increase in equivalence ratio [20]. The similar effect of equivalence ratio on three different reactant temperature is observed.

Figure 13 indicates the mass fraction of OH with different equivalence ratios at 300, 400, and 500 K reactant temperature. It indicates that at lower lean equivalence ratio, the production of OH is less as compared to rich mixture condition. At lower equivalence ratio up to 0.6, OH molecules leisurely increase with increases

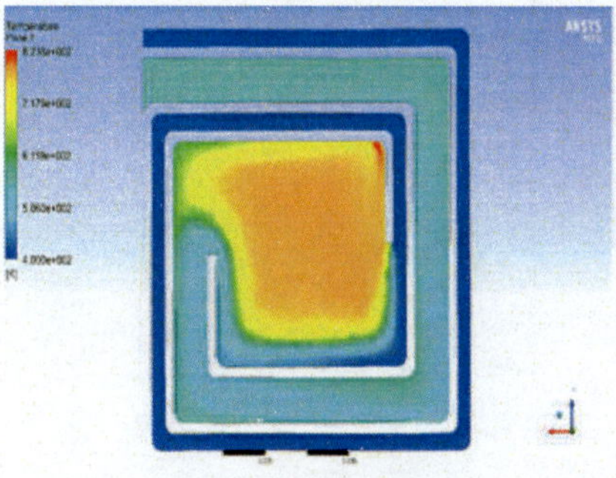

(a) $\phi = \mathbf{0.132}$

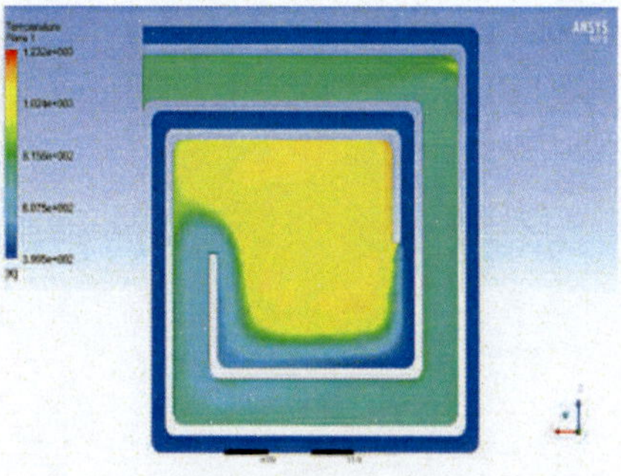

(b) $\phi = \mathbf{0.3}$

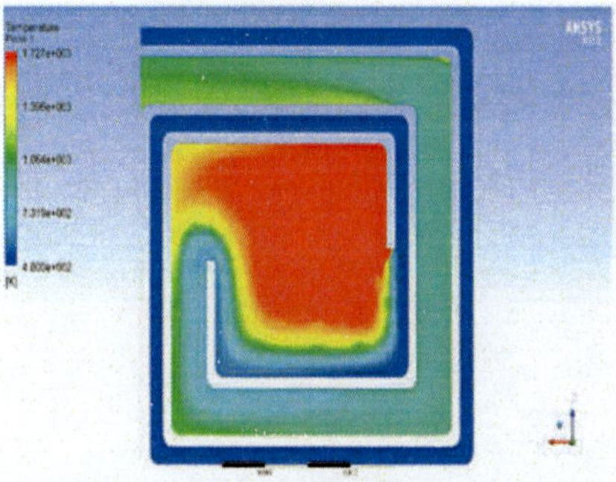

(c) $\phi = \mathbf{0.4}$

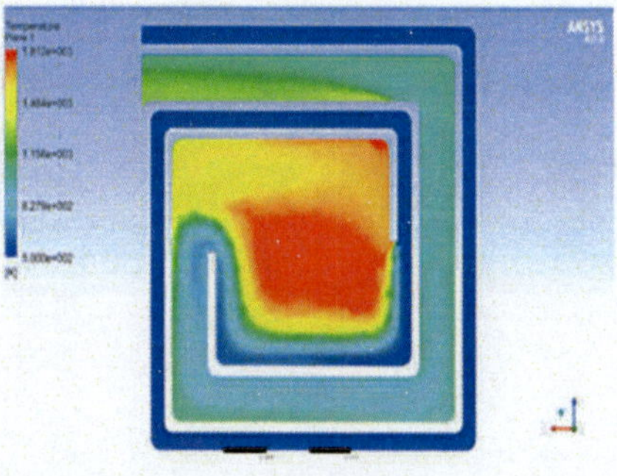

(d) $\phi = \mathbf{0.5}$

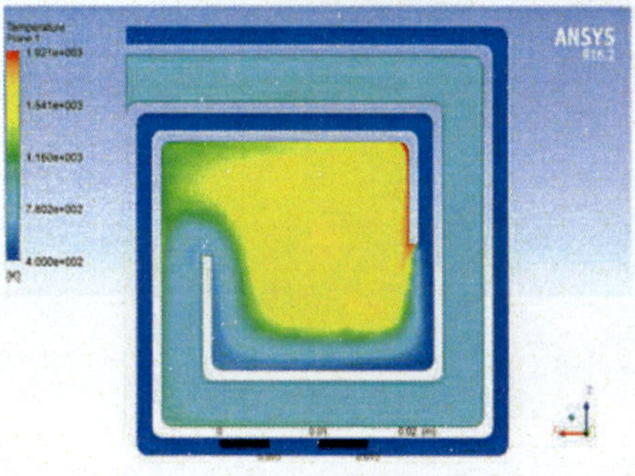

(e) $\phi = \mathbf{0.6}$

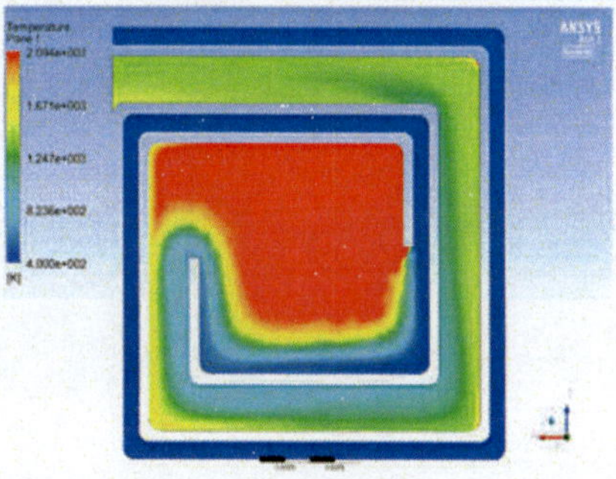

(f) $\phi = \mathbf{0.7}$

Fig. 6 The temperature distribution at 400 K reactant temperature with different equivalence ratio (ϕ)

equivalence ratio due to lower reaction between hydrogen and oxidizer with lower temperature. After the equivalence ratio 0.6, OH molecules rapidly increases with equivalence ratio due to the faster reaction at higher temperature. This suggests that higher residence time is required at lower equivalence ratio.

Figure 14 shows a mass fraction of NO with equivalence ratio at different reactant temperature. The mass fraction of NO is less in lower lean equivalence ratio due to

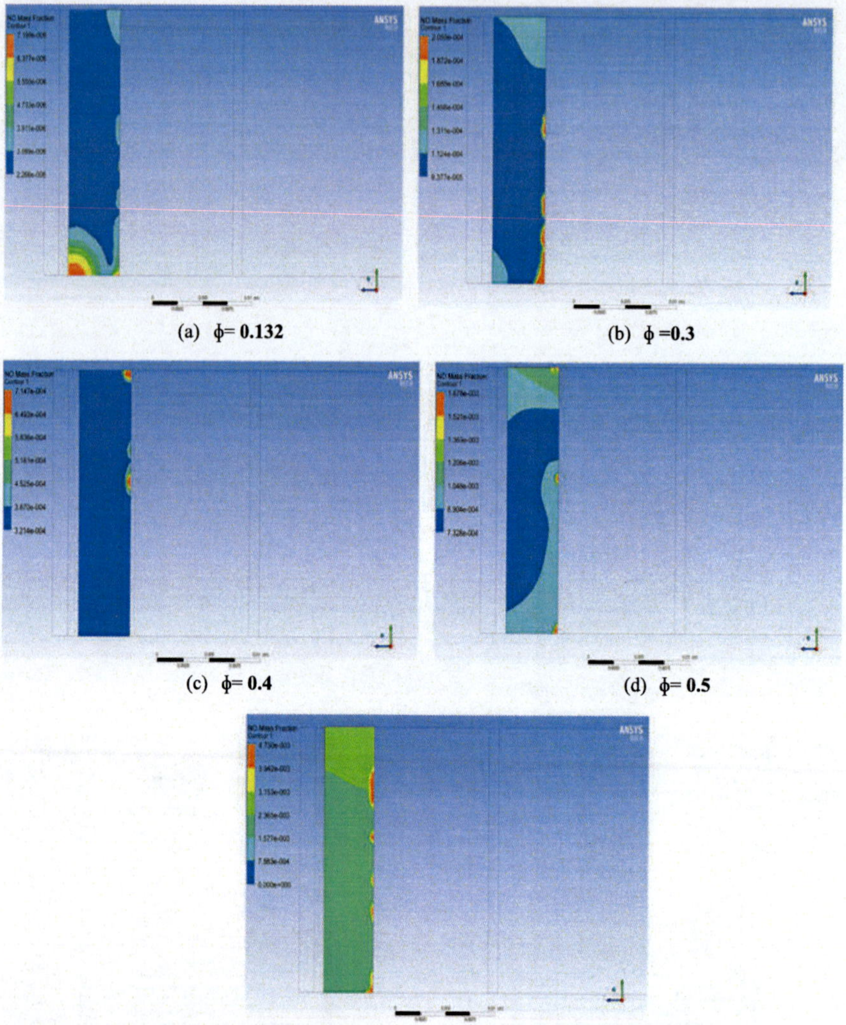

(a) ϕ= 0.132 (b) ϕ =0.3

(c) ϕ= 0.4 (d) ϕ= 0.5

(e) ϕ =0.7

Fig. 7 NO species at 400 K reactant temperature with different equivalence ratio (ϕ)

lower reaction between N_2 and oxidizer. After equivalence ratio 0.5, NO molecules rapidly increase with an increase in the equivalence ratio [21].

After a stoichiometric situation, the less amount of oxidizer is available due to this reason the NOx formation is less in the rich mixture than higher lean mixture condition. NOx increases with increasing with the flame temperature in the premixture combustor [20]. It means a lower lean equivalence ratio is a suitable condition for designing near the green micro combustor.

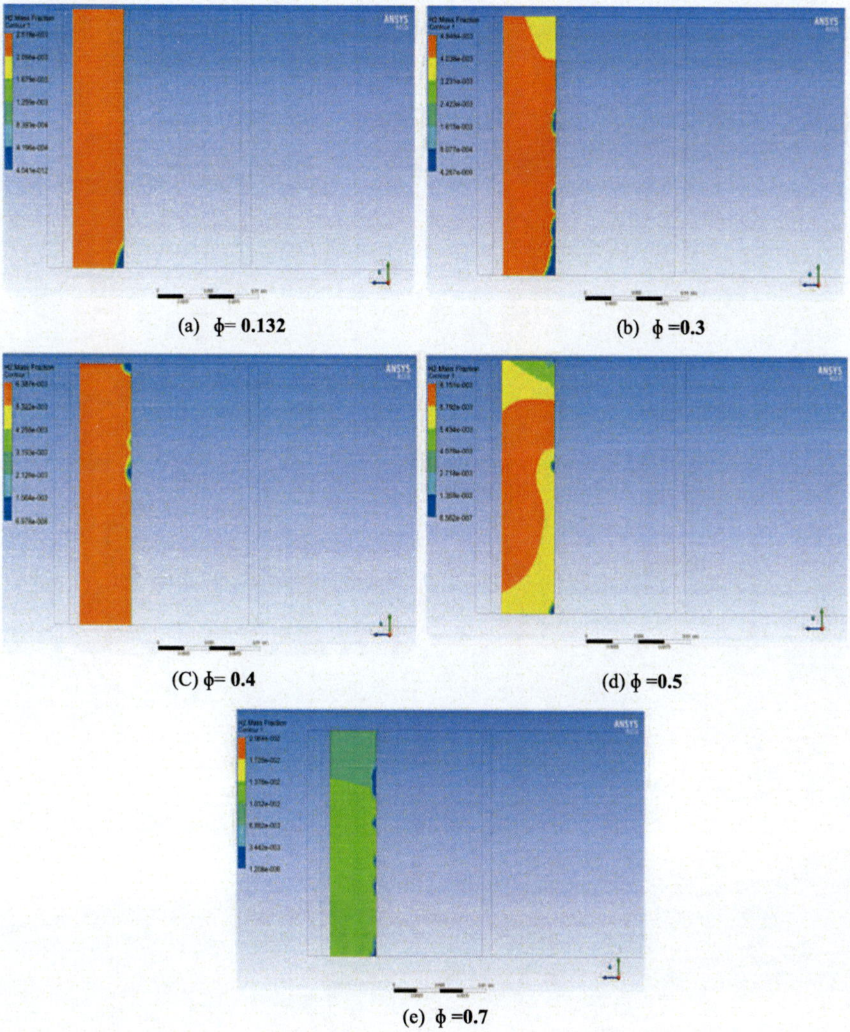

Fig. 8 H_2 species at 400 K reactant temperature with different equivalence ratio (ϕ)

5 Conclusions

Numerical study was made for micro combustion with the help of the proposed modified O'Conaire mechanism with varying reactant temperature and equivalence ratio. The following conclusion has been emerged.

1. Stable combustion process is possible in premixed micro combustor.
2. Combustion process is completed in micro combustor with lower equivalence ratio at lower reactant temperature.

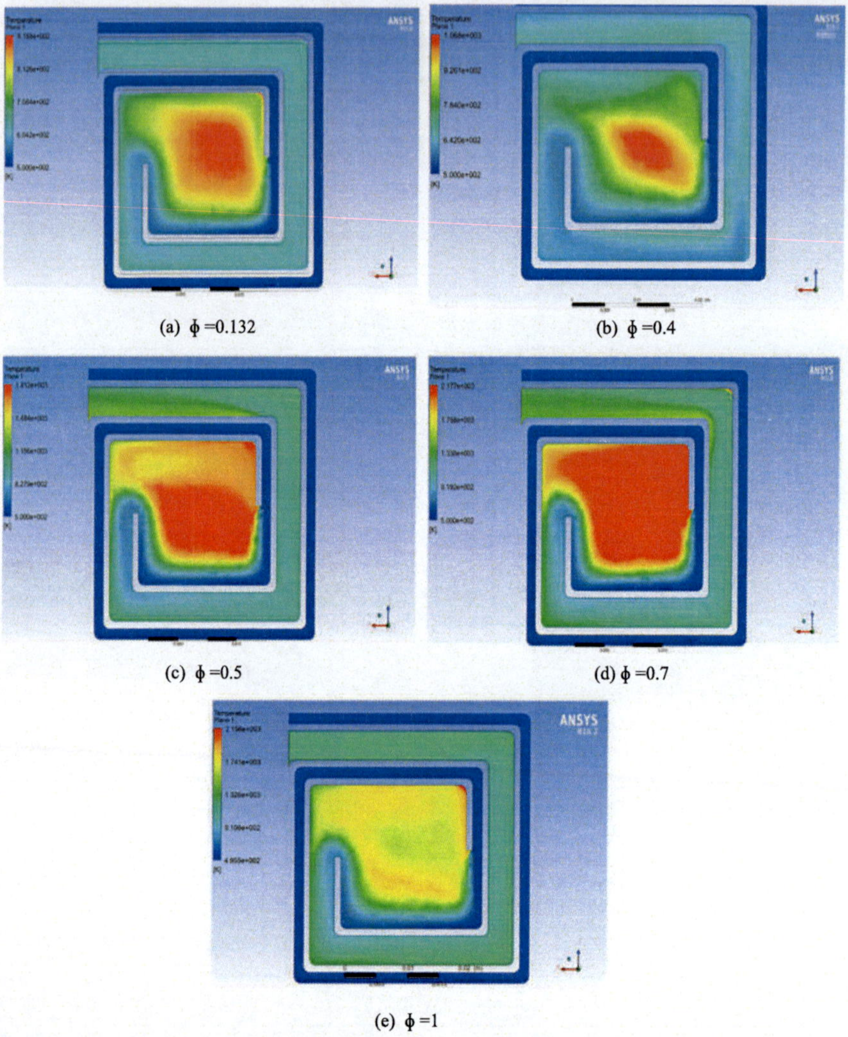

Fig. 9 The temperature distribution at 500 K reactant temperature with different equivalence ratio (ϕ)

3. NO_x formation is lower in below 0.5 equivalence ratio. Less emission after combustion in micro combustor with lower lean equivalence ratio.
4. OH formation is also less in lower reactant temperature at lower lean condition which suggests more residence time for combustion.

All above observations are useful to design and fabricate the micro combustion chamber with hydrogen/air premixed mixture. The micro combustor working at lower equivalence ratio with lower reactant temperate gives stable combustion with lower emission and it is the best alternative of batteries in micropower source devices.

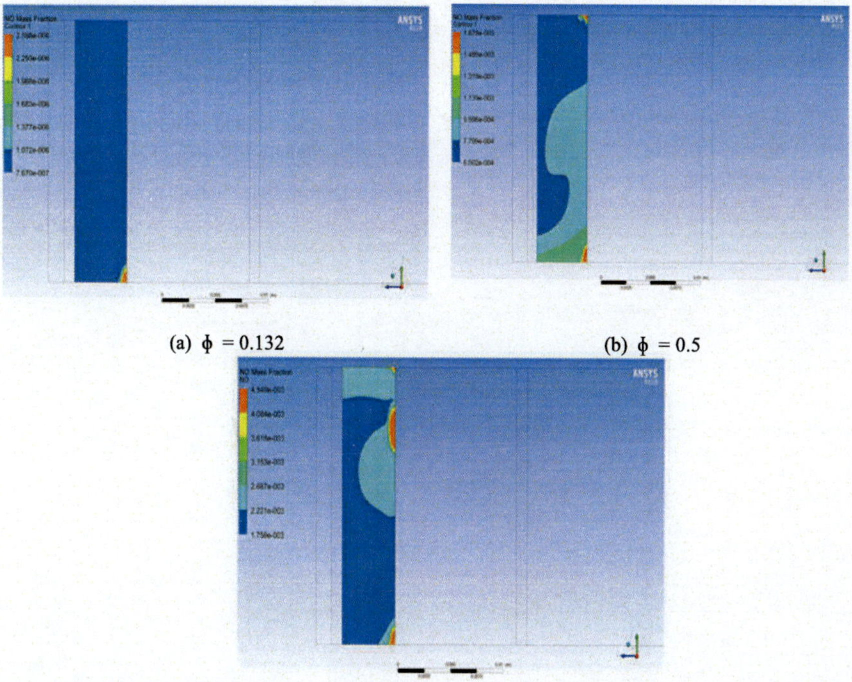

(a) ϕ = 0.132 (b) ϕ = 0.5

(c) ϕ =0.7

Fig. 10 NO species at 500 K reactant temperature with different equivalence ratio (ϕ)

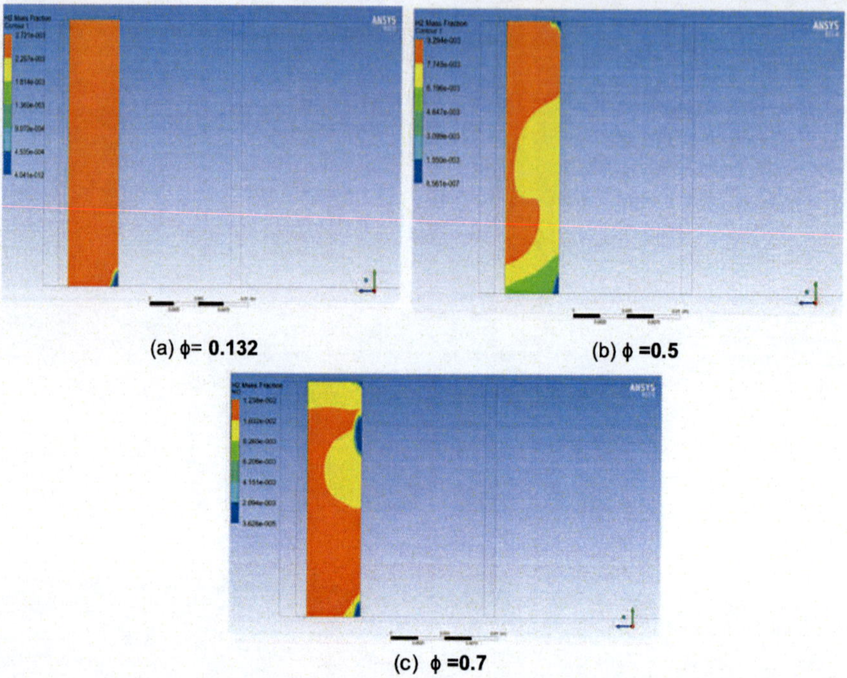

Fig. 11 H_2 species at 500 K reactant temperature with different equivalence ratio (ϕ)

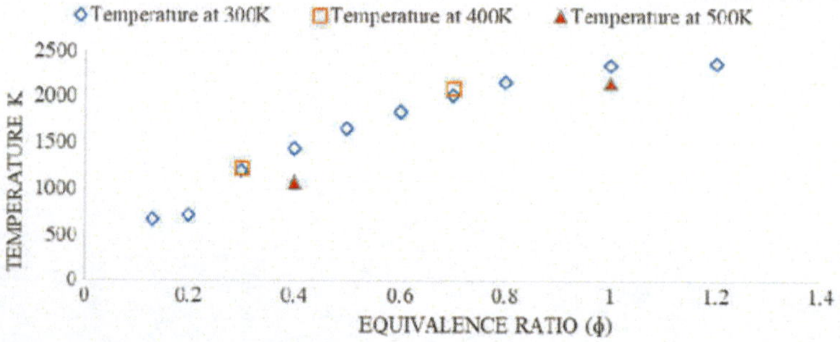

Fig. 12 Temperature versus equivalence ratio

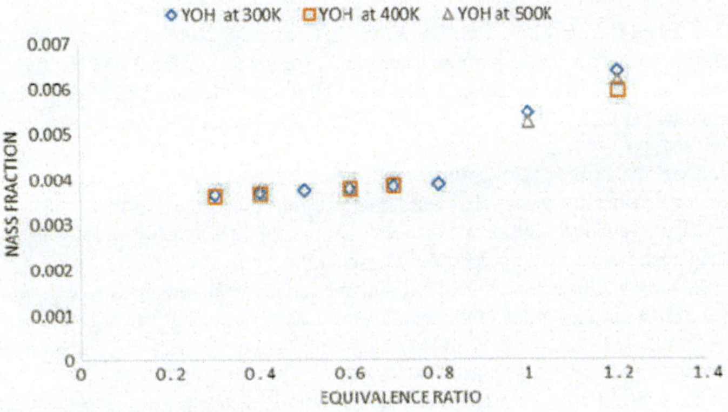

Fig. 13 Mass fraction (OH) versus equivalence ratio

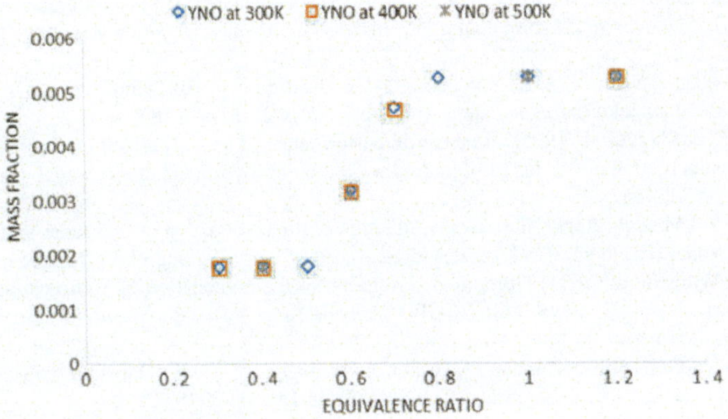

Fig. 14 Mass fraction (NO) versus equivalence ratio

References

1. Dunn-Rankin D, Leal EM, Walther DC (2005) Personal power systems. Prog Energy Combust Sci 31:422–465
2. Kaisare NS, Vlachos DG (2012) A review on micro combustion: fundamentals, devices and applications. Prog Energy Combust Sci 38:321–359
3. Epstein AH, Jacobson SA, Protz JM, Frechette LG (2000) Shirt button-sized gas turbines: the engineering challenges of micro high speed rotating machinery. In: Proceedings, 8th Int'l symposium on transport phenomena and dynamics of rotating machinery, Honolulu, HI
4. Fernandez-Pello AC (2002) Micro power generation using combustion: issues and approaches. Proc Combust Inst 29:883–899
5. Sitzki L, Borer K, Schuster E, Ronney PD, Wussow S (2001) Combustion in micro scale heat recirculating burners. In: The third Asia-Pacific conference on combustion. Seoul, Korea, pp 11–14
6. Ju Y, Maruta K (2011) Micro scale combustion: technology development and fundamental research. Prog Energy Combust Sci 37:669–715

7. Chao Y-C, Chen G-B, Hsu C-J, Leu T-S, Wu C-Y, Cheng T-S (2004) Operational characteristics of catalytic combustion in a platinum micro tube. Combust Sci Technol 176:1755–1777
8. Maruta K, Takeda K, Sitzki L, Borer K, Ronney PD, Wussow S et al (2001) Catalytic combustion in microchannel for MEMS power generation. In: Third Asia-Pacific conference on combustion, Seoul, Korea, pp 1–4
9. Chen J, Song W, Xu D (2017) Optimal combustor dimensions for the catalytic combustion of methane-air mixtures in micro-channels. Energy Convers Manage 134:197–207
10. Marbach T, Agrawal A (2006) Heat-recirculating combustor using porous inert media for mesoscale applications. J Propul Power 22:145–150
11. Li J, Li Q, Wang Y, Guo Z, Liu X (2016) Fundamental flame characteristics of premixed H_2–air combustion in a planar porous micro-combustor. Chem Eng J 283:1187–1196
12. Li J, Li Q, Shi J, Liu X, Guo Z (2016) Numerical study on heat recirculation in a porous micro-combustor. Combust Flame 171:152–161
13. Chou S, Yang W, Li J, Li Z (2010) Porous media combustion for micro thermo photovoltaic system applications. Appl Energy 87:2862–2867
14. Taywade UW, Deshpande AA, Kumar S (2013) Thermal performance of a micro combustor with heat recirculation. Fuel Process Technol 109:179–188
15. Khandelwal B, Deshpande AA, Kumar S (2013) Experimental studies on flame stabilization in a three step rearward facing configuration based micro channel combustor. Appl Therm Eng 58:363–368
16. Yang W, Chou S, Shu C, Li Z, Xue H (2002) Combustion in micro-cylindrical combustors with and without a backward facing step. Appl Therm Eng 22:1777–1787
17. Yoshida S, Shigeta M (1995) Combustion performance of lean premixed type combustors for a hydrogen fuelled micro gas turbine. In: Proceeding of Yokohama international gas congress, Japan, pp 347–352
18. Hua J, Wu M et al (2005) Numerical simulation of the combustion of hydrogen-air mixture in micro scaled chambers. Part I: fundamental study. J Chem Eng Sci 60:3497–3506
19. Hua J, Wu M et al (2005) Numerical simulation of the combustion of hydrogen-air mixture in micro scaled chambers. Part II: CFD analysis for a micro-combustor. J Chem Eng Sci 60:3507–3515
20. Fernández-Galisteo D, Sánchez AL, Linan A, Williams FA (2009) One-step reduced kinetics for lean hydrogen–air deflagration. Combust Flame 156:985–996
21. Das LM (1996) Hydrogen-Oxygen reaction mechanism and its implication to hydrogen engine combustion. Int J Hydrogen Energy 21(8):703–715

Ronak R. Shah was born in Piplod village in Gujarat, India in 1976. He received the B.E. degree in mechanical engineering in 1997 and M. Tech. in 2006. He obtained his Ph.D. degree in mechanical engineering from CHARUSAT, Changa in 2018. Currently, he is working as an Assistant Professor in Mechanical Engineering Department, ADIT, New V.V. Nagar, Gujarat. He has guided 10 PG students. Dr. Shah has authored more than 15 publications in different journals and conferences. His fields of interest in Renewable Energy, Energy Conservation, Refrigeration, and air conditioning, combustion and Power Plant Engineering. He is a registered member of different societies including ISTE, The Combustion Institute, and ISHRAE.

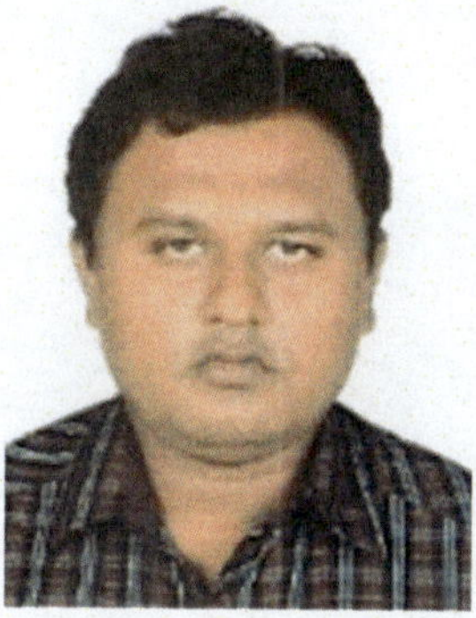

Digvijay B. Kulshreshtha was born at Ahmedabad, India on January 14, 1978. He graduated from S. V. M. Institute of Technology, Bharuch. He received his Masters in Turbomachines in 2003 and Ph.D. in Mechanical Engineering in 2009 from S. V. National Institute of Technology, Surat Digvijay Kulshreshtha is currently serving as Associate Professor in Mechanical Engineering Department at C K Pithawalla College of Engineering and Technology, Surat, India. He was previously Research Assistant at MHRD Sponsored Research Project on Gas Turbine Combustion. His current research areas are Combustion, Reacting Gas Dynamics, and Thermodynamics. He specializes in Analytical, Numerical Solutions, and Experimentation of combustion processes in gas turbine combustion chambers and spray dynamics. He directs anchor research on different types of gas turbine combustion chambers, atomization of liquid fuels, and premix fuel nozzles for gaseous fuels. Dr. Kulshreshtha has authored more than 70 publications in different journals and conferences of international repute. He has two designs registered. He has guided three PhD students and more than 10 PG students. He received N. K. Memorial Prize for Best Paper Published in The Institution of Engineers (I), Mechanical Engineering Division Journal. He is a registered member of different societies including Institute of Engineers (India), The Combustion Institute, Society of Automotive Engineers, and American Society of Mechanical Engineers.

Review of Laminar Burning Velocity of Methane–Air Mixtures at High Pressure and Temperature Conditions

Robin John Varghese, Harshal Kolekar, Swetha Lakshmy Hariharan, and Sudarshan Kumar

1 Introduction

The techniques like carbon-capture and lean premixed mode of combustion have helped achieve a reduction in carbonaceous and NO_X emissions, respectively [1]. Integrated gasification and combined cycle (IGCC) employs these techniques to the environment-friendly generation of power. The fuel used to run the gas turbines in IGCC power plants can be coal or natural gas. Abundant reserves of natural gas can help meet the soaring power demands.

Methane alone can be used as a fuel in IC engines or along with some other fuel ($H_2 + CH_4$, $H_2 + CO + CH_4$). For employing any fuel to produce power, understanding the combustion characteristics is inevitable. The fundamental combustion characteristics that the research scientists invariably focus on are ignition delay, flame speed, reaction rates. Having a thorough knowledge of the blow-off and flashback limits is necessary from the safety perspective for the use of fuels in combustion devices.

Various kinetic schemes are available, which can predict the elementary combustion characteristics, including laminar burning velocity (LBV) and ignition delay using simulations. These predictions come handy when applying a new blend in an IC engine. To make the predictions and hence the use of those chemical kinetic schemes worthy, validation becomes inevitable. The LBV and ignition delay times are vital in the validation of detailed reaction chemistry.

Thc most widely used techniques for measurement of LBV are spherically expanding flame method [2], heat flux method [3], Bunsen flame method [4] and counterflow flame method [5]. All the methods (except for heat flux method) employs

R. J. Varghese (✉) · H. Kolekar · S. L. Hariharan · S. Kumar
Department of Aerospace Engineering, Indian Institute of Technology Bombay, Powai, Mumbai 400076, India
e-mail: robinjohnvarghese@gmail.com

A. K. Gupta et al. (eds.), *Advances in IC Engines and Combustion Technology*, Lecture Notes in Mechanical Engineering,
https://doi.org/10.1007/978-981-15-5996-9_52

stretched flames and extrapolate the burning velocities from experiments to zero stretch. In contrast, the heat flux method interpolates the obtained burning velocities to zero heat loss. The heat flux method is regarded as the most accurate method for the measurement of burning velocities since they employ flat adiabatic flames. However, the measurement of burning velocities using heat flux method is limited by the measurable range (60 cm/s initially, extended to 80 cm/s by Goswami et al. [3]) and hence is not suitable for LBV determination at high-temperature conditions.

The development of the present method initially depended on fundamental studies on diverging channels [6, 7]. They investigated the flame structure of CH_4-air flames in a diverging channel. Following the studies, Akram et al. [8–10] validated the diverging channel method for LBV measurement of various fuel–air mixtures at high temperatures (350–650 K) and atmospheric pressure (1 atm). Along with various gaseous fuels, the method has been employed for LBV measurement of liquid fuels such as methanol [11], ethanol [12], propanol [13], butanol [11] and methyl formate [14].

The heated diverging channel technique for the measurement of LBV was extended to studies at elevated pressures in the current study. The present study focusses on the LBV measurement of CH_4-air mixtures at high mixture temperature and pressure conditions simultaneously. There are a few studies that experimentally determined the LBV at elevated pressures and temperatures. However, not studies that measure burning velocities simultaneously at high temperature and pressure conditions. The primary objective of the present study is to experimentally determine LBV of CH_4-air mixtures at high mixture temperatures ($350-650\ K$) and elevated pressures ($1-5$ atm). The LBV measured using the heated diverging channel is compared with the available literature and recent kinetic schemes (Aramco 2 [15, 16] and GRI Mech 3.0 [17]).

2 Methodology

Figure 1 shows the representation of the experimental facility developed for the measurement of LBV simultaneously at high temperature and pressure conditions. The experimental facility comprises a diverging mesoscale channel, infrared heater, thermocouple and an ignition device. The facility developed for the measurement of burning velocities simultaneously at elevated temperatures and pressures was housed in a 40 L cylindrical pressure vessel. The pressure vessel is maintained at the specified pressure (say 3 atm) with a pressure relief valve mounted on the top of the vessel. The relief mechanism was calibrated for accuracy at 3 atm, and the indicated pressure is accurate up to ± 0.05 atm. The gas supply system comprises of fuel cylinders and MFCs. A viewing window (made of toughened glass) helps visualization of the stabilized flame. The infrared heater is kept inside the pressure vessel and helps heat the diverging channel. The temperature profile in the axial and transverse directions of the channel at different inlet velocities and preheat conditions were measured

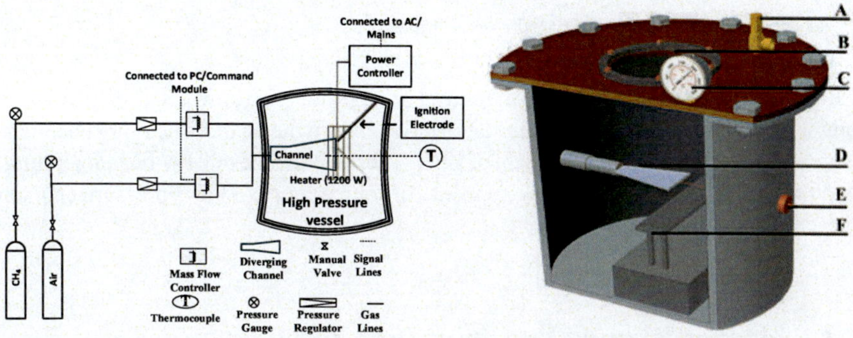

Fig. 1 Left: representation of the high-pressure facility, Right: sectional view; A—Pressure relief valve, B—Glass window, C—Pressure gauge, D—Diverging channel, E—Thermocouple, F—Infrared heater

beforehand. The traverse mechanism mounted with a thermocouple can move in the axial direction for measurement of temperature.

Mesoscale channels refer to those with the characteristic length similar to the quenching distance of the fuel–air mixture being investigated. The divergence angle (10°) helps in providing a range of flow velocities along the axis due to the decreasing cross-sectional area. The propensity of flashback can be limited with the help of divergence. A positive temperature gradient with respect to the flow velocity with the help of an infrared heater placed 2 cm below and a horizontal overlap of 2 cm from the exit of the channel. The temperature in the transverse direction was found to be uniform (maximum deviation of ±3 K) at a given axial position. The metered flow from the MFCs at a given equivalence ratio and inlet velocity are ignited at the channel exit using the ignition electrode. The flame gradually moves in the direction of the incoming fuel–air mixture and stabilizes at a position where the LBV equals the flow velocity.

The distance from the exit of the channel to the anchored position of a stationary flame is calculated from the flame images captured. The area of the flame can be calculated based on the geometry of the rectangular diverging channel. The temperature of the unburnt gas before the flame is measured beforehand. The burning velocity is calculated using a reorganized form (Eq. 1) of the continuity equation.

$$S_u = U_{\text{inlet}} \times \left(\frac{A_{\text{inlet}}}{A_f}\right) \times \left(\frac{T_u}{T_{u,o}}\right) \tag{1}$$

3 Results and Discussion

LBV of CH_4-air mixtures were measured at high temperatures (350–650 K) and pressure (1–5 atm) conditions. The stabilized flames in the heated diverging channels were employed for the measurement of LBV. The dependence of LBV on temperature and the determination of power exponents of temperature (α) and pressure (β) are presented.

3.1 Laminar Burning Velocity at High Temperatures

The rate of increase of the burning velocity with the mixture temperature (α) and the burning velocity at ambient conditions ($S_{u,o}$) are determined based on the power-law correlation. These two values characterize the LBV at various elevated temperatures for a particular equivalence ratio. A rise in the burning velocity with higher mixture temperature is visible in Fig. 2. The rate of increase of burning velocity as a power-law function similar to Eq. 2 is 2.26, 1.91 for $\phi = 0.7$, 1.1, respectively. The increase in burning velocity at elevated temperatures is due to the increase in thermal diffusivity and the enhanced reaction rate at higher initial temperatures. Also, a higher mixture temperature translates to an increased enthalpy of reactants and results in higher adiabatic flame temperature compared to lower mixture temperatures—these result in higher energy release and, in turn, higher burning velocities.

The following equation helps determine LBV of CH_4-air mixtures at elevated temperatures.

$$S_u = S_{u,o} * \left(T_u / T_{u,o}\right)^{\alpha} \tag{2}$$

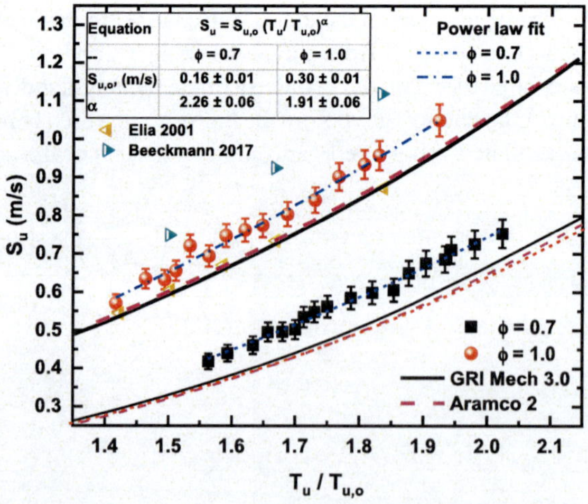

Fig. 2 LBV at elevated temperatures for two different equivalence ratios ($\phi = 0.7$, 1.0) and comparison with the literature [18, 19]

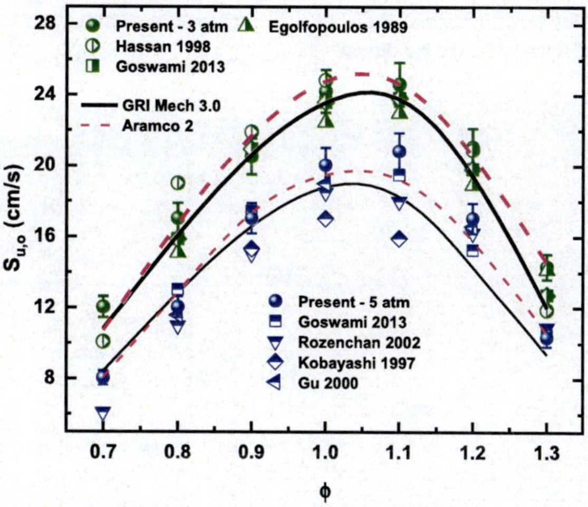

Fig. 3 LBV of CH_4-air mixture at various pressures (3, 5 atm). Symbols: [2, 3, 20] Kinetic simulations: lines

3.2 *Laminar Burning Velocities at Elevated Pressures and 300 K*

Figure 3 shows the comparison of LBV of CH_4-air mixtures at 300 K and high pressures (3, 5 atm) with the available literature and the predictions of two kinetic schemes (Aramco 2 [15, 16] and GRI Mech 3.0 [17]). Burning velocities decrease with an increase in pressure for all the equivalence ratios. The comparison with the available literature is in reasonable agreement with the current results. The predictions using Aramco 2 and GRI Mech 3.0 are in perfect agreement with the current LBV measurements at lean ($\phi \leq 0.9$) and rich ($\phi \geq 1.1$) mixtures.

3.3 *Temperature Exponent*

Power exponents of temperature (α) are a derived quantity and are a function of ϕ and P_u. It signifies the rate with which the burning velocity increases with temperature. The temperature exponent is lowest at an equivalence ratio for which the burning velocity is maximum. The temperature exponents derived from the determined LBV are compared with available literature and numerical predictions of kinetic schemes. The temperature exponents from the kinetic schemes with a similar temperature range to the experiments (300–650 K) show excellent agreement with the present experiments. Various curves with different upper and lower limits are shown in Fig. 4, to ascertain the effect of temperature range on the determination of power exponents of temperature (α). The temperature exponents should be compared with only those that have a similar temperature range for determination.

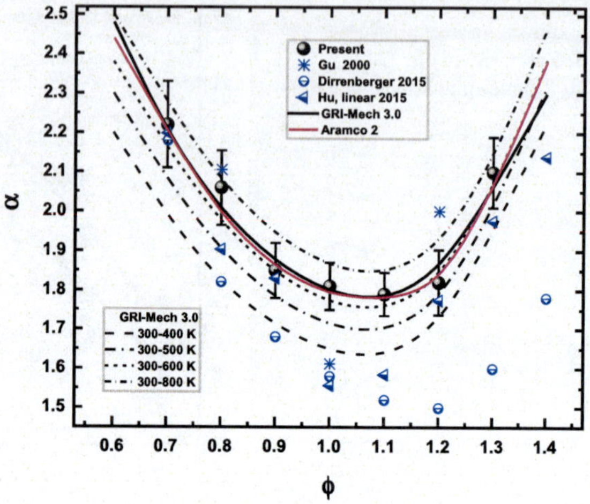

Fig. 4 Comparison of the change in power exponents of temperature (α) with temperature range

3.4 *Pressure Exponent*

Figure 5 shows the pressure exponent (β) at various ϕ (0.7–1.4) at ambient temperatures (300 K). Similar to the discussion on temperature exponent (3.2), the pressure exponents are a function of temperature and the pressure range used for its determination. The pressure range employed in the present experiments is 1–5 atm. The variation of follows a similar curve as to that of burning velocities. The pressure exponents derived from the current LBV measurements agree with the simulations of the Aramco 2 reaction scheme. Goswami et al. [3] and Dirrenberger et al. [21] results also match with the current measurements within the range of equivalence

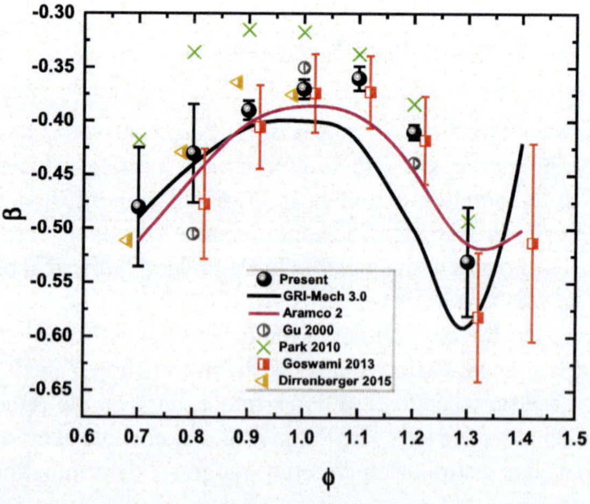

Fig. 5 Pressure exponents of CH_4-air mixtures at various equivalence ratios

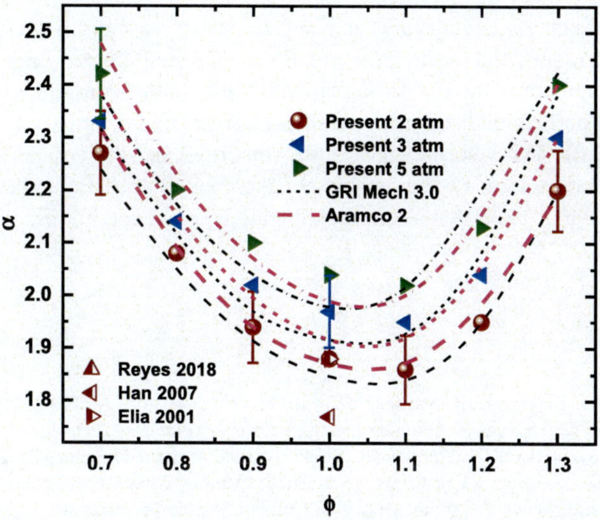

Fig. 6 The power exponents of temperature (α) for the CH_4-air mixture at different pressures (2, 3, 5 atm) and their dependence on pressure at different equivalence ratios

ratio ($\phi = 0.7{-}1.2$). The pressure exponents reported by Park et al. [5] are significantly different from the present measurements. The strained flames obtained by Park et al. [5] in counterflow stagnation flames and subsequent linear extrapolation to zero strain rate might be the reason for the discrepancy.

3.5 *Variation of Temperature Exponent with Mixture Pressure*

Figure 6 shows the power exponents of temperature (α) at different ϕ for elevated pressures ($P_u = 1{-}5$ atm). The temperature exponents linearly increase with an increase in pressure. The general variation of temperature exponent (α) with equivalence ratio (ϕ) remains the same with a minimum value at $\phi = 1.1$ for different pressures. For stoichiometric mixtures, the temperature exponents reported by Elia et al. [18] and Reyes et al. [22] agree well with the present results at 2 atm.

4 Conclusions

LBV of CH_4-air mixtures were determined at high mixture temperatures (350–650 K) and pressures (1–5 atm) using the externally heated diverging channel technique. The burning velocities varied non-monotonously over a range of ϕ

(0.7–1.3) with the maximum value at $\phi = 1.1$. The decrease in LBV with an increase in the inlet pressure and an increase in LBV at elevated temperatures were visible in the measurements. The results match well with recent literature at high mixture temperatures and simulations using two detailed reaction chemistry (Aramco 2 and GRI Mech 3.0). The determined burning velocities and the comparison with the available literature indicate the technique developed is accurate and can be further extended to higher pressures resembling engine relevant conditions.

References

1. Varghese RJ, Kolekar H, Kumar S (2019) Int J Hydrogen Energy 44/23:12188
2. Gu XJ, Haq MZ, Lawes M, Woolley R (2000) Combust Flame 121(1):41
3. Goswami M, Derks SCR, Coumans K, Slikker WJ, de Andrade Oliveira MH, Bastiaans RJM, Luijten CCM, de Goey LPH, Konnov AA (2013) Combust Flame 160(9):1627
4. Natarajan J, Lieuwen T, Seitzman J (2007) Combust Flame 151(1–2):104
5. Park O, Veloo PS, Liu N, Egolfopoulos FN (2011) Proc Combust Inst 33(1):887
6. Khandelwal B, Kumar S (2010) Appl Therm Eng 30(17–18):2718
7. Alekseev VA, Christensen M, Konnov AA (2015) Combust Flame 162(5):1884
8. Akram M, Saxena P, Kumar S (2013) Energy Fuels 27(6):3460
9. Akram M, Saxena P, Kumar S (2012) In ASME 2012 GasTurbine India Conference (American Society of Mechanical Engineers, 2012) pp 581–586
10. Akram M, Kishore VR, Kumar S (2012) Energy Fuels 26(9):5509
11. Katoch A, Asad M, Minaev S, Kumar S (2016) Fuel 182(2016):57
12. Katoch A, Millan-Merino A, Kumar S (2018) Fuel 231:37
13. Katoch A, Chauhan A, Kumar S (2018) Energy Fuels 32(5):6363
14. Kumar R, Katoch A, Singhal A, Kumar S (2018) Energy Fuels 32(12):12936
15. Metcalfe WK, Burke SM, Ahmed SS, Curran HJ (2013) Int J Chem Kinet 45(10):638
16. Zhou CW, Li Y, O'Connor E, Somers KP, Thion S, Keesee C, Mathieu O, Petersen EL, DeVerter TA, Oehlschlaeger MA (2016) Combust Flame 167:353
17. Smith GP, Golden DM, Frenklach M, Moriarty NW, Eiteneer B, Goldenberg M, Bowman CT, Hanson RK, Song S, Gardiner WC Jr (1999) GRI-Mech 3.0
18. Elia M, Ulinski M, Metghalchi M (2001) J Eng Gas Turbines Power 123(1):190
19. Beeckmann J, Hesse R, Bejot F, Xu N, Pitsch H (2017) Assessment of the approximation formula for the calculation of methane/air laminar burning velocities used in engine combustion models. Tech Rep, SAE Technical Paper
20. Ogami Y, Kobayashi H (2006) JSME Int J Ser B: Fluids and Thermal Eng 48(3):603
21. Dirrenberger P, Le Gall H, Bounaceur R, Glaude PA, Battin-Leclerc F (2015) Energy Fuels 29(1):398
22. Reyes M, Tinaut FV, Horrillo A, Lafuente A (2018) Appl Therm Eng 130:684

Spray Characterization and Structure Analysis in a Model LPP Atomizer

Shirin Patil and Srikrishna Sahu

1 Introduction

The injection of liquid jet into crossflow is widely studied in literature due to its enormous industrial applications. It is employed in Lean Premixed Pre-vaporized (LPP) combustor for gas turbines which has gained popularity for its capability for NO_x reduction. This is achieved by combustion of homogeneous mixture of fuel and air at equivalence ratios close to the lean blowout limit that ensures localized regions of high temperature are always avoided in the combustor. Several studies are available in the literature which report spray characterization using different optical techniques [1–7]. Pei-Kuan et al. [6] employed Phase Doppler Particle Analyzer (PDPA) technique to measure droplet size, axial velocity, and volume flux in a spray plume. The droplet velocity was found to be uniform throughout the spray region. The sauter mean diameter (SMD) was found to increase with increase in transverse distance from the orifice and also decrease with crossflow air velocity. Becker and Hassa [1] measured spray properties for a kerosene jet injected in crossflow at elevated pressure and found weaker effect of crossflow velocity on SMD compared to that at atmospheric pressure. Tambe and Jeng [5] considered the injection of liquid jet into the swirling crossflow. The radial penetration of spray increased with increase in swirl strength and spray continued to expand further downstream due to the centrifugal forces in crossflow. However, the measurements of droplet characteristics alone are not enough for understanding on unsteadiness in sprays which is important for flame stability and optimum engine performance. Only handful studies have addressed this issue for instance, Batarseh et al. [8] reported the frequency associated with an

S. Patil (✉) · S. Sahu
Indian Institute of Technology, Madras, India
e-mail: shirin3151@gmail.com

S. Sahu
e-mail: ssahu@iitm.ac.in

A. K. Gupta et al. (eds.), *Advances in IC Engines and Combustion Technology*, Lecture Notes in Mechanical Engineering,
https://doi.org/10.1007/978-981-15-5996-9_53

airblast spray using proper orthogonal decomposition (POD) technique and found them to be equal to frequencies obtained by applying slotting technique to LDV data. The spray frequency was found to strongly depend on atomizing air velocity and atomizer geometry.

The present paper aims to experimentally study liquid jet atomization in a model twin-jet cross-stream airblast atomizer in the absence and presence of annular air swirl. The atomizer comprises of two radially opposite orifices on a central tube for liquid injection. The atomizing air flows in an annular space around the central tube. The atomizer operating conditions are selected such that any wall filming due to jet-wall interactions are avoided before the jet breakup process. The spray produced by the atomizer is visualized using laser sheet imaging and droplet measurements are taken using PDPA technique (Artium Technologies). The POD technique is applied to study spray unsteadiness and dominant structures in spray of cross-stream airblast atomizer.

2 Experimental Method

2.1 Cross-Stream Airblast Atomizer

A schematic of the twin-jet cross-stream airblast atomizer is shown in Fig. 1a. It consists of an outer cylindrical acrylic tube of 45 mm outer diameter and 39 mm inner diameter. The concentric inner cylindrical acrylic tube has 10 mm outer diameter and 6 mm inner diameter. Water enters the central tube from the top and is injected radially through opposite holes each 1 mm diameter. The holes are situated 5 mm from the end of the central tube. The end of the inner tube is kept at 10 mm above to that of the end of outer tube.

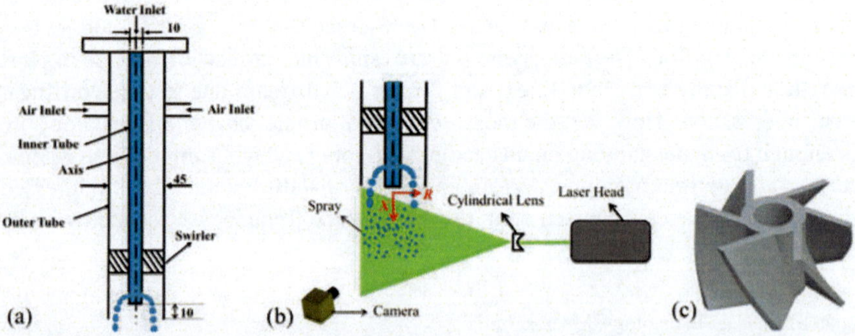

Fig. 1 **a** Schematic of the twin-jet cross-stream airblast atomizer. (All dimensions are in mm). **b** Optical arrangement for laser sheet imaging. **c** 3D printed axial swirler

The atomizing air enters the outer tube via four radial inlets and passes through the annular gap in a direction normal to the direction of liquid exit. An in-house 3D-printed axial air swirler as shown in Fig. 1c (swirl vane angle of 45 deg, swirl number, $S = 0.74$) is placed 40 mm above the orifice to impart tangential momentum to crossflow. For the experiments in the absence of swirling air, an annular holder with straight vanes (vane angle 0°) was used. Earlier authors have done study on primary breakup characteristics of twin-jet atomizer in Patil and Sahu [9].

2.2 *Optical Arrangement for Laser Sheet Imaging*

A single-pulse Q smart laser (532 nm) has been used to form laser sheet using cylindrical lens as shown in Fig. 1b. Images are captured with PCO Pixelfly Camera (14 bit, 1040 × 1392 pixels^2) using Nikon lens (50 mm focal length; f/1.8D). The exposure time of camera was kept at 30 μs to capture the instantaneous spray image. The field of view for imaging is 150 mm × 200 mm such that the spatial resolution is 150 μm/pixel.

2.3 *Injector Operating Conditions*

The non-dimensional numbers for studying liquid jet breakup are defined below.

$$\mathrm{We}_g = \frac{\rho_g U_g^2 d}{\sigma} \tag{1}$$

$$\mathrm{MFR} = \frac{\rho_l U_l^2}{\rho_g U_g^2} \tag{2}$$

where We_g is aerodynamic Weber number, MFR is liquid-to-air momentum flux ratio, ρ_l is density of liquid water jet, U_l is liquid jet velocity, d is diameter of jet, μ_l is dynamic viscosity of water, ρ_g is density of gas, *i.e.*, air, U_g is velocity of gas, σ is surface tension for water–air interface. The flow rates of air and water in to the atomizer were varied over a wide range of We_g (40–140) and MFR (3–7) that ensure the liquid jets never hit the outer tube prior to their breakup, and the jet atomization always occurs within the annular region between the inner and outer acrylic tubes. The injector operating conditions and the corresponding non-dimensional numbers are summarized in Table 1.

Table 1 Operating conditions for cross-stream airblast atomizer

Case	U_l (m/s)	U_g (m/s)	Re_l	We_g	MFR
1	4.2	49.1	5323.8	38.9	6.4
2	4.2	68.7	5323.8	76.0	3.3
3	5.3	68.7	6654.7	76.0	5.1
4	5.3	80.1	6654.7	103.3	3.8
5	6.4	91.9	7985.7	136.2	4.1

3 Result and Discussion

3.1 Spray Characterization Using PDPA

Figure 2 shows the droplet characteristics for swirling and non-swirling crossflows at $X = 50$ mm downstream from the injector and $R = 20$ mm. It can be observed in Fig. 2a that SMD reduces with increase in We_g for swirling and non-swirling crossflows as earlier reported by Pei-Kuan et al. [6] and Takao and Nagai [4]. SMD decreases from 130 to 60 μm as We_g increases from 40 to 140 and it is observed to decrease almost linearly for swirling and non-swirling crossflows. In other words, as We_g increases by 25%, SMD decreases by 5%. It is generally expected that with the introduction of swirling crossflow, atomization quality improves and SMD reduces as compared to non-swirling crossflow but it can be clearly seen in Fig. 2a that SMD is almost similar for swirling and non-swirling crossflows at axial station of $X =$ 50 mm. Droplet axial velocity (U) increases linearly with increase in We_g as can be seen in Fig. 2b. U increases from 20 to 50 m/s as We_g increases from 40 to 140. As the axial velocity of gas U_g increases, the drag force acting on the droplet increases which further increases the droplet velocity. Due to the increase in drag force on the surface of droplets, it tends to breakup, i.e., secondary atomization increases and hence SMD reduces as seen in Fig. 2a. The primary breakup regimes change from bag, multimode to surface breakup modes as We_g is increased [10–12]. Droplets are stripped off from the liquid jet surface in surface breakup mode thereby generating very small droplets and hence SMD tends to reduce as seen in Fig. 2a.

Figure 2c shows the droplet tangential velocity (V) normalized with local axial droplet velocity (U). As it is expected due to the swirling crossflow, tangential velocity is imparted to the droplets while for non-swirling crossflow, droplets have negligible tangential velocity as observed in Fig. 2c. The tangential velocity is up to 40% of axial velocity and as V/U is constant over a range of We_g, it implies that V increases with We_g. It is noted that the swirler used in the present study had swirl number ($S = 0.74$) while we measured $V/U = 0.4$. It might be because of the fact that at the end of atomizer, crossflow is allowed to expand and it is no more restricted by the outer tube and hence the swirl number is decreasing. Radial velocity of droplets (W) normalized with U (see Fig. 2d) remains constant at $W/U = 0.6$ for a range of We_g for swirling crossflow while $W/U = 0.2$ for non-swirling crossflow. It means that

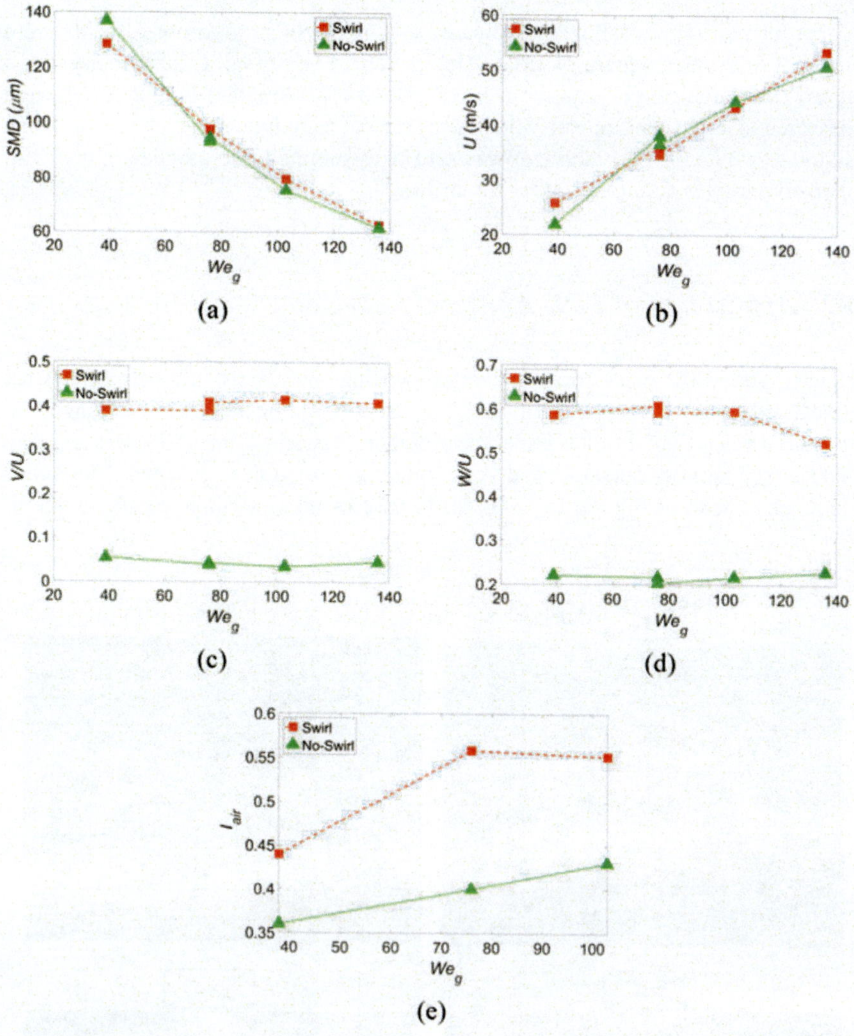

Fig. 2 Spray characterization at $X = 50$ mm, $R = 20$ mm **a** *SMD* versus We_g. **b** *U* versus We_g. **c** *V/U* versus We_g. **d** *W/U* versus We_g. **e** I_{air} versus We_g

W increases with We_g and hence spray is expected to expand radially outwards and void is expected to form in central region of the atomizer. It is noted that as tangential and radial velocities in case of non-swirling crossflow are negligible, hence droplets are not expected to spread radially outwards, and instead droplets will be accumulated near the central region of the atomizer axis. Moreover, the turbulent intensity of air, I_{air} (based on axial velocity) is between 36 and 43% for non-swirling crossflow while for swirling crossflow I_{air} is between 44 and 55% when We_g is varied from 38 to 103 at $X = 50$ mm and $R = 20$ mm as can be seen in Fig. 2e. I_{air} is almost 25%

higher for swirling crossflow as compared to non-swirling crossflow. I_{air} is the ratio between root mean square of axial velocity (U_{rms}) and mean axial velocity (U_{air}) of air. The statistical uncertainty in I_{air} is around 1%. Note that U_{rms} and U_{air} were measured using Laser Doppler Velocimetry (LDV) technique in which airflow alone (liquid flow rate into the atomizer was zero) was supplied into atomizer along with olive oil particles acting as tracers for airflow.

3.2 Application of POD Analysis

Figure 3 shows the mean spray image for swirling and non-swirling crossflows for Case 2 and Case 5. As discussed in Sect. 3.1, there is significant increase of tangential and radial velocity of droplets under the presence of swirling crossflow; hence, spray is dispersed radially outwards and void region is formed near the central region of atomizer as seen in Fig. 3a ($S = 0.74$). In case of non-swirling crossflow, due to

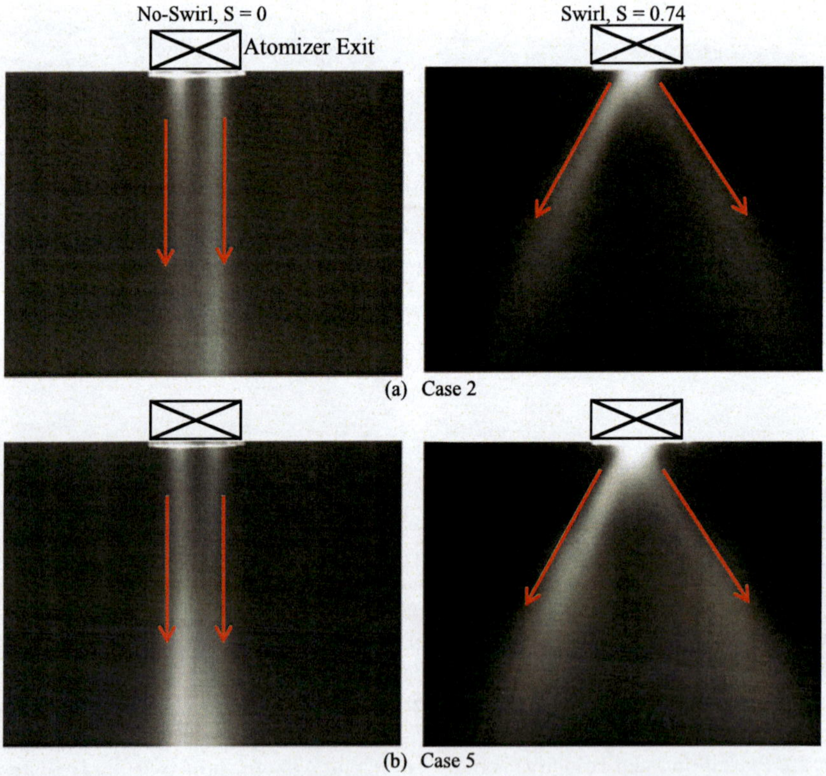

Fig. 3 Mean spray images for **a** Case 2 ($We_g = 76$, MFR = 3.3). **b** Case 5 ($We_g = 136$, MFR = 4.1)

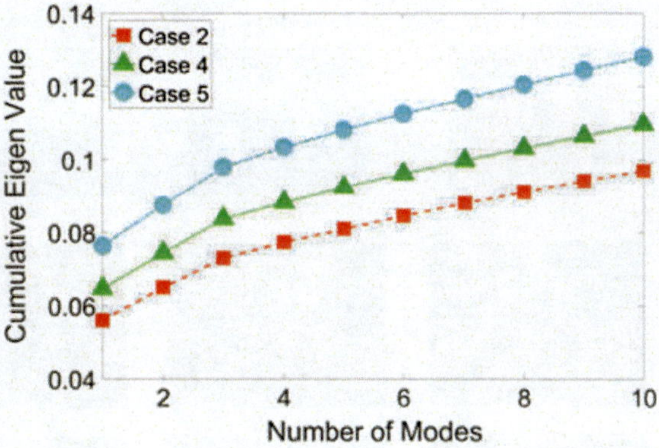

Fig. 4 Cumulative eigen value versus number of modes for Case 2 ($We_g = 76$, MFR = 3.3), Case 4 ($We_g = 103$, MFR = 3.8) and Case 5 ($We_g = 136$, MFR = 4.1)

the negligible tangential and radial velocity of droplets, spray is restricted to central region of atomizer as seen in Fig. 3a ($S = 0$). The droplet number density increases as indicated by the increase in intensity of mean image in Fig. 3b as the We_g is increased from Case 2 ($We_g = 76$) to Case 5 ($We_g = 136$). The number of droplets increases and the void region in the central region of atomizer is no more visible in Fig. 3b ($S = 0.74$) possibly due to the formation of stronger recirculation zone in the central region of spray.

Figure 4 shows the variation of cumulative eigen value with the number of modes for Cases 2, 4, and 5. Eigen value represents the amount of energy present in that particular mode and in the context of spray images, energy means intensity of the pixels in an image. It can be observed that eigen value contribution for mode 1 increases as the We_g is increased from Case 2 ($We_g = 76$), Case 4 ($We_g = 103$) to Case 5 ($We_g = 136$). It implies that the pixel illumination or intensity of image which indicates the number density of image is increased as the We_g is increased. It can be noticed that the cumulative eigen value contribution till ten modes is about 8, 10, and 13% for Case 2, Case 4, and Case 5, respectively. It means that as the We_g is increased, higher energy is captured within few modes hence first few modes are sufficient to demonstrate the dominant features on spray while for lower We_g, more number of modes will be required to study dominant features of spray.

Figure 5 (First row) shows the contour plots of first four POD modes for Case 2. The first POD mode shows the mean structure of the spray as depicted by the mean image as shown in Fig. 3. The second and third POD modes show the breakup of ligaments and depict the region of concentrated droplet number density below the atomizer exit. The fourth POD mode shows the flapping of spray structure. The similar POD features of breakup and flapping of liquid jet were observed in context of liquid jet in crossflow [13]. It implies that the memory of breakup and flapping features is still retained in the spray structure downstream of the atomizer exit. The dominant structures and features shown by the first four POD modes can be understood by looking at the instantaneous spray images corresponding to the minimum negative coefficient ($\alpha_{min} < 0$) and maximum positive coefficient ($\alpha_{max} > 0$) of each POD mode as shown in Fig. 5 (second row) and Fig. 5 (third row),

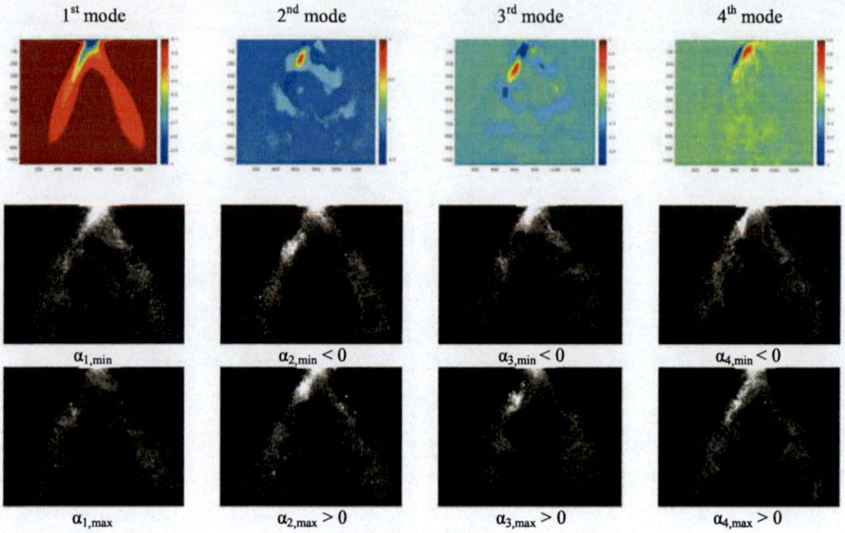

Fig. 5 First row: Contour plots for POD modes for Case 2 ($We_g = 76$, MFR = 3.3), Second row: Minimum negative coefficient for corresponding modes in first row, Third row: Maximum positive coefficients for corresponding modes in first row

respectively. It is noted for first mode that all the coefficients were negative. The droplet count is higher for $\alpha_{1,min}$ while droplet count is observed less for $\alpha_{1,max}$. It shows that first POD mode shows the fluctuations of droplet number at the exit of the atomizer. The second and third modes together represent shedding of ligaments further downstream of the exit of the atomizer. The region of concentrated droplet number density is away from the atomizer exit for $\alpha_{2,min} < 0$ while the chunk of concentrated droplets is closer to atomizer exit for $\alpha_{2,max} > 0$ (see Fig. 5). The vice versa is observed for third mode, i.e., the chunk of concentrated droplet is closer to atomizer exit for $\alpha_{3,min} < 0$ while it is away from the atomizer exit for $\alpha_{3,max} > 0$. Such features were observed for the liquid jet itself by Marco and Soteriou [14] in context of liquid jet breakup in crossflow where they reported that such alternate features depicted travelling waves on the liquid jet surface. It is clearly evident (from the alternate modal features of second and third mode) that the upstream liquid jet breakup process strongly affects the downstream spray characteristics. The spray structure can be seen to be tilted toward left side (near the atomizer exit) for $\alpha_{4,min} < 0$, while the spray structure is tilted toward right side for $\alpha_{4,max} > 0$ (see Fig. 5) which depicts the flapping behavior of spray structure.

Figure 6 (First row) shows first four POD modes for Case 5. Similar to Fig. 5 (Case 2), first POD mode shows the mean structure of the spray. Due to smaller size of droplets, the tangential and radial velocity are smaller (compared to Case 2), so centrifugal forces are negligible and hence such smaller sized droplets would be trapped in the central region of atomizer as the SMD reduces from Case 2 to Case 5 as shown earlier in Fig. 2 and hence more droplets are trapped near the central

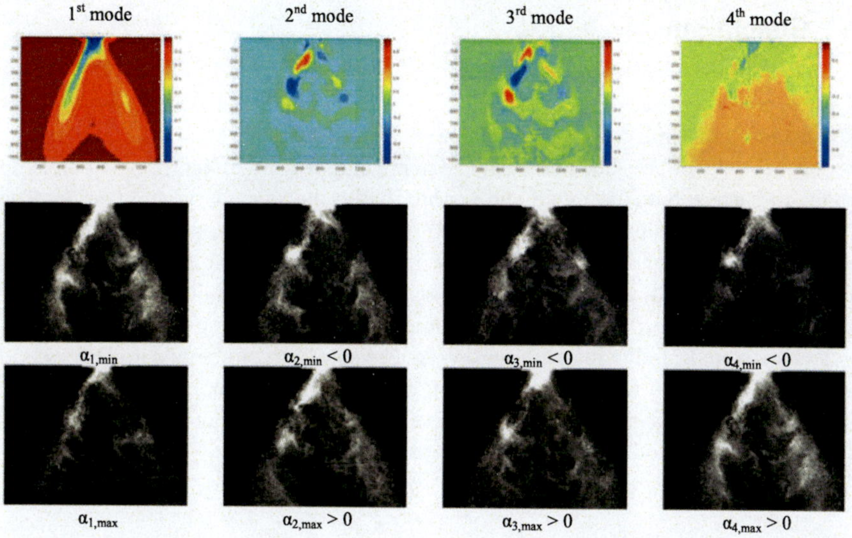

Fig. 6 First row: First four POD modes for Case 5 ($We_g = 136$, MFR = 4.1), Second row: Minimum negative coefficient for corresponding modes in first row, Third row: Maximum positive coefficients for corresponding modes in first row

region of atomizer. It is due to this reason that the void region was present in the central region of atomizer for Case 2 (see Fig. 5, first mode) while the central region of atomizer is filled with droplets for Case 5 (see Fig. 6, first mode). The fluctuations of droplet number is represented by $\alpha_{1,\text{min}}$ and $\alpha_{1,\text{max}}$ in Fig. 6 similar to Case 2 in Fig. 5. The alternate feature of the region with concentrated droplet number density is shown by second and third mode in Fig. 6 similar to Fig. 5. The fourth POD mode for Case 5 as shown in Fig. 6 showed the fluctuations in the droplet count while the flapping behavior of the spray structure is now shifted to fifth mode (not shown here). It means that for higher We_g, fluctuations in droplet count become dominant feature of the spray. The droplet count is observed less for $\alpha_{4,\text{min}} < 0$ while droplet count is significantly high and additionally, more droplet clusters can be seen for $\alpha_{4,\text{max}} > 0$. The fourth POD mode brings out very important feature of spray fluctuations which is otherwise cannot be captured by PDPA technique. Even though flow rates of water and air into the atomizer were maintained constant, significant difference in droplet number is observed from one instant to another which is due to different types of instabilities such as Kelvin-Helmholtz, Rayleigh-Taylor instabilities which govern the atomization process. These instabilities in spray may further cause spatially non-uniform mixing of fuel–air mixture which can lead to the formation of pockets with rich equivalence ratio and hence giving rise to the NO_x formation. The temporal variation in fuel–air mixing may lead to combustion instabilities inside gas turbine combustor [15]. In addition, due to the turbulent crossflow and interaction of turbulent eddies with droplets, preferential concentration of droplets may appear in some of the region in spray.

4 Conclusion

Optical measurements of spray characteristics are reported for a model twin-jet cross-stream airblast atomizer for LPP application. It was found that local *SMD* in the spray reduces for higher We_g due to the change in breakup regimes of jet from bag breakup mode to surface breakup mode and further due to the secondary atomization. The *SMD* close to the injector exit was similar for swirling and non-swirling crossflows. The axial velocity of droplets increased with We_g and it was similar for swirling and non-swirling crossflows. The tangential and radial velocity of droplets was significantly higher for swirling crossflow as compared to non-swirling crossflow and both velocities were found to increase with We_g. The spray structure indicated by second and third modes showed alternate region of concentrated droplet number density. The flapping of spray structure was observed in fourth mode. For higher We_g, spray, fluctuations are found to be more dominant.

References

1. Becker J, Hassa C (2002) Breakup and atomization of a kerosene jet in crossflow at elevated pressure. Atomization Sprays 12(1–3):49–68
2. Becker J, Heitz D, Hassa C (2004) Spray dispersion in a counter-swirling double-annular air flow at gas turbine conditions. Atomization Sprays 14(1):15–36
3. Elshamy OM, Tambe SB, Cai J, Jeng S-M (2007) PIV and LDV measurements for liquid jets in crossflow. In: 45th AIAA Aerospace Sciences Meeting and Exhibit, Reno, Nevada
4. Inamura T, Nagai N (1997) Spray characteristics of liquid jet traversing subsonic airstreams. J Propul Power 13(2):250–256
5. Tambe SB, Jeng S (2008) A study of liquid jets injected transversely into a swirling crossflow. In: 21st annual conference on liquid atomization and spray systems (April):1–12
6. Pei-Kuan W, Kirkendall KA, Fuller RP (1998) Spray structures of liquid jets atomized in subsonic crossflows. J Propul Power 14(2):173–182
7. Zheng Y, Marshall AW (2011) Characterization of the initial spray from low weber-number jets in crossflow 21(7):575–589
8. Batarseh FZ, Gnirß M, Roisman IV, Tropea C (2009) Fluctuations of a spray generated by an airblast atomizer. Exp Fluids 46(6):1081–1091
9. Patil S, Sahu S (2019) Breakup dynamics and near nozzle spray fluctuations in a twin-jet crossflow airblast atomizer. Atomization Sprays 29(3):217–250
10. Mazallon J, Dai Z, Faeth GM (1999) Primary breakup of nonturbulent round liquid jets in gas crossflows. Atomization Sprays 9:291–311
11. Sallam KA, Aalburg C, Faeth GM (2004) Breakup of round nonturbulent liquid jets in gaseous crossflow. AIAA J 42(12):2529–2540
12. Wu P-K, Kirkendall KA, Fuller RP, Nejad AS (1997) Breakup processes of liquid jets in subsonic crossflows. J Propul Power 13(1):64–73
13. Patil S, Sahu S (2018) Breakup dynamics in a twin-jet crossflow airblast atomizer. ICLASS 2018. In: 14th triennial international conference on liquid atomization and spray systems, Chicago, IL, USA
14. Arienti M, Soteriou MC (2009) Time-resolved proper orthogonal decomposition of liquid jet dynamics. Phys Fluids 21(11):1–15
15. Huang Y, Yang V (2009) Dynamics and stability of lean-premixed swirl-stabilized combustion. Prog Energy Combust Sci 35(4):293–364

Design of Free-Piston Linear Generator

Aditya Purkar, P. R. Dhamangaonkar, and K. Muralidharan

Abbreviations

FPE	Free-Piston Engine
FPLG	Free-Piston Linear Generator
CFD	Computational Fluid Dynamics
FSI	Fluid Structure Interaction
RLC circuit	Resistance Inductor Capacitor circuit
TDC	Top Dead Centre
BDC	Bottom Dead Centre
ε*	Damping Ratio

Nomenclature

F_p	Pressure Force (N)
F_p	Pressure Force (N)
F_f	Friction Force (N)
F_e	Electromagnetic Force (N)
m	Mass of Piston (kg)
R	Resistance (Ω)
φ	Magnetic Flux (Wb)
e_g	EMF inside Generator (V)

A. Purkar (✉) · P. R. Dhamangaonkar
College of Engineering, Pune, India
e-mail: adityapurkar5@gmail.com

K. Muralidharan
MTU India Pvt. Ltd. (EARC), Pune, India

A. K. Gupta et al. (eds.), *Advances in IC Engines and Combustion Technology*,
Lecture Notes in Mechanical Engineering,
https://doi.org/10.1007/978-981-15-5996-9_54

k_g	Generator Constant (Wb/m)
f_{mep}	Mean Effective Friction Pressure (N/m^2)
S	Stroke Length (m)
rps	Revolutions per Seconds
F_{mep}	Mean Effective Pressure Force (N)
F_{comb}	Combustion Force (N)
K	Stiffness of Spring (N/m)
C	Damping Constant (Ns/m)
∂_n	Natural Vibrating Frequency (s^{-1})
∂_d	Damped Natural Frequency (s^{-1})

1 Introduction

A lot research has been happening in the field of alternative power sources for propulsion systems, off-high way and hybrid electric vehicle (HEV) applications due to need of highly efficient devices. In the process of development of such devices, many methods have been devised to generate and provide the electricity by converting any form of energy into electrical form; general trend is to convert mechanical energy of into electricity which is usual method in conventional generators. But providing primary energy to generators involves mechanical losses which are dependent on factors such as mass, inertia and friction between involved parts. Further, converting this motion into electricity generation involves generation loss. So primarily, these are two losses lead the process towards higher inefficiency. To minimise this inefficiency, we should focus on minimising the losses in generator sets.

The free-piston engine concept was first patented by Pescara [1] in the 1940s as an air compressor. Many companies worldwide tried to study and design free-piston engines (as air compressors). Design of a dual piston free-piston engine generator (FPLG),a crank-less linear dynamic system, was presented by Sandia National Laboratory powered with homogeneous charge compression ignition (HCCI) gave electrical power output 40 kW with a thermal efficiency of 56% and low emissions [2]. At Czech Technical University, work has been done on precise motion control to serve steady operation. The prototype was running with a frequency of 27 Hz having 9 compression ratio; the output power was 650 W approximately [3]. Mikalsen and Roskilly gave many papers on operation of free-piston engines having gas bounce chamber, with the help of simulations, and discussed the influences of some of the parameters on the performance of the concept over a wide operating range [4–6]. Some researchers studied methods for piston motion management in FPLG because of effective engine management is the main requirement for this novel system to be commercially possible [7, 8]. One of such systems is shown in Fig. 1.

Natural frequency of the oscillation system was obtained from their model. A simulation program was developed in MATLAB/Simulink to unravel these mathematical equations; also, the simulation results showed that the motion of FPLG was a

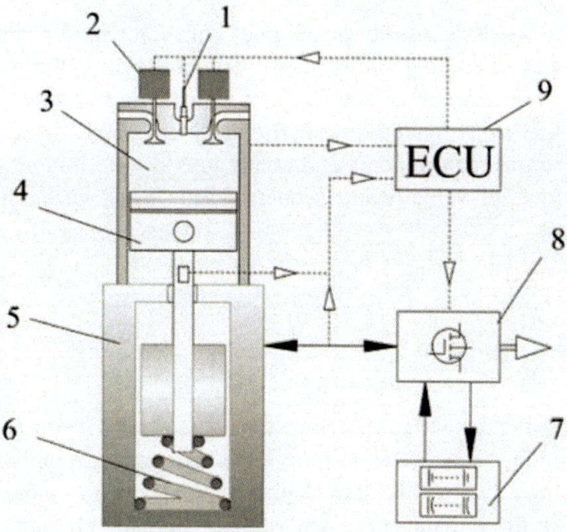

Fig. 1 1.spark plug, 2.electromagnetic valve, 3.combustion chamber, 4.free piston, 5.linear generator, 6.return spring, 7.battery pack, 8.convertor, 9.engine control unit Spring assisted free-piston engine [9]

forced vibration system with variable damping constant and stiffness. Fluid structure interaction (FSI) was used to analyse a multiphysics interaction in the paper paves a similar working in which a physics model is applied to the fluid domain, wherever one or several solid structures intermingle with an interior or encompassing fluid flow [10].

More of the notable literature is focused towards parametric investigation in free-piston engine using either springs or gas chambers as rebounding devices and thermal analysis using one-dimensional numerical modelling [8]. Some researchers also worked on the generator structure which can convert linear motion of piston into electromagnetically induced voltage in electrical circuit [11].

Based on principle of working for FPE and elaboration of linear electric generation concept, coupling of these two concepts, an efficient device can be achieved. Of course, operating parameters are needed to be analysed and tweaked according to the optimum and required performance. Design of free-piston linear generator (FPLG) involves a lot of variables as explained in Sect. 1 and we do not know which parameter needs to be changed that will satisfy our requirement for now. Use of numerical simulations is the preferred way to analyse such complex system. A proper CFD software with capability of analyse spring force and combustion simultaneously should be preferred. To use FPLG as a portable electricity generation system, questions that arise after reviewing present work are, what happens when instantaneous reciprocation velocity is given as input to generators, what is the minimum velocity that can be allowed for piston so that minimum voltage flows according the load requirement, do we really need to design a totally different engine with some assumed geometrical parameters or can we use conventional engines with some modifications, how much fuel is required per oscillating cycle, does every stroke of piston requires combustion, what will be efficiency of total system considering a stable output, remain unanswered. FPLG system should be particularly designed considering minimum output voltage output requirements and continuous or intermittent supply of electricity and reciprocation velocity and frequency. This thesis work is an attempt to fill up these gaps.

2 Research Approach

Different response is expected from free-piston engine with a spring as compared with conventional engine, hence detailed in-cylinder analysis is required. Combustion, electromagnetic and friction and inertia force cause dynamic imbalance on piston, and depending on the imbalance, piston velocity varies [11]. And, this velocity is input to generator which will induce voltage in windings connected to automotive battery, Output can be measured when a load is connected to battery. Approach is explained in Fig. 2.

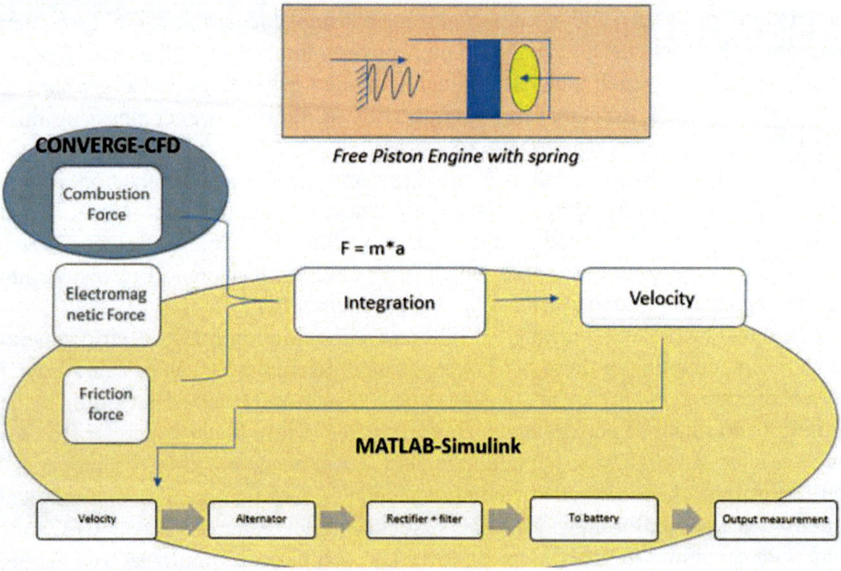

Fig. 2 Research approach

2.1 Mechanics of FPLG

We kept conventional single cylinder diesel engine geometry into a free-piston engine; hence, operation of intake, exhaust valves and suction and scavenging phenomenon experienced here are same as regular diesel engine. Refer Fig. 3 for pictorial representation of FPLG. Piston is lined with a permanent magnet material and windings are wound on cylinder liner. These windings are connected to a battery through a rectifier circuit for seasoning the voltage. After appropriate starting mechanism and fuel injection timing, trapped air fuel mass experiences high pressure and temperature, When the increased temperature matches autoignition temperature of diesel, the air fuel mixture combusts. And, combustion forces the piston towards spring side. Spring absorbs kinetic energy of piston and converts into its potential energy. Depending upon the energy storing capacity, spring deflects and slows the piston down to a complete stall. Now, spring has potential energy stored inside it, and hence, it tries to give it back, of course depending upon the efficiency, remaining energy is given back to piston, this potential energy is transferred into piston again in the form of kinetic energy, but this time momentum is damped, i.e. piston experiences reverse motion towards TDC again.

This time, if there is combustion, same process will repeat, but if there is no fuel injection, the working process at TDC will change. Though fuel injection is not there, air will still be trapped with intake valves closing during piston motion, and air is getting compressed so to increase its internal energy by absorbing kinetic energy of piston.

When all the kinetic energy gets absorbed and piston is at a complete stall, air releases its energy again into piston motion (not all the internal heat energy is transferred back in doing work, according to Carnot's theorem). Piston gains momentum in opposite direction, i.e. towards BDC again but with damped momentum as compared with the previous motion. This kinetic energy is again absorbed by springs, and depending on the efficiency of springs, part of energy is given back to piston (and hence damping the motion again).

Slowly, piston momentum dies, and it comes to stall unless there is fuel injection and combustion in between these oscillations. During no combustion condition,

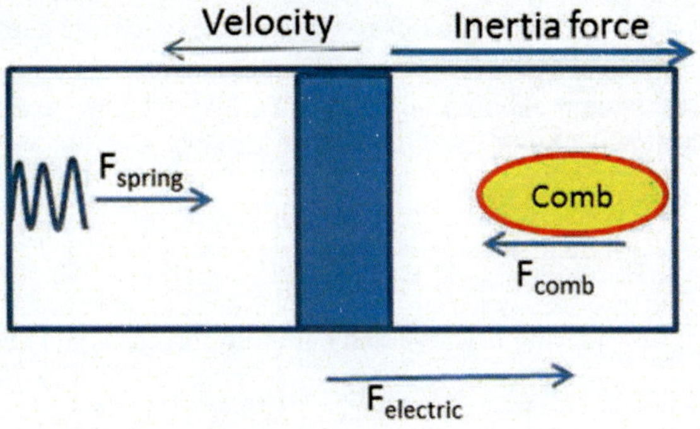

Fig. 3 Pictorial representation of FPLG

piston can be imagined acting between two springs (parallel system of springs) with dampening effect. This process may be imagined as working of air springs in case of air suspensions.

There is magnetic flux acting inside the liner area because the piston is lined with a permanent magnet. Hence, whenever piston is moving inside the liner, magnetic flux is getting cut and voltage is getting induced inside the windings. As flux cut is inducing a voltage is electrical circuit, an opposing force acts on the piston. This force is called as electromagnetic force. Also, during reciprocation, piston rings experience a friction force which is opposing the motion. Hence, the dampening effect is increased by the friction force and electromagnetic force.

2.2 Mathematical Formulation

Assumptions before modelling:

(a) Process is assumed to start when piston is moving towards TDC.
(b) Perfect, one-dimensionally variable spring: Energy release assumed to take place only in one direction, so that we can formulate energy absorbing and releasing physics properly.
(c) Same and constant compression and expansion coefficients for gases: This assumption is important when air is behaving like spring. Coefficient of expansion and compression would play major role in damping the piston motion and considering different coefficients every time would be tedious, and hence, the assumption is taken.
(d) Constant volume heat rejection and addition: Combustion force is not an impact force but considering time varying (i.e. CAD varying) force whilst considering the timing of whole reciprocation cycle timings, a confusion in timings arose. Hence, combustion force is assumed to act as a constant force independent of crank angle degrees. Value of this force is taken as average value (mean effective value) of time-dependent force.

In the FPLG, referring Fig. 4, the motion of piston is governed by equivalent of unbalanced forces acting on it. These forces include combustion chamber pressure force F_{p}, the frictional force F_{f} and the electromagnetic force F_{e}. Let x denote the position of the moving part, t be the time and m be the mass of the piston assembly. The system should obey Newton's Second Law, and hence, piston motion can be designated by:

$$m\frac{\mathrm{d}^2x}{\mathrm{d}^2t} = F_{\mathrm{p}} + F_{\mathrm{f}} + F_{\mathrm{e}} \tag{1}$$

Considering resistive and damping forces, Eq. (1) becomes

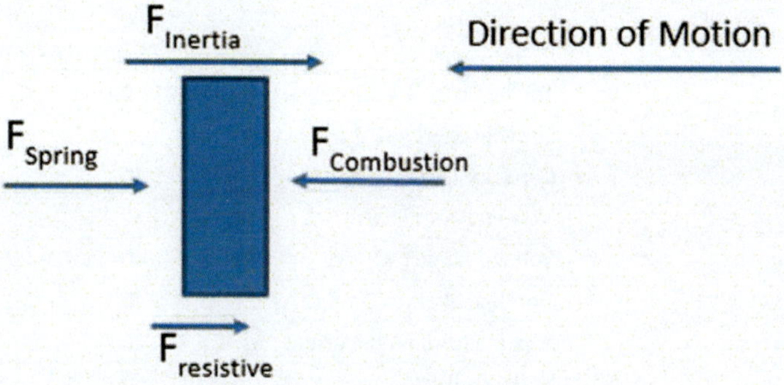

Fig. 4 Free body diagram of piston (after combustion)

$$m\frac{\mathrm{d}^2x}{\mathrm{d}t^2} + c\frac{\mathrm{d}x}{\mathrm{d}t} + kx = (F_{\mathrm{combustion}}) \tag{2}$$

where 'm' is mass of piston (kg), 'c' is damping constant (N s/m) and 'k' is the stiffness of spring used (N/m)

Equation (2) was solved using differential equation methods and solution came as,

$$x = \mathrm{e}^{-w_n t}(c_1 \cos w_\mathrm{d} t + c_2 \sin w_\mathrm{d} t) + F_{\mathrm{mep}} \times \frac{m}{K}$$

where $w_\mathrm{n} = \left(\sqrt{\frac{K}{m}}\right)$ = Natural vibrating frequency of the system.

And, $w_\mathrm{d} = w_\mathrm{n}(\surd(1 - \varepsilon^2)$ = Damped natural frequency.

K: Stiffness of spring, m: Mass of piston, ε: Damping ratio.

c_1, c_2 = Constants to be calculated from initial conditions.

This differential equation can be solved analytically. Numerical and analytical solutions were compared just to be sure that analysis was going in correct direction.

2.2.1 Battery Selection

A sample battery is selected to measure output of system.

It would get charged/discharged depending on the voltage supplied to it from generator. Table 1 shows battery datasheet referred.

2.2.2 Calculation of Minimum Threshold Velocity Requirement

Charging state of battery is decided on its nominal voltage.

When voltage more than nominal voltage is supplied, battery is charged and vice versa. As explained previously, voltage induced in linear generator depends

Table 1 Battery datasheet

Model	Capacity 2 kWh
Total energy (kWh)	2
Useable energy (kWh)	1.84
Capacity (Ah)	40
Nominal voltage (V)	51.2 V
Voltage range (V)	40–58.4 V
Nominal charge current (A)	40
Nominal discharge current (A)	40
Peak discharge current (A)	60

on oscillation velocity of reciprocating piston. Hence, there is a minimum value of reciprocation velocity below which if piston is allowed to reciprocate, voltage induced is not enough to charge the battery. This can be calculated when generator structure (i.e. number of turns, spacing, pitch, diameter of winding wire etc.) and battery nominal voltage are known. For this study, a Ph.D. report by William R. Cawthorne from West Virginia University is used [12]. The generator contained 296 number of turns; output was 120 volts and 488 W power with velocity of 1.4 m/s.

$$e_g = -\frac{\partial \phi}{\partial t} = -\frac{\partial \phi}{\partial x}.\frac{\partial x}{\partial t} = -kg.\dot{x}.\left(k_g \text{is generator constant}\right)$$

120 (volts) = 296 * k_g * 1.4 (m/s)

$k_g = 0.289$ (Wb/m)

Keeping same generator structure for our case gives,

40 (volts) = 296 * 0.289 * (minimum velocity)

So, ***Minimum velocity required = 0.467*** m/s

As justified previously, we are observing the effects when spring is placed inside the cylinder which is behaving similarly as with the case of spring at outside of cylinder liner. Stiffness of spring is calculated by considering all the kinetic energy of free piston is absorbed by spring within total displacement equal to length of stroke.

Potential energy in spring = Kinetic energy in free piston

$$\frac{1}{2} \times Kx^2 = F_{\text{comb}} \times x$$

$$\frac{1}{2} \times K(0.21)^2 = 270558.56 \times 0.21$$

$K = 2576.748$ kN/m.

Damping ratio was chosen to be 0.5 (underdamped system). Point to observe here is that piston is reciprocating, and the amplitude of reciprocation is decreasing as time is passing. Along with the position (i.e. displacement), velocity can also be plotted if displacement equation is analytically differentiated.

2.3 Numerical Modelling

The free-piston engine has to be modelled with the new methodology to study about the free-piston motion. To understand the piston motion, simulations must be performed on free-piston engine in three-dimensional CFD software. Proposed design also consists of spring that is going to take part in defining piston motion; hence, fluid structure interaction (FSI) module was used to incorporate spring force effects. Hence, CONVERGE CFD software was chosen to analyse the system properly which can use physical models of turbulence, FSI, dynamic meshing, combustion, spray models, etc.

For the system modelling and analysis of linear generator and battery charging discharging circuits, MATLAB–Simulink and especially Simscape module is chosen.

Prepared models are described in the following subsections.

CONVERGE generates the mesh at run time. A body fitted volume mesh is generated using cut cell technique. The mesh has orthogonal hexahedral interior cells, which are ideal for the solver algorithm.

- Base grid size used is 4 mm. And grid scale used was of scale 2.

$$\text{Scaled grid} = \text{dx\_base}/\left(2^{\text{grid scale}}\right)$$

- Fixed embedding is the grid control technique used to refine the grid only in specific locations in the domain where a finer resolution is critical to the accuracy and precision of the solution. When simulating sprays, an area of fixed embedding by the nozzle to resolve the complex flow behaviour should be added. Fixed embedding allows the rest of the grid to remain coarse to minimise simulation time. Scale was taken as 2.
- The AMR technique in CONVERGE CFD tool enabled local mesh refinement according to temperature, velocity and species gradients. AMR is used to automatically refine the grid based on fluctuating and moving conditions such as temperature and velocity. The scale of AMR for velocity and temperature was kept as 3 which made the grid size of AMR as 0.5 mm.
- A 3D CFD model of crank shaft engine was developed using CONVERGE CFD coupled with detailed chemical kinetics. The SAGE detailed chemistry has used for the combustion model with the 48 species and 152 reactions. Table 2 summarises the engine specifications, including the engine speed which is 1900 rpm.

Initial conditions for the simulations, including the equivalence ratio and combustion chamber pressure, temperature were taken from the experimental data.

Table 2 Engine specifications

Bore	170 mm
Stroke	210 mm
Connecting rod length	364 mm
Fuel	Diesel 2 (standard diesel)
Engine speed	1900 rpm
Max intake and exhaust valve lift	11.45 mm

Fig. 5 3D CFD model, view 1

Figures 5, 6 and 7 show views of the CFD model. The computational domain included three distinct regions: intake, exhaust and in-cylinder region. Each of which was initialized separately. The CFD model described above was validated against experimental data.

To analyse piston position and velocity, boundary conditions of fluid structure interaction were applied. In CONVERGE, springs were modelled but physical spring is not visible in the model. For our model, spring is fixed at both ends. One end is fixed at cylinder head and another end is attached to piston. Spring is compressed/tensed by piston when the situation comes and is deflected back to its original position due to dynamic imbalance of force across it. Faces were flagged as shown in Table 3.

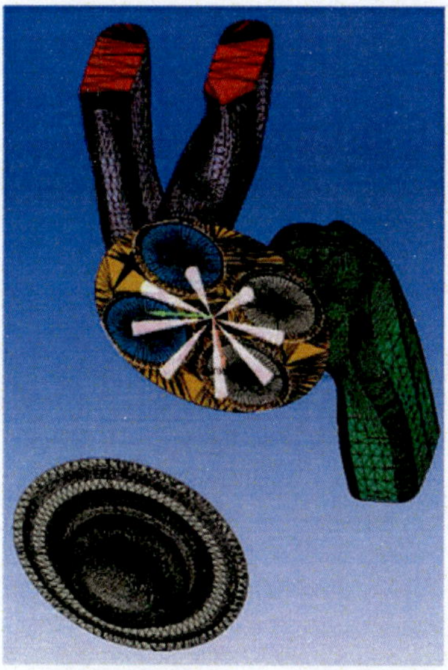

Fig. 6 CFD model view 2

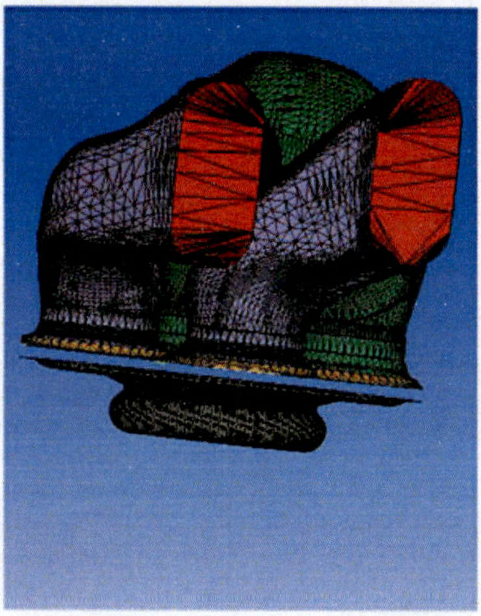

Fig. 7 CFD model view 3

2.4 Explanation of Battery Charging Circuit

Figure 8 shows the MATLAB–Simulink circuit. Voltage output is periodic as well as both positive and negative in nature. Therefore, alternator output will also be of two polarities. But, for battery to get charged, voltage to be supplied should be of positive polarity only. Full wave rectifier was used in the circuit to convert the output in single

Table 3 Boundary condition setup

Boundary name	Boundary type, condition
Piston	Wall, FSI
Cylinder liner	Wall, Stationary
Cylinder head	Wall, Stationary
Intake port	Wall, Stationary
Intake valve top	Wall, Translating
Intake valve bottom	Wall, Translating
Inflow	Inflow
Exhaust port	Wall, Stationary
Exhaust valve top	Wall, Translating
Exhaust valve bottom	Wall, Translating
Outflow	Outflow

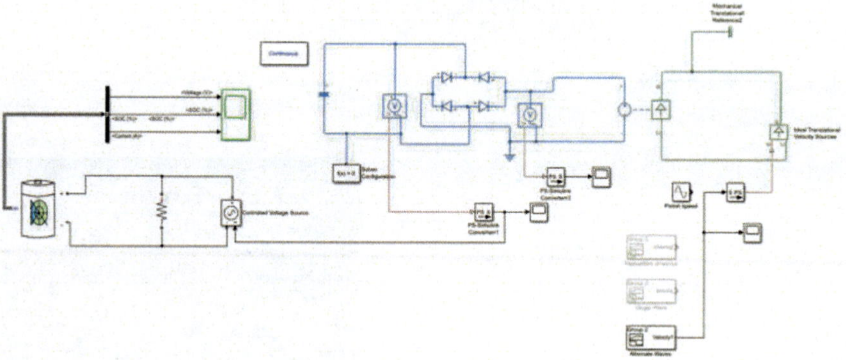

Fig. 8 Battery charging circuit

polarity, i.e. positive output only. Rectifier converts the AC voltage to pulsating DC, which is then smoothened by filter circuit.

3 Results

Figure 9 shows comparison between analytical (plotted using MATLAB) and computational (CONVERGE CFD) results and computational results show rapid dampening as compared. Piston position trend could be seen reciprocating and with damped amplitude. Similar trend was observed with piston velocity, maximum when combustion was there and then damping its amplitude with respect to time. We can see that the exact values did not match in simulation and analytical results. Referring Sect. 2.2,

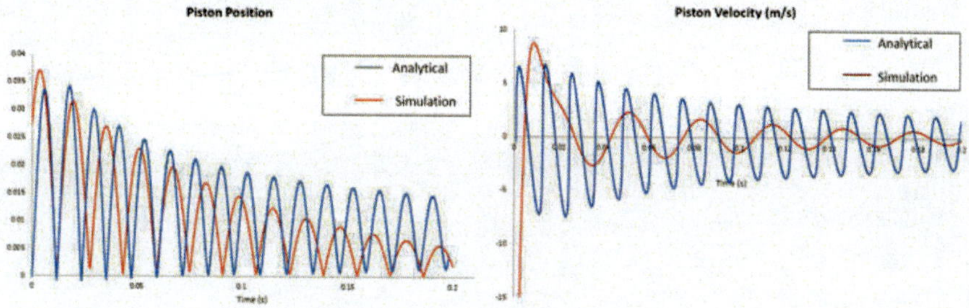

Fig. 9 Variation trend similarity between computational and analytical method

damping factor 'c' was assumed to be including just the friction and electromagnetic force but, the gas energy absorption–release process also plays important role in resisting piston motion; hence, more damping acts on the piston in reality. Also, there were assumptions about (a) the value and way combustion force acts on piston; (b) no loss whilst energy transfer between kinetic energy of piston and potential energy of springs. Considering this gas energy exchange process and others during formulating the problem analytically would involve complicated gas compressibility relations and so our system got too complex to solve analytically. Hence, system was very simplified with the explained assumptions so that we could solve analytically. But, when simulation was carried out using CFD, we could integrate model with different physics related to real gas, combustion and energy exchange phenomenon. Hence, in simulation, a real like situation was possible to analyse; therefore, the difference is observed simulation and analytical.

4 Discussion

The velocity was given as input to the Simulink circuit. Different studies were carried with different velocity input profiles. For simplicity, velocity in form of a sin wave was given as input and circuit model was calibrated with reference values. After successful calibration, obtained velocity trends were given as input. As explained earlier, minimum velocity of 0.467 m/s was required for charging condition of battery. Hence, velocity below this value was considered as zero (non-productive). As shown in Fig. 9, the minimum velocity is met after six cycles of reciprocation when only single combustion cycle is given. So mathematically speaking, engine should be fired at this point so that again similar velocity profile gets repeated. Two of the studies are shown below in Figs. 10 and 11.

To simulate simultaneous charge and discharge condition, a resistance was added as load. In first study, only one combustion cycle was given and obtained velocity curve was directly given to generator circuit. As long as velocity was above the threshold value, battery was getting charged but when velocity dropped, battery started showing continuous discharge, and discharged fully after some time.

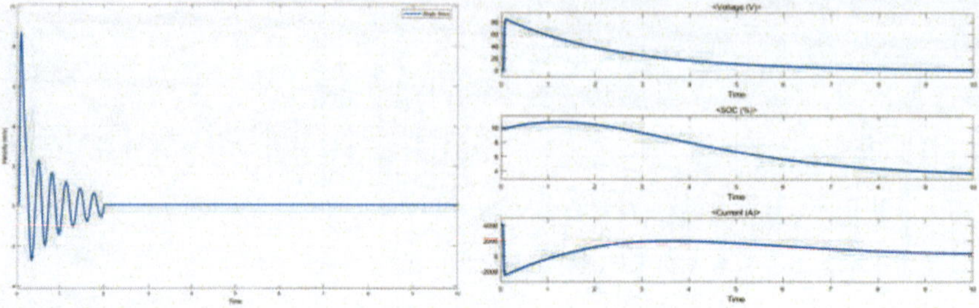

Fig. 10 Study 1 results

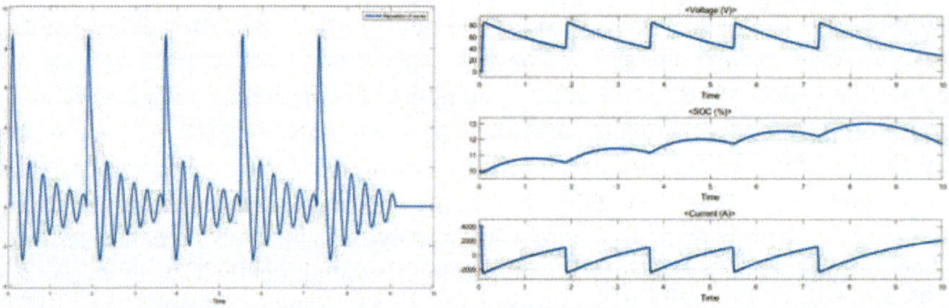

Fig. 11 Study 2 results

If engine could be fired intermittently, voltage can be maintained above required value and that too with lesser fuel unlike continuous firing condition. This logic was applied during study 2 as shown in Fig. 11. Engine was supposed to be fired when velocity drops below the minimum value; hence, increasing the velocity value above the threshold which in turn will increase induced voltage and hence can charge the battery.

With reference to Fig. 9, repeating combustion after six reciprocation cycles is recommended after every six cycles, velocity behaviour will repeat itself. Simulation of such repeated combustion was done, and results were obtained as seen from Fig. 11, voltage, current are fluctuating, state of charge has been increased by 1.7%.

Efficiency of system:

Overall efficiency of the system was calculated considering fuel input to engine and total battery energy stored due to charging during five combustion cycles.

Efficiency of the system was calculated to be **49.28%**.

Fuel requirement calculation:

For the simulation time, fuel was injected 5 times and 1.7% of usable battery capacity was charged.

Hence, battery capacity per kg of fuel can be calculated as,

$$= \frac{\frac{1.7}{100} \times 1.84}{5 \times 0.0005534} \text{ kWh/kg of fuel}$$

Fuel efficiency of the FPLG system came out as **0.1 litres per kWh.**

As experiment data of conventional generator with similar conditions of input and output was not available, direct comparison was not possible. But, general thumb rule for diesel engine generators is 0.4 l of diesel per kWh is used [13] and proposed system performing approximately at 0.1 L diesel per kWh hence, more efficient.

5 Conclusions

In this work, the literature review on free-piston engine has done to study about free-piston engine, its modelling methodologies. Fluid structure interaction methodology has been validated with the analytical calculations of Newton's motion equation. Simulation setup for DI combustion has been done and these results were validated with available experimental data.

- FPLG system is feasible and showing fair efficiency. This system can be developed as a portable unit that can be used as range extender for battery electric vehicles from efficiency point of view.
- Fluid structure interaction methodology has given the good results for Newton's motion equation compared with the analytical calculations. This methodology paves the new way to model the free-piston engine
- The operating frequency of free-piston engine is high compared with the crankshaft engine. This will give the more electric power output from the free piton linear generator.
- Starting mechanism needs to be studied. How much initial motor force to be invested in starting the engine and how many reciprocations are required by piston to reach the required temperature and pressure conditions will affect overall efficiency of FPLG.

Acknowledgements I would like to thank faculty of College of Engineering, Pune and staff of MTU India Pvt. Ltd., EARC, Pune for extending their support in this research.

References

1. Pateras PR (1941) Motor compressor of the free piston type
2. Goldsborough SS, Van Blarigan P (2010) A numerical study of a free piston IC engine operating on homogeneous charge compression ignition combustion. SAE Tech Pap Ser 1. https://doi.org/10.4271/1999-01-0619
3. Feng H, Song Y, Zuo Z et al (2015) Stable operation and electricity generating characteristics of a single-cylinder free piston engine linear generator: Simulation and experiments. Energies 8:765–785. https://doi.org/10.3390/en8020765
4. Mikalsen R, Roskilly AP (2007) A review of free-piston engine history and applications. Appl Therm Eng 27:2339–2352

5. Jia B, Mikalsen R, Smallbone A, Roskilly AP (2018) A study and comparison of frictional losses in free-piston engine and crankshaft engines. Appl Therm Eng 140:217–224. https://doi.org/10.1016/j.applthermaleng.2018.05.018
6. Jia B, Smallbone A, Feng H et al (2016) A fast response free-piston engine generator numerical model for control applications. Appl Energy 162:321–329. https://doi.org/10.1016/j.apenergy.2015.10.108
7. Mikalsen R, Roskilly AP (2010) The control of a free-piston engine generator. Part 1: fundamental analyses. Appl Energy 87:1273–1280. https://doi.org/10.1016/j.apenergy.2009.06.036
8. Kosaka H, Akita T, Moriya K et al. (2014) Development of free piston engine linear generator system part 1—investigation of fundamental characteristics. SAE Tech Pap Ser 1:. https://doi.org/10.4271/2014-01-1203
9. Xu Z, Chang S (2010) Prototype testing and analysis of a novel internal combustion linear generator integrated power system. Appl Energy 87:1342–1348. https://doi.org/10.1016/j.apenergy.2009.08.027
10. Goza A, Colonius T (2018) Modal decomposition of fluid–structure interaction with application to flag flapping. J Fluids Struct 81:728–737. https://doi.org/10.1016/j.jfluidstructs.2018.06.014
11. LI Q-F, Xiao J, Huang Z (2008) Parametric study of a free piston linear alternator. Int J Automtive Technol 11:111–117
12. Cawthorne WR, Clark N, Klein R, Nutter R (1999) Optimization of Brushless Permanent Magnet Linear Alternator for Use With a Linear Internal Combustion Engine
13. University of Calgary (2015) Diesel generator fuel efficiency. In: Energy education. Diesel generator

Prediction of Heat Losses in Scramjet Vitiator

Rocky Simon Pinto, S. Gagana, T. Sree Renganathan, S. M. D. Hamid Ansari, and Thiruchengode Mahalingam Muruganandam

Nomenclature

Pr	Prandtl's number,
Re	Reynold's number,
Nu	Nusselt's number,
L_e	Effective beam length,
V	Volume of the chamber,
A	Area of the chamber,
U	Velocity,
L	Length of the chamber,
h	Convection coefficient,
K	Thermal conductivity,
T_{ad}	Adiabatic temperature of the gas
T_w	Temperature of the wall,
T_O	Total temperature

R. S. Pinto (✉) · S. Gagana · T. Sree Renganathan · S. M. D. Hamid Ansari · T. M. Muruganandam
National Centre for Combustion Research and Development, Indian Institute of Technology Madras, Chennai, India
e-mail: aerok12939@gmail.com

S. Gagana
e-mail: gagana.satya@gmail.com

T. Sree Renganathan
e-mail: thanusr4@gmail.com

S. M. D. Hamid Ansari
e-mail: hameedansari789@gmail.com

T. M. Muruganandam
e-mail: murgi@ae.iitm.ac.in

A. K. Gupta et al. (eds.), *Advances in IC Engines and Combustion Technology*, Lecture Notes in Mechanical Engineering,
https://doi.org/10.1007/978-981-15-5996-9_55

P_O Chamber pressure
A^* Throat area,
Q_R Heat loss due to radiation,
Q_C Heat loss due to convection,
μ Dynamic viscosity,
ϑ Kinematic viscosity,
R Specific gas constant,
$\dot{m}$ Mass flow rate,
γ Specific heat ratio,
ε_g Emittance,
ε_w Emissivity,
σ Stefan–Boltzmann Constant

1 Introduction

Vitiation of air is one of the many ways to preheat the air used for studying supersonic combustion [1]. The supersonic combustion-based shock-induced ignition is largely dependent on the total temperature achieved at the preheater [2]. A hydrogen–oxygen-based vitiation is used by [3] for studying supersonic combustion. The mixture after vitiation consists considerable amount of water vapour [1]. The pressure of water vapour leads to huge radiation losses. The jet at the exit of hydrogen–oxygen burner entering the preheater forms large recirculation zone in the preheater. The jet from individual burner entering the preheater causes convection to occur from the gas to the surface of the preheater [1]. These losses are not linear with increasing hydrogen mass flow rate as there is a strong coupling between the convective and radiative heat losses [4].

Due to heat transfer losses, it is important to predict the final total temperature after losses in order to design experiments to achieve desirable total temperature for a given equivalence ratio. Gövert et al. [5] predicted numerically the heat losses of a confined premixed jet flame. Broatch et al. [6] predicted the heat losses in an internal combustion engine. Costa et al. [7] predicted the heat losses in rotatory engine. Ince et al. [8] studied the heat loss effects on turbulent flow in a cavity, and found that convection heat loss does significantly reducing the total temperature. Tsuji et al. [9] found the heat transfer due to convection near the wall using the mean velocity and mean temperature of the hot gases. Saha et al. [10] studied the heat transfer behaviour of a vertical wall subjected to sudden heat rise and found that convection heat transfer increases rapidly as the system reaches a steady state. Pilkington et al. [11] found that convective heat transfer losses could not be predicted accurately using RANS simulation. The Liftoff model and the Kandlikar model can be used to predict the heat fluxes on smooth surfaces [12]. Predicting radiative heat transfer losses is quite difficult as we need to predict the emittance of the gas [13]. Also, neglecting radiation-based heat transfer losses significantly affects the predicted temperature

[13]. The weighted sum of grey gases model (WSGGM) used by Fordoei et al. [14] gives an accurate prediction of emittance of gases required for heat transfer losses. A better way of predicting emittance is to use Hottel's emissivity charts, which give emittance of a gas containing water vapour at 1 atm pressure [15]. For elevated pressures [16] give approximate value of emittance for total pressures up to 40 bar based on the water vapour partial pressure.

In this paper, an easier and versatile method is discussed for predicting the convective and radiative heat losses. These predictions can be used to calculate the total temperature after all the losses, and help in choosing the correct mass flow rates to achieve the desired total temperature for studying supersonic combustion. These predicted temperatures are validated with experimental results.

2 Experimental Setup

The preheater and the supersonic combustor are designed at National Centre for Combustion Research and Development (NCCRD), IIT, Madras. The experimental setup shown in Fig. 1 consists of a preheater in which air was heated by a hydrogen-based vitiation heater with oxygen replenishment. Two spark igniters are used for ignition. The mass flow rate for each of the gases is electronically controlled using a pressure feedback system which includes a pressure regulating valve with mass flow metre and a pressure transducer, along with a choked orifice. Downstream of the preheater, a contoured nozzle, is used to accelerate the flow to a Mach number of 2.5.

The preheater showed in Fig. 1 consists of 8 burners of which 6 burners are used to vitiate the incoming air. Figure 2 shows the expected lateral flow field by the hot jet from the burners. Since the velocity of the hot gases is very high compared to the

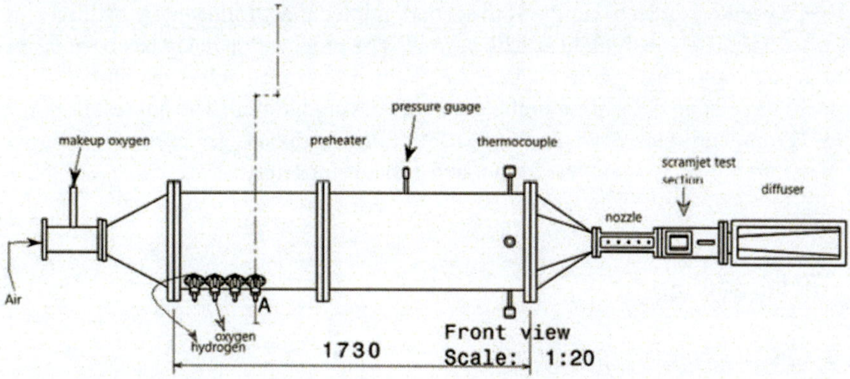

Fig. 1 Experimental setup

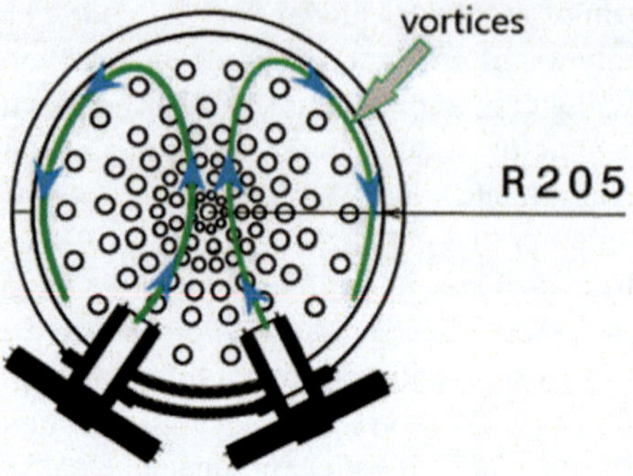

Fig. 2 Preheater cross-section with vortices

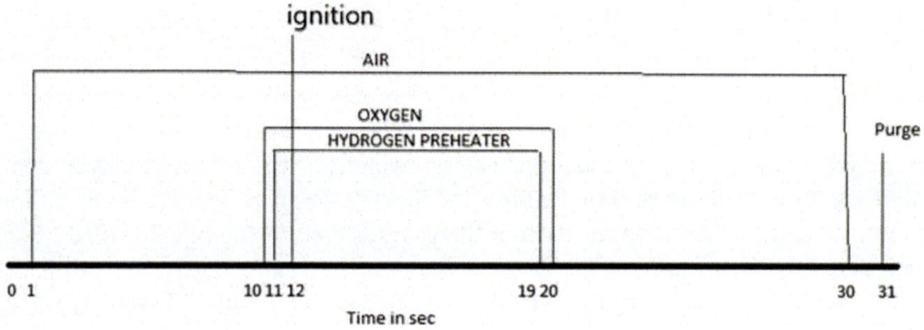

Fig. 3 Timing diagram

axial velocity of air, there will be lateral recirculation zones which help in mixing of the hot gases with air.

The experiment timing diagram is shown in Fig. 3. Air starts at 1 s and runs for $t =$ 29 s. Oxygen was injected at the 10th second, immediately followed by hydrogen at 11th second. They are ignited at 12th second. The propagation delay has been taken into account and thus ignition happens just when hydrogen enters the burner. This ensures there is no detonation in the preheater during ignition. The hot test lasts for 10 s. Hydrogen is shut down first at $t = 19$ s followed by oxygen at $t = 20$ s. Purge cycles start at $t = 31$ s and last for about 5 s to ensure safety.

3 Methodology

3.1 Convection Heat Losses

The velocity of the combustion products at the exit of an individual burner is significant enough to cause convective heat losses. The jet at the exit of the burner crosses the incoming air normally and impinges at the top of the preheater. This jet forms two large recirculation zones as shown in Fig. 2. The fluid elements at the periphery of the recirculation zone flow over the inner wall of the chamber which can be approximated as flow over a flat plate. Based on this approximation, the convective heat loss can be calculated as follows.

$$Q_{\mathrm{C}} = 2hA(T_{\mathrm{ad}} - T_{\mathrm{w}}) \tag{1}$$

$$h = \frac{N_{\mathrm{u}}\, K}{L} \tag{2}$$

The area considered is the product of preheater's half perimeter and the width of the preheater that experiences the hot jet exhausting the burners.

For 5 e5 < *Re* < 1 e7,

$$Nu = 0.037 Re^{0.8} Pr^{0.33} \tag{3}$$

For *Re* < 5 e5,

$$Nu = 0.0332 Re^{0.5} Pr^{0.33} \tag{4}$$

$$Re = \frac{UL}{\vartheta} \tag{5}$$

$$\mathrm{Pr} = \frac{\mu C_P}{K} \tag{6}$$

The velocity used in (5) is found using the density after combustion at the exit of the burner. All the properties used in the calculations are found using GASEQ for the film temperature, which is the average of the adiabatic temperature and the initial wall temperature.

3.2 *Radiation Heat Loss*

Along with convective heat loss, radiation heat loss also occurs. The same was calculated using the following equations.

$$Q_{\mathrm{R}} = \sigma\left(\varepsilon_{\mathrm{g}} T_{\mathrm{ad}}^4 - \varepsilon_{\mathrm{w}} T_{\mathrm{w}}^4\right) A_{\mathrm{h}} \tag{7}$$

The emittance of the mixture used in (7) is a function of adiabatic temperature of the combustion, chamber pressure, and partial pressure of water vapour in the combustion product [14–16]. The partial pressure of water vapour and adiabatic temperature was found using GASEQ. The emittance also depends on the product of effective beam length and partial pressure of water vapour [14–16]. The effective beam length is calculated using (8)

$$L_e = 3.6\frac{V}{A} \tag{8}$$

3.3 *Total Temperature Prediction*

The total temperature is calculated in two ways: one, after the heat losses and other using the chamber pressure and the mass flow rates.

Based on Heat Losses.

The total temperature after heat losses is predicted as follows

Initially, the adiabatic temperature is acquired for the given mass flow rates of air, oxygen, and hydrogen. The mixture properties are found using GASEQ for this temperature. Velocity used for convection is found using the density of mixture at the burner exit after combustion, and the mass flow rate in the burner. The heat loss is found from (9), where Q_A is the actual heat release.

$$Q_L = Q_A - Q_C - Q_R \tag{9}$$

The total temperature after losses is calculated using (10).

$$T_{OL} = T_{ad} + \left(\frac{Q_L}{C_{Pmix}\dot{m}}\right) \tag{10}$$

Based on Mass Flow Rates

The chamber pressure and mass flow rates from the experimental test are used to calculate the total temperature. The total temperature is calculated using (11).

$$T_o = \left(\frac{P_o A^*}{\dot{m}}\right)^2 \frac{\gamma}{R}\left(\frac{2}{\gamma+1}\right)^{\frac{\gamma+1}{\gamma-1}} \tag{11}$$

4 Results and Discussion

Experimental tests were done for varying hydrogen mass flow rates. The chamber pressure, hydrogen injection pressure, and oxygen injection pressure varying with time are shown in Fig. 4. The pressure in the chamber is denoted as 'air' in the figure.

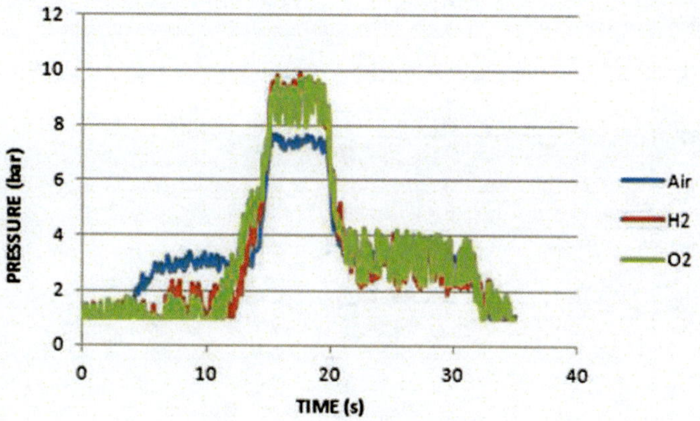

Fig. 4 Chamber pressure and injection pressure variation with time

Table 1 Comparison of predicted total temperature and mass flow rate-based total temperature

Test number	T_L (K)	T_N (K)
1	829	755
2	934	867
3	844	749
4	1069	1002
5	1095	1010

The airflow stabilizes after $t = 6$ s. The pressure rises to 7.5 bar at the 12th second due to ignition in the preheater.

The total hot test is conducted only for 10 s as seen in the figure. The predicted total temperature along with mass flow rate-based total temperature is shown in Table 1.

The predicted temperature (Tl) and mass flow rate-based temperature (Tn) are compared in Fig. 5 for five tests. It can be seen that both are matching within the range of 35–95 K. An S-type thermocouple was used to measure the temperature (Tt) for test 5, and it gave a value of 1044 K which closely matches with predicted total temperature (1095 and 1010 K). Further tests will be carried out to achieve a total temperature of 1500 K.

5 Conclusion

This paper describes a method for predicting total temperature of a scramjet vitiator. The hydrogen–oxygen–air vitiator was inside a chamber which acted as a settling chamber for the scramjet. The hot gases from the hydrogen–oxygen burner mixed with the air but the actual temperatures were lower than expected. This paper outlined

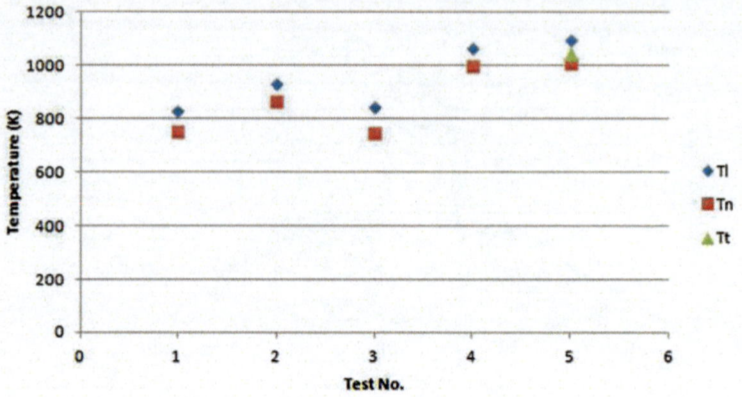

Fig. 5 Predicted temperature for various test and its comparison with mass flow rate-based temperature

a method for quickly estimating the temperature achievable by taking into account convective and radiative losses. The convective losses were found by approximating as a hot gas flowing over a plate radiative losses were accounted for by using emittance based on water vapour concentration. It was found that the predictions were within 95 K of the actual values which was again confirmed by the thermocouple.

Thus, this method can be used for predicting final temperature of the preheated air in any preheater of this type. The actual temperature required for scramjet is around 1550 K and it is not yet achieved as of now due to limitations in the flow control rig design which are being modified now. A total temperature of 1500 K will be achieved in the subsequent tests and supersonic combustion will be studied.

References

1. Pellett GL, Bruno C, Chinitz W (2002) Review of air vitiation effects on scramjet ignition and flameholding combustion process. In: Joint Propulsion Conference AIAA
2. Liang J-H, S W-T (2014) The vitiation effects of water vapor and carbon dioxide on the autoignition characteristics of kerosene. Acta Mechanica Sinica, 485–494
3. Tomioka S, Kohchi T, Masumoto R, Izumikawa M, Matsuo A (2009) Supersonic combustion with supersonic injection through diamond-shaped orifice. In: Joined propulsion conference and exhibit
4. Modest DHMF (2016) Radiative heat transfer in turbulent combustion systems: theory and applications. Springer International Publishing
5. Gövert S, Mira D, Zavala-Ake M, Kok JBW, Vázquez M, Houzeaux G (2017) Heat loss prediction of a confined premixed jet flame using a conjugate heat transfer approach. Int J Heat Mass Transfer 107:882–894
6. Costa TJ, Nickerson M, Littera D, Martins J, Shkolnik A, Shkolnik N, Brito F (2016) Measurement and prediction of heat transfer losses on the XMv3 rotary engine. SAE Int J Engines 9(4):2368–2380

7. Ince NZ, Launder BE (1995) Three-dimensional and heat-loss effects on turbulent flow in a nominally two-dimensional cavity. Int J Heat Fluid Flow 16:171–177
8. Tsuji T, Nagano Y (1988) Characteristics of a turbulent natural convection boundary layer along a vertical flat plate. Int J Heat Fluid Flow 31(8):1723–1734
9. Saha SC, Patterson JC, Lei C (2010) Natural convection boundary-layer adjacent to an inclined flat plate subject to sudden and ramp heating. Int J Thermal Sci 49(9):1600–1612
10. Broatch A, Margot X, Garcia-Tiscar J, Escalona J (2019) Validation and analysis of heat losses prediction using conjugate heat transfer simulation for an internal combustion engine. SAE Technical Paper 2019-24-0091
11. Pilkington A, Rosic BA, Tanimoto KB, Horie SB (2019) Prediction of natural convection heat transfer in gas turbines. Int J Heat Mass Transfer 141:233–244
12. Cao Z, Wu Z, Sundén B(2019) Heat transfer prediction and critical heat flux mechanism for pool boiling of NOVEC-649 on microporous copper surfaces. Int J Heat Mass Transfer 141:818–834
13. Tien C (1968) Thermal radiation properties of gases, advances in heat transfer, vol 5. Academic Press, New York, pp 253–324
14. Ebrhimi Fordoei E, Mazaheri K (2019) Numerical study of chemical kinetics and radiation heat transfer characteristics on the temperature distribution in the oxy-fuel combustion. Heat Mass Transfer 55(7):2025–2036
15. Hottel H, Sarofim A (1967) Radiative transfer, McGraw-Hill
16. Alberti M, Weber R, Mancini M (2016) Re-creating Hottel's emissivity charts for water vapor and extending them to 40 bar pressure using HITEMP-2010 data base. In: Combustion and flame, pp 141–153

CH-PLIF in Horizontal Slab PMMA Laminar Flame

Poorva Shrivastava, Deepika Ram, and Thiruchengode Mahalingam Muruganandam

1 Introduction

Flame spread is an area of acute research and requires careful attention to evolve the methods for fire safety. In order to develop effective and efficient fire safety measures, a behavioural study of the causes pertaining to flame spread is demanded. This leads to finding out the outward reasons for fires occurring in buildings, transports and different places. Almost all these fires occur due to the presence of solid fuel in the vicinity which constitutes complex heat and mass transfer interactions.

Polymethylmethacrylate (PMMA) is a versatile material that is useful in the above-listed applications. It is a polymer used in electronic instruments, appliances, household goods, glass, lenses, furniture, etc. Combustion of polymer is a phenomenon involving complex chemical and thermal reactions which necessitate a comprehensive study of fire behaviour and constituting chemical species getting produced during the reaction.

Development of non-interference and non-intrusive techniques made it simpler to obtain the required data for classifying the fire behaviour. Laser diagnostics' one such technique can be used to determine velocity and temperature fields, heat release rate, etc. and thus effectively defines the thermal and chemical structure of the flame. CH planar laser-induced fluorescence is a technique based on the finding the concentration of CH* radical in the flame which marks the location of primary reactions zones of flame.

P. Shrivastava · D. Ram · T. M. Muruganandam (✉)
National Centre for Combustion Research and Development, Indian Institute of Technology Madras, Chennai, India
e-mail: murgi@ae.iitm.ac.in

A. K. Gupta et al. (eds.), *Advances in IC Engines and Combustion Technology*, Lecture Notes in Mechanical Engineering,
https://doi.org/10.1007/978-981-15-5996-9_56

2 Literature Review

PMMA (polymethylmethacrylate) is a popular choice for fire research and diagnostics as it is a great substitute for materials in building construction due to its long-chain polymers. PMMA is also used as it is non-charring and does not flow like a liquid during combustion.

PLIF (planar laser-induced fluorescence) is a non-intrusive, non- interference diagnostics technique which helps in determining CH, OH, or PAH (polycyclic aromatic hydrocarbons) concentrations. There have been a lot of studies over the past years on the flame speed, flame propagation and burning behaviour of PMMA.

Pello and Williams [1] work was on mechanisms by which laminar flames spread over flat surfaces of PMMA in directions ranging from downwards to horizontal. Following this, Orloff et al. [2] studied the 2D upward flame spread and subsequent steady turbulent burning.

Apte et al. [3] studied the concurrent fire spread on horizontal PMMA and found that the flame propagates in two successive modes: boundary layer, where the pyrolysis mass loss flux is constant and the second, plume mode, happens sooner with lower wind velocities, is marked by a rapid increase with flame–fuel surface heat flux and pyrolysis flux which are independent. Tewarson and Ogden [4] studied the fire behaviours in PMMA slabs and cylinders. They developed a new technique to calculate the pyrolysis height and flame spread rate from chemical and convective heat release rates using conventional flammability apparatus. Gollner et al. [5] studied the upward flame propagation on a long lab of PMMA. The experimental and numerical results for auto-ignition and pilot ignition of PMMA in cone calorimeter by Tsai et al. [6] and the sample width and thickness effects have been studied by Jiang et al. [7].

Introduction of diagnostics techniques produced some qualitative works in the last decade, and Aldén [8] gave an overview of all the possibilities and limitations in laser diagnostic techniques for the understanding of combustion processes. Hebert et al. [9] employed laser-induced incandescence (LII) to measure soot concentration for vertically burning PMMA slab. Chen et al. [10] worked on bench-scale lateral flame spread tests on PMMA with various widths experimentally and theoretically. They found that the convective heat feedback was dominant in heat transfer for smaller widths less than 8 cm. With the increase in widths, convective heat transfer was weakened but the radiant heat transfer was enhanced. Rosell et al. [11] studied the simultaneous PLIF measurements OH, CH, CH_2O, and toluene so as to understand the structure of a turbulent premixed methane/air jet flames in the flamelet and thin-region zones regime. Sarma et al. [12] worked on PMMA slabs with non-planar surfaces in the form of steps. Time-dependent % mass loss, surface regression, OH*, CH* chemiluminescence, and PLIF of soot and PAH were determined.

Recently, Korobeinichev et al. [13] experimentally studied the flame spread rate, temperature distribution, and spatial variation of the species concentration over the surface of horizontally placed slabs of PMMA species in still air. The effect of ullage and structure of the flames on vertical slab has been studied by Ranga et al. [14, 15]

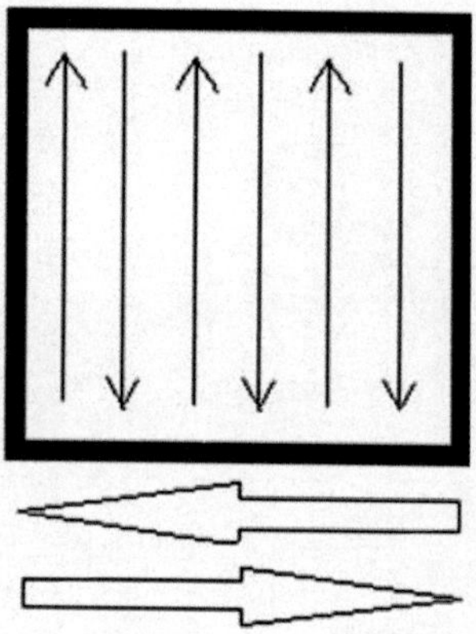

Fig. 1 Pattern for igniting PMMA surface

Significant efforts have been put towards studying the heat transfer interactions and soot emissions; however, the complexity of the problem prevented a thorough understanding of the phenomenon. Present work aims to fill the gaps by studying the CH concentrations during the burning of PMMA.

3 Experimental Setup

3.1 PMMA Arrangement

A square block of PMMA measuring 28 × 28 × 5 mm is used for combustion. To allow the regression of surface, it is coated with silicone grease on sides and wrapped in aluminium foil all through except for top surface. For extra precaution, it is kept in a square aluminium container of sides 5 mm more than that of PMMA slab and of thickness 1 mm less than that of the same. In order to bring in the equal distribution of air all around the slab, it is placed on top of the metal breadboard which allows the air to flow from the bottom as well through the holes present in it. The slab is ignited uniformly with the help of butane blowtorch in a regular pattern (Fig. 1).

3.2 PLIF Setup and Calibration

For PLIF experimentation, Nd: YAG (neodymium-doped yttrium aluminium garnet) laser (Quantel Qsmart 850) generating 50 μs pulses at 355 nm (3ω) with 5 Hz repetition rate is used. Maximum laser pulse energy is 100 mJ. Laser sheet is formed

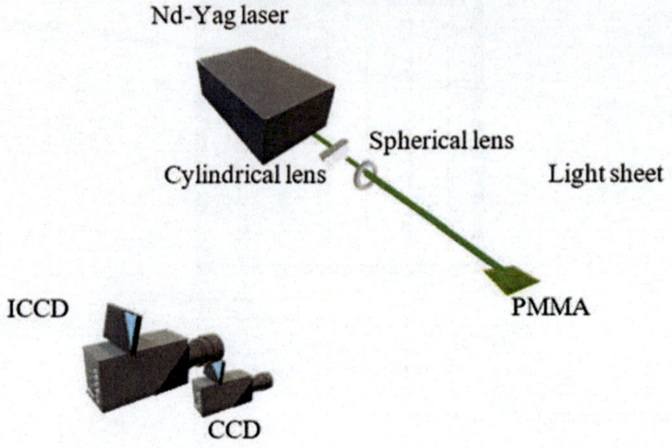

Fig. 2 Schematic of experimental setup

using two lenses viz cylindrical and spherical which are installed at little distance from the laser output.

An intensified CCD (Princeton Instruments PIMAX 3) is placed perpendicular to the laser sheet to capture the 2D CH*-PLIF images. A 50 mm zoom lens with aperture f/8 is used to record the flame signal. Resolution of the studied field is 0.36 mm/pixel on a maximum of 1024 × 1024 pixels. Camera is synchronised with the laser pulse using gate delay of 34 ns. Band pass filter of 430 ± 5 nm is used to eliminate the surplus signals such as OH and PAH. Gate width of 500 ns and intensifier gain of 95× are chosen to allow a good signal-to-noise ratio. Camera is focussed on the centre-plane of the PMMA. Total 100 frames are acquired at the rate of 5 fps.

Simultaneous flame image is captured using a CCD camera (PCO Pixelfly) installed with a 50 mm lens placed perpendicular to the laser sheet below the ICCD (Fig. 2) and focussed at the centre-plane as well. Aperture is kept at f/8. Resolution of the studied field is 0.17 mm/pixel on a maximum of 1392 × 1040 pixels. Camera is triggered using the q-switch of the laser. It has an intrinsic delay of 5.6 μs which allows capturing the full flame signal ahead of ICCD. Schematic for the experimental setup is given in Fig. 2 which shows the location and positioning of laser and cameras.

PMMA surface is ignited for 60 s evenly using a blowtorch. Data is acquired at 90, 240 and 380 s after the ignition. Additionally, an image is acquired in between the runs to make sure that flame alone does not give any signal to ICCD.

3.3 *Post-Processing*

Post-processing of data is done using MathWorks MATLAB. All the images have been subtracted with the reference background image for both ICCD and CCD. 100

flame emission images followed by a series of 100 images PLIF and PAH signal are recorded and averaged over time. They have been multiplied with a single factor each to get a good display. In doing this, ICCD images tend to show unwanted salt and pepper noise which is corrected using a median filter of size 3. Further, they are being cropped in order to visualise the flame clearly.

4 Results and Discussion

Simultaneous images of flame emission and CH-PLIF are acquired at different time intervals for a PMMA slab of 28 × 28 × 5 mm. Images are acquired at 5 Hz and averaged over 20 s (100 images). Data is acquired at 90, 240 and 380 s after the ignition at the centre-plane of the flame. Results for the first two instances are displayed after post-processing in Fig. 3a–d.

It can be seen that the flame has a simple inverted conical shape like a simple diffusion flame. The flame had a small swaying motion due to slight cross flow in the room and thus was averaged over 20 s duration. The CH-PLIF images confirm that the flame is hollow, and the region of flame luminosity matches with the CH-PLIF region.

It can be observed that the flame grows in height from 90 to 240 s tending towards more sharp-pointed conical shape. This is due to larger entrainment setup by the well-established flame around itself. The CH-PLIF image shows that the flame has a slight unsymmetry, and thus, one leg of the flame is longer than the other. It is evident that the CH-PLIF region is within the luminous region of the flame. This suggests that the CH radical is dominant in the same region as the soot particles, which give the luminosity.

At $t = 380$ s, the height of the flame reduces, and the flame is very shallow in shape. It was blocked by the burnt silicone grease ash at the viewing side; thus, no flame was found at this time. Thus, the data is not presented for this time instance.

Normalised intensity profiles of the flame emission and CH-PLIF data are plotted simultaneously at three different heights ($h = 38$, 50, and 75% of the flame height) at $t = 90$ s. These are presented in Figs. 4, 5 and 6. The red curves correspond to flame emission intensity variation, and the blue curves give the CH-PLIF intensities. It can be seen for the 38% case that, the intensity plot has two peaks, suggesting that there are two legs of the flame. It can also be seen that the intensity is unsymmetric which shows that the flame shape is unsymmetric.

Figure 5 shows the intensity plots at a height of 50% of flame height. It can be seen again that there are two peaks in the CH-PLIF intensity, saying that the flame is still having a hollow region. However, the flame luminosity is not having a clear gap in the middle. This is possibly because of the three-dimensional flame being projected into the camera. It may also be because of soot being high in concentration in the core of the flow and the whole middle region giving black body emission. However, it is again seen that the CH field is in the same region as the luminous zone; however, the CH does not fill the whole volume.

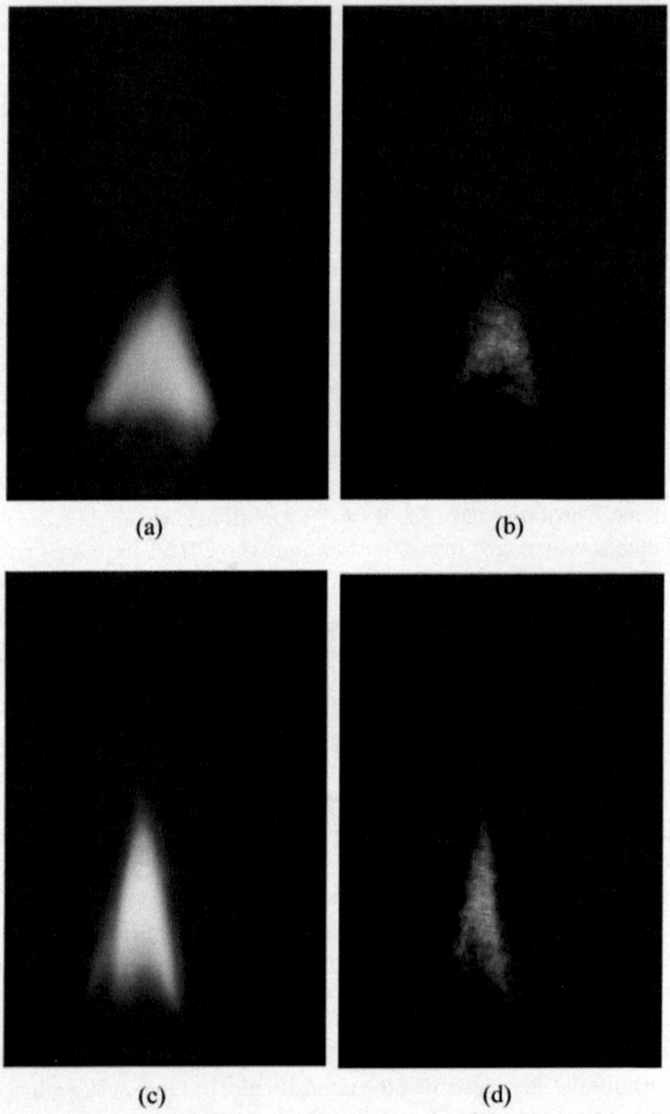

Fig. 3 **a** and **c** Flame emission at 90 and 240 s, **b** and **d** CH-PLIF at 90 and 240 s

Figure 6 shows the CH-PLIF and flame luminosity intensity plots at a height of 75% of the flame height. It can be seen now that the CH-PLIF peaks have merged suggesting that both the sides of the flame are starting to overlap at around this height. The flame image intensity is also having a single peak that coincides with this.

These observations confirm that the CH field is within the flame luminosity region and is thinner than the region of soot luminosity. It is also observed that the CH field is more closer to the outer edge of the luminous region. It is well known that the soot is formed in the fuel-rich side of the flame, and the black body emission from the flame gives the luminosity of the flame. These together suggest that the CH radical is dominant closer to the flame on the rich side of the flame. This observation can play an important role in validating various models for fire simulation.

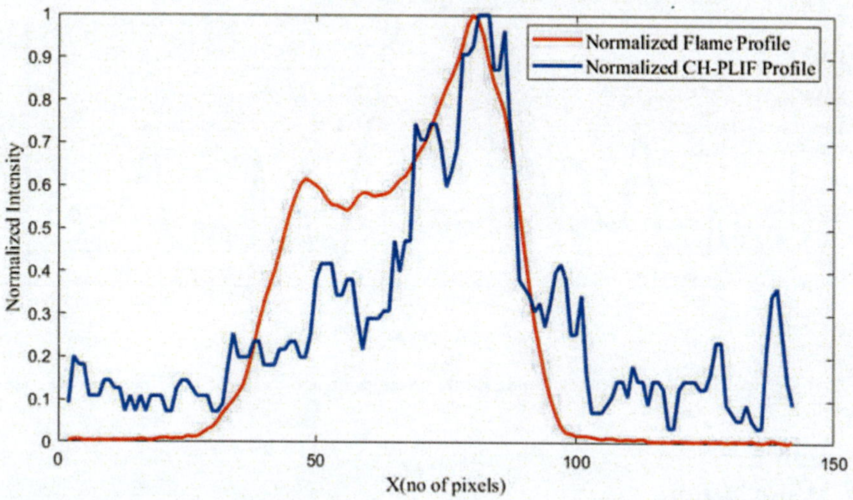

Fig. 4 Normalised flame intensity profile for flame emission and CH-PLIF at 38 per cent of flame height at 90 s

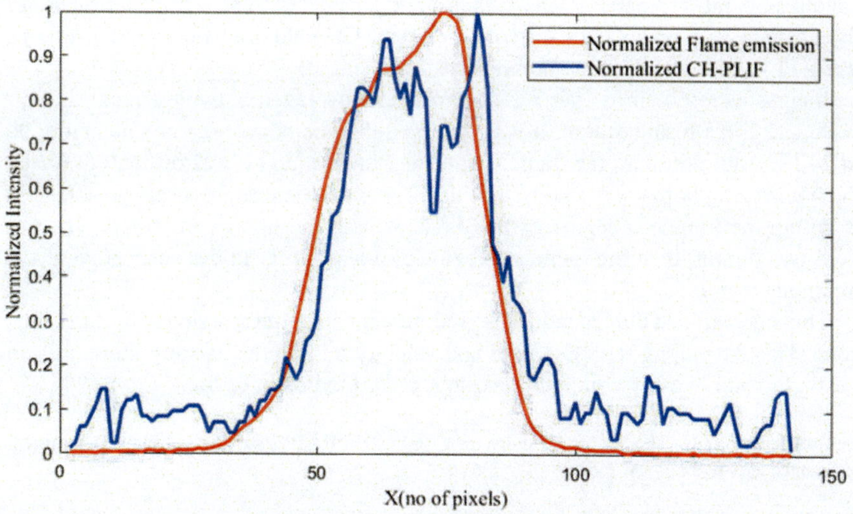

Fig. 5 Normalised flame intensity profile for flame emission and CH-PLIF at 50 per cent of flame height at 90 s

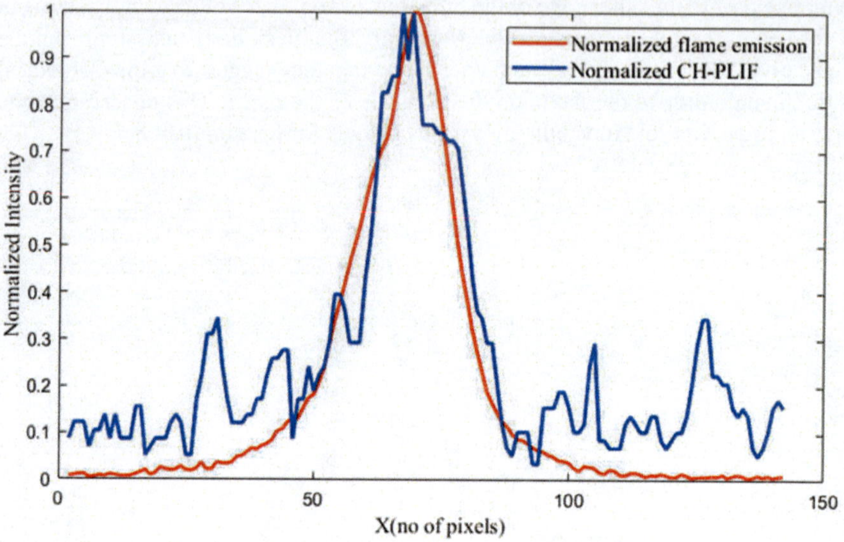

Fig. 6 Normalised flame intensity profile for flame emission and CH-PLIF at 75 per cent of flame height at 90 s

5 Conclusion

Prediction of fire is still considered an art because of the numerous tuning coefficients in the models used. This is because of very scarce data in the open literature on any canonical flame geometry. PMMA flames are considered to be a good surrogate for fire on many substances. This work measured the CH field in a laminar PMMA flame using PLIF, along with flame luminosity.

Images were obtained over a 20 s period and averaged at two different times, t = 90 and 240 s from ignition. It was observed that the flame became taller from 90 to 240 s. And after that, the flame diminished in size, and it was difficult to image the flame using both the methods. The data presented here suggest that the CH field is within the luminous region of the flame at both the times considered. The CH field was thinner than the luminous region and was closer to the outer edge of the luminous region.

This suggests that the CH radical is predominant in the outer region of the luminous zone. This data along with OH field and velocity field in the existing literature can be used to validate any computational model for prediction of fire.

Acknowledgements The National Centre for Combustion R&D is supported by the Department of Science and Technology, India.

References

1. Pello FA, Williams FA (1975) Laminar flame spread over PMMA surfaces. Symp Int Combust Proc 15:217–231. https://doi.org/10.1016/S0082-0784(75)80299-2
2. Orloff L, de Ris J, Markstein GH (1975) Upward turbulent fire spread and burning of fuel surface. Proc Combust Inst 15:183–192. https://doi.org/10.1016/s0082-0784(75)80296-7
3. Apte VB, Bilger RW, GreenJ AR, Quintiere G (1991) Wind-aided turbulent flame spread and burning over large-scale horizontal PMMA surfaces. Combust Flame 85(1–2):169–184. https://doi.org/10.1016/0010-2180(91)90185-E
4. Tewarson A, Ogden SD (1992) Fire behavior of polymethylmethacrylate. Combust Flame 89:237–259. https://doi.org/10.1016/0010-2180(92)90013-F
5. Gollner MJ, Huang X, Cobian J, Rangwala AS, Williams FA (2013) Experimental study of upward flame spread of an inclined fuel surface. Proc Combust Inst 34:2531–2538. https://doi.org/10.1016/j.proci.2012.06.063
6. Tsai TH, Li MJ, Shih IY, Jih R, Wong SC (2001) Experimental and numerical study of autoignition and pilot ignition of PMMA plates in a cone calorimeter. Combust Flame 124:466–480. https://doi.org/10.1016/S0010-2180(00)00219-4
7. Jiang L, Miller CH, Gollner MJ, Sun JH (2017) Sample width and thickness effects on horizontal flame spread over a thin PMMA surface. Proc Combust Inst 36:2987–2994. https://doi.org/10.1016/j.proci.2016.06.157
8. Aldén M, Bood J, Li Z, Richter M (2011) Visualization and understanding of combustion processes using spatially and temporally resolved laser diagnostic techniques. Proc Combust Inst 33:69–97. https://doi.org/10.1016/j.proci.2010.09.004
9. Hebert D, Coppalle A, Talbaut M (2013) 2D soot concentration and burning rate of a vertical PMMA slab using Laser-Induced Incandescence. Proc Combust Inst 34:2575–2582. https://doi.org/10.1016/j.proci.2012.06.096
10. Chen X, Liu J, Zhou Z, Li P, Zhou T, Zhou D, Wang J (2015) Experimental and theoretical analysis on lateral flame spread over inclined PMMA surface. Int J Heat Mass Transf 91:68–76. https://doi.org/10.1016/j.ijheatmasstransfer.2015.07.072
11. Rosell J, Bai XS, Sjoholm J, Zhou B, Li Z, Wang Z, Pettersson P, Li Z, Richter M, Alden M (2017) Multi-species PLIF study of the structures of turbulent premixed methane/air jet flames in the flamelet and thin-reaction zones regimes. Combust Flame 182:324–338. https://doi.org/10.1016/j.combustflame.2017.04.003
12. Sarma S, Chakraborty A, Manu NM, Muruganandam TM, Raghavan V, Chakravarthy SR (2017) Spatio-temporal structure of vertically spreading flame over non-planar PMMA surfaces. Proc Combust Inst 36:3027–3035. https://doi.org/10.1016/j.proci.2016.06.151
13. Korobeinichev O, Gonchikzhapov M, Tereshchenko A, Gerasimov I, Shmakov A, Paletsky A, Karpov A (2018) An experimental study of horizontal flame spread over PMMA surface in still air. Combust Flame 188:388–398. https://doi.org/10.1016/j.combustflame.2017.10.008
14. Ranga RHR, Korobeinichev OP, Harish A, Raghavan V, Kumar A, Gerasimov IE, Gonchikzhapov MB, Tereshchenko AG, Trubachev SA, Shmakov AG (2018) Investigation of the structure and spread rate of flames over PMMA slabs. Appl Therm Eng 130:477–491. https://doi.org/10.1016/j.applthermaleng.2017.11.041
15. Ranga RHR, Korobeinichev OP, Raghavan V, Tereshchenko AG, Trubachev SA, Shmakov AG (2019) A study of the effects of ullage during the burning of horizontal PMMA and MMA surfaces. Fire Mater 43:241–255. https://doi.org/10.1002/fam.2692

Numerical Analysis on Effect of Radiation in Laboratory-Scale Hybrid Rocket Motors

G. Krishna Prasad and Amit Kumar

Nomenclature

A, A_s	Pre-exponential factor for gas-phase reaction and pyrolysis of solid fuel
C	Species concentration
C^*	Characteristic velocity
$D_{i,m}$	Diffusivity of species i in the mixture
$D_{o,}$ D_i	Final and initial port diameter
D_ω	Cross-diffusion term temperature
E_a, E_{as}	Activation energy in gas-phase reaction and pyrolysis of solid fuel
$\widetilde{G}_k$, G_ω	Production of κ and ω
h, h_i	Sensible enthalpy of mixture and species
H_R, H_P	Heat of reaction for gas-phase reaction and pyrolysis of solid fuel
$\overrightarrow{J_j}$	Diffusion flux of species j
k	Turbulent kinetic energy
k_r	Reaction rate constant
L	Fuel length grain
P	Pressure
r_b	Regression rate
s	Ray direction
D_j	Diffusivity of species j
R_o	Radius of combustion chamber
R_u	Universal gas constant
T_s, T_∞	Surface temperature and initial temperature of fuel
v_x, v_r, v_z	Axial, radial and tangential velocity

G. K. Prasad · A. Kumar (✉)
National Centre for Combustion Research and Development, Department of Aerospace Engineering, Indian Institute of Technology Madras, Chennai 600036, Tamil Nadu, India
e-mail: amitk@ae.iitm.ac.in

A. K. Gupta et al. (eds.), *Advances in IC Engines and Combustion Technology*, Lecture Notes in Mechanical Engineering,
https://doi.org/10.1007/978-981-15-5996-9_57

x, r, z	Cylindrical coordinates in axial, radial and tangential direction
Y_i	Mass fraction of specie 'i'
Y_k, Y_ω	Dissipation of κ and ω
Γ_k, Γ_ω	Effective diffusivities of κ and ω
λ_{eff}	Effective thermal conductivity
λ_s	Thermal conductivity of the solid
μ	Molecular viscosity
ρ	Density
ω	Specific dissipation rate
W, W_i	Global reaction rate and consumption rate of species i
$\alpha_a, \varepsilon_e, \rho_s$	Absorption, emission and scattering coefficients
I	Radiation intensity

1 Introduction

Hybrid rocket motors have significant advantages over conventional liquid and solid rocket motors in mechanical simplicity, denser fuels, economical in costs and better safety. In large hybrid rocket motors, regression rate is less than the laboratory scale ones because the size scaling changes the location of the flame with respect to the surface and radiative heat transfer tends to offset this decrease [1]. Studies on performance of the hybrid rocket motors have usually ignored the effects of radiation [2–5], but radiation heat transfer may be a prominent mode of heat transfer as the temperature and size of the system increases substantially. Hence, the aim of the present study is to bring the effect of radiation in hybrid rocket motor performance.

2 Numerical Model and Solution Methodology

2.1 Geometry and Computational Domain

The schematic of hybrid rocket motor combustion chamber is shown in Fig. 1. Invariance in azimuthal direction is assumed to reduce the computational domain to a 2D axis-symmetric configuration. The computational domain comprises of a combustion chamber followed by a convergent nozzle.

A hybrid rocket motor with fuel grain length $L = 130$ mm is considered. The solid fuel grain has an external outer diameter of $D_o = 60$ mm and initial port diameter of $D_p = 20$ mm. The combustion chamber diameter considered in the computational domain is at an arbitrary instant of burning of solid fuel $D_i = 40$ mm. The convergent–divergent nozzle at the end of the combustion chamber has a throat diameter, $D_t = 17$ mm.

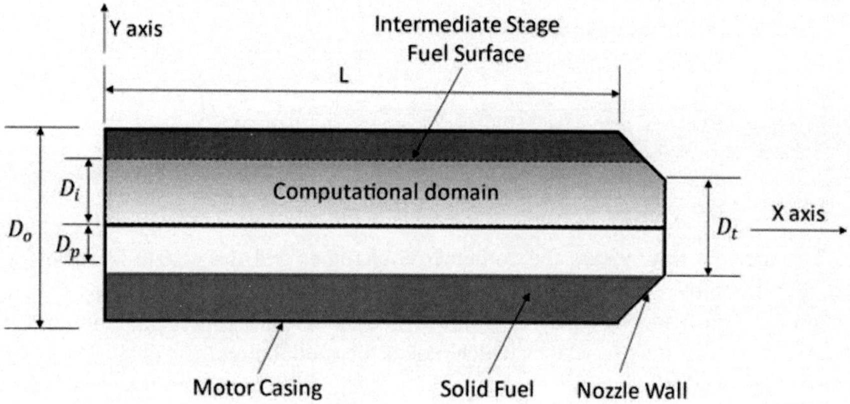

Fig. 1 Schematics of a single-port hybrid rocket motor

2.2 *Governing Equations of the Problem*

The transport processes inside the combustion chamber of a hybrid rocket can be appropriately expressed by the fundamental equations of flow, namely continuity, momentum, energy and species. The governing equations for 2D cylindrical coordinates are presented below.

Continuity equation

$$\frac{\partial \rho}{\partial t} + \frac{\partial(\rho v_x)}{\partial x} + \frac{\partial(\rho v_r)}{\partial r} + \frac{\rho v_r}{r} = 0$$

Axial momentum equation

$$\frac{\partial(\rho v_x)}{\partial t} + \frac{1}{r}\frac{\partial(r\rho v_x v_x)}{\partial x} + \frac{1}{r}\frac{\partial(r\rho v_r v_x)}{\partial r} = -\frac{\partial p}{\partial x} + \frac{1}{r}\frac{\partial}{\partial x}\left[r\mu\left(2\frac{\partial v_x}{\partial x} - \frac{2}{3}(\nabla.\vec{v}\,)\right)\right]$$
$$+ \frac{1}{r}\frac{\partial}{\partial x}\left[r\mu\left(\frac{\partial v_x}{\partial r} + \frac{\partial v_r}{\partial x}\right)\right]$$

Radial momentum equation

$$\frac{\partial(\rho v_r)}{\partial t} + \frac{1}{r}\frac{\partial(r\rho v_x v_r)}{\partial x} + \frac{1}{r}\frac{\partial(r\rho v_r v_r)}{\partial r} = -\frac{\partial p}{\partial r} + \frac{1}{r}\frac{\partial}{\partial r}\left[r\mu\left(2\frac{\partial v_r}{\partial r} - \frac{2}{3}(\nabla.\vec{v}\,)\right)\right]$$
$$+ \frac{1}{r}\frac{\partial}{\partial x}\left[r\mu\left(\frac{\partial v_x}{\partial r} + \frac{\partial v_r}{\partial x}\right)\right]$$

$$-2\mu\frac{v_r}{r^2} + \frac{2}{3}\frac{\mu}{r}(\nabla.\vec{v}\,) + \rho\frac{v_z^2}{r}$$

Tangential momentum equation

$$\frac{\partial(\rho v_z)}{\partial t} + \frac{1}{r}\frac{\partial(r\rho v_x v_z)}{\partial x} + \frac{1}{r}\frac{\partial(r\rho v_r v_z)}{\partial r} = \frac{1}{r}\frac{\partial}{\partial x}\left[r\mu\left(\frac{\partial v_z}{\partial x}\right)\right]$$
$$+ \frac{1}{r^2}\frac{\partial}{\partial r}\left[r^3\mu\frac{\partial}{\partial x}\left(\frac{v_z}{r}\right)\right] - \rho\frac{v_r v_z}{r}$$

The reactive flow inside the combustion chamber and the nozzle is turbulent; hence, a turbulence model appropriate for the present problem is required for the numerical simulation. The SST k–ω turbulence model uses k–ω formulation in the inner parts of the boundary layer which makes the model directly usable all the way down to the wall through the viscous sub-layer. The SST formulation also switches to a k–ε behaviour in the free-stream, and thereby avoids the common k–ω problem that the model is too sensitive to the inlet free-stream turbulence properties. Thus, in the present study, SST k–ω was used to predict turbulence.

SST k–ω equations

$$\frac{\partial(\rho k)}{\partial t} + \nabla.(\vec{v}(\rho k)) = \nabla.(\Gamma_k \nabla k) + \widetilde{G}_k - Y_k$$

$$\frac{\partial(\rho\omega)}{\partial t} + \nabla.(\vec{v}(\rho\omega)) = \nabla.(\Gamma_\omega \nabla\omega) + G_\omega - Y_\omega + D_\omega$$

Energy equation

$$\frac{\partial(\rho E)}{\partial t} + \nabla.(\vec{v}(\rho E + p))$$
$$= \nabla.\left(\lambda_{\text{eff}}\nabla T - \sum_j h_j \vec{J}_j + (\overline{\overline{\tau}}.\vec{v})\right) + H_R$$

Radiative transport equation

$$\frac{\mathrm{d}I}{\mathrm{d}s} = \varepsilon_e I_b - \alpha_a I - \rho_s I - \int_0^{4\pi} \rho I \frac{\varphi(\theta_i, \theta_e)}{4\pi} \mathrm{d}\theta_i$$

Species transport equation

$$\frac{\partial(\rho Y_i)}{\partial t} + \nabla.(\rho\vec{v}Y_i) = -\nabla.\left(\vec{J}_j\right) + W_i$$

2.3 Gas-Phase Reaction Kinetics

While for an accurate analysis of the combustion process inside the hybrid rocket motor, a detailed chemistry model may be required. The increase in computational costs with detailed chemistry is very time consuming and expensive. Hence, to reduce the computational costs, a global chemistry model with the thermodynamic effects of multi-species is incorporated. This is expected to yield appropriate prediction of the temperature field, heat and mass transfer processes. Considering HTPB pyrolysis produces butadiene as the main product, the global reactions is taken as,

$$C_4H_6 + 3.5O_2 \rightarrow 4CO + 3H_2O$$

$$CO + 0.5O_2 \rightarrow CO_2$$

The reaction rate W_r is given by a second-order reaction as

$$W = k_r C_{\text{fuel}} C_{\text{oxidizer}} = k_r C_{C_4H_6} C_{O_2}$$

where k_r is given by Arrhenius law $k_r = \text{A}e^{-E_a/R_u T}$

Here, A = 2.474E+11 and E_a = 1.2637E+08 J/kmol.

2.4 Fuel Pyrolysis Kinetics

The solid fuel pyrolytic reaction was modelled by the zeroth-order Arrhenius equation. The mass flux of fuel released due to the pyrolysis is specified by

$$\dot{m} = \rho_s r_b = k_r = \text{A}e^{-E_{as}/R_u T_s}.$$

2.5 Boundary Conditions

The various boundaries are shown in Fig. 2. In the present work, the fuel inlet is described as an interface where the mass and energy were conserved across the interface and mass flux is predicted. The reduced energy balance at the interface is given by

$$\lambda_{\text{eff}}\left(\frac{\partial T}{\partial y}\right)_g - \lambda_s\left(\frac{\partial T}{\partial y}\right)_s + \rho_s r_b H_p + q_{\text{rad}} = 0$$

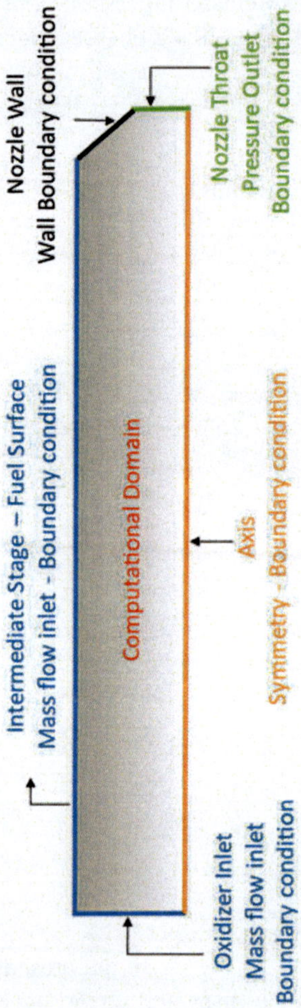

Fig. 2 Boundary conditions in the combustion chamber computational domain

The mass balance at the interface is given by

$$\dot{m} = \rho_s r_b = -\rho v$$

2.6 Numerical Method of Solution and Model Validation

The governing equations along with the boundary conditions were solved numerically in double precision. A density-based unsteady solver was used [6]. Spatial discretization for convective terms was implemented using second-order upwind scheme, while second-order central difference scheme was used for other terms in the governing transport equations. A second-order implicit formulation was used for temporal discretization. Second-order discretization schemes were opted as the velocity profiles and the local r_b along the axis did not change appreciably between the second and third-order discretization schemes. The discretized governing equations were iterated with the time step of 0.001 s and 50 iterations per time step. The converged solution for the laboratory-scale motor was achieved typically in around 1200-time steps. Convergence was ensured by the condition that the scaled residuals reduce were below a value of 1E−5 for all transport equations and did not change with time.

The numerical model for the reacting flow inside a hybrid rocket motor was adopted from Ph.D. thesis of C. Palani Kumar [7] which did not account for the radiation model. In order to validate the radiation model, the problem of a finite axis-symmetric cylindrical enclosure containing absorbing–emitting medium with constant temperature and constant absorption coefficient is simulated by using the Ansys Fluent to solve the radiative transport equation by the discrete ordinates method with S_6 approximation and were compared with the exact solutions presented by Dua and Cheng [8].

The problem under consideration is a cylindrical enclosure with radius of 1 m and height 2 m is cold ($T_w = 0$ K) and black ($\varepsilon = 1$), and the enclosed medium is hot as $T_g = T_{ref}$ and has three different absorption coefficients of $\kappa_a = 0.1, 1.0, 5.0\,m^{-1}$. The non-dimensional wall heat flux $\left(\frac{q_w}{\sigma T_{ref}^4}\right)$ on the circumferential surface area of the cylinder computed numerically using Ansys Fluent was in accordance with exact solution with very small margin of error. The comparison of numerical results with the exact solution is shown in Fig. 3.

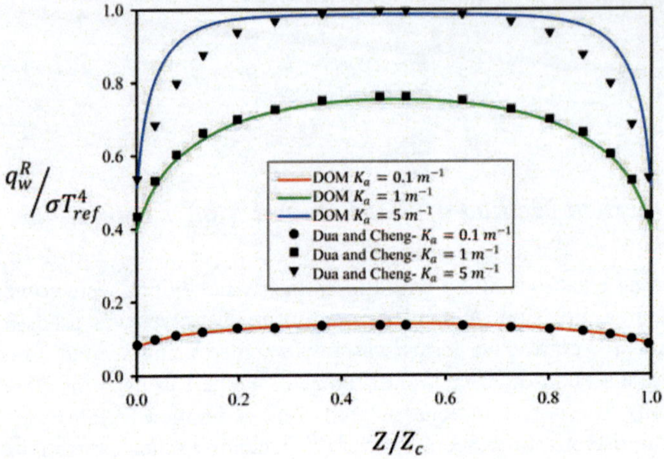

Fig. 3 Comparison of exact solution of non-dimensional radiative heat flux with the present numerical model

3 Results and Discussion

Simulations were carried out on a small laboratory-scale single-port hybrid rocket motor with port diameter of 40 mm and fuel grain length of 130 mm. Two independent cases were simulated with and without radiative heat transfer in the combustion chamber.

3.1 Hybrid Motor Geometry Details

A small hybrid rocket motor (of laboratory scale) with port diameter of 40 mm and fuel grain length of 130 mm was considered as the base motor geometry. The nozzle throat diameter was 17 mm. Note that the choice of throat diameter does affect the chamber pressure, and the pressure has negligible effect on burn rate r_b of the hybrid rocket motor. The average inlet oxidizer mass flux was maintained at 159 kg/m^2 s.

The mesh is made of structured quadrilateral grids which are clustered near fuel surface and oxidizer inlet to predict flame stabilization point precisely and resolve the steep gradients. The grid independence study for the motor was performed by Palani Kumar [7] and the grid size adapted in the combustion chamber is 355 × 80.

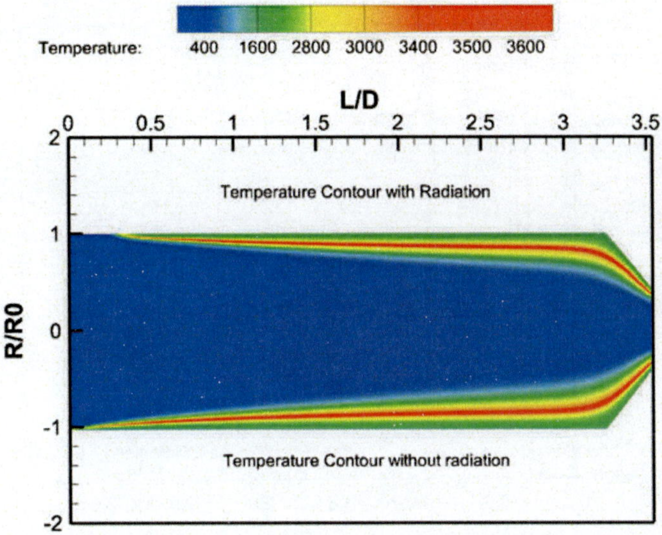

Fig. 4 Temperature contours variation along the length of the motor

3.2 *Effect of Radiation*

The radiation heat transfer is included in the present study governed by the radiative transport equation. Figure 4 shows the temperature contours in the motor with (top) and without radiation (bottom). It can be observed that the flame begins on the solid fuel surface at X/D = 0.3 in the case of radiation, whereas it begins at the beginning of the solid fuel surface in the case without radiation. One can also note that the flame indicated by the temperature contour of 3000–3500 K is narrower in the case with radiation.

The variation in temperature of solid fuel surface and the gas adjacent to the fuel surface lengthwise of the motor is plotted in Fig. 5. The overall trend in variation of the temperature for the same is shown in the inset of Fig. 5. It can be observed that the surface temperature of the case with radiation is considerably higher in the beginning of motor compared to the case without radiation. The emitted radiation heat flux from the flame sheet is absorbed by the fuel surface which increases its temperature. However, as the flame moves away from fuel surface downstream as shown in the contours of Fig. 4, the absorbed radiation heat flux reduces and radiative heat emitted from fuel surface cools down the fuel surface. Hence, the surface temperature falls below the surface temperature of case without radiation at about X/D = 1.9. Also, as the flame emits radiation heat, its temperature also reduces, and thereby reducing the gas temperature near the surface. This causes the gas temperature to fall below the gas temperature of case without radiation at about X/D = 2.3.

The temperature along the radial direction is plotted in Fig. 6a at three different locations along the length of the motor. One can note that the temperature profile

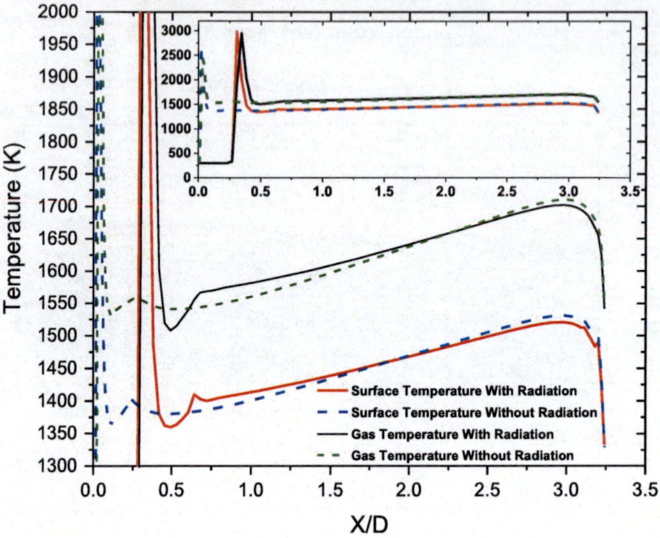

Fig. 5 Comparison of fuel surface and gas temperature with and without radiation model

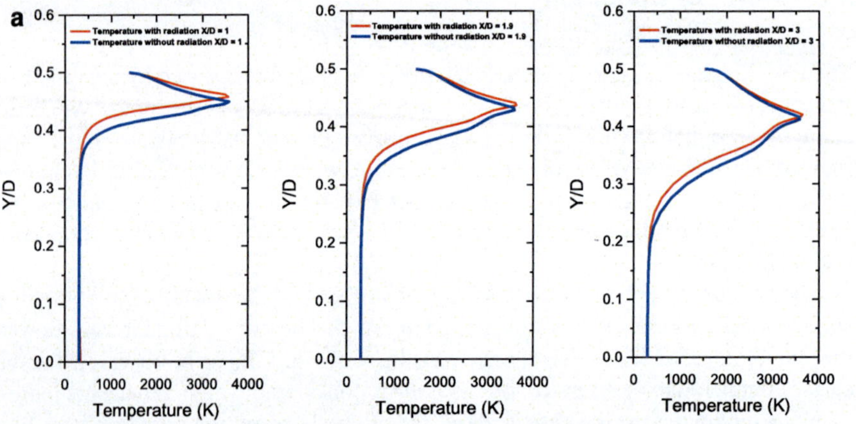

Fig. 6 **a** Comparison of temperature profiles along radial direction with and without radiation model, **b** variation of radiative absorption/emission per unit volume along the radial direction for various X/D, **c** comparison of velocity profiles along radial direction with and without radiation mode

of the flame increases in width as the flow moves downstream in the motor. It is also observed the flame temperature of the case with radiation is slightly more in the downstream locations. However, the flame temperature profile width is marginally more for the case without radiation.

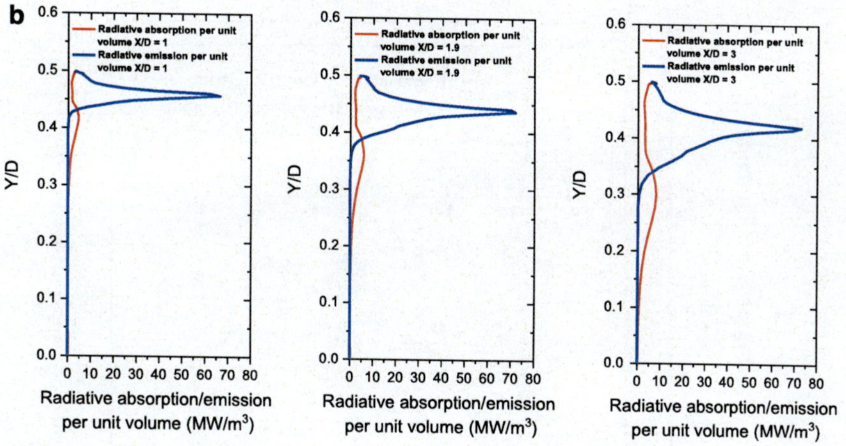

Fig. 6 (continued)

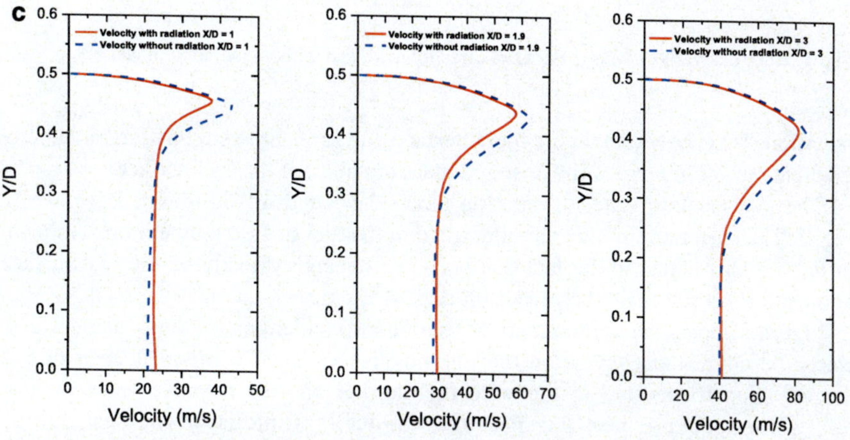

Fig. 6 (continued)

The variation of radiative absorption/emission per unit volume along the radial direction is plotted in Fig. 6b at three different locations along the length of the motor. It can be observed that the flame emits radiatively along the length of the motor and the participating gases also absorb a fraction of this emitted radiation. Since the motor under consideration is a laboratory-scale motor, the small volume of the combustion chamber causes the actual emitted and absorbed radiation heat to be small.

The velocity along the radial direction is shown in Fig. 6c at the same locations as temperatures along the length of the motor. The velocity without radiation is slightly more at all three locations because the flame does not loose energy by emission due to

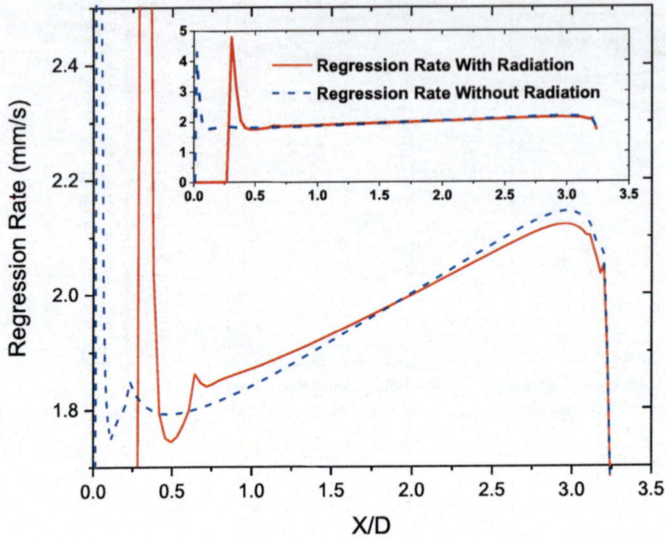

Fig. 7 Burn rate variation along the axial direction with and without radiation model

radiation. This shows that at the fuel interface, the gases have slightly larger velocity gradients which increases the heat and mass transport at the fuel surface.

The average burn rate r_b variation along the length of the motor is shown in Fig. 7. The burn rate follows a similar trend of the fuel surface temperature as shown in Fig. 5. This is due to the fact that burn rate depends directly on the fuel surface temperature as stated by the zeroth-order of Arrhenius equation.

Figure 8 shows the components of heat fluxes incident on the fuel-gas interface. Here, we adopt the convention that the positive value of component heat flux is the heat flux transferred to the solid fuel surface and the negative value of heat flux is considered as heat flux lost from the fuel-gas interface. The heat flux by convection is the primary source of heat transfer into the fuel surface. Therefore, the regression rate and the surface temperature have a similar trend lengthwise of the motor. The convective heat flux decreases initially as flame standoff distance from the fuel surface is very small. As the combustion products increase in quantity downstream, the fuel vapour needs to diffuse for a longer distance to mix with the oxidizer and attain stoichiometric composition. Flame standoff distance from the fuel surface, therefore, increases which in turn reduces heat transferred to the fuel surface and thus the regression rate. On the other hand, continuous addition of products to local mass flux in the combustion chamber increases the convective heat flux downstream. This results in decreasing followed by increasing trend of regression rate and heat flux components. The radiation heat flux variation is shown in the inset of Fig. 8. Though the radiation heat flux is significantly less compared to other heat flux components, it is responsible for the marginal changes in the surface and gas temperatures and average burn rate along the hybrid motor length.

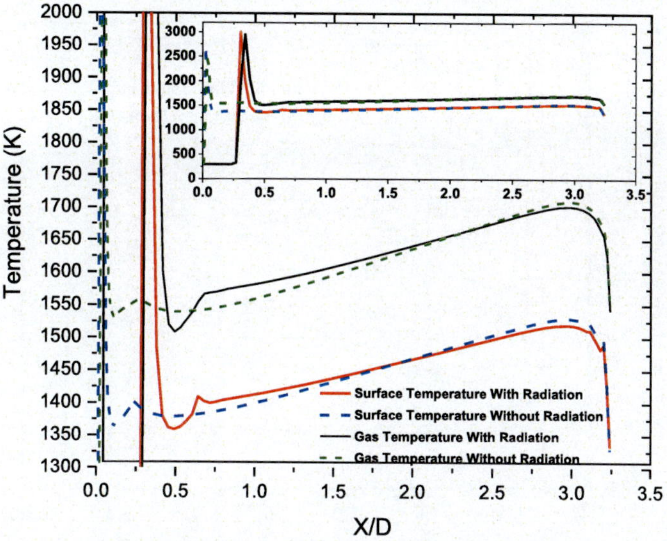

Fig. 8 Components of heat fluxes on the fuel surface for the motor with the radiation model

The variation of net heat flux incident on the surface of the fuel is shown in Fig. 9a. It can be observed that the average net heat flux transferred to the fuel surface interface and its components convective and conductive heat flux follows a similar trend in accordance with the surface temperature and burn rate. The radiative heat emission from flame to the fuel surface results in cooling of the gas which corresponds to a lower convective heat flux transferred to the fuel surface in the tail end of the motor. Similarly, the emission of radiative heat from the fuel surface results in a lower conductive heat flux in the tail end of the motor and absorption of radiative heat from the flame results in higher conductive heat flux close to head of the motor. However, the motor under consideration is a very small laboratory-scale motor, and the effect of radiation is expected to be more prominent in larger scale motors.

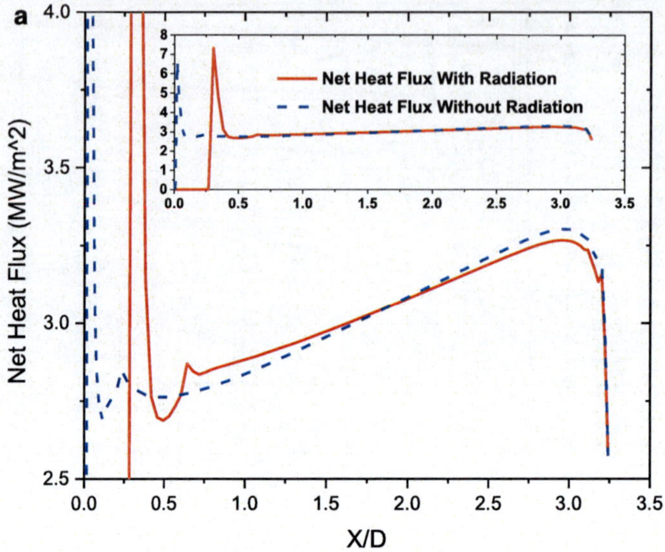

Fig. 9 **a** Net heat flux variation along the axial direction for the motor with and without radiation, **b** convective heat flux variation along the axial direction for the motor with and without radiation, **c** conductive heat flux variation along the axial direction for the motor with and without radiation

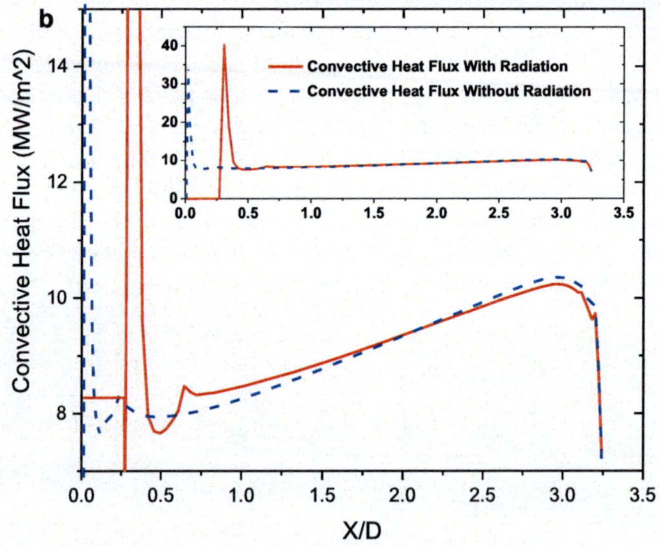

Fig. 9 (continued)

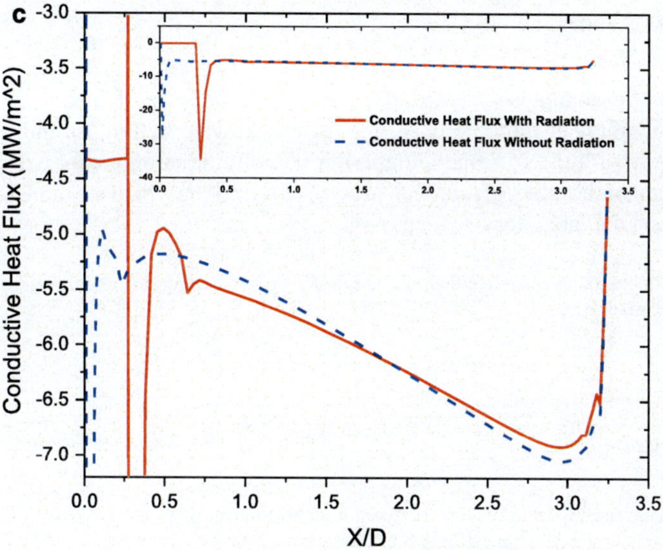

Fig. 9 (continued)

4 Conclusions

The effect of radiative heat transfer in hybrid rocket motor combustion chamber was studied numerically studied by comparing it the same motor without the radiation modelling. The following conclusion may be drawn from this study

1. The inclusion of radiation model has a marginal increase in the fuel surface temperature at the beginning of the motor but decreases beyond a certain X/D location and falls below the case without radiation model.
2. The burn rate also follows a similar trend to fuel surface temperature with marginal increase at the beginning of the motor and decrease beyond a certain X/D location and fallen below the case without radiation model.
3. The velocity decreased in the motor with radiation model in comparison with the one without radiation model. Hence, the gases have slightly smaller velocity gradients which decrease the heat and mass transport at the fuel surface.
4. The radiative heat flux at the fuel surface is significantly small in comparison with the other components of heat fluxes. But it has a significant effect on convection heat flux and diffusive heat flux. But the loss in convective heat flux is gained by the diffusive heat flux.
5. The net flux transferred into the fuel surface is in accordance with the surface temperature and burn rate profiles proving the heat and energy balance at the fuel and gas interface.

5 Future Work

The effect of radiation is expected to increase with the increase in the size of motor and configuration of motor. Hence, for further studies, scaling up the motor and comparing it with the corresponding motor without radiation model will provide a better insight of the effect of radiation in large scale motors which are more practically relevant than the laboratory-scale motors.

Acknowledgements Authors are thankful to DRDO for sponsoring this research under the aegis of CoPT, IIT Madras.

References

1. Venkateswaran S, Merkle CL Size scale-up in hybrid rocket motors. In: AIAA 34th aerospace sciences meeting and exhibit, Reno, NV, 15–18 Jan 1996
2. Chidambaram PK, Kumar A (2012) A numerical study on the regression rate of hybrid rocket motors using a combination of enhancement techniques. In: 48th AIAA/ASME/SAE/ASEE Joint Propulsion Conference & Exhibit, July 2012
3. Kumar CP, Kumar A (2013) Effect of diaphragms on regression rate in hybrid rocket motors. J Propuls Power (in press). https://doi.org/10.2514/1.b34671
4. Kuo KK, Chiaverini MJ (2007) Challenges of hybrid rocket propulsion in the 21st century. In: Chiaverini MJ, Kuo KK (Eds) Fundamentals of hybrid rocket combustion and propulsion. AIAA progress in astronautics & aeronautics, vol 218, AIAA, Reston, VA 2007, pp 593–638
5. Kuo KK, Lu Y-C, Chiaverini MJ, Johnson DK, Serin N, Risha GA, Merkle CL, Venkateshwaran S Fundamental phenomena on fuel decomposition and boundary layer combustion processes with applications to hybrid rocket motors. In: Semi-annual progress report, NASA-CR-201843, 1996
6. ANSYS-Fluent Software Release Version No. 18.1, 2017
7. Chidambaram PK Numerical study on hybrid rocket motors incorporating regression rate enhancement techniques, Doctor of Philosophy Thesis, Indian Institute of Technology, Madras
8. Dua SS, Cheng P (1975) Multi-dimensional radiative transfer in non-isothermal cylindrical media with non-isothermal bounding walls. Int J Heat Mass Transfer 18:245–259

Characterization of Engine Combustion Flames Using Inverse Abel Transform

Shashikant Verma and Rajneesh Kaushal

Abbreviations

IAT	Inverse Abel Transform
AT	Abel Transform
DD	Direct Discretization
NO	Nestor–Olsen
GUI	Graphical User Interface
GUIDE	Graphical User Interface Development Environment

1 Introduction

1.1 Chemiluminescence Phenomena

The phenomenon of chemiluminescence in a chemical reaction refers to the emission of light due to the chemical reaction. In a chemical reaction, if an excited intermediate species is formed, it reached its ground state accompanied by the emission of photons. This phenomenon is employed to identify the concentration of species in combustion systems. A typical hydrocarbon-based fuel burns with the production of CH*,

S. Verma
Department of Mechanical Engineering, National Institute of Technology Kurukshetra, Kurukshetra, Haryana 136119, India

R. Kaushal (✉)
Department of Mechanical Engineering, National Institute of Technology Kurukshetra, Kurukshetra 136119, India
e-mail: rajneesh@nitkkr.ac.in

A. K. Gupta et al. (eds.), *Advances in IC Engines and Combustion Technology*,
Lecture Notes in Mechanical Engineering,
https://doi.org/10.1007/978-981-15-5996-9_58

OH^*, C_2^* and CO_2^* radicals and other weaker radicals. CH* and OH* radicals are considered to be the flame and heat release markers in premixed flames and are used extensively in combustion diagnostics. The concentration field of a given species can be obtained by viewing the reacting zone through an appropriate optical band-pass filter. However, such an image contains the concentration of the entire volume, as each pixel captures the integrated light from the entire volume passing through its line of sight. In order to resolve an axisymmetric integrated line of the sight data distribution to one at a given azimuthal plane, a mathematical transformation such as inverse Abel transform can be used.

1.2 Inversion Function

Inverse Abel transform (IAT) is a mathematical function which takes a 2D projection and reconstructs a slice of the cylindrically symmetric 3D distribution. IAT also plays an essential role in analyzing the projection of plasma plumes and flames.

When the source is assumed to be cylindrically symmetric and optically thin, the observed lateral data can be converted to radial distribution profile. The Abel transformation (AT) is given by:

$$F(x) = 2\int_{x}^{L} \frac{f(r)r\,\mathrm{d}r}{\sqrt{r^2 - x^2}} \tag{1}$$

The IAT of a system is given as:

$$f(r) = \frac{-1}{\pi}\int_{r}^{L} \frac{F'(x)\mathrm{d}(x)}{\sqrt{x^2 - r^2}} \tag{2}$$

where

$F(x)$ Abel transform
$f(r)$ circularly symmetric function
x distance from origin
r radius
L maximum limit of radius.

The significant difficulties in implementing IAT are the discontinuity and its dependence on the first derivative of the original function. The latter is particularly challenging when used with experimental data, which is discrete and can be noisy. To remove these difficulties, several methods are available. Presently, a direct discretization method (DD) and Nestor–Olsen (NO) method implemented in MATLAB are used to calculate the IAT.

Difference between the numerical and analytic approach:

Analytical methods fit the experimental data to a function and allow direct integration of function and avoid the discretization problems [1–3]. Some functions, such as Gaussian function, are used to serve the purpose.

Numerical methods convert the Abel inversion equation (Eq. 2) to a summation to allow processing on a discrete set of data. It is a fast way of solving the integral equation.

2 Methodology

The IAT was operated on data points obtained from a flame image obtained from open literature [4–8]. The algorithm which was generated tested initially in MATLAB on data points obtained through a different trial function such as linear function, exponential function, and logarithmic function. It consists of two steps:

First, the discretized Abel transform was performed on the trial function to generate the data points which treated as data points of a flame image. Second, IAT was performed on previously obtained data points to get the original function. The algorithm seems to provide satisfactory results. Further, the method was implemented on actual data of a flame image.

2.1 *Algorithm for Discretized Abel Transform*

The following algorithm for Abel transform was implemented on trial function ($f\_r = \log(r)$) for 51 axisymmetric data points, as shown in Fig. 1.

```
% Finding abel transform numerically
r = (.2:.2:10.2);
x = r;
Abel_transform = zeros(1,50);
 for i = 1:50
  Y = (2.*f_r(i:51).*r(i:51))./sqrt((r(i:51)+.019).^2 - x(1,i).^2);
  Abel_transform(1,i)= trapz(r(i:51),Y);
 end
```

Fig. 1 Abel transform implemented in MATLAB

2.2 Algorithm for Inverse Abel Transform

There are two methods of implementing the IAT on the given data points. The following algorithms are to be applied on a $2n + 1$ data point discrete distribution $I(x)$ with axial symmetry. The first method is known as direct discretization (DD) reported by Alvarez et al. [9], and the second method is Nestor–Olsen (NO) reported by Blades et al. [10].

The DD is based on a direct discretization of (Eq. 2). This method addresses the singularity by modifying a term suitably to avoid division by zero. It is given as:

$$f\_r = \frac{-1}{\pi} \sum_{i=j}^{N-1} \frac{I(i+1) - I(i)}{\sqrt{\left(x_i + \frac{\Delta x}{2}\right)^2 - r_j^2}} \tag{3}$$

where

$(2N + 1)$ discrete data points
$I(i)$ intensity of data points.

The Nestor–Olsen method uses a numerical algorithm which is easy to compute, and the IAT is given as follows:

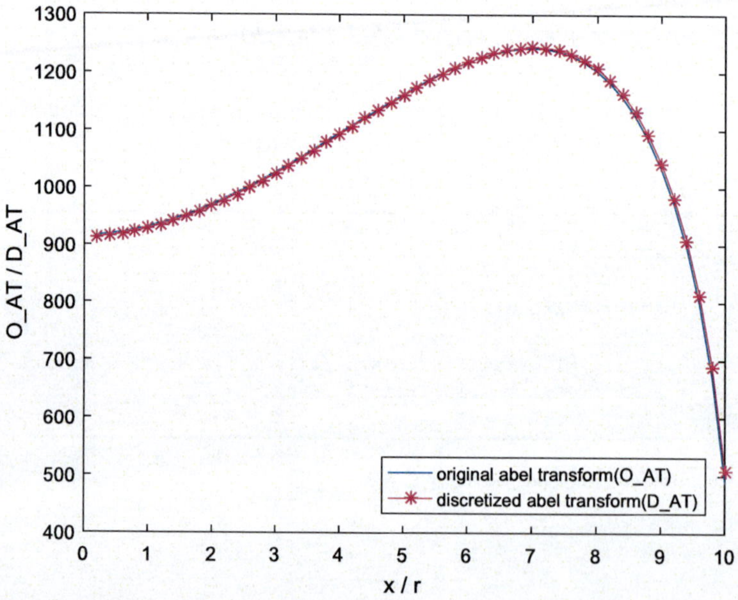

Fig. 2 Abel transform of $f(r) = (r^2 + 2r)$

$$I(r) = \frac{-2}{\pi a} \sum_{n=k}^{N-1} M(x) B(k, n) \quad (4)$$

where

$M(x)$ measured lateral intensity
a increment in value of x

$$B(k, n) = -A(k, k) \quad \text{for } n = k$$
$$B(k, n) = A(k, n-1) - A(k, n) \text{ for } n \geq k + 1$$

$$A(k, n) = \frac{\sqrt{n^2 - (k-1)^2} - \sqrt{(n-1)^2 - (k-1)^2}}{(2n-1)}$$

where k and n are the integer position indices for radial and lateral intensity.

3 Results and Discussion

3.1 *Validation of the Algorithm*

The algorithm was generated to solve the Abel transformation numerically, and inverse Abel transformation was performed on a different trial function.

3.1.1 Abel Transformation

Abel transform function was executed analytically and numerically, and the results were compared. Four functions were taken as a trial function.

The first function is a linear function ($f(r) = (r^2 + 2r)$), where r is the radius and its Abel transform was plotted in original and discretized form as shown in Fig. 2. Analytically, Abel transform was represented by (O_AT), and numerically, it was described by (D_AT).

Second function is exponential function ($f(r) = \exp(r)$), and Abel transform is plotted in original and discretized form as shown in Fig. 3.

Third function is $\left(f(r) = \sqrt{r}\right)$ and Abel transform is plotted in original and discretized form as shown in Fig. 4.

Fourth function is logarithmic function ($f(r) = \log(r)$) and Abel transform is plotted in original and discretized form as shown in Fig. 5.

It can be observed from the graphs (Figs. 2, 3, 4 and 5) that there is a minimal deviation between the actual and discretized Abel transform.

3.1.2 Inverse Abel Transform

Both methods of IAT were executed discretely and compared with the actual function values. The results were compared. IAT for trail function is shown in the subsequent figures. IAT for the function $f(r) = \left(r^2 + 2r\right)$ is shown in Fig. 6. The original function is represented by (F_r), and IAT is represented by (F_r1) corresponding to DD method and by (F_r2) corresponding to NO method.

IAT for the function ($f(r) = \exp(r)$) is shown in Fig. 7.

IAT for the function $f(r) = \sqrt{r}$ is shown in Fig. 8.

IAT for the function $f(r) = \log(r)$ is shown in Fig. 9.

It can be observed from the graphs that there is some deviation of both the methods of inversion from the actual values. The deviation varies for different trail functions at different intervals. For some intervals, direct discretization gives satisfactory results, while for another range Nestor–Olsen method gave better results. For example, in case of the exponential function (between interval 0.2 and 10.2), NO method gave better results at intervals 4.2–9.2 and DD method gave better results at range 0.2–4.2 and 9.2–10.2. However, for most of the functions, DD method proved to be optimum when compared with NO method. It was also observed that a significant deviation between both ways occurs at the start and end points.

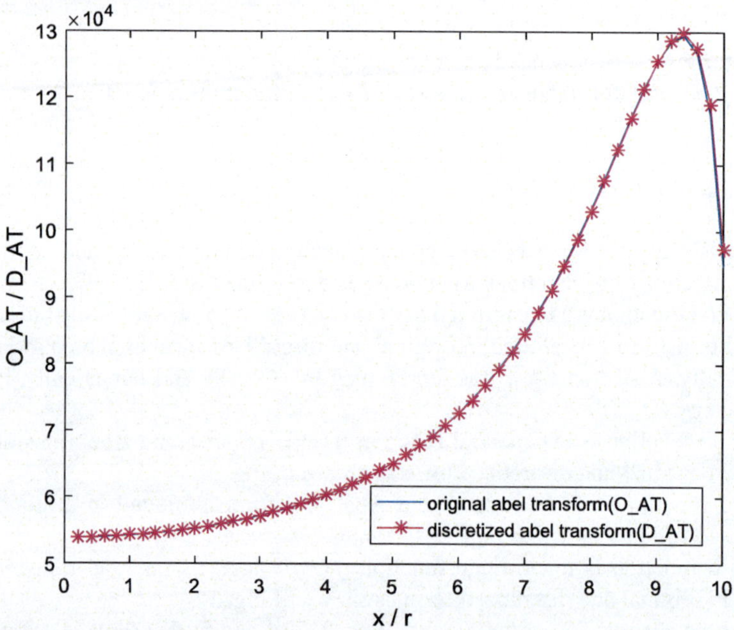

Fig. 3 Abel transform of $f(r) = \exp(r)$

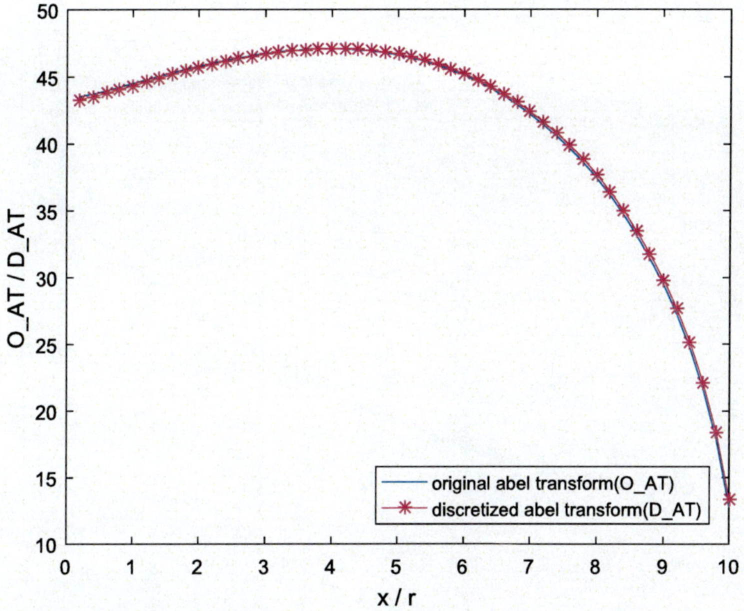

Fig. 4 Abel transform of $f(r) = \sqrt{r}$

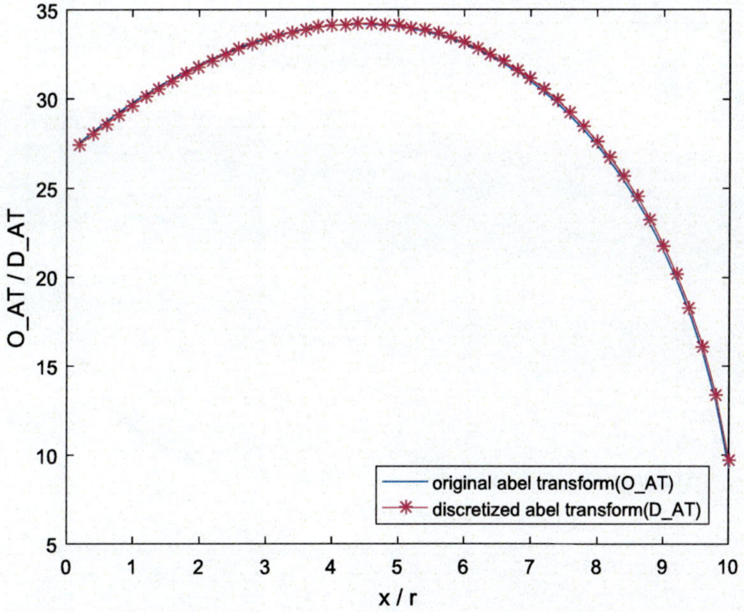

Fig. 5 Abel transform of $f(r) = \log(r)$

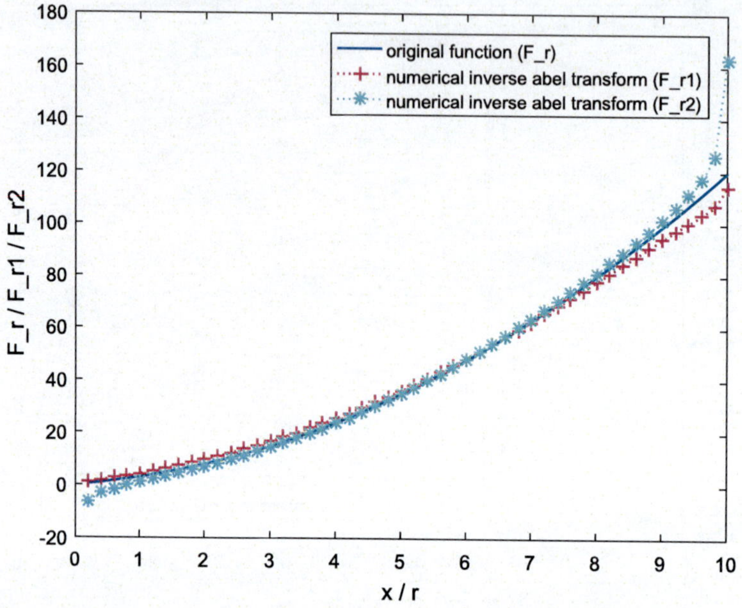

Fig. 6 IAT for $f(r) = (r^2 + 2r)$

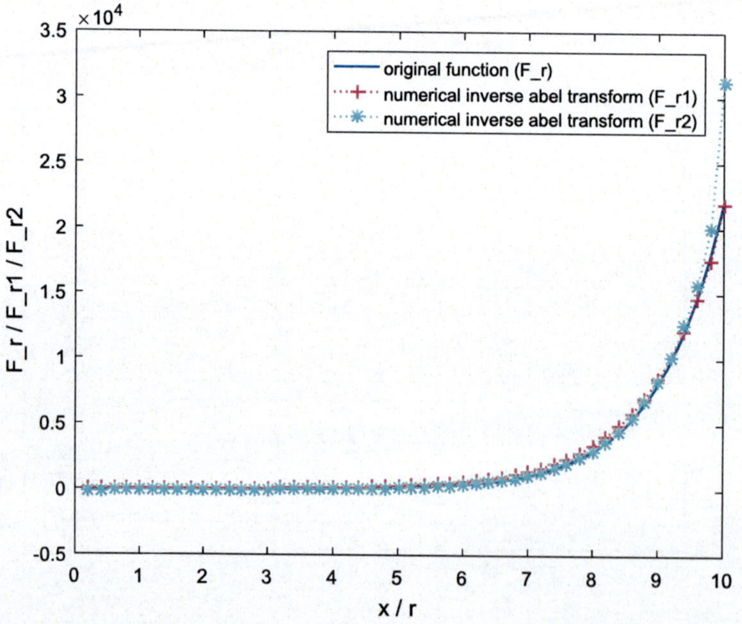

Fig. 7 IAT for $f(r) = \exp(r)$

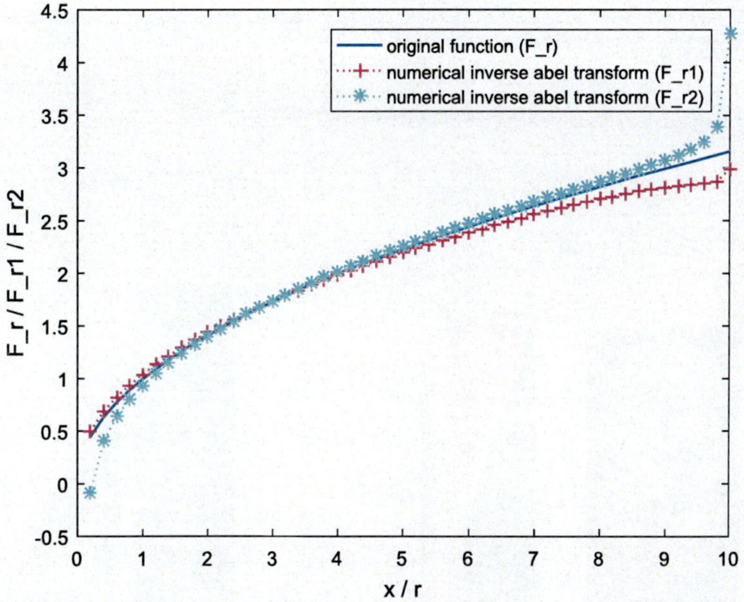

Fig. 8 IAT for $f(r) = \sqrt{r}$

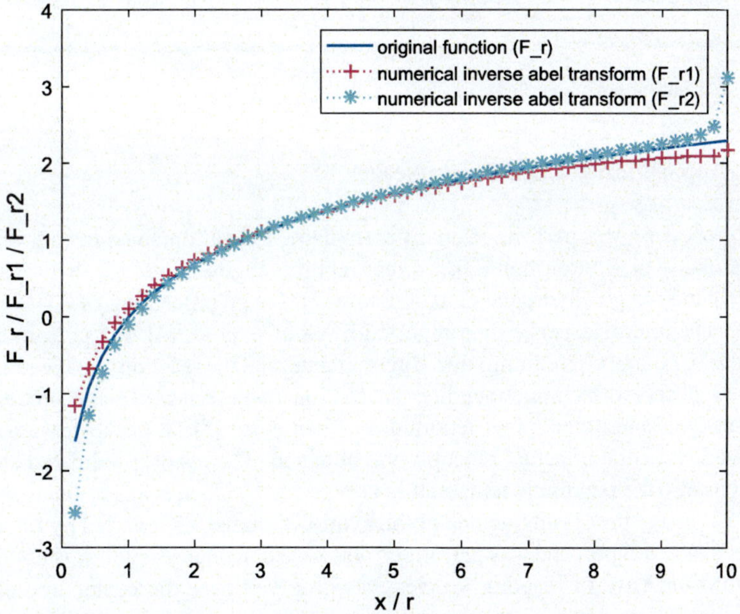

Fig. 9 IAT for $f(r) = \log(r)$

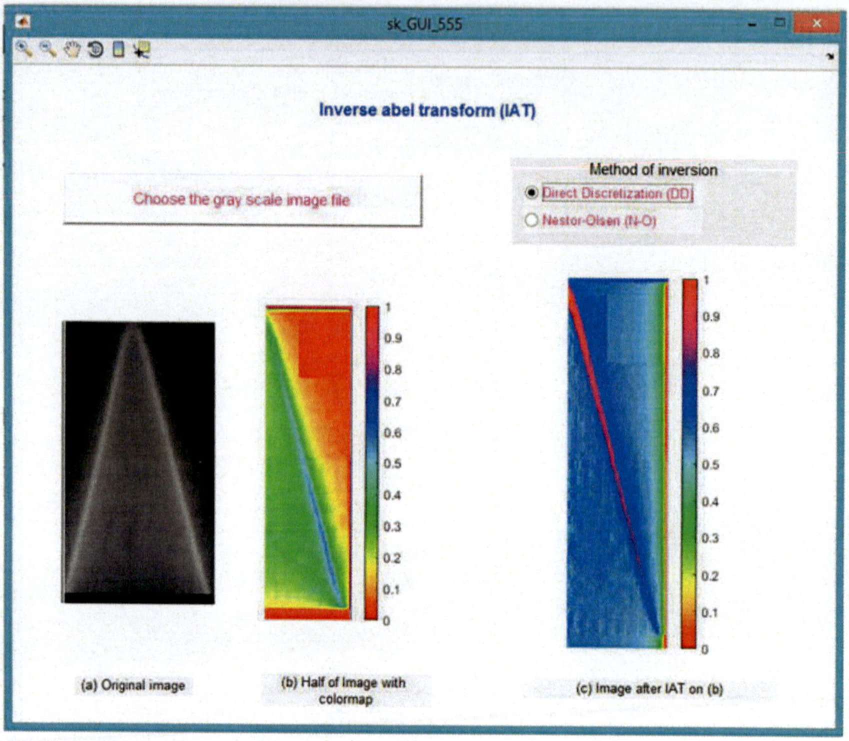

Fig. 10 IAT on Bunsen burner flame image

3.1.3 Application to Chemiluminescence

Image processing was performed on different flame images obtained from literature such as image of Bunsen flame [4], direct ignition engine flame [5], hydrocarbon flame [6] and oxygen/hydrogen coaxial flame [7]. Image processing of CH*, CO_2* and C_2* chemiluminescence data reported by Souflas et al. [4] was performed. It consists of CH_4 and C_3H_8 air mixture Bunsen flame, and the reaction zone or reacting edge was observed by implementing the IAT on flame image. Image processing on direct ignition engine flame reported by Chen et al. [5], at peak pressure, was performed and zone of actual reaction was obtained. The result is validated as that region has also the maximum temperature as reported by Chen et al. [5] and is shown in Fig. 12. Image processing on one of the hydrocarbon flames reported by Lin et al. [6] of coal gas and propane was performed and zone of actual reaction was obtained. Maximum intensity of reacting species was observed near the center line of the flame image. Processing of the flame image reported by Kim et al. [7], for OH* chemiluminescence, was performed to obtain the concentration of OH*. The image was of oxygen and hydrogen coaxial flame at fully opened nozzle. It was observed that the region of maximum concentration of OH* occurs near the nozzle. The result

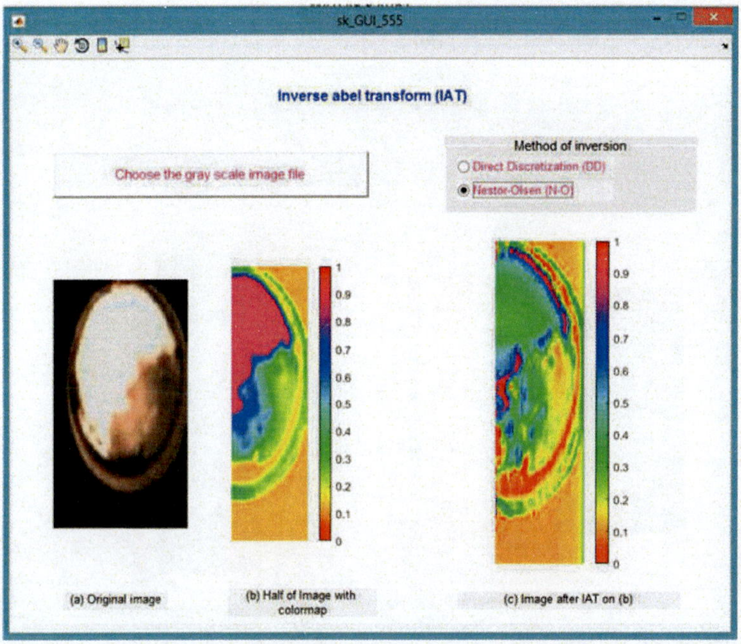

Fig. 11 IAT on diesel engine flame image

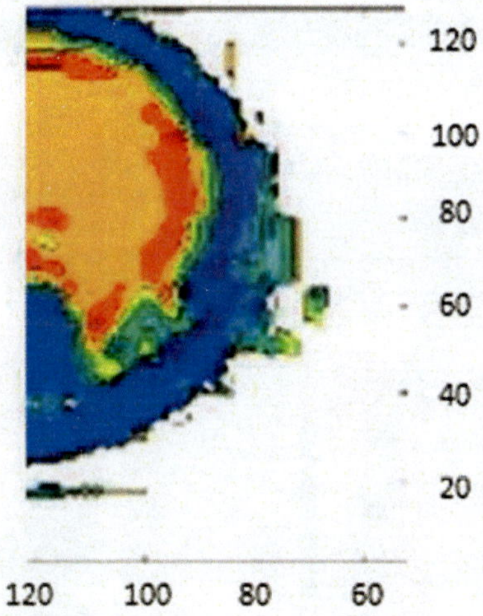

Fig. 12 Temperature field image at highest pressure [5]

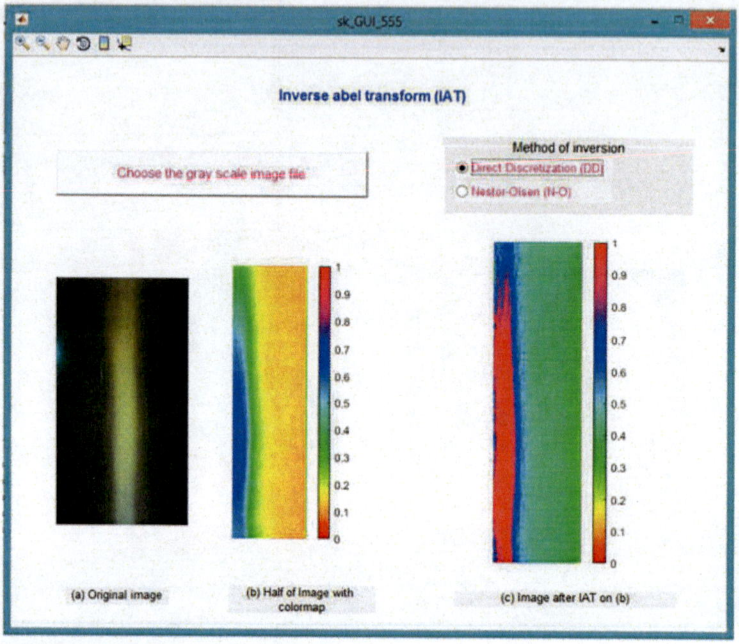

Fig. 13 IAT on hydrocarbon flame image

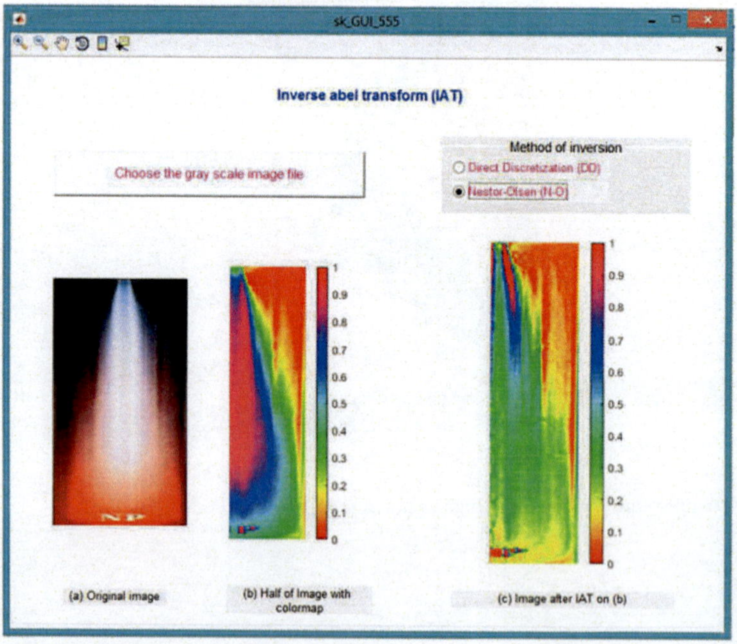

Fig. 14 IAT on oxygen and hydrogen coaxial flame image

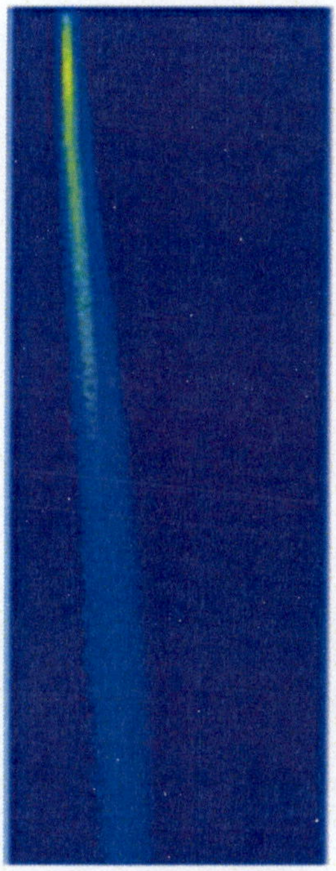

Fig. 15 OH* distribution of flame image at fully opened nozzle

is validated as that region has nearly same OH* distribution as reported by Kim et al. [7] and is shown in Fig. 15.

Processing of the image reported by Tong et al. [8], for CH* chemiluminescence, was performed to obtain the azimuthal distribution of CH* relative concentration. Implementation of IAT on the flame image was done, and it was observed that the region of maximum concentration of CH* occurs at the outer edges of the flame, which is the region where intensity is maximum.

On analyzing the inverted image by two methods, it was observed that the number of different zones of the flame image obtained with the help of DD method is more as compared to the NO method.

3.1.4 MATLAB-Based GUI for Inverse Abel Transform

GUI was generated in MATLAB using GUIDE. GUI is a user interface that includes graphical elements such as icons and buttons. The significant advantage of GUI is that they make operation more intuitive and easy to learn and use. The user does not require any knowledge of programming. The GUI was generated for finding the IAT, and it allows the user to choose an image and the method for finding the inverse Abel transform. Depending upon the method selected by the user either DD or NO,

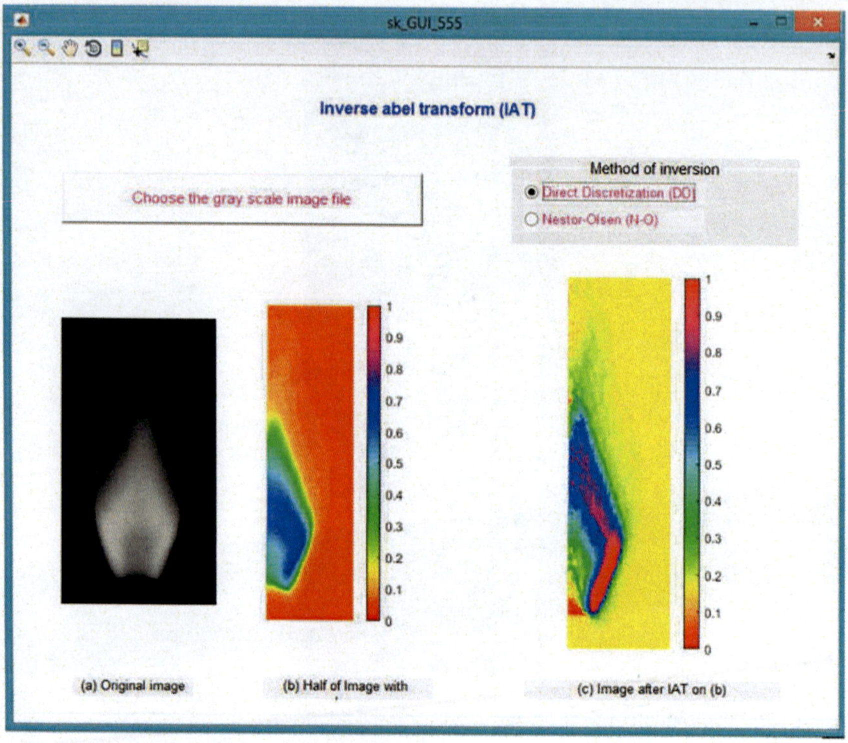

Fig. 16 IAT on combustion flame image (CH_4–O_2)

it will show the IAT of the previously selected image. At the interface, the user can see the actual chosen image, half of the image with color-map and image after Abel inversion. The GUI for the IAT of different flame images is shown in Figs. 10, 11, 12, 13, 14, 15 and 16.

Developed GUI for the image of Bunsen burner is shown in Fig. 10.

Developed GUI of the image of internal combustion engine is shown in Fig. 11.

GUI developed for hydrocarbon flame is shown in Fig. 13.

Developed GUI for oxygen and hydrogen coaxial flame is shown in Fig. 14.

Developed GUI for methane–air combustion flame image is shown in Fig. 16.

4 Conclusion

The present work involves the implementation of a mathematical transformation called inverse Abel transform for the purposes of analyzing chemiluminescence data of reacting flow fields. Two numerical methods, namely direct and Nestor–Olsen methods, were implemented in MATLAB. The methods are first validated for known one-dimensional functions and finally applied to a two-dimensional chemiluminescence data. A MATLAB-based function, including one with a GUI, is developed for this purpose. The results showed that the region of maximum radical concentration is captured adequately for different combustion flame images.

Acknowledgments The authors would like to gratefully acknowledge the help provided by Dr. Saptarshi Basu, Professor, Mechanical Engineering Department, IISc Bangalore and Dr. Keerthi, Postdoctoral Researcher, Mechanical Engineering Department, IISc, Bangalore.

References

1. Freeman MP, Katz S (1963) Determination of a radiance-coefficient profile from the observed asymmetric radiance distribution of an optically thin radiating medium. JOSA 53(10):1172–1179
2. Minerbo GN, Levy ME (1969) Inversion of Abel's integral equation bymeans of orthogonal polynomials. SIAM J Numer Anal 6(4):598–616
3. Piessens R (1973) Calculation of the radial distribution of emitters in a cylindrical source. Comput Phys Commun 5(4):294–298
4. Souflas K, Psarakis EZ, Koutmos P, Egolfopoulos FN (2019) Low cost image processing of Bunsen flame photography for estimation of flame speeds. Combust Sci Technol 191(7):1123–1138
5. Chen H, Hou Y, Wang X, Pan Z, Xu H (2019) Characterization of in-cylinder combustion temperature based on a flame-image processing technique. Energies 12(12):2386
6. Lin J, Zhang X, Liu K, Zhang W (2019) Emissivity characteristics of hydrocarbon flame and temperature measurement by color image processing. Energies 12(11):2185
7. Kim TY, Kim YH, Ahn YJ, Choi S, Kwon OC (2019) Combustion stability of inverse oxygen/hydrogen coaxial jet flames at high pressure. Energy 180:121–132
8. Tong Y, Liu X, Wang Z, Richter M, Klingmann J (2018) Experimental and numerical study on bluff-body and swirl stabilized diffusion flames. Fuel 217:352–364
9. Alvarez R, Rodero A, Quintero MC (2002) An Abel inversion method for radially resolved measurements in the axial injection torch. SpectrochimicaActa Part B: Atomic Spectrosc 57(11):1665–1680
10. Blades MW, Gary Horlick (1980) Photodiode array measurement system for implementing Abel inversions on emission from an inductively coupled plasma. Appl Spectrosc 34(6):696–699

Numerical Simulation

Numerical Investigation of Split Injection Strategy on Performance and Emission Characteristics of Diesel Engine

Ankit Kesharwani and Rajesh Gupta

1 Introduction

Due to better thermal efficiency and economic feasibility, diesel engine is widely used in the field of transportation and engineering applications [1–3]. However, the emission from the diesel engine such as nitrogen oxide (NO_x) and soot caused many serious air pollution problems [4, 5]. This attracts the researcher interest as to resolve the serious challenge like low efficiency and high emission from the diesel engine. Various control measures and regulation aimed at reducing exhaust emission in many countries all over the world. It is very well explained in the literature that the process of combustion can be enhanced by adequately controlling the air-fuel mixing inside the combustion chamber followed by emission. There are various strategies which are proposed to reduce the emission of diesel engine [6, 7]. The most effective strategy is split injection in which injection of fuel inside the combustion chamber is divided into more than one part—the part having low injected mass is called pilot injection and the one having high injected mass is called main injection. The split injection strategy proves effective result in increasing the area and decreasing the concentration of air/fuel mixture [8, 9]. The split injection strategy targeted on improving economy of fuel and control emission of direct injection diesel engine The duration between the pilot injection and main injection is called dwell time, if it is too short then the fuel injected in the main injection would penetrate in the zone formed by pilot injection resulting in the similarity of split injection to single injection, whereas if it is too long then the fuel injected in main injection may not burn properly results in enhancement of the soot emission and negative effect on thermal efficiency [10, 11].

A. Kesharwani (✉) · R. Gupta
Department of Mechanical Engineering, MANIT BHOPAL, Bhopal, India
e-mail: kesharwaniankit93@gmail.com

R. Gupta
e-mail: dr.gupta.rjp@gmail.com

A. K. Gupta et al. (eds.), *Advances in IC Engines and Combustion Technology*, Lecture Notes in Mechanical Engineering,
https://doi.org/10.1007/978-981-15-5996-9_59

The other strategy widely used recently to reduce emission is by manipulating the combustion chamber design and piston bowl geometry. It is proved by various studies that there is less influence in-cylinder flow during the initial stage of compression and intake stroke by manipulating the piston bowl design but it effectively enhances the in-cylinder flow during the initial stage of expansion stroke near TDC [12, 13]. A faster and an effective burning rate can be expected with an elevated vortex inside the combustion chamber, as it causes high intensity of turbulence at later stage of compression stroke. The design of piston bowl helps effectively in the formation of swirl; squish and turbulence inside the combustion chamber thereby improve air-fuel mixing. The circular motion of air about the cylinder axis is generally initiated by properly designed inlet port. The interaction of circular motion of air about and perpendicular to cylinder axis contributes to the formation of turbulent field at the end of compression stroke [14].

Literature surveyed yet, it is obvious that the optimization of the piston bowl design coupled with the injection ratio (mass of fuel injected into pilot to main injection) is the common concern of most of the investigation that focused on the improvement of engine performance with subsequent decrement in engine emissions. In the considered piston bowl design, emission like NO_x, soot as well as unburned HC can be controlled reasonably to low value with suitable combination of piston design and injection ratio. It noted that there was no optimum piston design that results in minimum emission level, since some parameter tends to reduce NO_x while some reduce soot emission. Therefore, it needs to understand the relationship among all these parameter. The effect of piston design with injection ratio has been examined numerically using AVL FIRE software. The computational study is performed in single-cylinder engine. Efforts been made to attain most favourable piston design with injection ratio so as to optimize the performance of engine and emission characteristics.

2 Model Used and Validation

The CFD code AVL FIRE is used in the current study. That is developed on the basis of finite volume method. The combustion process in diesel engine is being modelled by ECFM-3Z. K-epsilon model is chosen for turbulence inside the combustion chamber and heat transfer through cylinder wall. The spray behaviour of fuel is modelled by using WAVE break-up model. The evaporation of fuel particle and wall spray interaction is modelled by using Dukowicz and Walljet models, respectively. NO_X formation during combustion is estimated by Extended Zeldovich model. The prediction of Soot is made by using Kennedy-Hiroyasu-Magnussen model. To justify the model used and numerical steadiness of the codes, the result obtained from simulations is plotted against experimental results taken from literature [15]. Figure 1 shows the plot of pressure and rate of heat release variation with crank angle for experimental and simulation result; the simulation results are well agreed with the experimental results at least quality wise. However, quantity wise the figure shows

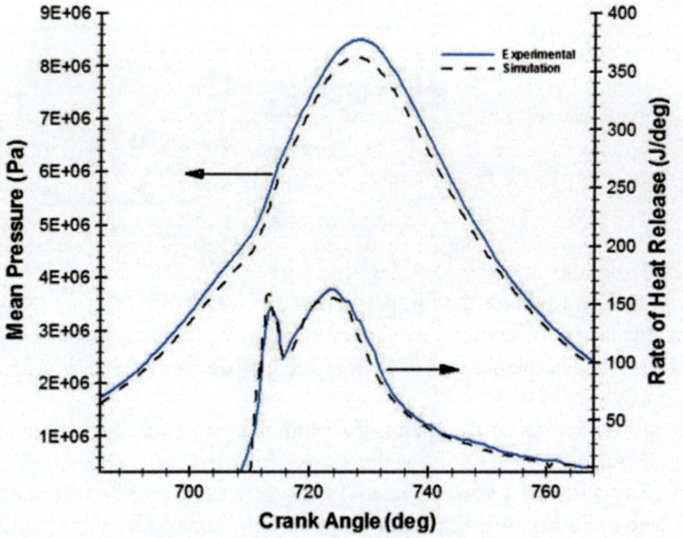

Fig. 1 Variation of pressure and heat liberated with crank angle

the considerable variation for the period of expansion stroke. While the peak value of heat release exhibits good agreement, mesh independency test is performed for the validation case by considering three different grid size having number of cells as 26,092, 30,524 and 48,312, respectively. It appears from Fig. 2 that the deviation

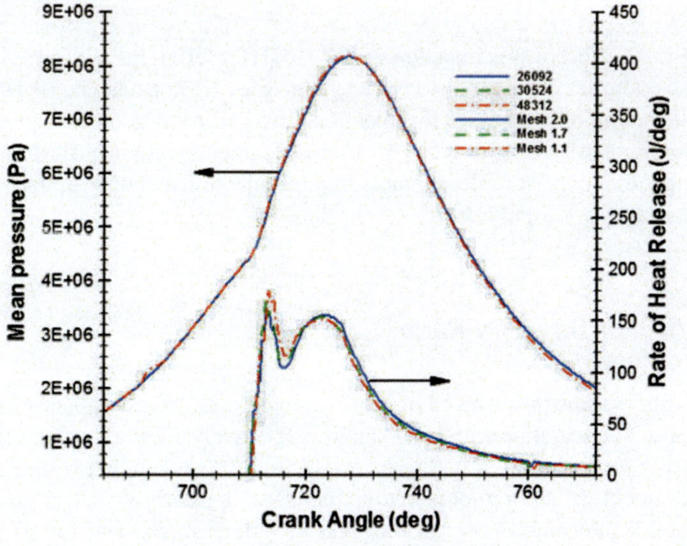

Fig. 2 Mean pressure and heat release variation with crank angle

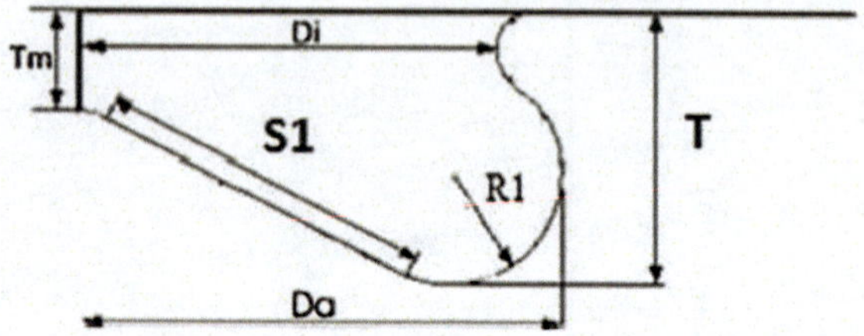

Fig. 3 Typical outline of piston bowl with represented parameter

of pressure and heat release data in prediction are well within the acceptable limit. Therefore, for other simulations mesh size with 30,524 cells is considered at TDC and slight variation in the number of cells is acceptable for other cases with different injection ratios.

In this study, a four-stroke, naturally aspirated, water-cooled, direct injection diesel engine is used with piston bowl design as shown in Fig. 3. The design parameter values are shown in Table 4 which result in six different piston bowl configurations D1–D6 as shown in Fig. 4. Table 2 shows the engine specification and initial conditions. Engine speed, compression ratio, load on engine and mass of fuel injected remain constant for all cases. The split injection strategy used in the present study is double injection in which certain quantity of mass is injected inside the chamber in pilot injection followed by the main injection. Fuel injected in pilot injection at 38° BTDC followed by 20° dwell then main injection at 16° BTDC (Tables 1, 2 and 3).

3 Result and Discussion

To examine the effect of split injection strategy in DI diesel engine with three injection ratios along with six different piston bowl geometries are considered. Table 4 shows the considered number of cases for injection ratio and bowl geometry, give overall eighteen combination. R1(P5M95) ("P" stands for pilot injection and "M" stands for main injection), i.e. 5% of injected mass in pilot injection and 95% of mass in main injection, same for R2 and R3.

3.1 Effect of Injection Ratio

Figures 5 and 6 show the effect of injection ratio on mean pressure, heat release and CO mass fraction variation with crank angle, respectively. Figure 5 reflects the effect of increasing the injection of fuel in pilot injection; it is noted that with increase in amount of injected fuel in pilot injection, the effect is clearly visible at 16° BTDC; that is, for 5% pilot injection mass the pressure is lowest, whereas for 10 and 15% pilot injection it is intermediate and maximum, respectively. The same effect can also be seen in case of heat release; that is, as the pilot injection mass increases, the

Fig. 4 Cosidered piston bowl design

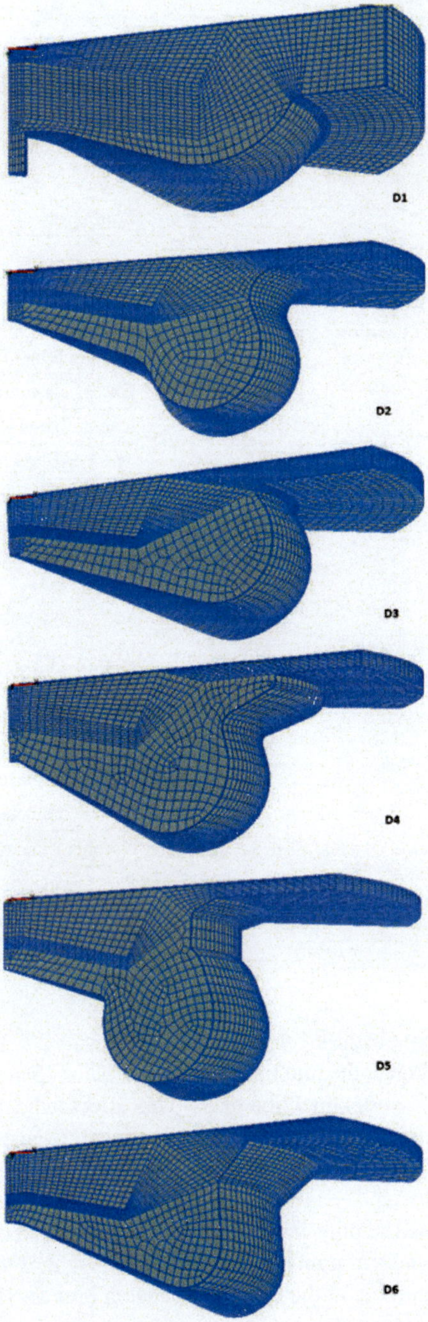

Table 1 Sub-models for calculation

Turbulence	k–ε–f
Wall interaction	Walljet1
Break-up	WAVE
Evaporation	Dukowicz
NO	Extended Zeldovich
Soot	Kennedy–Hiroyasu–Magnussen

Table 2 Initial conditions for simulation

Parameter (units)	Value
Suction temperature (K)	333
Excess air coefficient	1.8
Engine speed	2100
Mass injected (mg)	100
Suction pressure (bar)	1.19
Pilot injection interval and main injection (CA)	20
Swirl ratio	1

Table 3 Specification of test engine

Cylinder bore (mm)	132
Length of stroke (mm)	145
Connecting rod length (mm)	262
Compression ratio	13.5
Nozzle diameter (mm)	0.195
Number of nozzle holes	7
Spray angle (°)	153°
Ratio of pilot injection (%)	5, 10, 15
Rail pressure (MPa)	140

rate of heat release also increases at 20° BTDC the two peaks can be noticed from Fig. 5, the maximum first peak is for 15% pilot injection mass, and it is minimum for 5% pilot injection mass. This effect in heat release rate helps in reducing the ignition delay, and the second peak moved towards left as the pilot injection mass increases.

Figure 6 depicts the consequence of injection ratio on CO mass fraction; that is, it can be seen from figure that the CO mass at the end of expansion process is minimum as the pilot injection mass increases, whereas Fig. 7 depicts that of injection ratio on swirl number and it represents that as the pilot injection mass increases, the swirl number at the end of expansion process is improved that implies the better air-fuel mixing thereby improving combustion process.

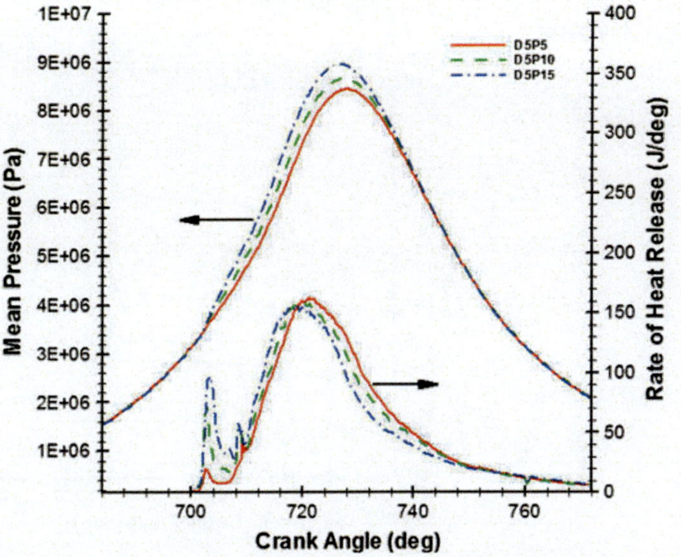

Fig. 5 In-cylinder pressure and heat liberated variation with crank angle

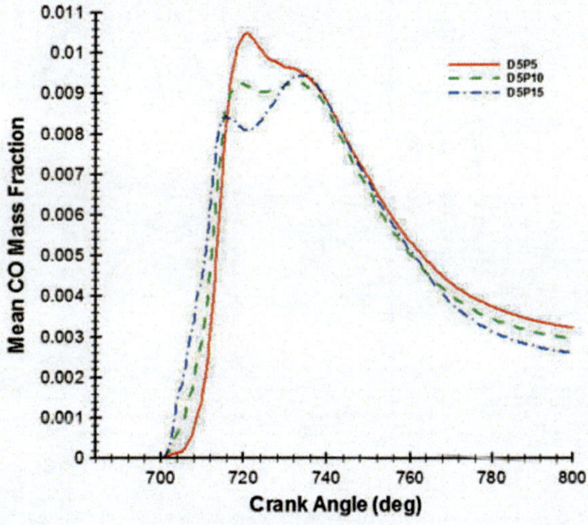

Fig. 6 CO mass fraction variation with injection ratio

3.2 *Effect of Bowl Geometry*

Figure 8 shows the effect of mean cylinder pressure for six different geometries and injection ratios P5M95. The cylinder pressure versus crank angle data over the compression and expansion strokes of the engine operating cycle can be used to obtain quantitative information on the progress of combustion. The engine had a

Fig. 7 Effect of injection ratio on swirl number

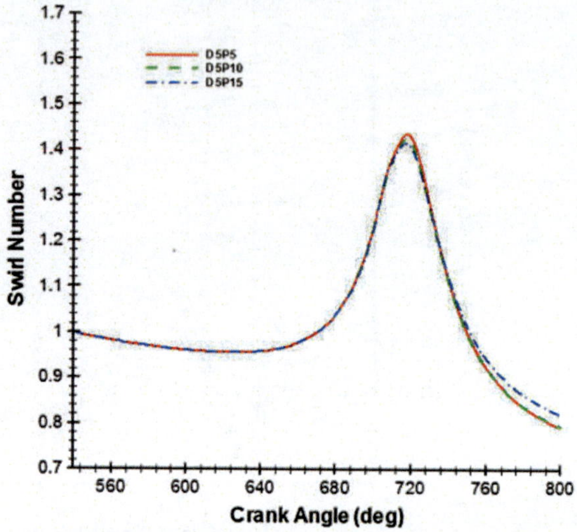

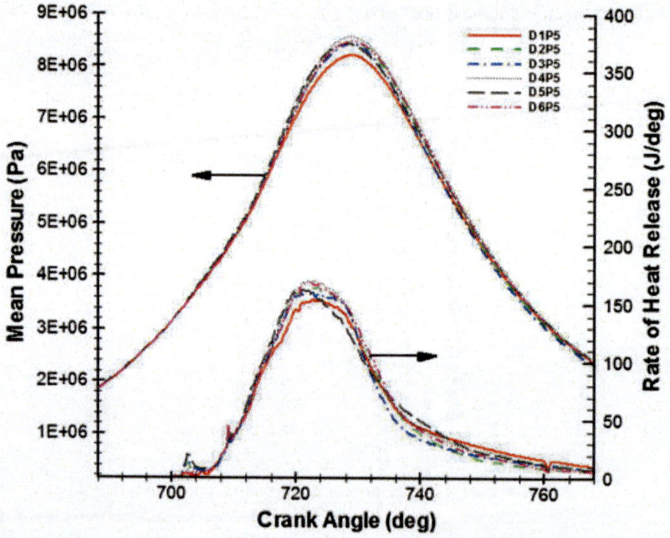

Fig. 8 Effect of geometries on mean pressure and rate of heat release for P5M95

central fuel injection through a seven-hole nozzle into the combustion chamber. The rate of fuel injection can be obtained from fuel line pressure, cylinder pressure and nozzle geometry. In general, for all three injection ratios mean pressure profile is similar to that of the conventional diesel fuel. However, it can be seen that the peak pressure is maximum in "D4" case for the three injection ratios. It can be noted that ignition delay is approximately 10° (identified by the change of slope of the P–Θ curve). From figures, the pressure curve initially seems to be increasing rapidly for a few crank angle degrees, then slowly attains the peak value about 7° after TDC. It can be observed from these figures that the ignition delay tends to reduce as the pilot injection ratio increases, and further, the peak pressure not only tends to shift towards TDC but also the magnitude of peak pressure also increases; that is, the peak pressure is maximum for injection ratio P15M85 for all the six configurations. Figure 8 also shows the effect of geometries on rate of heat release for injection ratio P5M95. It can be seen from figures that the maximum rate of heat release is for "D4" configuration for injection ratio P5M95 whereas it is also observed that for injection ratio P10M90 and P15M85, the peak for heat liberation is for "D4" and "D3", respectively. The heat release diagram shows negligible heat release until towards the end of compression when a slight loss of heat during the delay period (which is due to heat transfer to the walls and to fuel vaporization and heating) is evident. In case of P15M95, 15% of fuel is injected at early stage of compression which thereby increases the duration of ignition delay, the increased ignition delay improves the air-fuel mixture and gives scope to accumulate more evaporated fuel before the start of combustion, thereby enhances the combustion rate.

Both physical and chemical processes must take place before a significant fraction of chemical energy of the injected fuel is released. The fraction of fuel which is being injected during pilot injection helps to the pre-combustion reaction of the fuel which leads to auto-ignition and provides better thermal atmosphere for the main injection to burn. Chemical delay is more effective for duration of the ignition delay period. If the amount of fuel collected in the combustion chamber during ignition delay is much, then the high heat release results in rapid rise of pressure which causes diesel knock. Figure 9 shows the effect of geometries on NO mass fraction for injection ratio P5M95. It can be seen from figure that at the end of combustion, the NO mass fraction is lowest for "D1" configuration for injection ratio P5M95, whereas for P15M85 it is maximum for "D5" design. Exhaust emission is the by-product of the combustion of fuel; the complete combustion of fuel is not practically observed. Hence, unwanted gases such as carbon monoxide and unburnt hydrocarbon are emitted. In addition, high temperature that occurs in the combustion chamber promoted an unwanted reaction between nitrogen and oxygen resulting in the formation of oxide of nitrogen called as NO_x. It is well known that it is very difficult to reduce both soot and NO_x production simultaneously during combustion process. Many emission reduction technologies tend to increase soot emission while reducing NO_x and vice versa. In split injection, the fuel is injected into two pulses which leads to the reduction in ignition delay. This results in the greater fraction of combustion to occur in the expansion stroke. Since the majority of the NO_x is formed

Table 4 Configuration used for study

Injection ratios	Geometry D1	D2	D3	D4	D5	D6
R1(P5M95)	D1R1	D2R1	D3R1	D4R1	D5R1	D6R1
R2(P10M90)	D1R2	D2R2	D3R2	D4R2	D5R2	D6R2
R3(P15M85)	D1R3	D2R3	D3R3	D4R3	D5R3	D6R3

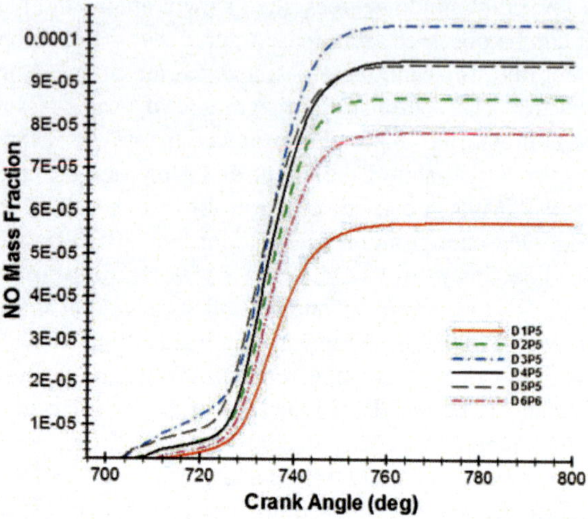

Fig. 9 NO mass fraction for P5M95

in the premixed combustion, the net amount of NO_x produced in split injection is reduced. Figure 10 shows the effect of geometries on CO mass fraction for injection ratio P5M95. It can be seen from figure that the lowest CO mass fraction at the end of combustion process is for "D4" configuration in all the three injection ratios. Carbon monoxide (CO) emissions from internal combustion engine are controlled primarily by fuel/air equivalence ratio. From the figures, it can be seen that during pilot injection the CO mass fraction increases rapidly till TDC after that during the main injection period the CO mass fraction starts decreasing towards the end of the combustion process. As the injected mass in the pilot injection increases beyond 5%, we can figure out from the graphs that the CO concentration increases till TDC but

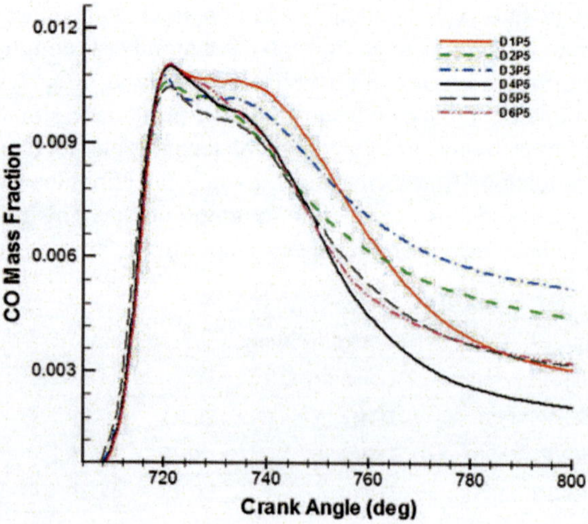

Fig. 10 CO mass fraction for P5M95

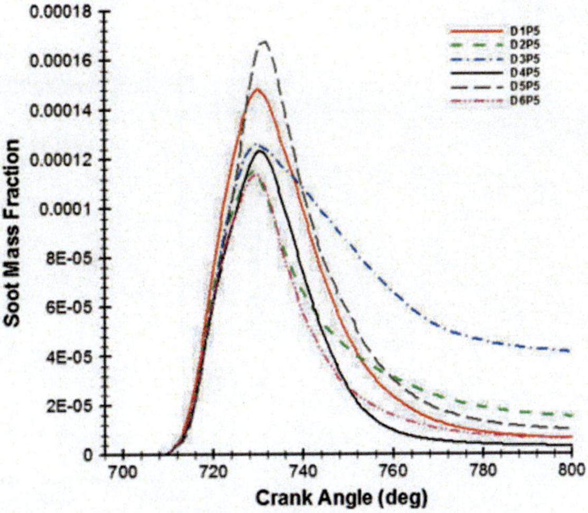

Fig. 11 Soot mass fraction for P5M95

decreases rapidly towards the end of the combustion process provided it is least in the P15M85 case. Figure 11 shows the effect of geometries on soot mass fraction for injection ratio P5M95. It can be seen from figures that the lowest soot mass fraction at the end of combustion process is for "D4" configuration for all injection ratios. It can also be figured out from the graphs that at the beginning of the pilot injection the soot mass fraction increases gradually and it is maximum at about 10° after TDC and it starts decreasing towards the end of the combustion process. As the injected mass in the pilot injection increases beyond 5%, then the peak of the soot mass fraction shifts towards the TDC.

4 Optimization

By examining the results of all the eighteen cases, an optimization study is carried out to determine the optimum trade-off between NO, CO and SOOT mass fraction. Figures 12 and 13 show the CO, NO and SOOT mass fraction at the end of expansion process. From these figures, eight cases are being selected on the basis of minimum of these parameters, i.e. closer to origin. Hence, the comparison of performance characteristics for these eight cases is made to find the case with optimum emission levels (Fig. 14).

From emission point of view, D1R1 is the best case as it has low value of NO mass compared to other selected cases shown in Fig. 15. But from performance point of view, D4R1 considered as the best case because it has high indicated mean effective pressure shown in Fig. 14 as well as it has minimum soot which is about 50% less than the previous case, i.e. D1R1. Also, D4R1 case CO mass amount is low compared to other selected cases; therefore, it is considered as the optimized case.

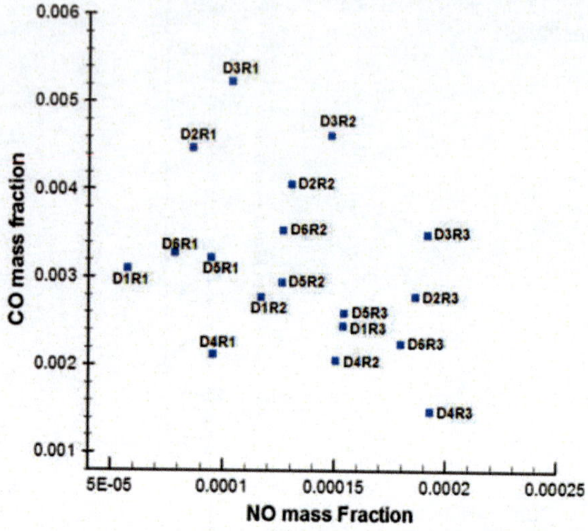

Fig. 12 CO and NO mass fraction for all cases

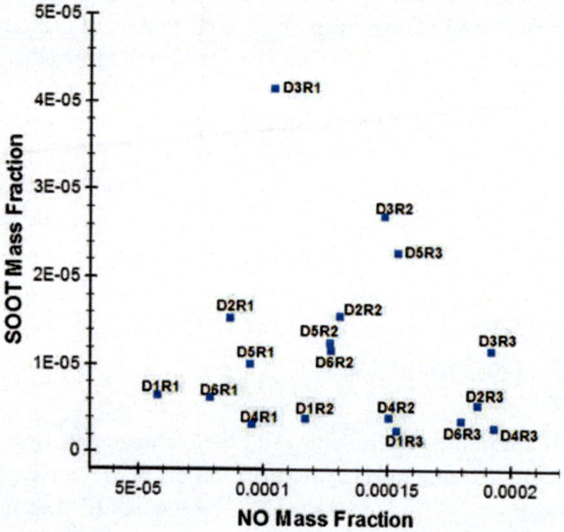

Fig. 13 SOOT and NO mass fraction for all cases

5 Conclusion

The effect of split injection on piston bowl geometry with different injection ratio in a single-cylinder DI diesel engine is numerically investigated by using AVL FIRE code. In the present investigation, a total of eighteen configurations were considered with six-piston bowl designs and three injection ratios. Finally, an optimization analysis is done to figure out the best combination of piston bowl design and injection ratio for better emission and performance. From the above study, the following conclusions are made:

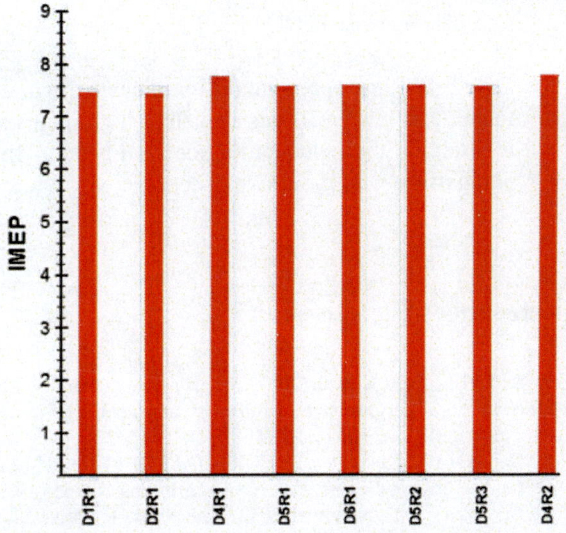

Fig. 14 IMEP for selected cases

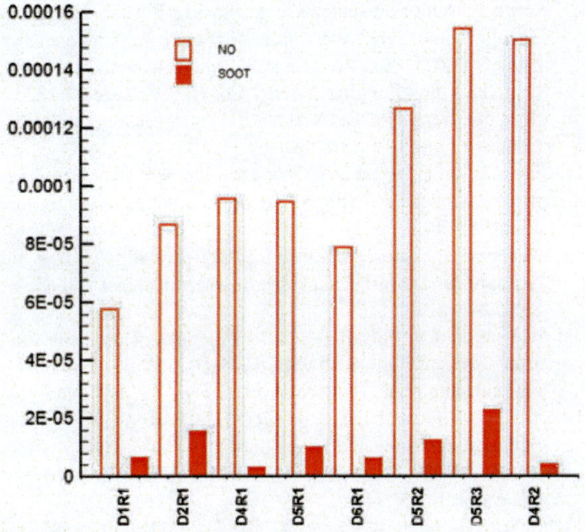

Fig. 15 SOOT and NO mass fraction for selected cases

(1) By injecting some amount of mass into pilot injection, there is an increase in ignition delay but that amount of fuel gets chemically ready to burn which results in effective burning of mass injected in main injection due to better thermal atmosphere provided by pilot injection mass thereby improving the combustion process.

(2) The results show that as the injection ratio increases, the soot and CO mass fraction decreases and the NO mass fraction increases due to high combustion

rate. The increase in NO mass fraction can be reduced by proper after-treatment of exhaust gases.

(3) Case D4R1 is proposed to the optimum configuration resulting in better performance and low emissions as well as high indicated mean effective pressure as it results in the reduction of soot and NO_x to 16% and 10%, respectively, in comparison with the single injection.

References

1. Su L, Li X, He X et al (2015) Experimental research on the diffusion flame formation and combustion performance of forced swirl combustion system for DI diesel engines. Energy Convers Manage 106:826–834
2. Gao H, Li X, Xue J, Bai H, He X, Liu F (2016) A modification to the WAVE breakup model for evaporating diesel s pray. Appl Therm Eng 108:555–566
3. Berggren C, Magnusson T (2012) Reducing automotive emissions—the potentials of combustion engine technologies and the power of policy. Energy Policy 41:636–643
4. Ntziachristos L, Mellios G, Tsokolis D et al (2014) In-use versus type-approval fuel consumption of current passenger cars in Europe. Energy Policy 67:404–411
5. Zhang J, Zhang H, Yang K et al (2014) Performance analysis of regenerative organicRankine cycle (RORC) using the pure working fluid and the zeotropic mixture over the whole operating range of a diesel engine. Energy Convers Manage 84:282–294
6. Song JO, Song CL, Tao Y et al (2011) Diesel soot oxidation during the late combustion phase. Combust Flame 158(3):446–451
7. Plamondon E, Seers P (2014) Development of a simplified dynamic model for apiezoelectric injector using multiple injection strategies with biodiesel/diesel fuel blends. Appl Energy 131:411–424
8. Mangus M, Kiani F, Mattson J (2015) Investigating the compression ignition combustion of multiple biodiesel/ULSD (ultra-lowsulfur diesel) blends via common-rail injection. Energy 89:932–945
9. d'Ambrosio S, Ferrari A (2015) Potential of multiple injection strategies implementing the after shot and optimized with the design of experiments procedure to improve diesel engine emissions and performance. Appl Energy 155:933–946
10. Tow TC, Pierpont DA, Reitz RD (1994) Reducing partiulates and NO_x emissions by using multiple injections in a heavy duty D.I. diesel engine. SAE paper 940897
11. Reitz R (1987) Modeling atomization processes in high-pressure vaporizing sprays. Atom Sprays Tech 3:309–337
12. Payri F, Benajes J, Margot X, Gil A (2004) CFD modeling of the in-cylinder flow indirect-injection diesel engines. Comput Fluids 33:995–1021
13. Heywood JB (2011) Internal combustion engine fundamental, 2011 edn. McGraw-Hill, New York
14. Brandl F, Reverencic I, Cartellieri W, Dent JC Turbulent air flow in the combustion bowl of a DI diesel engine and its effect on engine performance. SAE Paper 790040
15. Li X, Gao H, Zhao L et al (2016) Combustion and emission performance of a split injection diesel engine in a double swirl combustion system. Energy 114:1135–1146

Multiple Optimizations of Engine Parameters of Single-Cylinder Four-Stroke Direct Injection Diesel Engine Operated on Dual Fuel Mode Using Biodiesel-Treated and Untreated Biogas Combination

V. S. Yaliwal, S. R. Daboji, K. N. Patil, M. K. Marikatti, and N. R. Banapurmath

1 Introduction

Reducing harmful tail pipe gases and toxic emissions from the internal combustion (IC) engines is a major problem for the world. In addition, reduction in underground oil reserves and a substantial rising of fossil fuel prices in world market leading to lowered life standard. Hence, it is led to widespread search for alternative or interchangeable fuels [1, 2]. For a sustainable development, biofuels play a very important role because biofuels derived from biomass and other renewable sources have several benefits compared to fossil fuels. Biofuels derived from biomass are renewable and biodegradable in nature. Further, they address food and energy safety, overseas exchange saving and also address ecological and socioeconomic issues more appropriately.

In present days, a number of precise standards and regulations have been initiated in India and is same for rest of the world. Mainly, it is intended to promote novel and eco-friendly solutions in power generation sectors, and particularly in industry and agriculture applications. In this regard, Government of India has taken many facilitating policies toward renewable energy technologies. Further, several policies were favored by considerable subsidies provided by central and state governments for various organizations which are engaged in power generation and agriculture

V. S. Yaliwal (✉) · S. R. Daboji · K. N. Patil · M. K. Marikatti
Department of Mechanical Engineering, S.D.M. College of Engineering and Technology, Dharwad, Karnataka 580002, India
e-mail: vsyaliwal2000@rediffmail.com

N. R. Banapurmath
Department of Mechanical Engineering, B.V.B. College of Engineering and Technology, KLE Technological University, Hubli, Karnataka 580031, India

A. K. Gupta et al. (eds.), *Advances in IC Engines and Combustion Technology*, Lecture Notes in Mechanical Engineering,
https://doi.org/10.1007/978-981-15-5996-9_60

applications. India is made up of more than 600, 000 villages; nearly 1800 villages still do not have access to electricity; and more than 62,000 villages are still waiting to be wired. Biogas is a flexible energy source, and it finds wide application at both domestic and industrial. Nearly about 980 million tones per year of cattle dung is available in India and has a potential of generating 6.38×10^{10} m^3 of biogas. In addition, 350 million tones of manure can be produced. India is already installed about 4.5 million domestic biogas plants. In view of this, India is much more concentrating on power generation through renewable sources. It is because transmission of electricity is not feasible in hilly and remote places due to ecological and unfavorable conditions. Hence, the use of locally available renewable power source such as biogas provides an advantage for distributed power production compared to grid extension. Such that even hilly and remote villages can also get access to affordable electricity. In this context, several researchers were conducted exhaustive research on the power generation using renewable fuels in providing electricity to all such mentioned areas [1–4]. In view of this, the promotion of biogas-based power generation is currently being encouraged in India. This is due to the fact that India is dependent more on agriculture and rank second in the world for harnessing thermal energy using biogas. It is observed that power generation using renewable energy sources contribute about 5% of the total power generation. Among all the renewable energy sources, biogas derived from anaerobic digestion seems to be one of the most promising solutions [5]. Further, the use of biogas in diesel engines offers fuel flexibility, acceptable thermal efficiency with lower exhaust gas emissions, and easy conversion of existing diesel engines by retaining higher compression ratio (CR) and with minor hardware modifications [6, 7].

Renewable fuels such as biodiesels, producer gas, and biogas present a significant role in achieving energy demand. Biofuels derived from biomass and other organic waste have nearly similar properties as that of fossil fuels; hence, such fuel may be used for transport and electricity generation purposes more conveniently. Utilization of biofuels for diesel engine applications has been reported by several investigators. Use of biodiesels for diesel engine applications resulted in acceptable performance and emission levels. Moreover, several researchers have focused considerable attention on the development of alternative fuel sources with emphasis on biofuels and biogas in particular [5, 8]. Recently, several researchers have been conducted experiments using various biofuels and blends operating on single fuel method. Lower performance with amplified smoke, hydrocarbon (HC)/carbon monoxide (CO) emission levels and decreased nitric oxide (NO_x) emissions have been described in the literatures [9–14]. However, Patel et al. reported that the heat release rate curve for neat biodiesel was nearly similar to diesel operation. Therefore, combustion characteristics of diesel engine with blended fuels resulted in slightly different heat release rates. Effect of spray characteristics on the engine performance, knocking, and vibration caused by the use of biodiesels has been reported. Biogas operation in dual fuel engine resulted in prolonged ignition delay due to reduction of oxygen concentration and increased charges in polytropic index. This fact significantly affects the combustion characteristics, tail pipe emissions, engine performance, and fuel consumption.

Biodiesel and its blends operation with and without gaseous fuels resulted in differences in combustion characteristics compared to diesel operation due to differences in the physicochemical properties. Hence, they can lead to combustion with more complexity because it involves two fuels with different properties and is burnt simultaneously inside the engine cylinder. Therefore, fuel combination combustion in a dual fuel engine is a result of three stages [15]. Biogas is obtained by anaerobic digestion of cow dung and any other organic matter in a digester. They serve as an assured renewable fuel for both SI and CI engine application. Most existing dual-fueled engines are operated either on gaseous type fuels with injected fuel or with liquid fuel alone in a diesel engine. Engine operation with liquid and gaseous fuels requires small amount of liquid fuel with an in-cylinder injection system. However, diesel engines cannot be operated with gaseous fuel alone without the injection of a small amount of pilot fuel because biogas will not ignite under the prevailing conditions of temperature and pressure. In such engines, the combustion was regulated by allowing gaseous fuel along with air through suitable mixing chamber or venture via inlet manifold. Several investigators reported that decreased engine output, others reported unchanged or increased power output. Significant research work on biogas-fueled diesel engine and its influence on combustion behavior have been described in the literature. References [16–18] presented unexplored feedstocks for biogas generation and assessed progress, potential challenges, perspectives and future opportunities of anaerobic membrane bioreactors and presented various technological solutions for optimizing biogas conversion and operational conditions. Dual fuel engines are recognized for better pilot fuel saving with decreased smoke and nitric oxide (NO_x) emission levels. Researchers have stated that decreased performance with increased HC, CO and decreased smoke and NO_x levels under biogas dual-fueling [5, 8]. Further, several investigators have reported that raw biogas resulted in poor performance due to impurity constituents such as carbon dioxide (CO_2), hydrogen sulfide (H_2S). However, methane (CH_4)-rich biogas resulted in improved performance with higher pilot fuel savings ranging from 20 to 87% [19–21]. Himsar et al. reported that the efficiency of dual-fueled engine is affected to a great extent by biogas flow rate and methane concentration. Composition of methane and carbon dioxide in biogas depends on the feedstock type and loading rate, atmospheric conditions, water content, and type of enzyme [22]. Main combustible component of biogas is methane and hydrogen. However, the percentage of hydrogen is very less. Biogas is a high octane number fuel and has considerable amount of CO_2. In addition, biogas is a high octane number gaseous fuel and use of a high CR for biogas-fueled dual fuel engine results into superior performance without engine knocking.

Further, the performance of biogas-fueled dual fuel engine greatly affected by air-fuel mixing and combustion chamber design. In view of this, suitable carburetors for biogas operation have been developed [23]. This will greatly assist the biogas utilization commercially. In addition, to operate diesel engine on 100% biogas needs lot of conversion, and hence, the conversion of diesel engine takes more cost for removal of the conventional fuel injection system, placing a suitable spark plug in place of the nozzle injector, modifying intake manifold system, changing compression ratio, modifying the combustion chamber, etc. [24] reported that upgraded biogas behaves similar to that of compressed natural gas and can result better engine performance.

2 Present Work

The review of literature suggests that the consequence of injection parameters and compression ratio on the combustion characteristics of biogas-fueled engine has been less investigated. Hence, this area still needs detailed study. In this context, main attention was concentrated on the maximum substitution of biogas and improving efficiency of biogas operated dual fuel engine with decreased emission levels. In this work, experimentations were carried out with a single-cylinder direct injection (DI) CI engine functioning on dual fuel mode using biodiesel of HOME and biogas. In the first phase, optimum injection timing (ranging from 19 to 27 deg. bTDC) and injection pressure (ranging from 200 to 260 bar in steps of 20) were determined. Further in order to examine the consequences of compression ratio, diesel engine was operated with varied compression ratio from 15 to 17.5. Finally, combustion characteristics obtained functioning on dual fuel mode using HOME and biogas (treated and untreated biogas) combination were compared with baseline data and analyzed.

3 Properties of Fuels

Biogas consists of methane and carbon dioxide, both constitute nearly about 95%, and rests are organic/non-organic traces of variable compositions. Moreover, quality of biogas depends on the sources of origin. Compressed biogas was readily available in cylinder and purchased from local industry. Properties of fuels used such as HOME and biogas used in the present work were computed in the college laboratory and are presented in Table 1. Composition of biogas was compared with literature [25].

4 Experimental Setup

In this present work, Kirloskar, single-cylinder, four-stroke, direct injection compression ignition (CI) engine test rig was used. Figure 1 shows the gas flow diagram for duel fuel engine. Figure 2 shows photographic views of mixing chamber. The specification of the compression ignition (CI) engine is given in Table 2. Electrical loading was used for engine loading. Exhaustive experiments on the diesel engine have been conducted, and data were recorded at different loads and operated at 1500 rev/min constant speed. Biogas was generated by anaerobic digestion of cow dung and organic material using a digester and is directly admitted into the combustion

Table 1 Properties of pilot liquid fuel and biogas composition [25]

Sl. No.	Properties	Pilot fuel properties		Biogas composition		
		Diesel	HOME	Composition	TBG	UTBG
1	Viscosity @ 40 °C (cSt)	3.6–3.9	5.6	Methane	42–72%	52–71%
2	Flash point, °C	56	163	Carbon dioxide	28–58%	28–41%
3	Calorific value, kJ/kg	45,000	36,010	Hydrogen sulfide	12–20,000 ppm	4–8000 ppm
4	Specific gravity	0.830	0.870	Oxygen	0.06–2.5%	0.75–2.5%
5	Density, kg/m^3	830	890	Nitrogen	0.012–4.8%	0.01–0.25%
6	–	–	–	Hydrogen	1–3%	2–4%
7	–	–	–	Water vapor	Remaining	Remaining

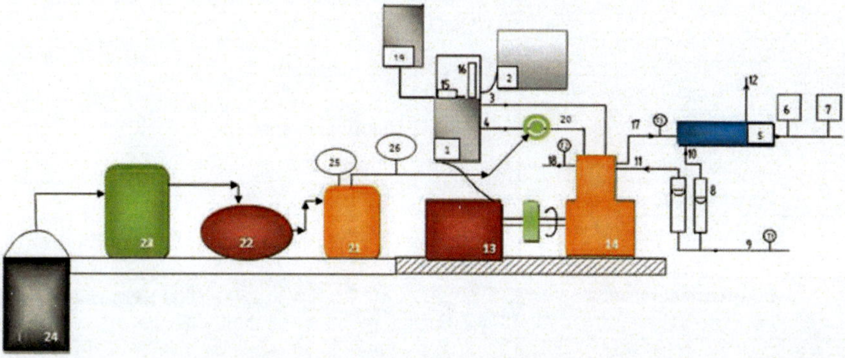

Fig. 1 Experimental setup for duel fuel engine. Experimental setup: 1—control panel, 2—computer system, 3—diesel flow line, 4—air flow line, 5—calorimeter, 6—exhaust gas analyzer, 7—smoke meter, 8—rotameter, 9, 11—inlet water temperature, 10—calorimeter inlet water temperature, 12—calorimeter outlet water temperature, 13—dynamometer, 14—CI engine, 15—speed measurement, 16—burette for fuel measurement, 17—exhaust gas outlet, 18—outlet water temperature, 19—fuel tank, T1—inlet water temperature, T2—outlet water temperature, T3—exhaust gas temperature, 20—air-gas mixing chamber, 21—storage tank at 3 bar, 22—compressor, 23—storage tank, 24—digester, 25—pressure gauge, 26—gas flow meter

chamber during suction of the engine. The flow rate air and gas was measured using a calibrated single tube manometer and gas flow meter, respectively. To prevent flow disturbance of biogas and air mixture, a suitable mixing chamber was used. The biogas flow was not restricted and mixed with air using suitable mixing chamber shown in Fig. 3. Mixing chamber was built up by Dr. R. C. Hallikeri, Dairy Science division, Agricultural University, Dharwad, India. It is developed in order to have a

Fig. 2 Carburetor used for biogas induction

Table 2 Specification of experimental test rig

Sl No	Parameters	Specification
1	Machine supplier	Apex Innovations Pvt. Ltd., Sangli. Maharastra State
2	Engine type	Single-cylinder four-stroke water-cooled direct injection TV1 CI engine with a displacement volume of 662 cc, compression ratio of 17:1, developing 5.2 kW at 1500 rev/min TV1 (Kirolsker make)
3	Software used	Engine soft
4	Nozzle opening pressure	200–225 bar
5	Governor type	Mechanical centrifugal type
6	Cylinder diameter (Bore)	0.0875 m
7	Stroke length	0.11 m
8	Combustion camber	Open chamber (direct injection) with hemispherical cavity
9	Eddy current dynamometer	Model: AG—10, 7.5 KW at 1500–3000 RPM and water flows through dynamometer during the use

capability to keep the required air-fuel ratio with varying load and pulsating gas flow conditions and ensured smooth operation. Fuel consumption during dual fuel operation has been measured on volumetric basis. Engine speed was kept constant in both versions of the injected fuels (diesel/HOME) operation by adjusting the governor speed. Engine injection timing (IT), injection pressure (IP), and compression ratio (CR) were at 230 bTDC, 205 bar, and 17.5. However, in the present work, pilot fuel IT, IP, and CR were varied from 19 to 270 bTDC, 205–260 bar, and 15–17.5, respectively.

The exhaust five gas analyzer and HARTRIDGE smoke meter were periodically calibrated and used to measure composition of exhaust gas emission levels. In the beginning of the experimentation, both were changed to ON position and permitted to stable condition before recording data. The cooling water temperature was held at 70 °C. The instrument used to measure the smoke works on the principle of comparative basis. Exhaust gas analyzer measure regulated emission levels such as

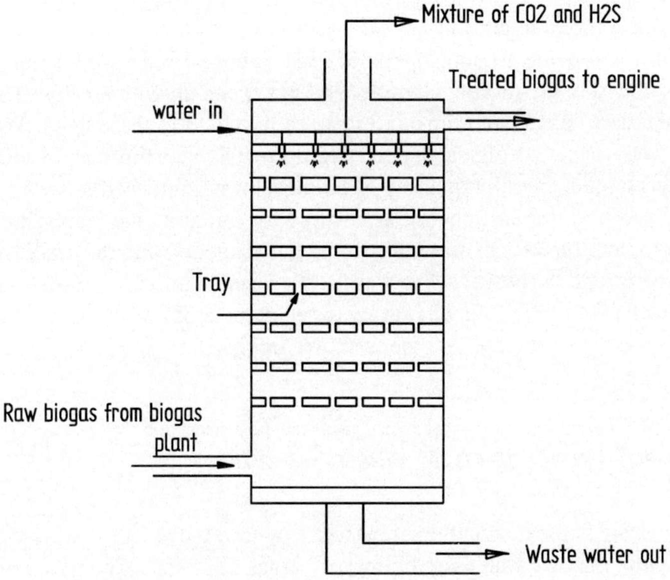

Fig. 3 Water scrubbing for biogas treatment

HC, CO, and NO_x. It works on the basis of non-dispersive infrared technology, and it has various sensors for measuring an individual gas and pyroelectric detectors. Measured emission levels were transmitted to voltage signals proportional to light intensity. Finally, it is converted from analogue or digital signal and thereby provides the actual data. The injection pressure for HOME-biogas operation was varied in the range of 200–260 bar (in steps of 20 bar). Finally, the results obtained with HOME-biogas operation were evaluated in comparison with diesel-biogas operation. The specification of the biogas digester and compression ignition (CI) engine is given in Tables 2. For each load, five measurements were recorded to agreeing the correctness of the data generated and averaged out data was considered for analysis.

4.1 Gaseous Fuel Supply System

Raw or untreated biogas (UTBG) was obtained from a 37.2 m^3 fixed dome biogas plant and has 6 m in diameter and 6 m height with a capacity of generating a biogas of 25 m^3 per day. Biogas was further transferred to water scrubber column through a 3/4th inch flexible hose pipe. Purified or treated biogas (TBG) was then admitted to suitable compressor operated by 0.5 h.p electric motor. Uniform supply of quality gas to the dual-fueled engine is ensured, and it depends on the flow conditions, pressure drop across the flow line. Flow rate of gas significantly depends on the design of engine and operating circumstances. Suitable mixing chamber or carburetor was used for combining air and gas, and pulsating gas flow conditions were taken care, besides it provide smooth operation with minimal pressure losses. The flow meter was united to the mixing chamber and gas speed up with air flow. During the engine operation, biogas and air mixture were provided to the engine. The supply of biogas was regulated physically to obtain greatest substitution. In this present work, for each load, the pilot quantity of biodiesel was controlled in such a way that minimum quantity of their injection is ensured for initiating the combustion of fuel combinations and further ensuring maximum gas induction into the intake manifold. During experiment, the biogas valve was fully opened in order to substitute more gas and the quantity of liquid fuel was varied to maintain constant speed. Carburetor for air-bio gas used in a duel fuel engine is shown in Fig. 2.

4.2 *Biogas Purification by Water Scrubbing*

Water scrubbing is most economical easiest way to remove CO_2 and H_2S from the biogas. Figure 3 shows water scrubbing for biogas treatment. The major advantage is without specialized equipments and chemicals; higher conversion with amplified purity has been recorded. Methane-enriched gas leads to higher specific output and thermal efficiency caused by the faster burning rate of methane. In normal biogas, nearly 30% of CO_2 by volume presence is observed. Moreover, the solubility also depends on the pressure of the water. More dissolubility for CO_2 than methane was noticed with higher pressure of water. Figure 3 illustrates the simple arrangement of water scrubbing [26].

5 Results and Discussion

This section discusses the outcome of the work on the combustion of the HOME–biogas dual-fueled engine. Comparative assessment of dual fuel engine by suitable fuel and engine modification is a unique feature of this study. In this present work, the effect of IT, IP, and CR on the combustion and tire out emissions of diesel engine functioning on the application of two fuels using diesel/HOME-biogas mixture at maximum load condition was investigated. In this context, a concise overview of preliminary engine testing is summarized in the following subsequent sections.

5.1 *Effect of Injection Timing*

This section presents the consequences of IT on the performance of diesel/biodiesel-biogas-fueled dual fuel engine. Injection timings were varied from 19 to 270 bTDC, while biogas flow rate was kept constant. Results are reported in the following subsequent section.

5.1.1 Performance Characteristics

Figure 4 presents influence of IT on the brake thermal efficiency (BTE) for various fuel combinations. Increased BTE was noticed for diesel-biogas dual fuel method of operation than HOME-biogas operation at 80% load. Biogas contains methane and being universal uncontaminated burning gaseous fuel having considerably greater energy content. In this present work, injected pilot fuel and its properties can role a significant responsibility on the dual fuel engine efficiency. HOME is a low volatile fuel which in turn results into complexity during fuel vaporization and atomization. In addition, its lower energy content than diesel fuel together leads to decreased thermal efficiency. Decreased intake air during dual fuel operation is also accountable for the observed trend, which will further lower the volumetric efficiency. At enhanced magnitude of pilot fuel, biogas utilization improves leading to increased thermal efficiency due to augmented ignition sources and also it leads to amplified combustion temperature which lowers the gaseous fuel escaping considerably from the combustion.

Advancing the IT from 19° to 27 deg. bTDC improved engine performance was noticed. Therefore, dual fuel process with IT of 27 deg. bTDC and use of treated biogas (TBG) resulted in enhanced thermal efficiency due to increased time for mixing of air-gaseous fuel mixture with pilot injected fuel. It is observed that for the same TBG induction, diesel-based operation resulted in 8.22% increased thermal efficiency compared HOME-based operation. Pilot fuel injection timing advancing enhances the thermal efficiency with normal running of the engine and lowers the delay period resulting into amplified performance, but this may be inclined to increased fuel consumption. Increased thermal efficiency with TBG could be due to the presence of more methane in the biogas during combustion. Furthermore,

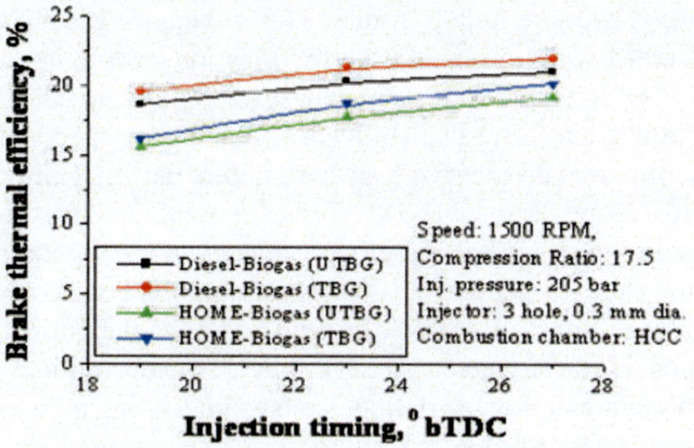

Fig. 4 Effect of injection timing on the BTE

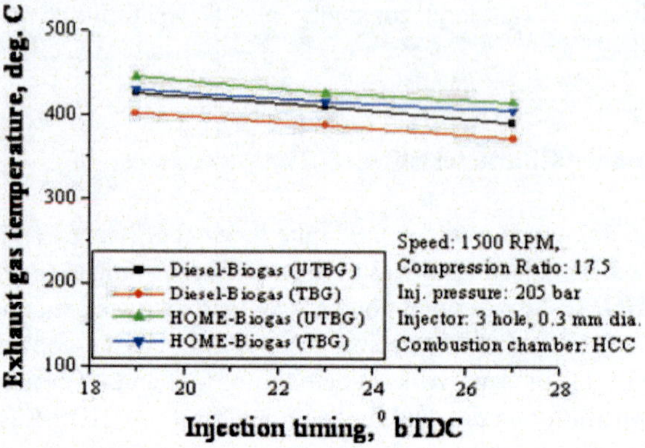

Fig. 5 Effect of injection timing on the EGT

thermal efficiency increases with advancing of injection timing. This may be due to the use of biogas requiring more time for physical mixing with air. TBG in the present work consists of larger content of methane. This in turn leads to increased pilot fuel saving. Hence, it is expected that the specific fuel consumption reduces with increasing methane concentration. However, the presence of carbon dioxide in biogas called as untreated biogas (UTBG) that affects lower heating value of a biogas leading to lower the performance caused by the reduced burning characteristics. Hence, decreased combustion rate with UTBG is bound to take place due to the participation of carbon dioxide during combustion. At 80% load, BTE for diesel-TBG, diesel-UTBG, HOME-TBG, and HOME-UTBG operation at 27° bTDC injection timing were 22.01, 21.01, 19.2, and 20.12%, respectively.

Effect of IT on the temperature of exhaust gas EGT is presented in Fig. 5. For the same operating conditions, diesel-biogas operation provides decreased EGT compared to HOME-biogas operation. This may be due to differences in the pilot injected fuel. Compared with 19 and 230 bTDC injection timing, 270 bTDC injection timing gave lower EGT. It is characterized to the fact that better oxidation of the fuel combination caused by more time is available for mixing air-gas mixture with pilot fuel droplets. Therefore, appropriated mixture formation leads to improved combustion during premixed and rapid combustion phase during diesel/HOME-biogas operation. Better air utilization due to availability of advancing the injection timing results better mixing of pilot fuel droplets with air-gas mixture during ignition delay before impinging of pilot fuel on the piston bowl. Further, it is observed that significant variations in the combustion have been observed with a dual fuel operation when TBG was used. For the same diesel and TBG combination, EGT was reduced by 3.8% compared to the operation with UTBG. It could be due to the presence of methane during combustion. However, increased EGT with UTBG operation is mainly due to the participation of carbon dioxide during combustion leading to increase diffusion combustion phase. It indicates lowered combustion rate considerably. At 80% load, EGT's obtained for diesel-TBG, diesel-UTBG, HOME-TBG, and HOME-UTBG operation at IT of 27 deg. bTDC were 402, 416, 390 °C, and 372 °C, respectively, at IT of 27 deg. bTDC

5.1.2 Emission Characteristics

Figure 6 presents the disparity in smoke heights at 80% load for dual fuel operation. Inducted fuel being same, the smoke levels were decreased for diesel-based dual fuel operation than HOME-based operation. Low volatile nature of biodiesel due to higher viscosity and presence of heavier fatty acid molecular structure is accountable for the increased smoke levels. However, for the same fuel combination, advancing the IT from 19 to 27 deg. bTDC lowers the smoke levels compared to the operation with 19 and 230 bTDC injection timing. This is because of better utilization of available oxygen during combustion with pilot fuel and biogas leading to better oxidation. In addition, better mixture formation caused by the more time availability is also responsible for the observed trend. Therefore, enhanced combustion was succeeded significantly and attributes to lower smoke levels. For all injection timings tested, smoke emission was further decreased with use of TBG induction. This may be due to the induction of TBG in the engine cylinder during dual fuel process; the pilot fuel consumption is lowered leading to decrease the exhaust gas opaqueness. Influence of methane concentration present in the TBG on the smoke opacity is presented in Fig. 6. It is noticed that diesel-based dual fuel operation, induction of TBG lowers smoke opacity by 18.75% compared to the operation with UTBG. Similar trend was obtained for HOME-based dual fuel operation. Hence, the presence of methane in the biogas leads to decreased smoke opacity in the tail pipe exhaust due to reduced consumption of the carbon-based pilot fuel consumption [7]. In addition, it may be due to superior combustion caused by the presence of methane in the biogas. However, the dual fuel operation with UTBG leads to increased smoke levels due

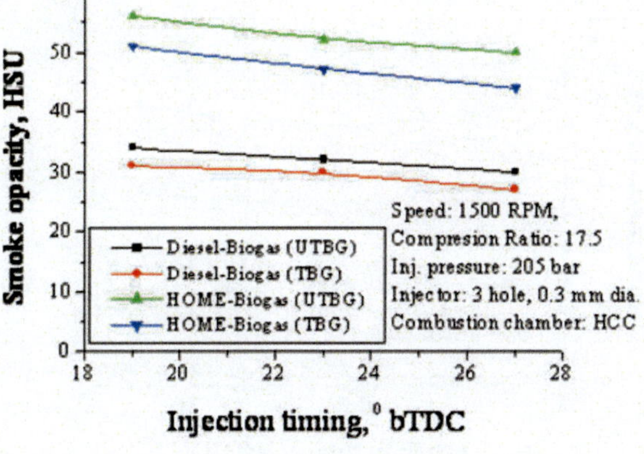

Fig. 6 Effect of injection timing on the smoke opacity

to the participation of carbon dioxide during combustion, which in turn indicates lowered combustion rate considerably. Hence, it is concluded that, for the inducted fuel, diesel/HOME-based dual fuel process decreases smoke levels with advancing pilot fuel injection timing. It indicates lowered combustion rate considerably. At 80% load, smoke levels obtained for diesel-TBG, diesel-UTBG, HOME-TBG, and HOME-UTBG operation at IT of 270 bTDC were 26, 32, 50 HSU, and 42 HSU, respectively, at IT of 270 bTDC.

The HC and CO variations with respect to timing of injection for the selected fuel combinations are presented in Figs. 7 and 8 at 80% load. HC and CO levels are the outcome of partial burning of hydrocarbon fuel. It depends mainly on the air-fuel ratio relative to chemically correct proportion. Amplified HC and CO levels were bound to occur for biodiesel-biogas combination than diesel-biogas dual fuel process. This may be attributed to inferior properties of biodiesel compared to fossil fuel. Both HC and CO levels were diminished considerably, when IT was advanced

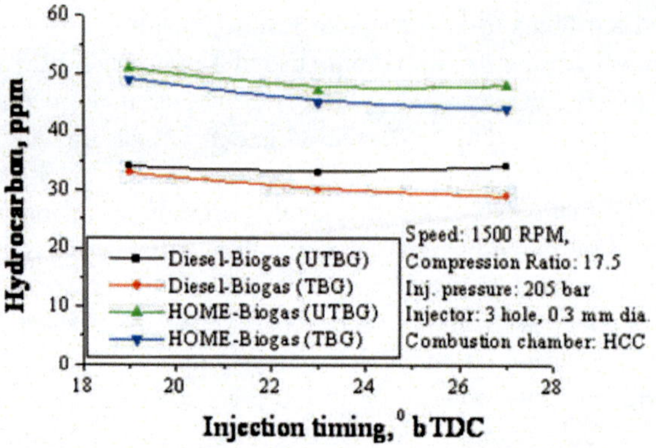

Fig. 7 Effect of injection timing on the HC emissions

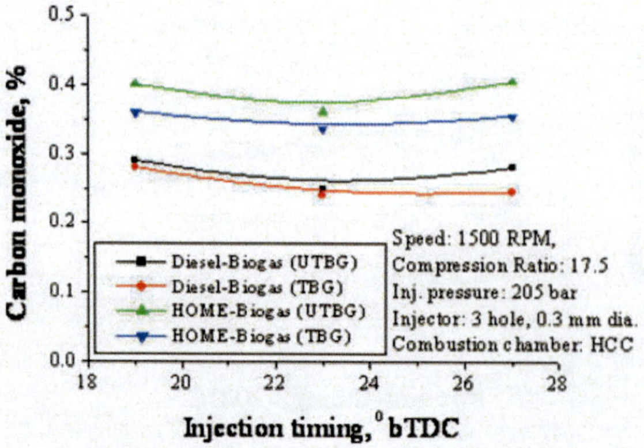

Fig. 8 Effect of injection timing on the CO emissions

and it may be caused by superior combustion taking place inside the engine leading to amplified thermal efficiency. Advancing of IT provides increased heat release during uncontrolled combustion caused by availability of more time for mixing the fuel combination. Advancing fuel IT results in elongated delay period which in turn leads to lowered UBHC and CO emissions. Longer delay period allows better spray characteristics through a flame propagation region. Further increased premixed zones build up greater combustion pressure and temperature leading to decrease the carbon-based emissions. It is also noticed that CO concentration in the tail pipe emission for diesel-based operation at all IT's lower than HOME-based operation. This may be due to the fact that dual fuel process under diesel, the stoichiometric flame velocity is much greater compared to the operation with HOME. Both HC and CO emissions depend on the oxygen availability. But during dual fuel operation oxygen supply is lowered due to the addition of biogas during induction. Oxygen deficiency during combustion is not powerfully affected by injection parameters and compression ratio. When the biogas was introduced inside the engine cylinder, the oxygen percentage is reduced [7]. Furthermore, CO emission was not affected by the presence of methane in the biogas. This may be due to the fact that the oxygen concentration in the air is not affected by methane concentration. In dual fuel engine when methane gas increases, automatically pilot fuel consumption reduces. Therefore, total energy content in the engine cylinder almost remains same leading to same CO emission. But the type of pilot fuel and presence of impurities in the biogas significantly affects the combustion more negatively. Thus, the CO emission is found to be greater with biodiesel and untreated biogas combination [7].

However, for diesel-based dual fuel operation induction of TBG lowers both HC and CO levels by 14.8 and 14.2% compared to the operation with UTBG. Similar trend was noticed for the HOME-based operation. This may be due to the use of improved properties of biogas leading to better combustion. However, dual fuel operation with UTBG results into greater HC and CO levels due to partial combustion caused by the participation of carbon dioxide during combustion. It indicates condensed combustion rate significantly. It is also observed that for the same inducted fuel, HC and CO levels of diesel/HOME-based operation decrease both emission levels with advancing pilot fuel IT. At 80% load, HC levels obtained for diesel-TBG, diesel-UTBG, HOME-TBG, and HOME-UTBG operation at IT of 27 deg. bTDC were 29, 34, 44, and 52 HSU, respectively, at IT of 270 bTDC. Similarly, CO levels at 80% load obtained for diesel-TBG, diesel-UTBG, HOME-TBG, and HOME-UTBG operation at IT of 27 deg. bTDC were 0.24, 0.28, 0.34%, and 0.41%, respectively, at IT of 27 deg. bTDC.

NO_x levels in the exhaust are produced during combustion depending on availability oxygen and nitrogen. Absence of any one gas leads to affect the NO_x production negatively. However, dual fuel engine uses biogas, and hence, part of oxygen is replaced by biogas leading to lower air intake. Hence, it reduces oxygen concentration in the engine cylinder, and as a result, the formation of NO_x was reduced significantly. Figure 9 presents the NO_x levels with respect to injection timing for selected fuel combinations at 80% load. NO_x emissions were greater at advanced IT (27 deg. bTDC) due to enhanced combustion rates caused by the increased delay

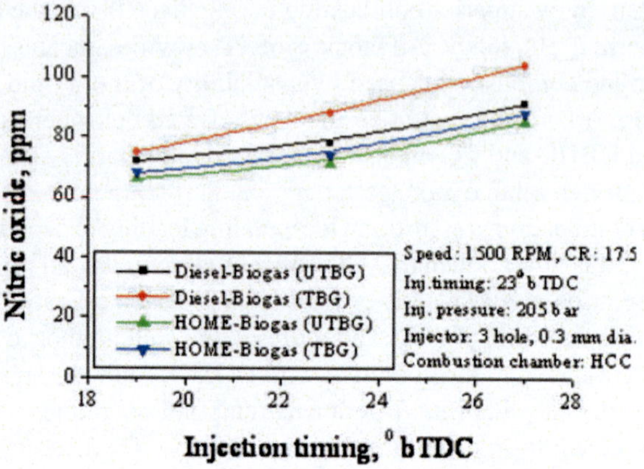

Fig. 9 Effect of injection timing on the NO_x emissions

period resulted in amplified heat release. Dual fuel operation with TBG and UTBG induction, biodiesel fueled operation gave 8–16% decreased NO_x levels compared to diesel-based operation. Advancing the injection timing results into greater combustion temperature in flame zone, which in turn results into increased combustion pressure. Further, investigation showed that for the same pilot injected fuel and use of TBG as inducted fuel showed increased NO_x levels due to enhanced combustion rate. This may be due to the participation of combustible gas during combustion. However, decreased NO_x levels for the dual fuel process has been noticed when UTBG was added. This is due to the fact that the occurrence of carbon dioxide in the biogas leads to retard the burning of fuel combination. At maximum load, NO_x levels obtained for diesel-TBG, diesel-UTBG, HOME-TBG, and HOME-UTBG operation at IT of 27 deg. bTDC were 104, 91, 88 ppm, and 82 ppm, respectively, at IT of 27 deg. bTDC.

5.2 *Effect of Injection Pressure*

This part of discussion presents the results obtained on the study of IP on the diesel/biodiesel-biogas-fueled dual fuel engine performance. Injection pressure was varied from 205 to 240 bar. Results are reported in the following subsequent section.

5.2.1 Performance Characteristics

Figure 10 provides brake thermal efficiency for different injectors opening pressure (IOP) at 80% load. Spray characteristics significantly affect engine performance. Investigation showed that biodiesel operation favors increased injection pressure.

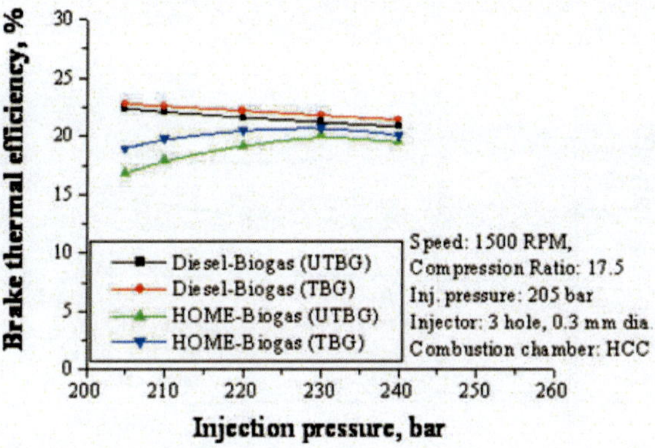

Fig. 10 Effect of injection pressure on the BTE

For both TBG and UTBG fueled dual fuel operation and at identical injection pressure, diesel-based operation resulted in considerably amplified thermal efficiency than HOME-based operation. It may be owing to inappropriate mixing of air-biogas mixture with biodiesel droplets caused by low volatility character of biodiesel. Lower energy content of HOME along with lower flame speed and higher auto-ignition temperature of biogas further adjoins to this trend. But investigation showed that, increased injection pressure from 205 to 230 bar, HOME-biogas operation provides 3.6% increased thermal efficiency compared to the operation at 205 bar. This may be caused the improved evaporation of high viscous fuel, enhanced fuel penetration, and reduced wall wetting at 230 bar injection pressure. However, decreased thermal efficiency was noticed at 240 bar injection pressure. At 240 bar injection pressure maximum fuel evaporate and droplets may accumulate on the compressed gas due to reduced relative velocity of droplets. In addition at decreased injection pressure (less than 230 bar), the hydraulic flow rate of injected pilot fuel decreases resulting into larger droplets which in turn leads to reduced thermal efficiency due to difficulty in evaporation and improper air-fuel mixing. The thermal efficiency at 230 bar injection pressure for diesel-TBG, diesel-UTBG HOME-TBG, and HOME-UTBG operation were found to be 21.8, 21.12, 20.89%, and 20.6%, respectively, at 80% load.

5.2.2 Emission Characteristics

It is well identified that during dual fuel process, oxygen deficiency is bound to occur. However, utilizing available oxygen is utmost necessary. This is possible by adopting proper injection strategy, altering engine design and operating conditions. In this context, in the present work injection pressure was varied from 205 bar to 240 bar and its effect on emissions has been discussed in the following subsequent section.

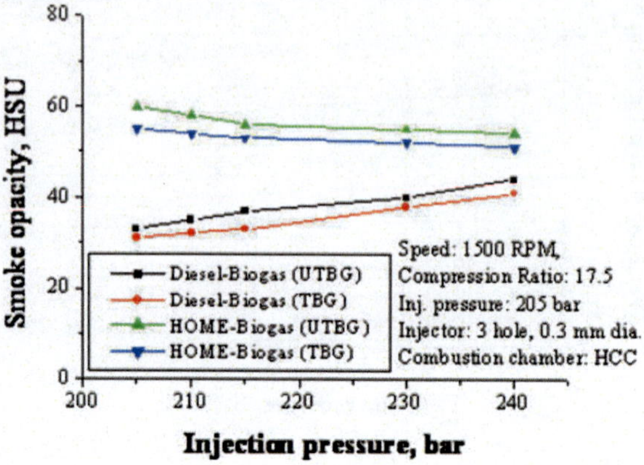

Fig. 11 Effect of injection pressure on the smoke opacity

Figure 11 presents the effect of injection pressure (IP) on the smoke levels for various selected fuel combinations. For the same operating conditions, dual fuel operation with TBG and UBTG as introduced fuel during suction and diesel as injected fuel, dual fuel operation provided decreased smoke levels compared to HOME operation. Partial combustion due to decreased oxidation caused by the insufficient oxygen and low volatility characteristic of biodiesel may be accountable for this trend. However, at a 230 bar IP, HOME-based dual fuel operation provided decreased smoke levels compared to the operation with other injection pressures. From the results, it is noticed that 5.6% reduced smoke levels when injection pressure was varied from lower IP to 230 bar. Reduced droplets size and increased surface area of droplets leading to better utilization of available oxygen thereby combustion rate was enhanced significantly. Further, another possible reason may be that the flame developed by combustion may cover entire cylinder area. However, at a 240 bar IP, combustion rate may retard due to reduced air-fuel mixing rates leading to reduced combustion efficiency. Further, dual fuel operation with lower IP (less than 230 bar), size of the fuel droplets may be larger leading to difficulty in mixing of pilot fuel with air. This feature may slow down the combustion rate compared operation with 230 bar. It is also observed that decreased soot oxidation is possible with diesel fuel with higher IP (more than 205 bar) due to reduced relative velocity of droplets, reduced lower mixing rates and decreased diesel droplets penetration into fully compressed air-biogas mixture The smoke opacity at 230 bar injection pressure for diesel-TBG, diesel-UTBG HOME-TBG, and HOME-UTBG operation were 38, 40, 51 HSU, and 56 HSU, respectively, at maximum load.

Figures 12 and 13 present the influence of IP HC and CO emissions. For the same operating conditions, dual fuel process with TBG and UTBG and diesel as pilot injected fuel present reduced levels of emission than HOME-based operation over entire pressure range. It may be due to partial burning of fuel because the existence of greater molecular mass of fatty acids and reduced burning of biodiesel. Investigation showed that HOME-biogas (TBG) operation showed that HC and CO levels were

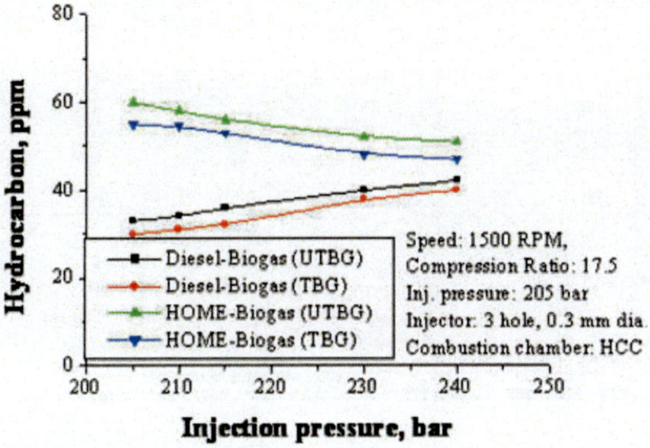

Fig. 12 Effect of injection pressure on the HC emissions

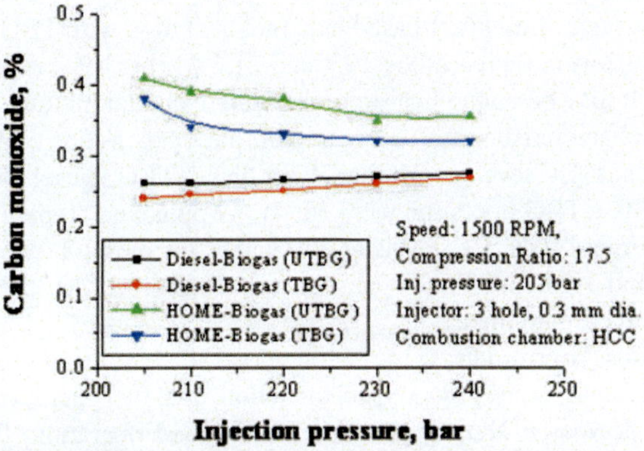

Fig. 13 Effect of injection pressure on the CO emissions

condensed by 12.6% and 15.8%, respectively, when IP was enhanced from 205 to 230 bar due to reduced partial combustion caused by better mixing rates during HOME-biogas (TBG) operation. Increase in combustion pressures and temperature due to appropriate consumption of available oxygen leading to improved combustion at an injection pressure of 230 bar. Also, the production of tiny droplets caused by greater IP (230 bar) results into rapid flame propagation and which may lead to burn entire air-fuel mixture. This fact may reduce HC and CO emissions at an IP of 230 bar compared to the operation with other IP's. However, for the identical fuel combination the HC and CO levels were higher at lower IP (less than 230 bar) due to inappropriate fuel and air mixing resulting into condensing of the oxidation process. Reduced oxygen and decreased hydraulic flow rate of pilot fuel is accountable for this trend. Further at an increased IP (more than 230 bar), the pilot fuel droplets momentum was decreased due to decreased density of injected fuel droplets leading to accumulate on the compressed gas. This may lead to incomplete mixing air and fuel

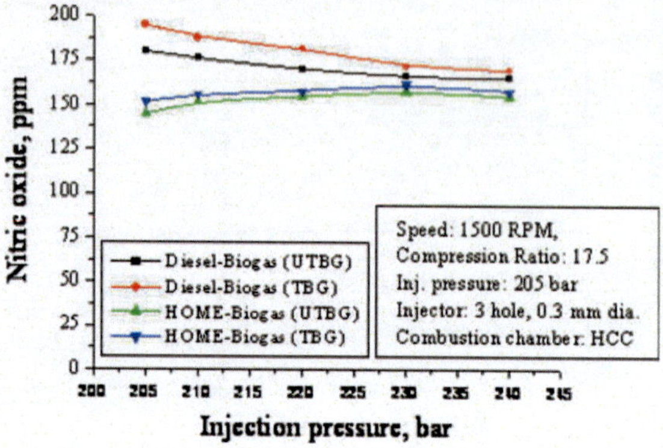

Fig. 14 Effect of injection pressure on the NO_x emissions

combination. As far as diesel fuel-based dual fuel operation with TBG is concerned, HC and CO levels were increased by 19.1 and 21.2% when IP was increased from 205 to 230 bar. It may be due to inappropriate united air-fuel mixture caused by the improper spray characteristics with decreased droplets penetration of diesel fuel into the compressed gas. HC levels at 230 bar IP for diesel-TBG, diesel-UTBG HOME-TBG, and HOME-UTBG operation were 38, 41, 47 ppm, and 50 ppm, respectively, at 80% load. Similarly, the CO levels at 230 bar IP for diesel-TBG, diesel-UTBG HOME-TBG, and HOME-UTBG operation were 0.26, 0.28, 0.33%, and 0.36%, respectively, at maximum load.

As it is obvious that from Figs. 14 that increased IP significantly influences NO_x emission levels. For the same gaseous fuel induction, dual fuel operation with HOME provides 6.1% decreased NO_x levels than diesel-based operation. Poorer quality of biodiesel and reduced temperature inside the engine cylinder is accountable. In addition, increased diffusion combustion compared to premixed combustion is also affecting the NO_x levels. For HOME-TBG fueled operation with an IP of 230 bar resulted in 5.6% increased NO_x levels than the operation with lower IP tested. This may be owing to the fact that engine operation with higher injection pressure allows improved spray characteristics of HOME; that is, reduced biodiesel droplet's size leads to better fuel penetration with adequate dispersion. This is the reason why in-cylinder pressures were found to be higher at 230 bar IP with biodiesel-based operation. However, NO_x levels were still lower at lower IP due to production of larger droplets leading to incomplete mixing rates. Also at higher injection pressure (240 bar) lowers the droplet size leading reduced the mixing rate caused by the reduced momentum of biodiesel droplets. In addition, for the gaseous fuel induction, diesel-based operation provides decreased NO_x emission levels at an injection pressure higher than 205 bar due to decreased rate of combustion during premixed combustion phase. Fuel penetration due to lowered relative velocity of diesel droplets is accountable for this trend. NO_x levels at 230 bar injection pressure for diesel-TBG, diesel-UTBG HOME-TBG, and HOME-UTBG operation were 173, 166, 161 ppm, and 156 ppm, respectively.

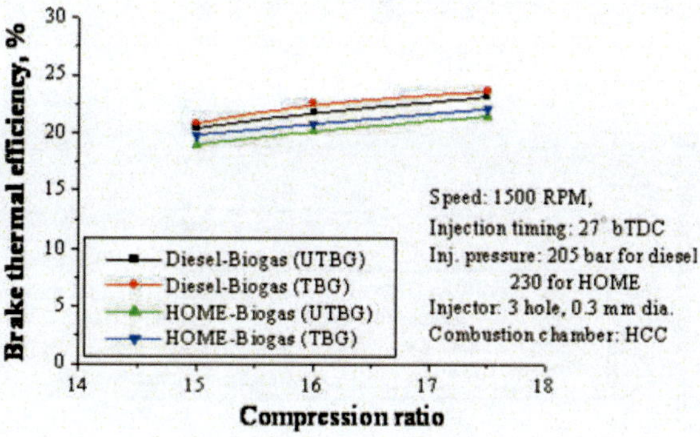

Fig. 15 Effect of compression ratio on the BTE

5.3 Effect of Compression Ratio

Influence of compression ratio (CR) on the combustion behavior of diesel/biodiesel-biogas-fueled dual fuel process has been presented in this section. In this investigation, the CR 15, 16, and 17.5 effect were analyzed.

5.3.1 Performance Characteristics

Variations in thermal efficiency with CR are shown in Fig. 15. It is evident from Fig. 15 that, for the same gaseous fuel induction (TBG or UTBG), efficiency was greater for both diesel and HOME-based dual fuel operation at a CR of 17.5 compared to the operation at lower CR. At a CR of 17.5, HOME-based operation provided 4.3% decreased thermal efficiency than diesel-based operation due to variations in the fuel physicochemical properties. However, for the TBG induction and at increased CR, the thermal efficiency of the HOME fueled dual fuel engine can be brought very near to the diesel-based operation. This may be due to the existence of methane (TBG) in the working mixture leading to increase combustion temperature with an increased CR. In addition, it leads to a greater flame temperature and burning of the fuel mixture with a reduced residual gas concentration. Further high ignition quality like gasoline of gaseous fuel may perform better at advanced compression ratio. On an average, for the HOME-based operation with TBG, thermal efficiencies were enhanced by 10.4% when the CR was increased from 15 to 17.5, respectively. It was reduced by 5.8% when the CR was raised from 16 to 17.5, respectively.

5.3.2 Emission Characteristics

As far as environmental perspective, it is essential to control the emissions from the diesel engine. Magnitude of emissions directly reflects the combustion behavior that

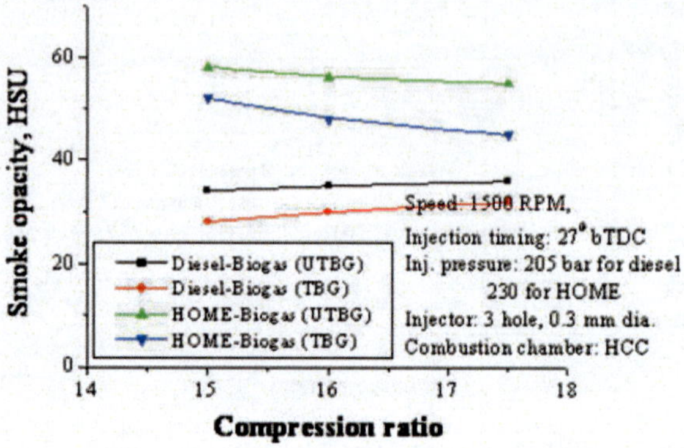

Fig. 16 Effect of compression ratio on the smoke emissions

taking place in engine cylinder. For both pilot fuels, emission levels for the UTBG and TBG fueled dual fuel operation were determined under steady-state condition. Emission levels were computed by using well-calibrated instruments. The various emissions with dual fuel operation are summarized in the succeeding paragraph.

Figure 16 presents the influence of CR on the smoke levels. It is noticed that for the same inducted fuel, smoke levels from both diesel and HOME-based dual fuel operation resulted in reduced smoke levels at CR of 17.5. This may be due to fact that the flame developed by the combustion of fuel combination arrive whole area of the cylinder caused by the increased combustion pressure and temperature leading to better combustion. Investigation showed that for the identical CR and gaseous fuel induction, diesel-based operation provides 19.4% decreased smoke levels than HOME-based operation. This is due to retarded combustion rate of biodiesel and burning of elevated molecular weight fatty acids. Also, incomplete mixing and lower oxidation caused by the differences in the fuel properties significantly ruled the combustion. Smoke levels for diesel-based operation at 15, 16, and 17.5 CR were 33, 30, and 28 HSU, respectively, and for HOME-based operation at 15, 16, and 17.5 CR were 52, 48, and 44 HSU, respectively, at 80% load. Therefore, higher CR favors the combustion of high octane number gaseous fuels. On an average, for the HOME-based operation with TBG, smoke opacity was lowered by 13.4% when the CR was enhanced from 15 to 17.5, respectively. It was reduced by 6.5% when the CR was raised from 16 to 17.5, respectively.

Influence of CR on the HC and CO levels at different CR for the selected fuel combinations is presented in Figs. 17 and 18 for 80% load. For the same gaseous fuel induction, results showed that HOME-based dual fuel operation provides amplified HC and CO levels than diesel-based operation. This may be due to the actuality that the combustion pressure and temperature were greater for diesel-based operation caused by the increased flame propagation leading to improved combustion. Further oxygen availability during pre-flame combustion reaction dominates the quality of combustion. Therefore, the utilization of oxygen during diesel operation is better compared to HOME operation. Because differences in a spray characteristics caused

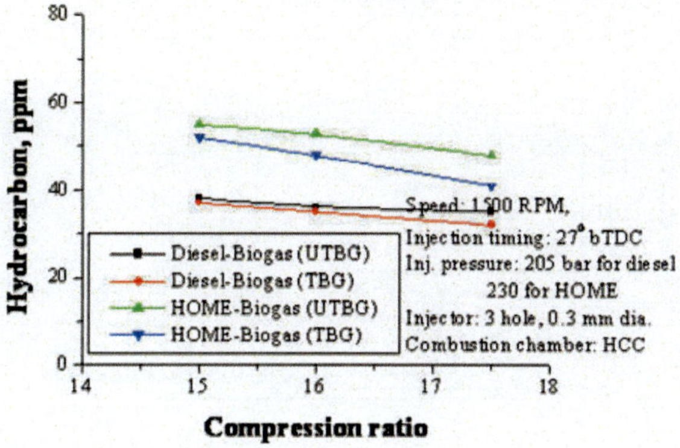

Fig. 17 Effect of compression ratio on the HC emissions

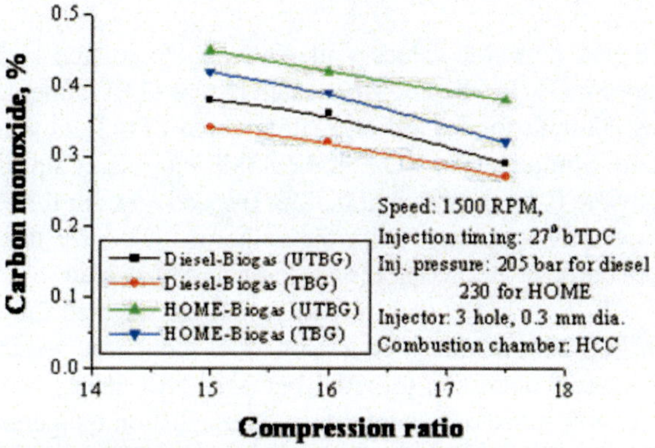

Fig. 18 Effect of compression ratio on the CO emissions

by variations in the fuel properties may be accountable for this trend. However, it is noticed that with higher CR (17.5), HOME-based operation provided lesser HC and CO emissions than the operation with lower CR (15 and 16). For the HOME-treated biogas (TBG) operation at various CR's, the HC emission levels were lowered by 21.2% when the compression ratio was varied from 15 to 17.5, respectively. Similarly, CO levels were lowered by 14.5% when the CR was elevated from 16 to 17.5, respectively. This is because at higher compression ratio, the combustion pressure and temperature increases due to increased density of compressed gas leading to increased flame propagation, and hence, better burning of entire fuel mixture is possible. At a CR of 15, the combustion pressure and temperature is lower resulting into lowered oxidation process.

Further, it is noticed that for same injected fuel, dual fuel operation with TBG provided in reduced HC and CO levels in the exhaust than the operation with UTBG due to better burning caused by the presence of combustible methane gas in the fuel

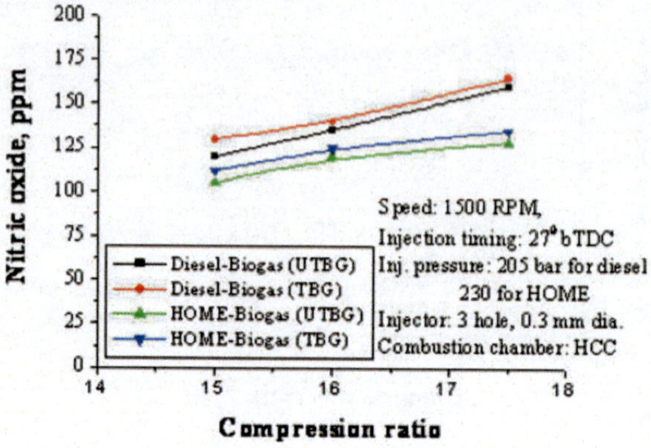

Fig. 19 Effect of compression ratio on the NO_x emissions

combination. The HC emission values at the CR's of 15, 16, and 17.5 were 52, 48, and 41 ppm, respectively, for the HOME-treated biogas (TBG) operation compared with the value of 32 ppm for the diesel-treated biogas (TBG) combination at 17.5 CR and 80% load. Similarly, the CO emission values at the compression ratios of 15, 16, and 17.5 were 0.42, 0.39%, and 0.32%, respectively, for the HOME-treated biogas (TBG) operation compared with the value of 0.27% for the diesel-treated biogas (TBG) operation at 17.5 compression ratio and 80% load.

The disparities in NO_x levels with respect to CR for selected fuel combinations are presented in Fig. 19. For the same gaseous fuel induction, results showed that HOME-based dual fuel operation provide decreased NO_x levels than diesel-based operation. It may be because of lower premixed combustion temperature caused by the variations in the fuel properties. Poor physicochemical properties of HOME lead to reduced flame temperature leading to reduced combustion. Further inappropriate mixing of fuel combination caused by the variations in the spray characteristics may be responsible. Results showed that oxygen utilization is more appropriate with diesel-based double-fueled operation than HOME-based operation. On the other hand, it is seen that with HOME-based dual fuel operation showed improved emission levels, i.e., decreased HC, CO, and marginally increased NO_x emissions with higher CR (17.5) than the operation with lower CR (15 and 16). For the HOME-treated biogas (TBG) operation at different CR, the HC levels were diminished by 21.2% when the CR was varied from 15 to 17.5, respectively. In the same way, CO emission levels were declined by 14.5% when the CR was elevated from 16 to 17.5, respectively. This is because at higher CR, the combustion pressure and temperature increases due to increased density of compressed gas leading to increased flame propagation and hence better burning of entire fuel mixture is possible. At a lower CR, the combustion pressure and temperature is lower resulting into freezing of the oxidation process.

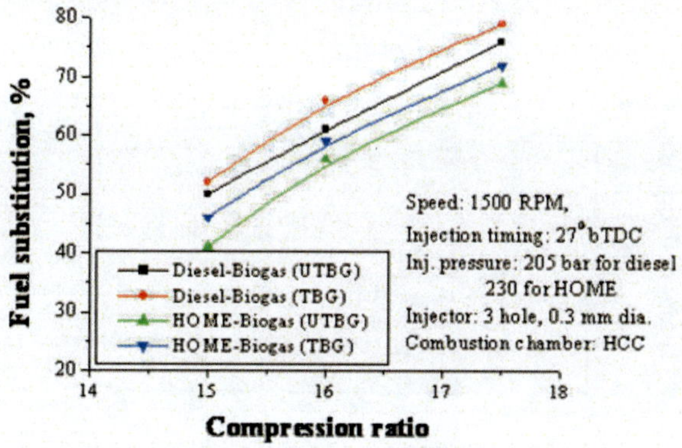

Fig. 20 Effect of compression ratio on the fuel substitution

5.3.3 Fuel Substitution

Figure 20 demonstrates the substitution of fuel at various CR's for different fuel combination at 80% load. Fuel substitution is significant during dual fuel process, but it depends on the pilot fuel properties, engine design, and operating modes. Fuel substitution values were greater at increased compression ratio and which in turn leads to increased efficiency and declined specific fuel consumption. It means that little quantity of fuel is consumed at a higher compression ratio. This feature permits a large quantity of biogas for achieving similar power output. Values of fuel substituted in percentage at a compression ratio of 17.5 were 79 and 72% for the diesel and HOME-based dual fuel operations with TBG as inducted fuel at the 80% load.

5.3.4 Combustion Characteristics

Combustion of liquid and gas combination in an engine is significantly dissimilar than conventional CI engine combustion rate during premixed combustion phase, fuels properties, design of engine and parameters of operating can affect the combustion behavior. In addition, spray characteristics with mixture strength quality, penetration, and dispersion play a main role while combustion. Various combustion characteristics are explained in the following subsequent section.

The variations in ignition delay (ID) with CR for a variety of fuel combinations are shown in Fig. 21. Increased CR decreases ID for fuel combinations tested. It is because increase in CR amplifies peak cylinder gas temperature caused by the amplified fuel quantity being burnt. Delay period was determined based on the static IT. For the same gaseous fuel induction, increased ignition delay has been noticed for HOME fuel-based operation than diesel operation at 80% load. Type of liquid fuel and gaseous fuel during the operation resulted in variations in the ID. This could be

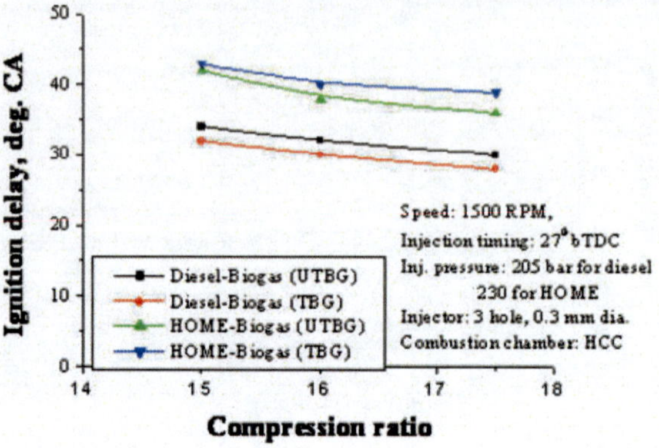

Fig. 21 Effect of compression ratio on the ignition delay

attributed to variations in the liquid fuel and gas composition. During the combustion, diesel was ignited earlier leading to burn biogas. This is possible due to the fact that diesel has lesser self ignition temperature than biogas which in turn leads to reduced ID for diesel-based dual fuel operation. However, for ignition delay for dual fuel operation with TBG was lesser than the operation with UTBG. Increased flame velocity caused by the presence of methane combustible gas leading to rapid burning of the fuel combination may be accountable. However, increased compression ratio (17.5) favors the biogas combustion due to its high octane number. But higher ignition delay has been noticed for lower compression ratios due to decreased density of gas mixture, decreased combustion temperature and deficient mixing of fuel combinations. Commonly, delay period was greater for HOME and UBTG-based dual fuel operation. It may be due to larger viscosity and lesser energy content of HOME, and the presence of carbon dioxide and other impurities in the biogas could be responsible.

Figure 22 presents the influence of CR on the combustion duration (CD) for different fuel combinations. It was determined as the time period between the combustion start and 90% cumulative release of heat. For the same inducted fuel, CD was greater for HOME-based dual fuel operation than the operation with diesel fuel at 80% load. HOME has a higher viscosity with greater molecular weight of fatty acids, and hence, HOME-based dual fuel operation provides increased combustion duration, this in turn lowers thermal efficiency. Further, it is seen that decreased CD was noticed with TBG-fueled dual fuel operation because of improved premixed combustion caused by optimum utilization of air. Presence of TBG in the fuel combination leads to rapid burning due to higher flame velocity along with HOME is mainly accountable. A critical issue in decreasing compression ratio below 17.5 is that the combustion duration was found to be longer due to decreased flame velocity caused by the reduced combustion temperature. However, dual fuel operation with UTBG resulted in a second peak in the later stages of combustion. This may be due to the reduced mixture strength and improper utilization of available air.

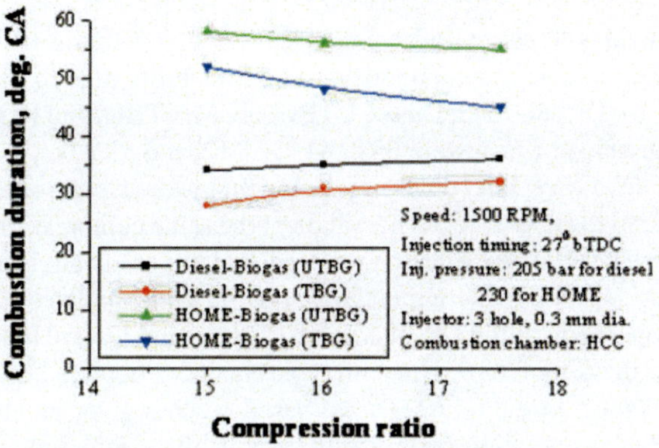

Fig. 22 Effect of compression ratio on the combustion duration

Variations in peak pressure with respect to crank angle for the selected fuel combinations at 80% load are presented in Fig. 23. From the figure (BTE), it is seen that cylinder pressure was amplified with increase in CR's. It may be due to use of high octane number gaseous fuel along with pilot fuel and it provides better results with increased compression ratio. Further, it is noticed that the combustion of fuel was comparatively better during premixed combustion stage with higher compression ratio. For the fixed compression ratio and same inducted gaseous fuels, HOME-based operation provided in decreased cylinder pressure than diesel-based operation caused by the fact that incomplete burning of the biodiesel caused by the significant variations in the fuel properties. Further, it is distinguished that dual fuel operation

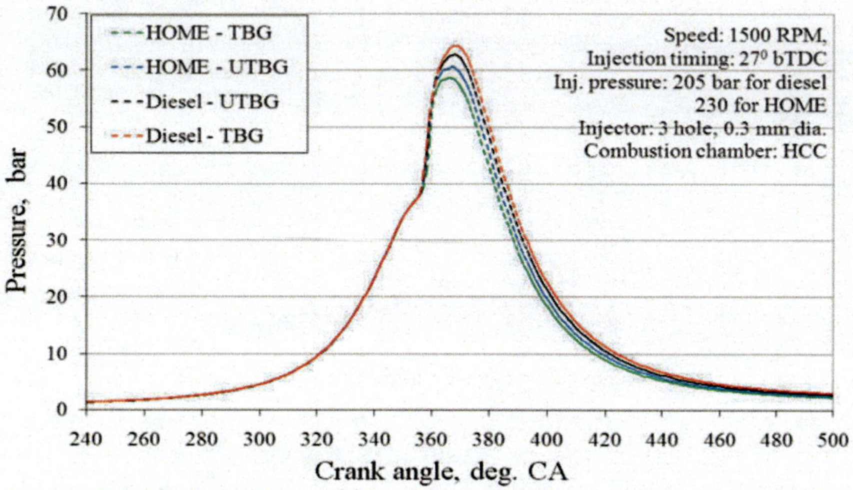

Fig. 23 Cylinder pressure verses crank angle diagram for various fuel combinations at a compression ratio of 17.5 and 80% load

with UTBG provides lesser cylinder pressure than the operation with TBG due to presence of other chemical compounds and inferior properties in the UTBG. The pressure values for Diesel-TBG, Diesel-UTBG, HOME-TBG, and HOME-UTBG at the compression ratios of 17.5 were 66.8, 64.4, 61.12, and 57.2 bar, respectively. On an average, for the diesel-TBG dual fuel operation, the peak pressure was enhanced by 8.5% compared to HOME–TBG operation at the same compression ratio of 17.5.

Figure 24 presents the heat release rate (HRR) versus crank angle at 17.5 CR and 80% load for the selected fuel combinations. For the same inducted fuels operation, HOME-based operation at the CR of 15 and 16 showed a lesser HRR compared to the operation at the CR of 17.5. However, consequences obtained for the CR of 15 and 16 are not shown in the figure. Diesel-based process gives greater rate of heat release than HOME operation owing to the result of a decreased second peak in the diffusion combustion phase obtained at the compression ratio of 17.5. Further, for the same pilot injected fuel, improved combustion during premixed combustion phase has been noticed for TBG fueled dual fuel engine. It may be characterized with the presence of purest form of combustible gas in the engine cylinder which in turn has better burning characteristics because of methane gas. It is evident from the figure, double fueled operation with base fuel (diesel) and TBG, the cylinder pressure and temperature increases, due to more burning of fuel near very close to the top dead center. The results disclose that thermal efficiency and energy release rate significantly depend on load, density of mixture of air and biogas, compression ratio, pilot fuel injection parameters, purity of biogas, i.e., methane concentration. From the results, it is seen that larger methane concentration in the biogas provides increased thermal efficiency.

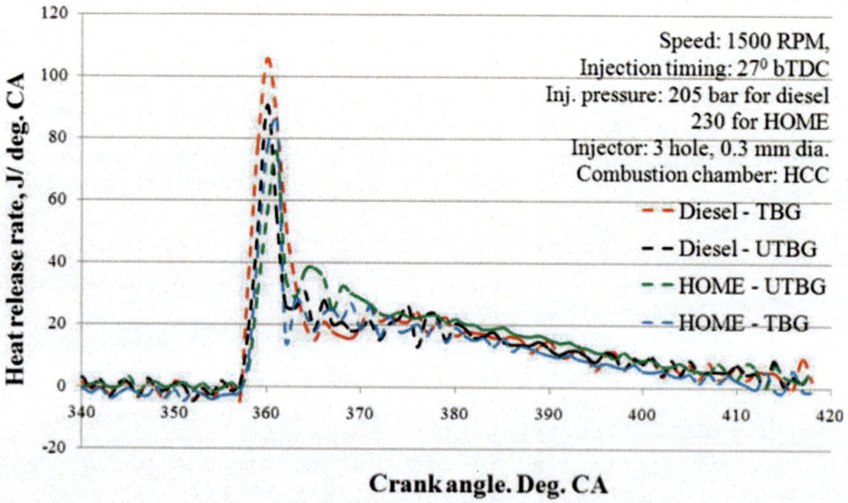

Fig. 24 HRR verses crank angle diagram for various fuel combinations at a compression ratio of 17.5 and 80% load

As the combustion starts, heat release rate achieved positive values and is followed by the diffusion combustion phase. Overall fuel combination combustion is governed by the mixture strength. It is seen that the engine operation with biodiesel and UTBG, combustion provide second peak during diffusion combustion phase. With this fuel combination, combustion was initiated later by a few degrees of crank angles than the diesel-TBG operation. This may be due to extensive ID.

6 Conclusions

Engine design and operating parameters strongly influences combustion. Method adopted in this present work is successful in achieving the goals of the conversion process and a reduction of emissions when optimum engine parameters were used.

- Experimental investigations showed that HOME-TBG operation provided improved thermal efficiency with an IT of 270 bTDC, fuel IP of 230 bar, and CR of 17.5.
- At an advanced IT diesel-based dual fuel operation with TBG provided in 8.22% increased thermal efficiency compared to HOME-based operation. Lower smoke, HC and CO levels were obtained with advanced injection timing due to better mixing rates caused by the more time availability.
- With all optimized IP, on an average, HOME-TBG operation provides 3.6% increased thermal efficiency and 5.6% reduced smoke levels when IP was varied from lower IP to 230 bar. Further, HC and CO levels were reduced by 12.6 and 15.8% when IP was changed from 205 bar to 230 bar, For same fuel combination at an IP of 230 bar resulted in 5.6% increased NO_x levels than the operation with lower injection pressure.
- At CR 17.5 and HOME-based operation with TBG, resulted in 4.3% decreased thermal efficiency compared to diesel-based operation due to variations in the fuel physicochemical properties of pilot fuel. As far as emission levels were concerned smoke, HC and CO levels were reduced by 19.4%, 21.2% and 14.5%, respectively, at 80% load when the CR was changed from 15 to 17.5. Further, about 79 and 72% fuel saving was achieved for diesel and HOME-based dual fuel operations with TBG at the 80% load. For the same gaseous fuel induction, increased ID, CD, lower HRR has been noticed for HOME fuel-based double-fueled operation compared to diesel-based operation at 80% load.
- It is seen that HOME-biogas combination resulted in smooth engine operation. Hence, their extensive utilization for power generation applications will cover the way for the country energy safety.
- In addition, the use of hydrogen with proper nozzle geometry and matched combustion chamber for diesel engine along with higher compression ratio results into better combustion.

On the whole, experimental observations showed that treated biogas (TBG) performed better compared to untreated biogas (UTBG) operation. HOME and biogas (TBG/UTBG) can be used as fuels in a diesel engine. Further, they do not require major engine alterations in the diesel engine. HOME-biogas process with maximum operating limitations resulted in generally better combustion with decreased emissions.

References

1. Banerjee R (2006) Comparison of options for distributed generation in India. Energy Policy 34:101–111
2. Ravindranath NH, Somashekar HI, Dasappa S, Jayasheela Reddy CN (2004) Sustainable biomass power for rural India—case study of biomass gasifier for village electrification. Spec Sect: Appl S&TRural Areas Curr Sci 87(7):932–941
3. Nouni MR, Mullick SC, Kandpal TC (2007) Biomass gasifier projects for decentralized power supply in India: a financial evaluation. Energy Policy 35:1373–1385
4. Gupta A, Saini RP, Sharma MP 2010 Steady-state modeling of hybrid energy system for off grid electrification of cluster of villages. Renew Energy 35:520–535
5. Senthilkumar K, Vivekanandan S (2016) Investigating the biogas as secondary fuel for CI engine. Int J Appl Environ Sci 11(1):155–163. ISSN 0973-6077
6. Banapurmath NR, Yaliwal VS, Kambalimath S, Hunashyal AM, Tewari PG (2011) Effect of wood type and carburetor on the performance of producer gas–biodiesel operated dual fuel engines. Waste Biomass Valoriz 2(4):1–11
7. Ambarita H (2017) Performance and emission characteristics of a small diesel engine run in dual-fuel (diesel-biogas) mode. Case Stud Therm Eng 10:179–191
8. Duc PM, Wattanavichien K (2007) Study on biogas premixed charge diesel dual fuelled engine. Energy Convers Manag 48:2286–2308
9. Kruczynski SW, Orlinski P (2013) Combustion of methyl esters of various origins in the agricultural engine. Indian J Eng Mater Sci 20:483–491
10. Mohsin R, Majid ZA, Shihnan AH, Nasri NS, Sharer Z (2014) Effect of biodiesel blends on engine performance and exhaust emission for diesel dual fuel engine. Energy Convers Manag 88:821–828
11. Ashraful AM, Masjuki HH, Kalam MA, Rizwanul Fattah IM, Imtenan S, Shahir SA, Mobarak HM (2014) Production and comparison of fuel properties, engine performance, and emission characteristics of biodiesel from various non-edible vegetable oils: a review. Energy Convers Manag 80:202–228
12. Kalam MA, Mosarofa MH, Zulkiflia NWM, Teohab YH, How HG (2016) Assessment of performance, emission and combustion characteristics of palm, jatropha and Calophyllum inophyllum biodiesel blends. Fuel 181:985–995
13. Dhole AE, Yarasu RB, Lata DB (2016) Investigations on the combustion duration and ignition delay period of a dual fuel diesel engine with hydrogen and producer gas as secondary fuels. Appl Therm Eng 107:524–532
14. Punnarapong P, Sucharitakul T, Tippayawong N (2017) Simulation of producer gas combustion in a premixed burner for ceramic firing process. Energy Procedia 138:622–627
15. Sahoo BB, Sahoo N, Saha UK (2009) Effect of engine parameters and type of gaseous fuel on the performance of dual-fuel gas diesel engines e a critical review. Renew Sustain Energy Rev 13:1151e1184
16. Chena C, Guo W, Ngo HH, Lee D-J, Tung K-L, Jin P, Wang J, Wu Y (2016) Challenges in biogas production from anaerobic membrane bioreactors. Renew Energy 98:120–134
17. Hagos K, Zong J, Li D, Liu C, Lu X (2017) Anaerobic co-digestion process for biogas production: progress, challenges and perspectives. Renew Sustain Energy Rev 76:1485–1496
18. Budzianowski WM (2012) Sustainable biogas energy in Poland: prospects and challenges. Renew Sustain Energy Rev 16(1):342–349

19. Srichat A, Suntivarakorn R, Kamwilaisak K (2017) A development of biogas purification system using calcium hydroxide and amine solution. Energy Procedia 138:441–445
20. Jaber MB, Couvert A, Amrane A, Le Cloirec P, Dumont E (2017) Hydrogen sulfide removal from a biogas mimic by biofiltration under anoxic conditions. J Environ Chem Eng 5(6):5617–5623
21. Kapoor R, Subbarao P.M.V, Vijay VK, Shah G, Sahota S, Singh D, Verma M (2017) Factors affecting methane loss from a water scrubbing based biogas upgrading system. Appl Energy 208:1379–1388
22. Bari S (1996) Effect of carbon dioxide on the performance of biogas/diesel duel-fuel engine. Renew Energy 9:1007–1010. https://doi.org/10.1016/0960-1481(96)88450-3
23. Sridhar G, Sridhar HV, Dasappa S, Paul PJ, Rajan NKS, Mukunda HS (2005) Development of producer gas engines
24. Angelidaki I, Treu L, Tsapekos P, Luo G, Kougias PG (2018) Biogas upgrading and utilization: current status and perspectives. Biotechnol Adv 36(2):452–466
25. Wierzbicki S (2012) Biogas as a fuel for diesel engines. J KONES Powertrain Transp 19(3)
26. Walozi R, Nabuuma B, Sebiti A (2016) Application of low pressure water scrubbing technique for increasing methane content in biogas. Univ J Agric Res 4(2):60–65. https://doi.org/10.13189/ujar.2016.040206

Bond Graph Based Modelling of Four-Cylinder In-Line Engine and Study of the Various Engine Characteristics Parameters

Rajmeet Singh, Joypreet Singh, and Tarun Kumar Bera

Nomenclature

C_v	Specific heat at constant volume
C_{pm}	Specific heat capacity of the body
C_{di}	Coefficient of discharge of inlet valve
C_{do}	Coefficient of discharge of outlet valve
d	Width of piston
$J_{flywheel}$	Polar moment of inertia of flywheel
J_c	Polar moment of inertia of cylinder
J_p	Polar moment of inertia of piston
k_{bend}	Stiffness
k_{p}, k_{pad}	Stiffness of pad
k_r	Upper structure stiffness
l_c	Length of connecting rod
l_{cg}	Distance between point E_c and centre of gravity of cylinder
l_p	Distance between point E_P and centre of gravity of piston
l_{pg}	Distance between point E_P and centre of gravity of piston rod
M_b	Mass of block
M_f	Mass of flow
M_c	Mass of cylinder
M_p	Mass of piston
$Pist_a$	Piston area
P_{atm}	Atmospheric pressure
q_{hc}	Heat of combustion

R. Singh (✉) · J. Singh · T. K. Bera
Mechanical Engineering Department, Thapar Institute of Engineering and Technology, Patiala 147001, India
e-mail: rajmeetsng@gmail.com; rajmeet.singh@bbsbec.ac.in

A. K. Gupta et al. (eds.), *Advances in IC Engines and Combustion Technology*,
Lecture Notes in Mechanical Engineering,
https://doi.org/10.1007/978-981-15-5996-9_61

r, R	Damping
R'	Gas constant
r_c	Radius of crank
r_{pad}	Damping of pad
R_d	Equivalent rubber damping coefficient
t	Time
T_{atm}	Atmospheric temperature
V_{ia}	Valve inner area
V_{oa}	Valve outer area
γ	Ratio of specific heats
λ_{ba}	Overall heat transfer coefficient between the body and the environment
λ_{gb}	Overall heat transfer coefficient between the gas and body
μ	Transformer modulus

Subscripts

c	Cylinder
atm	Atmospheric
f	Friction
p	Piston/Slider
1, 2	Contact point number

1 Introduction

In the last two decades, a rapid improvement in the field of the power computing results great improvement in many technical areas especially in the vehicle sector. The cost reduction is the prime issue in the automobile industries. Earlier the dynamics of the vehicle engine was a mainly focused area in the manufacturing of the vehicles [1–3]. Later, the thermodynamic aspect and after affect were discussed [4]. The modelling of the engine based on the set of governing equations and thermodynamics relations is mostly used [5, 6]. The study of the engine in the area of dynamics and thermodynamics gives some important facts that helped to develop new engines. This research work helps to lessen the large amount of the experiment work for validation of the new design of the engine. The results of study such as dynamically balanced crankshaft, thermal effect on the crankshaft, behaviour of the performance of the engine, etc., and many more helps in the modification of existing or new design of the engine. The mathematical model of single-cylinder spark ignition engine is developed for studying the various fuel injection strategies [7].

The modelling of any complex system is required to predict the system behaviour before fabricating the actual system. The modelling through bond graph (BG) is adopted in this work as this technique does not require dynamic equations for modelling. Only kinematic equations are sufficient to model through it. This modelling procedure is very compact in nature as this is bi-directional in nature, i.e., signals flow in both the directions through bonds [8, 9]. For the present system, i.e., four-cylinder engine mounted on hydraulic mounts, the components of the system lie in different energy domains, i.e., mechanical (cylinder), thermal (combustion process) and hydraulic (mounts). So, BG modelling is suitable for this type of system (multi-energy domain). If the several BG variables are required to be presented, multi-bond graph is an option for it [10]. Different control algorithms can be implemented using BG [11, 12]. Pathak et al. have developed overwhelming control scheme for the space robot [13]. The dynamics equations of reciprocating machine are taken into consideration while modelling of the four-cylinder in-line engine. Bond graph (BG) approach is used for modelling the engine as per the dynamics and thermodynamics equations using SYMBOLS Shakthi software [14]. The single-cylinder-piston model is developed first and then extended to the four-cylinder engine model. The 1-3-4-2 firing order is considered for the simulation of the engine model.

The structure of paper is follows: In Sect. 2, the motion equation and thermodynamics of the engine model are discussed, also piston cylinder model based on prismatic joint is discussed, and the BG model of the engine is developed. The BG model of engine is developed to generate an incorporated model of the mechanical, thermodynamics and hydraulic systems. In Sect. 3, the results are presented and discussed. The conclusions are drawn in Sect. 4.

2 Engine Model

2.1 Thermal Model

The engine used for modelling and simulation under this research work is the four-cylinder in-line diesel engine application to marine having low speed (20 rad/s) and high torque (2500 N m). The capacity of the engine used in this research paper is 50 KW. [Capacity (kw) = torque (2500 N m) X speed (20 rad/s)]. Four-stroke in-line engine is considered and shown in Fig. 1. At any given instant, the pressure

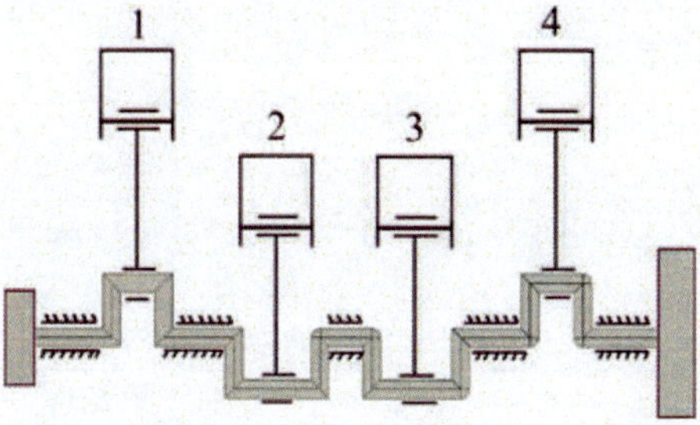

Fig. 1 A four-cylinder in-line engine

and temperature is uniform inside the cylinder. The motion is constrained to vertical plane only. A cam mechanism is used for the actuation of valves. The gas properties are calculated from the known values of Δ, V and m. The flow variables ($\dot{m}\dot{V}$ and $\dot{\Delta}$) and the corresponding effort variables are required for the development of the bond graph model. Obviously, sometimes, pseudo BG (temperature and heat flow rate are effort and flow variable, respectively) is better for representing thermodynamic systems using C-field element [15, 16].

The effort variables (output from C-field) can be computed using Eq. (1) as shown below [17]:

$$T(t) = \frac{\int_0^t \dot{\Delta}(\tau)\mathrm{d}\tau + \Delta_0}{\left(\int_0^t \dot{m}(\tau)\mathrm{d}\tau + m_0\right)C_\mathrm{v}},\ P(t) = \frac{R\left(\int_0^t \dot{\Delta}(\tau)\mathrm{d}\tau + \Delta_0\right)}{\left(\int_0^t \dot{m}(\tau)\mathrm{d}\tau + m_0\right)C_\mathrm{v}} \tag{1}$$

The engine produces the mechanical work output which is used to rotate the crank shaft. The external force acting on the piston slide is

$$F_\mathrm{P} = P_\mathrm{atm}A_\mathrm{P} - P\,A_\mathrm{P} - R_\mathrm{P}\dot{x}_\mathrm{P} - \mu_\mathrm{T}\tau_\mathrm{C} \tag{2}$$

The angular velocity $\dot{\theta}_1$ of crankshaft can be expressed as

$$\dot{\theta}_1 = \mu_\mathrm{T} V_\mathrm{P} \tag{3}$$

where

$$\mu_\mathrm{T} = \frac{1}{r_\mathrm{c}}\left[\frac{(r_\mathrm{c}/l_\mathrm{c})\sin 2\theta_1}{2\left[1-(r_\mathrm{c}/l_\mathrm{c})^2\sin^2\theta_1\right]^{1/2}} + \sin\theta_1\right]^{-1} \tag{4}$$

μ_T is the transformer modulus defined in Eq. (4), and T_C is the load torque. The critical pressure ratio is given by

$$P_\mathrm{Cr} = \left(\frac{2}{\gamma-1}\right)^{\gamma/(\gamma-1)} \tag{5}$$

The airflow rate $\dot{m}_a$ is given by

$$\dot{m}_\mathrm{a} = \frac{A_{\mathrm{v_i}}C_{\mathrm{d_i}}P_\mathrm{o}}{\sqrt{T_\mathrm{o}}}\sqrt{\frac{2\gamma}{R'(\gamma-1)}\left[\mathrm{P_r^{2/\gamma}} - \mathrm{P_r^{(\gamma-1)/\gamma}}\right]} \tag{6}$$

The discharge rate through valve is given by

$$\dot{\mathrm{m}}_\mathrm{o} = \frac{\mathrm{A_{v_o}C_{d_o}P_o}}{\sqrt{\mathrm{T}}}\sqrt{\frac{2\gamma}{R'(\gamma-1)}\left[\mathrm{P_r^{2/\gamma}} - \mathrm{P_r^{(\gamma-1)/\gamma}}\right]} \tag{7}$$

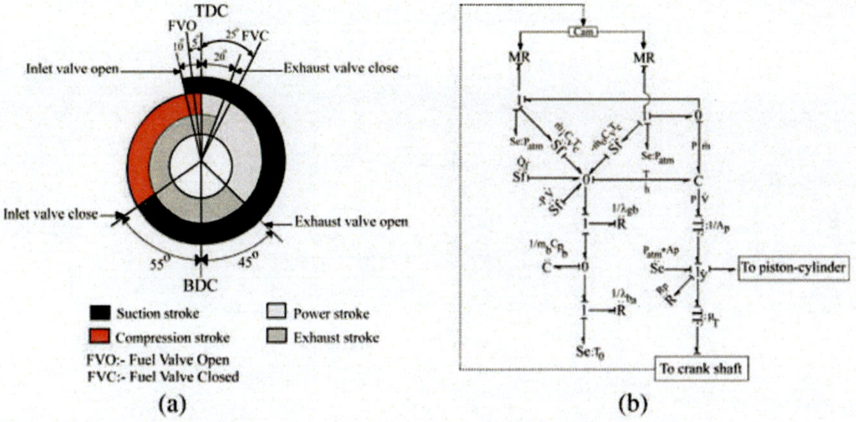

Fig. 2 **a** Valve timing diagram and **b** bond graph model of single cylinder

Figure 2a, b shows the valve timing diagram and model of engine. The mass flow is represented by modulated resistance (MR) element. The suction model in the left and discharge model in the right can be represented by MR elements using Eqs. (6) and (7), respectively. The crankshaft model is connected to *C*-field. The force acting on the piston is represented by Eq. (2). The engine output, i.e., torque, is provided to the crankshaft.

2.2 *Cylinder and Piston Model*

A schema of a planar prismatic member is shown in Fig. 3. The velocities of the ends

Fig. 3 Schematic diagram of the prismatic link

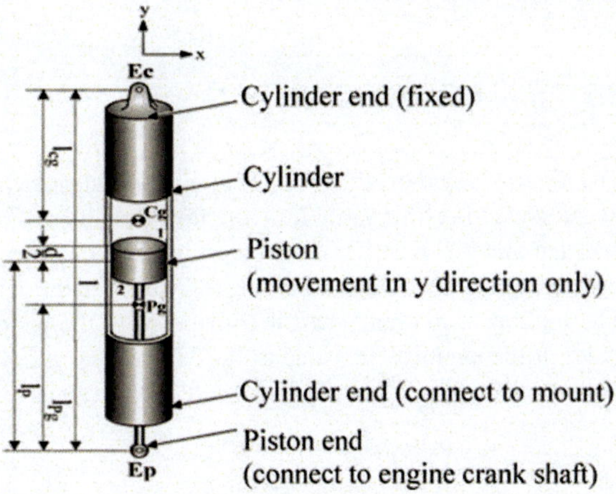

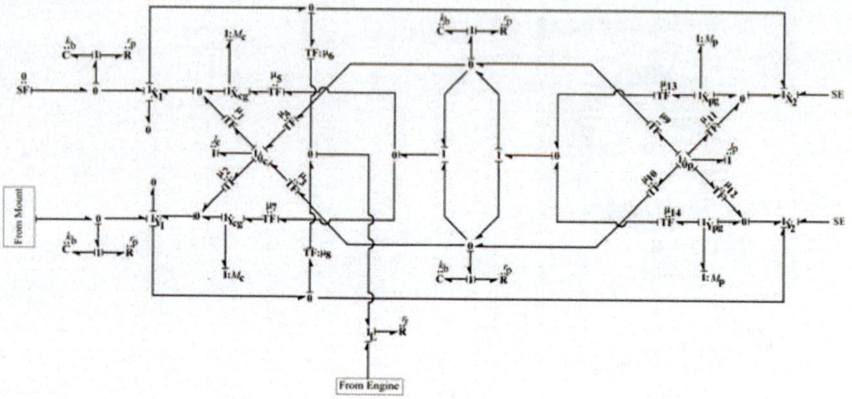

Fig. 4 Piston–cylinder BG model

of the prismatic member can be represented with linear velocities of their centroid and the angular velocities of the system about the axis that is perpendicular to the plane. The change of length of prismatic member is given as

$$\dot{l} = \left(\frac{x_1 - x_2}{l}\right)(\dot{x}_1 - \dot{x}_2) + \left(\frac{y_1 - y_2}{l}\right)(\dot{y}_1 - \dot{y}_2) \tag{8}$$

The BG model of prismatic link is shown in Fig. 4. This is framed using kinematic equations. M_P, M_c and J_P, J_c, are the piston and cylinder masses and mass moment of inertias, respectively. These are described by inertia elements (*I*). The joint forces can be expressed by sources of efforts (SE). The model also considers the contact stiffness and damping values. The total weight of the piston assembly is 2 kg, which consists of weight of piston, gudgeon pin, piston rings, etc. The heavy weight pistons were used in the marine diesel engines.

2.3 *Bond Graph Model of Engine*

The detailed bond graph model of engine is illustrated in Fig. 5. The *C*-field is used to calculate the effort variables from flow variables of heat transfer port, mechanical port and material port. The BG consists of four-cylinder in-line engine with cylinder–piston arrangement. Each mounting is attached to the cylinder end in vertical position. The motion from crank shaft is distributed to the piston depending upon the firing order of the engine. The value of the TF element is considered in this model as 0.5, because the one rotation of crank shaft is equivalent to twice the piston movement.

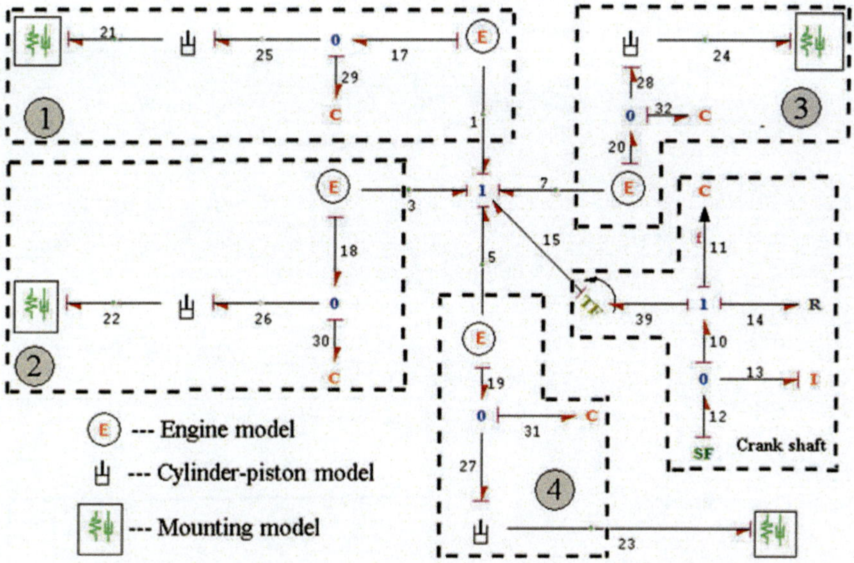

Fig. 5 Four-cylinder engine with mount BG model

3 Results and Discussion

The dynamics of the four-cylinder in-line engine is investigated in this paper. The engine characteristics results are given and discussed in Sect. 3.1.

3.1 *Engine Characteristics*

Table 1 shows the simulation parameters for engine model. Figure 6 illustrate the pressure (N/m^2) developed in the engine as per the firing order 1-4-3-2. Figure 6 depicts that cylinder 1 generates pressure first then cylinder 4 and cylinder 3, at last pressure is developed in cylinder 2. After 1 s, the pressures in all cylinders maintain a constant level between approximately 60–70 bar.

The pressure–volume diagram for all cylinders is shown in Fig. 7. The cylinder 1 develops the pressure of 64 bar. The suction stroke starts first as the inlet valve in open state, and after closing of inlet valve, the compression stroke takes place, and pressure inside the cylinder starts to increase up to 45 bar.

When the combustion stroke starts, the pressure inside the cylinder rapidly increases to 60 bar. Then expansion/power stroke takes place in which pressure reduces as force exerted on the crank shaft by the piston generated. During exhaust stroke, the un-burnt substances start to move out from cylinder through exhaust valve opening due to pressure difference. Figure 8a shows the trend of piston movement

Table 1 Parameter values for simulations

Engine parameters			
$k_p = 10^5$ N/m	$V_{oa} = 0.008$ m^2	$M_b = 10$ kg	$\lambda_{gb} = 1$ Wm^{-2}/K
$C_v = 720$ Jkg^{-1} K^{-1}	$C_{pm} = 459.8$ Jkg^{-1} K^{-1}	$J_{flywheel} = 20$ kg m^2	$\lambda_{ba} = 25$ Wm^{-2}/K
$k_{bend} = 10^5$ N/m	Pist$_a = 7.822e^{-3}$ m^2	$R' = 287$ Jkg^{-1} k^{-1}	$M_f = 3 \times 10^{-4}$ kg
$r_c = 0.06$ m^2	$l_c = 0.185$ m	$P_{atm} = 10^5$ N/m^2	$q_{hc} = 4.73 \times 10^7$ Jkg^{-1}
$C_{di} = 0.2$	Via $= 0.008$ m^2	$T_{atm} = 300$ K	$\gamma = 1.41$
Cylinder–piston parameters			
$M_c = 10$ kg	$M_p = 2$ kg	$k_{pad} = 10^5$ Ns/m	$r_{pad} = 10^3$ Ns/m
$l_{cg} = 0.06$ m	$J_p = 0.1$ kg m^2	$l_p = 0.1$ m	$l_{pg} = 0.05$ m
$J_c = 5$ kg m^2	$k_{bend} = 10^5$ N/m	$d = 0.09015$ m	

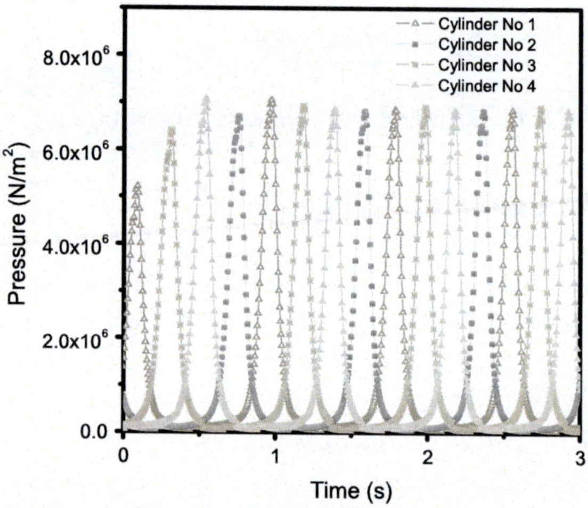

Fig. 6 Pressure developed in cylinders

inside the cylinders. As depicted in Fig. 8a that piston of cylinders 1 and 4 starts to move from BDC to TDC, and in similar pattern, the piston of cylinders 3 and 2 displaces from TDC to BDC. The total displacement travel by pistons of all cylinders is 0.12–0.24 m. The torque developed by the mass is always greater than produced by gas during power stroke. Figure 8b illustrated the torque developed by gas and mass. The deviation in crank speed and torque developed in engine start-up are presented in Fig. 8c. Figure 8d shows the relation between the crank speed and secondary force generated by piston in the four-stroke engine. The frequency of crank speed is double the frequency of secondary force.

The speed and acceleration of the crank is calculated for various values of J (mass moment of inertia of flywheel) as shown in Fig. 9. The speed of the crank increases as the value of J increases and vice versa as shown in Fig. 9a. On contrary, the acceleration of the crank decreases as value of J increases as presented in Fig. 9b.

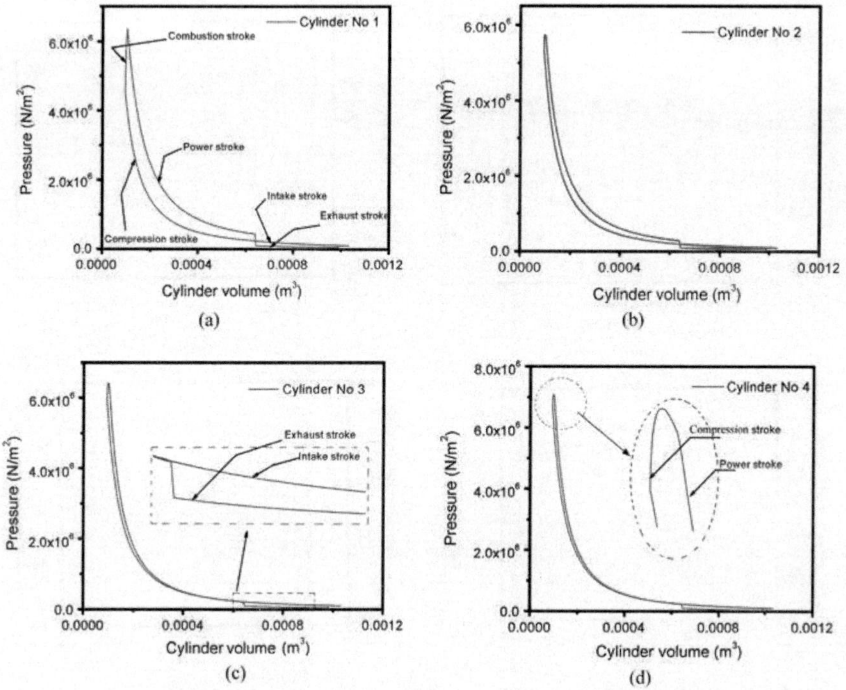

Fig. 7 Pressure–volume diagram of **a** cylinder no 1, **b** cylinder no 2, **c** cylinder no 3, and **d** cylinder no 4

4 Conclusions

The modelling and simulation of four-cylinder in-line diesel engine application to marine is done in this paper. Firstly, the four-cylinder in-line engine model is discussed. The motion and thermodynamics equations of the engine model are discussed. The engine bond graph model is developed and simulated for various conditions such as pressure developed in cylinders as per firing order 1-3-4-2 sequence, crank speed versus torque developed, speed and acceleration of the crank for different values of J (mass moment of inertia of flywheel), piston stroke length of all cylinders, pressure–volume diagram etc. The piston–cylinder model based on prismatic joint was discussed, and the BG model of the same was developed. The thermodynamics and mechanical port was considered while modelling the BG model of engine. The bond graph technique reduces the system modelling complexities of very complex models like vehicle engine which was not easily handled by other simulation software's. For the future scope, this work will be extended by validating the simulation results with real-time four-cylinder in-line diesel engine test bench apparatus during running conditions.

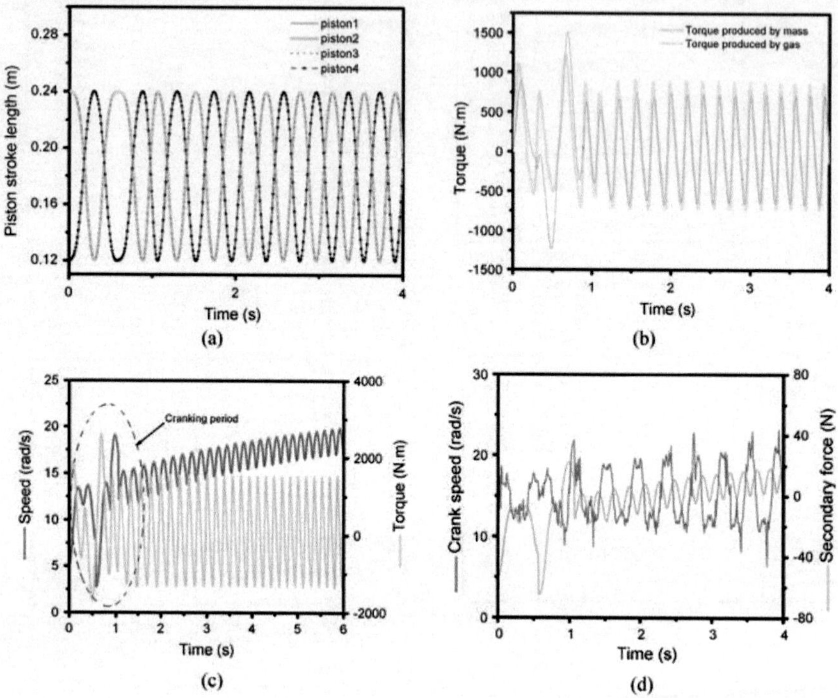

Fig. 8 **a** Piston stroke displacement of all cylinders, **b** torque produced by gas and mass, **c** crank speed and torque at engine start-up and **d** comparison of frequency between crank speed and secondary force

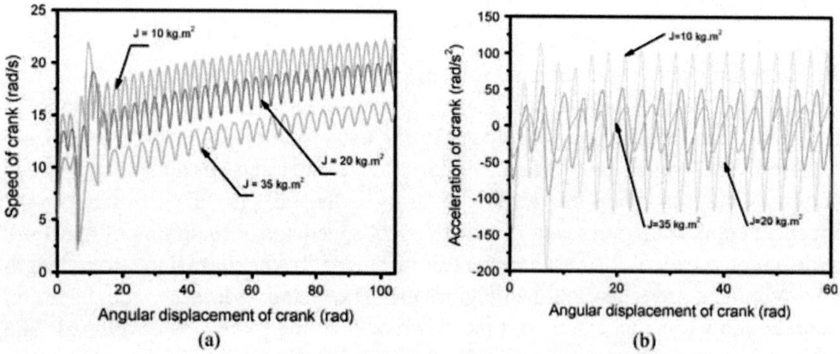

Fig. 9 **a** Speed of the crank with different values of J in Kg m^2 and **b** acceleration of the crank with different values of J in Kg m^2

References

1. Denman HH (1991) Exact solution for the rigid slider-crank mechanism with gas pressure. Mech Mach Theory 26(2):415–420
2. Hsieh SR, Shaw SW (1993) The dynamic stability and nonlinear resonance of a flexible connecting rod: continuous parameter model. Nonlinear Dyn 4:573–603
3. Shiao Y, Pan CH, Moskwa JJ (1994) Advanced dynamic spark ignition engine modelling for diagnostics and control. Internat J Veh Des 15:578–596
4. Heywood JB (1988) Internal combustion engine fundamentals. McGraw-Hill, New York
5. Dinca L, Aldemir T, Rizzoni G (1999) Fault detection and identification in dynamic systems with noisy data and parameter/modelling uncertainties. Reliab Eng Syst Saf 65:17–28
6. Lyshevski SE (2000) Energy conversion and optimal energy management in diesel-electric drive trains of hybrid-electric vehicles. Energy Convers Manage 41:13–24
7. Thompson S, Duan SY (1991) Modelling, parameter selection and simulation of a single-cylinder four-cycle engine. Proc Inst Mech Eng, Part I: J Syst Control Eng 205:49–57
8. Karnopp DC, Margolis DL, Rosenberg RC (2000) System dynamics, modeling and simulation of mechatronic systems. Wiley, New York
9. Borutzky W (2004) Bond graphs-a methodology for modelling multidisciplinary dynamic systems. SCS Publishing House, Erlangen, San Diego
10. Samantaray AK, Ould-Bouamama B (2008) Model-based process supervision. Springer, London
11. Gawthrop PJ (1995) Physical model-based control: a bond graph approach. J Franklin Inst 332(3):285–305
12. Granda JJ (2002) The role of bond graph modeling and simulation in mechatronics systems: an integrated software tool: CAMP-G, MATLAB–SIMULINK. Mechatronics 12:1271–1295
13. Pathak PM, Kumar RP, Mukherjee A, Dasgupta A (2008) A scheme for robust trajectory control of space robots. Simul Model Pract Theory 16(9):1337–1349
14. Mukherjee A (2006) Users manual of SYMBOLS Shakti. High Tech Consultants, STEP, Indian Institute of Technology, Kharagpur
15. Louca LS, Stein JL, Rideout DG (2001) Generating proper integrated dynamic models for vehicle mobility using a bond graph formulation. In: Proceedings of the 2001 international conference on bond graph modeling (ICBGM'01), Phoenix, 33(1):339–345
16. Karnopp D (1985) Modelling and simulation of adaptive vehicle air suspens ions with pseudo bond graphs in CAMP and ACSL. In: Proceedings of 11th IMACS world congress on system simulation and scientific computation, Oslo, Norway
17. Bera TK, Samantaray AK, Karmakar R (2011) Bond graph modeling of planar prismatic joints. Mech Mach Theory 49:2–20

Numerical Study of Deflagration to Detonation Transition in 2D and Axisymmetric Detonation Tube with Obstacles Using OpenFOAM

Udit Vohra, Rajpreet Singh, Vinayak Bassi, T. K. Jindal, and Amarjit Singh

1 Introduction

Due to severe industrial accidents that have occurred which were directly or indirectly linked to combustion phenomenon of the fuel, various researches are being done on the different modes of combustion namely deflagration and detonation. These modes of combustion are marked by the significant increase in the pressure and temperature in confined volume. A large number of changes occur in the thermodynamic (pressure, temperature, etc.) and gas-dynamic properties of the combustion wave as it propagates away from the ignition source, transforming reactants into products, producing and sustaining energy for further propagation of the combustion wave. There are two types of combustion waves, namely deflagration and detonation wave.

Deflagration wave travels through the reaction mixture at low subsonic speed, and the heat is released at a slower pace. This wave travels via diffusion of heat and mass through the reactants ahead of it. There is an increase in the temperature and pressure states as the wave propagates. Detonation phenomenon of the combustion is the one in which the detonation wave travels at supersonic speed with sharp increase in thermodynamic states of the flow. The shock wave which is very thin and the reaction zone which may be thicker are coupled together in the detonation wave and both of them traverse throughout the medium at supersonic speed. For equivalent energy of both deflagration and detonation wave, the latter produces higher pressure, temperature and density at higher speed [1, 2].

Although, till date two theories, namely Chapman–Jouguet theory (CJ Theory) and Zeldovich-von Neumann-Doering theory (ZND theory) are used for modelling

U. Vohra (✉) · R. Singh · V. Bassi · T. K. Jindal · A. Singh
Aerospace Department, Punjab Engineering College, Chandigarh, India
e-mail: vohra.udit@gmail.com

A. K. Gupta et al. (eds.), *Advances in IC Engines and Combustion Technology*, Lecture Notes in Mechanical Engineering,
https://doi.org/10.1007/978-981-15-5996-9_62

detonation waves. The former theory assumes equilibrium downstream of the detonation wave to determine its chemical equilibrium detonation properties. Three conservational laws, conservation of mass, momentum and energy, are applied across the detonation wave which is treated to be of infinitesimal thickness 1D reaction zone [2, 3]. The strong and weak detonation solutions are the two solutions that exist for a detonation wave at some wave speed. Chapman hypothesis stated that for a given fuel mixture, its detonation velocity will be the minimum velocity which is the velocity corresponding to the point of convergence of strong and weak solution [1]. The theoretical detonation properties predicted by CJ theory is also referred to as CJ state for a given explosive mixture.

According to ZND theory, detonation wave comprises of the very thin shock wave region followed by induction zone whose thickness depends on the shock strength, and this zone is further succeeded by the reaction zone where the thermodynamic properties like pressure and temperature reach CJ state. ZND theory predicts similar end states of the reaction as CJ theory, thereby providing a sort of validation to CJ theory and is also able to predict the steep rise in pressure (30–100 times than initial pressure) of strong detonation known as von Neumann spike [2]. Since most of the detonation phenomenon occurs in the confined space, hence there is deviation of detonation properties from the theoretical data of explosive mixture provided by CJ and ZND theory. Hence, the detonation velocity obtained for a confined space detonation will differ from equilibrium CJ theoretical value [1].

The point to ponder is that how can detonation be achieved. The most frequently used techniques are direct initiation, shock focusing and deflagration to detonation transition (DDT) [4]. The most commonly used technique in pulse detonation engines, deflagration to detonation transition (DDT) occurs due to the fact that deflagration wave after initiation is quite unstable and has a tendency to accelerate to some critical velocity that is the consequence of boundary interactions, after which detonation transition is observed [1, 2]. The detonation wave may attain velocities higher than CJ state detonation velocity due to the increased temperature and pressure in the reaction zone which accelerates the detonation wave. This wave is called as over-detonation wave and is less stable to CJ detonation wave. While propagating, over-detonation wave tends to attain CJ state [2, 5]. PDE removes the need of all these by providing propulsive thrust with the help of high-frequency detonations [6]. It has shown better efficiencies and thrust generation then conventionally used aero-propulsive engines [6]. If the detonation frequency of the engine is achieved to be of the order of 250 Hz, it will resemble to a constant thrust engine [7]. PDE cycle comprises of filling fuel mixture, detonation initiation, wave propagation and product expansion, exhaustion of gases and finally purging out the exhaust [8]. One of the bigger challenges currently for PDE researchers and experimentalists is optimising the length of the tube so that DDT occurs at a shorter distance from ignition and a sustainable detonation wave propagates through the tube. To achieve this challenge, several researchers have come with usage of DDT enhancement devices, which tend to shorten DDT time and distance [6]. Some of the most commonly used DDT enhancement devices are Shchelkin spiral, orifice plates, circumferential grooves, spiral grooves, etc. Shchelkin spirals are differentiated on the basis of namely two parameters, its blockage ratio and pitch, where former being the parameter of spiral diameter and thickness. Pitch is actually the distance between two consecutive helical turns in the spiral. The smaller the pitch of the Shchelkin spiral closer will be its helical turns. The purpose of this study is to find the effect of these parameters on deflagration to detonation transition performance.

The very significant advantage of hydrogen-based fuels is that its product after combustion is mainly water which has no harmful impact on the environment. Although hydrogen combustion with air might produce NO_x gases, but researches are going on to reduce the production of NO_x. Hydrogen with air has a total of 19 chemical reactions as reported by Smirnov, and the reaction kinetics for the same has been evaluated using the ZND model [8]. This study thus uses hydrogen-air based stoichiometric fuel mixture for deflagration to detonation transition studies. The numerical simulation of deflagration to detonation requires very high grid resolution to observe all the microscopic flow details [9–11]. Various publications can be found, which uses very high grid resolution of at least 20 computational cells per half reaction length to sufficiently resolve both length scales associated with chemistry and physical diffusion to obtain the required results [12, 13]. This is a challenge to achieve in large computational domains, and it is important to include the physical diffusion terms for detonation wave solutions while performing the numerical simulation for solving reactive Navier–Stokes equations [13].

Computational complexity for the 'direct' simulation of deflagration to detonations is not tractable, and therefore, the 'direct' simulations of DDT cannot be applied to large-scale engineering simulations. The current study is aimed at using a CFD solver capable of simulating both deflagration to detonation and the transition between both the regimes. This CFD solver named 'ddtfoam' was developed by Florian Ettner and two of his colleagues [14] to simulate DDT in explosion tubes in 2D without resolving the microstructure of the flow in the CFD grid. In their implementation of the developed 'ddtfoam' solver, they investigated the influence of concentration gradients on DDT formation [15]. The current study deals with homogeneous mixtures, whereas the influence of concentration gradients has come into the focus of research [16, 17]. It is found that ddtFoam can predict various modes of DDT in obstacle-laden macro-shock tube of 30 mm width. Vashishtha et al. (2018) have also used this solver for numerical study of detonation-driven miniature shock tubes [18]. Different configurations of obstacles representing blockages are used with varying parameters and their combinations, to observe the formation of DDT. The solver is validated by its developer through experimental data [15]. Further, detonation equilibrium properties are validated with CJ values of hydrogen-air mixture provided by NASA Chemical Equilibrium with Applications (CEA) code [2, 19].

In this study, the emphasis was on the optimisation of the currently existing researches on detonation and PDE. Computational modelling of Shchelkin spiral was done in the form of circular blockages where the blockage offered by the circular blockage was evaluated by the same formula used to calculate blockage ratio of Shchelkin spiral. Pitch was marked by the distance between two consecutive circular blockages. The effect of variation in these parameters on DDT were further studied and reported.

2 Numerical Modelling

The deflagration to detonation transition is a complex phenomenon including fame instabilities, interaction with turbulence, thermodynamic phenomenon and gas-dynamic phenomenon. To numerically simulate the flow at very high Reynolds number with all these phenomena, it requires a huge amount of computational resources.

A reliable DDT model does not have to resolve all details of the flow as concluded by Thomas [20]. But features such as correct turbulent burning rates, local density increases due to shocks and the capability of giving rise to detonations as a result of blast waves are considered as necessary for an accurate DDT model. The open-source CFD platform OpenFOAM [21] has been chosen for the current study. OpenFOAM can deal with unstructured grids and sophisticated geometries. The model developed by Ettner [15] is made available freely online to the scientific community.

The solver used in the current study, i.e., ddtFoam, is a density-based code which solves the unsteady, compressible Navier–Stokes equations [15]. The numerical simulation of compressible flows at high Mach number requires a scheme which is much better in shock capturing as compared to the standard schemes [22]. PISO scheme sometimes predicts a wrong shock location and displays overshoots at discontinuities. This scheme is highly dissipative. This is a very critical issue for transonic reacting flows which includes auto-ignition caused by shocks. The scheme used in ddtFoam solver to determine the convective fluxes is HLLC, which is very accurate and robust. This scheme shows advantage in terms of computational efficiency [15].

HLLC scheme does not work well in low Mach number flows. So ddtFoam uses PISO scheme to initiate the numerical simulation and then switches to the HLLC scheme once the flow starts developing due to combustion. The chemkin thermodynamic database is used for properties of reacting hydrogen-air mixture, and Sutherland correlation is used to determine the molecular transport coefficients. The approach to model the combustion is via a reaction progress variable denoted by c. This variable can take value either 0 or 1 and determines the behaviour of the mixture. $c = 1$ specifies that it is a completely burned mixture, whereas $c = 0$ is for unburned mixture. The quantity $\tilde{c}(x, t)$ is Favre-averaged density-weighted probability of encountering burned gas at a particular instance of time and space. The transport equation for the reaction progress variable c is given as

$$\frac{\partial(\bar{\rho}\tilde{c})}{\partial t} + \frac{\partial\left(\bar{\rho}\tilde{c}\tilde{u}_j\right)}{\partial x_j} = \frac{\partial}{\partial x_j}\left(\bar{\rho}\, D_{\text{eff}}\frac{\partial\tilde{c}}{\partial x_j}\right) + \bar{\omega}_{c,\text{def}} + \bar{\omega}_{c,\text{ign}} \qquad (1)$$

where the tilde is used for Favre averaging and overbar for Reynolds averaging. There are two source terms. The one with the subscript *c*,def accounts for deflagration, and the other term with the subscript *c*,ign accounts for detonative combustion. The RANS version of Weller combustion model extended by a factor $0 \leq G \leq 1$ is used to model the deflagrative source term. Quenching of turbulent flames is considered by this model [15] through the relation given by:

$$\bar{\omega}_{c,\mathrm{def}} = \bar{\rho}_{\mathrm{u}} s_{\mathrm{T}} |\nabla \tilde{c}| G \tag{2}$$

$$s_{\mathrm{T}} = \xi s_{\mathrm{L}} \tag{3}$$

ρ_{u} denote the density of the unburned mixture and s_{T} is the turbulent burning speed which is modelled as the product of laminar burning speed s_{L} and a flame wrinkling factor ξ. A transport equation for the flame wrinkling factor is also solved [15]. An approximate relation between laminar flame speed, pressure and temperature used in the ddtFoam solver is given by:

$$s_{\mathrm{L}} = s_{\mathrm{L},0} \left(\frac{T}{T_0} \right)^{\alpha} \left(\frac{p}{p_0} \right)^{\beta} \tag{4}$$

The values of α and β used in the current study are 1.75 and −0.2, respectively [15].

Another source term, i.e., detonative source term $\bar{\omega}_{c,\mathrm{ign}}$, accounts for auto-ignition effects. The auto-ignition delay time is governed by the local temperature T, pressure p and mixture composition. Mixture composition is described using the mixture fraction f_{H}, which is the amount of hydrogen that would be present if the cell was completely unburned. The ddtFoam solver uses a table of t_{ign} as a function of temperature, pressure and mixture fraction using Cantera [23] and the O Conaire mechanism [24]. There is an upper limit for the application of the O Conaire mechanism which states it cannot be used for pressure above 87 atm and temperatures above 2700 K. The ddtFoam solver uses the values of auto-ignition time delay from the t_{ign} table in each computational cell. The grid size used for the current study is 1 mm. This is a very coarse grid. But this has been validated by the Ettner, and his colleagues [15], that this solver works well for 1 mm grid size. They have validated the results by comparing the numerical results with the experimental data. ‘ddtFoam’ implements a sub-model that increases the accuracy of auto-ignition modelling on coarse grids. This sub-model was introduced due to the fact that auto-ignition can be triggered by shock-induced heating. If there is model that predicts auto-ignition based on the average temperature and average pressure, then it will lead to incorrect results. This model leads to separate calculation of ignition time delay across the shock in computational cell. The advantage of this model is that it can be applied to the entire computational domain even if there is no shock formation in some regions [15].

3 Computational Domain

Computational domain is modelled for 1 m length PDE tube and 50 mm diameter. The computational domain consists of half the tube (25 mm) in Y-direction and 1000 mm length in X-direction. Conventionally, circular obstacles (Shchelkin spiral), rectangular and square obstacles are utilised for experimental purposes [25]. The

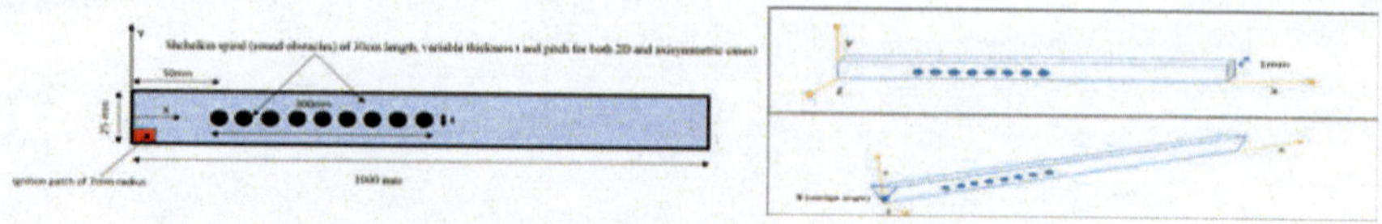

Fig. 1 Computation Domain for 2D and axisymmetric simulations

Shchelkin spiral is modelled as circular blockages in tandem for 2D simulation as well as axisymmetric simulations, where t is thickness of the obstacle and blockage ratio is proportional to thickness. Length of the obstacle domain is also kept constant, i.e., 30 cm. The ignition is modelled at the centre of tube with 7 mm radius patch. When $t = 5.5$ mm, BR = 22%; and when $t = 11$ mm, BR = 44%.

Different Shchelkin spiral (round obstacle) of various pitches is placed at 50 mm from the ignition source. It offers blockage corresponding to its thickness and pitch. The domain is discretised by using structured grid for all the parts of computational domain except for the round obstacles. To perform 2D and axisymmetric simulation in OpenFOAM, the single grid in z- and θ-direction, respectively, are generated (Fig. 1).

3.1 Initial and Boundary Conditions

2D Simulations

In two-dimensional simulation, the faces of single grid, normal to Z-directions are given as empty boundary condition in mesh generation method; hence, no solution in Z-direction is calculated. Further, adiabatic no slip wall boundary condition is applied at wall near the ignition point and on the top surfaces. Outlet face of the tube is given patch boundary condition, and lastly, the faces normal to the Y-direction are given symmetric boundary condition. The solution is symmetric about this face.

Axisymmetric Simulations

In case of axisymmetric simulations, the angular planes are given the wedge boundary condition. The wall near ignition point and the top wall are given adiabatic no slip wall boundary condition. Outlet face of the tube is given patch boundary condition.

Ignition is modelled as patch near the centre with radius of 7 mm. The tube is filled with hydrogen-air stoichiometric mixture at 1 atmospheric pressure and 293 K temperature. It is also assumed that the mixture is homogeneously distributed along X- and Y-directions in the tube without any gradient, at given pressure and temperature. Initially, pressure-based solver, pddtFoam, is used for initialising the flame and flow from the rest for 1 ms. Further, density-based solver, ddtFoam, is used by mapping the solution from pddtFoam.

3.2 Grid Independency Test

See Fig. 2.

For the grid independency test, the simulations are performed on straight tube (without obstacle) assuming all the appropriate boundary condition. The hydrogen-air stoichiometric mixture is filled with pressure of 101,325 Pa and 293 K temperature. Three uniform grid sizes of 0.5, 1 and 2 mm are studied for computations for 8 ms time for straight tube without obstacles. The flame tip locations along *X*-axis and surface-averaged pressures at outlet have been monitored. The flame tip locations have been calculated based on maximum axial extent of reaction progress variable at each time step. Both 2D and axisymmetric cases without obstacle are analysed for grid independency test varying the grid size as mentioned above. The flame tip accelerates towards the outlet and reaches at the end of the tube at approximately 5 ms time as shown in flame tip location versus time plot. The pressure at the outlet remains closed for all the grid sizes up to 5 ms time, and further deviates as reaction products goes out from the tube. Hence, the grid sizes 0.5, 1 and 2 mm behave in similar fashion up to the time flame reaches outlet. For further cases, 1 mm grid size has been used for 2D simulations.

In case of axisymmetric case tube without obstacle, for 2 mm grid size, the DDT transition occurs at approximately 2.5 ms. However, for finer grid, there is no DDT transition. Hence, 2 mm grid simulation for axisymmetric case may not be correct. Further, simulations for 0.5 and 1 mm in axisymmetric case show close agreement in case of flame tip location and pressure at the outlet. Hence, for all the further simulations, 1 mm grid size is chosen.

4 Results and Discussions

The computational study began with extensive literature review of the detonation and DDT enhancement devices, especially Shchelkin spiral. It was seen that detonation wave comprises of a very strong shock wave, which is accompanied by expansion trail which contains reactions products at very high temperature and pressure [18]. The computational domain was set up for two different numerical studies, viz 2D

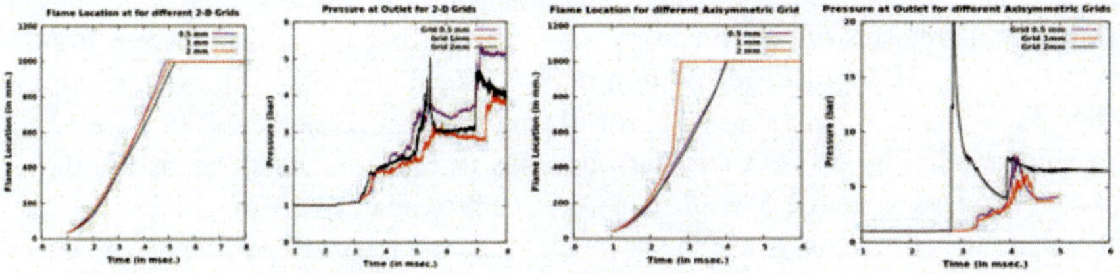

Fig. 2 Flame tip location and surface-averaged pressure for 2D and axisymmetric simulations

and axisymmetric, by varying the parameters of obstacles such as pitch and blockage ratio. Spiral was modelled as circular obstacle system. Initial conditions, length of the obstacle system and starting position of the obstacle system are kept same for all the cases. Every case is simulated for a minimum of 4 ms after the ignition of H_2-air mixture. The thought behind doing axisymmetric simulation was that it will resemble 3D tube more closely than 2D simulation. Also, the former has three-dimensional interactions as compared to 2D simulation, where the flow interaction is only considered in *X*- and *Y*-direction. Due to the increase in wall interactions in axisymmetric simulations, it is assumed that DDT will occur earlier in comparison to same 2D simulation case. Further, effect of variable pitch parameter on DDT is studied for a fixed blockage ratio for both 2D and axisymmetric simulation. DDT transition occurs earlier in cases where the obstacles are placed close to each other, i.e., pitch is small. Since, DDT is marked by sudden pressure and velocity rise, the flame tip location versus time and pressure–time graphs are plotted for all the cases. In addition to that, temperature–time graph and *X*–*t* plots for both temperature and pressure are plotted. The detonation velocity and pressure results for all the cases are validated with the CJ state detonation velocity and pressure for hydrogen-air mixture provided by NASA CEA codes [19]. The results differed slightly with the CJ state values probably because of the boundary interactions in the confined space. Also, the computational results are approximate as compared to the experimental results, and further, the flame speed is calculated by differentiating time step which is in itself an approximation. For instance, if we choose a smaller time step, then the results will converge towards the CJ state values. The probable reason that the detonation velocity exceeds CJ velocity is that, since local flame speed depends on the available fuel and pressure–temperature at particular location, so local flame speed calculated computationally may exceed CJ speed value which is about 1970 m/s. Although, ddtFoam solver have been tested and validated for DDT for macro-tubes through experimental results and the pressure, temperature and flame tip plots results thus obtained are in accordance with the trends [15].

4.1 2D Versus Axisymmetric Simulations

Three different cases were modelled and are studied for both 2D and axisymmetric simulations. Blockage ratio and pitch parameters of the obstacle system were chosen arbitrarily. Three sets of combinations were chosen, that are 44% blockage-20 mm pitch (Case I), 22% blockage- 24 mm pitch (Case II) and 22% blockage- 15 mm pitch (Case III) and were simulated for 2D and axisymmetric cases. In Case I, for 2D simulation (Fig. 3) with circular obstacles in tandem, the slope of the flame tip plot changes near the first obstacle, i.e., approximately 53 mm, location, and the slope further increases, which shows that the obstacles offer further increase in flame speed. The transition to detonation occurs when the slope is maximum in the flame tip location plot which is approximately at 251 mm location and at roughly 2.03 ms time in 2D simulation for Case I. Comparing it with its axisymmetric (Fig. 4)

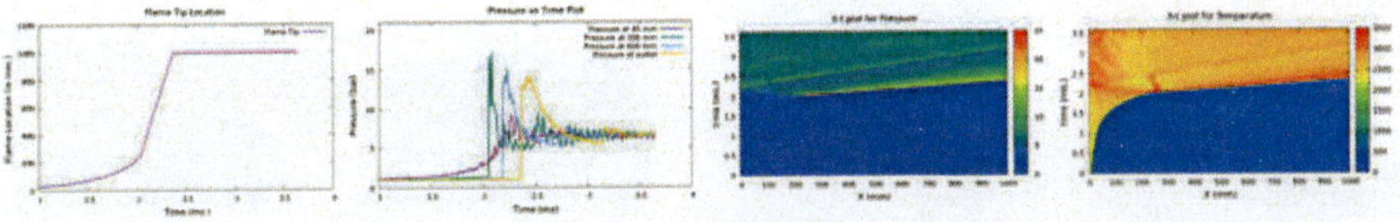

Fig. 3 Flame tip location versus time, surface-averaged pressure versus time at different locations, *x–t* plot for pressure and *x–t* plot for temperature plotted for 2D simulation (BR = 44%, pitch = 20 mm)

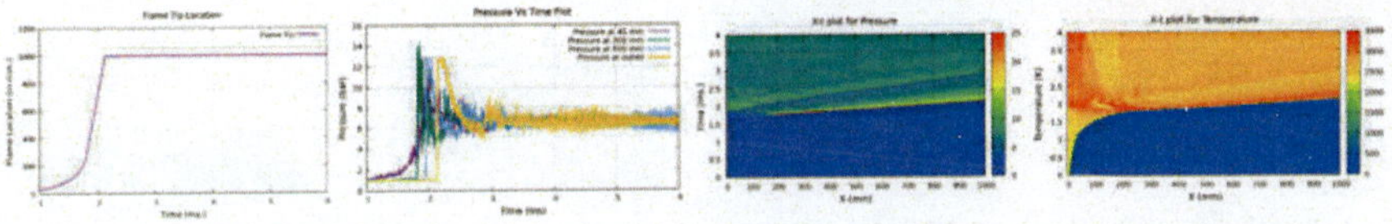

Fig. 4 Flame tip location versus time, surface-averaged pressure versus time at different locations, *x–t* plot for pressure and *x–t* plot for temperature plotted for axisymmetric simulation (BR = 44%, pitch = 20 mm)

counterpart, first increase in flame tip speed is marked at 42 mm and further transition to detonation is seen just about at 233 mm location and at 1.79 ms time. It shows that for axisymmetric simulation, DDT occurs earlier with respect to the ignition point. From the *X–t* plot for pressure and pressure–time plot, the pressure peak rise is almost equal for both the simulations which is approx. 22 times the initial pressure which further approaches CJ state pressure value (less than 16 times initial pressure) [19]. This further shows that a transition to detonation from deflagration has occurred along the tube. The detonation velocity is calculated using differentiating time step from flame tip location plot and pressure–time plot which averages out to be approximately 2200 m/s for 2D case and 2050 m/s for axisymmetric simulation. This value is quite close to the CJ detonation velocity for hydrogen-air mixture. The *X–t* temperature plot for the tube shows that there is rise in the temperature, since combustion takes place and the peak temperature is seen to be about 3200 K for both the simulations and reduces thereafter tending towards CJ state temperature value. The surface-averaged pressure and temperature plots for different locations, 45, 300, 600 mm and outlet have been plotted which shows that pressure and temperature rise is earlier for axisymmetric simulation than 2D simulation. The flame reaches the exit at 2.36 and 2.13 ms for 2D and axisymmetric simulations, respectively.

Similar trends were seen for Case II (Figs. 5 and 6) with blockage 22%- and 24-mm pitch. As after ignition, deflagration wave interacts with the first obstacle placed at 50 mm, and this positive feedback mechanism with flame and turbulent flow field causes first increase in the slope of flame tip plot for both the simulations. Although axisymmetric geometry considers 3D wall interactions with the flow field has sudden increase in the slope of the flame tip plot at just less than 305 mm and at 2.02 ms from ignition point as compared to its analogue where it occurs roughly at 330 mm and at 2.33 ms, marking the transition to detonation. Also, *X–t* plot for pressure shows that

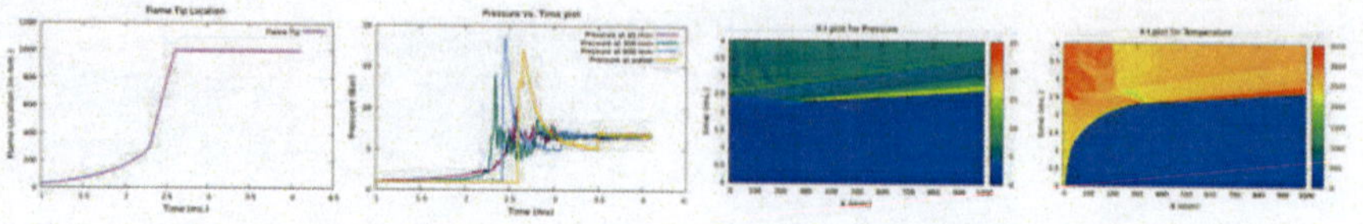

Fig. 5 Flame tip location versus time, surface-averaged pressure versus time at different locations, x–t plot for pressure and x–t plot for temperature plotted for 2D simulation (BR = 22%, pitch = 24 mm)

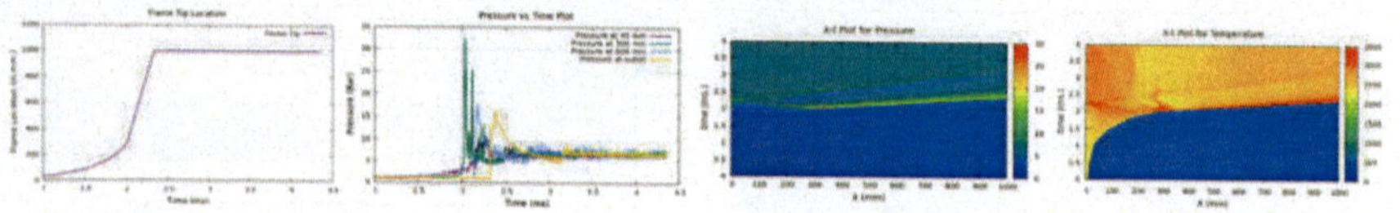

Fig. 6 Flame tip location versus time, surface-averaged pressure versus time at different locations, x–t plot for pressure and x–t plot for temperature plotted for axisymmetric simulation (BR = 22%, pitch = 24 mm)

the peak pressure rise for this location is approximately 32 and 25 times the initial pressure and the surface-averaged pressure reduces thereafter to the range of 16–18 bar for axisymmetric and 2D simulations, respectively. The detonation velocity calculated by flame tip and pressure–time plot averages out to be about 2005 m/s for both the simulations. The rise in temperature marking the combustion process as seen in X–t plot of temperature shows that the peak temperature reaches roughly 3400 and 3900 K at the location of detonation transition for 2D and axisymmetric simulations, respectively. This surface-averaged temperature further decreases with the wave propagation, nearby to CJ temperature value limits which is roughly around 2800 K for H_2-air mixture. These results further strengthen our assumption that axisymmetric simulations consider 3D interactions and do resemble the actual tube more realistically than 2D simulations, and hence, DDT will occur slightly earlier in them.

And for the last Case III (Figs. 7 and 8), with same blockage as Case II, but a lesser pitch, i.e., 15 mm, shows same trend in the results, hence further solidifying our results. In both the simulations, the ignited mixture shows gradual increase in the slope of flame tip versus time plot at nearly 50–51 mm, thus highlighting the presence

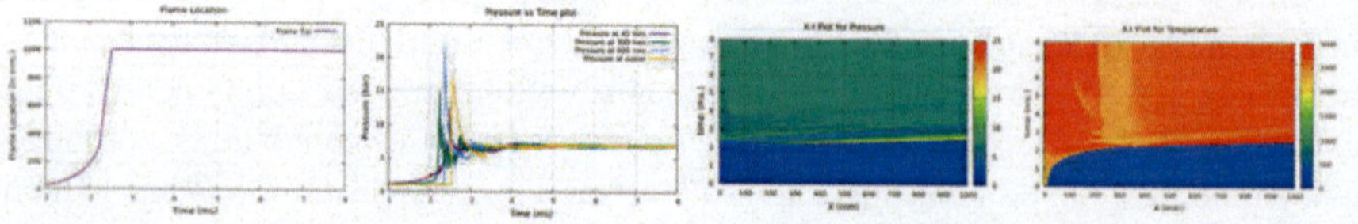

Fig. 7 Flame tip location versus time, surface-averaged pressure versus time at different locations, x–t plot for pressure and x–t plot for temperature plotted for 2D simulation (BR = 22%, pitch = 15 mm)

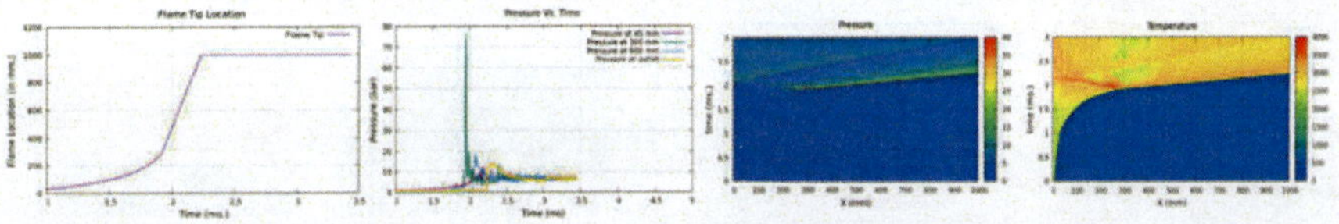

Fig. 8 Flame tip location versus time, surface-averaged pressure versus time at different locations, *x–t* plot for pressure and *x–t* plot for temperature plotted for axisymmetric simulation (BR = 22%, pitch = 15 mm)

of obstacles. After this due to further wall interactions because of the presence of blockage, the turbulence in the flow medium increases, which causes deflagration wave to reach its critical limit and transit into a detonation wave at 291 mm and 1.92 ms for axisymmetric simulation, whereas it transits at 320 mm location and 2.24 ms for 2D simulation. After studying *X–t* plot and pressure–time plot for 2D simulation, spike pressure was obtained to be approximately 25 times the initial pressure. Although when it was seen for axisymmetric simulation, a bit unexpected peak was seen about 75 time the initial pressure.

This value of spike differs from all the other results, and the reason might be some computational approximation uncertainty, but the graph further shows that surface-averaged pressure thereby approaches CJ state detonation pressure value. The detonation velocity calculated using flame tip and pressure plot comes out to be roughly 2001 and 2216 m/s for 2D and axisymmetric case. The *X–t* plot for temperature shows that the temperature peak attained during the 2D and axisymmetric case comes out to be 3300 K and 3800 K, respectively. Case III also shows quite similar trend for axisymmetric and 2D simulations comparisons. The assumption was thereby true that axisymmetric which resembles more to the actual 3D case shows detonation earlier compared to 2D simulations. Also, while evaluating Case II and Case III, the possibility of comparing DDT properties on the basis of pitch variations came into picture. If the blockages are closer to each other, there is a possibility of them to increase wall interactions of the flow with respect to time, which further increases turbulence in the deflagration phase to start the onset of detonation.

4.2 Effect of Pitch Variations

The effect of pitch variation was compared with the aim to find how the location and time of DDT change with change in pitch parameters of obstacle system. Three different cases having same blockage ratio (22%) but different pitch (24, 15 and 10 mm) are studied for 2D simulations. The results for 2D simulation of pitch 24 and 15 mm are discussed above. For 10 mm pitch (Fig. 9), the first increase in the flame tip plot was seen at, as expected, 50 mm which is the point of flame interaction with first obstacle. Further, the transition to detonation occurs at 302 mm which is roughly 18 mm and 28 mm earlier from 15 and 24 mm pitch obstacle system, respectively. The

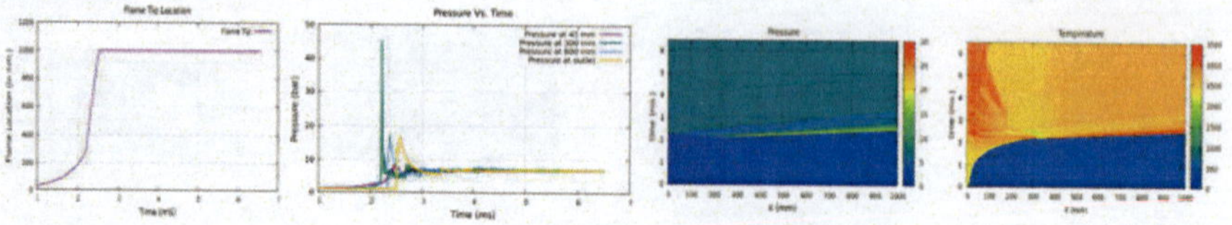

Fig. 9 Flame tip location versus time, surface-averaged pressure versus time at different locations, *x–t* plot for pressure and *x–t* plot for temperature plotted for 2D simulation (BR = 22%, pitch = 10 mm)

detonation velocity calculated for 10 mm case comes out to be 2300 m/s, having slight deviation from CJ velocity for H_2-air. The peak of the pressure rises as observed from *X* to *t* plot for pressure is roughly at 300 mm marking the DDT, and this peak is about 40 times the initial pressure. The impact of closeness of obstacles is evident from the results. Keeping all the parameters constant but varying pitch of the obstacle, system shows that presence of obstacles closer to each other has a considerable effect in DDT location. Although the difference in pitch of the various obstacle system was less, but for a bigger variation of pitch values, a more significant difference in DDT length can be observed. Pitch, hence, is an important parameter for length optimisation of the obstacle system and tube.

5 Conclusion

A numerical investigation was carried out to study the deflagration to detonation transition phenomenon in detonation tube with the presence of DDT enhancement devices for 2D and axisymmetric geometric configurations. The use of ddtFoam solver for the numerical study of H_2-air stoichiometric fuel mixture gave the results which were as per the results validated by Florian etner et al. for 2D simulations. For axisymmetric cases, the results showed similar trends, and further, detonation properties were compared with NASA CEA codes given for hydrogen-air fuel mixture. It was seen that use of axisymmetric computational domain provides a better insight for real 3D tube since it considers all three-dimensional interactions. This was further proved as due to the additional interactions along the *z*-axis showed that DDT occurs roughly 20–30 mm earlier than 2D simulation. Hence, an emphasis on axisymmetric computational domain can be given while modelling a case for DDT transition along with 2D domain for better look-out of the case. The pressure plots verified the occurrence of detonation phenomenon since at the location of detonation there was a spike in pressure observed. Also, these detonation properties approached CJ state detonation properties. The slight deviation would be due to the reason that CJ theory eludes the effect of boundary interactions and also numerical simulations find approximate solution unlike experimental results. Further, the effect of closeness of blockages in obstacle system also had an impact in detonation location. Closer the blockages, i.e., smaller the pitch of obstacle medium, it was seen that DDT occurs at earlier time and location from ignition. The interaction of deflagration wave with obstacles closer to each other increases the turbulence and causes the wave to reach its critical limit to transit into detonation wave. Lastly, in couple of cases, unexpected spike in pressure was observed. The reason of this uncertainty can be a part of future studies which can be validated with experimental results. Also, finding an empirical formula about how pitch variations of obstacle systems affect DDT phenomenon in a tube, is one of the future works. There is still a possibility of simulating a 3D tube, and if successful, those results can have more real relation with actual tube.

References

1. LeeJohn HS (2008) The detonation phenomenon. Cambridge University Press, NY, USA
2. Panicker PK (2008) The development and testing of pulsed detonation engine ground demonstrators. The University of Texas at Arlington
3. Kuo KK (2005) Principles of combustion, 2nd edn. Wiley
4. Chander S, Jindal TK (2012) Integration challenges in design and development of pulse detonation test rig. Int J Adv Res Electr, Electroni Instrum Eng 1(4). ISSN: 2278-8875
5. Li J, Lai WH, Chung K (2006) Tube diameter effect on deflagration to detonation transition of propane-oxygen mixtures. Shock Waves 16(2):109–117
6. New TH, Panicker PK, Chui KF, Tsai HM, Lu FK (2006) Experimental study on deflagration-to-detonation transition enhancement methods in a PDE. In: American Institute of Aeronautics and Astronautics, AIAA-2006-7958
7. Jindal TK, Dogra BA (2017) Development of PDE based power generation configuration. In: International conference on science, technology, engineering and mathematics, ICSTEM
8. Maciel EC, Marques CST (2019) 2-D simulation with OH* kinetics of a single-cycle pulse detonation engine. J Appl Fluid Mech 12(4):1249–1263. https://doi.org/10.29252/jafm.12.04.29593
9. Oran ES, Khokhlov AM (2000) Deflagrations, hot spots, and the transition to detonation. Philos Trans R Soc A: Math, Phys Eng Sci 358(1764):3539–3551
10. Oran ES, Gamezo VN (2007) Origins of the deflagration-to- detonation transition in gas-phase combustion. Combust Flame 148(1–2):4–47
11. Gamezo VN, Ogawa T, Oran ES (2008) Flame acceleration and DDT in channels with obstacles: effect of obstacle spacing. Combust Flame 155(1–2):302–315
12. Hwang P, Fedkiw RP, Merriman B, Aslam TD, Karagozian AR, Osher SJ (2000) Numerical resolution of pulsating detonation waves. Combust Theory Model 4(3):217–240
13. Singh S, Powers JM, Paolucci, S () Detonation solutions from reactive Navier-Stokes equations. In: AIAA-99-0966
14. Ettner F, Sattelmayer T (2013) ddtFoam. http://sourceforge.net/projects/ddtfoam
15. Ettner F, Vollmer KG, Sattelmayer T (2014) Numerical simulation of the deflagration-to-detonation transition in inhomogeneous mixutres, vol 2014. Article ID 686347, 15 pages http://dx.doi.org/10.1155/2014/686347
16. Vollmer KG, Ettner F, Sattelmayer T (2012) Deflagration-todetonation transition in hydrogen/air mixtures with a concentration gradient. Combust Sci Technol 184:1903–1915
17. Ettner F, Vollmer KG, Sattelmayer T (2010) Numerical investigation of DDT in inhomogeneous hydrogen-air mixtures. In: Proceedings of the 8th international symposium on hazards, prevention and mitigation of industrial explosions, Yokohama, Japan, 2010
18. Vashishtha A, Rao SMV (2018) Numerical study of detonation driven miniature shock tubes using OpenFOAM. In: National symposium on shock waves, Chandigarh, India
19. McBride BJ, Gordon S (2008) Chemical equilibrium with applications (CEA) code, CEA2 with CEAgui, Version 1.0. http://www.grc.nasa.gov/WWW/CEAWeb/. Accessed July 2008

20. Thomas G (2012) Some observations on the initiation and onset of detonation. Philos Trans R Soc A: Math, Phys Eng Sci 370(1960):715–739
21. OpenFOAM (2012) User guide, version 2.1.1. http://www.openfoam.org
22. Issa RI (1986) Solution of the implicitly discretised fluid flow equations by operator-splitting. J Comput Phys 62(1):40–65
23. Goodwin D (2009) Cantera: an object oriented software toolkit for chemical kinetics, thermodynamics and transport processes. http://code.google.com/p/cantera
24. Conaire MÓ, Curran HJ, Simmie JM, Pitz WJ, Westbrook CK (2004) A comprehensive modeling study of hydrogen oxidation. Int J Chem Kineti 36(11):603–622
25. Soni SK, Singh A, Sandhu M, Goel A (2013) Numerical simulation to investigate the effect of obstacle on detonation wave propagation in a Pulse Detonation Engine Combustor. In: ICERTSD 2013

Numerical and Simulation Approach for Design of Variable Valve Actuation Mechanism on Single-Cylinder Diesel Engine

Ashish Jain, E. Porpathamn, and Sukrut S. Thipse

1 Introduction

Diesel engine is widely used in commercial vehicles. There are less than 1% market share of other fuel than diesel particularly for commercial vehicles. The reason is very clear—better low-end torque, fuel efficiency and comparatively relaxed norms. However, these vehicles are major source of environmental pollution, particularly NO_x emission. Old vehicles, unmonitored driving practices, overloading, unavailability of vehicle scrapping policy, improper maintenance and poor inspection technique are the major reason of higher pollution from these type of vehicles. Regulation needs to be introduced along with stringent monitoring system to control this pollution.

The market share of LCV is increased from 43% in FY 2006 to 58% in FY 2016 and growing at the rate of 8–10% YoY basis. However, controlling pollution will remain a problem until proper scrapping policy and upgradation of old vehicle at lower cost for better emission control.

While VVA systems have been successfully implemented in gasoline engines to reduce throttling losses without much difficulties; however, there is a restriction in its application to diesel engine due to mechanical constraint of lower clearance between piston and valves on account of high compression ratios. In the recent past, it has been concluded that the VVA systems have increased the volumetric efficiency; however, research is still going on to verify its effect on the emissions at various engine operating conditions.

A. Jain (✉) · S. S. Thipse
The Automotive Research Association of India, Pune 411038, India
e-mail: jainashish1985@yahoo.com

E. Porpathamn
Vellore Institute of Technology University, Vellore 632014, India

A. K. Gupta et al. (eds.), *Advances in IC Engines and Combustion Technology*,
Lecture Notes in Mechanical Engineering,
https://doi.org/10.1007/978-981-15-5996-9_63

With contradicting requirements like stringent government regulations for emissions and need of increased performance, variable valve actuation has gained importance as a topic of research in the recent past. Parvate et al. [1] studied the effects of various valve lift strategies on single-cylinder diesel engine through computer simulation and described VVA to reduce NO_x emissions with early inlet valve opening (EIVO) and late exhaust valve closing (LEVC). Using simulation studies, they also found that while EIVO and LEVC reduce NO_x emissions, LIVO and LEVO were also useful to reduce unburnt hydrocarbon emissions. EEVO and EEVC also helped to reduce the NO_x emissions; however, they are detrimental to the engine performance. Pierik et al. [2] have developed a mechanical VVA system in which output cam lift can be varied from 0 to 9 mm using input cam. Further, they had verified this mechanism on a standard baseline European four-cylinder diesel engine and achieved BSFC improvement of 12% at idle speed, 7–10% at mid-load and 0–3% at high load. They observed that the HC emissions degrade, while the NO_x emissions improve with VVA. Ojeda [3] analyzed the effect of EIVC on the performance of a 6.8L V8 engine. The VVA mechanism was controlled, using an electro-hydraulic mechanism, and performs the EIVC according to the engine requirement using a solenoid valve. The tests showed reduction in soot above 90% with fuel efficiency improvements of 5% and NO_x levels below 0.2 g/bhp-h. Gurney et al. [4] carried out 1D and 3D numerical simulation approaches to study the effect of EIVC and LIVC on a four-valve engine. They concluded that the EIVC and LIVC reduce the volumetric efficiency of the engine and drastically reduce the soot emissions during tests. Lancefield et al. [5, 6] build a model to simulate and predict engine performance using VVA mechanism with cyclically increased and decreased angular speed of the cam. They have observed that both EIVC and LEVO decreased the BSFC of the engine by approximately 2.3% during simulation. The results also showed improvements in torque with EIVCC increasing torque by 6.3–8.2% and LEVO by 8.6–12.6%.

This paper describes the design of a VVA mechanism carried out with the help of multibody dynamic analysis and static structural analysis. Present work is design, development and implementation of VVA mechanism on LCV diesel engine for emission reduction. A simplified mechanism of cam shift is developed to increase the engine performance and emission control. Design/developed mechanism's performance is compared to the baseline mechanism to check the actual benefits.

2 Variable Valve Actuation

The valvetrain mechanism in the considered engine is pushrod type which is shown in Fig. 1. It consists of valve which is operated by camshaft with the help of pushrod and rocker arm.

Therefore, a mechanism is developed for the baseline engine mechanism considering the available space and restrictions in the engine, and a secondary camshaft is designed which is placed above the rocker arm. The secondary camshaft is initially

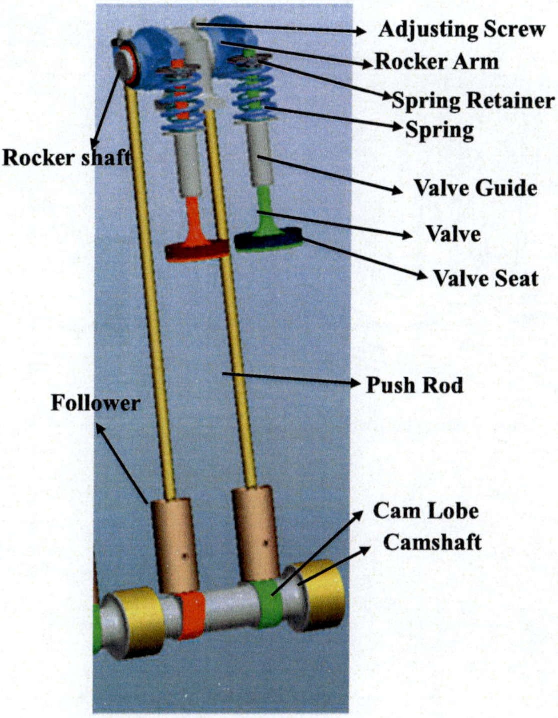

Fig. 1 Pushrod operated valvetrain mechanism

driven by motor for mechanism validation. The new camshaft designed has provision for three cam lobes; however, initially two cam lobs are used for mechanism validation. The key reason for implementing the two lobe options is to validate the performance and emission improvement in baseline vehicle as vehicle was not meeting the desired emission limit as shown in Fig. 2, and vehicle simulation result shows vehicle is majorly operating in mid-to-higher engine-speed zone (Fig. 3). Another reason for selection of only two-lobe camshaft is the space constraint in baseline engine. Hence, it is decided initially to analyze the effects in baseline engine, and then, it can be adopted in other engines. The first lobe's lift is same to baselinc engines lift, i.e., 7.6 mm, and brief engine specifications are given in Table 1. The other lobe has different lift as 8.3 mm based on 1D simulation results. When the cam rotates in normal working condition, the lift remains same as baseline lift. As the engine speed increases, the mechanism has to actuate and shift the cam so that other lift starts working. The actuation is done by a solenoid valve which gets signal from the RPM sensor and shift accordingly to vary the lift of the valve. The schematic layout of the modified system is shown in Fig. 4. Reason for specific selection for two cam lobes is decided based on the vehicle and engine modeling of the baseline emission-failed vehicle.

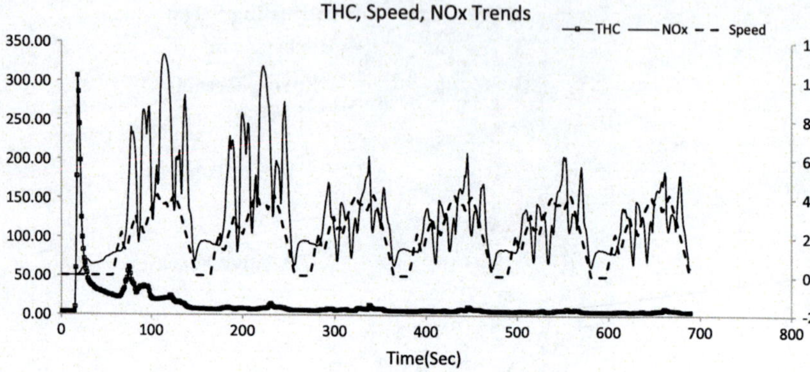

Fig. 2 Vehicle baseline emission performance at IDC

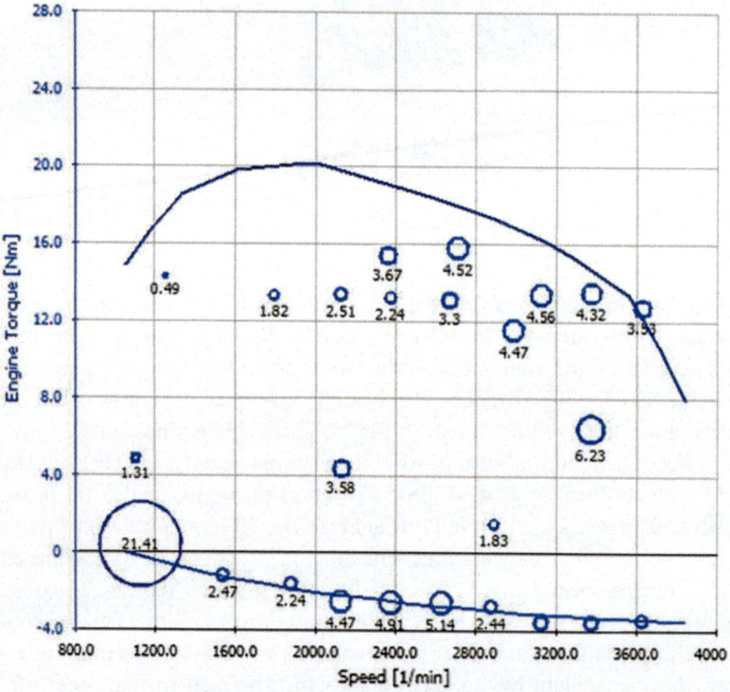

Fig. 3 Vehicle operating zones and speed w.r.t torque

Table 1 Engine specification

Type	Four stroke, air cooled, single cylinder, CI engine
Fuel	Diesel
Number of cylinders	One
Bore/stroke ratio	1.15
Volume	0.4 L
Compression ratio	19:1
Rated power	5.2 kW @ 3600 rpm
Rated torque	18 Nm @ 2200–2400
Max valve lift	7.6 mm
Start of injection	6 ± 1° CAD BTDC
Low idle rpm	1050
Fly up rpm	3960

3 Design Methodology

The methodology used in this paper is shown in Fig. 5 which explains the procedure followed to design, analyze and simulate the baseline and modified VVA system as shown in Fig. 6.

3.1 Simulation of Baseline Engine Mechanism

Baseline engine has pushrod-actuated valvetrain mechanism as shown in Fig. 4, and this baseline model of the pushrod-actuated valvetrain is modeled in 3D CAD design software. This CAD model in then imported in simulation software (ADAMS) for multibody dynamics simulation. Its kinematic parameters are studied and analyzed during initial trails based on type of joints and degree of freedom of each component. Four types of joints used for the analysis are translation, revolution, fixed and motion joints, and degree of freedom of each is decided based on type of joints. All the simulations are carried out at rated engine rpm; hence, camshaft speed is 1800 rpm, i.e., half the rated engine speed. Each analysis is carried out with end time of 10/30 s at step size of 0.001 s. The baseline engines' lift, velocity and acceleration data w.r.t time are shown in Figs. 7, 8 and 9.

Baseline system configuration has good lift and velocity profile; however, the acceleration curve is irregular. Some corrective measures are also taken in the analysis part in constraints and degree of freedom; however, there is minimum improvement in the profile. As we are more concerned about modified VVA, and the lift data is from cam measurement, hence we have not touched this further.

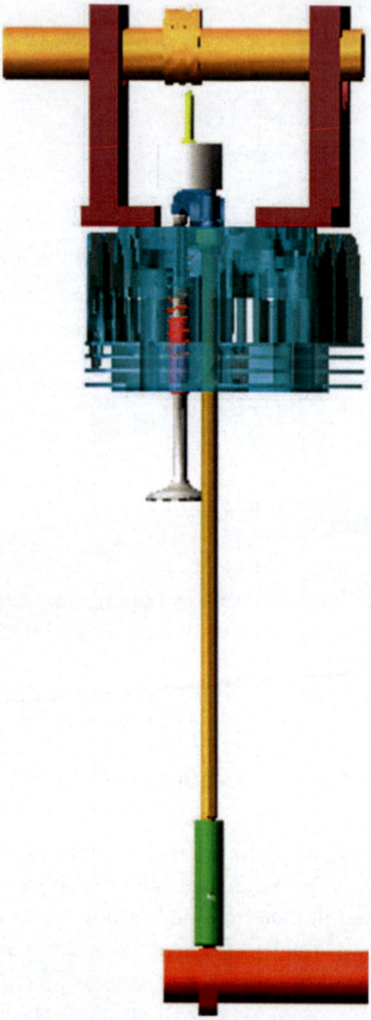

Fig. 4 Proposed VVA valvetrain mechanism

3.2 *Numerical Analysis*

To actuate the new mechanism, we need to know the force required for its actuation and structural design validation of the components. As the rated engine speed is 3600 rpm, spring stiffness (K) is 22.54 N/mm and preload is 226 N known to us. Therefore, spring force of the system is calculated as follows.

$$F = K * x$$
$$= 22.54 * 7.6$$

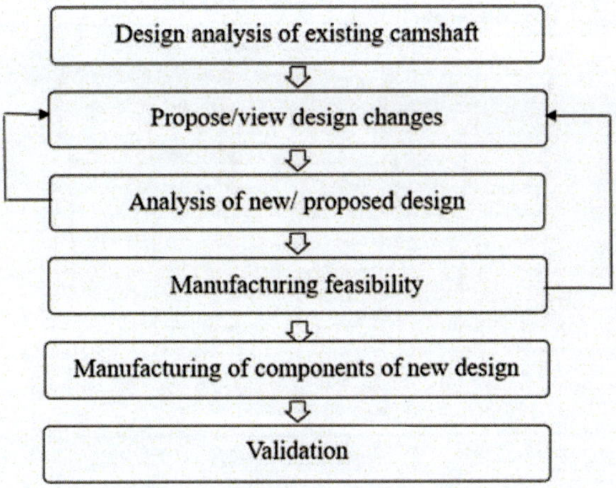

Fig. 5 Flowchart of methodology

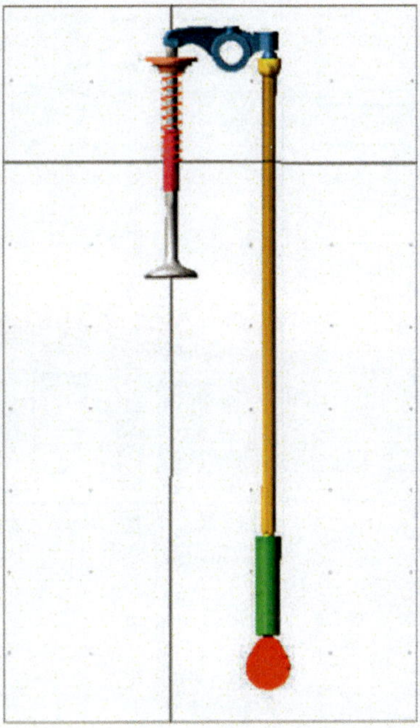

Fig. 6 ADAMS simulation model of baseline valvetrain model

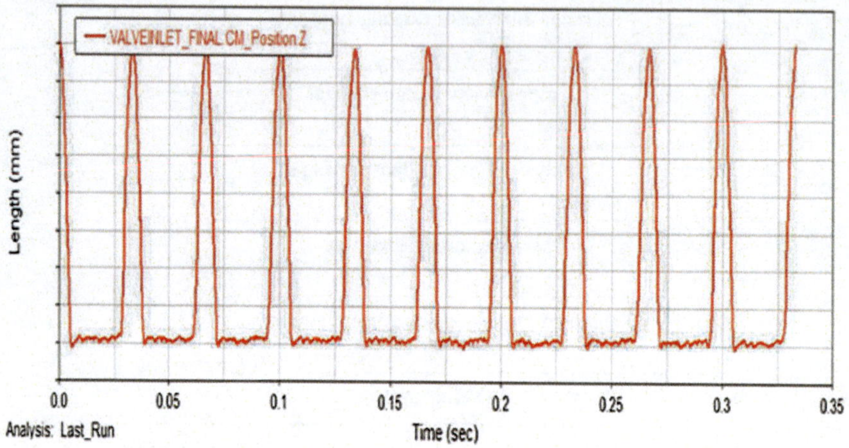

Fig. 7 Lift length versus time for baseline valvetrain system

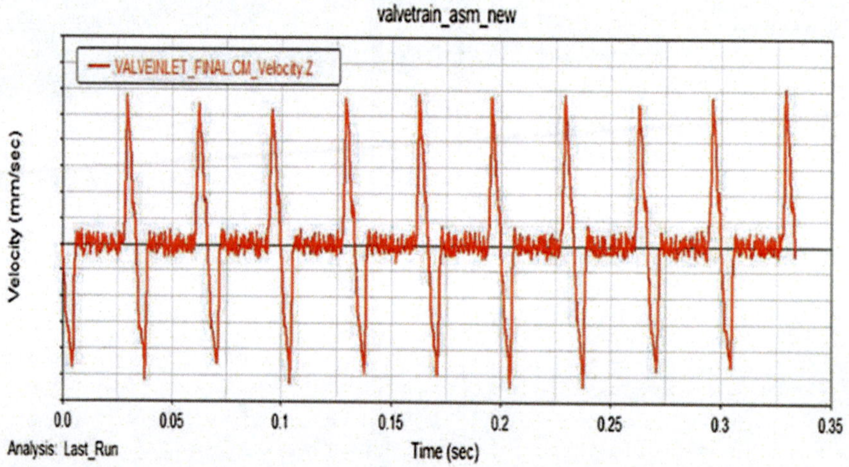

Fig. 8 Velocity versus time for baseline valvetrain system

$$= 173.1N \tag{1}$$

where x is the valve lift off the baseline valvetrain system.

Therefore, the total force F can be calculated by adding the force and the preload. Therefore,

$$F = 226 + 173.3 = 399.3\ \mathrm{N}$$

Now, to calculate the actuation force for the new mechanism,

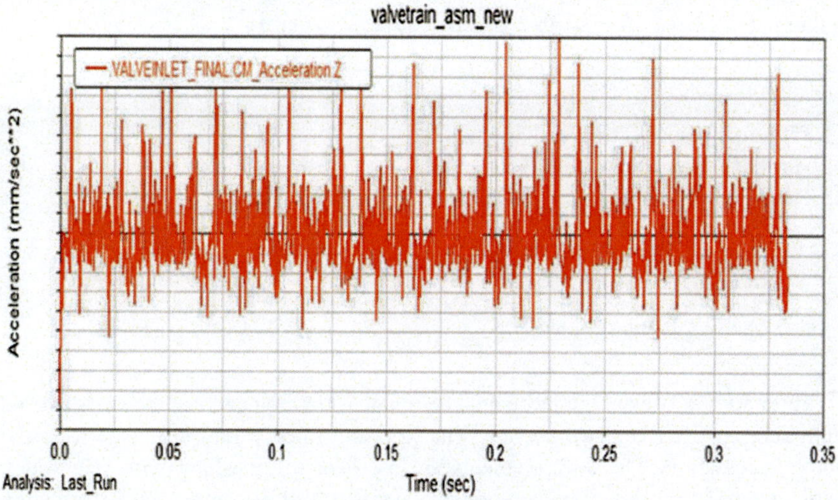

Fig. 9 Acceleration versus time for baseline valvetrain system

$$\text{Engine speed} = 3600\,\text{rpm} = 60\,\text{rps}$$
$$\text{Time for one rotation} = (1/60) = 0.016$$

Angle of dwell = 260°.

$$\begin{aligned}\text{And time for change} &= (260/360) * 0.016\\ &= 0.0115\,\text{s}\end{aligned}$$

Therefore, time for change is 11.5 ms
As the distance required to travel is 4 mm. within time of 11.5 ms
Hence, velocity is

$$V = (4/11.5) = 0.035\,\text{m/s}$$

Now from equation of motion, we know

$$S = ut + 1/2at^2 \ldots \tag{2}$$

As there is no initial movement, hence

$$\begin{aligned}&u = 0\\ &S = \frac{1}{2}at^2\\ &a = 2S/t^2 \ldots\end{aligned}$$

$$= (2 * 4 \times 10^{-3})/(11.5 \times 10^{-3})^2$$
$$= 60.49\ \text{m/s}^2 \quad (3)$$

Therefore, acceleration = 60.49 m/s^2.
Mass of the camshaft is 1.7 kg.
Therefore, force required is calculated as follows

$$F = m * a$$
$$= 1.7 * 60.49$$
$$= 102.8\ \text{N}$$

The static analysis of the individual components of the new mechanism is carried out using ANSYS 16.0 workbench. The load and force values are taken from the above calculation. The tetra elements are used for the meshing with size which varies from 2 to 5 mm. The components are analyzed for stress and deformation under loading conditions, and results are discussed.

The tappet is analyzed for compressive stress. The maximum stress due to a force of 400 N was found to be 80.937 MPa. This stress is very much less than the yield stress. The deformation is also calculated for compressive load of 400 N, and total deformation is 0.0062486 mm which is very less, and this component is safe for the mechanism. The results of the stress and deformation for tappet are shown in Figs. 10 and 11.

Stress and deformation analysis is also carried out for the camshaft mounting bracket. The stress value is found to be 35.987 MPa which is less than the yield stress of the material, and total deformation value of the cam bracket mount is 0.021284 mm which is very less and is allowable range for the proto-testing of the mechanism. The

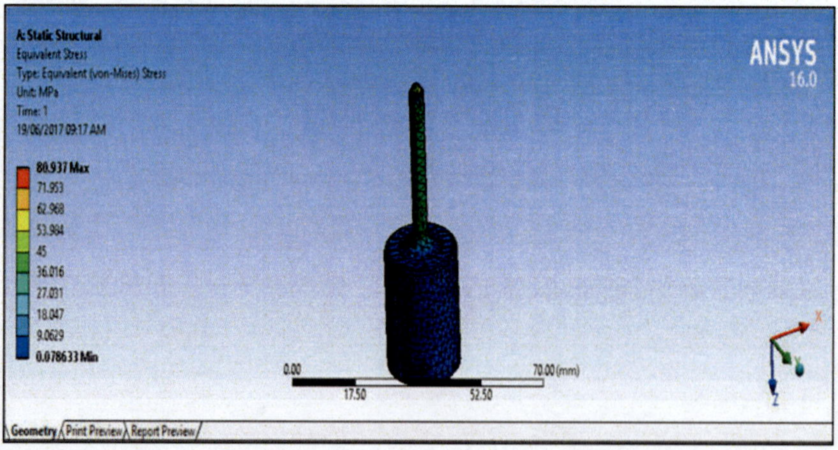

Fig. 10 Stress for tappet

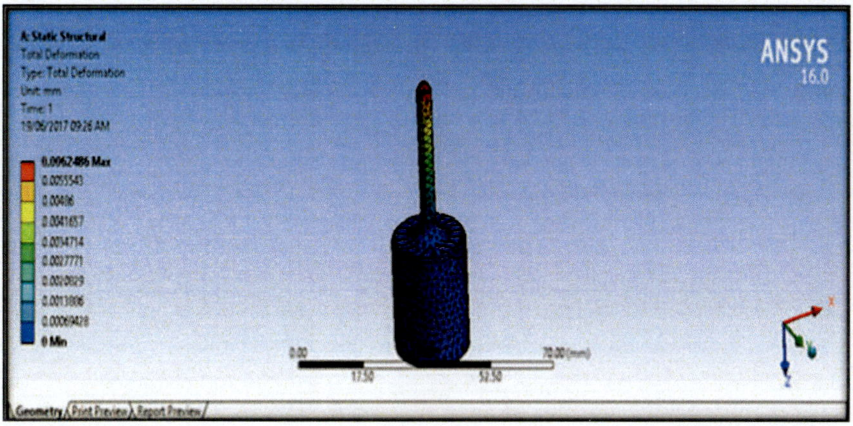

Fig. 11 Total deformation for tappet

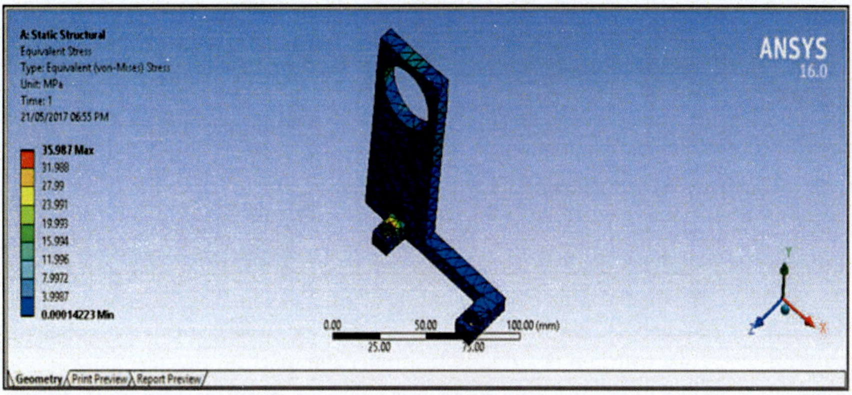

Fig. 12 Stress for cam bracket mount

results of the stress and deformation for cam bracket mount are shown in Figs. 12 and 13.

3.3 *Results for Structural Analysis*

The structural analysis of the individual valvetrain components is carried out for the load of 400 N which is the total force coming on the system. Table 2 shows the combined results of the components.

The stress values are very less compared to the yield stress of the material; hence, the components are safe.

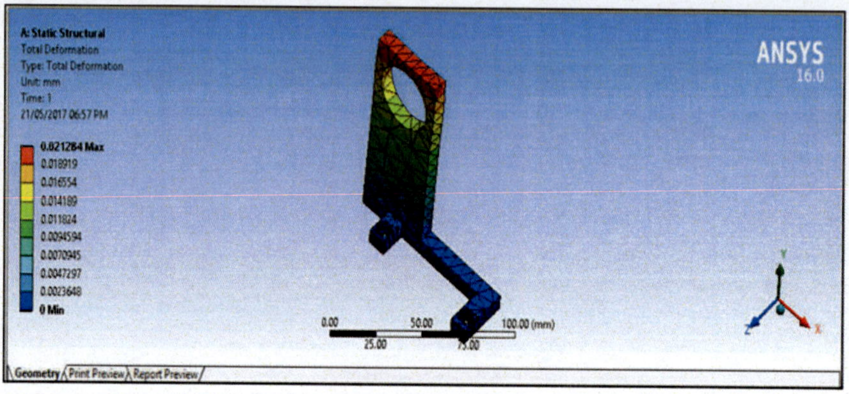

Fig. 13 Total deformation for cam bracket mount

Table 2 Maximum values of stress and deformation

Sr. No.	Components	Stress (MPa)	Deformation (mm)
1	Tappet	80.937	0.0062486
2	Cam bracket mount	35.987	0.021284

3.4 *Simulation of Proposed VA Mechanism*

3D design model of modified VVA system is shown in Fig. 14. The system is mounted on cylinder head with the help of two mounting bracket in which camshaft is mounted with help of bearing. On the left side, some additional overhand is provided for coupling the setup with motor drive with rubber dampers during mechanism and mounting of pulley to run the setup in belt drive under operating condition. All the analysis carried out on modified VVA system is also carried out at rated engine rpm, i.e., camshaft speed of 1800 rpm. Initially, some redundancies are found in the model to match the lifts as the complete system is running with two camshafts; however, model is modified and this problem is resolved.

Similar procedure is applied for the analysis of this mechanism. As there are two different lift profiles, two simulations are carried out for each lift profile. The first simulation is carried out for baseline lift of 7.6 mm, and the second simulation is carried out for modified lift of 8.3 mm. These simulations are carried out to check the lift and working mechanism.

Lift, velocity and acceleration curve of the modified VVA system with base lifts of 7.6 mm are shown in Figs. 15, 16 and 17 respectively. Even though with the same lift, the velocity and acceleration curves show the improvements. This shows the validation of the VVA system with baseline lift data; however, the baseline pushrod system needs to be fine-tuned for same profile as some variation is observed in acceleration profile. When the same system is tested for cam/valve lift of 8.3 mm, the lift,

Fig. 14 ADAMS simulation model of baseline valvetrain model

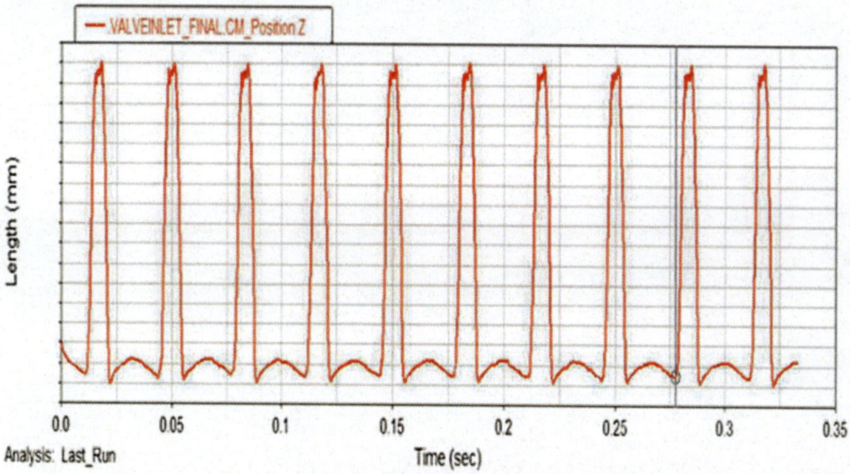

Fig. 15 Lift length versus time for VVA system with baseline lift

velocity and acceleration of this system are higher than the baseline valvetrain system and baseline VVA lift system. The reason for this higher velocity and acceleration in VVA system with higher lift is higher lift with same or slightly higher duration. As the higher lift needs to be completed within same or slightly higher duration creates sharp rise and fall in the cam, velocity and acceleration are also reflecting this rise and

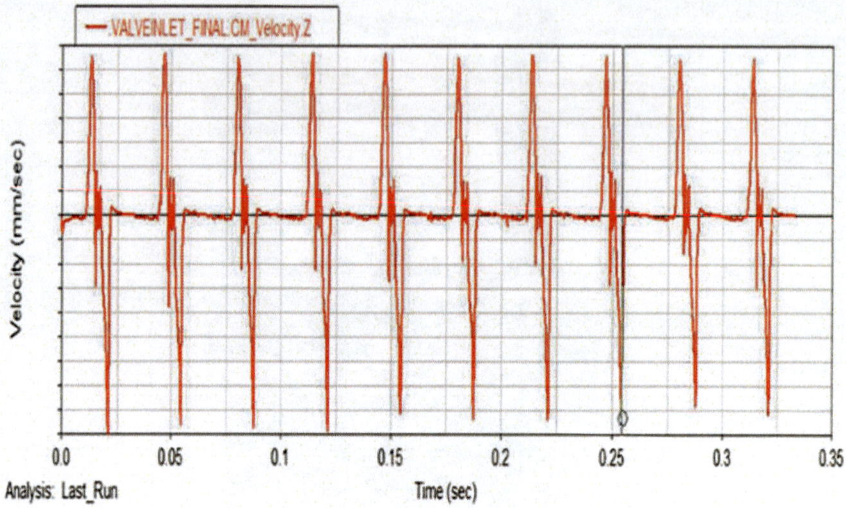

Fig. 16 Velocity versus time for VVA system with baseline lift

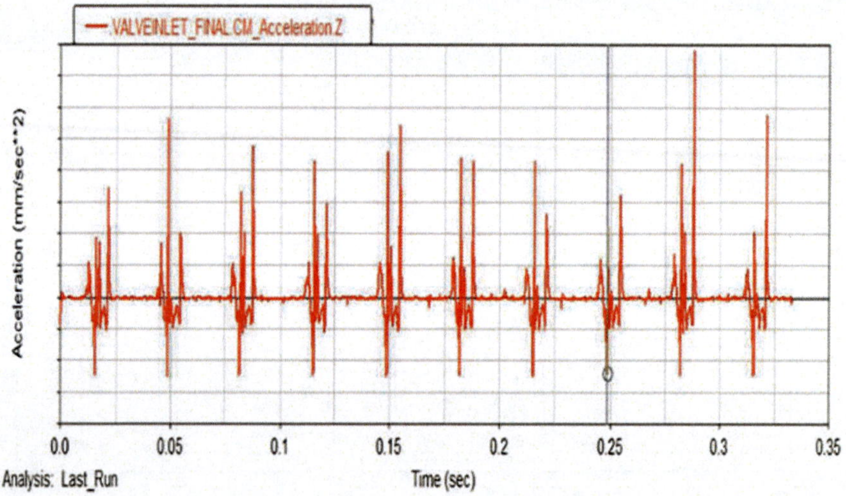

Fig. 17 Acceleration versus time for VVA system with baseline lift

fall. This can be clearly seen in Figs. 18, 19 and 20 respectively for higher lift cam lobe. The same can also be seen while comparing both the cam profile. Figures 21, 22 and 23 show the comparison of lift, velocity and acceleration of VVA system with baseline lift and modified lift.

The lift for the VVA system is higher than the baseline valvetrain system, and for the same duration, the additional force will be required for higher valve opening,

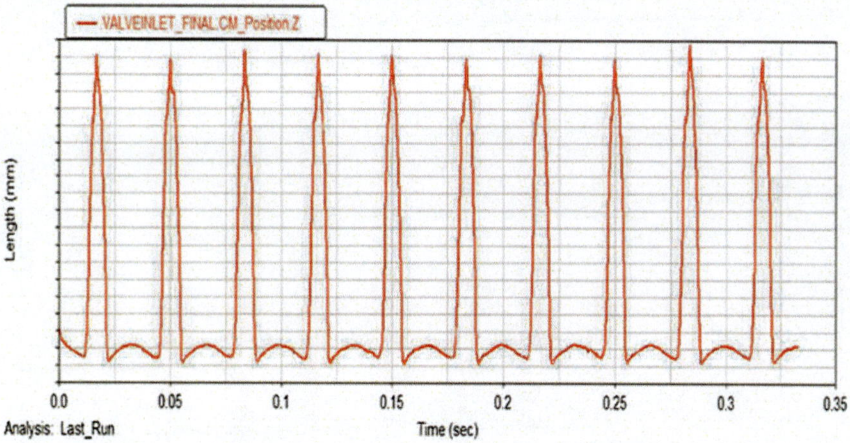

Fig. 18 Lift length versus time for VVA system with modified lift 8.3 mm

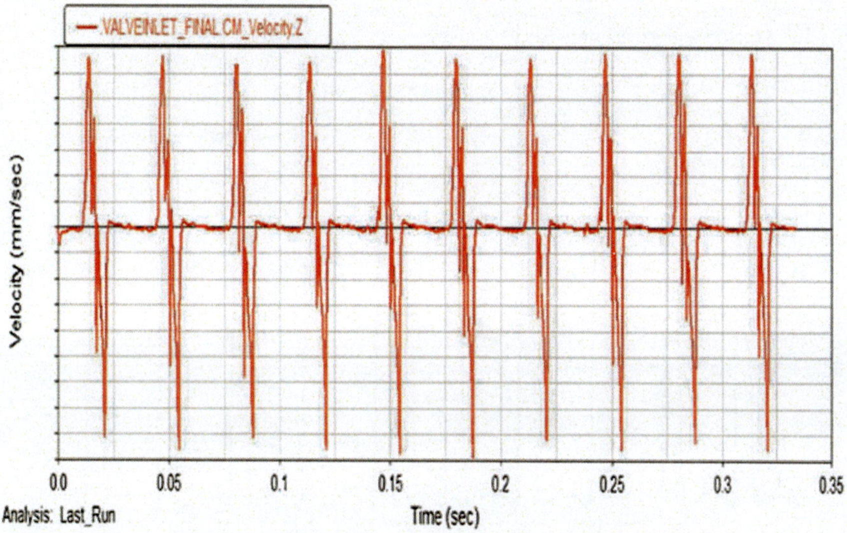

Fig. 19 Velocity versus time for VVA system with modified lift 8.3 mm

as time required for the higher opening is same. This leads to higher velocity and acceleration. The same is reflecting in Figs. 22 and 23, where the valve velocity and acceleration for VVA system are higher than the baseline system. Some spikes are also visible in acceleration curves which show during valve closing region, the profile needs to be tuned; however, this phenomenon is not visible in all closing condition. The same mechanism when adapted to the baseline engine and performance is analyzed.

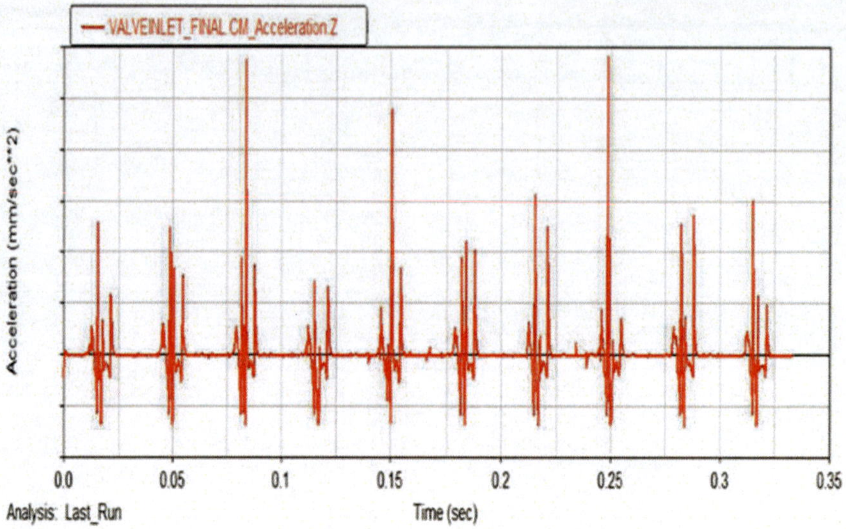

Fig. 20 Acceleration versus time for VVA system with modified lift of 8.3 mm

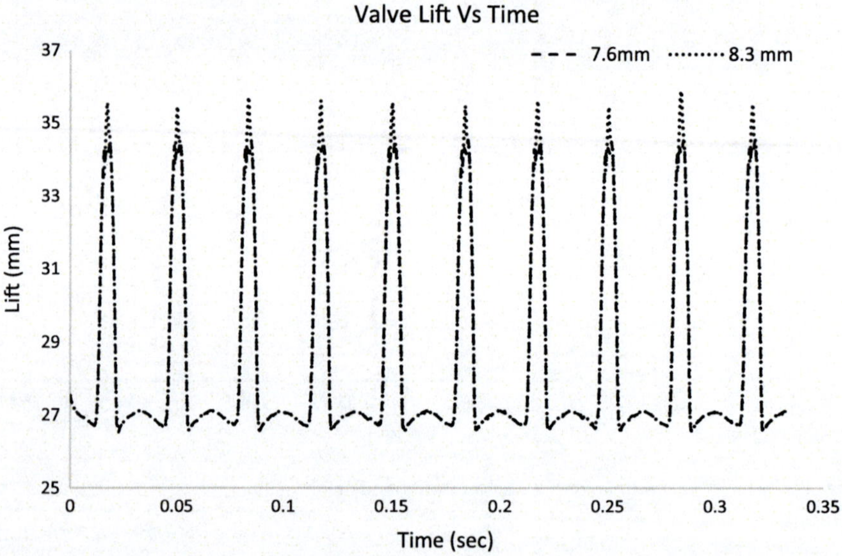

Fig. 21 Comparison of lifts for baseline versus modified lift for VVA

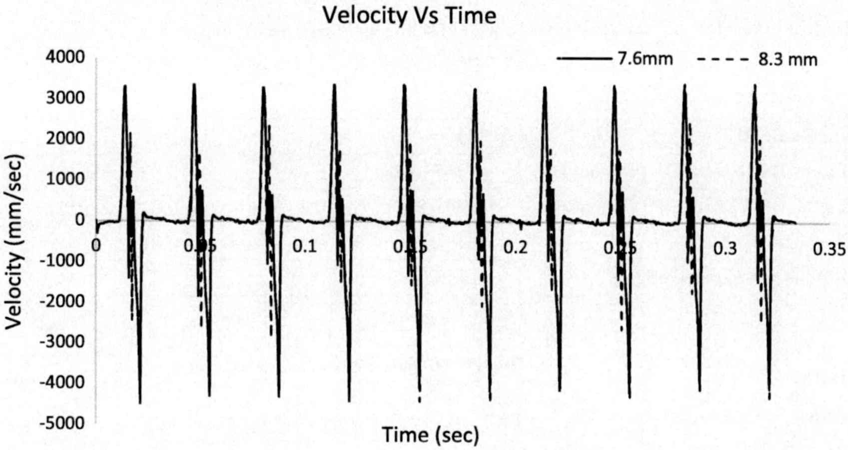

Fig. 22 Comparison of velocity for baseline versus modified lift for VVA

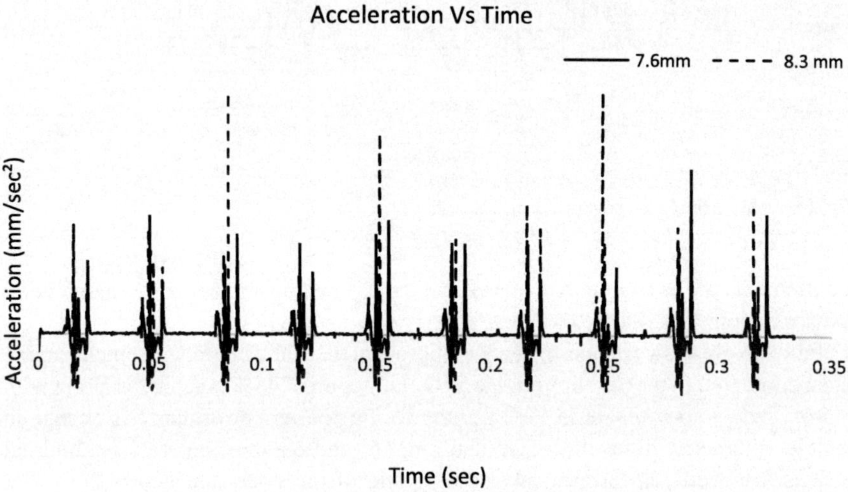

Fig. 23 Comparison of acceleration for baseline versus modified lift for VVA

3.5 Vehicle Performance with VVT

The performance of the upgraded engine with adaptation of VVA mechanism is measured at chassis dynamometer with Indian driving cycle. The new performance shows improvements in NO_x, hydrocarbon emission. The engine is meeting the existing emission norms and can be taken further for upcoming emission norms; the limits of the same are mentioned in Table 3. Cylinder head and piston of the upgraded engine are modified along with valve timing and actuation mechanism to cater the

Table 3 Vehicle emission performance with baseline and upgraded engine

	CO	HC + NO_x	PM
	g/km		
Limits BSIII	0.5	0.5	0.05
Limits BS IV with DF (without DF)	0.345 (0.38)	0.38	0.0354 (0.0425)
Limits BS VI with DF (without DF)	0.2 (0.22)	0.2	0.0208 (0.025)
Baseline results	0.141	0.56	0.047
Results of upgraded engine	0.158	0.31	0.0331

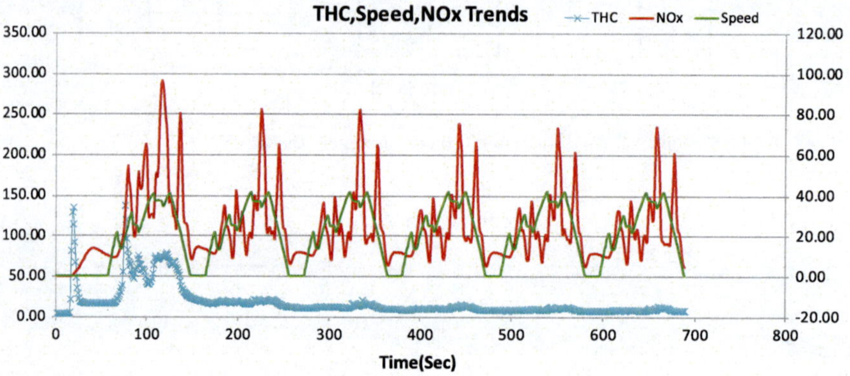

Fig. 24 Vehicle final emission performance at IDC

requirements of the engine. All values of the measured data in chassis dynamometer test are in ppm (Fig. 24).

Table below shows the performance improvement of the upgraded vehicle engine when compared to the baseline engine. NO_x reduction of 44% is achieved along with approx. 29% improvement in PM. Reason for this better performance is change in vehicle volumetric efficiency, chemical kinetics and combustion strategy. Internal EGR is achieved with proper valve timing and of the system, which is the major reason for NO_x reduction. However, change in cylinder head swirl has helped in reduction of PM emission along with quantity of fuel injected.

4 Conclusions

Design analysis of VVA mechanism is carried out under dynamic condition, and structural analysis of the component of the mechanism is also carried out under operating load conditions. Following are the conclusion from the detailed analysis of the complete system:

(1) Mechanism analysis of baseline valvetrain system is carried out with 7.6 mm lift under dynamic load condition with rated engine rpm; however, acceleration profile of the system is irregular.
(2) Mechanism analysis of the modified VVA with baseline lift of 7.6 mm is carried out and found to be good. Based on acceleration profile, comparison of baseline system needs to be further tuned; however, cam profile cannot be altered as it is already proved on engine.
(3) Structural analysis of the component of the modified VVA system is carried out for stress and deformation under operating condition load and found to be safe for the required actuation force applied.
(4) Mechanism analysis of the modified VVA with modified lift of 8.3 mm is carried out, and velocity and acceleration of the modified lift system are higher compared to the baseline system as lift is higher for same or slightly higher duration. Hence, the operating load condition will also be higher.
(5) Velocity and acceleration profile of the modified lift system is higher while comparing with baseline system, the reason being higher lift and forces due to higher lift.
(6) Vehicle's upgraded performance shows 44 and 29% improvement in NO_x and PM emissions, respectively. The major reason for the this is better combustion optimization strategy along with gas dynamics.

Further variability can be added in the engine to meet upcoming emission norms. This can be in the form of after treatment devices, mechanical boosting or both along with combustion optimization.

References

1. Parvate-Patil GB, Hong H, Gordon B (2004) Analysis of variable valve timing events and their effects on single cylinder diesel engine. SAE Technical Paper Series 2004-01-2965 25–28 Oct 2004
2. Pierik RJ, Burkhard JF (2000) Design and development of a mechanical variable valve actuation system. SAE Technical Paper Series 2000-01-1221 6–9 March 2000
3. De Ojeda W Effect of variable valve timing on diesel combustion characteristics. 2010 SAE International, 2010-01-1124, 04/12/2010
4. Gurney D, Mitcalf J, Warth M Integrated simulation, analysis and testing of a variable valve train for passenger car diesel engines. 2012 SAE International, 2012-01-0829, 04/16/2012
5. Lancefield T, Methley I The application of variable event valve timing to a modern diesel engine. SAE Technical Paper Series 2000-01-1229
6. Lancefield T (2003) The influence of variable valve actuation on the part load fuel economy of a modern light-duty diesel engine. SAE Technical Paper Series 2003-01-0028 3–6 March 2003

Determination of Stretch-Corrected Laminar Burning Velocity and Selection of Accurate Analytical Model for Burned Gas Mass Fraction Using Constant Volume Method

Anup Singh, Vikas Jangir, Anjan Ray, and M. R. Ravi

1 Introduction

The laminar burning velocity is an intrinsic property of a fuel–air mixture that affects many aspects of combustion behaviour. It is, for example, an important parameter for analysis and performance predictions of various SI engines. It is defined as the velocity, relative to the unburned gas, with which a planar, one-dimensional flame front travels along the normal to its surface. Laminar burning velocity values are also important for tuning and validation of chemical mechanisms.

There are various methods to determine laminar burning velocity. Of the propagating flame methods, constant pressure and constant volume spherically expanding flame methods are of major significance. The key difference is the range of pressures and temperatures in which tests can be conducted. In constant volume method, velocity can be determined for a range of pressures from a single experiment. In constant pressure method, flame images are captured using high-speed photography during the pre-pressure period, i.e. the duration over which the pressure in the vessel does not rise significantly from its initial value. Using the flame radius-time history, the laminar burning velocity and stretch are computed and extrapolated to obtain unstretched laminar burning velocity. In the present work, the constant volume method is used in which the premixed CH_4–air mixture is ignited at the centre of the spherical vessel. A spherical vessel is preferred over cylindrical one since the flame remains spherical in the former, while in the latter, it gets distorted. It is important to get the accurate flame surface area to know the flame stretch. During the combustion in a spherical vessel, the flame propagates from the centre towards the wall of the vessel accompanied by a pressure rise which results in a temperature increase of the unburned gas. If these changes in pressure and temperature, as well as the position of

A. Singh · V. Jangir · A. Ray (✉) · M. R. Ravi
Indian Institute of Technology Delhi, New Delhi, India
e-mail: raya@mech.iitd.ac.in

A. K. Gupta et al. (eds.), *Advances in IC Engines and Combustion Technology*, Lecture Notes in Mechanical Engineering,
https://doi.org/10.1007/978-981-15-5996-9_64

the flame front can be accurately determined, the method becomes extremely versatile and laminar burning velocity over a range of pressure and temperature values can be obtained using a single experiment. However, stretch correction at each condition requires a systematic sequence of experiments, by varying the initial pressure and temperature for each experiment.

While most of the research in the past has been done on the constant pressure method, very less focus has been given to the constant volume method. This is because constant pressure method allows correction for stretch correction in a more straightforward manner when compared to constant volume method. Nathan Hinton et al. [1] compared laminar burning velocity measurements in constant pressure method and constant volume method, in which latter required the more complex analysis. In another study by Chen et al. [2], results showed flame stretch had significant impacts on the accuracy of flame speed determination. Thus, the current authors have focused on calculating the unstretched laminar burning velocity by employing a simpler and accurate method of stretch correction, from the constant volume method based on constant volume combustion simulation employing detailed chemical kinetics mechanisms and detailed transport properties. Most of the simulations and experiments in literature are available for the standard initial condition ($T_0 = 300$ K, $P_0 = 1$ bar). In this research, the laminar burning velocity is also calculated for elevated temperature and pressure conditions.

Another issue in focus here is the relation between burned gas mass fraction and pressure. This relation is necessary in determining the laminar burning velocity once we get the pressure time trace from experiments, especially when flame radius is not obtained by imaging. Many analytical models have been discussed in the literature based on some assumptions. In the present work, three analytical models have been discussed and compared with the results of detailed numerical simulation.

2 Method of Stretch Correction

During the combustion of gases in a constant volume vessel, the unburned gas gets compressed continuously, and therefore, its temperature and pressure keep on increasing with time. Considering no heat loss, the compression process can very well be approximated as isentropic compression process. This means that if a constant volume combustion experiment is started with an initial pressure (P), initial unburned gas temperature (T), and initial specific volume (v) at any point along an isentrope $Pv^{\gamma} =$ constant through, say, 1 bar and 300 K, then subsequent states of the unburned gas during combustion also lie on the same isentrope. This enables us to achieve a desired state (P_{02}, T_{02}) on this curve by starting the experiment at any lower state (P_{01}, T_{01}) on this isentrope, given by Eq. (1).

$$\left(\frac{T_{02}}{T_{01}}\right)^{\gamma} = \left(\frac{P_{02}}{P_{01}}\right)^{\gamma-1} \tag{1}$$

Thus by setting one state (P_{02}, T_{02}) as the target condition, several experiments could be done by taking different initial conditions. In every experiment, the target condition would be achieved but at different radii and hence different stretch rates. Then the stretched laminar burning velocity could be obtained at this state but with different values of stretch rates from different experiments. The stretched laminar burning velocity (S_u) is calculated according to mass conservation and the relation is given by Eq. (2).

$$S_u = -\frac{1}{A_f \rho_u}\frac{dm_u}{dt} = \frac{1}{A_f \rho_u}\frac{dm_b}{dt} = \frac{1}{r_f^2 * \rho_u}\frac{d}{dt}\left(\int_0^{r_f} r^2 \rho dr\right) \quad (2)$$

In Eq. (2), A_f is the flame curvature surface area, ρ_u is the instantaneous density of unburned gas, m_u and m_b are the mass fraction of unburned gas and burned gas, respectively, and r_f is the flame radius. In numerical simulations, the flame is tracked by taking the mean value of the temperatures at grid points where the temperature value is just less than and just greater than the $T_{isotherm}$ value. $T_{isotherm}$ is one of the input parameters which is the monitor temperature whose value is assigned as 650 K. During the simulation, it will always take the value of temperature 650 K greater than the unburned gas temperature at any instant of time. In experiments, flame radius could be obtained by high-speed photography. If the experiment measures only pressure as a function of time, a relation between pressure time history and burned gas mass fraction history is required to evaluate laminar burning velocity from experiments. The stretch rate experienced by a spherically expanding flame is given by Eq. (3).

$$\text{Stretch rate}(k) = \frac{1}{A}\frac{dA}{dt} = \frac{2 * (dr_f/dt)}{r_f} \quad (3)$$

Here, A is the flame surface area, and r_f is the flame radius.

If the stretched laminar burning velocities are plotted against stretch rates, then the trend obtained could be extrapolated to get the laminar burning velocity at zero stretch.

3 Models for Determination of Burned Gas Mass Fraction

As discussed before, laminar burning velocities can be obtained from the pressure time data in a constant volume combustion experiment provided we can derive a relation between pressure history and burned gas mass fraction history. In the literature, various analytical relations have been derived between pressure and burned gas mass fraction. It is easier to determine the pressure time trace than the flame radius-time trace from the experiments. In the past researches, various analytical relations have been derived between pressure and burned gas mass fraction. Therefore, the relation between pressure (p) and mass fraction burned (x) is crucial to solve the differential equation as derived by Luijten et al. [3] in laminar burning velocity (S_u), pressure (p), burned gas mass fraction (x) and time (t).

Three analytical models available in the literature have been discussed in this section.

3.1 Lewis and Von Elbe Linear Model [4]

This relation by Lewis and Von Elbe [4] is still widely used due to its appealing simplicity. It is derived by assuming the burned and unburned gas specific heat ratio values to be constant and equal. The linear approximation may introduce error when γ_u and γ_b differ, which will always be the case due to difference in both composition and temperature. The linear relation has been used frequently owing to its simplicity and reasonable accuracy, for obtaining laminar burning velocities.

$$x = \frac{p - p_i}{p_e - p_i} \tag{4}$$

Here, p_e: pressure at the end of complete combustion, p_i: initial pressure.

3.2 Luijten's Two-Zone Model [3]

In a two-zone model [3], at any time the space is divided into two zones, i.e. the burned and unburned zones. Both have uniform temperatures and compositions. Conservation of specific volume and internal energy is done to get the expression for burned gas mass fraction as given by Eq. (5). Here, the assumption of isentropic compression of unburned gas is taken.

$$x = \frac{p - p_i f(p)}{p_e - p_i f(p)} \tag{5}$$

Here

$$f(p) = \left(\frac{\gamma_b - 1}{\gamma_u - 1}\right) + \left(\frac{\gamma_u - \gamma_b}{\gamma_u - 1}\right)\left(\frac{p}{p_i}\right)^{(\gamma_u - 1)/\gamma_u} \tag{6}$$

This will approach the linear relation if $\gamma_u = \gamma_b$.

For calculation of f(p), γ_b and γ_u are taken as 1.24 and 1.39, respectively, in the literature [3].

3.3 Nagy Relation [5]

During constant volume combustion, an intermediate situation has been assumed where a mass fraction x has burned and the total pressure p at that moment is uniform throughout the vessel. Thermodynamically, this intermediate state might be achieved by considering two subsequent processes. First, a fraction of gas mixture mass x is burning at constant volume, followed by pressure equilibration. In the first step, pressure and temperature of the burned gas mixture rise at constant volume, while the unburned gas mixture remains at the initial temperature and initial pressure p_i. In the second step, the unburned gas mixture is compressed by the expanding burned gases. Both expansion and compression are assumed to be adiabatic. Hence, the expression (7) is derived.

$$x = \frac{p^{1/\gamma_u} - p_i^{1/\gamma_u}}{p_e^{1/\gamma_b} p^{(1/\gamma_u - 1/\gamma_b)} - p_i^{1/\gamma_u}} \tag{7}$$

γ_b = Burned gas specific heat ratio, γ_u = unburned gas specific heat ratio.
If $\gamma_u = \gamma_b = 1$, then this relation will also approach the linear relation in Eq. (4).

3.4 Burned Gas Mass Fraction Calculated by Numerical Simulation

Burned gas mass fraction (x) at any instant of time is defined as the mass of the gas mixture that has burned divided by the total mass of gas mixture. It is calculated according to Eq. (8).

$$x = \frac{m_b}{m_o} = \frac{\int_0^{r_f} 4\pi r^2 \rho dr}{\rho_0 * \left(\frac{4}{3}\pi R^3\right)} \tag{8}$$

Here, ρ_0 is the initial unburned gas density, R is the inner radius of vessel, m_b and m_0 are the instantaneous mass of burned gas and initial mass of total mixture, respectively. The integral expression in Eq. (8) is calculated using a numerical integration procedure on the computed results at any instant. Trapezoidal rule is used in the present work.

4 Numerical Simulation

Unsteady numerical simulations of outwardly propagating spherical laminar flames were carried out using RUN-1DL software, developed by Rogg and co-workers [6].

The calculations for unstretched (plane) flames were carried out using the steady, one-dimensional laminar premixed flame code CHEMKIN [7]. The latter computations were performed to compare with the unstretched laminar burning velocities obtained by stretch correction to spherical flame laminar burning velocity. Detailed finite rate chemistry using GRI 3.0 mechanism consisting of 5 elements (C, H, O, N and AR), 53 species and 325 homogenous reactions is computed in both above codes. For all simulations, the vessel radius was taken to be 6.2 cm. The initial temperature at the centre of the vessel is taken as 2500 K. The input parameters CURV = 1.05, GRAD = 0.1 and N.G. (initial number of grids) = 121 were taken to have a fine adaptive 1-D grid to simulate the combustion process completely without any error. The fuel is taken as CH_4–air mixture at equivalence ratio 1.

5 Results and Discussion

5.1 *Extrapolation to Get Laminar Burning Velocity (S_u) at Zero Stretch*

To plot the variation between laminar burning velocity and stretch rate, different values of stretch rates are required. First of all, the target condition is set at which the unstretched laminar burning velocity value has to be known, and then the simulations are carried out by taking different initial conditions. The initial conditions are chosen such that flame stretch in the range 40/s to 220/s is covered by the data. For the standard condition ($T_0 = 300$ K, $P_0 = 1$ bar), the stretch variation is not so much; hence, only four different initial conditions are sufficient to extrapolate the burning velocity to zero stretch. After the simulation, values of flame radius (r_f), flame speed (S_b) and stretched laminar burning velocity (S_u) at the target condition are tabulated in Table 1.

The plot shown in Fig. 1 gives the extrapolated value of laminar burning velocity at zero stretch which is 36.675 cm/s. The unstretched laminar burning velocity calculated by CHEMKIN is 37.824 cm/s. The difference between both the unstretched velocities is approximately 3%.

Table 1 Details of simulation runs for target condition 1 bar, 300 K for stoichiometric CH_4–air mixture

P_0 (bar)	T_0 (K)	r_f (cm)	k (/s)	S_u (cm/s)
0.75	282	3.702	115.359	35.405
0.8	287	3.427	130.122	35.201
0.85	292	3.107	149.260	34.982
0.9	297	2.707	177.332	34.704

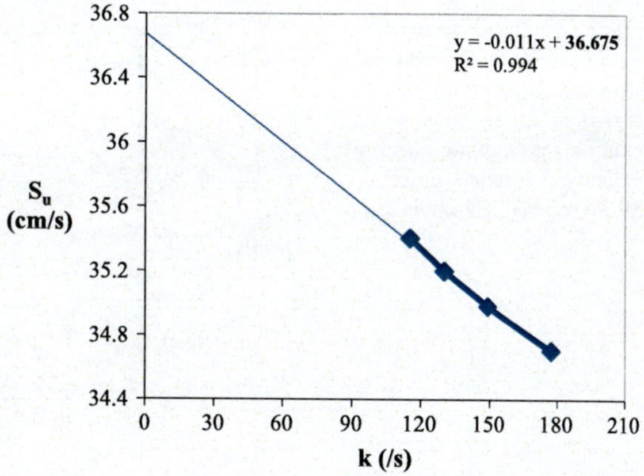

Fig. 1 Stretched laminar burning velocities obtained from simulations plotted against stretch rates for the target condition 1 bar, 300 K. The extrapolated laminar burning velocity at zero stretch is 36.675 cm/s (CH_4–air, $\phi = 1$)

For the elevated pressure and temperature conditions, the range of stretch rate is larger than that shown in Fig. 1. To cover the whole range of stretch rates, seven values of stretch rate are taken for uniform distribution.

Laminar Burning Velocity (S_u) Versus Uniformly Distributed Stretch Rates (k) Here, the stretch rate values are equally spaced in between seven points. By running simulations at various initial conditions, equally distributed stretch rates are not obtained. First, simulations are carried out at different initial conditions, and then, the curve fitting is done between pressure and stretch rate to get the initial pressure condition at which simulation could be carried out to get the desired stretch rate. Stretch rate is varied from 40/s to 220/s with a gap of 30/s.

From Figs. 2, 3 and 4 and Table 2, it is clear that the laminar burning velocity values obtained through extrapolation by taking the uniform distribution of stretch rates give values closer to the CHEMKIN values.

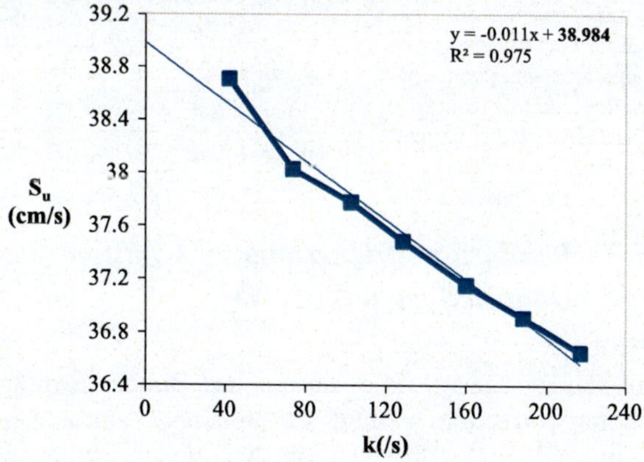

Fig. 2 Stretched laminar burning velocity (S_u) versus stretch rate (k) for the target condition 2 bar, 364 K. The extrapolated laminar burning velocity at zero stretch is 38.984 cm/s (CH_4 air, $\phi = 1$)

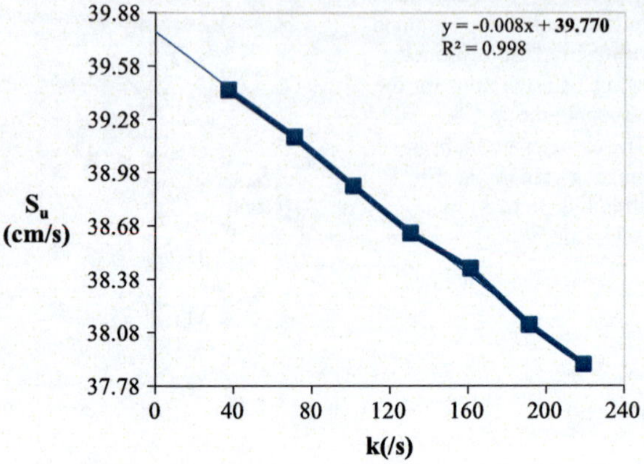

Fig. 3 Stretched laminar burning velocity (S_u) versus stretch rate (k) for the target condition 3 bar, 406 K. The extrapolated laminar burning velocity at zero stretch is 39.770 cm/s (CH_4-air, $\phi = 1$)

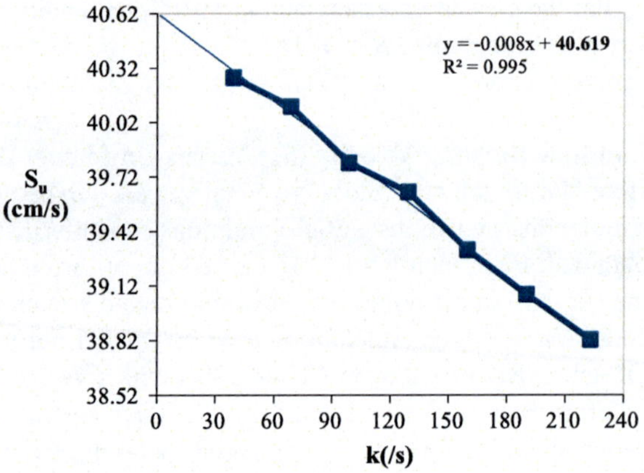

Fig. 4 Stretched laminar burning velocity (S_u) versus stretch rate (k) for the target condition 4 bar, 460 K. The extrapolated laminar burning velocity at zero stretch is 40.619 cm/s (CH_4-air, $\phi = 1$)

Table 2 Stretch-corrected laminar burning velocity (S_u^0) values in comparison with laminar burning velocity values obtained from CHEMKIN (CH_4-air, $\phi = 1$)

Target condition	S_u^0 (RUN1DL)(cm/s)	S_u^0 (CHEMKIN) (cm/s)
2 bar, 364 K	38.984	40.002
3 bar, 406 K	39.770	41.278
4 bar, 460 K	40.619	42.111

5.2 *Validation of Unstretched Laminar Burning Velocity (S_u^0) Obtained from RUN1DL*

To validate the values of unstretched laminar burning velocity obtained from the stretch correction method, the obtained values from RUN1DL simulations are compared against the CHEMKIN software results and experimental results from literature. Experiments were carried out using the constant pressure method (CPM)

Table 3 Comparison of S_u^0 (cm/s) values obtained from RUN1DL with CHEMKIN

Target condition	S_u^0(cm/s) (RUN1DL)	S_u^0(cm/s) (CHEMKIN)	Per cent relative difference
1 bar, 300 K	36.675	37.824	3.04
2 bar, 364 K	38.984	40.002	2.54
3 bar, 406 K	39.770	41.278	3.65
4 bar, 460 K	40.619	42.111	3.54

Table 4 Comparison of S_u^0 (cm/s) values obtained from RUN1DL with experimental results

Method	Target condition	S_u^0(cm/s)
RUN1DL [6]	$P = 1$ bar, $T = 300$ K	36.675
CPM [8]	$P = 1$ bar, $T = 301$ K	36.8
CPM [9]	$P = 1.012$ bar, $T = 298$ K	35.7569
HFM [10]	$P = 1.012$ bar, $T = 298$ K	35.92
HFM [11]	$P = 1.012$ bar, $T = 298$ K	36.3 ± 0.9
HFM [11]	$P = 0.993$ bar, $T = 298$ K	36.1 ± 0.9

[8, 9] and heat flux method (HFM) [10, 11]. In the table below, the difference between the unstretched laminar burning velocities (obtained through extrapolation by taking uniformly distributed stretch rates) and CHEMKIN is shown by per cent relative difference. The difference is relative to the CHEMKIN value.

From Tables 3 and 4, it is validated that the value of unstretched laminar burning velocity (S_u^0) obtained after the stretch correction from the numerical simulations (RUN1DL) comes closer to the value obtained from CHEMKIN software calculations and the experimental results. Thus, the method can be taken as validated.

5.3 *Deviation of Analytical Results from the Simulation Results*

To have a clear picture of the deviation of burned gas mass fraction obtained from various analytical relations, a non-dimensional parameter for pressure rise is introduced. The difference of various analytical burned gas mass fraction values from the one obtained from numerical simulation is plotted against fractional pressure rise.

$$\text{Fractional pressure rise} = \frac{p - p_0}{p_e - p_0} \tag{9}$$

Here, p: instantaneous pressure in a vessel, p_0: initial pressure, p_e: end pressure

From Fig. 5 it could be clearly seen that the burned gas mass fraction obtained from the Nagy's relation deviates most from the RUN1DL one, while the Luijten's

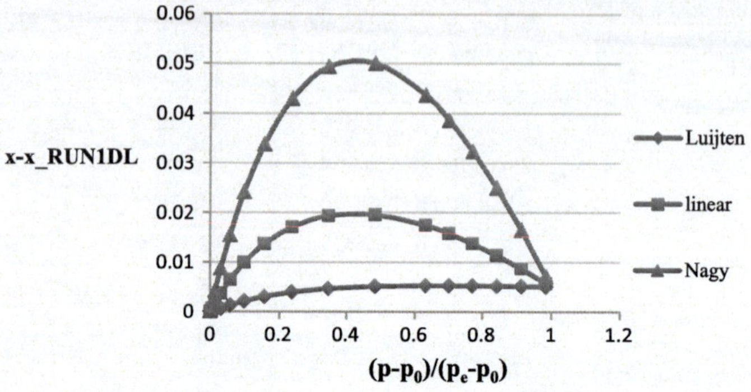

Fig. 5 Difference in burned gas mass fraction obtained from analytical models and RUN1DL simulations plotted against fractional pressure rise for $T_0 = 300$ K and $P_0 = 1$ bar

two-zone model shows the least deviation. This is due to the assumption taken while deriving the expression. The linear model simplifies the expression by taking the specific heat capacity ratio values for burned gas and unburned gas equal to unity. This is wrong as the burned gas and unburned gas have a vast difference in their temperature so the difference in the values of specific heat capacity ratios. Nagy's relation was proven to violate integral energy conservation as discussed in [3].

5.4 Comparison of S_u Values Calculated from Analytical Models with RUN1DL

In this section, the laminar burning velocity calculated with the help of the different analytical burned gas mass fraction relations (Eqs. 4, 5, and 7) is compared with simulation result. Analytical relations between burned gas mass fraction and pressure are used to calculate the laminar burning velocity by differential Eq. (10) as derived by Luijten et al. [3].

$$S_u = \frac{R_v}{3} * \left[1 - \left(\frac{p_0}{p}\right)^{1/\gamma_u} * (1 - x)\right]^{-2/3} * \frac{p_0}{P}^{1/\gamma_u} * \frac{dx}{dp} * \frac{dp}{dt} \quad (10)$$

From Fig. 6, it is clear that Luijten's burned gas mass fraction model gives a similar trend as shown by the numerical simulation and the values are closer to simulation results. The linear model shows larger differences at initial stages and is closer to RUN1DL results only at larger flame radii, where the stretch rate is relatively lower.

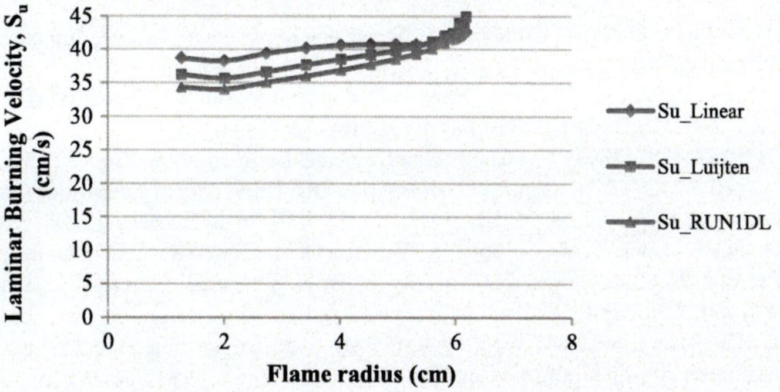

Fig. 6 Laminar burning velocity obtained from linear model, Luijten's model and RUN1DL plotted against flame radius for the initial condition $T_0 = 300$ K and $P_0 = 1$ bar (CH_4–air, $\phi = 1$)

6 Conclusions

1. The stretch correction method of extrapolating the stretched laminar burning velocities at a given target condition, [obtained through simulations at different initial conditions related with isentropic relation (1)] to zero stretch gives values of laminar burning velocities having relative difference under 4% from the CHEMKIN values.
2. The Luijten's two-zone analytical model closely predicts the burned gas mass fraction and is better than linear and Nagy models.
3. As the Luijten's two-zone model for burned gas mass fraction deviates least from the RUN1DL, the same thing could be seen for laminar burning velocity calculated with the use of this relation. Thus, Luijten's model is found to be more accurate than the linear model for calculating the laminar burning velocity. It is recommended that the pressure time trace obtained from the experiment should use Luijten's model to get burned gas mass fraction, and the laminar burning velocity should be calculated using the differential Eq. (10) for best results.

References

1. Hinton N, Stone R, Cracknell R (2018) Laminar burning velocity measurements in constant volume vessels-reconciliation of flame front imaging and pressure rise methods. Fuel 211:446–457. https://doi.org/10.1016/j.fuel.2017.09.031
2. Chen Z, Burke M, Ju Y (2009) Effects of compression and stretch on the determination of laminar flame speeds using propagating spherical flames. Combust Theor Model 13(2):443–464. https://doi.org/10.1080/13647830802632192
3. Luijten CCM, Doosje E, de Goey LPH (2009) Accurate analytical models for fractional pressure rise in constant volume combustion. Int J Therm Sci 48:1049–1258. https://doi.org/10.1016/j.ijthermalsci.2008.12.020

4. Lewis B, von Elbe G (1961) Combustion, flames and explosions of gases, 2nd edn. Academic Press, New York
5. Nagy J, Conn J, Verakis H (1969) Explosion development in a spherical vessel. Technical Report RI 7279, US Department of Interior, Bureau of Mines
6. Rogg, B. RUN-1DL (1991) The Cambridge Universal Laminar Flame Code. Technical Report CUED/ATHERMO/TR39, Department of Engineering, University of Cambridge, England, UK
7. R. J. Kee, F. M. Rupley, J. A. Miller, M. E. Coltrin, J. F. Grcar, E. Meeks, H. K. Moffat, A. E. Lutz, G. Dixon Lewis, M. D. Smooke, J. Warnatz, G. H. Evans, R. S. Larson, R. E. Mitchell, L. R. Petzold, W. C. Reynolds, M. Caracotsios, W. E. Stewart, P. Glarborg, C. Wang, and O. Adigun (2000) CHEMKIN Collection, Release 3.6, Reaction Design, Inc., San Diego, CA
8. Gu XJ, Haq MZ, Lawes M, Woolley R (2000) Laminar burning velocity and Markstein lengths of methane–air mixtures. Combust Flame 121:41–58. https://doi.org/10.1016/S0010-2180(99)00142-X
9. Rozenchan G, Zhu DL, Law CK, Tse SD (2002) Outward propagation, burning velocities, and chemical effects of methane flames up to 60 ATM. Proc Combust Inst 29:1461–1470. https://doi.org/10.1016/S1540-7489(02)80179-1
10. Mayuri Goswami, Sander C.R.Derks, Kris Coumans, Willemyn J.Slikker, Marcelo H.de Andrade Oliveira, Rob J.M.Bastiaans, Carlo C.M.Luijten, L. Philipus H.de Goey, Alexander A.Konnov (2013) The effect of elevated pressures on the laminar burning velocity of methane + air mixtures. Combust Flame 160:1627–1635. https://doi.org/10.1016/j.combustflame.2013.03.032
11. Dirrenberger Patricia, Le Gall Hervé, Glaude Roda Bounaceur Pierre-Alexandre, Battin-Leclerc Frédérique (2015) Measurements of laminar burning velocities above atmospheric pressure using the heat flux method-application to the case of *n*-pentane. Energy Fuels 29:398–404. https://doi.org/10.1021/ef502036j

A Three-Dimensional Numerical Model to Predict the Performance of a Microcombustion-Based Thermoelectric Generator

B. Aravind, Karan Hiranandani, and Sudarshan Kumar

1 Introduction

The development of a portable power source is governed by two essential factors: a high energy density and an environment-friendly nature. The advancements in the field of microcombustion research [1] have led to a greater demand for thermoelectric power generators due to the high energy density of hydrocarbon fuels [2, 3]. Though hydrocarbon fuels offer energy densities as high as ~100 times that of conventional lithium batteries [4], the use of thermoelectric power generators (TEG) for standalone power applications is limited in practical systems due to the low conversion efficiency of TEGs [5]. However, a combustion-driven thermoelectric generator, having an efficiency as low as 5%, is considered to be ~10 times more efficient than the conventional batteries. In addition to this, the bulky volumes and low energy densities of electrochemical batteries make them unviable for portable applications [6]. Thus, it is more advantageous to adopt combustion-based sources powered by the high energy density hydrocarbon fuels.

Power generation by using thermophotovoltaic (TPV) cells [7] and thermoelectric generators (TEG) [8] has started receiving immense interest from the research community due to the various advancements in the field of microcombustion during

B. Aravind (✉) · S. Kumar
Department of Aerospace Engineering, Indian Institute of Technology Bombay, Powai, Mumbai, Maharashtra 400076, India
e-mail: aravindbpillai69@gmail.com

S. Kumar
e-mail: sudar4@gmail.com

K. Hiranandani
Department of Mechanical Engineering, Manipal Institute of Technology, Manipal Academy of Higher Education (MAHE), Manipal, Udupi 576104, India
e-mail: karan.hiranandani97@gmail.com

A. K. Gupta et al. (eds.), *Advances in IC Engines and Combustion Technology*, Lecture Notes in Mechanical Engineering,
https://doi.org/10.1007/978-981-15-5996-9_65

the past few decades. TEGs are found to be more efficient and economical than the TPVs due to their enhanced conversion efficiencies and compact sizes. Therefore, there are several research studies being carried out on thermoelectric materials and various thermal management techniques are being adopted for heat sources to ensure optimum utilization of TEGs for portable power applications. By using a 250 g butane cartridge for a duration of 5 h, Shimokuri et al. [9] used a vortex combustor to obtain a conversion efficiency of 3.01%. Aravind et al. [10] designed a dual microcombustor-based TEG to obtain a power output of 4.52 W with an overall efficiency of 4.66%.

Numerical studies on the TEG effect through FEA solvers have been carried out in the past decade [11]. Meng et al. [12] modeled a TEG and obtained a power output of 0.13 W with an efficiency of 0.87 W numerically. Chen et al. [13] successfully simulated the TEG model in a CFD environment by assuming constant temperatures across the module. Although Meng et al. [14] coupled a microcombustor with a thermoelectric module numerically, they did not investigate the thermal characteristics of the combustor in depth. Most of the aforementioned studies do not account for the non-uniform temperature profile obtained across the hot side of the thermoelectric module due to the presence of a combustion-based heat source.

In the present study, three-dimensional modeling of an integrated microcombustor-thermoelectric power generator has been carried out. Although numerous studies can be found on the numerical modeling of microcombustors and TEGs individually, designing an integrated system numerically is of great challenge and is of interest for optimizing the real micro power generator systems. The study derives its novelty from being the first such model to simulate an integrated microcombustor-thermoelectric power generator in a CFD environment and analyze the thermal output of the combustor in 3D.

2 Computational Specifications

2.1 Computational Domain and Governing Equations

The computational domain of the integrated system, shown in Fig. 1, is symmetric about the XY and XZ planes. The axial flow direction of the fuel-air mixture is along the x-axis of the system, the height of the system (13.5 mm) is along the y-axis, and the width of the system (20 mm) is along the z-axis. The combustor has two rectangular backward-facing steps at distances of 5 mm and 25 mm from the inlet at combustor widths of 4 mm and 6 mm, respectively. A heat recirculation cup is placed 2 mm away from the combustor exit to recirculate the exhaust gases and thus, preheat the reactant mixture. It also serves as a housing for the thermoelectric module, which is placed on the mounting plate. A premixed CH_4–air mixture is passed through the 2-mm inlet and ignited at the end of the combustor through a 2-mm temperature patch of 2400 K. The governing equations for the combustion reactions taking place in the

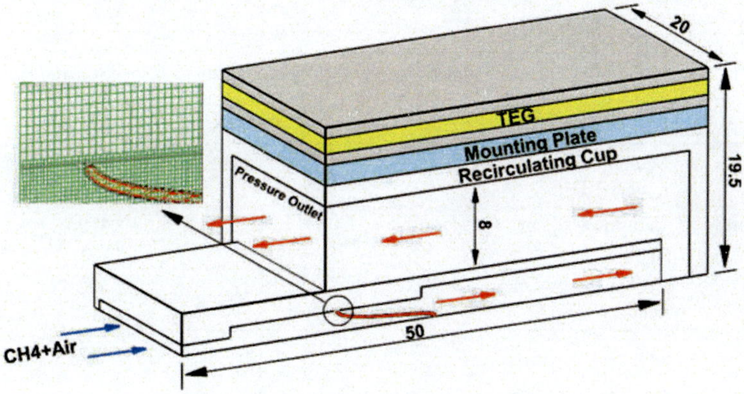

Fig. 1 Computational domain of the integrated thermoelectric generator

microcombustor [15] are solved by making the following assumptions: (a) The fluid is a continuum (Knudsen number <1), (b) no Dufour effects, (c) work done by pressure and viscous forces is neglected, and (d) flow is assumed to be incompressible. Chen et al. [13] summarized the governing equations for the thermoelectric effect. The equations are modified to account for the nonuniform temperature profile across the hot side of the module:

$$\nabla J_c = 0 \tag{1}$$

$$\nabla q = -\nabla[k\nabla T] = \mathrm{mc}\frac{\partial T}{\partial t} + J_c^2\rho + [\nabla\alpha]T J_c \tag{2}$$

$$\nabla V = -\alpha\nabla T - \rho J_c \tag{3}$$

where J_c, q, k, T, ρ, α, and V denote the current flux, thermal field, thermal conductivity, temperature, electrical resistivity, Seebeck coefficient, and the open-circuit voltage, respectively.

2.2 *Boundary Conditions and Numerical Modeling*

The premixed CH_4–air mixture is passed through the inlet at an ambient temperature $T_a = 300$ K, which is simulated by the GRI 3.0 mechanism [16]. The mixture velocity is varied from $v_{in} = 1$ to 7 m/s for $\phi = 0.8$ and 1.0. Due to a low Knudsen number (~10–5), a no-slip boundary condition is applied on all the interior walls. The walls of the combustor that are exposed to the ambient are assigned a mixed boundary condition, with a standard heat loss coefficient (20 W/mK^2) and emissivity (0.9), to account for the heat loss. The exhaust gas outlet is assigned a Neumann boundary

Table 1 Material specifications

Material	Cell zone	Density (kg/m^3)	Thermal Conductivity (W/m K)	Specific Heat (J/kg K)
Bi_2Te_3	Semiconductor material TEG	10,922	Variable [14]	200
Ceramic	Ceramic walls of module	4300	18	419
Steel	Combustor and cup walls	8030	16.27	502.48

condition and the k–ε turbulence model is adopted. The properties of the materials used in the integrated system are listed below in Table 1. However, the properties of the semiconductor material vary significantly with temperature, and to account for this variation, their polynomial relations [14] are used in the calculations made to solve the thermoelectric model.

The simulations are carried out on a general-purpose CFD software ANSYS FLUENT 16.2 [17]. A steady-state simulation was performed by using spatial integration, wherein the diffusion terms were solved by the central difference scheme, and the convective terms by the second-order upwind scheme. While recent studies have employed standard FEM solvers to simulate the thermoelectric effect [11], neither do they account for a non-uniform temperature profile across the module ends, nor do they have a model that is coupled with a combustor. The current study accounts for the heat and mass transfer through the combustor while simultaneously solving the thermoelectric effect by accounting for the variable temperature profile obtained across the hot side of the module.

The thermoelectric model is integrated into the CFD environment by manually integrating an ANSI-C code into the software [13]. User-defined scalars are used to define the source terms and are solved with the help of user-defined functions. The code can then be interpreted or compiled in the CFD software, before hooking it with the existing CFD model. The scalars and functions used for this model are summarized in Table 2.

2.3 *Grid Independence Study and Validation of TEG Models*

In order to verify grid independence, simulations are carried out for different grids ranging from 0.5 million cells to 7.6 million cells. Figure 2a compares the mass fraction of OH along the inner, stepped combustor wall for different grids. It can be observed that there is negligible difference between the values obtained for grids having 7.6 million cells and 4.1 million cells. Thus, a grid consisting of 4.1 million cells is adopted for the computational simulations in the present work.

Table 2 User-defined scalars

Scalar	Mathematical formula	Source term	Diffusivity
UDS_A	∇UDS_ A $= -\rho J$	0	$\frac{1}{\rho}$
UDS_B	∇UDS_ B $= \alpha$	0	1
UDS_C	$\nabla UDS\_C = -\alpha \nabla T$	$\frac{\partial\alpha\left(\frac{\partial T}{\partial x}\right)}{dx} + \frac{\partial\alpha\left(\frac{\partial T}{\partial y}\right)}{dy} + \frac{\partial\alpha\left(\frac{\partial T}{\partial z}\right)}{dz}$	1
UDS_D	∇UDS_ D $= \frac{\partial\alpha\left(\frac{\partial T}{\partial x}\right)}{dx}$	0	1
UDS_E	∇UDS_ E $= \frac{\partial\alpha\left(\frac{\partial T}{\partial y}\right)}{dy}$	0	1
UDS_F	∇UDS_ F $= \frac{\partial\alpha\left(\frac{\partial T}{\partial z}\right)}{dz}$	0	1
Energy	–	$\frac{1}{\rho}$(UDS_ A)^2 − ∇UDS_ B$\frac{T}{\rho}\nabla$UDS_ A	k

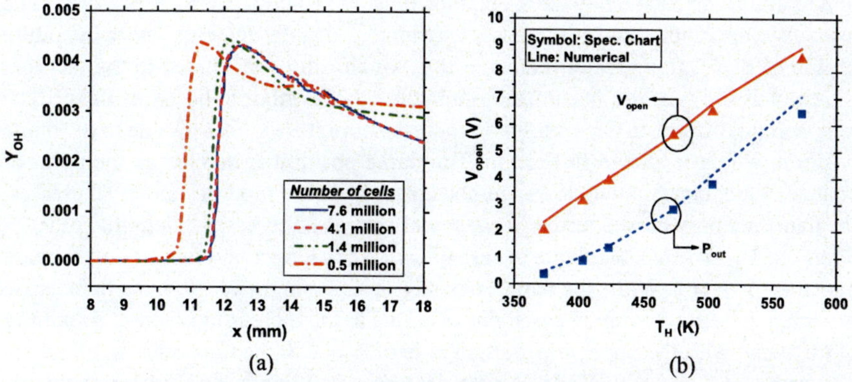

Fig. 2 **a** Grid independence study and **b** validation of TEG model

Before analyzing the thermoelectric output, accuracy of the thermoelectric module has to be verified. Simulations were carried out on the thermoelectric module individually. For comparison purposes, constant temperature boundary conditions were imposed on the hot and cold sides of the module. With T_c = 323 K, simulations were carried out for different temperatures on the hot side of the module. Figure 2b shows a comparison between the numerical values and the values on the manufacturer's specification sheet [18] for the open-circuit voltage and the output power for different values of the hot side temperature. Due to a negligible difference between the numerical and specified values, the accuracy of the model was confirmed.

3 Results and Discussions

The thermal characteristics and the thermoelectric output of the integrated system are analyzed for different mixture velocities and equivalence ratios. Variation of different parameters such as wall temperature and conversion efficiency is plotted and discussed.

3.1 *Thermal Characteristics*

Since the temperature difference across the module determines the output power generated, it is important to analyze the temperature distribution across the hot side of the module. Figure 3a shows the variation of the average temperature of the hot side of the module with inlet velocities for different equivalence ratios. Maximum average temperatures of 551 and 549 K are obtained for $\phi = 1.0$ and $\phi = 0.8$, respectively. The average temperature steadily increases with increasing inlet velocity. The temperature profile of $\phi = 1.0$ is greater than $\phi = 0.8$, which can be attributed to the fact that the combustion reaction becomes more intense as the mixture becomes richer. The temperature difference decreases with increasing mixture velocities due to the flame position, which is shown in Fig. 3b. The flame position is defined as the point of contact of the flame tip with lower combustor wall. As the mixture velocity increases, the flame not only moves further downstream, but also increases in length. For $v_{in} = 5$ m/s and $v_{in} = 7$ m/s, the flame in case of $\phi = 0.8$ is longer and further downstream than it is for $\phi = 1$. While the flame is located at ~32 mm for $\phi = 0.8$, the flame tip is observed to be at ~27 mm in case of $\phi = 1$. Due to the increasingly closer proximity of the flame with the upper cup wall, in case of $\phi = 0.8$, the temperature difference between $\phi = 1$ and $\phi = 0.8$ in Fig. 1a decreases with increasing mixture velocity.

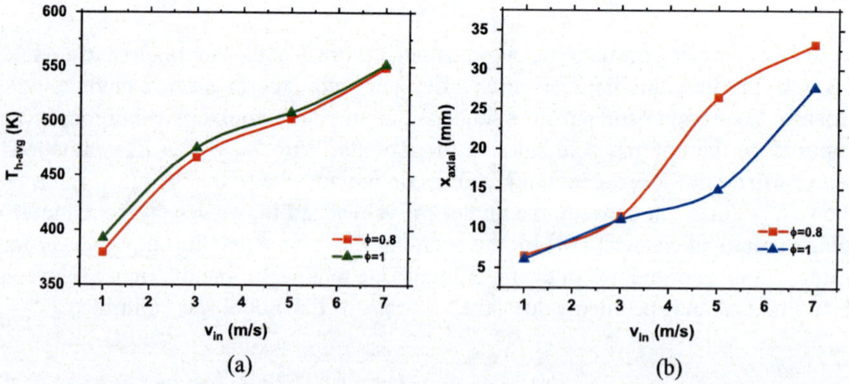

Fig. 3 Variation of **a** average temperature of the hot side of the module and **b** flame position with inlet velocity for different equivalence ratios

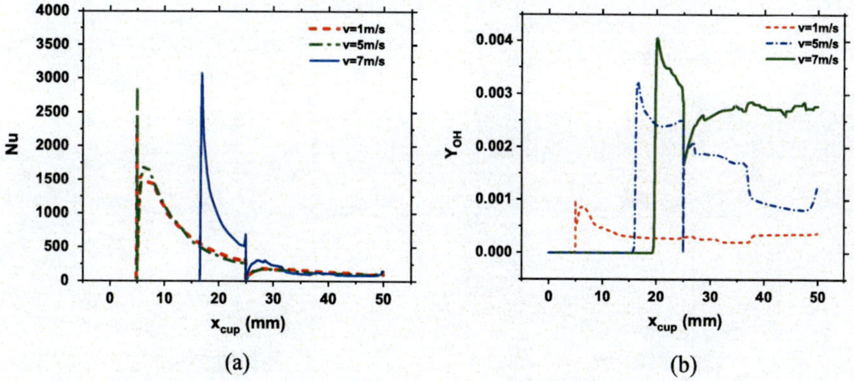

Fig. 4 Variation of **a** Nusselt number along the inner, stepped wall of the combustor and **b** mass fraction of OH along the lower wall of the combustor

Thus, it can be concluded from Fig. 3 that the flame in case of $\phi = 1$ is more intense and shorter than in case of $\phi = 0.8$.

In an attempt to study the performance of the microcombustor, the intensity of the combustion reaction was analyzed with the help of two parameters. Variation of Nusselt number is shown along the inner stepped wall of the combustor for different mixture velocities in Fig. 4a. It is observed that the peak values of Nu increase with increasing mixture velocities. Since the Nusselt number is a measure of the heat transferred to the combustor wall, heat transferred to the wall is maximum in case of $v_{in} = 7$ m/s. These readings are confirmed by the values obtained in Fig. 4b, which shows the variation of the mass fraction of OH along the lower combustor wall. The mass fraction of OH is a measure of the intensity of the combustion reaction [19] and has a maximum peak value for $v_{in} = 7$ m/s. Figure 4 conclusively proves that the intensity of combustion increases with increasing mixture velocity. It is interesting to note that the peak value of Nu denotes the point of maximum heat transfer to the stepped combustor wall, which also corresponds to the point where the flame root is anchored on the stepped wall. Correspondingly, the maximum values of Y_{OH} denote the flame position, which is the point of contact between the tip of the flame and the lower combustor wall.

Figure 5 shows the contours of temperature (upper half) and velocity (lower half) for different mixture velocities. Thc contours of HCO superimposed with the contours of velocity to depict the location of the flame. The flame moves downstream with increasing velocity with an increase in the flame length. There is a significant rise in the velocity of the gases after combustion, with exhaust gas velocities as high as 19 m/s obtained for $v_{in} = 7$ m/s.

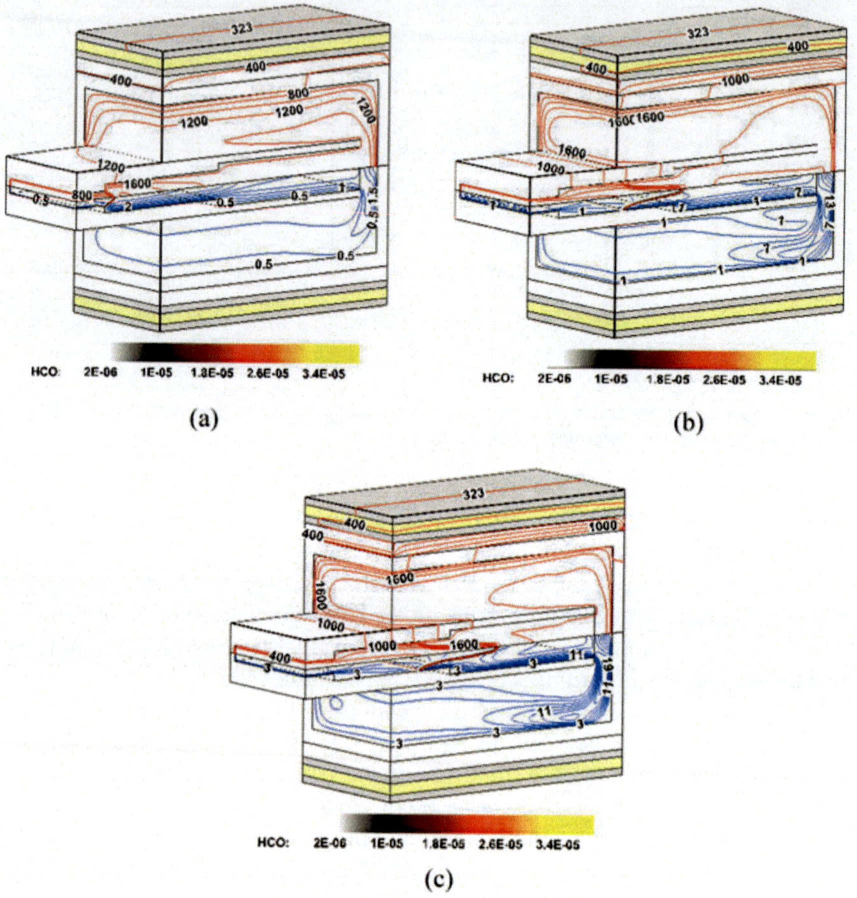

Fig. 5 Contours of temperature (upper half), velocity and HCO (lower half) for **a** $v_{in} = 1$ m/s, **b** $v_{in} = 5$ m/s, **c** $v_{in} = 7$ m/s

3.2 *Thermoelectric Characteristics*

Different parameters were studied to analyze the electric output of the system. Variation of the open-circuit voltage and the output current with the average temperature of the hot side is shown in Fig. 6a. Maximum voltage and current of ~7.1 V and ~1.1 A were obtained numerically. The difference between the numerical and specified values can be attributed to the non-uniform temperature distribution along the hot side of the module. While the specified values are obtained for constant temperatures, the numerically obtained values are varied against the average temperature of the hot side of the module. Figure 6b shows the variation of the output and the overall conversion efficiency of the module with the average temperature of the hot side of the module. The output power increases with increasing temperature, which

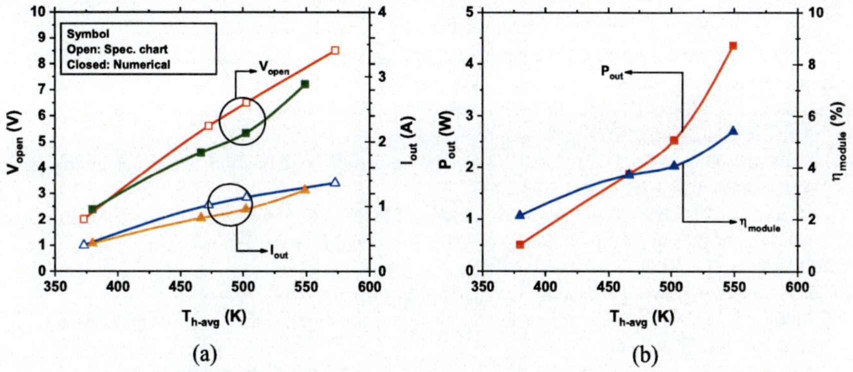

Fig. 6 Variation of **a** open-circuit voltage and current and **b** output power and conversion efficiency with average temperature of the hot side of the module

is in accordance with the readings obtained in Fig. 6a. The conversion efficiency is also reported to increase with increasing temperatures, with a maximum efficiency of 5.39% obtained for a power output of 4.35 W at $v_{in} = 7$ m/s.

4 Conclusion

In an attempt to simulate the thermoelectric effect in a CFD environment, the present study integrates the thermoelectric model with the pre-existing combustion model. An optimized combustor is designed to maximize the thermal output of the integrated system. Following this, the TEG model is coded into the software and verified by comparing the obtained values against the values given in the specification sheet. The integrated system's thermal characteristics and thermoelectric output are then analyzed by studying different parameters. The integrated system produced a maximum output power of 4.35 W at an efficiency of 5.39%, which is a significant improvement upon prior experimental TEGs [10]. The successful integration of the code allows the possibility to design combustors with better outputs without having to a create an experimental setup from scratch.

References

1. Weinberg F et al (2002) On thermoelectric power conversion from heat recirculating combustion systems. Proc Combust Inst 29(1):941–947
2. Riffat SB, Ma X (2003) Thermoelectrics: a review of present and potential applications. Appl Therm Eng 23(8):913–935
3. Ismail BI, Ahmed WH (2009) Thermoelectric power generation using waste-heat energy as an alternative green technology. Recent Pat Electr Electron Eng (Former Recent Pat Electr Eng)

2(1):27–39
4. Walther DC, Ahn J (2011) Advances and challenges in the development of power-generation systems at small scales. Prog Energy Combust Sci 37(5):583–610
5. Stevens JW (2001) Optimal design of small ΔT thermoelectric generation systems. Energy Convers Manag 42(6):709–720
6. Fernandez-Pello AC (2002) Micropower generation using combustion: issues and approaches. Proc Combust Inst 29(1):883–899
7. Alipoor A, Saidi MH (2017) Numerical study of hydrogen-air combustion characteristics in a novel micro-thermophotovoltaic power generator. Appl Energy 199:382–399
8. Jiang L et al (2011) Experimental study of a plat-flame micro combustor burning DME for thermoelectric power generation. Energy Convers Manag 52(1):596–602
9. Shimokuri D (2018) Autonomous power system using small scale vortex combustor. J Phys: Conf Ser. IOP Publishing
10. Aravind B, Khandelwal B, Kumar S (2018) Experimental investigations on a new high intensity dual microcombustor based thermoelectric micropower generator. Appl Energy 228:1173–1181
11. Gao X et al (2012) Numerical model of a thermoelectric generator with compact plate-fin heat exchanger for high temperature PEM fuel cell exhaust heat recovery. Int J Hydrogen Energy 37(10):8490–8498
12. Meng F, Chen L, Sun F (2011) A numerical model and comparative investigation of a thermoelectric generator with multi-irreversibilities. Energy 36(5):3513–3522
13. Chen M, Rosendahl LA, Condra T (2011) A three-dimensional numerical model of thermoelectric generators in fluid power systems. Int J Heat Mass Transf 54(1–3):345–355
14. Meng L, Li J, Li Q (2017) A miniaturized power generation system cascade utilizing thermal energy of a micro-combustor: design and modelling. Int J Hydrogen Energy 42(27):17275–17283
15. Srivastava N et al (2019) Numerical investigations on behaviour bifurcation of premixed H2-air flames in mesoscale tubes. Combust Theory Model 1–25
16. Smith GP (1999) GRI-Mech 3.0. http://www.me.berkley.edu/gri_mech/
17. Fluent A (2015) Theory guide. Ansys Inc.
18. Thermonamic electronics (Jiangxi) Corp. Ltd; http://www.thermonamic.com/TEP1-1264-1.5-English.pdf
19. Zhang Y et al (2019) Effects of inlet parameters on combustion characteristics of premixed CH_4/Air in micro channels. J Energy Inst 92(3):824–834

Numerical Investigation on Combustion Characteristics of Premixed H_2/Air in Stepped Micro-Combustors

Gannena K. S. Raghuram, B. Aravind, E. V. Jithin, Sudarshan Kumar, and Ratna Kishore Velamati

1 Introduction

Recent technology developments in the field of nano- and micro-fabrication have ushered the miniaturization of various devices in the field of micro-mechanical, communicational, image sensing, biomedical and many other applications [1, 2]. These devices require efficient power sources capable of providing power from several milliwatts to hundreds of watts. However, the energy densities of the existing conventional electrochemical batteries are very low and unable to serve the purpose for long duration due to low energy densities and long recharging times. Even the most advanced lithium-ion batteries have energy densities as low as ~0.20 kWh/kg [3, 4] as compared to 33.3 kWh/kg of hydrogen fuel [5]. The availability of reliable micro-power generators significantly enhances the functionality of these miniature devices, in which micro-combustion plays a significant role [6]. Hence, design of micro-combustors and their development at a wider operational range have propelled the attention of researchers over the past few years [3].

Several studies were conducted to analyse the dynamics of premixed H_2/air flames in micro-scale tubes till date [4–10]. The backward-facing step leads to a flow expansion resulting in generation of low-velocity recirculation zone near the step [11]. This helps in improving the flame stabilization characteristics in these micro-combustors. Even though there are some numerical studies on the stepped micro-combustors with k-ε turbulence model [12–15], a detailed comparison on the effect of turbulence

G. K. S. Raghuram · E. V. Jithin (✉) · R. K. Velamati
Department of Mechanical Engineering, Amrita School of Engineering, Amrita Vishwa Vidyapeetham, Coimbatore, India
e-mail: jithineved@gmail.com

B. Aravind · S. Kumar
Department of Aerospace Engineering, Indian Institute of Technology Bombay, Powai, Mumbai, Maharashtra 400076, India

A. K. Gupta et al. (eds.), *Advances in IC Engines and Combustion Technology*, Lecture Notes in Mechanical Engineering,
https://doi.org/10.1007/978-981-15-5996-9_66

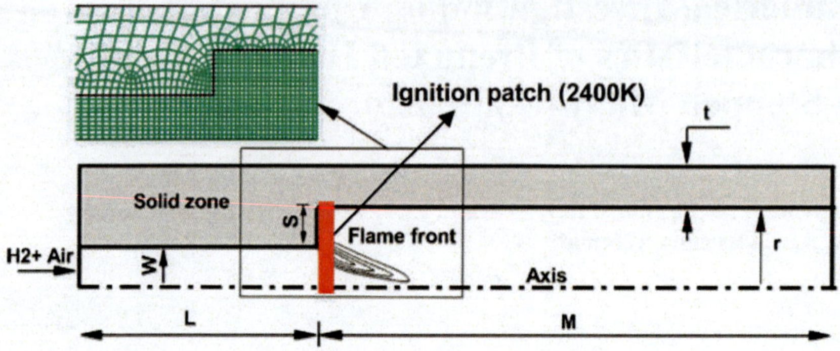

Fig. 1 Schematic of computational domain

model for predicting the actual flame dynamics in the microscale channels over the laminar model is yet to be explored. In the present study, the importance of the inclusion of the turbulence model for modelling the reacting flow in micro-combustors and its comparison with a laminar flow model has been discussed. Hydrogen was selected as the target fuel for the present investigations because of its short reaction time, smaller quenching distance, high heating value and faster diffusion velocity. The present study aims at numerically analysing the effect of inlet velocity and step ratio of the micro-combustor on combustion and thermal characteristics of the combustor operating with hydrogen fuel.

2 Numerical Methodology

2.1 Computational Domain and Grid

The computational domain is shown in Fig. 1. Stainless steel micro-combustor having total length (M) of 15 mm with an inlet of 7 mm is considered. The inner diameter (w) before the step is kept constant at 1 mm, and diameter (r) after the step has been varied by 2, 2.5, 3 and 3.5 mm. The step ratio of the combustor is delineated as $\beta = r/w$. The mesh is prepared using GAMBIT. The original grid size is $\Delta x = 50\ \mu\text{m}$, $\Delta y = 50\ \mu\text{m}$.

2.2 Governing Equations

The following assumptions are made, as the chemical reaction and flow governing equations are hard to solve in its detailed form. The combustion is assumed to be steady state. Dufour effect is neglected [14]. Radiation of the gaseous mixture and gravity effects are neglected [15] and (e) incompressible flow owing to low Mach number of the flow.

2.3 Boundary Conditions

The inlet velocity (v), inlet temperature (300 K) and the mixture mole fractions were defined at combustor inlet. At combustor exit, ambient pressure condition was specified. The turbulence intensity is specified to be 5% at the inlet of the combustor. Combined heat loss due to both natural convection and thermal radiation is provided at combustor outer wall by considering heat transfer coefficient, emissivity of the solid surface as 15 W/m^2 K and 0.85, respectively.

2.4 Grid Independence

Initial grid for the computational domain consists of 9876 cells. Once a stabilized flame is achieved in the computational domain, the region in the vicinity of the reaction zone is refined further to obtain a finer grid. The total number of cells with the refined grid is 33,816. Figure 2a, b, respectively, shows the variation of static temperature and mass fraction of OH species mass fraction at different mesh densities, for the turbulent model with $\beta = 3$, $\phi = 0.6$, $v = 8$ m/s. The figures show that the profiles of static temperature and OH mass fraction in finer and coarser grids match well. Thus, the adapted grid used in the model was found to be reliable to capture all the essential characteristics.

3 Result and Discussions

The burning characteristics of premixed H_2/air mixtures in stepped micro-combustors were examined by considering various inlet velocities such as 8, 12, 24 and 32 m/s, at equivalence ratio (ϕ) 0.6. The effect of step ratio on the outer wall temperature is studied by considering different step ratio combustors such as $\beta = 2$, 2.5, 3 and 3.5.

3.1 Effect of Flow Modelling

The numerical model is validated by comparing its mean outer wall temperature with the experimental results reported by Li et al. [16] under same operating conditions. Figure 3 presents the comparison between numerical and experimental results with a step ratio of micro-combustor, $\beta = 2$, $\phi = 0.6$ and equivalent velocity $u_e = v/(\beta)^2$,

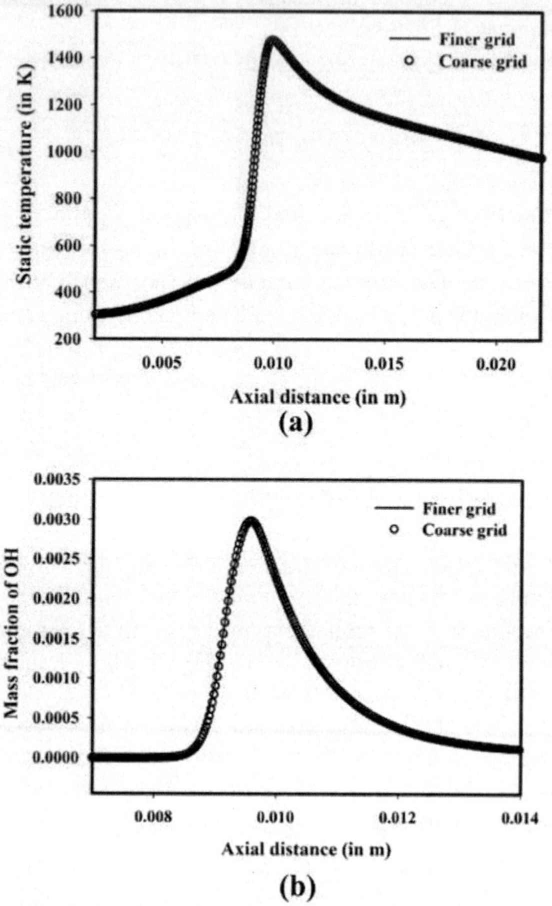

Fig. 2 Comparison of **a** static temperature and **b** OH mass fraction with coarse and finer grid, for turbulent model at $\beta = 3$, $\phi = 0.6$, $v = 8$ m/s along the axial direction

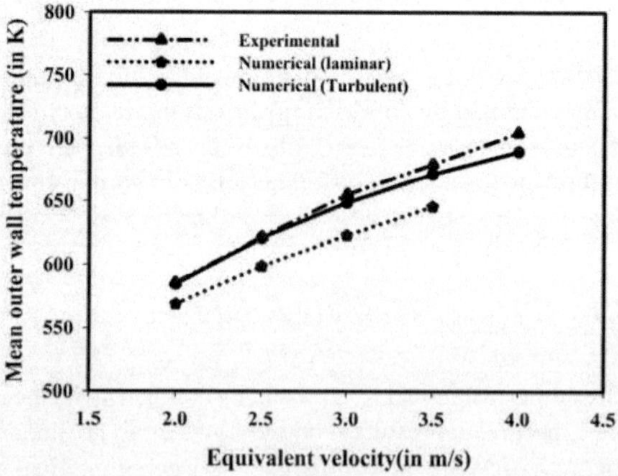

Fig. 3 Graph showing computational results of the mean To_w, at $\beta = 2$, $\phi = 0.6$, along with the experimental result

where v is the inlet velocity and β is the step ratio. The graph shows that the variation of mean To_w with experimental and turbulent numerical model is very small. The maximum relative error between the turbulent numerical predictions of mean wall temperature with that of experimental result was found to be 2.1%. The variation between the experimental and numerical results increases as the equivalent velocity increases. Therefore, the numerical simulations are limited to lower velocities in the present study. Overall, the turbulent numerical model realizable k–ε appears to predict the experimental behaviour way better compared to the laminar flow model. Similar conclusions have also made by Yilmaz et al. [17].

Figure 4a shows the comparison of the static temperature and mass fraction of OH for laminar and turbulent models at $v = 8$ m/s, $\phi = 0.6$ and $\beta = 2$ along the axial direction. The static temperature predicted with turbulent model near the step is slightly higher than that predicted by laminar model. Also, the static temperature from turbulent model decreases in the downstream and becomes lower than the static temperature predicted by laminar model.

This is perhaps due to enhanced mixing characteristics predicted by turbulent model, which increases the intensity of the sensitive chemical reactions. The mass fraction of OH for laminar and turbulent models also follows a similar trend. Figure 4b reveals that the magnitude of the flow velocity predicted using turbulent model is more than the laminar model, and it follows a pattern similar to the outer wall temperature. The enhanced mixing characteristics predicted using turbulent flow model increase the heat transfer coefficient and in turn result in increased outer wall temperature of the combustor.

3.2 *Effect of Step Ratio on Combustion Characteristics and Wall Temperature*

Figure 5a shows the variation of heat flux along the inner wall for different step ratios. An increase in the step ratio, β, causes the flow velocity to be decreased. The gap between the flame and wall also increases. As a result of these, the convection heat loss to the wall from the gas is reduced. Due to this, the heat flux transferred to the inner wall decreases. This has a direct effect on To_w of the combustor. Figure 5b shows the variation of OH mass fraction along the axis for different step ratios. Maximum OH mass fraction is observed for $\beta = 2$ and decreases with an increase in the step ratio, β. The peak value of OH mass fraction shifts to a downstream location with increase in step ratio. This is perhaps due to the reason that an increase in step ratio, β, shifts the reaction zone to a downstream location due to increased size of the recirculation zone as given in Fig. 6. It signifies that the intensity of chemical reaction decreases with an increase in step ratio, β. This results in reduced heat release rate as clear from Fig. 6.

Figure 7a shows the outer wall temperature at different step ratios for $\phi = 0.6$ and $v = 12$ m/s. The outer wall temperature decreases with an increase in step ratio. As

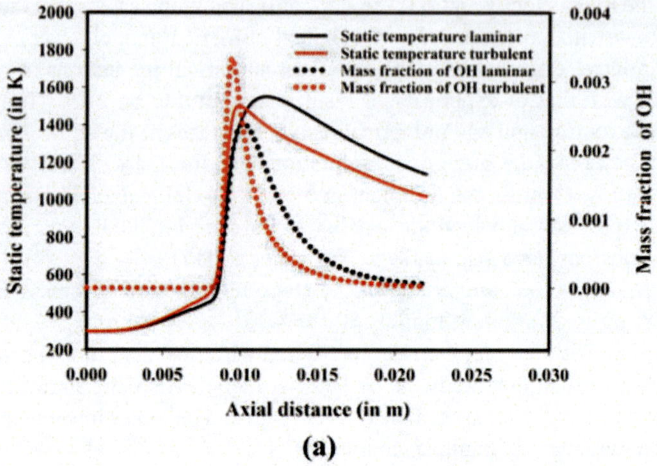

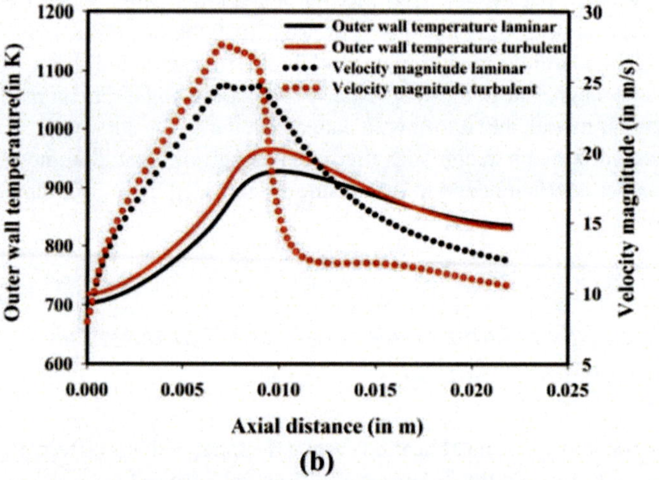

Fig. 4 Comparison of **a** static temperature and OH mass fraction, **b** outer wall temperature and velocity magnitude along the axis, for laminar and turbulent flow models at $v = 8$ m/s, $\beta = 2$, $\phi = 0.6$

the step ratio increases, the shear zone length increases, which increases the thermal resistance in radial direction. Due to this, smaller amount of heat is conducted towards the solid wall. Figure 7b depicts the variation of mean To_w with the variation of outer diameter of the micro-combustor at $\phi = 0.6$, for $v = 12$, 24 and 32 m/s. There is a linear decrease in the mean To_w with an increase in the step ratio of the combustor. The percentage reduction in mean wall temperature with respect to the increase in step ratio is found to be high at lower inlet velocity, compared to that at higher inlet velocity conditions.

Fig. 5 **a** Variation of heat flux along the inner wall and **b** OH mass fraction along the axis for different step ratios at $\phi = 0.6$ and $v = 12$ m/s

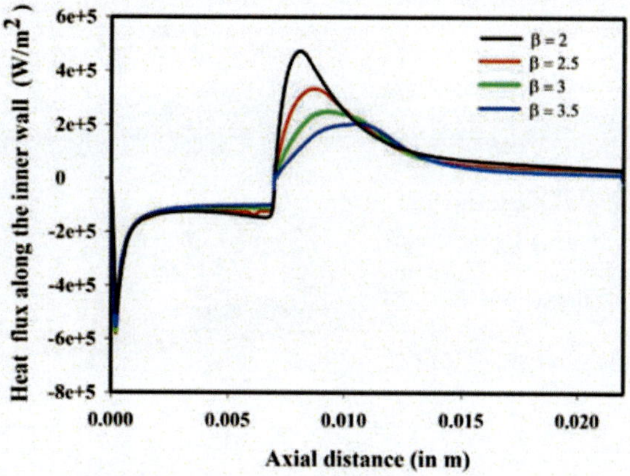

(a)

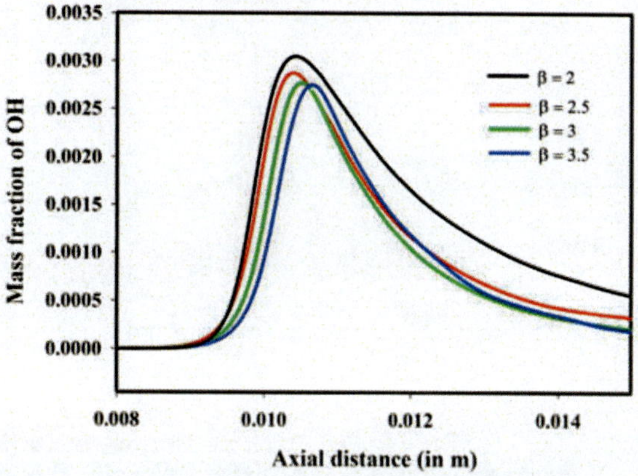

(b)

4 Conclusions

Burning properties of H_2/air mixtures in a single-step micro-combustor were investigated numerically by considering detailed reaction mechanism for H_2/air mixtures. The effect of inlet velocity, turbulent/laminar flow model and step ratio on combustion characteristics and To_w of the combustor were investigated. The mean To_w increases nonlinearly with increase in mixture velocity due to increase in the flame length and flame moving outside the combustor. Mean To_w increases nonlinearly with an increase in mixture equivalence ratio. An increase in step ratio from 2 to 3.5 at $\phi = 0.6$ and $v = 12$ m/s reduces the outer wall temperature of the combustor linearly.

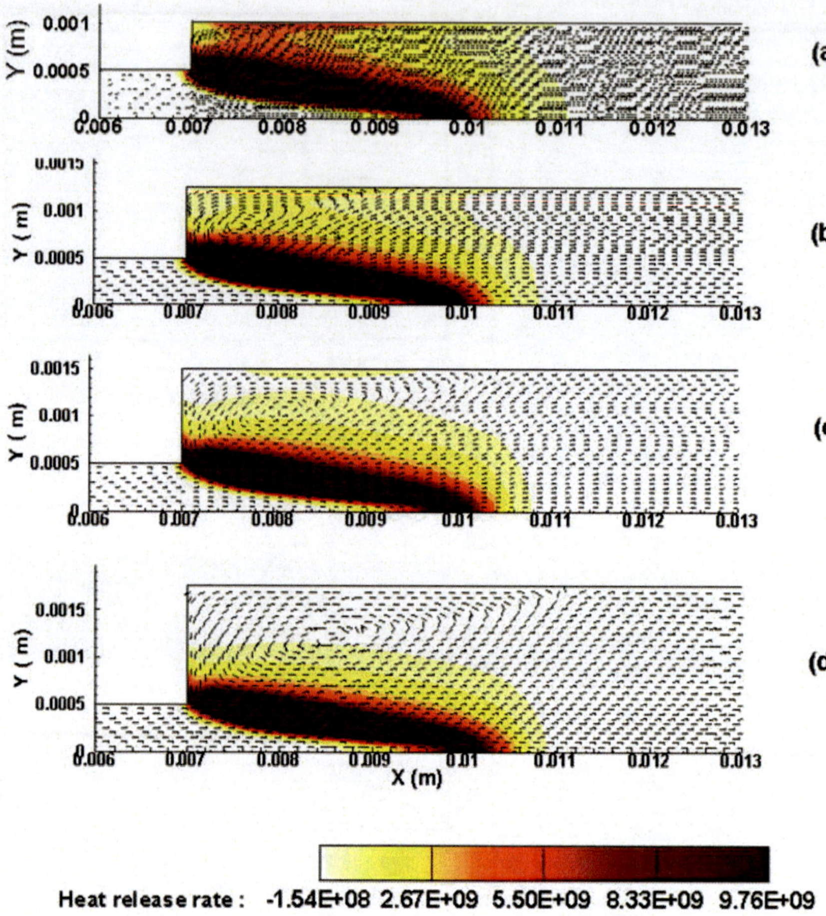

Fig. 6 Heat release rate field overlaid with velocity vectors at different step ratios **a** $\beta = 2$, **b** $\beta = 2.5$, **c** $\beta = 3$ and **d** $\beta = 3.5$, at $\phi = 0.6$, $v = 12$ m/s

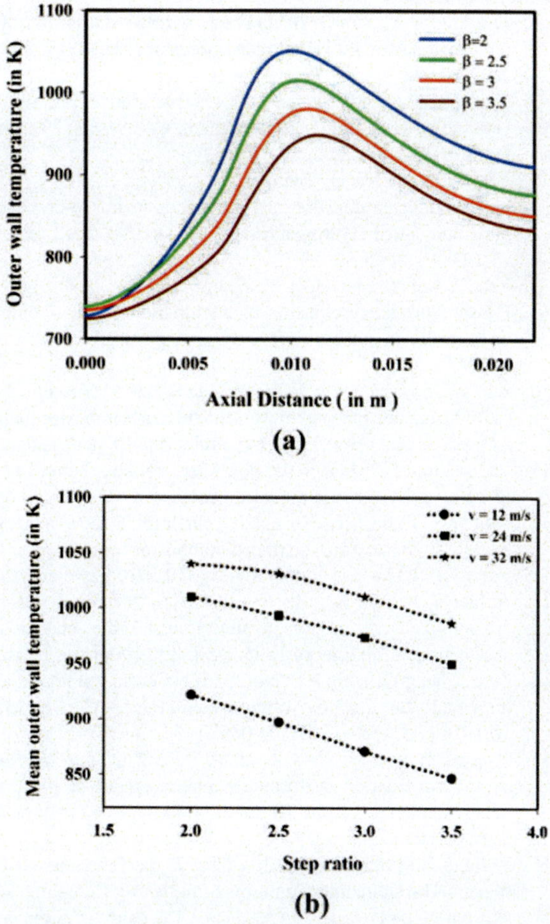

Fig. 7 **a** Variation of outer wall temperature along the axis for different step ratios at $\phi = 0.6$, $v = 12$ m/s, **b** variation of mean outer wall temperature with step ratio at $\phi = 0.6$, $v = 12$ m/s, 24 m/s and 32 m/s

References

1. Ju Y, Maruta K (2011) Microscale combustion: technology development and fundamental research. Prog Energy Combust Sci 37(6):669–715. https://doi.org/10.1016/j.pecs.2011.03.001
2. Walther DC, Ahn J (2011) Advances and challenges in the development of power-generation systems at small scales. Prog Energy Combust Sci 37(5):583–610. https://doi.org/10.1016/j.pecs.2010.12.002
3. Maruta K (2011) Micro and mesoscale combustion. Proc Combust Inst 33(1):125–150. https://doi.org/10.1016/j.proci.2010.09.005
4. Wan J, Yang W, Fan A, Liu Y, Yao H, Liu W, Du Y, Zhao D (2014) A numerical investigation on combustion characteristics of H_2/air mixture in a micro-combustor with wall cavities. Int J Hydrogen Energy 39(15):8138–8146
5. Aravind B, Velamati RK, Singh AP, Yoon Y, Minaev S, Kumar S (2016) Investigations on flame dynamics of premixed H_2–air mixtures in microscale tubes. RSC Adv 6(55):50358–50367. https://doi.org/10.1039/C6RA05539G

6. Jejurkar SY, Mishra D (2011) Flame stability studies in a hydrogen–air premixed flame annular microcombustor. Int J Hydrogen Energy 36(12):7326–7338. https://doi.org/10.1016/j.ijhydene.2011.02.098
7. Nair A, Kishore VR, Kumar S (2015) Dynamics of premixed hydrogen-air flames in microchannels with a wall temperature gradient. Combust Sci Technol 187(10):1620–1637. https://doi.org/10.1080/00102202.2015.1059326
8. Yan Y, He Z, Xu Q, Zhang L, Li L, Yang Z, Ran J (2019) Numerical study on premixed hydrogen/air combustion characteristics in micro-combustor with slits on both sides of the bluff body. Int J Hydrogen Energy 44(3):1998–2012. https://doi.org/10.1016/j.ijhydene.2018.11.128
9. Yan Y, Liu Y, Li L, Cui Y, Zhang L, Yang Z, Zhang Z (2019) Numerical comparison of H_2/air catalytic combustion characteristic of micro-combustors with a conventional, slotted or controllable slotted bluff body. Energy 189:116242. https://doi.org/10.1016/j.energy.2019.116242
10. Zuo W, Jiaqiang E, Lin R, Jin Y, Han D (2018) Numerical investigations on different configurations of a four-channel meso-scale planar combustor fueled by hydrogen/air mixture. Energy Convers Manage 160:1–13. https://doi.org/10.1016/j.enconman.2018
11. Rajasekaran J (2011) On the flow characteristics behind a backward-facing step and the design of a new axisymmetric model for their study. University of Toronto, Canada
12. Akhtar S, Khan MN, Kurnia JC, Shamim T (2017) Investigation of energy conversion and flame stability in a curved micro-combustor for thermo-photovoltaic (TPV) applications. Appl Energy 192:134–145. https://doi.org/10.1016/j.apenergy.2017
13. Akhtar S, Kurnia JC, Shamim T (2015) A three-dimensional computational model of H_2–air premixed combustion in non-circular micro-channels for a thermo-photovoltaic (TPV) application. Appl Energy 152:47–57. https://doi.org/10.1016/j.apenergy.2015.04.068
14. Su Y, Cheng Q, Song J, Si M (2016) Numerical study on a multiple-channel micro combustor for a micro-thermophotovoltaic system. Energy Convers Manage 120:197–205. https://doi.org/10.1016/j.enconman.2016.04.088
15. Jiaqiang E, Liu H, Zhao X, Han D, Peng Q, Zuo W, Meng T, Qiu R (2018) Investigation on the combustion performance enhancement of the premixed methane/air in a two-step micro combustor. Appl Therm Eng 141:114–125. https://doi.org/10.1016/j.applthermaleng.2018.05.101
16. William F (1985) Combustion theory: the fundamental theory of chemically reacting flow system. The Benjamin/Cummings Publishing Company, Menlo Park, CA
17. Norton DG, Vlachos DG (2004) A CFD study of propane/air microflame stability. Combust Flame 138(1):97–107. https://doi.org/10.1016/j.combustflame.2004.04.004

Numerical Investigation on Flame Dynamics of Premixed Hydrogen–Air Flame in a Sudden Converging–Diverging Microscale Tube

Akshit Yadav, B. Aravind, and Sudarshan Kumar

1 Introduction

The development of small-scale devices like drones, robots, micro-air vehicles, and other various electronics devices is receiving strong demand and great attention from the research community. There has been a growing interest in the development of micro-power generation devices due to their simple structure and non-moving parts. Combustion of hydrocarbons fuels is more intriguing because of its high energy density, which is almost 20–50 times higher than the most advanced Li-ion batteries [1]. Physics of micro-combustion is quite different from that of large-scale combustion. In micro-combustion, there is an increased effect of flame–wall interaction and molecular diffusion. Also, surface-to-volume ratio increases drastically which entails more heat loss and thus causes the flame to quench easily. The principle on which the micro-power generators work is the conversion of chemical energy into the radiant energy. This is the foremost component to ponder over because this affects the overall efficiency of these devices [2–4].

A series of investigations in understanding the flame dynamics in microscale combustors have been summed up below. Numerical studies of Kim and Maruta outline the effect of flow velocities profile, wall heat transfer coefficient, and tube

A. Yadav
Department of Mechanical Engineering, National Institute of Technology Kurukshetra, Kurukshetra, Haryana 136119, India
e-mail: akshit_11610697@nitkkr.ac.in

B. Aravind (✉) · S. Kumar
Department of Aerospace Engineering, Indian Institute of Technology Bombay, Powai, Mumbai, Maharashtra 400076, India
e-mail: aravindbpillai69@gmail.com

S. Kumar
e-mail: sudar4@gmail.com

A. K. Gupta et al. (eds.), *Advances in IC Engines and Combustion Technology*, Lecture Notes in Mechanical Engineering,
https://doi.org/10.1007/978-981-15-5996-9_67

diameters on laminar flame propagation velocities [5]. Effects of inlet velocity and mixture composition on wall heat losses and flame temperature were investigated by various researchers. The term micro-combustion is usually used when the characteristic length scale is of the order as that of the flame quenching distance [1, 6]. The importance of the thermal and chemical properties of the wall of the channel was studied by Kim et al. [7] by carrying out experiments on the flame behavior. They observed that chemical quenching is influential when temperature is higher than 1000 K and at very low pressure. Detailed numerical investigations have been carried out with various wall heat transfer conditions which demonstrate that the flame behavior is non-monotonic in nature [8–10]. Moreover, the progression in flame shape from a concave to convex flame has been studied for different wall conditions by Singh et al. [11] for a premixed methane–air mixture. In addition to this, Aravind et al. [8] studied the transition behavior of hydrogen–air flames in different tube diameters and thermal wall conditions.

Understanding flame dynamics in microscale tubes is fundamental for the future development of micro-combustors for power generation. Although many advancements have taken place in the past few decades in this field which have also led some to inventions [12], there is significant potential to design even more efficient micro-combustors. The present study is aimed primarily toward fast-propagating flames in tubes. Therefore, due to very short time period of propagation of flame in the mesoscale tubes, the involvement of conjugate heat transfer [13, 14] remains irrelevant. However, this phenomenon is of great importance and has been studied satisfactorily earlier.

The objective of the present study is to realize the transient flame behavior in a converging–diverging tube with different thermal boundary conditions. This study will further help visualize the dynamics and flame behavior in microscale tubes and would further contribute in designing of micro-combustors. Hydrogen–air stoichiometric mixture is used in the numerical investigation due to its simplicity in modeling as it involves only nine species and 21 reactions.

2 Computational Details

2.1 Computational Domain

The details of computational domain considered in the present work are shown in Fig. 1. It consists of a tube of length 110 mm and diameter of 2 mm with contraction channel (C–D) in the middle of length 10 mm with 1 mm diameter. The contraction channel diameter is chosen close to the flame quenching distance of hydrogen–air which is 0.74 mm and that makes the contraction ratio of $D/d = 2$. Because of axis symmetric nature of tube, half of the geometry has been solved in the present study.

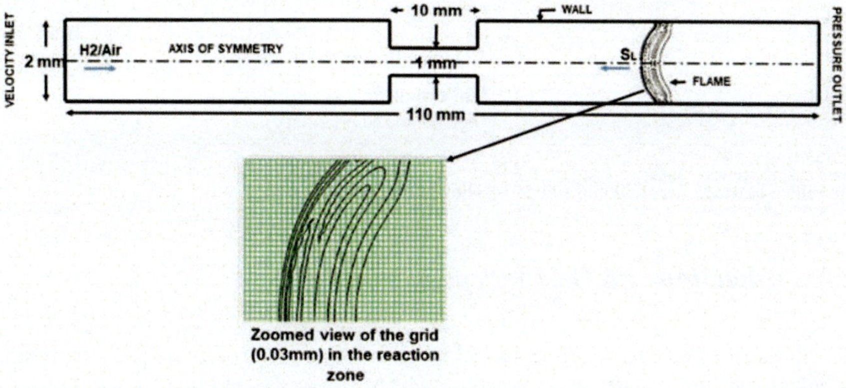

Fig. 1 Schematic of computational domain and grid

2.2 *Chemical Reaction Modeling*

In the present study, the governing equations for mass, momentum, energy, and species conservation were solved using finite volume method solver ANSYS Fluent 15.0 [15] because of its ability to integrate both chemical kinetics and fluid dynamics. The chemical kinetics information was imported into the Fluent solver in the CHEMKIN format. The chemical reaction mechanism proposed by Li et al. [16] is used to model the hydrogen combustion which consist of 9 species and 21 reactions.

2.3 *Boundary and Initial Conditions*

A uniform velocity is provided at the inlet from where stoichiometric proportions of premixed hydrogen–air enter the channel and ambient pressure of 1 atm is assigned at the outlet. Velocity of $V_{in} = 0.3$ m/s (at 300 K) is taken for all simulations which is small enough to make the relative velocity of flame propagation quite high in the microscale tube due to which heat transfer in the wall can be neglected in such small period of time. Species transport model involves only volumetric reactions, and therefore, inner wall of the tube is considered chemically inert. Heat transfer coefficient of the wall is varied from 300 to 1000 W/m^2 K. A rectangular patch of width 2 mm with 2400 K temperature is provided at the exit for igniting the mixture. For this transient simulation, SIMPLE algorithm is used to solve the computation, and laminar flow is considered in the model. Second-order implicit transient formulation is used to solve unsteady terms with the time step of 10^{-5} s. The chemistry integration was implemented using ISAT integration method. A summary of various boundary conditions is shown in Table 1.

Table 1 Details of boundary conditions

Inner wall	Chemically inert with no slip
Inlet	Uniform velocity distribution of 0.3 m/s with 300 K (normal to the boundary)
Outlet	Neumann boundary
Interface between the fluid and solid	The method of 'coupled'

2.4 Validation and Grid Independence Study

To validate the present numerical model, laminar flame speed of hydrogen/air mixture is calculated numerically and compared with the experimental values for different equivalence ratios as shown in Fig. 2. To obtain a freely propagating flame, the thermal wall boundary of the present domain is changed to a symmetry boundary. This allows the stretch-free, adiabatic flame to propagate upstream. Aravind et al. [8] have taken a similar approach for the validation purpose. It can be concluded from Fig. 2 that there is minimal fluctuation between the present computational methodology and the different experimental procedure. This ensures the validity of the present model. Further, a detailed grid independence study was conducted for grids ranging from 20 to 50 μm. Since there was considerable difference between the results obtained for the 30, 40, and 50 μm cases and negligible fluctuation was observed in the results obtained for the 20 and 30 μm grids, the latter was chosen for computational purposes. The average flame front thickness is of 0.3 mm for hydrogen–air mixture, so the grid size taken in the domain sufficiently captures the reaction and flame propagation features. Since the flame is unsteady, computation data was stored after every 0.5 ms to ensure that details of region of interest could be studied more accurately.

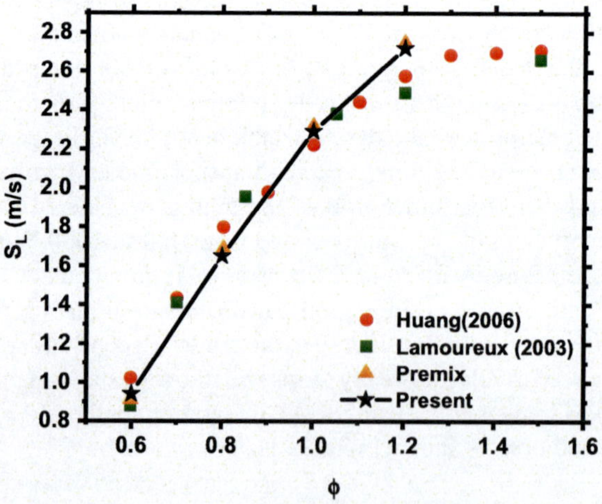

Fig. 2 Validation of laminar burning velocity

3 Result and Discussion

3.1 *Effect of Wall Heat Transfer Coefficient on Flame Shape Angle*

The effect of wall heat transfer coefficient (h) on the flame shape angle has been analyzed in this section. The flame shape in the present work is determined by using volumetric heat release rate contours. In this work, the flame angle (θ) is delineate as the angle between the line normal to wall from axis and the line joining the point of contact of VHRR contour (corresponding to $0.4 \times \text{VHRR}_{max}$) and wall to the foremost point of the contour on axis. In addition to that, the convex shape of the flame toward the unburnt mixture (inlet) is taken as the positive flame shape angle and the concave shape toward the mixture is taken as negative. The variation of the flame shape angle follows a non-monotonic trend in a straight micro-tubes as already predicted by Singh et al. [11] and Aravind et al. [8] recently. However, flame dynamics in different tube geometries such as stepped tubes and converging–diverging tubes need to be investigated in detail to realize the underlying physics behind the flame propagation and flashback characteristics. This perhaps helps in designing optimized combustors and to take care of the safety concerns associated with engine tubing. A converging–diverging (CD) channel has been considered, and flame shape dynamics are studied in the present study. The thermal wall coupling and flow hydrodynamics are the key factors affecting the flame shape change in micro-tubes. Among these two, wall coupling has more effect on flame dynamics as micro-flames are prone to extinction due to high surface heat loss to heat generation ratio.

The variation of flame shape angle with different wall heat transfer coefficients is shown in Fig. 3. After the ignition, flame evolves its shape and forms concave and convex shapes for h = 300 W/m^2 K and 1000 W/m^2 K, respectively, and propagates

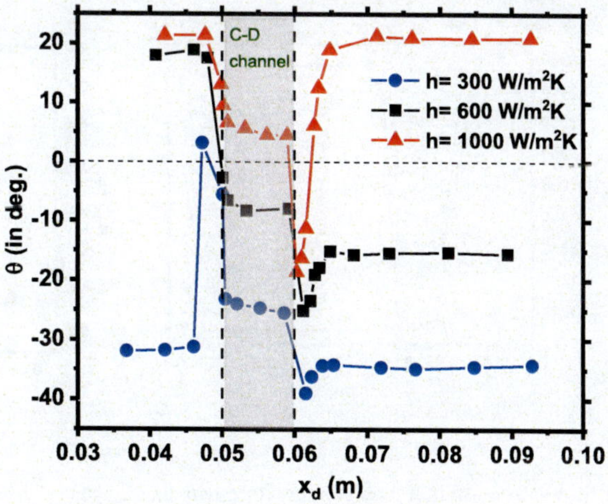

Fig. 3 Variation in flame shape angle along the tube

toward the upstream direction. The initial convex shape of the flame at higher h values relates to the significant convective heat loss from the tube. The laminar flame speed will be less near the wall boundary compared to the center line for higher h values. Therefore, the flame readjusts its shape to the convex to match the laminar flame speed for achieving its stable position. Further, it is also noted that the HRR contours shift toward the wall boundary that pushes the center of the flame upstream side via strong thermal diffusion and thus makes a convex shape. On the contrary, for lower h values, the hydrodynamic effect dominates over the heat loss and leads to the formation of concave shape. Therefore, the flame propagates with a concave shape from the downstream end which is similar to the typical parabolic velocity profile in any pipe flow. However, there is an abrupt change in the flame shape which is noted at the entrance and exit of the convergent–divergent (C–D) channel. It can be understood from Fig. 3 that as the wall heat transfer coefficient increases from 300 to 1000 W/m^2 K, flame shape changes from concave to convex from the point of contraction at the channel entrance. Heat loss would be higher in the reaction zone for higher h values. The flame temperature is the direct manifestation of heat content in the reaction zone, i.e., increase in heat loss would result in lowering of flame temperature. It can be observed from Fig. 4a that there is a sudden drop in the flame temperature in the contraction channel due to an abrupt change in the surface-to-volume ratio. It is observed that the initial and final flame shapes remain almost same. However, unpredictable behavior of flame shape angle is observed for $h =$ 600 W/m^2 K, especially at the C–D exit. This perhaps due to the transition of h value as predicted by Aravind et al. [8] earlier. A thorough investigation on intermediate h values is required to draw a proper conclusion on this variation.

Further, a detailed investigation on the ratio of heat loss to heat generation (H_{loss}/H_{gen}) is carried out and shown in Fig. 4b. The heat generation depends on the type of fuel, equivalence ratio, combustor volume, etc. It is clear from Fig. 4b that a steep increase in H_{loss}/H_{gen} is observed when the flame approaches the contraction channel. This is probably because of the increased surface area near the entrance

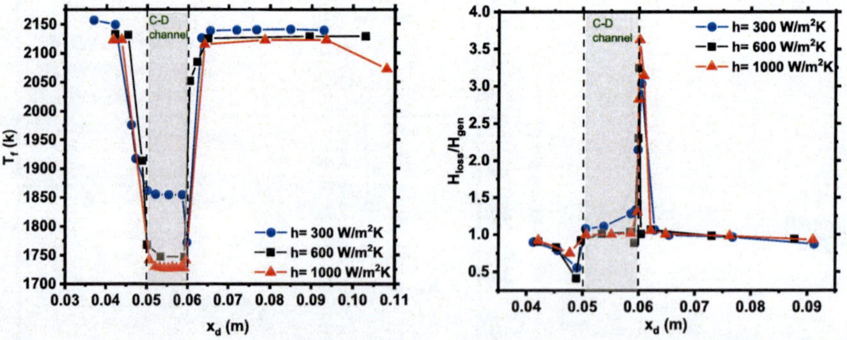

(a) (b)

Fig. 4 Variation in **a** flame temperature with axial distance, **b** H_{loss}/H_{gen} ratio along the tube of the contraction channel. This has also affected the flame shape substantially in the region within a very short period of time. And apart from it, if we observe the flame dynamics at the exit of contraction channel, this heat loss to heat generation ratio drops a little for a considerable part of channel length. This might be due to sudden exposure to the larger mass of unburnt fuel–air mixture and decrease in surface-to-volume ratio. This results in more heat generation and lesser heat loss.

3.2 *Effect of Wall Heat Transfer Coefficient on Flame Shape Angle*

In the present study, flame tip displacement (x_d) relates to the physical location of flame at different time intervals which is defined as the axial point corresponding to the maximum mass fraction of HO_2. The HO_2 remains for a very short period of time, so it can provide a quite precise location of the flame in the tube. The flame propagation velocity (S_L) is obtained by calculating the slope of the flame tip displacement—time graph. Flame propagation speed depends upon the heat content of the flame, i.e., more heat content implies more energy of the propagating flame. The variation of the flame tip displacement and the corresponding propagation velocity along the length of channel is shown in Fig. 5a, b, respectively. The variation of flame propagation velocity with different wall heat transfer coefficients can be explained with respect to the flame shape as discussed in the previous section. There is an abrupt change in the flame tip displacement—time graph observed for $h = 600$ W/m^{2}K and 1000 W/m^2 K cases near the C–D section which can relate to the change in mass flux as shown in Fig. 5a. However, there is only a slight change in the slope for $h =$ 300 W/m^2 K, which is perhaps due to the reduced heat loss. A detailed analysis is

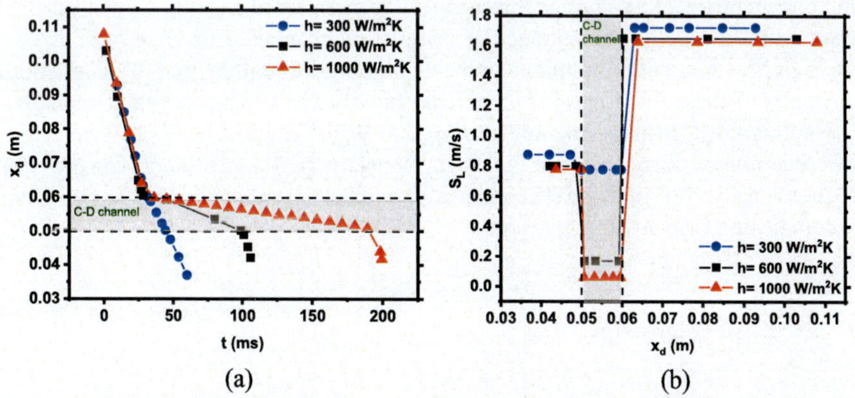

Fig. 5 Variation in **a** flame position with time, **b** flame propagating velocity with axial distance

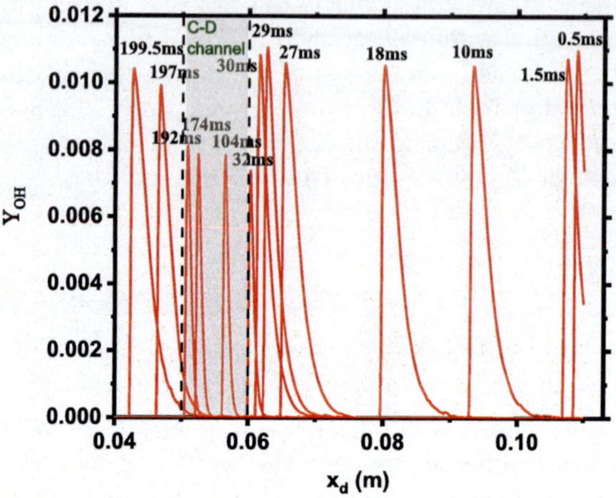

Fig. 6 Temporal variation OH mass fraction along the tube loss for $h = 1000$ W/m^2 K

required for the low wall heat transfer coefficient values to draw a solid conclusion for this peculiar variation.

It is noted that the propagation velocity decreases with an increase in the h values as shown in Fig. 5. This can be attributed to the fact that the heat loss is more at higher h values. In the contraction channel, propagation velocity for $h = 1000$ W/m^2 K is quite less (0.058 m/s approx.) due to greater heat loss as a result of the increased surface-area-to-volume ratio. The variation of the OH mass fraction at different time steps along the tube is shown in Fig. 6. The mass fraction of OH represents the reaction intensity. It can be observed from Fig. 6 that at the time of ignition, OH mass fraction is as high as 0.011. Then, as the flame starts moving upstream, mass fraction of OH decreases initially and follows a constant trend before the C–D entrance and then again increases as the flame approaches the contraction channel. As the flame approaches nearer to the contraction channel, the intensity of reaction suddenly drops to a minimal value till the entrance of contraction channel. This is perhaps due to the high heat loss and minimum combustion volume. From the exit of contraction channel, OH mass fraction starts to increase rapidly and remains constant onwards.

Contours of OH mass fraction for $h = 300$ W/m^2 K and $h = 1000$ W/m^2 K at different time steps are shown in Fig. 7a, b, respectively. These contours can provide a better visualization of flame shape transition and help in understanding the variation of combustion intensity along the axis.

4 Conclusion

A detailed study of transient numerical simulations has been carried out to study the effect of various wall heat transfer conditions on the flame dynamics of premixed hydrogen–air mixture in a sudden converging–diverging microscale tubes. It has

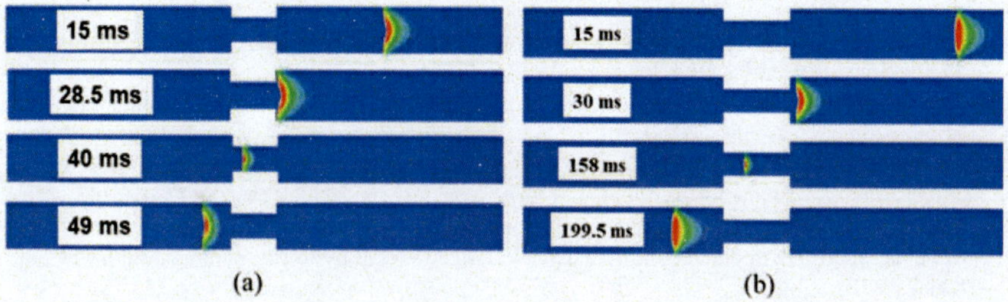

Fig. 7 Contours of OH mass fraction at different time steps for **a** $h = 300$ W/m^2 K, **b** $h = 1000$ W/m^2 K

been observed that there is a significant effect of wall heat transfer coefficient on the reaction zone at various locations in the stepped microscale tube. Flame shape angle, heat release rate, flame temperature and propagation velocity changes rapidly near the end points of contraction channel due to immediate change in the flame surrounding in the very short time period. Also heat loss to heat generation ratio plays a vital role in understanding the flame dynamics at such small scale particularly in case of studying the flame propagation. The flame shape change near the end points of contraction channel is extensively determined by heat loss to heat generation ratio and mass flux redistribution of unburnt particle of hydrogen–air mixture in the region.

References

1. Ju Y, Maruta K (2011) Microscale combustion: technology development and fundamental research. Prog Energy Combust Sci 37(6):669–715
2. Tang A et al (2015) A comparative study on combustion characteristics of methane, propane and hydrogen fuels in a micro-combustor. Int J Hydrogen Energy 40(46):16587–16596
3. Jiang D, Yang W, Teng J (2015) Entropy generation analysis of fuel lean premixed CO/H_2/air flames. Int J Hydrogen Energy 40(15):5210–5220
4. Li Y-H, Hong JR (2018) Power generation performance of hydrogen-fueled micro thermophotovoltaic reactor. Int J Hydrogen Energy 43(3):1459–1469
5. Kim NI, Maruta K (2006) A numerical study on propagation of premixed flames in small tubes. Combust Flame 146(1–2):283–301
6. Ju Y, Xu B (2005) Theoretical and experimental studies on mesoscale flame propagation and extinction. Proc Combust Inst 30(2):2445–2453
7. Kim KT, Lee DH, Kwon S (2006) Effects of thermal and chemical surface–flame interaction on flame quenching. Combust Flame 146(1–2):19–28
8. Aravind B et al (2016) Investigations on flame dynamics of premixed H_2–air mixtures in microscale tubes. RSC Adv 6(55):50358–50367
9. Singh AP et al (2015) Numerical investigations of unsteady flame propagation in stepped microtubes. RSC Adv 5(122):100879–100890
10. Akram M, Kumar S (2011) Experimental studies on dynamics of methane–air premixed flame in meso-scale diverging channels. Combust Flame 158(5):915–924
11. Singh AP et al (2017) Effect of wall thermal boundary conditions on flame dynamics of CH_4-air and H_2-air mixtures in straight microtubes. Combust Sci Technol 189(1):150–168

12. Jejurkar SY, Mishra DP (2009) A review of recent patents on micro-combustion and applications. Recent Pat Eng 3(3):194–209
13. Kim NI et al (2007) Development and scale effects of small Swiss-roll combustors. Proc Combust Inst 31(2):3243–3250
14. Lee MJ et al (2010) Scale and material effects on flame characteristics in small heat recirculation combustors of a counter-current channel type. Appl Therm Eng 30(14–15):2227–2235
15. Fluent A (2009) 12.0 Theory guide. Ansys Inc. 5(5)
16. Li J et al (2004) An updated comprehensive kinetic model of hydrogen combustion. Int J Chem Kinet 36(10):566–575

Numerical Investigation on the Effect of Wall Preheating on Flame Stability of Stepped Microcombustor

Chilanka P. Panicker, Iram Maqbool, B. Aravind, and Sudarshan Kumar

1 Introduction

From the past few decades, there is a considerable rise in the rate of research and development toward diminishment of various small-scale electrochemical devices such as smartphones, mobiles, drones, and laptops. Most of these devices rely on conventional chemical batteries as a source of electric power. Therefore, use of such devices for long duration is limited owing to the low energy density [1, 2]. In addition to that, short recharging time, limited recharging cycles, environmental concerns, and burdensome weight make it further problematic [3, 4]. Microcombustion-based power sources can be considered as reliable power sources due to their increased power density, which leads to reduction in weight and increased lifetime [1]. Presently used advanced Li-ion batteries have an energy density two time lesser than that of hydrocarbon fuels [5]. The operating range is lesser for hydrocarbons, thus reducing the emission of pollutants from the microcombustor. Therefore, small-scale power generating devices that can be operated at very high efficiencies would offer a good alternative to conventional electrochemical batteries. The major issue associated with

C. P. Panicker · I. Maqbool
Amity Institute of Aerospace Engineering, Amity University, Sector-125, Noida, Uttar Pradesh 201313, India
e-mail: chilanka1998@gmail.com

I. Maqbool
e-mail: irammqbl@gmail.com

B. Aravind (✉) · S. Kumar
Department of Aerospace Engineering, Indian Institute of Technology Bombay, Powai, Mumbai, Maharashtra 400076, India
e-mail: aravindbpillai69@gmail.com

S. Kumar
e-mail: sudar4@gmail.com

A. K. Gupta et al. (eds.), *Advances in IC Engines and Combustion Technology*, Lecture Notes in Mechanical Engineering,
https://doi.org/10.1007/978-981-15-5996-9_68

the microcombustors is a high surface-to-volume ratio which results in poor flame stabilization. Adsorption of radicals on inner combustion walls and recombination lead to radical quenching, which ultimately results in the reduction of homogenous chemistry. Flow recirculation and heat recirculation are common strategies adopted for the flame stabilization in microcombustors. The use of backward-facing steps [6], bluff bodies, cavities, etc., helps in flow recirculation, while heat recirculation can be achieved by performing excess enthalpy combustion through Swiss-roll combustors, use of recirculating cup, etc. [7, 8].

Maruta et al. [9] have experimentally investigated a heated channel of 2.0 mm diameter with a positive temperature gradient along the direction of fluid flow and studied the propagation characteristics of a premixed methane–air mixture. They achieved unstable pulsating flame propagation modes at moderate flow rates and a stable flame mode at very high and low flow rates. Characteristic features of a microcombustion with methane–air mixture combustion were investigated by Li et al. [10]. Pan et al. [11] performed an experimental study on single backward-facing stepped microcombustor, and it was discovered that flame stabilization is affected by the nozzle to combustor diameter ratio. Effect of recirculating cup on operational regime of combustor was reported by Ronney et al. [12]. Yan et al. [13] performed parametric study for combustor with recirculating cup. Kim and Kwon examined the characteristics of flame propogation in a two-staged microcombustor with premixed propane–air mixtures computationally and experimentally [14]. They showed the importance of heat recirculation in stabilization of the flame as well as the rise in flame stability limits. Khandelwal et al. [6, 15] have further revealed analogous studies on the effect of flow and heat recirculation on flame stability, flame dynamics, and flame structure by using backward-facing stepped combustors. Zarvandi et al. [16] presented different ways to enhance flame stabilization by activating certain chemical reactions that will have an effect on the combustion reaction. Peng et al. [17] compared different inlet shapes for backward-facing stepped microcombustors and analyzed the power output and radiation efficiency for all configurations. Schoegl et al. [18] developed an analytical model to predict the combustion characteristics in externally heated microtubes.

In the present study, various factors affecting flame stabilization in a backward-facing three-stepped combustors are investigated numerically. Methane–air is used as a fuel because it is widely used in various research studies and experiments. It is also accessible in the form of CNG which is used for many industrial and automobile applications. Although our research group has already studied the enhancement of flame stability in backward-facing stepped combustors through experiments, a detailed study on various factors affecting the flame stability needs to be explored numerically. Therefore, a two-dimensional model of a three-stepped backward-facing step with a heat recirculating cup has been developed on a CFD platform. The model has been successfully validated by using preliminary experiments. The influence of heat recirculation on flame stabilization by varying the cup lengths has been studied. This study indeed helps in developing an optimized combustor that can be used as a heating source for power applications.

2 Computational Details

2.1 Computational Domain

The computational details of the microcombustor used in the current study are shown in Fig. 1. Half the geometry has been solved due to the axisymmetric nature. A cylindrical microcombustor consisting of three backward-facing steps located at 40, 50, and 65 mm from the inlet with respective diameters of 2 mm, 3 mm, and 6 mm is considered. Methane–air mixture is fueled from the 2-mm-diameter inlet. The thickness of the microcombustor wall is 1 mm, and the total length of the combustor is 75 mm. To understand the effect of heat recirculation, a heating cup is placed 2 mm away from the outlet of the combustor. Various heating cup lengths (HC_L) have been considered such as (a) 10 mm, (b) 25 mm, and (c) 50 mm to examine preheating effect on the flame stabilization. The combustor material used is quartz, with a density of 2320 kg/m^3, thermal conductivity of 1.38 W/m K, and specific heat of 700 J/kg K. Surface reactions are not considered due to chemically inert interior surfaces. Throughout the simulations, the equivalence ratio is taken 1.0 and the heat transfer coefficient of external wall of combustor is also taken constant as 20 W/m^2 K. Similarly, such boundary conditions are given at the inlet for specifying the inlet velocities and the mole fraction of reactants. The combustion reaction is initiated by a temperature patch of 2400 K, specified at the combustor exit. Volumetric reactions were considered and stiff chemistry solver was also enabled to deal with the reaction mechanism.

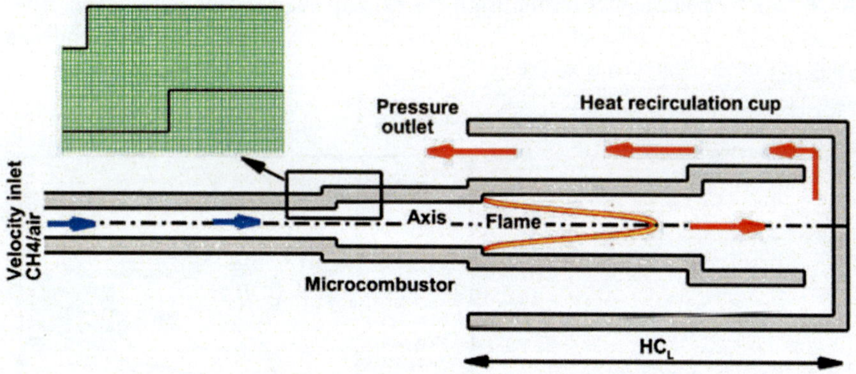

Fig. 1 Schematic diagram of the stepped microcombustor with heat recirculation cup

2.2 Numerical Solution Methodology

A two-dimensional numerical steady-state simulation was performed by using general-purpose CFD solver Fluent 15.0 [19]. Pressure-based solver and steady-state simulations have been carried out in the current study. A coupled wall boundary condition is adopted for the inner walls to solve the conjugate heat transfer. The thermodynamic and transport property of fuel–air mixture was imported from CHEMKIN database. In this numerical study, the simulations were done with the help of methane–air CHEMKIN mechanism which consists of 27 species and 159 reactions [20]. The SIMPLE algorithm (pressure-velocity coupling) is used to carry out steady-state simulations. Second-order discretization is used for better accuracy. ISAT integration method is used to perform chemistry integration. To initiate the combustion of mixture, a 2-mm patch of temperature 2400 K was given near the end of the combustor.

2.3 Grid Independence Study and Model Validation

The two cases of the combustor with and without recirculation are considered to verify the utility of the above numerical model. Preliminary experiments are carried out on a combustor-recirculation cup configuration made up of quartz at an inlet velocity of 5 m/s. The temperature on combustor wall (no HC) and cup wall (HC) is measured using a K-type thermocouple at various locations. Interestingly, for both the cases, the variation of numerical results from the experimental results is less than 10%, which asserts the validity of the present model (Fig. 2). Also, a grid independence study has been conducted for three different grids of 15, 20, and 50 μm for the heat

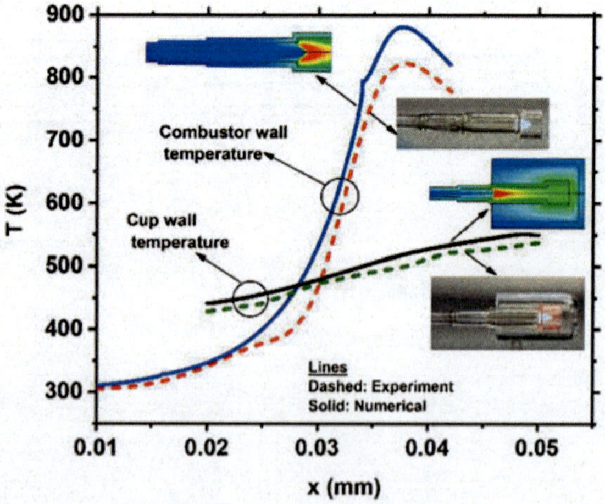

Fig. 2 Numerical model validation

recirculation case, and the grid corresponding to 20 μm, with 47,229 nodes, was found to be satisfactory.

3 Results and Discussions

3.1 Effect of Heat Recirculation Cup

In this section, flame stability has been compared for cases of with the cup and without the cup. All the simulations are carried out with a stoichiometric ratio (ϕ) of 1.0, and a heat transfer coefficient of the external wall of combustor is taken as 20 W/m^2 K.

The variations of flame front position with and without heat recirculation cup with inlet velocity are shown in Fig. 3. It is witnessed that the heat recirculation cup helps in shifting the flame towards upstream. This is perhaps due to the significant preheating of reactants. At lower inlet velocities, flame stabilization occurs near the first and second step for both the cases of with and without heat recirculation. For the no recirculation case, further increase in velocity beyond 3 m/s causes the flame front to move toward the third step and eventually blow out at 5.2 m/s. On the contrary, for the case with heat recirculation, due to the significant preheating of the reactant mixture, the flame stabilizes near the first step even for 10 m/s.

This flame stabilization phenomenon is essentially an interplay between inlet mixture velocity and the laminar flame speed. The preheating of the reactant increases the laminar flame speed, which pushes the flame more upstream. Although the inlet velocity also increases due to the conjugate heat transfer, the increase in laminar flame speed dominates over it. Therefore, at the second step, flame is stabilized

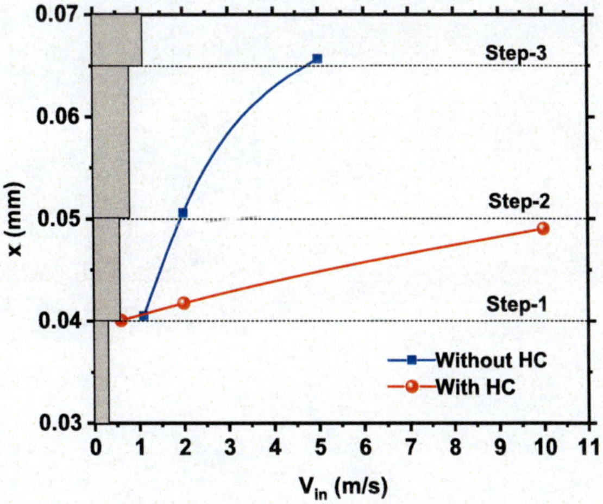

Fig. 3 Variation of flame position with inlet velocity

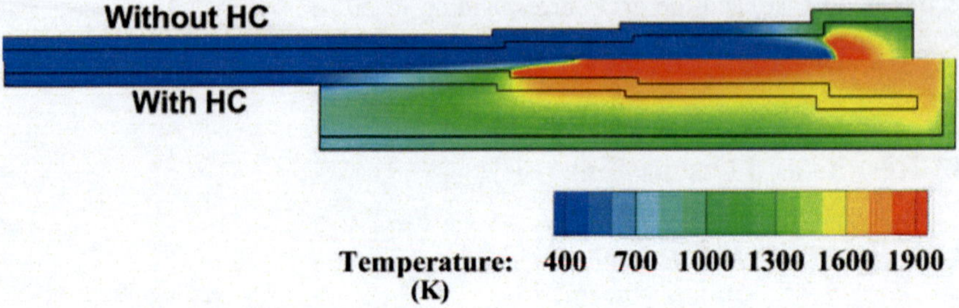

Fig. 4 Variation of temperature for (upper half) without recirculation and (lower half) with recirculation case

even at very high inlet velocity for the case with recirculation cup. Figure 4 shows the contours of static temperature at 5 m/s for both the cases of a combustor with and without heat recirculation cup. From this figure, it is observed that the flame stabilization takes place at the first and third steps of the combustor with and without HC, respectively. This represents the effect of preheating on the flame stabilization. The variation of temperature along the axis for the two cases is shown in Fig. 5. This figure shows a sudden increase in the temperature near the first step of the combustor for the heat recirculation case, whereas for the combustor without heat recirculation, the temperature rise is at a further distance. This sudden rise in temperature indicates the location of the flame front. The peak temperature reported for the HC case is approximately 200 K greater than that obtained in the no HC case due to excess enthalpy. This can help in better flame stabilization.

Fig. 5 Variation of inner combustor wall temperature with axial distance

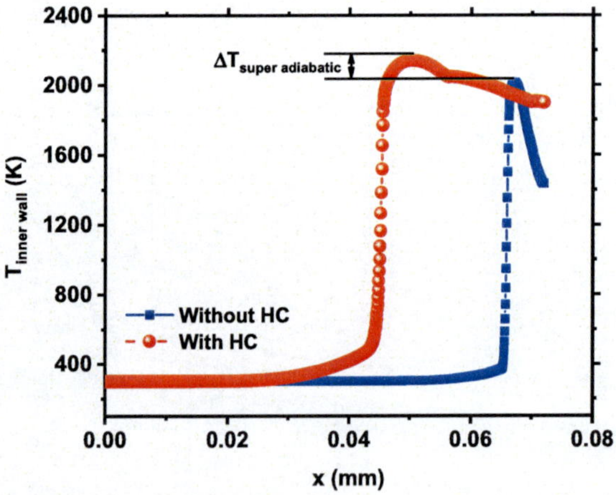

3.2 *Effect of Heat Recirculation Cup Length*

In this section, the effect of different cup lengths on the flame stability in a microcombustor is studied. Heat recirculation cup lengths of 10, 25, and 50 mm are considered for the analysis. Flame stabilization for each case is compared at different inlet velocities.

Figure 6a shows the variation of the lower flame stability limit for different HC_L. Lower flame stability limit is the minimum velocity at which a flame can be stabilized in a combustor. It is observed that the lower flame stability limit decreases with an increase in the recirculation cup length. Lower flame stability limits of 0.6, 0.9, and 1.1 m/s are observed for $HC_L = 50$ mm, 10 mm, and no recirculation case, respectively. This variation is due to the phenomenon of preheating with the help of heat recirculation cup. The variation of flame position with different inlet velocities for various heat recirculation cup lengths is shown in Fig. 6b. This figure shows that as the recirculation cup length increases, the flame shifts upstream with an increase in inlet velocity. In general, flame stability can be achieved by the flow recirculation method and heat recirculation method. Flow recirculation is achieved through the implementation of multiple backward-facing steps which enhances the residence time of the reactant, whereas the use of heat recirculation cup helps in reducing the chemical time through enthalpy exchange. At low velocities (<2 m/s), the flame stabilizes near the first step for all the cup lengths. However, at high inlet velocities (=10 m/s), the flame stabilizes at the first and third steps for 50 and 10 mm lengths of the recirculation cup, respectively. As the heat recirculation length increases, the laminar flame speed also increases due to rise in preheating temperature. So, the increase in laminar flame speed shifts the flame upstream.

The variation of preheating temperature along the axial distance for different HC_L is shown in Fig. 7. It can be identified from Fig. 7 that the reactant preheating temperature increases with increase in recirculation cup length for all the velocities

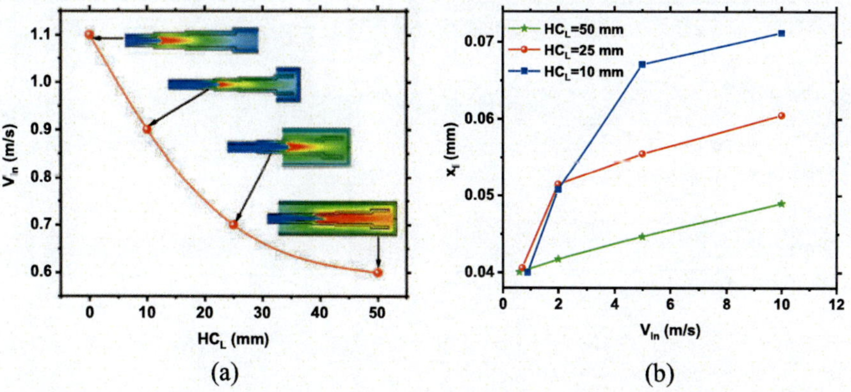

Fig. 6 Variation of **a** lower flame stability limit, **b** flame position for different heat recirculation cup lengths

Fig. 7 Variation of temperature profile along the axial distance for different heat recirculation cup lengths

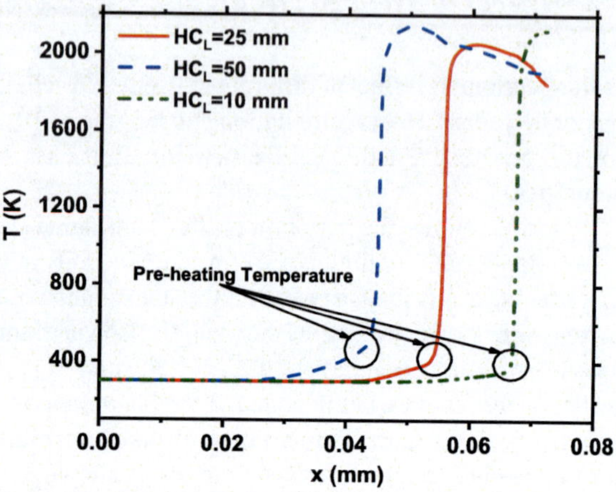

investigated. However, it is also noted that this effect is stronger for higher inlet velocities. Initially, the reactant, at a constant temperature of 300 K, flows through the combustor, and then, a gradual rise in reactant temperature is observed due to wall preheating through conjugate heat transfer. Further, a sudden surge in the temperature profile is seen which indicates the location of the flame.

The variation of the temperature for different lengths of the heat recirculation cup for different velocities is illustrated in Fig. 8. The temperature on the heat recirculation cup has shown to have a gradual rise from the combustor exit to cup exit direction. The temperature of the recirculation cup wall increases with an increase in inlet velocity. There is a considerable difference in the cup wall temperature between $HC_L = 50$ mm and $HC_L = 25$ mm cases, especially at higher velocities. This is perhaps caused by the enhanced residence time of the exhaust gases inside the cup.

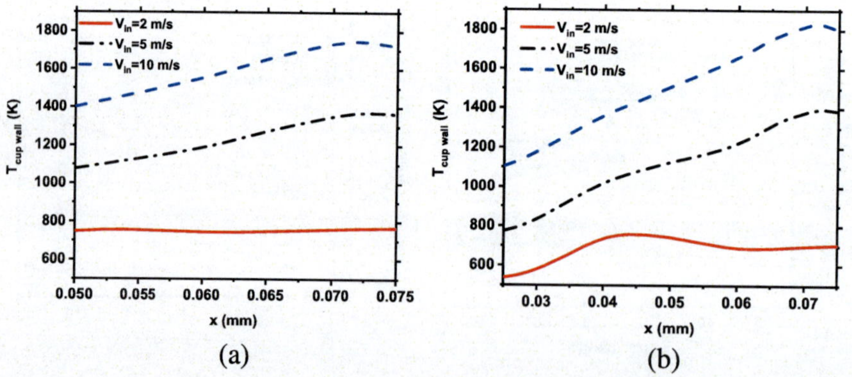

Fig. 8 Variation of cup wall temperature along the distance from cup front **a** for a combustor with heat recirculation cup of 25 mm, **b** for a combustor with recirculation cup of 50 mm

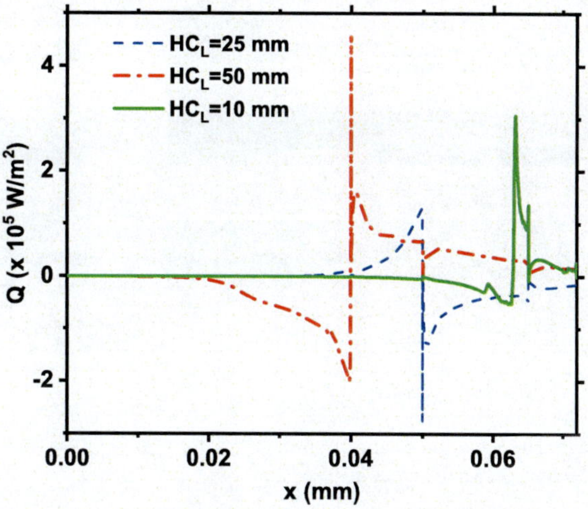

Fig. 9 Variation of heat flux along the interior wall of combustor for various heat recirculation cups

The variation of heat flux for different cup lengths is shown in Fig. 9. Heat flux is the amount of heat transmitted per unit area per unit time from or to a body. The figure shows a rapid increase in the heat flux at different steps for combustors with different recirculation cup lengths. For the 50-mm recirculation cup, the sudden increase in heat flux is formed at the first step, whereas for the 25- and 10-mm recirculation cup lengths, the rapid heat flux is observed at the second and third step, respectively. The sudden increment in the heat flux gives the location of the flame front in the combustor. The figure shows both positive and negative variations of heat flux along the interior wall of the combustor. Negative heat flux implies that the direction of the heat transmission is from fluid to wall, and positive heat flux refers to heat transmission from the wall to the fluid. So, at the location of stabilized flame, the temperature of wall is lesser than the fluid temperature. Thus, heat transfer takes place from fluid to wall. Hence, a negative heat flux is generated.

Variation of the temperature profile for different heat recirculation lengths is studied for 2 and 10 m/s velocities in Fig. 10. It can be noticed that at same velocities, flame in $HC_L = 50$ mm case shifts more toward the inlet. This can be attributed to the enhanced contact time of the exhaust gases with the inner wall of the cup. Also, it can be seen that the flame is strong in case of high velocities for longer cup lengths because the temperature of the flame is more when compared with the shorter heat recirculation cups.

4 Conclusion

In the present study, flame stabilization mechanism in a backward-facing three-stepped microcombustor with heat recirculation cup is investigated in detail. Initially, the performance of the microcombustor without heat recirculation cup is analyzed.

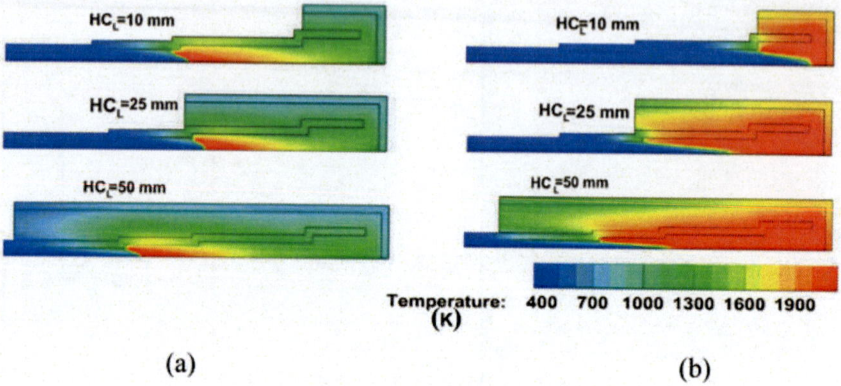

Fig. 10 Variation of temperature profile for heat recirculation cup lengths of 10 mm, 25 mm, and 50 mm, respectively, at **a** $V_{\text{in}} = 2$ m/s, **b** $V_{\text{in}} = 10$ m/s

Various parameters such as flame position and wall temperature are noted. Further, a heat recirculation cup is mounted on the top of the combustion and showed a positive impact on the flame stabilization due to the mixture preheating effect. It is noticed that an increase in temperature of 200 K is achieved with the help of heat recirculation. In the second stage, a thorough study on the effect of heat recirculation cup length on flame stabilization is performed. It is observed that the preheating temperature rises with the increase in the cup length due to the enhanced conjugate heat transfer. At low velocities ($\leq$2 m/s), the flame stabilizes at the first step for all the cases. On the other hand, flame is found to stabilize at the first, second, and third step for 50 mm, 25 mm, and 10 mm, respectively, for higher velocities (>5 m/s). The variation of cup wall temperature also indicates high temperature for longer heat recirculation cups due to the increase in residence time of exhaust gases in those long recirculation cups. For 10 m/s, maximum temperature of 1748.8 and 1831.8 K is observed for a recirculation cup length of 25 mm and 50 mm, respectively. A rapid rise in heat flux is observed at the location of stabilized flame.

References

1. Fernandez-Pello AC (2002) Micropower generation using combustion: issues and approaches. Proc Combust Inst 29(1):883–899
2. Yadav S et al (2014) A prototype micro-thermoelectric power generator for micro-electromechanical systems. Appl Phys Lett 104(12):123903
3. Aravind B et al (2018) Compact design of planar stepped micro combustor for portable thermoelectric power generation. Energy Convers Manag 156:224–234
4. Aravind B, Khandelwal B, Kumar S (2018) Experimental investigations on a new high intensity dual microcombustor based thermoelectric micropower generator. Appl Energy 228:1173–1181

5. Kumar S, Maruta K, Minaev S (2007) Experimental investigations on the combustion behavior of methane–air mixtures in a micro-scale radial combustor configuration. J Micromech Microeng 17(5):900
6. Khandelwal B, Deshpande AA, Kumar S (2013) Experimental studies on flame stabilization in a three step rearward facing configuration based micro channel combustor. Appl Therm Eng 58(1–2):363–368
7. Kim NI et al (2005) Flame stabilization and emission of small Swiss-roll combustors as heaters. Combust Flame 141(3):229–240
8. Ahn J et al (2005) Gas-phase and catalytic combustion in heat-recirculating burners. Proc Combust Inst 30(2):2463–2472
9. Maruta K et al (2005) Characteristics of combustion in a narrow channel with a temperature gradient. Proc Combust Inst 30(2):2429–2436
10. Li J, Zhong B (2008) Experimental investigation on heat loss and combustion in methane/oxygen micro-tube combustor. Appl Therm Eng 28(7):707–716
11. Pan J et al (2007) Effects of major parameters on micro-combustion for thermophotovoltaic energy conversion. Appl Therm Eng 27(5–6):1089–1095
12. Ronney PD (2003) Analysis of non-adiabatic heat-recirculating combustors. Combust Flame 135(4):421–439
13. Yan Y et al (2015) Numerical study of the geometrical parameters on CH4/air premixed combustion in heat recirculation micro-combustor. Fuel 159:45–51
14. Kim K, Kwon O (2008) Studies on a two-staged micro-combustor for a micro-reformer integrated with a micro-evaporator. J Power Sources 182(2):609–615
15. Khandelwal B, Sahota GPS, Kumar S (2010) Investigations into the flame stability limits in a backward step micro scale combustor with premixed methane–air mixtures. J Micromech Microeng 20(9):095030
16. Zarvandi J, Tabejamaat S, Baigmohammadi M (2012) Numerical study of the effects of heat transfer methods on $CH_4/(CH_4 + H_2)$-AIR pre-mixed flames in a micro-stepped tube. Energy 44(1):396–409
17. Peng Q et al (2019) Combustion characteristics and thermal performance of premixed hydrogen-air in a two-rearward-step micro tube. Appl Energy 242:424–438
18. Schoegl I, Sauer VM, Sharma P (2019) Predicting combustion characteristics in externally heated micro-tubes. Combust Flame 204:33–48
19. Fluent A (2009) 12.0 Theory guide. Ansys Inc. 5(5)
20. Singh AP et al (2015) Numerical investigations of unsteady flame propagation in stepped microtubes. RSC Adv 5(122):100879–100890

Computational Analysis of Intake Manifold Design Variants on Induction Swirl of Single-Cylinder Diesel Engine

Dhananjay G. Thombare, Vivek V. Ghare, and S. A. Dunung

1 Introduction

Small internal combustion engine is widely used for many power generation and agricultural applications over the years because of its reliability. The engine intake and compression process influence incoming air flow pattern and in-cylinder swirl. The incoming air flow pattern and in-cylinder swirl during these processes affect mixing of fuel with air during injection. High turbulent kinetic energy of an air increases the mixing of fuel with air and improves vaporization process. The swirl created inside the engine cylinder is helping to break the molecular chain of the fuel and mix up fuel effectively with air. The process of improving the swirl motion in-side cylinder improves the separation, evaporation and diffusion of fuel droplets. Good mixing of fuel improves rate of fuel vaporization and ultimately improves the combustion performance.

Many investigators have studied the effect of intake manifold design on in flow air characteristics like pressure, velocity, swirl ratio and turbulent kinetic energy. Researchers have tried various methods to improve air flow inside the cylinder. Air in-cylinder flow characteristics depend on intake manifold size, shape and geometry of combustion chamber and valve profile. Robert et al. [1] have analysed the effect of tumble and swirl air motion developed inside the engine cylinder on fuel vaporization.

D. G. Thombare (✉) · V. V. Ghare
Automobile Engineering Department, Rajarambapu Institute of Technology, Islampur, Shivaji University, Kolhapur, 415414, Sangli, Maharashtra, India
e-mail: dgtrit@gmail.com; dgtrit@yahoo.co.in

V. V. Ghare
e-mail: royaldr.v@gmail.com

S. A. Dunung
Research & Development Department, Rocket Engineering Corporation Pvt. Ltd., Shiroli MIDC, Kolhapur 416005, India

A. K. Gupta et al. (eds.), *Advances in IC Engines and Combustion Technology*, Lecture Notes in Mechanical Engineering,
https://doi.org/10.1007/978-981-15-5996-9_69

They have made analysis by using visualization techniques. Paul and Ganesan [2] have theoretically investigated the effect of helical- and spiral-shaped manifolds on air motion and flow pattern in a diesel engine by using CFD tool. Helical spiral manifold has shown promising improvement. Suleiman et al. [3] have investigated design variants of an intake manifold. A surge tank with taper edges and bell mouth design of intake has improved the flow coefficient by 6%.

Krishna et al. [4] studied the effect of intake manifold orientation on air flow characteristics. They have carried out visual inspection with the help of particle image velocimetry. These results show that the manifold of 60° inclination gives better results. Saad and Bari [5] have improved the air swirl to maximize the air fuel mixing by using guide vanes and tumble devices for viscous fluid. Different devices were tested with ANSYS-Fluent software to optimize the guide vane design. From the simulation results, the 35° twist angle model gives better results than base engine HINO model. Rahiman et al. [6] investigated the effect of helical and spiral manifold configuration on the motion of air inside the engine cylinder by using CFD tool. Flow patterns of these manifold configurations were studied, results of experiment show that spiral-shaped and helical-spiral-shaped manifolds have higher swirl than a conventional inlet manifold. The engine volumetric efficiency was observed to be higher for helical manifold. Perini et al. [7] developed new intake manifold having two-throttle restricted ducts with same shape. The throttle valve design was fitted between the cylinder head and intake runner. The shape of the throttle body changed to helical port throttle and tangential port throttle. The new intake design gives swirl ratio 2.2 which improved flow quantity and tangential velocity profile. Also flow field in a cylinder is given better effect on local mixture temperature. Raj Kumar et al. [8] have used three different types of vane profile installed at downstream of an intake manifold. From the experiment, they observed that the performance of an engine is improved. About 9.24% higher brake thermal efficiency is obtained by guide vane-fitted manifold. Holkar et al. [9] have predicted and analysed the flow-through intake manifold and inlet port system using CFD. Flow is highly three-dimensional, and it is depended on valve lift and port size. The simulation studies made by Perini et al. [7], Saad and Bari [5] and Sun et al. [10] have shown that the simulation methodology used for in cylinder simulation of air flow is valid and results of simulation are matching to the actual results.

Finding of various researchers' shows that the design of intake manifold of its size, shape and geometry influences the flow characteristics of incoming air flow in the cylinder. The in-cylinder air flow characteristics influence swirl of incoming air. The use of guide vanes in the intake manifold close to intake valve induce swirl motion to incoming air and continues this swirl inside the cylinder. The position of the guide vanes should be inside the intake manifold close to the intake port. The research carried out is limited to theoretical study and that too for large diesel engines; however, effect on small constant speed engines is out of research focus. So there is a need to investigate effect of design variants of guide vanes on small constant speed engine performance and emission. In this research work, the effect of guide vanes and its design variants on a small single cylinder diesel engine is carried out by engine simulation using computational fluid dynamics software. The

Table 1 Technical specification of the engine selected for investigation

Engine parameters	Details
Engine model	8 HP, Comet single cylinder, naturally aspirated
Speed	2600 rpm
Stroke and bore	80 and 87.5 mm
Intake diameter	41 mm
Compression ratio	16.7:1
Exhaust valve open	35°30′ BBDC
Exhaust valve close	4°30′ ATDC
Inlet valve open	4°30′ BTDC
Inlet valve close	35°30′ ABDC
Piston type	Hemispherical type
Start of injection	27° BTDC

designed and optimized guide vanes are manufactured as per requirement, and the experimental investigations are carried out on engine test set-up. The optimized guide vanes with modified intake manifold are tested for engine performance like BSFC, thermal efficiency, power and emission. The theories of combustion [11, 12] also shows that the induction swirl can have positive influence on emission reduction.

2 Base Model of Diesel Engine

The geometry of the Comet single-cylinder diesel engine with the engine specifications is generated as base model. The technical specifications of the engine are given in Table 1. Both valves are set as solid domains as they are available in real domain, and the other domains are set as fluid domains to compute the fluid flow. ANSYS-Mesh is set to generate tetrahedral cell to allow very fine mesh with moving mesh for simulation. By doing the tetrahedral mesh, the complicated shape is finely covered by elements to get better simulation results.

3 Guide Vane Model

The work is aimed to develop air swirl motion by placing the guide vanes inside the intake manifold. An intake manifold is designed with guide vane devices, and engine flow simulation is carried out to analyse air characteristics. The guide vanes are designed as dimension given in Table 2. The guide vanes outside dimensions are kept constant only the number of guide vanes, and angle of inclination is changed.

Table 2 Specification of guide vane device

Particulars	Specification	
Intake port radius (R)	14.5 mm	
Number of vanes (n)	Configuration a	Two guide vanes with 180° to each other
	Configuration b, c	Three guide vanes with 90° and 120° to each other
	Configuration d, e	Four guide vanes with 90° and 180° to each other
Vane length (L)	$3 \times R = 43.5$ mm	
Vane height (H)	$0.2 \times R = 2.9$ mm	

Figure 1 illustrates the sample design with model numbers a, b, c, d, e used for analysis.

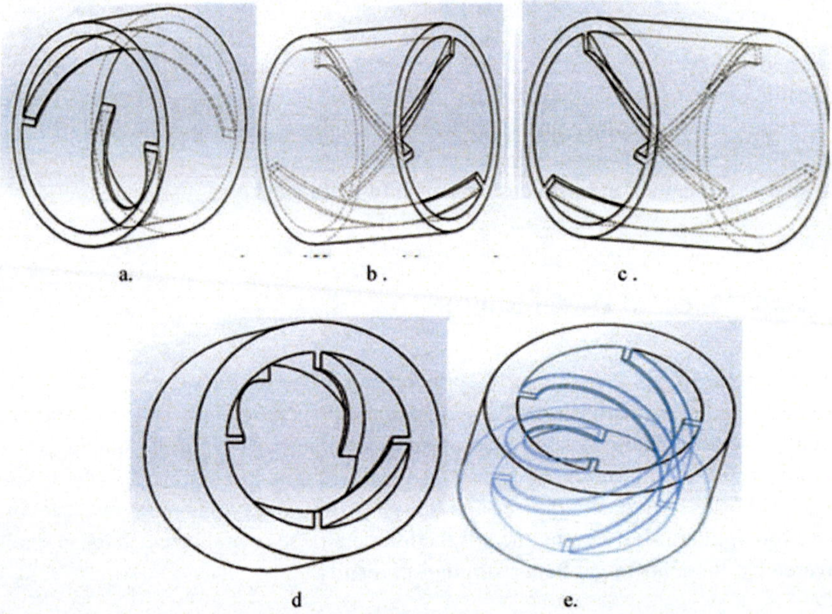

Fig. 1 Developed guide vane model used in simulation. **a** Two guide vanes with 180° helix angle. **b** Three guide vanes with 120° helix angle. **c** Three guide vanes with 90° helix angle. **d** Four guide vanes with 90° helix angle. **e** Four guide vanes with 180° helix angle

4 Simulation Results

With the help of guide vanes inside the intake manifold of the engine, simulation is carried out. The simulation result shows that there is an improvement in air flow characteristics. The simulation results for four guide vanes having 180° helix angle show good results than other guide vanes devices. From the simulation results, maximum swirl ratio achieved as 0.8, whereas for base model, it was 0.58. Similarly, tumble ratio for guide vane is computed as 0.4, whereas for base model it was 0.38 (Fig. 2).

The turbulent kinetic energy (TKE) is required to break the fuel droplets effectively. The maximum turbulent kinetic energy of 120 kg m/s^2 is achieved for guide vane model, whereas for base model it was attained to 95 kg m/s^2. The velocity of air improves the flow rate of incoming air inside the engine cylinder. So maximum air velocity achieved for guide vane is 16 m/s, whereas for base manifold it is 14 m/s. The simulation result shows that the four guide vanes having 180° helix angle devices give better results than base model results.

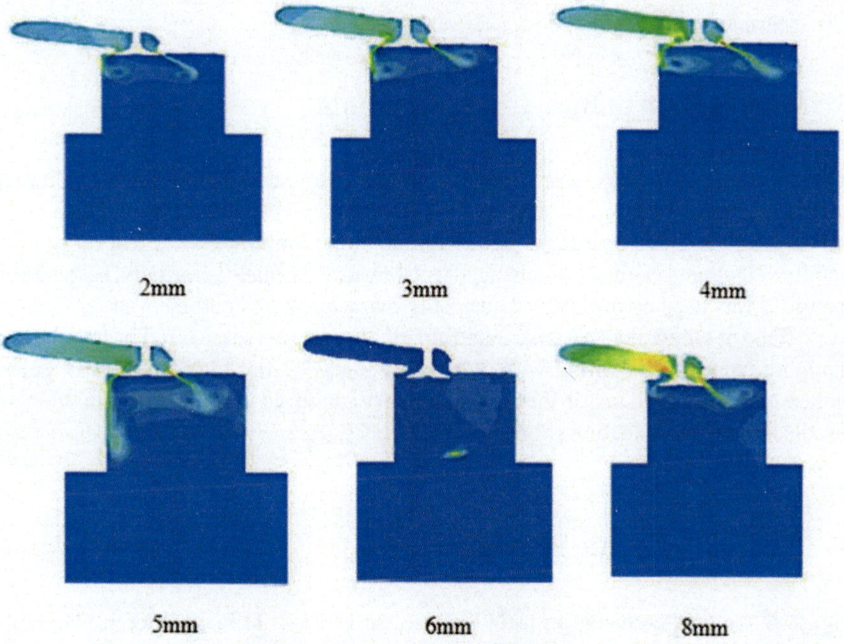

Fig. 2 Step-wise flow field development inside engine cylinder

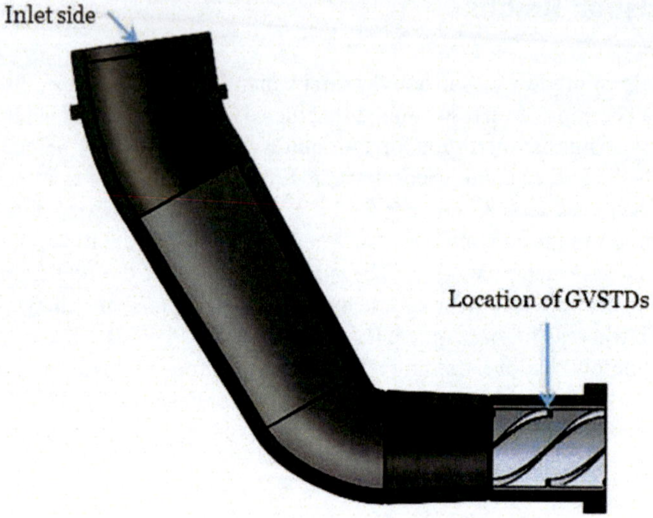

Fig. 3 Modified intake manifold with guide vane-fitted

5 Modification in Base Intake Manifold

The inclination of intake manifold reduces the mass flow rate of air inside the cylinder, which has direct effect on air flow quality. The research carried out by Krishna et al. [4] on effect of intake manifold inclination on induction flow shows that 60° intake manifold inclination gives promising effects on air flow quality. Initially, base intake manifold was used elbow (90°) shape. This shape has high coefficient of resistance. So, instead of elbow shape, intake manifold inclination was set as 60°. Figure 3 shows the modified intake manifold with guide vane placed. After modification, all guide vane models are fitted inside the modified intake manifold to carry out simulation as per discussed above section (Figs. 4 and 5).

6 Turbulent Kinetic Energy

Figure 6 shows the change in turbulent kinetic energy (TKE) with crank angle at peak engine speed with base model and manifold with guide vanes. It is observed that the inlet manifold configuration influences the level of fluid turbulence in the cylinder. Turbulence approaches to peak value at maximum valve open situation.

The variation of turbulent kinetic energy may be due to amount of air induced through the inlet manifold. The dissipation of kinetic energy is account of increased fluid motion. The maximum turbulent kinetic energy achieved for four guide vanes

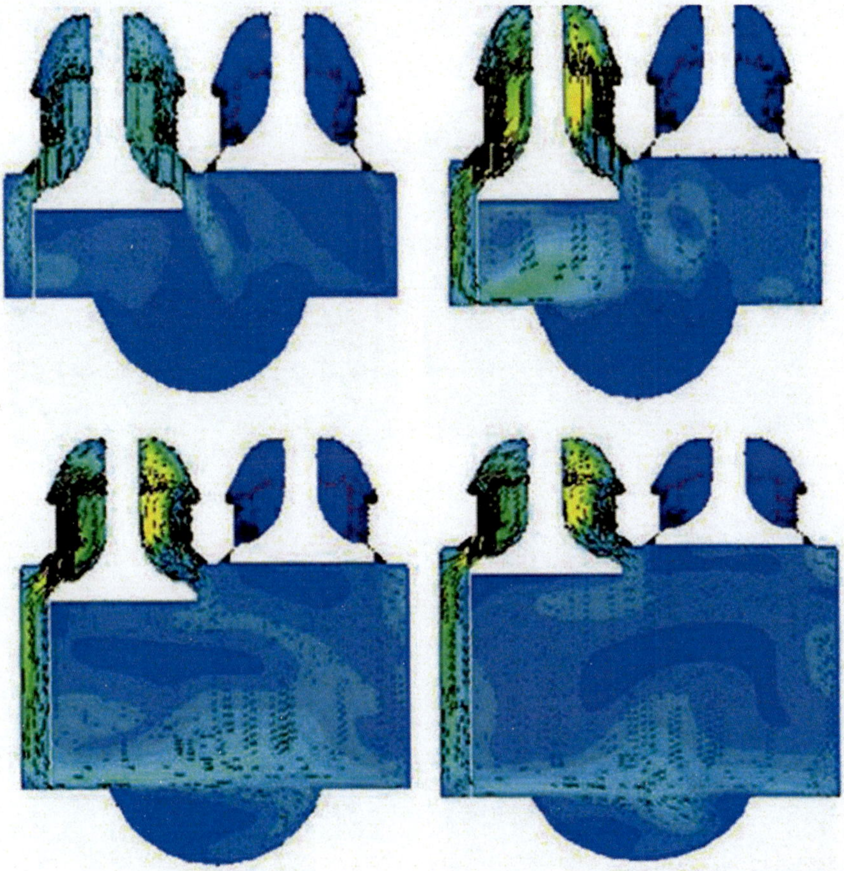

Fig. 4 Flow of air through modified manifold with guide vanes during suction

having 180° helix angle (e) is 145 kg m/s^2, whereas for base model it gives 100 kg m/s^2.

7 Swirl-and-Tumble Ratio

Swirl ratio is defined as ratio of circumferential air speed in the cylinder to the axial speed of air flow in the cylinder. Figure 7 shows the swirl ratio with respect to crank angle for all guide vane models and modified model. The swirl ratio is increasing as valve lift is increasing. The swirl ratio is increasing as valve lift is increasing. When the valve lift is at maximum, the swirl ratio is at peak value. After that again reduces linearly. The four guide vanes having 180° helix angle (e) give maximum swirl ratio

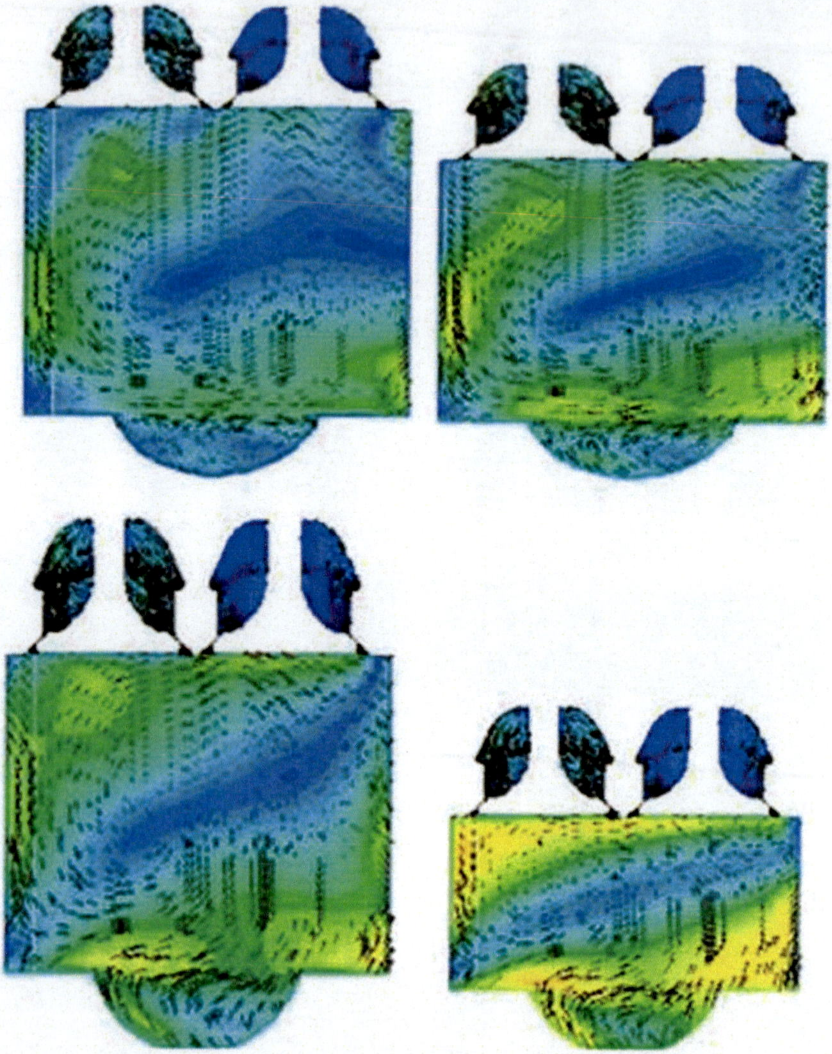

Fig. 5 Flow of air through modified manifold with guide vanes during compression

1.35, whereas base model gives only 0.9 values. By placing the guide vanes device inside the intake manifold, it improves swirl-and-tumble motion of inducting air.

Figure 8 shows the effect of guide vane and its arrangement on tumble ratio. Four guide vanes with 90° helix angle give highest tumble ratio especially at 380° of crank angle.

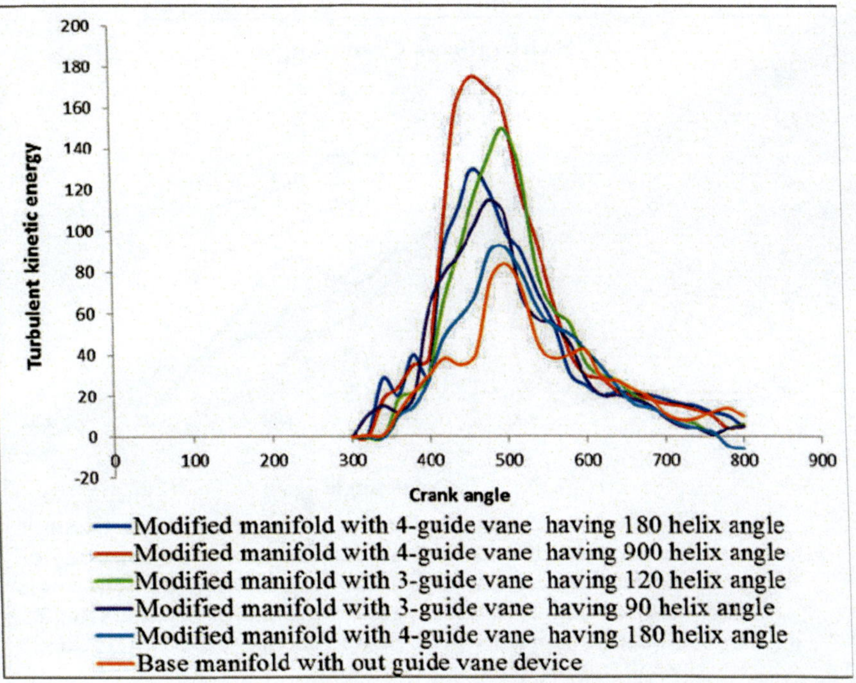

Fig. 6 In-cylinder turbulent kinetic energy for modified manifold with guide vanes models

8 In-Cylinder Velocity and Pressure

The maximum air velocities are plotted in Fig. 9. Wherein the velocity increased till 360° of crank angle and declined after 360°. This trend is occurring because of the movement of the piston who changes the cylinder volume. It is observed that the velocity of air and number of vane do not have linear relationship. It is observed from Fig. 9 that four guide vanes with 180° helix angle (e) configuration has the highest maximum velocities amongst all models. The maximum velocities are occurring after 330° of crank angle. Other guide vane models are not showing much of the rise in velocity. The in-cylinder pressure is plotted against crank angle in Fig. 10. Pressure inside the cylinder at the beginning of compression stroke goes below atmospheric pressure, and latter the pressure goes on increasing. Maximum in-cylinder air pressure is attained at 72 bar, whereas for base model it reaches to 65 bar.

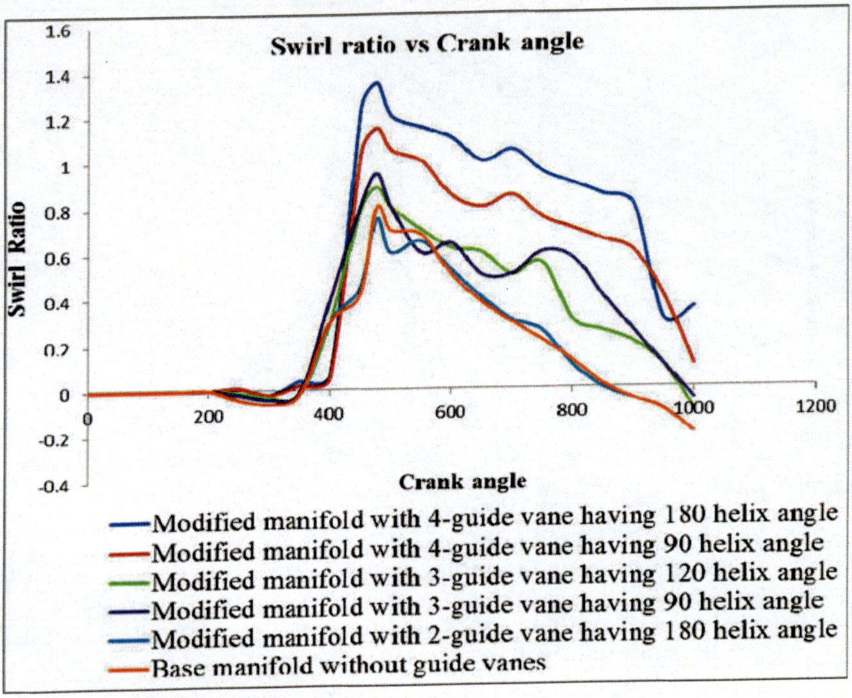

Fig. 7 Swirl ratio for modified intake manifold with guide vanes models

9 Optimum Guide Vane

The analysis of simulation results in terms of turbulent kinetic energy, in-cylinder velocity, swirl-and-tumble ratio shows that four guide vanes having 180° helix angle (e) have highest turbulent kinetic energy as well as highest swirl ratio. The high turbulent kinetic energy and high swirl ratio achieved after 347° of crank angle. As compared to base model, these four guide vanes having 180° helix angle (e) give best results.

Based on above results, four blades with 180° helix angle (e) are optimum design to achieve better in-cylinder air flow characteristics than other models.

10 Experimental Set-Up

To evaluate performance of designed guide vanes on engine experimentally, a test set-up is developed in the laboratory. Figure 10 shows the schematic diagram of various instruments used for measurement of parameters on engine. A volumetric fuel measurement system is used with a stopwatch to record the fuel consumption.

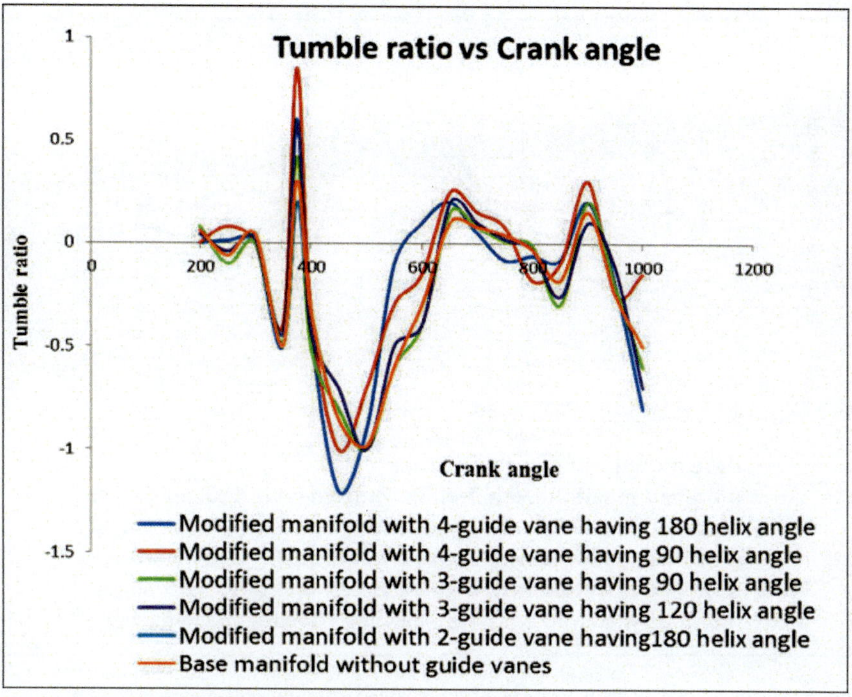

Fig. 8 Tumble ratio for modified intake manifold with guide vane models

Two thermocouples with accuracy of 0.05% is used to measure the exhaust gas temperature and engine oil temperature. A digital tachometer is used to record the engine speed. An AVL makes DiTEST1000 exhaust gas analyser used to measure CO, CO_2, O_2, HC and NO_X with an accuracy of ±0.05%, ±0.25%, ±0.02%, ±0.005 and ±0.025, respectively. The engine used for experimentation is of constant speed engine, and the load was varied from minimum to maximum of 20.8 kN at full power (Fig. 11).

11 Experimental Results and Discussion

a. **Brake-Specific Fuel Consumption (BSFC)**
The effect of guide vanes on brake specific fuel consumption is shown in Fig. 12. Brake-specific fuel consumption of the engine decreases with increase in load on engine. With respect to change in load, there is small variation in speed of engine. The base engine manifold has higher BSFC over modified system with guide vanes. The average fuel consumption of 0.263 kg/kW h is observed for base model, and for modified vane manifold, it is observed at 0.256 kg/kW h.

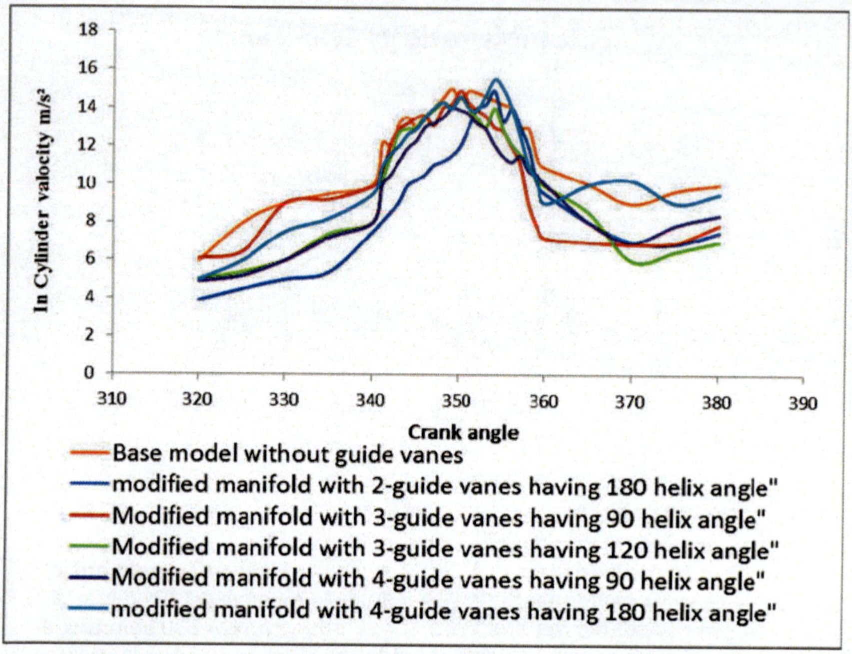

Fig. 9 Maximum in-cylinder velocity of air for modified manifold with guide vanes

This indicates that brake-specific fuel consumption is less for modified intake with guide vane model. The drop in fuel consumption may be mainly because of improvement in swirl which might have lead to better utilization of fuel and improvement in combustion of fuel. The variation of brake thermal efficiency for base model and modified model with vanes at different engine rpm is shown in Fig. 13. It is observed that the brake thermal efficiency is slightly improved for the modified model with vanes about 1–2%.

b. **Exhaust Gas Temperature**

Exhaust gas temperature for base model and for modified guide vane model is shown in Fig. 14. The exhaust gas temperature is indication of heat release during combustion. The exhaust gas temperature is higher for modified guide vane model which may be because of improved air fuel mixing and combustion. The exhaust temperature has effect of in-cylinder pressure during combustion which influences expansion work of engine. According to Fig. 14, the exhaust temperature at different load and speed is increased. For base model, exhaust gas temperature is observed from 388 to 395 °C, whereas for modified model with vanes is increased from 392 to 395 °C.

c. **Volumetric efficiency**

During the experimental work, the air flow to the engine is measured. It is observed that the volumetric efficiency is improved from 71 to 76%. This rise in volumetric efficiency is mainly because of modification in intake pipe from air

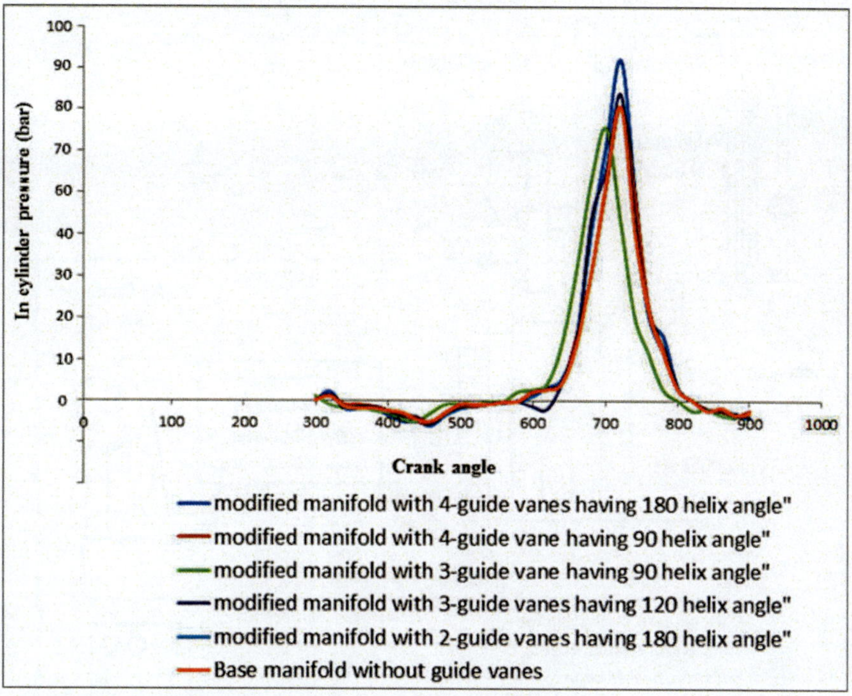

Fig. 10 Maximum in-cylinder pressure of air for modified manifold guide vanes models

cleaner to intake port. The fluid friction in the tube is reduced because of free flow of air in pipe. This has directly improved the mass flow rate of air to engine. As mass flow rate of air increases, the engine performance improves.

d. **HC Emission**

Increase in swirl level has shown significant improvement in the combustion process and exhaust emission. Figure 16 shows the variation of hydrocarbon emission for base model and modified model with guide vanes. Unburned hydrocarbon emission is caused by incomplete combustion of fuel and air mixture. Modified model with guide vanes has lower HC emission at all operating conditions of engine. The unburned hydrocarbons (UBHC) of guide vane model are between 4 and 10 ppm, whereas for base model it is varying from 14 to 18 ppm.

e. **CO emission**

Carbon monoxide occurs in the engine exhaust which is the result of incomplete combustion. In Fig. 17, the variation in carbon monoxide with respect to engine speed and load is shown. It is observed that as the load on engine increases, the CO emission increases. CO emission is low for low engine speed and high at high engine speed in both the cases; however, the level of CO emission is lower in modified model with guide vanes which are lower under all operating conditions. The CO emission for guide vane device model is 0.07, whereas for base model it is 0.14.

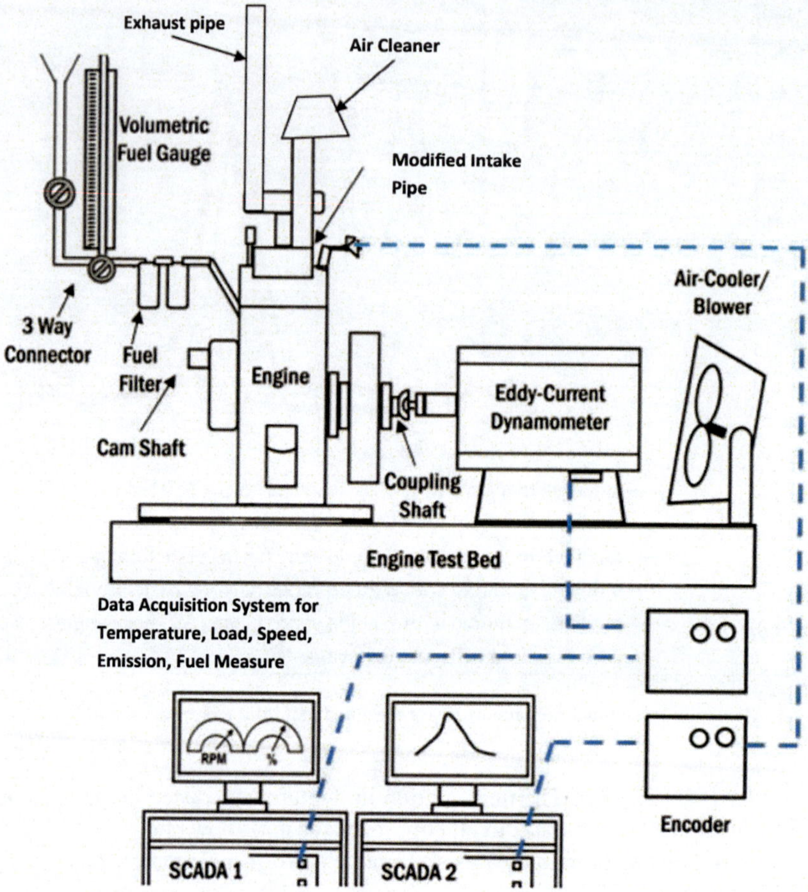

Fig. 11 Schematic diagram of engine test set-up with instrumentation

f. **NO_X emission**
Figure 18 gives the oxidation of nitrogen from the engine exhaust at different engine speeds and loads. NO_X results from reaction of nitrogen with oxygen at relatively high temperature. NO is major component in NO_x emission. As engine speed increases, the NO_X also increases. The NO_X emission for guide vanes model is about 230 ppm at constant load 20.8 Nm, whereas for same load with base model gives 320 ppm NO_X.

12 Conclusion

The results of simulation show that the four guide vane having 180° helix angle devices gives 1.35 swirl ratio, whereas for base model it is observed as 0.58. There

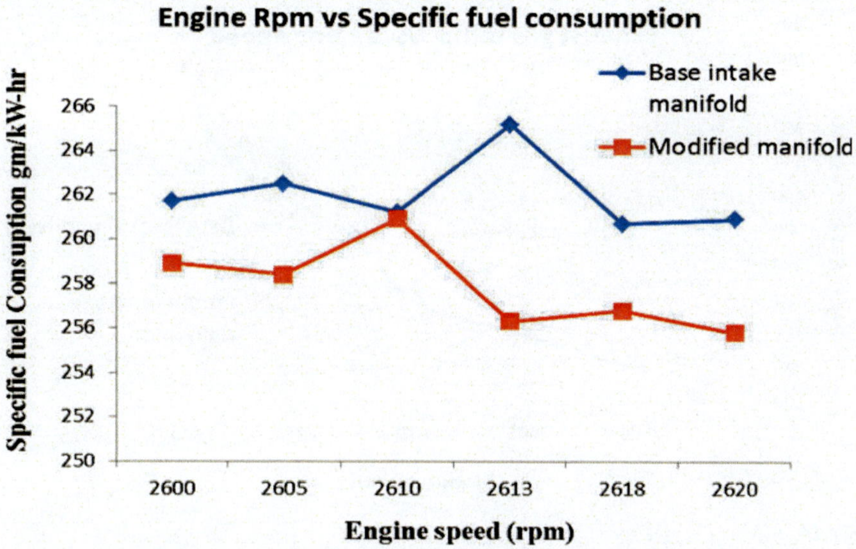

Fig. 12 Brake-specific fuel consumption of base model and modified model with vanes

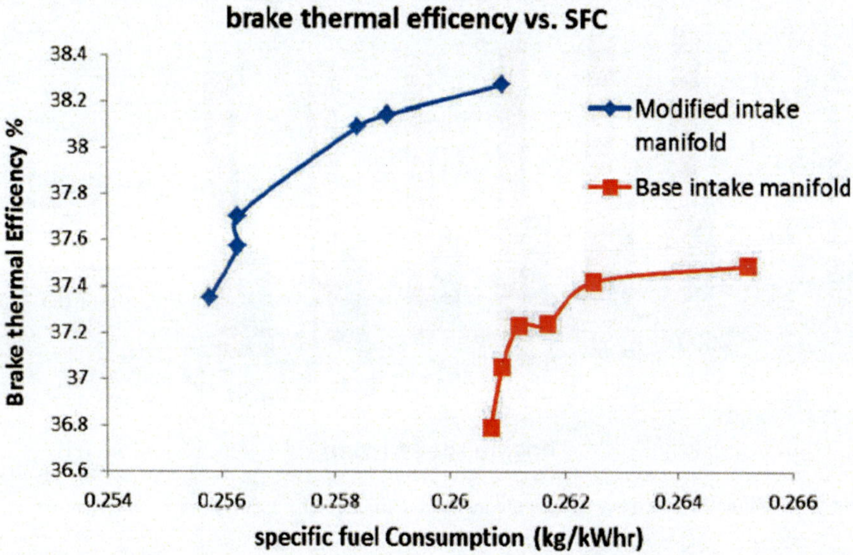

Fig. 13 BSFC versus brake thermal efficiency with and without modified manifold

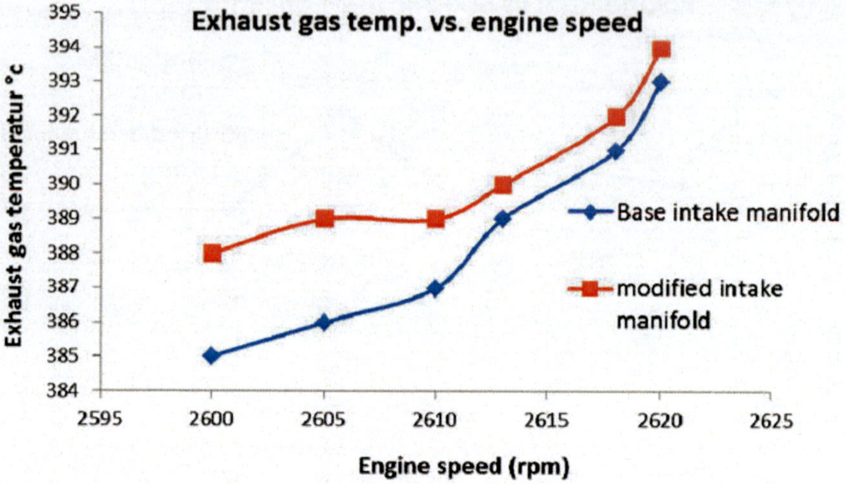

Fig. 14 Effect of engine speed on exhaust gas temperature

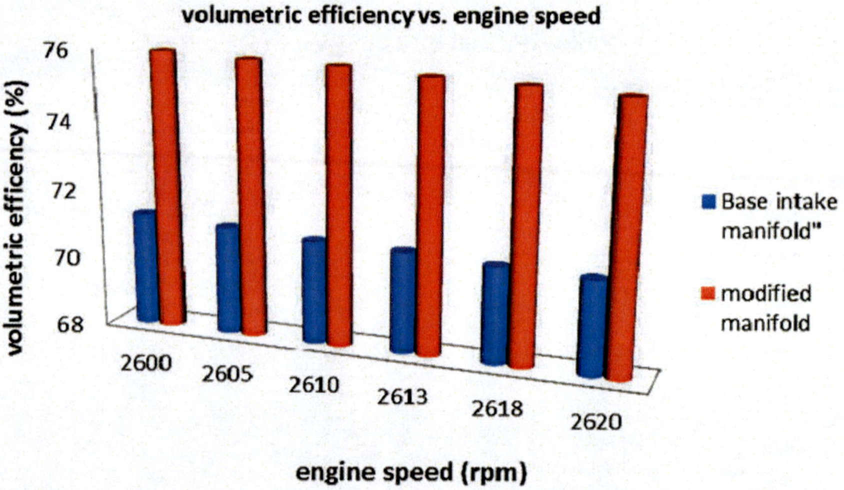

Fig. 15 Volumetric efficiency of base model and modified intake manifold with guide vanes

is significant increase in swirl ratio by use of guide vanes in the intake manifold. The experimental results show that the brake thermal efficiency of modified manifold with guide vane is increased by 1–2% as compared to base model. The exhaust gas temperature is achieved more for modified intake manifold over base model. The break-specific fuel consumption for base model is 0.263 kg/kW h, and for modified intake manifold with four guide vane gives 0.257 kg/kW h. This indicates that fuel consumption is reduced for modified manifold with guide vanes. Volumetric

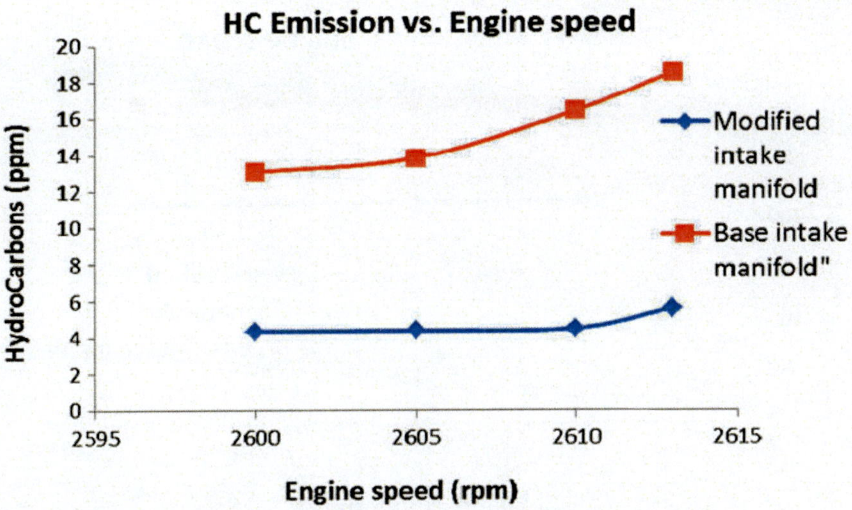

Fig. 16 Effect of engine speed on HC emission for both intake manifolds

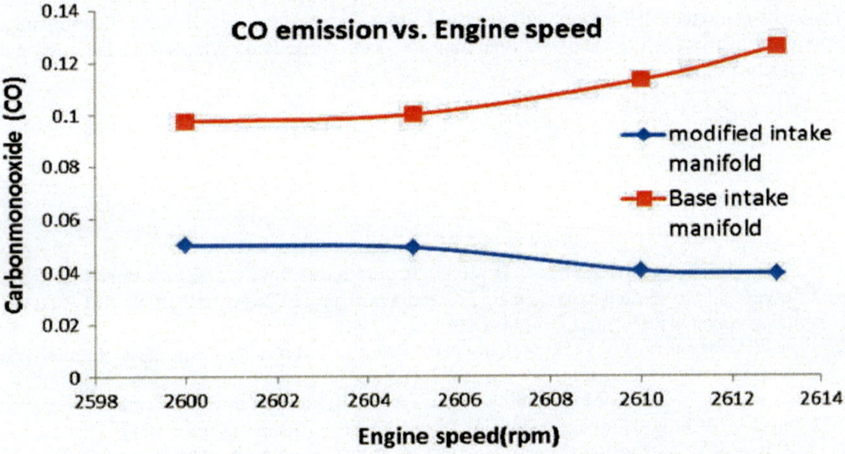

Fig. 17 Effect of engine speed on CO emission for both intake manifolds

efficiency of modified intake manifold is also increased by 4% than base intake manifold. Specific fuel consumption of fuel is reduced by 0.02%. The brake thermal efficiency increases by 0.78% than existing manifold engine. There is reduction in CO emission level as compared to base engine by 30%, whereas HC emission is reduced by 20% and NO_X is reduced by 28% in modified intake manifold than base model. To conclude, the use of guide vanes in intake manifold of small constant speed diesel engine is beneficial by all means.

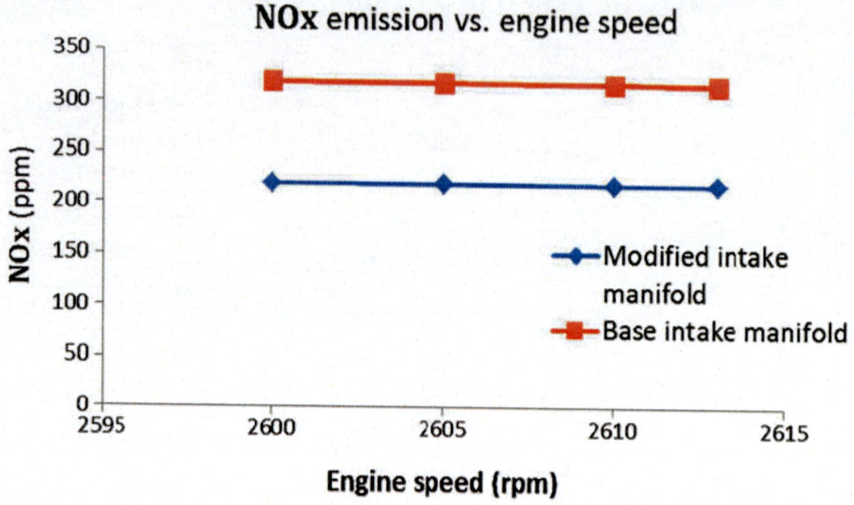

Fig. 18 Effect of engine speed on NO_X emission for both intake manifolds

Acknowledgements The authors are thankful to Management and Research and Development Department of Rocket Engineering Corporation Pvt. Ltd., for providing laboratory facilities and financial support.

References

1. Laramee RS, Schneider J, Hauser H (2004) Texture-based flow visualization on isosurfaces from computational fluid dynamics. In: Proceedings of joint EUROGRAPHICS—IEEE TCVG symposium on visualization, pp 134–142
2. Paul B, Ganesan V (2010) Flow field development in a direct injection diesel engine with different manifolds. Int J Eng, Sci Technol 2(1):80–91
3. Sulaiman SA, Murad SHM, Ibrahim I, Karim ZA (2010) Study of flow in air-intake system for a single-cylinder go-kart engine. Int J Autom Mech Eng (IJAME) 1(1):91–104
4. Krishna BM, Mallikarjuna JM (2011) Effect of engine speed on in-cylinder tumble flows in a motored internal combustion engine—an experimental investigation using particle image velocimetry. J Appl Fluid Mech 4(1):1–14
5. Saad I, Bari S (2013) Improving air-fuel mixing in diesel engine fuelled by higher viscous fuel using guide vane swirl and tumble device (GVSTD). SAE Technical Paper, pp 1–6
6. Rahiman A, Abdul Razak RK, Mohammad Samee AD, Ramis MK (2014) CFD analysis of flow field development in a direct injection diesel engine with different manifolds. Am J Fluid Dyn 4(3):102–113 12
7. Perini F, Miles PC, Reitz RD (2014) A comprehensive modeling study of in-cylinder fluid flows in a high-swirl, light-duty optical diesel engine. Comput Fluids 105:113–124
8. Kumar AR, Raju GJ, Reddy KH, Chakrapani M Experimental investigation of an airenhancement device on the performance of a compression ignition engine
9. Holkar R, Sule-Patil YN, Pise SM, Godase YA, Jagadale VS (2015) Numerical simulation of steady flow through engine intake system using CFD. IOSR J Mech Civ Eng (ISOR-JMCE) 12(1):30–45

10. Sun Z, Li X, Du W (2011) Research on swirler for intake induced swirl in DI diesel engine. In: Proceedings of computer distributed control and intelligent environmental monitoring (CDCIEM), international conference, pp 275–279
11. Pundir BP (2007) Engine emissions: pollutant formation and advances in control technology. Alpha Science International, Limited
12. John BH (2011) Internal combustion engine fundamentals. Mc-Graw-Hill Publications, New York, pp 345–353

Computational Fluid Dynamics of Co-axial Unconfined Isothermal Swirling Jets

Arun Pattanashetti and R. Santhosh

1 Introduction

Swirling flows are generally employed in most of the engineering applications such as cyclone separators, gas turbine combustors, and mixing enhancement of two different fluid streams. [1]. The application of the swirl jets in the gas turbine combustors is widespread because strong swirl creates recirculating flow which acts as flame anchoring mechanism near the burner exit [2]. The occurrence of these recirculation zones is due to a phenomenon called 'vortex breakdown', and the flow regime is characterized by the presence of negative axial velocity in the central region. Vortex breakdown generally occurs when swirl number exceeds a critical value. Swirl number is the dimensionless number used to characterize the intensity of swirl that is imparted to the flow field. It is defined as the ratio of the axial flux of angular momentum to the axial flux of linear momentum [3]. At high swirl number, an adverse pressure gradient is set up along the central axis resulting in the formation of the central toroidal recirculation zone (CTRZ). The presence of the large vortical motions within the CTRZ enables enhanced mixing of hot combustion products with incoming reactants and radical species leading to flame stabilization [3].

The co-axial swirling jets are vastly used in industrial furnaces, combustion chambers, cooling systems, and fluid mixing [4]. Werner J. A. Dahm et al. investigated the vortex structures and dynamics of their interactions for incompressible fluid in the near region of the round co-axial jet issuing into the ambient fluid [5]. Ryan B. Wicker et al. conducted an experiment on co-axial swirling jet with sufficient annular swirl for a single flow condition in order to investigate the effect of the flow field on particle motion [6]. Shiferaw Regassa Jufar et al. studied the effect of swirl on the

A. Pattanashetti · R. Santhosh (✉)
Department of Mechanical Engineering, Indian Institute of Technology Dharwad, Dharwad 580011, India
e-mail: santhoshr@iitdh.ac.in

A. K. Gupta et al. (eds.), *Advances in IC Engines and Combustion Technology*, Lecture Notes in Mechanical Engineering,
https://doi.org/10.1007/978-981-15-5996-9_70

flow and mixing characteristics of the acoustically excited co-axial jets and observed that better mixing was found in the near field region for low swirl number. Contrarily for high swirl numbers, better mixing was found in the far field [7].

Two well-known computational methodologies, namely Large Eddy Simulation (LES) and Reynolds Average Navier–Stokes (RANS), are used to predict the swirl type of complex flows. Past works have employed RANS numerical approach to predict NO_x emission in environmental burner and have been successful in obtaining numerical results that were used to optimize the combustor design [8]. In particular, the improvement of the performance of the tri-fuelled low-cost gas turbine has been done by analysing the effect of inner flow field on the flame structure employing RANS [9]. Sofiane Ouali et .al studied the effect of temperature on the formation of the NO_x emission in the context of low swirl burner [10]. Thus, it is seen that computationally less intense RANS (as compared to LES) is a useful technique to simulate swirling flow field.

Different turbulent models have been employed in the prediction of the turbulent flow field. For high swirl number flow fields, it is recommended to use the Reynolds stress turbulent model which predicts the flow field with high accuracy compared to one-equation and two-equation models [11]. RSM closes the unknown Reynolds stress terms in the RANS equation by solving the transport equation for Reynolds stresses [11]. Contrarily for low swirl number flow fields, $k-\varepsilon$ or $k-\omega$ are suggested as suitable models.

The aim of the present work is to employ RANS-based RSM model to simulate unconfined isothermal co-axial swirling jet and study the dependence of flow topology on operating swirl number. The flow conditions simulated here are the same as the flow cases studied (experimentally) in Santhosh et al. [3] which validates the observations reported in the present flow. We attempt to give further insights into the flow field by discussing vorticity and turbulent kinetic energy fields.

This paper is a work-in-progress study and reports numerical results (for the first time) of the dependence of RZ on swirl number.

2 Numerical Approach

2.1 Geometry and Outer Domain Preparation

The geometry of the swirl burner was created using 'SOLID WORKS' modelling software. The geometry consists of two major domains, i.e. co-axial burner and computational domain (Fig. 1a). Among the two, the burner geometry consists of a central pipe, annular pipe, and eight tangential swirl ports (to impart swirl motion to the annulus air) created on the surface of the annular pipe at an angle of 30° to the outward radial vector (Fig. 1b, c). The fluid employed in the present work is air, and its flow into each of the pipes (i.e. central pipe, co-annular pipe, and swirl ports) can be controlled independently. The entrainment of the swirled co-annular

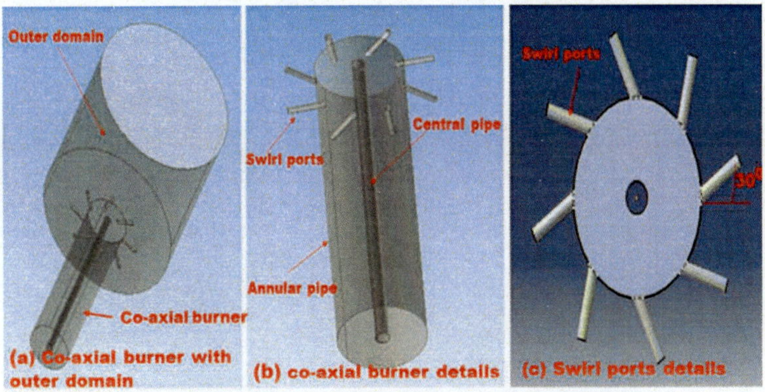

Parameter	Dimensions
Central pipe diameter	12
Annular pipe inner diameter	83.2mm
Swirl port diameter	6mm
Number of swirl ports	8
Angle of swirl port to radial outward	30^0
Outer domain diameter and length	280 & 300mm

Fig. 1 Details of co-axial burner with outer domain employed in the present study

air jet and central air jet into the quiescent fluid needs to be captured numerically. For this purpose, numerical outer domain of sufficient length 7 R (R = Radius of the co-annular burner) and width 3 R is used. Dimensions of the domain were fixed after successive trails to overcome the confinement effects.

2.2 Mesh Creation

ANSYS workbench geometry component is used to name the different parts of the geometry. The discretization of the domain is done using tetrahedron elements (unstructured meshing). In the region of interfacing between two domains, i.e. the interface between the burner geometry and outer domain, a necessarily fine meshing has been done in order to get accurate flow field data. The optimized fine meshing is done only for the outer zone which is the numerical zone of interest. Meshed geometry is shown in Fig. 2. It contains 1,460,000 number of elements and 135,000 nodes. The grid independence test has not been done in the present work but is envisaged as a work for the immediate future.

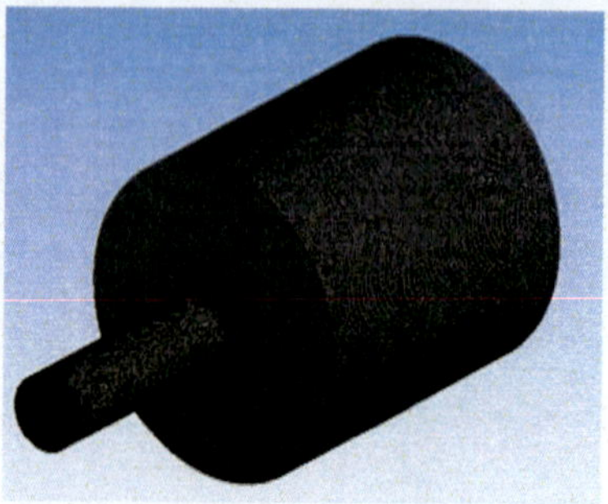

Fig. 2 Generated mesh for outer domain and co-axial burner

2.3 Governing Equations

The unsteady, incompressible 3D turbulent swirl flow fluid is analysed by solving the transport equation for $k - \varepsilon$, RSM, and Navier–Stokes equations. Following conservation equations [2] in indicial notation are considered.

Conservation of mass [12]

$$\frac{\partial u_i}{\partial x_i} = 0 \tag{1}$$

Conservation of momentum [12]

$$\frac{\partial \cup_i}{\partial t} + u_j \frac{\partial u_i}{\partial x_j} = -\frac{\partial p_i}{\partial x_i} + \nu \frac{\partial^2 u_i}{\partial x_j^2} \tag{2}$$

2.4 Turbulence Modelling and Validation

For the high swirl number flows ($S > 0.6$), ANSYS user guide [11] strongly recommends to employ the RSM turbulent model. This is because the RSM model predicts the non-isotropic nature of the flow field accurately as compared to $k - \varepsilon$ and $k - \omega$ models. Contrarily, for low swirl number ($S < 0.6$), $k - \varepsilon/k - \omega$ is found to be a suitable model. The validation of $k - \varepsilon$ and RSM turbulent models were done by comparing numerically simulated cases with experimentally determined axial velocity field (PIV technique) by Santhosh et al. [3]. Equations for $k - \varepsilon$ and RSM are written below.

Turbulent kinetic energy–dissipation equation [6]

$$\frac{\partial k}{\partial t} + u_j \frac{\partial \mathrm{k}}{\partial x_j} = \zeta_{ij} \frac{\partial u_i}{\partial x_j} - \varepsilon + \frac{\partial}{\partial x_j}\left[\left(\nu + \frac{\nu_t}{\sigma_\varepsilon}\right)\frac{\partial k}{\partial x_j}\right] \tag{3}$$

$$\frac{\partial \varepsilon}{\partial t} + u_j \frac{\partial \varepsilon}{\partial x_j} = C_{\varepsilon 1} \frac{\varepsilon}{k} \zeta_{ij} \frac{\partial u_i}{\partial x_j} - C_{\varepsilon 2} \frac{\varepsilon^2}{k} + \frac{\partial}{\partial x_j} \left[\left(\nu + \frac{\nu_t}{\sigma_\varepsilon} \right) \frac{\partial k}{\partial x_j} \right] \quad (4)$$

where ζ_{ij} is Reynolds stress tensor.

RSM Equation [11]

$$\frac{\partial R_{ij}}{\partial t} + u_k \frac{\partial R_{ij}}{\partial x_j} = D_{ij} + P\_ij + \pi_{ij} + \Omega_{ij} - \varepsilon_{ij} \quad (5)$$

First term in LHS	Rate of change of R_{ij}
Second term in LHS	Transport R_{ij} of due to convection
First term in RHS	Transport of R_{ij} due to diffusion
Second term in RHS	Rate of production of R_{ij}
Third term in RHS	Transport of R_{ij} due to turbulent pressure strain interaction
Fourth term in RHS	Transport of R_{ij} due to rotation
Fifth term in RHS	Dissipation of R_{ij}.

2.5 *Boundary Conditions*

The same inlet flow conditions as in the experimental study in Santhosh et al. [3] are applied for central pipe, annular pipe, and swirl ports. These are discussed in the next section. Since the co-axial flow is entraining into quiescent fluid, the atmospheric pressure condition is assigned to the outlet domain as outlet boundary condition and rest of the surfaces are assigned as wall.

3 Solution Procedure

The simulations were carried out using ANSYS Fluent software which uses an unstructured mesh and the finite volume method to solve the various transport equations. Second-order upwind scheme is used for convective terms to get velocity fields. The simple algorithm is used for pressure correction. For temporal discretization, the first-order implicit scheme is used. To analyse the unsteady nature of the flow field, simulations were run on transient mode, and the mass flow rate at the outlet was monitored continuously. Convergence criteria were set to 0.001 which were satisfied by all the set of equations except continuity, which was fluctuating between 0.001 and 0.01. The time step size of 0.0005 s was adopted which was sufficiently small, and the number of time steps chosen was 2000 to get the data for 1 s. To average the velocity and vorticity data, custom field function of velocity and vorticity was selected in the software, and data sampling for time averaging was done for the chosen custom field variables. The simulated data were exported to TECPLOT in order to plot various contour plots.

4 Test Cases

Different test cases shown in Table 1 are obtained by decreasing the central jet flow rate in steps while keeping the flow rate for annular and swirl ports unaltered. The Case 1 is called as Pre-VBB to denote Pre-Vortex breakdown, Case 2 is denoted by VBB (Vortex breakdown bubble), Case 3 is called 'Transition', and Case 4 is called CTRZ (Central Toroidal Recirculation Zone). The nomenclature is same as in [3]. The dimensionless numbers and their corresponding values are presented in Santhosh et al. [3].

The transition of recirculation structures in the flow field occurs as the Reynolds number (Re) gradually decreases. The Reynolds number (Re) is calculated as using [13].

$$\mathrm{Re} = \frac{UD}{\nu} \tag{6}$$

where U is the velocity obtained from volume flow rate, D—the diameter of the central pipe and ν the kinematic viscosity. The fluid considered is air at standard temperature and pressure.

Modified Rossby number is defined as [14],

$$\mathrm{Ro}_m = \frac{\Delta U_y}{U_\theta} \tag{7}$$

where U_θ is tangential velocity and ΔU_y is the axial velocity deficit between central and coflow jets.

$$\Delta U_y = U_c - U_a. \tag{8}$$

$$\emptyset = \frac{U_c}{U_a} \tag{9}$$

Table 1 Inlet conditions for different flow cases considered in the present study

Flow mode	Reynolds number (Re)	Swirl number (S)	Modified Rossby number (Ro_m)	Velocity ratio ($\emptyset$)
Pre-VBB (Case-1)	3604	0.592	3.15	7.15
VBB (Case-2)	1691	0.759	1.12	3.35
Transition (Case-3)	1127	0.788	0.33	1.92
CTRZ (Case-4)	563	0.801	0.02	0.98

Here, Ø is the velocity ratio between central and coflow jet.

Swirl number is defined as [15],

$$S = \frac{\int_0^{R_0} \rho U_y U_\theta 2\pi r^2 \mathrm{d}r}{R_0 \int_0^{R_0} \rho U_y^2 2\pi r \mathrm{d}r} \tag{10}$$

U_y and U_θ indicate the streamwise and tangential velocities, respectively.

5 Results

This section presents the time-averaged transition of the recirculation structures as the Reynolds number of central jet varies from 3607 to 563, while the swirl number is kept within $0.592 \le S \le 0.801$. Along with Reynolds number and swirl number, modified Rossby number (Eq. 7) is employed to explore the physics of transition of recirculation structure. In order to analyse and visualize the structures, various counterplots superimposed on streamlines are plotted.

5.1 Time-Averaged Axial Velocity Contours

Figure 3 represents the counters of the experimental (left side) and numerical (right side) time-averaged axial velocity. The annular pipe radius (R_0) is used as a normalizing factor for both abscissa and ordinate. The flow structure observed in Fig. 3a, b is known as pre-vortex bubble (pre-VBB) which is the first occurrence of the flow reversal in the central region of the inner jet when Re = 3607 (swirl number $S = 0.592$) and $\mathrm{Ro}_m = 3.15$. This flow mode consists of two different regions. First, high axial momentum of the central jet in the central region. Second, two counter-rotating eddies present on either side of the central jet. The characteristic dimension of these two rotating eddies (length and width) is mainly dependent on two factors; first, axial pressure gradient that sets along the central axis towards the burner exit due to the large axial velocity deficit between central and coflow jet. Second, centrifugal pressure gradient developed along radial direction due to tangential velocity component of the rotating particles. The resultant of these two pressure gradients is responsible for characterizing the dimension of the recirculation eddies [3]. This can be explained as the swirl imparted to the annulus air rolls up the intermediate shear layer (ISL-2) (Fig. 1b) which results in the increased penetration of the angular momentum in the radial direction [3]. Contrarily, the pressure deficit, i.e. $\left(\rho|\Delta U_y^2\right)$ in the inner shear layer (ISL-1) (Fig. 1b) due to the large entrainment of the central jet restricts the spread of the swirl influence towards the central axis [3]. Finally, the resultant of these two competing pressure gradients delineates the size of the recirculation structure. As the swirl number increases or Ro_m decreases, vortex

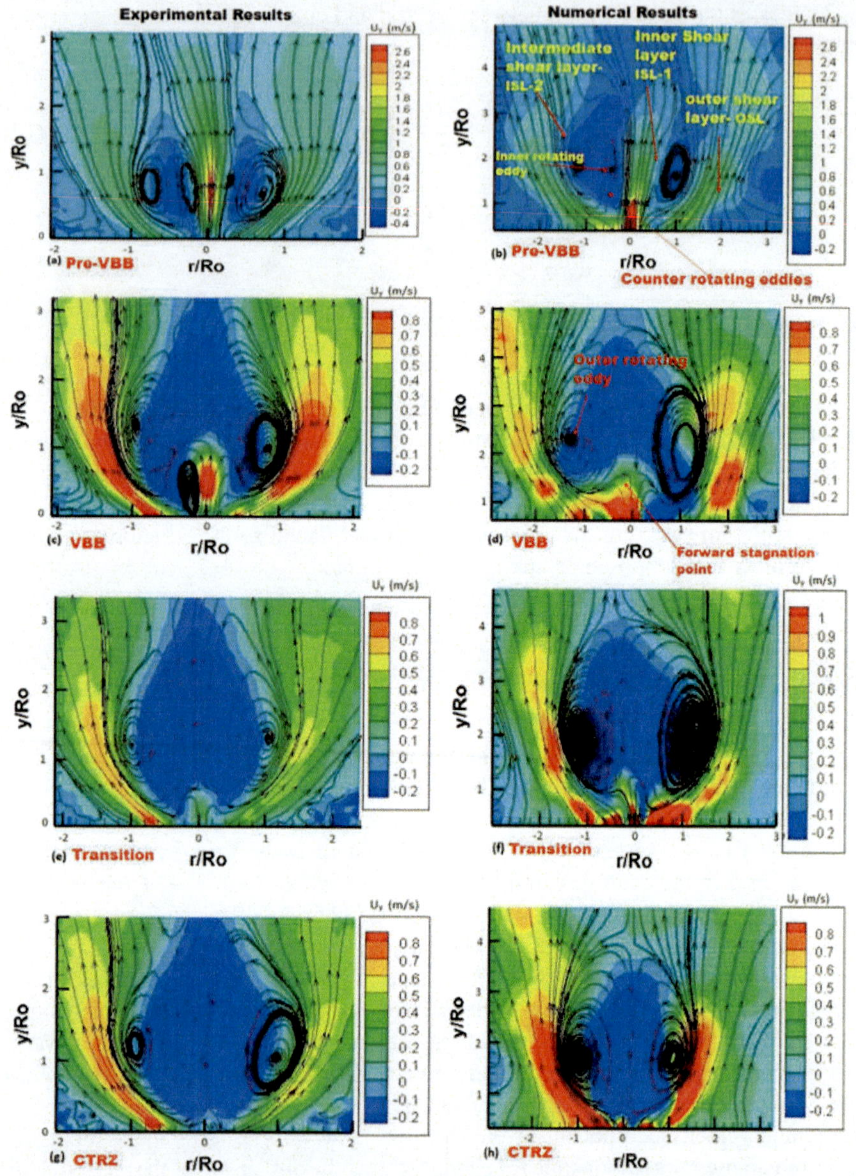

Fig. 3 Time-averaged velocity contours of experimental [3] and numerical results (present study) are placed side by side. Image **a**, **b** Pre-breakdown flow regime (Pre-VBB), **c**, **d** vortex breakdown bubble (VBB), **e**, **f** transition, **g**, **h** CTRZ

breakdown bubble (VBB) type of recirculation is observed (Fig. 3c, d). It is characterized by extensive amplification in the size of the outer recirculating eddy in the streamwise and radial direction. This remarkable increase in the size of the rotating structure implies the victorious penetration of the swirl influence in the transverse direction towards the inner jet [3]. On the other part, the inner recirculating eddies (Fig. 3b) get displaced in the upstream direction towards the exit of the burner. This results in the development of the forward stagnation point that restricts the axial penetration of the central jet shown in (Fig. 3d). Further decrease in the Ro_m (to a value of 0.33) causes the occurrence of the transition structure. In this, inner vortices are completely diminished, collapsed, and integrated with the outer eddies. Finally, further decrease in the Ro_m (to a value close to 0) results in the central toroidal recirculation zone leaving no traces of the inner vortices. CTRZ resembles the complete bubble-like structure within which hot combustion products, fresh reactants, and radical species reside and extensively mix due to large recirculating motions shown in Fig. 3g, h.

5.2 *Variation of Axial Velocity Along the Radial Direction*

Figure 4 represents the variation of time-averaged axial velocity along the radial direction for each of the four cases considered. The plots are drawn at an axial location which passes through the vortex core centre for each case. These distances are indicated (for each case) in Fig. 4 and are indicated by a red horizontal line in Fig. 5. The normalization factor for *Y*-coordinate is bulk mean velocity (Uo) which is defined as [3]

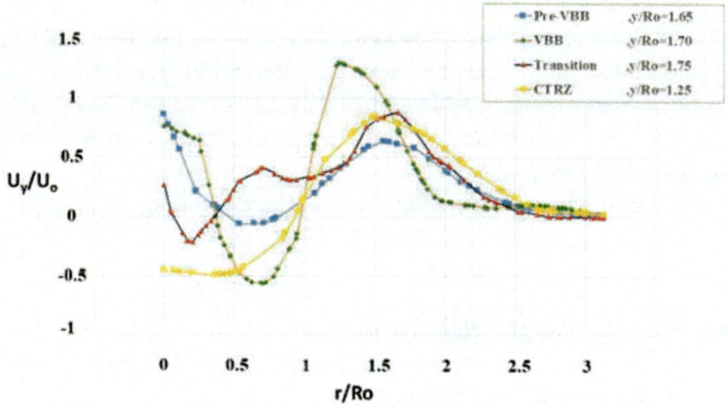

Fig. 4 Normalized time-averaged axial velocity distribution in the radial direction at an axial distance through the vortex core centre

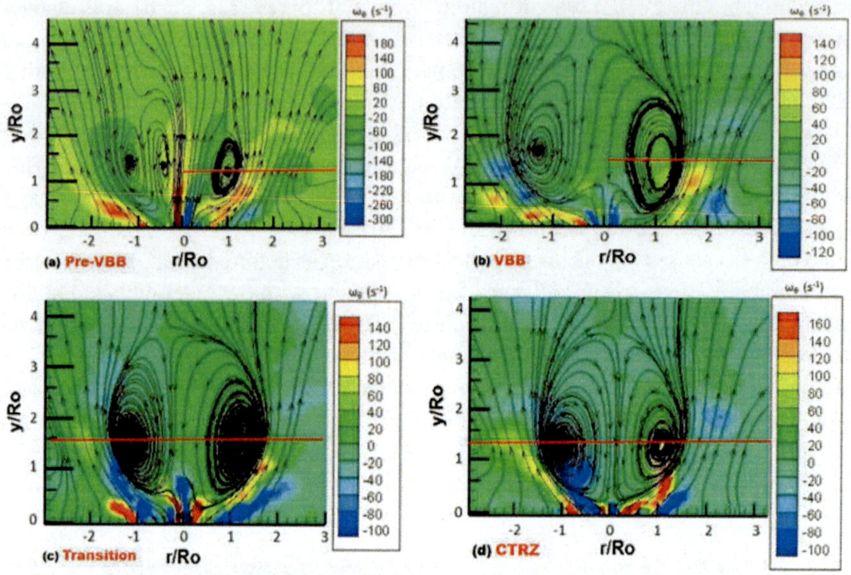

Fig. 5 Numerical results of time-averaged vorticity contours for the **a** Pre-breakdown flow regime, **b** partially penetrated vortex breakdown bubble, **c** transition and **d** CTRZ flow mode

$$Uo = \frac{A_c u_c + A_a u_a + A_s u_s}{A_c + A_a + A_s} \tag{7}$$

where $A_c u_c$, $A_a u_a$, $A_s u_s$ are area and velocity of central pipe, annular pipe, and swirl ports, respectively.

It is seen from Fig. 4 that due to the streamwise flowing central jet present in the Pre-VBB, the velocity is positive at the centreline ($r = 0$) and decreases to a value below zero (indicating the presence of recirculation zone). As the flow rate of the central jet is decreased to switch from Case 1 to Case 4 through Case 2 and Case 3, the axial velocity at the centreline gradually decreases. The streamwise velocity for CTRZ (Case 4) is negative at the centreline indicating the presence of recirculation. This plot further corroborates the transition described in the previous subsection.

5.3 *Time-Averaged Vorticity Contours*

Figure 5 represents the numerical result of time-averaged azimuthal vorticity contours for the different flow regimes. Since the axial velocity profiles for experimental and numerical are closely matching, so only numerical results are used to explain the variation of vorticity for different cases. For Case 1 (Pre-VBB), the azimuthal vorticity is

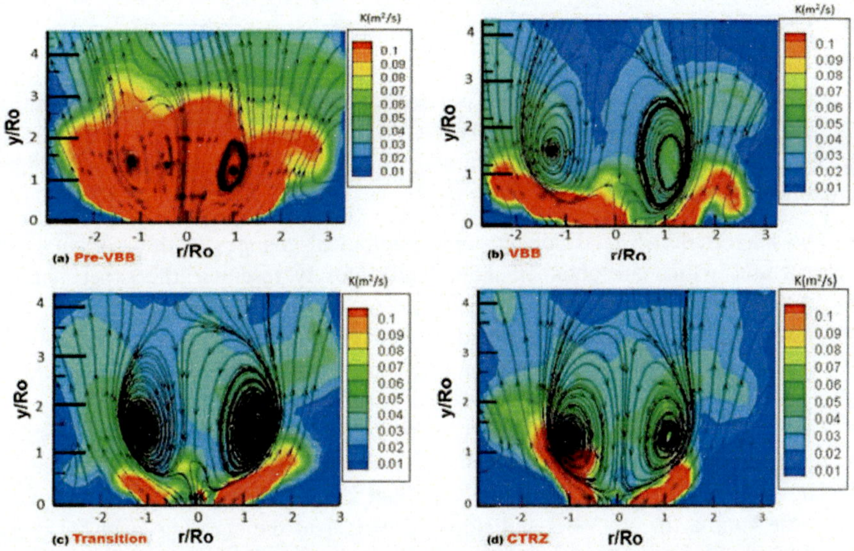

Fig. 6 Numerical time-averaged turbulent kinetic energy contours for the **a** Pre-breakdown flow regime, **b** partially penetrated vortex breakdown bubble, **c** transition and **d** CTRZ flow mode

found maximum at the central jet column, and as the swirl number increases, vorticity strength shifts and is found maximum near the shear layer between recirculation zone (RZ) and co-annular air.

5.4 Time-Averaged Turbulent Kinetic Energy (TKE) Contours

Figure 6 represents the time-averaged turbulent kinetic energy for all the cases considered previously. For the pre-VBB case (Fig. 6a), the concentration of the turbulent kinetic energy is found highest in the central region. This may due to the large fluctuation in the velocity components due to the presence of a shear layer between central jet and recirculation zones. As the flow Reynolds number decreases (Fig. 6b), the turbulent fluctuations slowly die out, and the maximum is now concentrated at the shear layer between the recirculation zone and quiescent surroundings. This leads us to conclude that the transition in the recirculation zone described before has considerable and interesting effects on the TKE in the interior of the flow field.

6 Conclusion

The numerical simulations are carried out to study transition in recirculating structure in swirl co-axial flow configuration. Two RANS turbulent models, $k - \varepsilon$ and RSM, are used to explore the nature of the turbulent flow field. The salient features of this investigation are

- The magnitude of axial velocity and topological changes of the recirculation structures which are obtained numerically closely resemble the experimental results.
- The magnitude of the vorticity found maximum in the interior region for pre-VBB case and shifts towards the shearing layer between rotating eddy and co-annular air as the transition proceeds.
- For the different turbulent models used to predict the turbulent flow field, RSM model is found as suitable in predicting the unsteady nature of the swirl flow field.
- The flow field contains maximum turbulent kinetic energy in the interior part of the bubble for pre-VBB and slowly directs near the shearing layer between rotating eddy and co-annular air as the transition proceeds.

References

1. Gupta A, Kumar R (2007) Three-dimensional turbulent swirling flow in a cylinder: experiments and computations. Int J Heat Fluid Flow 28(2):249–261
2. Saiki Y, Suzuki Y, Kasagi N (2011) Active control of swirling coaxial jet mixing with manipulation of large-scale vortical structures. Flow Turbul Combust 86:399–418
3. Santhosh R, Miglani A, Basu S (2013) Transition and acoustic response of recirculation structures in an unconfined co-axial isothermal swirling flow. Phys Fluids 25(8):083603
4. Santhosh R, Miglani A, Basu S (2014) Transition in vortex breakdown modes in a coaxial isothermal unconfined swirling jet. Phys Fluids 26:043601
5. Ouali S, Bentebbiche A, Belmrabet T (2015) Numerical simulation of methane-air equivalence ratio effect on premixed low swirl stabilized flame. J Bra Soc Mech Sci Eng 37(2):747–760
6. Werner JAD, Frieler CE, Tryggvason G (1992) Vortex structure and dynamics in the near field of a coaxial jet. J Fluid Mech 241:371402
7. Wicker RB, Eaton JK (2009) Structure of a swirling recirculating co-axial free jet and its effect on particle motion. Int J Multiph Flow 27:949–970
8. Cho CH, Baek GM, Sohn CH, Cho JH, Kim HS (2013) A numerical approach to reduction of NOx emission from swirl premix burner in a gas turbine combustor. Appl Thermal Eng 59:454–463
9. Krieger GC et al. (2012) A swirler stabilized combustion chamber for a micro-gas turbine fuelled with natural gas. J Bra Soc Mech Sci Eng 34(4):441–449
10. Jufar SR, Huang RF, Hsu CM (2013) Effects of swirl on flow and mixing of acoustically excited swirling double-concentric jets. Exp Thermal Fluid Sci 49:40–50
11. Fluent 6.3, theory and user guide
12. Kundu PK, Cohen IM (2002) Fluid mechanics. Academic Press, USA
13. Amirshaghaghi H, Zamaniyan A, Ebrahimi H, Zarkesh M (2010) Numerical simulation of methane partial oxidation in the burner and combustion chamber of auto thermal reformer. Appl Math Model 34(9):2312–2322

14. Giannadakis A, Perrakis K, Panidis Th (2008) A swirling jet under the influence of a coaxial flow. Exp Therm Fluid Sci 32:1548
15. Gupta AK, Lilley DG, Syred N (1984) Swirl flows. Abacus Press

Numerical Study on Sample Thickness Dependence of Fire Response Properties of Polymeric Materials (Charring and Non-charring) in Standard Cone Calorimeter Test

Vishal Srivastav, S. Sabarilal, and Amit Kumar

1 Introduction

In this section, an overview of the pyrolysis process is presented, following which the process of thermal degradation of charring and non-charring is described. In the following section, a brief description of the numerical modelling of both charring and non-charring material types is mentioned. Finally, model validation and results and the conclusions are presented.

Pyrolysis is a thermal decomposition of the solid fuels due to external heat flux. It is a thermochemical process. It plays a major role in the ignition of the solid fuels and also in the propagation of fire in the solid fuels; the rate of pyrolysis controls the time to ignition, heat release rate and the mass loss rate of the solid fuels. Solid fuels can be grouped into two namely charring and non-charring solid fuels. When subjected to external heat flux, solid fuels which give carbonaceous residues along with other products are known as charring materials and those which give very less or no char are non-charring solid fuels.

When the surface of solid fuel is exposed to the external heat flux, at elevated temperatures a series of physical and chemical processes follow that leads to the thermal decomposition of the charring and non-charring solid fuels. This process results in the formation of pyrolysate, which is a liquid mixture of different species and volatiles, which is a gaseous mixture of different species [2]. The rate of release of gaseous volatile decides the growth and spread of fire over the solid fuel. The fire, in turn, provides heat feedback to the fuel surface whose response to the feedback

V. Srivastav · S. Sabarilal · A. Kumar (✉)
National Centre for Combustion Research and Development, Department of Aerospace Engineering, Indian Institute of Technology Madras, Chennai, Tamil Nadu 600036, India
e-mail: amitk@ae.iitm.ac.in

V. Srivastav
e-mail: srvstvishal@gmail.com

A. K. Gupta et al. (eds.), *Advances in IC Engines and Combustion Technology*, Lecture Notes in Mechanical Engineering,
https://doi.org/10.1007/978-981-15-5996-9_71

further affect the fire growth and rate of propagation of fire over the rest of the fuel. Therefore, the study of the process thermal degradation of polymers will help one gain deeper insights into the growth and propagation of fire over polymeric substrates.

The cone calorimeter is a bench-scale instrument in fire testing [3]. In standard cone calorimeter test, a test sample is subjected to thermal degradation under prescribed radiant heat flux and fire response properties like time to ignition, peak mass loss rate, time to peak mass loss rate, average mass loss rate and burn out time are measured. The ASTM 1354 [1] recommends sample thickness of less than 50 mm for the test. However, for thinner samples conditions on the unexposed side of the sample should represent actual conditions in real-life application as the test results may be affected by these conditions. A study on the influence of sample size on cone calorimeter results is done by [4]. A study on the effect of thickness on flammability properties of the composites using cone calorimeter is done by [5]. Therefore, in this work, a numerical study is carried out to predict the effect of sample thickness on the fire response parameters of polymeric materials.

2 Numerical Modelling and Validation

There has been a tremendous amount of research in modelling of pyrolysis of the charring and non-charring polymers, Nicolas Bal in [6] has compared the results of models with different mechanisms and complexities. For the purpose of this work a 1D transient numerical model for pyrolysis was developed by Sabarilal and Kumar [7] based on conservation of energy, mass, and momentum have been used in this study. For more information on numerical modelling and its validation please refer to [7]. Here only an overview of the model is presented.

A fully implicit finite control volume approach was used for charring material to obtain the temperature in each control volume cell. In a numerical model for charring material, the dimensions of the fuel cells are constant while the density, specific heats and the mass fraction of char and virgin fuel were calculated at every time step as a function of temperature in each control volume cell. In non-charring as there is no char, there is no formation of any residues and the fuel is assumed to directly escape resulting in a reduction in the thickness of the fuel unlike in case of charring materials where the properties are varied keeping dimensions same.

A schematic of the model for charring fuel block and non-charring fuel block is shown in Fig. 1 [7]. In Fig. 1a there is a schematic representation of charring model where the heat flux enters from the surface of the fuel and the heat transfer inside the fuel happens by conduction. On the other hand in Fig. 1b schematic representation of non-charring fuel is shown. Here, one can see that heat flux enters through the surface of the fuel and then gets transferred into the solid by conduction as well as in-depth radiation. Further, heat gets reflected by the back boundary of the fuel resulting in further absorption of the heat flux by the cells. This transmitted heat flux along with absorption is calculated by solving Radiation Transfer Equation (RTE).

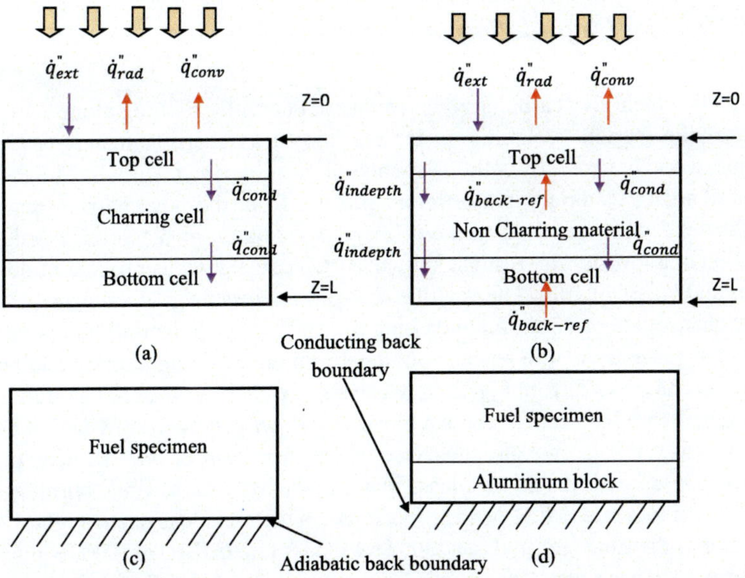

Fig. 1 **a** Schematic representation of cells and heat flux in charring fuel numerical model. **b** Schematic representation of cells and heat flux in non-charring fuel's numerical model. **c** Bottom fuel cell with adiabatic back boundary condition (or insulated). **d** Bottom fuel cell with conducting back boundary. (withback aluminium block)

Figure 1c illustrates the adiabatic back boundary condition used in modelling and Fig. 1 (d) an aluminium block is placed beneath the fuel sample for enforcing conducting back boundary condition.

3 Results and Discussions

In this section thermal degradation of charring and non-charring fuel blocks subjected to a radiant heat flux of 50 kW/m^2 is discussed for the effect of block thickness. Variation of key fire response properties with fuel thickness is studied to identify at what sample thickness these properties are invariant to fuel thickness. Effect of radiant heat flux magnitude on these invariant fire properties is also discussed. In the following sub-sections, first charring materials are discussed then non-charring materials.

3.1 Thermal Degradation of Charring Materials

Solid fuels which give carbonaceous residue or char after burning along with other products like gaseous volatile and tar are known as charring materials [2]. For example, wood or generally cellulosic materials are charring materials. The combustion of charring materials is a different process from the non-charring materials. The chemical processes lead to the primary degradation of the solids resulting in formation of a low-density residue known as char and a gaseous mixture of different species known as volatiles. The volatiles then due to pressure gradient passes through the hot char region or virgin fuel resulting in cracking or condensation, respectively [2]. As the gaseous volatile reaches out to surrounding oxygen flaming combustion occurs. As can be seen from Fig. 2, in pyrolysis of a charring material an initial peak was reached with high rate of change of MLR following which a low rate of change of MLR was observed which is because of the formation of the low-density char. Peak mass loss rate is the highest mass loss rate the fuel attains while burning. Time to peak mass loss rate (MLR) is the time at which the mass loss rate is highest. Time to ignition is the time required to attain a mass loss rate (MLR) of 0.003 kg/m^{2}s [8].

Figure 3 shows a mass loss rate (MLR) versus time for different thicknesses at the external heat flux of 50 kW/m^{2}s. This is because an initial peak is seen during the thermal degradation of charring materials which subsequently decay with time. The reaction front progresses deeper into the fuel leaving low-density char residue on top. This low-density char residue offers resistance to heat transfer into the fuel [9]. The second peak is usually observed towards the end of the pyrolysis process for thicker fuels and it is distinguishable at the higher heat fluxes. Figure 3a Shows plots of MLR versus time for different thicknesses of CPVC block. It may be noted from the figure that the peak MLR decreases with an increase in thickness. As in Fig. 3a peak MLR for 1 mm is higher than for 3 mm sample. In Fig. 3b for the case of CPVC with aluminium plate the peak MLR trend is different from that of Fig. 3a

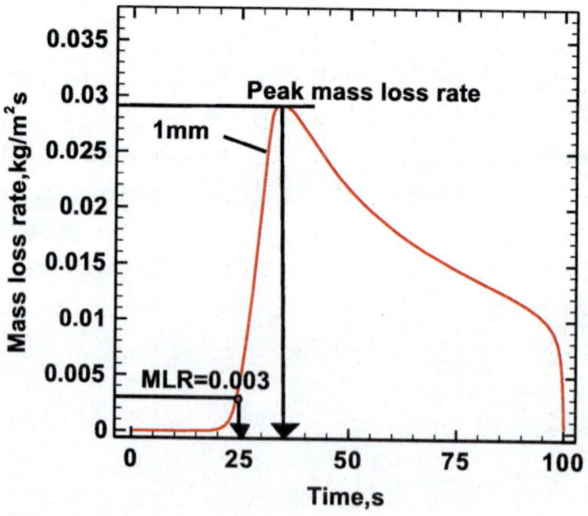

Fig. 2 Typical mass loss rate (MLR) curve for a charring material

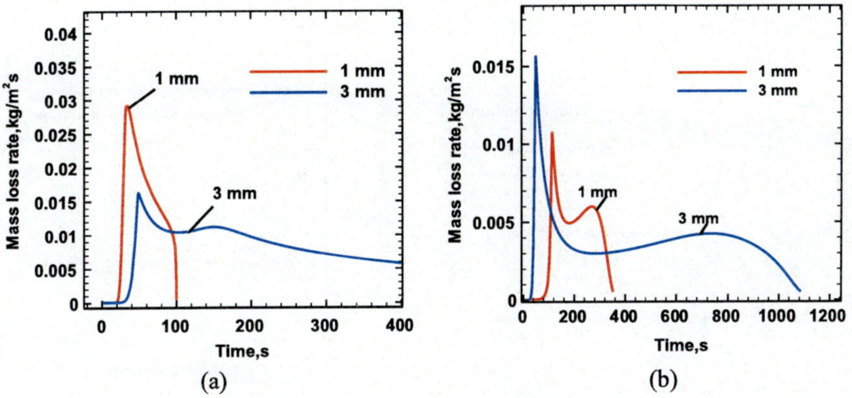

Fig. 3 **a** Plot of mass loss rate (MLR) versus time at 50 kW/m^2 for charring (CPVC) solid fuel without aluminium for different thicknesses. **b** Plot of mass loss rate (MLR) versus time at 50 kW/m^2 for charring (CPVC) solid fuel with aluminium for different thicknesses

where adiabatic boundary condition was considered. The peak MLR of 3 mm sample is higher than the peak MLR of 1 mm sample.

These opposite trends are noted because of the effect of back boundary adiabatic versus conducting block of aluminium [7]. Aluminium enhances the heat transfer resulting in more time taking place for reaching a peak MLR value. The plot of MLR versus time with aluminium bottom cases can be seen to have extended burn time compared to adiabatic cases of adiabatic back boundary.

A plot between peak mass loss rate, average mass loss rate, time to peak mass loss rate and time to ignition is shown for different thicknesses under an external heat flux of 50 kW/m^2. Figure 4a Shows plot between peak mass loss rate of CPVC (charring material) sample with bottom aluminium slab and without bottom aluminium slab for different thicknesses at heat flux of 50 kW/m^2. One can note that pyrolysis of CPVC without aluminium the peak mass loss rate first decreases and then becomes constant after 7 mm to be exact. When the bottom aluminium slab is present, the peak mass loss rate first increases rapidly and then becomes constant thereafter after a 7 mm thickness, the constant peak MLR value obtained in with aluminium and without aluminium case came out to be same, which indicates back boundary condition is no more influencing the property and hence peak mass loss rate has become independent of thickness.

Figure 4b, shows a plot between the time taken to peak mass loss rate versus thickness is shown for CPVC at 50 kW/m^2. It shows the time taken by the fuel sample of different thicknesses to reach peak mass loss rate. It is noted that time to peak mass loss rate changes steeply with thickness for thin fuels typically before 3 mm than for thicker fuels typically after 3 mm. For the burning of CPVC with aluminium case, the time to peak MLR is more than it is for without aluminium. As because of the aluminium plate, the heat transfer from the back surface is enhanced. Just like peak MLR, the constant value of time to peak MLR for with aluminium and

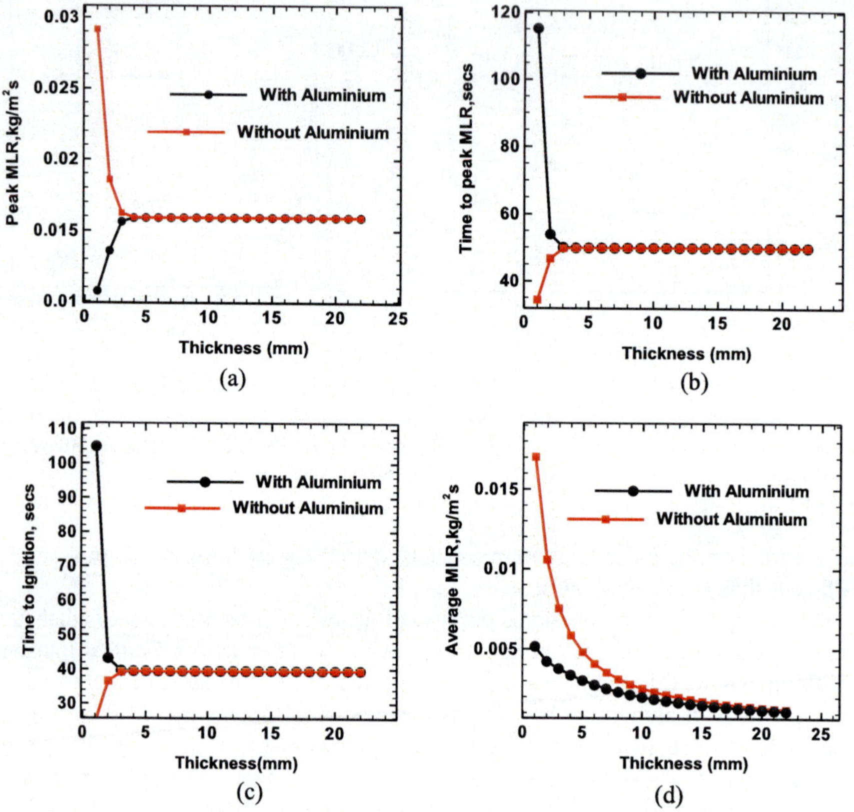

Fig. 4 Variation of various fuel properties with fuel thickness for adiabatic back boundary and conducting back boundary conditions. **a** Peak MLR, **b** time to peak MLR, **c** time to ignition, **d** average MLR

without aluminium becomes same after 4 mm and is equal to 49.8 s and hence does not change thereafter. This shows as the case of sample with aluminium is also giving the same value as of the sample without aluminium, the property is not being affected by the back boundary condition and hence has become independent of thickness.

In Fig. 4c, variation of time to ignition with fuel is shown in Fig. 4c. Time to ignition is defined as time to reach MLR of 3 g/m^{2}s. The trend is similar to that of time to peak MLR as seen in Fig. 4b. In Fig. 4d a plot between average mass loss rate for various thicknesses is shown. Average mass loss rate is the mass loss rate divided by the difference between time when pyrolysis started typically taken as when MLR is 0.001 and burnout time which is also taken as time at the end of burning when the MLR is 0.001 kg/m^{2}s. The average mass loss rate achieves a constant value for both cases, with aluminium and without aluminium. The thickness at which this property becomes independent of thickness is much larger (~20 mm here) compared to that for other properties (~3–4 mm).

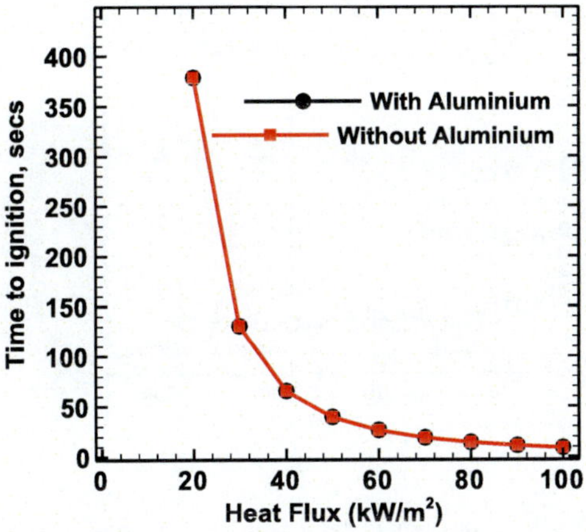

Fig. 5 Variation of thickness-independent time to ignition value with external radiant heat flux

As constant time to ignition value is reached at every heat flux and heat flux. The constant value of time to ignition was obtained at every heat flux and presented below in Fig. 5.

It may be noted here that time to ignition decreases with heat flux and changes steeply for lower heat fluxes and is less sensitive to higher heat fluxes. As the time to ignition becomes independent of thickness with aluminium and without aluminium values are equal.

3.2 Thermal Degradation of Non-charring Materials

Solid fuels which do not leave any residue or very less residue after undergoing thermal decomposition is known as non-charring materials. For example thermoplastic materials like Poly (Methyl methacrylate) (PMMA), polyethylene (PE), polypropylene (PP), etc. Usually, they burn completely and does not leave any residue unless fire retardants are added. As the burning of a non-charring material is a different process from the burning of a charring material. That's why the plot between MLR and time are different for both. Also, the back boundary and in-depth radiation heat transfer have more influence on the MLR for non-charring material. Figures 6 and 7 shows the plot between MLR and time for non-charring fuel without aluminium and with aluminium. Peak mass loss rate is the maximum MLR that is obtained in the burning and time to ignition criteria is same as for charring [8].

A plot of mass loss rate of non-charring fuel and time is shown in Fig. 7. It may be noted from the figure above that non-charring solid fuels show a certain period of steady-state burning for the thicker fuels [10] and follow by a peak at the end of the burning, unlike charring fuels which show a peak at the initial stage of burning. It has been shown that the peak at the end of burning is not a fuel property but a back boundary effect [7]. Figure 7b shows MLR versus time plot for PMMA with

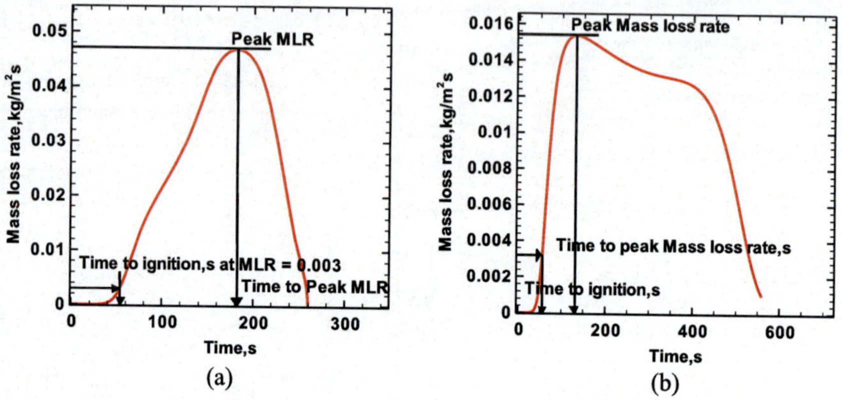

Fig. 6 **a** Variation of MLR with time for non-charring fuel block without aluminium. **b** Variation of MLR with time for non-charring fuel block with aluminium. Both plots are for 5 mm thick fuel block at external heat flux of 50 kW/m^2. Time to ignition is shown by arrow at 0.003 kg/m^{2}s MLR and time to peak MLR is shown by arrow at peak MLR value

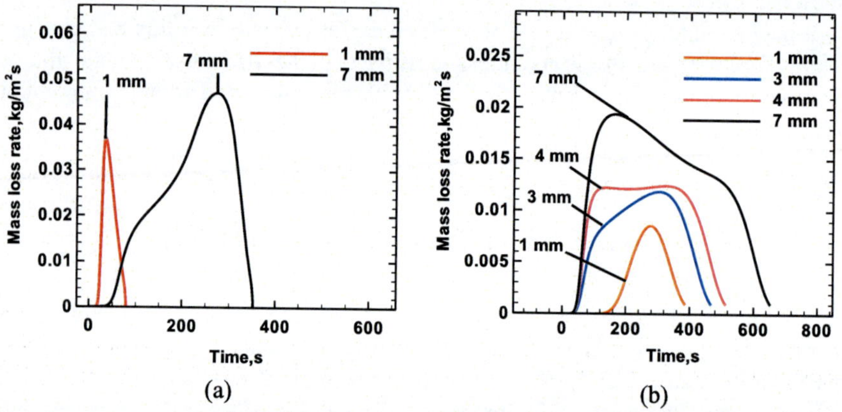

Fig. 7 **a** Variation of MLR with time for non-charring (PMMA) solid fuel without aluminium under external heat flux of 50 kW/m^2. **b** Variation of MLR versus time for non-charring (PMMA) solid fuel with an aluminium back plate under external heat flux of 50 kW/m^2

10 mm thick aluminium plate at the back of it. An initial a peak is observed for thin samples from 1 to 3 mm, it is noted that for thickness at 5 mm and onwards a peak is attained in the initial stage of burning as opposite to the burning of PMMA without aluminium where the peak was obtained at the end of burning. Hence the effect of aluminium is noticeable from Fig. 7 for non-charring.

A plot between peak mass loss rate, average mass loss rate, time to peak mass loss rate and time to ignition is shown in Fig. 8 for different thicknesses under external heat flux of 50 kW/m^2 for both cases of with aluminium backplate and without aluminium backplate. Figure 8a Shows MLR attains an almost constant value of

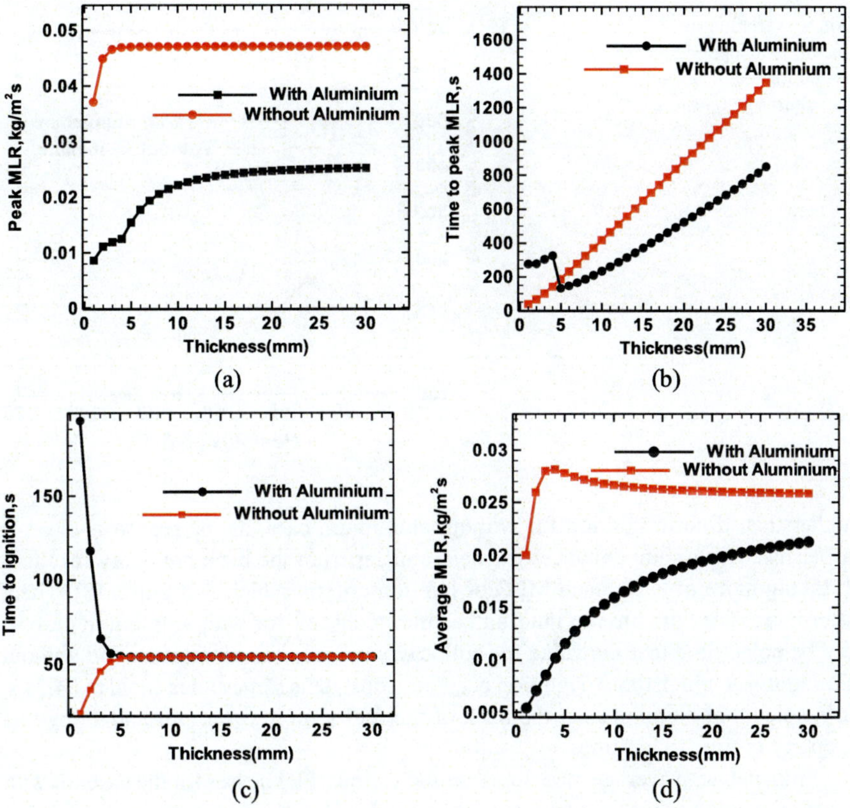

Fig. 8 Shows the variation of various fuel properties with thickness for both the cases of without aluminium slab (adiabatic back boundary) and with bottom aluminium slab (conducting back boundary). **a** Peak MLR, **b** time to peak MLR, **c** time to ignition, **d** average mass loss rate (MLR)

0.0470 after a thickness of 6 mm fuel sample. A similar trend is noted for the case of with aluminium where rate of increase in peak MLR decreases with thickness except for thin fuels, where the trend is different. It is noted that with aluminium MLR is less as compared to without aluminium. Unlike in charring where after the peak MLR becomes constant the value for both cases of without aluminium and with aluminium is equal. Figure 8b the time to peak mass loss rate increases with thickness unlike in charring where it attained a constant value. For the case of sample with aluminium time to peak mass loss rate first increases till 4 mm and then decreases instantly for 5 mm. The reason for this can be found by looking at Fig. 7b, where we can see a peak is forming at the initial stage of pyrolysis after 5 mm resulting in shifting of peak MLR to left side in plot.

In Fig. 8c time to ignition is shown for different thicknesses, it has been noted that time to ignition for PMMA without aluminium first increases till a8 mm thickness and then becomes constant. Time to ignition for PMMA with aluminium case is

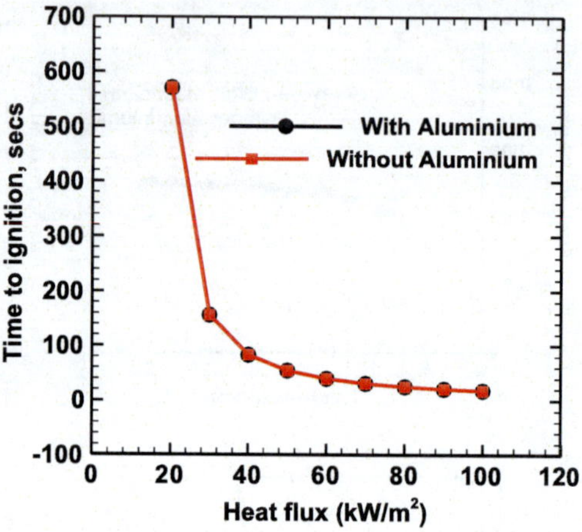

Fig. 9 Variation of thickness-independent time to ignition and heat flux for non-charring materials

higher than time to ignition for without aluminium case, as the reason is clear, it is because aluminium enhances the heat transfer from the back boundary resulting in taking more time to reach MLR of ignition criteria which is 3 g/m^2s [8]. Then, after increasing, the time to ignition becomes constant for with aluminium case. It can be noted here that the value for both cases of fuel with aluminium and without aluminium at which time to ignition becomes constant is same and is equal to 54.35 s. It is noted that aluminium at the back of the fuel sample is no more affecting this property of time to ignition.

A plot between average mass loss rate and various thicknesses for the cases of with aluminium and without aluminium is shown in Fig. 8d. It is noted that average MLR first increases and then like other properties rate of change is decreasing. It varies for thin fuels more steeply as compared to thicker fuels. The average MLR value for the case of with aluminium is less than that for the case of without aluminium.

A constant value of time to ignition reached was obtained for non-charring for both cases of with aluminium and without aluminium just like one done for charring (Fig. 9).

The trend here is similar to the trend obtained in charring case, time to ignition value decreases with an increase in heat flux. The time to ignition changes more steeply with an increase in heat flux at lower heat fluxes and is less sensitive to change in heat flux at higher heat fluxes.

4 Conclusions

For the charring materials (here CPVC) some fire response properties like peak mass loss rate, time to peak mass loss rate, time to ignition and average mass loss rate were noted to become constant after certain thickness for both the case of with aluminium and without aluminium under the sample. Hence it can be concluded here that while performing standard cone calorimeter test for these fire response properties, a thicker sample (3 and 5 mm) will give a better value of this properties with high accuracy in real-life response to fire situation without the effect of the back boundary.

For non-charring materials (here PMMA), fire response properties like mass loss rate, time to peak mass loss rate and average mass loss rate are noted to be different for with aluminium and without aluminium. So to extract these property proper back boundary conditions should be provided to simulate real-life response to fire. On the other hand, the property of time to ignition was seen to achieve a same constant value for both the case of with aluminium and without aluminium indicating back boundary of aluminium is no more affecting this particular property, hence standard cone calorimeter test can be performed at thicker sample where it becomes independent of thickness will give better result with good accuracy because it will be independent of back boundary effects.

For both charring and non-charring materials, it was also noted that time to ignition becomes constant for a prescribed radiant heat flux. Time to ignition decreased with increase in applied radiant heat flux. It was also noted that time to ignition value (independent of thickness) is sensitive (or prone to errors) at lower heat fluxes and less sensitive at higher heat fluxes (>40 KW/m^2). Hence to get time to ignition value performing the standard cone calorimeter test at higher heat flux will give better result with less error.

References

1. ASTM (2015) E-1354 Standard Test Method for heat and visible smoke release rates for materials and products using an oxygen consumption calorimeter. An Am Natl Stand
2. Di Colomba B (1993) Modeling and simulation of combustion processes of charring and non-charring solid fuels. Prog Energy Combust Sci 19:71–104
3. Babrauskas V (2016) The cone calorimeter. In: SFPE handbook of fire protection engineering, 5th edn
4. Lindholm J, Brink A, Hupa M (2012) Influence of decreased sample size on cone calorimeter results. Fire Mater
5. Fateh T, Kahanji C, Joseph P, Rogaume T (2017) A study of the effect of thickness on the thermal degradation and flammability characteristics of some composite materials using a cone calorimeter. J Fire Sci
6. Bal N, Rein G (2013) Relevant model complexity for non-charring polymer pyrolysis. Fire Saf J
7. Sabarilal S, Kumar A (2018) Effect of back boundary condition on pyrolysis of charring and non-charring materials. J Phys Conf Ser 1107(3)
8. Lyon RE, Quintiere JG (2007) Criteria for piloted ignition of combustible solids. Combust Flame 151(4):551–559
9. Schartel B, Hull TR (2007) Development of fire-retarded materials—interpretation of cone calorimeter data. Fire Mater
10. Carvel R, Steinhaus T, Rein G, Torero JL (2011) Determination of the flammability properties of polymeric materials: a novel method. Polym Degrad Stab